Klaus Günther / Wolfgang Schulz

Biophysical Theory of Radiation Action

Biophysical Theory of Radiation Action

A Treatise on Relative Biological Effectiveness

by

Klaus Günther
Wolfgang Schulz

with 80 Figures and 32 Tables

AKADEMIE-VERLAG · BERLIN
1983

Dr. sc. nat. Klaus Günther
Dr. rer. nat. Wolfgang Schulz

AdW der DDR

Zentralinstitut für Elektronenphysik
Zentralinstitut für Molekularbiologie

Erschienen im Akademie-Verlag, DDR-1086 Berlin, Leipziger Straße 3—4

Lizenznummer: 202 · 100/518/83
Printed in the German Democratic Republic
Gesamtherstellung: VEB Druckhaus „Maxim Gorki“, 7400 Altenburg
Lektor: Christiane Grunow
Einband: Annemarie Wagner
LSV 1315
Bestellnummer: 762 429 1 (6432)
DDR 68,— M

Preface

The science of radiation action on living systems has a rather long history. Practical problems in application of, and protection from, ionizing radiations have always been the main impetus to continue and extend research in this field, so that emphasis has been put on experimentally establishing radiation actions of practical interest rather than systematically developing a general theoretical framework to achieve comprehension of the underlying mechanisms. It is true, the early period until about the mid-50s was governed, indeed, by such kind of fundamental ideas which were widely accepted (hit principle and target theory), but thenceforth, when the simple models of target theory proved unsuitable in many cases, the development became divergent. At present, there is an obvious gap between the wide variety of facts described in terms of biology, biochemistry, and medicine and, on the other hand, the now well-established physics of radiation interaction with matter. This book has been written with the intention to help fill that gap.

Apart from the fundamental scientific aspects of the problem, its relevance to practice should be stressed. In contrast to the formerly dominant role of gamma- and X-rays, it is now a broad range of ionizing radiations that is increasingly important in the fields of radiotherapy and protection of men from radiation hazards. Due to this development, it gains high practical significance what has always been the central objective of radiation biophysics: to understand and to make predictable the varying relative biological effectiveness (RBE) of radiations. At present, in spite of several efforts made, there is no generally accepted theory of RBE.

In our opinion, this shortcoming in regard of RBE and the above-mentioned gap are much the same: to relate the biological effect to the primary physical events requires model building for which the crucial test is just the ability of a model to yield the right RBE for every radiation. By this powerful criterion, inadequate models or ideas about the mechanism of radiation action can largely be ruled out. This was convincingly demonstrated in the past when simple target models were evaluated consistently, in particular by D. E. Lea in his noted work "Actions of

Radiations on Living Cells". The theoretical approaches contained in LEA's book may be looked upon as the prototype of what will be found in this monograph.

Our primary objective was merely to give a systematic and comprehensive account of our own theoretical work about radiation action at the molecular and cellular levels, ending up in a definite predictive theory of RBE for mammalian cell killing and related effects. Thus, this monograph bears a great resemblance to an outsized original paper, as most of the results presented have not been published before. Nevertheless, the book has become broad in scope, mainly for two reasons. Firstly, the physical and mathematical fundamentals of the theory constitute a profound study in radiation stochastics including new concepts and methods. This part of the monograph could be entitled 'Theoretical microdosimetry'. Secondly, to demonstrate the capability of the theory, it was necessary and possible to discuss a good deal of modern radiation biology, as far as radiation quality is involved. This ranges from DNA breaks to gross tissue and tumour reactions. Numerous data are compendiously reviewed and systematically interpreted in terms of theory. Have a look at the section titles of chapter 7 to realize the topics covered.

With a view of enabling research workers of various engagement to make use of this book, we have aimed for making readable each chapter separately, at the expense of some redundancy. For example, the radiobiologist who wants to be informed only on the achievements of the theory in checking with experimental facts can start with chapter 7, eventually omitting subsection 7.2.1. Furthermore, the final procedures of application (mainly survival curve prediction) are simple and can be practised without having studied the theoretical background. For this particular purpose, chapter 9 has been written as a compendium of the required formulas and instructions with reference to tabulated parameters (appendix). On the other hand, potential readers with a general interest just in theory should start with chapter 1 where, first, existing approaches, strictly speaking, their rationale only, are analysed in a critical way and, second, our general model is introduced with regard to its basic principles and expected scope. To body forth, as a whole, the theory based on the latter model, chapter 2 has been included, addressing to a fairly wide circulation. It develops the subject systematically in a simplified version as to the underlying physics, thus keeping it within bounds and presenting, nevertheless, all the essential aspects, theorems, limitations, and perspectives of this theory. From here, one may take a leap direct to chapter 6 or chapter 7. In this way, people with only a limited training in mathematical physics and microdosimetry can easily make themselves familiar with the theory as regards both the essential principles and the applications.

The chapters not mentioned above deserve special comments. They would be superfluous in principle unless the strict evaluation of our model required a description of radiation stochastics in more detail than offered by customary microdosimetry. So it was necessary to start with first principles and to give a comprehensively systematic mathematical treatment of radiation fields and absorption events in both micrometre and nanometre regions, introducing new concepts, e.g., to take account of sites with inherent structure in molecular dimensions. The general and profound character of the approach, in spite of its special purpose in the framework of the considered theory of radiation action, will be appreciated, as we hope, by many workers engaged in radiation physics and theoretical fundamentals of microdosimetry. It is important to note that the formalism, which rests upon analytical procedures instead of Monte Carlo techniques, involves the calculation of the usual microdosimetric parameters, too. In this respect, a great number of numerical results are compiled and compared with the large body of corresponding experimental data in chapter 5. The intention is twofold: testing our approach and giving an abundant summary of calculated data for photons, fast neutrons, and monoenergetic ions (see appendix). Results for nanometre sites along with modern particle track simulation data are a topic of chapter 6.

It is a peculiarity of this monograph that it covers topics which may appear largely unrelated, at first glance. However, as the principal goal is developing and testing a systematic theory which is to build a bridge between physics and biology, the discussion of different areas is quite naturally justified and necessary but should not be confused with profoundly reviewing them like autonomous subjects. Hence, we cannot pretend that we have given an exhaustive and well-weighted account of the present state of affairs in the fields of radiation physics, microdosimetry, action models, and radiobiology. As to the experimental work discussed, emphasis has been placed on results and data rather than fundamentals and methods. In quoting literature we have confined ourselves to what seemed suitable and sufficiant for our purpose, and was easily available to us. We hope that this can be tolerated even though the most recent literature is only scarcely covered.

It is another shortcoming that this monograph is not exhaustive with respect to the theory's scope for possible application. In fact, the general idea of taking DNA lesions for the crucial primary radiation damage and the corresponding general formalism are likely to be valid for most of the biological effects observed on various kinds of objects. This may justify the title of the book, but actually the main attention has been focused on the impairment of the proliferative capacity of mammalian cells and tissues in regard of RBE. This topic being of particular importance, no doubt, other subjects are only touched on or totally omitted in order to keep the book within bounds. For example, this concerns an already accomplished special action model for bacteria as well as model building, in the framework of the general theory, with respect to other objects and endpoints such as viruses, phages, mutations, chromosome aberrations, division delay, cancerogenesis, etc. Furthermore, there are also limitations concerning the radiations considered. This is only a question of data calculation; the corresponding tables are restricted to photons, fast neutrons between 0.25 MeV and 20 MeV, and ions up to 20 MeV/nucl.. Extended calculations are under work in order to make practical application of the RBE theory possible for pions, high-energy neutrons and relativistic ions. We think that this will benefit the modern activities in high-LET radiotherapy as well as fundamentals of radiation protection in manned space research.

During the pretty long period of writing this book, SI units were declared to be obligatory. However, a corresponding revision of tables, figures, and the text has been forborne in the hope that this will hardly be realized as a serious defect, particularly as the imperative alterations would almost only refer to the unit of absorbed energy dose. We have used the older unit rad, throughout, whereas the SI unit is Gray = Joule/kg. The difference is by a factor of hundred: 1 Gy = 100 rad.

Finally, we should apologize to the reader for our English. Despite having made every effort to eradicate mistakes and uncustomary phrases, we are painfully aware of our very limited success in hushing up the German writer. We believe the book will nevertheless be readable, at least, and understandable with respect to the language, hoping that the blemishes make the English user smile rather than grieve him.

We are much obliged to our colleagues who have encouraged and supported this work in various ways. It was only in the kindly and stimulating atmosphere of the Department of Radiation Biophysics of the Central Institute of Molecular Biology that we were able to accomplish it. Most grateful we are to Prof. H. Abel, director of the department, who has steadily promoted our work. His valuable discussions, helpful criticism and stimulating encouragement merit our inmost gratitude. We are highly indebted to W. Leistner for his help in carrying out a great deal of calculations and data selection underlying the figures and tables in chapters 5 and 7. Our obligations to Mrs. Prof. L. Herforth are due to here patiently and thoroughly reading the whole manuscript and giving useful hints. The typing has exclusively been the work of Mrs. E. Homeyer, who has always taken so much care that she has deserved well of this book. Last not least, we thank the Akademie-Verlag, Berlin, for the beneficial co-operation.

K. Günther, W. Schulz

Contents

Chapter 1

Review of theoretical conceptions

1.1. Introduction

The injurious effect of ionizing radiations on living matter has been known since the beginning of this century, and attempts to understand these actions have constituted a good deal of early biophysics. Any theoretical considerations and model building require the phenomena to be described in *quantitative* terms with respect to both the acting physical agent (radiation) and the biological effect.

For the former the *radiation dose D* now defined as the energy absorbed per unit mass[1]) has been chosen. In principle, this is an arbitrary choice,[2]) but energy dose is the only quantity that (a) can be readily defined (and measured) for every ionizing radiation and (b) is the main determinant of the biological effect.

Different quantitative measures of *biological effects* are in use,[3]) but the most important one in regard to theoretical considerations is the rate of reacting (or non-reacting) biological units such as viruses, bacteria, and single eukaryotic cells. The proportion of units which do *not* show a certain well-defined reaction is generally denoted by the symbol S indicating 'survival' since the effect considered in most cases is loss of the capacity for unlimited proliferation.

Based on these definitions, an experimental dose-effect relation $S(D)$ is an input-output relation for the 'black box': biological object. Any theory aiming at a satisfactory understanding and prediction of $S(D)$ has to be based on both a proper description of the primary physical events (the input) and hypotheses regarding the chain of biological events that follow (the interior of the black box). It is the latter that raises principal problems since the

---

[1]) Rigorously, the dose at a given point has to be defined as the *expected value* of specific energy in a site round the point as the site diameter tends to zero (see section 1.4.1).

[2]) An imaginable alternative would be, e.g., the fluence of charged particles.

[3]) There are principally three types:
(i) quantifiable *degree of damage* to a single object (e.g., seedling root growth inhibition, cell division delay, scaled skin reaction, biochemical changes);
(ii) *mean number* of countable lesions (e.g., DNA breaks, chromosome aberrations);
(iii) *probability* of an object to react in a specified way (scored as a reaction frequency on a large population of objects).

hypotheses expressed as models cannot be verified by means of $S(D)$ alone. This signifies a lack of information: different models can equally well account for a measured dose-effect relation. Therefore, with very few exceptions, numerous efforts of this kind, starting with classical target theory, have finally yielded nothing but mathematical descriptions, rather than genuine explanations and predictions of $S(D)$, in terms of largely abstract conceptions and parameters such as number of hits, number of targets, repair rates, biological states with transition coefficients, etc.

A radical alternative consists in renouncing this kind of theorizing about dose-effect curves themselves, envisaging radiobiophysical theories that refer only to the question of how the relation $S(D)$ *changes* with changing sort, composition, and energy of radiation. In other words, the subject of such kind of theory is the *relative biological effectiveness*, RBE,[1]) of different ionizing radiations. The advantage is that the 'black box' may almost remain black while the variation of input (radiation quality) is a matter of physics only. In reality the things are not so simple, but conceptually it is very important to distinguish between the two aspects of theoretical approaches: bare dose-effect curves as such on the one hand, and RBE on the other hand. This has often not been done clearly enough, and a number of models suggested contain more assumptions restrictive to $S(D)$ than necessary if the question of RBE only is treated.

This first chapter, corresponding to the topic of the entire monograph, deals with approaches (or aspects of approaches) concerning the RBE problem. Even with this restriction to observe, completeness cannot be contemplated and would not be suitable in view of the numerous attempts published over a period of several decades. Instead, we will follow only the principal lines of development of this field of research for which also the common term *microdosimetry* has come in use, in a general sense. However, throughout the present monograph this vague denotation is not adopted, because a branche of science should be defined by fundamental principles, not by envisaged aims with the principles left open. Accordingly, in this book 'microdosimetry' signifies a *special* way of physical description of the random spatial patterns of energy deposition in matter without the implication of a definite RBE theory. In this restricted sense the concept of microdosimetry is reviewed in section 1.4.1, and it underlies a great part of what follows.

In discussing theoretical conceptions, it is presumed that the reader is familiar with the essential experimental facts, since comprehensively reporting the latter would have made this chapter too lengthy, inasmuch as a good deal of the material can be found in chapter 7 (for a survey see ALPER, 1979).

As $S(D)$ is a *probability*, namely the probability for each single object of a homogeneous population to survive a *given* dose of a *given* radiation (i.e., on well-defined conditions of experiment), the underlying processes of action must be of stochastic nature. It is a matter of fact that *two* kinds of stochastics are involved: *physical stochastics and biological stochastics* (HUG and KELLERER, 1966). To avoid any models or approximation concepts in advance, the following *general* characterizations may be given.

Physical stochastics means that despite specified dose and radiation a large variety of spatial patterns of radiation

---

[1]) The relative biological effectiveness is defined as the ratio of doses, $RBE = D_X/D$, that produce equal effects under equal conditions, where D refers to the radiation in question and D_X is the reference-radiation dose. The common usage is taking X-rays (mostly 250 kVp) for the latter, but no standard choice has been adopted in practice. Therefore, the reference radiation will be always indicated in cases where confusion may be possible.

energy deposition is engendered by quantum-mechanical chance. Varying elements of these patterns are: tracks of primary and secondary charged particles, along-track excitations and ionizations, and clusters of the latter. Physical stochastics obeys known physical laws, but a *complete* description in terms of random variables (position and intensity of all energy transfer points) is practically impossible (GÜNTHER, 1968b).

To explain what biological stochastics means, suppose that *identical* patterns of energy deposition have occurred in, from outward appearances, identical biological objects. In general, only part of the objects will show the reaction under consideration as a matter of chance. Molecular and cell biology can tell us a lot of reasons for this béhaviour, but on the other hand the assumption of 'identity' of biological objects is rather fictive. Never will two objects be identical in all respects, so that no clear distinction can be made between sheer biological stochastics and influence of inhomogeneity of a given population (biological variability). Nevertheless, a conceptual distinction should be made, and, in a more pragmatic way, let us speak of nothing but biological stochastics unless some kind of biological variability can be clearly specified (e.g., asynchrony of a cell population).

To separate physical stochastics from biological stochastics is not trivial. The prerequisite for this to be meaningful is that the physical stochastics does not essentially depend on the particular structure of the biological object. To give a counter-example, the distribution of energy deposition by U.V. radiation is governed by the sort and structure of the absorbing molecules (their electronic energy levels). By contrast, ionizing radiations consist of charged particles energetic enough for the binding energies of electrons in biological molecules to be of minor importance. As it were, the fast charged particles almost 'feel' nothing but a homogeneous 'sea' of free electrons with only the number per unit volume (electron density) being relevant. Therefore, physical stochastics can be treated as independent, approximately. On the other hand, it is only the physical stochastics that changes when varying radiation quality. Just for these reasons, theoretical approaches to RBE can in some way be separated from complete theories of dose-effect relations.

In the sequel we will repeatedly refer to the conception of physical and biological stochastics, since this will prove to be a good guide line for a unifying discussion of different theories or models, in each case asking about the special way in which both kinds of stochastics are taken into account.

1.2. Classical target theory

To begin with a restricting comment, the following critical discussion will only deal with the rationale of target theory in a rather abstract manner omitting details, variants, and interpretations as well as the historical development and the numerous applications and successes of target theory; inasmuch as there exist a number of excellent monographs. DESSAUER's book "Quantenbiologie" re-edited and supplemented by SOMMERMEYER (1964) is a comprehensive survey of the theory's development including the entire range of applications. The work of TIMOFÉEFF-RESSOVSKY and ZIMMER (1947) places emphasis on the analysis of genetic effects. The monograph by LEA (1946) is of particular value for the present purposes because of the efforts made to use a correct description of physical stochastics in the framework of simple models of action, which is most closely related to modern approaches. An instructive critical analysis of the fundamentals of target theory was given by HUG and KELLERER (1966).

1.2.1. General considerations

The leading idea of target theory was to directly connect two facts: the *shape* of dose-effect curves $S(D)$ and the *discontinuous* quantum-like absorption of radiation energy by matter. $S(D)$ is found to be always a gradually decreasing function, often of purely exponential character: $S(D) = \exp(-\alpha D)$ (inactivation of enzyme molecules, viruses, bacteria). Also if it has a shoulder, there does not exist a threshold dose below which $S(D) = 1$ would hold. This behaviour was attributed to the random occurrence of so-called *hits* (with a mean number proportional to dose) in one or several targets. With nothing more said this conception (hit principle) may be general enough; the crucial point, however, is the introduction of two specifying assumptions: (a) the number of hits in a target, and the number of targets damaged by hits, is assumed to determine *unambiguously* whether the object does or does not react (neglect of biological stochastics), and (b) hits are assumed, beforehand, to be *statistically independent* (to obey Poisson statistics) in any case, prior to deciding what kind of physical events are hits.

Numerous variants of such deterministic models can be, and have been, considered. The most common ones are the *single-target-multi-hit model*

$$S(D) = \sum_{k=0}^{m-1} \frac{(\alpha D)^k}{k!} e^{-\alpha D} \tag{1.1}$$

which assumes that m hits are necessary and sufficient as a trigger for the effect, and the *multi-target-single-hit model*

$$S(D) = 1 - (1 - e^{-\alpha D})^n \tag{1.2}$$

which assumes that the object possesses n targets each having to receive at least one hit for the effect to be produced. These models were primarily introduced with a view of explaining the *shape* of dose-response curves by the random nature of radiation absorption (physical stochastics). However, corresponding analyses of experimental dose-effect relations in terms of the parameters m or n and the hit probability α are quite formal in that they can never prove the validity of the particular model as to reflect anything real. Either the experimental data are anyway incompatible with the simple models (1.1) and (1.2), or the limited precision of date does not allow to make a distinction from other variants of models. In the former case, it has often been tried to attribute the discrepancies to biological variability, and to use, consequently, superpositions of the simple expressions with different parameters n, m, and α, with the drawback of a larger number of parameters and a complete loss of stringency (for examples see Fowler, 1966).

These difficulties could be overcome altogether if one would allow for biological stochastics (and biological variability) in the most general manner by changing Eqs. (1.1) and (1.2) into

$$S(D) = \sum_{k=0}^{\infty} P(k) \frac{(\alpha D)^k}{k!} e^{-\alpha D} \tag{1.3}$$

and

$$S(D) = \sum_{k=0}^{n} P(k) \binom{n}{k} e^{-(n-k)\alpha D}(1 - e^{-\alpha D})^k, \tag{1.4}$$

thereby giving up any theoretical assertion about $S(D)$ for a single given radiation: with $P(k)$ left arbitrary, the equivalent equations (1.3) and (1.4) (as $n \to \infty$) constitute nothing but a formal transformation of information — from $S(D)$ into $P(k)$. The only meaning of this generalization, considering $P(k)$ as a radiation-independent characteristic of the particular object and response, is that it could eventually be used as a starting-point of a theory of RBE, the main topic of the present monograph. This requires to specify 'targets' and 'hits'. Prior to that, the generalization (1.3), (1.4) is mentioned here (though it has never been used in the framework of target theory) in order to emphasize that the special models (1.1) and (1.2) are unessential to the RBE problem.

It is true that target theory does not leave it at forma lanalyses of dose-effect curves. The nature of targets and hits has for a long time been a subject of permanent discussions striving for a theoretical understanding of fitted values of the 'hit probability' α and its variation with radiation quality. However, satisfactory results warranting to speak of a genuine theory of RBE are obtained only for very small objects or reacting units (enzyme molecules, viruses, phages, gene mutations, chromosome breaks) which exhibit a pure exponential dose-effect curve for *every* radiation ($m = 1$ or $n = 1$ in Eq. (1.1) or Eq. (1.2)). Before entering into particulars, let us give a quite formal argument for this insufficiency.

It is a well-known empirical fact that sigmoidal dose-effect curves ($m > 1$ or $n > 1$) *change their shape* in dependence on radiation quality, becoming more nearly exponential for densely ionizing radiations (e.g., α-particles, fast neutrons). This cannot be accounted for by classical target theory unless m or n, in general $P(k)$, is allowed to be not the same for every radiation. As a logical consequence, it is impossible that 'hits' are identical physical events irrespective of radiation. *This circumstance hinders the development of a systematic theory of RBE*. This principal weakness of target theory is due to the unmotivated restriction to Poisson statistics; it is therefore that the equations contain no more than one parameter naturally connected with physical stochastics: α which is only a dose-modifying factor.

1.2.2. Single-hit reactions

In case the survival curve is exponential for photon radiations, there is no radiation-dependent change in shape, indeed, and therefore the above arguments cease to apply. One has to take $m = 1$ or $n = 1$, i.e.

$$S(D) = e^{-\alpha D}, \tag{1.5}$$

so that the so-called hit probability α is the only subject of theory.[1]) The simplest approach is assuming a spherical target of volume V and a single ionization in it to be sufficient for the object to react.

Saying that an ionization in the target is a hit would not be correct, however, because hits have to be statistically independent events, by definition. Actually, ionizations are clustered (2 to 3 ionizations per cluster, on the average), and these clusters, moreover, are localized along the charged particles's paths in mean consecutive distances which may well be smaller than the dimensions of the target. If these facts could be ignored (ionizations distributed at random), α would be equal to V times the mean number of ionizations per unit volume and unit dose:

$$\alpha = \frac{\varrho V}{W}, \tag{1.6}$$

where ϱ is the mass density and W the mean energy absorbed per ionization. As the latter (30–35 eV) does not appreciably depend on radiation, Eq. (1.6) would not account for the observed dependence of the logarithmic slope of the survival curves (1.5) on radiation quality.

The *correct definition of hits* in the framework of the single-ionization model must read: each passage of a particle which ionizes inside the target is a hit. This implies to evaluate the actual physical stochastics regarding the mean number per unit dose of so-defined (statistically independent) hits, which was done by Lea (1946) in good approximation by means of his 'associated volume method'. Round each single ionization as centre a sphere of volume V is imagined, considering the tracks of all charged particles forming the particular radiation field and taking account of ionization clusters and δ-tracks. These spheres which, in general, overlap to a greater or lesser degree will occupy a net volume which may be described as the associated volume. If V_{ass}

[1]) In other words, it is only in this case that a unique value of RBE independent of the effect level (or dose) considered can be assigned to each radiation.

signifies the associated volume per ionization, it is by the factor V_{ass}/V that the mean number of hits is smaller than the mean number of ionizations in the target:

$$\alpha = \frac{\varrho V_{ass}}{W}. \tag{1.7}$$

Eq. (1.7), i.e. the single-ionization model evaluated correctly, has been proved to be valid in a number of cases, e.g., inactivation of viruses and phages. Proof means, first, that the model yields a correct theory of RBE as an implication of obtaining *consistent* estimates of the target volume V from α values measured with *different* radiations. Second, the target volume thus obtained should correspond to a biological structure, mostly the whole virus or phage. An example is given in Table 1.1,

Tab. 1.1. Associated-volume-method analysis of the inactivation of dysentery phage S 13

	Ra γ-rays	X-rays (wave-length 1.5 Å)	α-rays (4 MeV)
Measured value of α^{-1} [krad]	580	990	3500
Calculated target diameter [nm]	15.5	15.9	16.3

which has been taken from LEA (1946). Note the good agreement of the target size estimates obtained despite of the α values varying by a factor of six, and the conformability to the actual virus diameter measured to be 16 nm. By contrast, simplified methods like Eq. (1.6) (hit = ionization) and another method based upon 'hit = *each* passage of a charged particle' are inadequate in general; they are only approximately valid for sparsely ionizing radiation (γ-rays) and densely ionizing radiation (α-rays), respectively. However, even in these limiting cases there is no complete coincidence with correct theory (associated volume method) due to the neglect of ionization clusters in the former, and δ-rays in the latter case. This emphasizes the necessity of a sufficiently correct treatment of physical stochastics. Therefore, several rough approaches will not be considered in this section, since this would lead to confusion regarding our main objective: to examine what can be learned from classical target theory in developing a modern theory of RBE.

Further evidence comes from comparing various viruses and phages. The general correlation thus found between inactivation dose ($D_0 = \alpha^{-1}$) and virus diameter (ranging from 20 to 200 nm) is striking. In keeping with the simplified approaches mentioned above, one finds approximately $\alpha \sim V$ for γ-rays and $\alpha \sim V^{2/3}$ (cross-sectional area) for α-rays. However, the quantitative analysis by means of the associated volume method does not always yield consistent results: the measured α values are often *smaller* than expected in theory, particularly for larger viruses, the internal structure of which may, hence, play a role. Thereby, which is a crucial test, the estimates of V become dependent on the type of radiation, which suggests modifications of the single-ionization model. These shall now be discussed in a general manner to elucidate the scope of hit theory in principle.

One can introduce a constant action probability p smaller than unity for each ionization by the definition: each passage of a particle that produces at least one *effective* ionization is a hit. This can easily be taken into account of the associated volume method (the proportion $(1-p)$ of ionizations is considered as absent, randomly). Another possibility is assuming n targets to exist of which *only one* must be hit to bring about inactivation.[1]) In this case, the measured α values have to be divided by n prior to being analysed as before. While from these generalizations

[1]) This should not be confused with the multi-target-one-hit-in-each model (1.2).

a decrease of effectiveness at growing ionization density of radiation (saturation effect) continues turning out, several-ionization models can yield an increase also (concentration effect). Numerous combinations of these models are imaginable (several targets differing in size, shape, and hit definition) and, which is important to emphasize, this general picture must be accomplished by taking account of *correlations* between hits in adjacent targets (e.g., two targets hit simultaneously by one ionizing particle). However, giving up the statistical independence of hits in several targets, one must define the hit as a specified event concerning the *totality* of targets, in general. This leads to the final consequence: return to the picture of one target only which may now have internal structure and should be, therefore, referred to as a *gross sensitive region* in which some complex class of spatial patterns of simultaneously produced ionizations (by one particle or several correlated ones) defines hits.

Of course, these considerations go by far beyond the efforts actually made in this field. Nevertheless, along similar lines a good deal of elementary single-hit reactions, particularly of enzymes and single genes in addition to viruses and phages, have been satisfactorily accounted for. The concentration effect has also been treated by several-ionization models (POLLARD *et al.*, 1955; HOLT, 1974). As a rule, the correctness of the description of physical stochastics (e.g., taking account of δ-rays) has proved crucial for the ability of the models to yield the right dependence of effectiveness on radiation quality. For a survey see HUTCHINSON and POLLARD (1961).

1.2.3. Multi-hit reactions

However complex the model may be, genuine single-hit reactions have in common the mere existence of hits which are both statistically independent and capable of being defined consistently for every radiation by classifying all imaginable absorption events in the gross sensitive region into hits and non-hits. Strictly speaking, this is nothing but an implication of the exponential shape of the dose-effect curves themselves. For any type of *cumulative action* (sigmoidal dose-effect curves), however, either of these properties of hits must be given up. The general arguments enforcing this alternative are put forward above (end of subsection 1.2.1). Accordingly, one can distinguish two attitudes concerning the way in which target theory should be reasonably applied to non-elementary reactions of more complex objects such as bacteria and eukaryotic cells. Some workers took the cumulative action of *statistically independent* hits and, hence, mathematical formalisms like Eqs. (1.1—2) for fundamental, while other scientists emphasized the original standpoint to identify a hit principally with some *elementary damage* to a single macromolecule.

The former way implies defining a hit, in general, as a somehow effective passage of a charged particle through the gross sensitive volume (GLOCKER, 1932), with the implication that more densely ionizing particles produce 'stronger' hits, so that the critical hit number m diminishes with increasing ionization density of the radiation (in the sequence of γ-rays, protons, α-particles, heavy ions). This picture can qualitatively account for the experimental findings, particularly for the well-established exponential survival curves of mammalian cells bombarded with neon or argon ions. In this case, each ion which goes across the cell nucleus is a hit that kills the cell (i.e. $m = 1$):

$$S = e^{-\sigma f}, \tag{1.8}$$

where f is the total fluence of ions per square centimetre and σ the cross-section of cell nucleus (for a modern special investigation see TODD, 1975). However, this approach is only explanatory but not predictive with respect to the RBE of varying radiations, apart from very rough estimations sparsely made (critical hit number as a function of radiation, see SOMMERMEYER, 1952).

The second way is most clearly asserted in the latest book by TIMOFÉEFF-RESSOVSKY *et al.* (1968). The hub of this treatise is the *hit principle* which is emphasized to be not yet a theory, thus rejecting the early formalisms of target theory, particularly in regard of cumulative actions. 'Hit principle' merely means that radiation action does *not* consist in a diffuse gradual injury to the entire biological object, but a usually *small* number of localized all-or-nothing molecular changes (hits), each of which is produced (directly or indirectly) by *one* ionizing particle only, underlie the biological effect. As a consequence, the primary process of hit production is governed by statistical laws, but not necessarily by Poisson statistics (the same particle can produce several hits). This implies that not only the occurrence of a hit in a single molecular target envisaged is a matter of physical stochastics, but also the statistics of hits to the object as a whole must be derived by means of physics. Making allowance, in addition, for biological stochastics (no critical hit number for the final effect to occur with certainty), this conception provides a very general framework of radiation action theories. An attempt of this kind is the topic of the present book. Therefore, in this very generalized sense, it may be considered as a further variant of target theory, if one likes to stress the conceptual equivalence: DNA = target, DNA-lesion = hit.

1.3. The LET concept

In the preceding section we have preliminarily used the term 'ionization density' for a crude characterization of radiation quality. The corresponding precise parameter now currently used is the *linear energy transfer* (LET designated as L) which is defined for a given charged particle of given energy as follows. LET means the quotient $\mathrm{d}E$ by $\mathrm{d}x$, where $\mathrm{d}x$ is the distance traversed by the particle and $\mathrm{d}E$ is the mean loss of energy which is locally absorbed:

$$L = \frac{\mathrm{d}E}{\mathrm{d}x}. \tag{1.9}$$

Some ambiguity exists concerning the restriction: 'locally absorbed'; it is the predominating usage to specify an energy cut-off, rather than a geometrical (radial) cut-off, by omitting collisions with energy transfer in excess of a cut-off energy δ, indicated by the notation L_δ. Accordingly, L_∞ means the unrestricted LET also called *stopping power*. LET or LET spectra (see below) can be theoretically calculated as well as approximately measured. For details the reader is referred to the ICRU Report 16 (1970) where the original literature is also quoted.

For arbitrary radiation fields to be described by LET, it is necessary to introduce LET spectra and associated mean values. A single value of LET (strictly speaking L_∞) is suitable for monoenergetic ion radiations only. Also in this case, δ-electrons must be taken into account as separate particles if L_δ is regarded. In general, the differential track segments of all particles at all energies involved have to be considered, thus yielding the so-called track-length spectrum, $t(L)$, of LET.[1] The dose spectrum $d(L)$, on the other hand, specifies the fraction of dose absorbed at the value L. It holds $d(L) = cLt(L)$, where c is the factor of renormalization. The associated mean values are designated as $\bar{L}_T$ and $\bar{L}_D$:

$$\bar{L}_T = \int_0^\infty \mathrm{d}L\,L\,t(L), \quad \bar{L}_D = \int_0^\infty \mathrm{d}L\,L\,d(L). \tag{1.10}$$

[1]) $t(L)\,\mathrm{d}L$ means the proportion of total track length where particles have an LET value between L and $L + \mathrm{d}L$.

LET is an important physical parameter which is needed in every theoretical description or model of radiation action. Target theory has also made use of it (e.g., DITTRICH, 1961). However, the LET *concept* to be discussed in this section should be understood more specifically as the attempt to somehow express the relative biological effectiveness of radiations in terms of LET *only* (spectrum or mean value), without referring to any model of the mechanism. Indeed, if one had a single parameter or spectrum consistently definable for every radiation, and if this parameter constituted a *complete* description of the physical properties of a radiation field (i.e., if it determined the physical stochastics in all respects), it would be clear from the outset that the biological effectiveness could only depend on that parameter, regardless of the mechanism. Of course, LET provides only an incomplete description in several respects. Therefore, it is the question whether LET is a 'good' parameter which happens to determine the RBE in sufficient approximation.

At first glance, LET *is* a good parameter, which has caused the constant interest in it up to the present. If one plots RBE versus a mean value of LET, usually $\bar{L}_D$, one obtains clear correlations of essentially two types depending on the biological response considered. The function $RBE(L)$ thus suggested is, as a rule, decreasing monotonously in the case of exponential responses (sometimes termed linear responses) such as virus inactivation, gene mutation, etc. For chromosome aberrations and mammalian cell killing, $RBE(L)$ first increases and then decreases, forming a maximum in the region of about 100 to 300 keV/μm.

Various attempts have been made to standardize functions $RBE(L)$ for particular classes of objects and effects (see, e.g., NAKAI and MORTIMER (1967), ANGLETON (1967)), starting mostly from experimental data for mono-LET radiations to set forth $RBE(L_\infty)$, and aiming at a prediction for radiations with broad LET spectra, especially fast neutrons, either by means of a mean value, $\bar{L}_D$ or $\bar{L}_T$, or by the tentative approach

$$RBE = \int_0^\infty \mathrm{d}L \, d(L) \, RBE(L). \tag{1.11}$$

For the purposes of radiation protection, it has been tried to estimate the relation $RBE(L)$ on the basis of measured relationships between RBE and mean values $\bar{L}_D$ for fast neutrons of varying energies. These efforts have lastly led to the definition of the so-called *Quality Factor*, QF, as a function of L_∞ accompanied by the recommendation to estimate the biologically equivalent dose, *DE*, of arbitrary (eventually mixed) radiation fields by $D\overline{QF}$, where D is the physical dose and $\overline{QF}$ has to be calculated, in analogy to Eq. (1.11), as the LET-integral over $QF(L)$. For a detailed discussion see ICRP (1966) and REGEL (1969).

Obviously, the LET concept cannot be regarded as a theory of RBE because it does not allow to derive RBE-LET relationships otherwise than from the experiment itself. Moreover, the special approach (1.11) as a procedure of merely generalizing experimental data lacks theoretical foundation. The crucial point is that Eq. (1.11) implies *additivity* of the biological effects of the single LET components, which is unwarranted save for true single-hit reactions under certain conditions.[1] The characteristic *cumulative* actions on eukaryotic cells would require to allow for

---

[1] The components of radiation must indeed act independently in an additive manner requiring absence of sublethal damage: a track segment in the sensitive region either causes the reaction or has no effect at all. Moreover, the object must be small enough so that the track segments in question can be considered as infinitely small, since the LET spectrum describes *differential* track segments. However, this does not remedy the fundamental limitations of the LET concept discussed below.

changing functions $RBE(L)$ in dependence on the considered level of effect, as a consequence of the LET-dependence of survival curve shape. Obviously this lacks both theoretical justification and practical usefulness (too many empirical relations),[1]) and such attempts were indeed unsuccessful. As a particularly instructive example we quote a paper by BEWLEY (1968).

BEWLEY (1968) started with a family of T1-cell survival curves measured over a wide range of single L_∞ values (track-segment irradiation with monoenergetic ions (BARENDSEN *et al.*, 1963; BARENDSEN, 1964; BARENDSEN *et al.*, 1966; TODD, 1966), see also chapter 7, Fig. 7.2 and Fig. 7.3). Based on calculated LET spectra of a number of different-energy neutron radiations, he constructed theoretical survival curves for neutrons, not only by means of Eq. (1.11), but also by a special procedure of composing the single-LET survival curves direct, using an appropriate subdivision of the neutron LET spectra. The ambiguity thereby involved is unessential[2]); the procedure can be regarded as the most general way to test the LET concept as such, avoiding any additional assumptions. BEWLEY's results (including variants, e.g., δ-ray cut-off: L_δ instead of L_∞) clearly prove the insufficiency of the LET concept: the obtained survival curves greatly deviate from the measured ones for fast neutrons with the same cells (BROERSE and BARENDSEN, 1966; BROERSE *et al.*, 1968).

According to BEWLEY's analysis of his results, the most obvious shortcomings of the LET concept in general are two facts. (i) Different particles at equal LET produce different δ-ray spectra, i.e. the microstructure of tracks cannot be specified by LET only. (ii) The LET spectrum, as it were, only describes what would be seen after 'hashing' the tracks into infinitely small segments, i.e. the diverse ranges of particles remain unspecified.

In summary, the complex physical stochastics does not, on principle, follow unequivocally from the reduced information represented by the LET spectrum, and this loss of information, unfortunately, proves to be not unimportant for the biological effectiveness. Thus, the LET concept, despite of its great value for a rough scaling of the biological quality of radiations and the importance for the requirements of radiation protection (evaluation of radiation hazards), is unsuitable to provide a general basis or heuristic guide line of RBE theories.

1.4. Microdosimetry

Microdosimetry has now grown into a well-established field of modern biophysical research aimed at an exact description of physical stochastics. In contrast to LET spectra, microdosimetric distributions *directly* display an *important aspect* of physical stochastics, namely the fluctuations of energy imparted to a single small volume of matter. From the formal point of view, these distributions for a set of diameters of the volume (mostly a sphere) constitute a similar *amount* of information as LET spectra for a set of cut-off energies, but it is a different *kind* of information, which is certainly more relevant to biological actions, though it is, of course, no exhaustive information on the peculiarities of physical stochastics associated with a given radiation.

The concept of microdosimetry was first introduced and to a great extent developed by ROSSI and his co-workers (ROSSI, 1960; BIAVATI *et al.*, 1963; ROSSI, 1968). The original idea was to assign direct significance to the pulse-height spectra obtained with gas-filled proportional counters (low pressure to simulate small regions of tissue), which had previously been used to derive approximate LET spectra.

---

[1]) These objections do not fully apply to the Quality Factor QF which is no RBE in the usual sense. QF is to estimate radiation hazards of men at *very small* doses, so that the crucial additivity at the cellular level is certainly met.

[2]) The joint action of radiation components in the same cell cannot be estimated exactly. $S(D)$ was obtained by assuming a consecutive reduction of survival averaged over all permutations in the sequence of components.

Along with the experimental techniques (e.g., wall-less counters), theoretical aspects have been developed regarding both precise definitions and methods of calculating microdosimetric parameters. Fundamental contributions in these respects have been made by Kellerer. In the framework of the present book, it is not intended to report the subject comprehensively. The following subsection, for the reader's convenience only, briefly summarizes the theoretical fundamentals of microdosimetry as far as needed for what follows, in a rather elementary manner. Further details as well as exact definitions and profound inquiries can be found in the pertinent literature (ICRU Report 19, 1971; Kellerer and Rossi, 1970; Kellerer, 1970; Kellerer, 1968a; Kellerer and Chmelevsky, 1975). Besides, chapter 5 reports a large number of experimental and theoretical results along with citations of a good deal of original papers.

1.4.1. Survey of microdosimetric quantities and relations

Consider a small volume V of spherical or any other shape in a uniformly irradiated medium. The energy ε deposited in that volume by irradiation with the dose D of an arbitrary radiation is a *random quantity*, i.e. it will fluctuate despite of fixed dose. Usually ε is replaced by z, the *specific energy*, which is defined as the quotient of ε by the mass of the volume, so that its expected value, $\bar{z}$, belonging to the normalized differential distribution $f(z; D)$ introduced to describe the fluctuations, is just the dose itself, by definition:

$$\bar{z} = \int_0^\infty \mathrm{d}z\, z\, f(z; D) = D. \tag{1.12}$$

The random value of z reached at a specified dose is always the sum of a varying number of varying increments during irradiation, each of which is due to an elementary process of energy deposition by a single incident particle and/or its secondaries. According to this definition, one speaks of *absorption events* or *energy deposition events*, which are statistically independent by definition.

The increment of specific energy imparted in *one* randomly chosen absorption event also fluctuates, which is described by the differential *event size distribution* $f_1(z)$ referring to a single absorption event. Its mean values,

$$\bar{z}_F = \int_0^\infty \mathrm{d}z\, z\, f_1(z), \quad \bar{z}_D = \bar{z}_F^{-1} \int_0^\infty \mathrm{d}z\, z^2\, f_1(z), \tag{1.13}$$

are of great importance. $\bar{z}_F$, the *frequency-mean*, determines the mean number, $D/\bar{z}_F$, of absorption events at dose D. The *energy-mean* $\bar{z}_D$ (also called 'dose-mean') is related to the variance of the distribution $f_1(z)$. It is $f_1(z)$ that is actually measured (pulse-height spectra from proportional counters), while the dose-dependent distribution $f(z; D)$ is the result of a *compound Poisson process* described as follows.

Due to the statistical independence of absorption events, their number n obeys a Poisson distribution with the mean value $D/\bar{z}_F$, and the superposition of increments from n events is described by the n-fold convolution of $f_1(z)$:

$$f(z; D) = \sum_{n=0}^{\infty} \frac{(D/\bar{z}_F)^n}{n!}\, \mathrm{e}^{-D/\bar{z}_F} f_n(z) \tag{1.14}$$

with $f_n(z)$ (distribution after exactly n events) defined by the self-explanatory recurrence formula (convolution integral)

$$f_n(z) = \int_0^z \mathrm{d}z' f_{n-1}(z')\, f_1(z - z'). \tag{1.15}$$

A practicable computational procedure was given by Kellerer (1968a).

The *moments* of the distribution $f(z; D)$ can be expressed by the moments of $f_1(z)$. The first moment, $\bar{z}$, is trivial; it must be equal to D according to Eq. (1.12). Let us outline the derivation of the second moment, $\overline{z^2}$, only, since the way is similar for all moments. First, by virtue of Eq. (1.14),

$$\overline{z^2} = \int_0^\infty \mathrm{d}z\, z^2\, f(z; D) \tag{1.16}$$

is reduced to the Poisson-distribution sum of

$$\begin{aligned}[\overline{z^2}]_n &= \int_0^\infty \mathrm{d}z\, z^2\, f_n(z) \\ &= \int_0^\infty \mathrm{d}z' f_{n-1}(z') \int_{z'}^\infty \mathrm{d}z z^2 f_1(z - z') \\ &= \int_0^\infty \mathrm{d}z' f_{n-1}(z') \int_{z'}^\infty \mathrm{d}z[(z - z')^2 \\ &\quad + 2(z - z')\, z' + z'^2]\, f_1(z - z') \\ &= \bar{z}_F \bar{z}_D + 2[\bar{z}]_{n-1} \bar{z}_F + [\overline{z^2}]_{n-1} \\ &= \bar{z}_F \bar{z}_D + 2(n-1)\, \bar{z}_F^2 + [\overline{z^2}]_{n-1}.\end{aligned}$$

This is again a recurrence formula, but it can be solved in closed form: $[\overline{z^2}]_n = n\bar{z}_F\bar{z}_D + n(n-1)\,\bar{z}_F^2$. Now the summation yields

$$\overline{z^2} = \bar{z}_D D + D^2. \tag{1.17}$$

This equation is of great importance because the linear term, $\bar{z}_D D$, is a direct measure of the fluctuations of specific energy.

A further point of interest concerns *mixed radiation fields* (e.g., fast neutrons contaminated with γ-rays). How are the total microdosimetric quantities to be expressed by those of the individual components? If $\mu_F^{(a)}$ is the probability that a randomly chosen absorption event is due to the component a, and if $f_1^{(a)}(z)$ denotes the event size spectrum pertaining to that component, the total spectrum and its frequency-mean (Eq. (1.13)) read, evidently:

$$f_1(z) = \sum_a \mu_F^{(a)} f_1^{(a)}(z), \quad \bar{z}_F = \sum_a \mu_F^{(a)} z_F^{(a)}. \tag{1.18}$$

Instead of $\mu_F^{(a)}$, one may introduce the more familiar *dose fraction*, $\mu_D^{(a)}$, of the a-th component by $\mu_F^{(a)} = \mu_D^{(a)} \bar{z}_F / \bar{z}_F^{(a)}$. This definition is easily seen to be consistent. Eq. (1.18) ensures the normalization $\sum \mu_D^{(a)} = 1$. From $\sum \mu_F^{(a)} = 1$, $\bar{z}_F$ turns out to be given by

$$\frac{1}{\bar{z}_F} = \sum_a \frac{\mu_D^{(a)}}{\bar{z}_F^{(a)}}, \tag{1.19}$$

while the total dose-mean derived from the second equation (1.13) by substituting the expression (1.18) for $f_1(z)$ reads

$$\bar{z}_D = \sum_a \mu_D^{(a)} \bar{z}_D^{(a)}. \tag{1.20}$$

Assuming the single event size spectrum $f_1(z)$ to be given by experiment, the above formalism is axiomatically complete. It is another question to discuss and describe the various stochastic factors that determine $f_1(z)$. With respect to simplified situations this can be done in terms of three types of distribution: chord length, LET, and straggling. In general, however, the totality of possible absorption events must be modelled by means of Monte Carlo techniques.

In the simplified case, the *spectrum of chord lengths*, $w(l')$, constitutes the necessary and sufficient geometrical description of the sensitive region considered. The expected value of l' is the *mean chord length*, $l = \overline{l'}$, which is equal to two thirds of the diameter d for a spherical region. For details see chapter 3, Eq. (3.25) and context. l has been used to define the so-called *lineal energy*, y, by

$$y = \frac{\varepsilon}{l} \tag{1.21}$$

as an alternative random variable of imparted energy. The relation between y and z is obvious (see chapter 5, Eq. (5.11)). The variable y is commonly used to describe single events: $f_1(y)$ rather than $f_1(z)$.

If particles of a given LET value L only traverse the reference volume along straight lines, and if we disregard the discontinuous random nature of energy transfer (straggling), $f_1(y)$ reflects nothing but the chord length spectrum $w(l')$ expressed as a function of $y = l'L/l$. Then it holds $\bar{y}_F = L$ and $\bar{y}_D = (\overline{l'^2}/\bar{l'}^2)\,L$ (cf. Eqs. (1.13)). If one has an LET spectrum, it can obviously be allowed for by averaging over $t(L)$, and the mean values read $\bar{y}_F = \bar{L}_T$ and $\bar{y}_D = (\overline{l'^2}/\bar{l'}^2)\,\bar{L}_D$ (cf. Eqs. (1.10) and context). These relations exhibit that lineal energy and LET are closely related quantities, but the full reduction of y-spectra to L-spectra and vice versa depends on neglecting a number of factors the consideration of which just constitutes the essence of microdosimetry. The most important factor is the so-called *energy straggling*.

Straggling results from the *discontinuous* energy transfer along charged particle tracks in form of separate collisions. The energy transfer E in a single collision is a random quantity described by the normalized *collision spectrum* $c(E)$ with the mean values

$$\delta_1 = \int_0^\infty dE\,E c(E),$$
$$\delta_2 = \frac{1}{\delta_1}\int_0^\infty dE\,E^2 c(E). \qquad (1.22)$$

In the region of large energy transfers E (production of δ-rays), it holds $c(E) \sim E^{-2}$ (Rutherford's scattering law), i.e. the spectrum has a long tail which gives rise to $\delta_2 \gg \delta_1$ (for details see KELLERER (1968a)).

Now the *straggling distribution* is introduced to be the spectrum of the sum of energy transfers within a small track segment[1]) of length l. It can be referred to as the event size distribution $f_1(y)$ in an idealized situation where particles of constant L_∞ perpendicularly traverse a layer of thickness l regarded as the sensitive volume, and in case that escape of δ-rays can be neglected.[2]) Upon these conditions, $f_1(y)$ results from a compound Poisson process which is completely analogous, mathematically, to the Poisson process underlying the distribution $f(z; D)$ (see above). The spectrum of the process is $l\,c(ly)$ (the analogon of $f_1(z)$ in Eqs. (1.14), (1.15)) and the mean number of collisions is lL_∞/δ_1 (the analogon of $D/\bar{z}_F$ in Eq. (1.14)). Consequently, the moments of the straggling distribution $f_1(y)$ are given by the analoga of expressions (1.12) and (1.17). Expressed as frequency-mean and dose-mean of lineal energy according to Eqs. (1.13), they turn out to be

$$\bar{y}_F = L_\infty, \quad \bar{y}_D = L_\infty + \delta_2/l. \qquad (1.23)$$

These equations demonstrate the influence of straggling. The broadening of the δ-function $f_1(y) = \delta(y - L_\infty)$ that would hold if there were no straggling is characterized by the *straggling parameter* δ_2 of the given radiation. Its influence on the fluctuations of energy imparted to small volumes in single events may exceed the influence of other random factors (chord length, LET) considered above.

1.4.2. The problem of constructing theory

A most direct implication of microdosimetry is the fact that a single irradiated cellular object must 'feel' the specific energy, z, imparted to it by chance, rather than the expected value D. Hence S, e.g. survival, has to be understood as the expected value of some function $S_0(z)$, the probability of survival for a given specific energy z:

$$S(D) = \int_0^\infty dz\,S_0(z)\,f(z; D). \qquad (1.24)$$

[1]) For what follows, the segment must be short enough so that the mean energy loss of the charged particle is small as compared with its kinetic energy. This ensures both statistical independence of collisions and constancy of the collision spectrum $c(E)$.

[2]) Energy deposition straggling identical with energy loss straggling.

This may be called the 'energy-only concept'. The region of reference, the *gross sensitive volume* (GSV), must be large enough to ensure that energy deposition outside it does not contribute to the effect (Rossi, 1964). Eq. (1.24) displays the superposition of physical stochastics described by the z-distribution, and biological stochastics described by the probability $S_0(z)$. For a hypothetical 'zero-LET radiation' characterized by $f(z; D) = \delta(z - D)$ one obtains $S(D) = S_0(D)$. Hence $S_0(D)$ may be regarded as the zero-LET dose-effect relation.

The relation (1.24) might already constitute a complete theory of RBE, assuming $S_0(z)$ to be independent of radiation. This would mean that the spatial *average* of energy density in the GSV is relevant enough to the final reaction. This is the only prerequisite, since the formalism itself is generally applicable: obviously, one can in each case find an $S_0(z)$ such that a given relation $S(D)$ for a particular radiation (e.g. X-rays) is exactly reproduced, and then $S(D)$ can be predicted for any other radiation, with the size of the GSV considered as an additional adjustable parameter. Ultimately, experience is the criterion of this concept.

Eq. (1.24) or what is equivalent to it (at least approximately, often without explicit reference to microdosimetry) has been discussed and evaluated by a number of authors (Rossi *et al.*, 1960, 1961; Dittrich, 1961; Biermann, 1963; Hug and Kellerer, 1966; Günther and Rossbach, 1966; Günther, 1968a, b; Kapultsevich *et al.*, 1971; Rossi, 1972; Chmelevsky *et al.*, 1972; Videnský *et al.*, 1973; Gubin *et al.*, 1974). These investigations show that in the crucial case of eukaryotic cell inactivation (sigmoidal survival curves) a satisfactory theory of RBE cannot be obtained in this way.[1) What the formalism describes correctly is the change of survival curve *shape* with growing LET (sigmoidal curves shading into exponential ones), but the slope at large doses turns out to decrease monotonously with growing LET. In other words, Eq. (1.24) is unable to yield RBE values greater than unity as $S \to 0$. A general mathematical proof is given in chapter 2 (section 2.3.3).

Conversely, if one confines to the initial slope, $S'(0)$, of survival curves (insert the expression (1.14) for this purpose),

$$-S'(0) = \frac{1}{\bar{z}_F} \int_0^\infty dz[1 - S_0(z)]\, f_1(z), \qquad (1.25)$$

there is no obvious dicrepancy since, by *free* choice of $S_0(z)$, giving up its above meaning, one can obtain various forms of dependence on radiation quality including an increasing RBE (up to a maximum) at increasing LET. This approach was adopted by several authors (e.g., Barendsen, 1966; Gubin *et al.*, 1977; Pohlit *et al.*, 1979), but it should be emphasized to be inconsistent from the theoretical point of view unless one deals with true single-hit reactions. In the latter case, realizing that Eq. (1.25) describes the effect of single absorption events, one can interpret $[1 - S_0(z)]$ as the probability of the energy z to produce a hit in an unstructured volume, and thus microdosimetry provides a physically correct basis to evaluate the hit probability α in dependence on radiation quality (see Eq. (1.5) and the discussion at page 6).

The rationale of Eq. (1.24) includes that the total specific energy z is produced simultaneously, i.e. the dose rate must be infinitely large. Hence, the difficulties might eventually be eliminated by considering the temporal sequence of events which produce the increments of z (Kellerer and Hug, 1963; Günther and Rossbach, 1966). Generally speaking, at high LET the same specific energy tends to consist of few large increments as compared with many small ones at low LET. Assuming some kind of recovery during irradiation, a corresponding time-dependent formalism (Booz, 1970; Payne and Garrett, 1975) in terms of the single event distribution $f_1(z)$ can, in principle, yield an increase of effectiveness at growing LET, but it implies a great influence of dose rate at low and medium LET (Günther,

---

1) This holds regardless of special models which may be used for $S_0(z)$. A step function ($S_0(z) = 1$ for $z < z_0$ and $S_0(z) = 0$ for $z > z_0$) has often been considered, but this is not essential to the RBE problem.

1968a; KAPULTSEVICH and PARFENOV, 1972), which is not observed in the range of already large dose rates.

Let us now draw the conclusions from Eq. (1.24) failing to predict RBE. First, it should be stressed that the integral expression is reasonable at any rate, since a population of irradiated objects can be imagined to be sampled by z, so that $S_0(z)$ can be defined as the surviving fraction for the z-sample. Then $S_0(z)$ must be allowed to depend on radiation quality, which means that it is not irrelevant to the cell whether it has received the same fixed energy z from one or another radiation. The reason is obvious: at low LET, the energy will be distributed rather uniformly over the GSV, while it is locally concentrated in a small number of particle tracks at high LET. However, this information is not given by $f(z; D)$, but it would be needed to take account of some *internal structure* of the GSV. For this reason, the scope of the microdosimetric concept as a general description of physical stochastics is limited on principle (see also PARETZKE, 1979).

In the case of eukaryotic cells, the GSV is almost certain to be the cell nucleus, and internal structure may be understood, in abstract terms, as the presence of demarcated targets or merely blurred regions of potential interaction of energy round every point. In either case, one can speak of *sites* that define another structural level, in addition to the GSV level, at which energy concentration is relevant. Now the tools of microdosimetry may be applied to treat hypothetical processes *in a site*, in terms of z-statistics, but the response probability of an individual cell will depend on what has happened in the *entire* GSV, by definition, i.e. in all sites simultaneously. Hence, looking only at a separate site, or sites on the average, one is forced to disregard the statistical correlations of energy deposition in adjacent sites. This may be a reasonable approximation, at least for radiations of not too high LET. In fact, the now most advanced RBE theory developed by KELLERER and ROSSI, which is discussed in the next subsection, constitutes an approach of this kind.

These considerations should not be understood to slight the fundamental importance of microdosimetry as *the* consistent concept to describe radiation quality. This description is, however, not detailed enough to serve as a general basis of RBE theories. Since any kind of a more sophisticated general description would hardly be practicable (several random variables),[1]) it is the opinion of the present authors that *one cannot but start with action models in which the GSV has internal structure*, and the treatment of physical stochastics must then be due adapted to the model, thus giving up the idea to define radiation quality once for all by a manageable number of physical parameters. Of course, the choice of the model is a matter of hazard with almost no chance unless one refers to what is already known from cellular and molecular radiobiology about targets, primary lesions, and radiation-induced processes. The present book is devoted to such an attempt.

1.4.3. The theory of dual radiation action

This theory of RBE, often signified by the abbreviation DRA, was suggested by KELLERER and ROSSI (1971). It has been said to be "so far the only approach that makes

---

1) An alternative is direct Monte Carlo simulation of particle tracks in all details (positions of ionizations and excitations), which has become possible in recent years (PARETZKE, 1974, 1979) on the basis of measured and calculated molecular interaction cross-sections. In this way any imaginable information on physical stochastics, in addition to the known microdosimetric parameters (KELLERER and CHMELEVSKY, 1975; CHMELEVSKY and KELLERER, 1977), can be obtained in principle, but it will be hard to guess what kind of evaluation is relevant to RBE.

direct use of microdosimetric parameters" (Booz, 1974). In a detailed treatise by KELLERER and ROSSI (1972) the theory has been described in all respects and variants, including profound discussions of the basic principles and possible interpretations as well as its limitations. The following discussion is only a very concise one with emphasis placed on the fundamental shortcomings of DRA theory.[1]) Also in this respect, we will not go into some details given elsewhere (GÜNTHER *et al.*, 1977a; PARETZKE, 1979).

Imagine an uniform infinite medium consisting of 'sensitive matter' which is uniformly irradiated. The basic assumption expressed by the term 'dual' is that radiation induces primary reactions which are of second order kinetics, i.e. the yield ϵ (number of 'dual lesions') is proportional to the *square* of energy density. Hence, $\epsilon = kD^2$ would hold if energy were uniformly distributed over the medium. Actually, the density of imparted energy varies abruptly from point to point, and a reasonable definition of this density function in space is only possible by averaging, at each point, over a finite region called the 'critical site'. The sites may also be considered as preformed structural entities in the cell, but this is irrelevant to the theory; in any case the site specifies the range of energy interaction in producing elementary lesions. The mere variation of energy density from site to site must increase the average yield ϵ as compared with kD^2. The probability to find the energy density z in a randomly selected site is given by the distribution $f(z; D)$. Thus, by the help of Eq. (1.17), one finds

$$\epsilon = \int_0^\infty dz k z^2 f(z; D) = k\overline{z^2} = k(\lambda D + D^2), \quad \lambda = \bar{z}_D. \tag{1.26}$$

This fundamental equation describes the *average* yield of dual lesions, and the essence of DRA theory is assuming this quantity (apart from some subsequent modifications) to be the only one on which a biological response can depend. In other words, if ϵ is used to replace dose as the measure of the *intensity* of irradiation, radiation quality does not any longer exist. It is implicitly contained in ϵ by virtue of the parameter $\lambda = \bar{z}_D$. No assertions are made on the relation between ϵ and a specifically scaled effect (e.g. survival rate), thus generally allowing for biological stochastics. For different radiations equal effect means equal ϵ, and this suffices to derive the RBE from Eq. (1.26). This concept has three fundamental aspects which need further comments:

(i) No consideration of the spatial distribution of dual lesions as a possible factor of radiation quality.

(ii) The hypothesis of dual action itself.

(iii) The constancy of the proportionality factor k with respect to radiation quality.

Point (i) marks the principal restriction involved in DRA theory since, in contrast to (ii) and (iii), there is no way to go further, in this respect, within the given theoretical framework. One is confronted with the correlation problem discussed in the preceding section. The avove-used picture of a 'sensitive medium' may be helpful to understand the point. The dual lesions will be distributed at random over the medium, but this randomness is no trivial one: it depends on radiation quality, in addition to the *mean* yield ϵ. The latter could be alone relevant if the biological response of a single reacting unit were due to the dual lesions produced in a large region of the medium.[2]) Actu-

[1]) Recently, the theory was generalized (KELLERER and ROSSI, 1978), but in a respect which is not related to the fundamental aspects in question.

[2]) This condition is certainly met if there is no reacting unit at all, as in the case of chromosome aberrations counted for a large number of irradiated cells. Hence, the average yield of aberrations may equal the average yield of dual lesions. The DRA theory has indeed sucessfully been applied to evaluate the RBE in this case (e.g., BRENOT *et al.*, 1974).

ally, the region to be considered is the GSV (cell nucleus) located with equal probability anywhere in the medium. The number of dual lesions (or 'dually' damaged sites) that happen to lie in the GSV is a random quantity, whatever, and the corresponding statistics would have to be at least approximately the same for all radiations to justify the principle of DRA theory. This may roughly hold for X-rays and fast neutrons, but not for high-energy heavy ions. In the latter case, at constant dose and growing LET, an increasing proportion of cells remains entirely unaffected (no particle traversal through the nucleus), and this is just the surviving fraction (an experimentally well-established fact). Hence each traversal produces overkill, i.e. a *saturating* amount of dual lesions (or damaged sites), so that the *mean* yield ϵ loses its significance with respect to the surviving fraction.[1])

The significance of dual action described by kz^2 in sites of the order of 1 µm diameter was originally inferred from fast neutron RBE-dose relationships for some effects in eukaryotic organisms. Dual damage is likely to be the main component at medium doses or for medium-LET radiations throughout. In general, however, a linear component, $k\lambda_0 z$,[2]) must be added:

$$\epsilon = k\overline{(\lambda_0 z + z^2)} = k(\lambda D + D^2), \qquad \lambda = \lambda_0 + \bar{z}_D. \tag{1.27}$$

Another modification concerns the possibility of site saturation at high LET, which has been evaluated for intra-track interaction only by the ansatz

$$\lambda = \frac{1}{\bar{z}_F}\int_0^\infty \mathrm{d}z z_0^2[1 - \mathrm{e}^{-(z/z_0)^2}]\, f_1(z) = \bar{z}_D^*, \tag{1.28}$$

which becomes the relation $\lambda = \bar{z}_D$ again if z is small as compared with z_0. In view of these modifications one speaks of *elementary lesions* rather than dual lesions.

The coefficient k should be constant if nothing but the total energy imparted to a site determines the kinetics of elementary lesion formation. However, if the submicroscopic structure of energy deposition in particle tracks inside the site also plays a role, k may involve an additional dependence on radiation quality, but there is no way to evaluate this dependence by means of microdosimetry. The problem is analogous to the non-equivalence of different-radiation specific energy in the GSV discussed in the foregoing section. Thus, acknowledging this dependence means giving up the predictive power of DRA theory.

The general RBE-dose relation of DRA theory is readily derived. If the subscript X refers to the reference radiation (usually X-rays) and no subscript is used for the radiation in question (e.g. fast neutrons), one has $RBE = D_X/D$ with D_X being related to D by equating ϵ_X to ϵ. This yields

$$RBE = \frac{2\dfrac{k}{k_X}\left(1 + \dfrac{D}{\lambda}\right)}{\dfrac{\lambda_X}{\lambda} + \sqrt{\left(\dfrac{\lambda_X}{\lambda}\right)^2 + 4\dfrac{k}{k_X}\left(1 + \dfrac{D}{\lambda}\right)\dfrac{D}{\lambda}}}. \tag{1.29}$$

The characteristic doses λ_X and λ, which are at least roughly equal to the dose-means of specific energy in sites, differ greatly for X-rays and densely ionizing radiations (e.g., in the case of $d = 1$ µm sites, $\bar{z}_D \approx 70$ rad for X-rays and $\bar{z}_D \approx 2000$ rad for 0.5-MeV neutrons). Ac-

[1]) Note that, at constant dose, ϵ increases with increasing LET.

[2]) In this form the ansatz was given by KELLERER and ROSSI (1972), which we follow here, but it can be accepted only as long as both λ_0 and k are looked upon as constant. If k is allowed to depend on radiation quality (see below), one has to realize that kz^2 describes the *squared* number of sublesions, so that an additional linear component must read $\sqrt{k}\,\lambda_0 z$. Hence, the following equations have to be modified by substituting $\lambda_0/\sqrt{k}$ for λ_0 if analyses with a variable k are to be performed.

cordingly, three dose regions for which simpler relations follow from Eq. (1.29) can be distinguished:

$$RBE = \frac{k}{k_X}\frac{\lambda}{\lambda_X} \text{ for } D \ll \frac{k_X}{k}\left(\frac{\lambda_X}{\lambda}\right)^2 \lambda \quad (1.30)$$

$$RBE = \sqrt{\frac{k}{k_X}\frac{\lambda}{D}} \text{ for } \frac{k_X}{k}\left(\frac{\lambda_X}{\lambda}\right)^2 \lambda \ll D \ll \lambda \quad (1.31)$$

$$RBE = \sqrt{\frac{k}{k_X}} \text{ for } D \gg \lambda. \quad (1.32)$$

Apparently most clear experimental evidence exists for various effects of *fast neutrons* on *higher organisms* studied within the *middle dose region* (1.31). It is this selected data that exhibits the $D^{-1/2}$-law of dose-dependence over a rather wide dose range, and the idea of DRA theory has thus been suggested. It is clear that, as long as the middle-range relation (1.31) applies, there is neither need to assume $k \neq k_X$ nor $\lambda_0 \neq 0$. Thus by curve fitting one obtains $\bar{z}_D$ for the given neutron energy and, therefrom, the diameter d of the site. This yields values of d somewhat in excess of 1 μm. Now one has to ask whether the corresponding predictions for large and small doses and for varying neutron energies are also compatible with experience.

If $\lambda_0 = 0$, the low-dose plateau (1.30) should be reached only at very small doses and should be as high as $\bar{z}_D/\bar{z}_{D,X} \approx 30$. Kellerer and Rossi (1972) present one example (mutations of *tradescantia* stamen hairs) where the curve extends to so small doses and flats out as it should. In another case (eye lense opacification) the curve *continues* to follow the $D^{-1/2}$-law down to similarly small doses reaching an RBE clearly *above* $\bar{z}_D/\bar{z}_{D,X}$.[1]) This forces to assume $k > k_X$; curve fitting is achieved by $\sqrt{k/k_X} = 4$ [2]) and a correspondingly smaller value of $\bar{z}_D$ (150 rad leading to $d = 3.5$ μm).

It is now generally accepted that k depends on radiation quality (Kellerer and Rossi, 1974), by evidence from the *high-dose region* (1.32). The large number of studies on cultured mammalian cell killing with fast neutrons (see chapter 7, section 7.4) clearly show that the RBE approaches a value above unity at high doses (survival curve slope significantly larger for neutrons than for X-rays). In addition, this important type of radiation action is incompatible with the hypothesis of a pure dual action in so far as the X-ray survival curves usually exhibit a substantial linear component of damage, i.e. a non-vanishing initial slope (Wambersie and Dutreix, 1976; Elkind, 1977). As a consequence, the low-dose plateau of RBE extends up to a rather large dose (> 50 rad), so that a $D^{-1/2}$-dependence region is often hardly found, and the plateau value is considerably smaller than $\bar{z}_D/\bar{z}_{D,X}$. This can be accounted for by a large cell-line-specific λ_0, but in this way, with λ_0 and k/k_X to be treated as adjustable parameters, DRA theory degenerates into a mere mathematical model to fit RBE-dose relations which is no longer predictive. If such fits are performed for several radiations, one may ask whether at least the parameter λ varies as it should in view of its meaning in theory: eventually modified dose-mean of specific energy in a site of constant diameter. Such an analysis (Günther *et al.*, 1977a) based on V79-cell survival curves for 11 neutron energies (Hall *et al.*, 1975a; Kellerer *et al.*, 1976) revealed that the variation of the fitted λ is *contrary* to a mean specific energy, while it is the variation of k/k_X that essentially reflects the pattern of neutron energy dependence of RBE. Goodhead (1977) arrived at similar results.

---

[1]) Much the same behaviour was found for the induction of mammary tumours in rats (Shellabarger *et al.*, 1974).

[2]) In the paper we refer to here (Kellerer and Rossi, 1972) one finds $k/k_X = 4$, but this is obviously a mistake; the original paper (Bateman *et al.*, 1972) reports $k/k_X = 16$ in accord with the fitted curve actually shown.

Hence, the DRA theory appears as an insufficient concept to understand and predict the RBE, at least in the important case of mammalian cell survival and related responses. The difficulties are, however, not so apparent if one confines to *low doses* (small effects). On this understanding it has been suggested to take the *modified dose-mean of specific energy* defined by Eq. (1.28) as a *relative measure* of radiation quality (particularly for comparing different-energy neutrons). To complete the definition, the 'threshold value' z_0 has been fixed by the help of track-segment irradiation data, in terms of lineal energy for convenience:

$$y^* = \frac{y_0^2}{\bar{y}_F} \int_0^\infty dy[1 - e^{-(y/y_0)^2}] f_1(y), \qquad (1.33)$$

$$y_0 = 125 \frac{\text{keV}}{\mu\text{m}}.$$

The recommended site diameter, 0.5 μm to 2 μm, is not critical within this range. The 'asterisked-y concept', which only states

$$RBE \sim y^* \qquad (1.34)$$

for small doses, can be slightly motivated by both Eq. (1.30) and Eq. (1.31) transformed into the equivalent expression for a constant level of effect (constant X-ray dose D_X). By giving up to predict the absolute RBE, neither λ_0 nor k/k_X is needed. One has $RBE \sim k\lambda$, and the theoretical argument for relation (1.34) seems to be $k = const.$ and $\lambda = \bar{z}_D{}^*$ (on condition that $\lambda_0 \ll \bar{z}_D{}^*$). However, this would be incompatible with the finding that high-dose data enforce a strong variation of k. Consequently, it is the entire product, $k\lambda$, that may *fortuitously* correlate with $\bar{z}_D{}^*$.

Thus, the relation (1.34) cannot be regarded as necessarily resulting from DRA theory. It rather constitutes an empirical rule, undoubtedly: a very useful one. Indeed, the asterisked y quite well reflects the dependence of effectiveness on neutron energy for a number of effects on higher organisms, including inactivation of cultured cells (Rossi, 1972; Kellerer and Rossi, 1972; Hall *et al.*, 1975a). y^* seems to be more relevant than any mean LET value or the integral (1.11), and it has the additional advantage of being readily measurable in any complex radiation field by a routine proportional counter equipment. Therefore, in recent time, this concept has been increasingly put to use in high-LET therapy to characterize the quality of different therapeutic beams of neutrons, π-mesons, etc. in the depth and the distance lateral from the beam axis.

y^*, a purely physical parameter without reference to the special biological object and the response considered, is likely to be the best compromise one can find to express *radiation quality* as a property of radiation as such. However, every concept of a single 'quality parameter' is insufficient in so far as the definite value of RBE, which depends on dose as well as the particular object and effect, should be reduced to *both* physical properties of radiation and relevant characteristics of the special reacting object, by what could be called a theory of RBE.

1.5. Gross sensitive volume with structure

It has been made plausible in the preceding section that one cannot obtain a satisfactory theory of RBE for effects on higher organisms, particularly cell killing, unless some kind of structure of the object is allowed for in such a manner that *correlated* energy concentrations below the level of gross sensitive volume (GSV) become significant. From Eq. (1.24) confronted with experimental RBE data it can be learned that models to be used must imply a *higher* effectiveness if a definite energy imparted to the GSV is *more concentrated* in a smaller part of the GSV (fewer particle tracks). Of course, these general hints hardly facilitate the choice of a suitable model. Therefore, we leave this abstract way of thinking and start with the idea, now generally adopted, that *molecular damage to the DNA* in the cell nucleus is at least the predominating factor of radiation injury to cells. Thus the substructure of the GSV (cell nucleus) is due to the DNA molecules in which discrete primary lesions are formed, and physical stochastics has to be evaluated with respect to both the production of DNA-lesions (yield per unit dose and unit length of DNA) *and* the statistics of lesions per cell nucleus.

The detailed elaboration and application of the *DNA-lesion theory of RBE* is the topic of this entire book, while in the present section only the fundamental principles as published a number of years ago (Günther and Schulz, 1972) will be reviewed. It should be emphasized that similar conceptions have been repeatedly suggested in the past (Munson *et al.*, 1967; Fowler, 1968; Neary, 1970; Burch, 1970; Oda *et al.*, 1971; Rossi, 1972; Barsukov and Malinovsky, 1974). However, if mathematical formulations have been attempted, special assumptions regarding the relation between primary lesions and biological endpoints (biological stochastics) have been made, and oversimplified treatment of physical stochastics has caused a further loss of generality. By contrast, our concept is as general as possible:[1]) it does not involve a definite model of dose-effect relations, since this proves unnecessary regarding RBE. On the other hand, this gives the theory a somewhat abstract appearance, but it is nevertheless felt unnecessary to repeat the large body of arguments and experimental evidence found in the literature in favour of (a) the mere idea that DNA-lesions (particularly double-strand breaks) dominate cell damage and (b) various aspects or hypotheses concerning relations between DNA-lesions, repair, recovery, modifying factors, and cell viability. Instructive surveys have been given, e.g., by Elkind and Ben-Hur (1974) and Barendsen (1974b). More recently, strong evidence was obtained by analysing the features of radiotoxicity of Auger emitters (Warters *et al.*, 1977; Hofer *et al.*, 1977; Warters and Hofer, 1977).

1.5.1. The model of DNA-lesion production

Numerous experimental investigations concerning the induction of DNA single- and double-strand breaks (which are accessible to observation by means of the sucrose gradient centrifugation technique) by ionizing radiations *under physiological conditions* (irradiation of intact cells) have shown that the yield is *proportional to dose* up to many krads. Therefore, the range of any interaction of absorbed energy on forming a lesion can be at most of nanometre order of magnitude,[2]) so that at normal dose each lesion is produced by one ionizing particle only (intra-track interaction). This is in keeping with findings indicating that there is no long-range radiation-induced radical attack on DNA in the cell (Chapman *et al.*, 1976; Frankenberg, 1976; Michaels and Hunt, 1978). This suggests a target-model description of DNA-lesion formation as follows.

(i) A cylindrical target volume is associated with the filamentous double-strand

---

[1]) It is, however, limited by disregarding a possible influence of superstructure of DNA (nucleosomes, supercoils, replicons, chromosomes). Generally speaking, the cumulative biological effect of a given number of DNA-lesions in the nucleus is assumed to be independent of the spatial arrangement of the lesions.

[2]) To draw this conclusion is a matter of microdosimetry: appreciable deviations from linearity must be expected at doses above the value $\bar{z}_D$ that refers to the size of the interaction region (cf. Eq. (1.26)). For direct experimental evidence as to the nanometre order of site size see Hofer *et al.* (1977).

DNA. Its diameter does not much exceed that of a DNA molecule (2 nm). Energy deposition outside the target has no effect.

(ii) An absorption event in the target (usually a single particle traversal) can produce a single DNA-lesion at the point affected.

(iii) The total number of ionizations produced in the event determines the probability that a lesion of the type considered will be formed.

With respect to the mean yield, s, of lesions per unit dose and unit length of DNA, the model can be expressed in microdosimetric notation by

$$s = \frac{1}{\bar{z}_F} \sum_{j=0}^{\infty} \lambda_j f_1(j). \qquad (1.35)$$

$f_1(j)$ is the event size distribution in terms of ionization number j rather than specific energy.[1] It refers to a randomly oriented cylindrical target volume of unit length, and $\bar{z}_F$ is the corresponding frequency-mean of specific energy (cf. Eq. (1.13)) which determines the mean number, $D/\bar{z}_F$, of events at dose D. λ_j denotes the probability that an event of j ionizations engenders the lesion. No *a priori* assumptions about this probability are made. Different lesion types will be described by different sets of λ_j. The mathematical evaluation of the model for arbitrary radiations is presented in chapter 3, and the results are confronted with experimental data on DNA strand breakage in chapter 6 (see also Fig. 2.6).

The model has much in common with single-hit target theory. In particular, the one-ionization model discussed in section 1.2.2 (LEA's associated volume method) is obtained by $\lambda_j = 1 - \delta_{0j}$; and allowance for an action probability p per ionization can be made by $\lambda_j = 1 - (1 - p)^j$. In these cases, s turns out to be smaller for radiations of higher LET (saturation effect), which corresponds to experimental findings for DNA single-strand breaks. Increase of effectiveness at increasing LET (concentration effect), as found for double-strand breaks, will be obtained with several-ionization models, e.g., $\lambda_j = 0$ for $j < n$ and $\lambda_j = 1$ for $j \geqq n$. Hence, in general, the dependence of λ_j on j will resemble a step function[2]), but the 'step' may be gradual. Generality in the evaluation of Eq. (1.35) can be maintained by using the ansatz

$$\lambda_j = \sum_{\nu} b_\nu \, e^{-\zeta_\nu j} \qquad (1.36)$$

with any finite number of terms. This expression constitutes no restriction, since every arbitrary function of j can be described in this way. Some restrictions by using special expressions of the form (1.36) containing a reasonably small number of parameters can be introduced later on, with the purpose of fitting experimental data.

Despite of its rather formal nature and the discretion in the choice and interpretation of λ_j, the model is more than a sophisticated mathematical construction, since its ability to fit the *absolute* yield of DNA

[1]) This has been done for several reasons. Taking imparted energy as the stochastic variable would also be an arbitrary choice, whereas there is some evidence for the ionizations to be dominant in producing molecular damage (LEA, 1946). Moreover, imparted energy is difficult to define and calculate for nanometre sites. While the distinction between ionizations and energy is not so important for micrometre sites, it has been suggested earlier (KELLERER and CHMELEVSKY, 1975) that microdosimetry based on the discrete variable 'number of ionizations' might be both the more definite and more relevant concept when going to very small sites. In consequence, recent numerical calculations of event size distributions for nanometre sites (CHMELEVSKY and KELLERER, 1977) do use this concept.

[2]) The step approaching the value 1 at large j on condition that the lesion type includes double-strand breaks is a reasonable assumption, since a sufficiently large number of ionizations concentrated in the target is likely to destroy the DNA completely at the point affected.

lesions in dependence on radiation quality is, nevertheless, not trivial. This will be demonstrated in chapter 6. However, it should be emphasized that the purpose of this model is not primarily to elucidate the mechanism of DNA-lesion formation.[1]) Its value for the theory of cellular radiation effects consists in relating the production of molecular lesions to *physical events* at the nanometre level, for which the mere ability to fit for observed yields of DNA breaks seems to be a sufficient criterion.

1.5.2. The theory of cellular radiation effects

Assuming DNA-lesions to be responsible for observable biological effects on cells implies that the probability of a cell to react according to the test stipulated (e.g. survival) depends on the actual DNA-lesions produced in the individual cell. The simplest assumption is to take the mere number, k, of lesions of some type, or the numbers $k_{\mathrm{I}}, k_{\mathrm{II}}, \ldots$ of several types, for the determinant of the reaction probability: $H(k)$ or $H(k_{\mathrm{I}}, k_{\mathrm{II}}, \ldots)$. Now, which is the crucial point, at a given dose these numbers will fluctuate from cell to cell, which has to be described by a statistical distribution $U_k(D)$ that changes with radiation quality. Hence, the dose-effect relation in the case of one lesion type reads:

$$S(D) = \sum_{k=0}^{\infty} H(k)\, U_k(D; \lambda_j, l_l, Y). \qquad (1.37)$$

$H(k)$, which describes the *biological stochastics*, is independent of radiation, by definition. Thus the formalism constitutes a theory of RBE resting upon the evaluation of the lesion number distribution $U_k(D)$ which depends on the size of the DNA-containing region (GSV) in addition to the constant parameters indicated: λ_j referring to the lesion type considered, l_l denoting the diameter of the cylindrical DNA targets, Y being the total length of DNA molecules in the nucleus. In this way the lesion production model as such, not only the mean yield s described by Eq. (1.35), is constitutive to the complex distribution $U_k(D)$. Its rather complicated evaluation is the topic of chapter 3.

Meaning of, and approaches to, the effect-probability function $H(k)$ are discussed in the next subsection, but independently of that it is imperative in any case to express $H(k)$ in the form

$$H(k) = \sum_{\mu=1}^{n} B_\mu\, \mathrm{e}^{-Z_\mu k} \qquad (1.38)$$

in order to be able to evaluate Eq. (1.37). Due to the free choice of n, B_μ, and Z_μ, the expression is flexible enough for approximating arbitrary functions $H(k)$ as precisely as needed. Some special models lead directly to this form of $H(k)$.

Eq. (1.37) may be looked upon as an approach which is closely related to, but remedies the shortcomings of, both Eq. (1.3) and Eq. (1.24). In the former equation an *invariable* Poisson distribution stands for U_k, but the mean number of hits, αD, *can* depend on radiation quality, in analogy to the expected value

$$\bar{k} = \sum_{k=0}^{\infty} k U_k(D; \lambda_j, l_l, Y) = sYD$$

$$= \frac{YD}{\bar{z}_F} \sum_{j=0}^{\infty} \lambda_j f_1(j). \qquad (1.39)$$

By contrast, in Eq. (1.24) the *realistic* distribution $f(z; D)$ depends on radiation quality like $U_k(D)$, but its expected value $\bar{z}$ does not; it is always equal to dose according to Eq. (1.12). These remarks are to elucidate the essence of the theory from the formal point of view. To take account

---

[1]) A somewhat more definite model, which is a simplified version of the present one, has been published recently (Obaturov, 1979): By restriction, from the outset, to single- and double-strand breaks, λ_j for both types of breaks is expressed by two constant probability parameters only, where j signifies the number of *primary* ionizations (i.e. ionization clusters). The rationale of the latter depends, however, on physical simplifications.

of at least these two kinds of dependence is the indispensable minimum requirement, and the lesion production model has been introduced as a plausible attempt to specify the stochastic variable to be used, k, and how its distribution and expected value depend on radiation quality. Thus, while the biological rationale of the approach may appear to be too simple,[1]) it can nevertheless be anticipated to pick up the main factors dominating RBE, and in consideration of its flexibility due to the adjustable parameters λ_j and l_l there should be sufficient latitude to account for the RBE in various cases (viruses up to mammalian cells). Furthermore, Eq. (1.37) by itself allows for a number of complicating factors, implicitly, as follows.

Notwithstanding that experimental situations usually involve *biological variability*, i.e. subpopulations of cells differing in radiosensitivity, there is no need to modify Eq. (1.37) as long as the variability concerns the response function $H(k)$ only. It is merely the interpretation that changes: $H(k)$ now represents the *average* behaviour of the population as a whole. In the important case of asynchronous cells an additional inhomogeneity exists concerning the content, Y, of DNA per cell. Eq. (1.37) can nevertheless be retained for the entire asynchronous population owing to the so-called *relation of equivalence* (for the proof see chapter 3): changing the parameter Y is equivalent to using a certain changed $H(k)$. Hence, the contributions to $S(D)$ of the subpopulations can be formally described by a common value of Y, so that again the differences are reducible to $H(k)$.

Finally, let us discuss the case that several different types of lesion have a share in the biological response. In the DNA one finds base damage, single- and double-strand breaks, cross-links. It has also been suggested that the same lesions may be of different quality if produced by different radiations, particularly regarding the ability to be repaired. However, in the present theory there is no place for a changing 'lesion quality', and repair is a matter of $H(k)$. Rather than speaking of lesion quality, one has to distinguish sub-types of different chemical constitution, and what eventually changes with radiation quality are the frequencies of these sub-types. Hence, in general, a number of lesion types described by individual sets of λ_j may contribute to the effect, and the corresponding generalization of Eq. (1.37) reads

$$S(D) = \sum_{k_{\text{I}}, k_{\text{II}}, \dots} H(k_{\text{I}}, k_{\text{II}}, \dots) \times U_{k_{\text{I}}, k_{\text{II}}, \dots}(D; \lambda_j^{\text{I}}, \lambda_j^{\text{II}}, \dots, l_l, Y). \tag{1.40}$$

Now two questions have to be asked. (a) Can this generalized theory be mathematically evaluated? (b) Is it possible formally to account for $S(D)$ by an equivalent one-lesion-type model given by Eq. (1.37)?

In detail these questions are dealt with in chapter 2 and chapter 3. Briefly, one arrives at the following results. The evaluation of Eq. (1.40) is analogous to Eq. (1.37). The final formula for the dose-effect-relation components $S(Z_\mu, D)$ arising from the terms of the expression (1.38) can be used to express the general $S(D)$, too, after introducing formal parameters Z_μ related to $H(k_{\text{I}}, k_{\text{II}}, \dots)$ and formal $\lambda_j^{(\mu)}$ composed of $\lambda_j^{\text{I}}, \lambda_j^{\text{II}}, \dots$ This statement implies the answer to question (b). In general, there is an *almost* equivalence in the sense that the totality of dose-effect relations (at varied radiation quality) arising from an arbitrary several-lesion-type model can be nearly obtained also from a one-lesion-type model referring to some formal combined λ_j. Under certain conditions (see chapter 2) the equivalence is complete. This case is met, e.g., if two types of lesion differ only in the efficiency of repair while the residual unrepaired lesions play the same role regarding the biological response in question. Then, a

[1]) See the footnote 1 at page 20.

linear combination $\lambda_j = \alpha_{\rm I}\lambda_j^{\rm I} + \alpha_{\rm II}\lambda_j^{\rm II}$, where $\alpha_{\rm I}$ and $\alpha_{\rm II}$ are proportional to the respective probabilities of non-repair, accounts exactly for the response to any kind of radiation.

1.5.3. Biological models of action and formal analysis

$H(k)$, in general $H(k_I, k_{II}, \ldots)$, is referred to as the *effect-probability function.* To make its meaning plausible, let us consider cell survival on condition that one lesion type only is responsible for the loss of reproductive capacity. Then, first of all, it is evident that $H(k)$, the survival probability at k lesions, must be a decreasing function:

$$H(k + 1) \leqq H(k). \tag{1.41}$$

If the lesions in a cell act *independently,* the following relation must be valid:

$$H(k + k') = H(k)\, H(k'), \tag{1.42}$$

which defines an exponential function:

$$H(k) = \mathrm{e}^{-Zk}. \tag{1.43}$$

In general, however, there will be some kind of *cumulative action* so that the effectiveness of each lesion is enhanced by the presence of others:

$$H(k + k') < H(k)\, H(k'). \tag{1.44}$$

The generalization of these relations in the case of more than one type of lesion is evident.

In analogy to hit theory, these two classes of models may be termed 'single-hit' and 'multi-hit' models, but important conceptual differences should not be overlooked. Only the single-hit model (1.43) yields exponential survival curves for *every* radiation, while in the general case the model by itself does not decide on the shape of survival curves. It is only for radiations of not too high LET (for which U_k in Eq. (1.37) resembles a Poisson distribution) that multi-hit models yield sigmoidal survival curves. Moreover, relation (1.44) does not imply a threshold hit number, i.e. $H(1) < 1$ is allowed so that the survival curves for any radiation can have a single-hit component (non-vanishing initial slope). As to the underlying biological mechanisms, there are multifarious possibilities, some of which shall be outlined.

A single-hit model will be obtained, e.g., in the following cases. (i) Each lesion, if not repaired within some time, inactivates the cell (then e^{-Z} means the probability of repair). (ii) Only a lesion in some sensitive part of DNA gives rise to the effect (then e^{-Z} means the insensitive proportion of DNA). (iii) Even in the case of more complex mechanisms, the mere inhomogeneity of the population can nearly lead to the expression (1.43) by averaging.

In dependence on the special object, the biological conditions, and the definition of the test, one can imagine lots of multi-hit models characterized by relation (1.44). (i) If the repair capacity is limited, the chance of each lesion to be repaired is decreased by the presence of further lesions. This type of cumulative action would have to be described by a kinetic or cybernetic model (LAURIE *et al.*, 1972; KAPPOS and POHLIT, 1972). (ii) In the case of partially replicated chromosomes as found, e.g., in proliferating bacteria, a *single* lesion may prevent colony formation *only if* located on the unreplicated part, while several lesions have an additional possibility to act by affecting *both* branches of the replication fork.[1]) (iii) A special

---

[1]) A generalized version of this model for bacteria was evaluated by the present authors. Both double- and single-strand breaks of DNA described by fitted λ_j's were considered relevant, and it was assumed that single-strand breaks can be repaired with some definite probability. The latter was the only adjustable parameter, while chromosome length, nucleoide diameter and replication state (or the distribution of states) were taken from experimental data for both *E. coli* and *B. subtilis.* The theory yielded good quantitative agreement with experimental results on in-

model of yeast cell inactivation developed by KAPULTSEVICH (1972) may serve as another example here. The stochastics of cell division after irradiation is assumed to be governed by a constant division probability P which is a decreasing function of the number k of 'hits' the parental cell has received: $P_k = (1 - \alpha)^k$.[2) The treatment of division stochastics yields

$$H_m(k) = 1 - 2 \sum_{n=2}^{m} \frac{(2n-3)!}{(n-2)!\,n!} \times P_k^{\,n-2}(1 - P_k)^n \qquad (1.45)$$

for the probability to find a colony containing more than m cells.

These examples are only to illustrate what a biological model of action could be, and the limited significance of such hypothetical models. It can be seen from the fundamental equation (1.37) that both the shape of a single survival curve for a given radiation and the dependence on radiation quality are governed by nothing but the *abstract* survival-probability function $H(k)$ which can be readily obtained by curve fitting. Therefore, it makes little use to theorize on biological models of action as long as one's interest concerns only the radiation-quality-dependence of survival curves. If, however, one is examining the dependence of survival curves on biological factors or the time factor (dose rate), post-irradiation conditions etc., it is indispensable to construct biological models of action, and one has a better chance, then, to find out the right model. For these reasons, no attempt is made in the present book to propose a definite inactivation model for eukaryotic cells, inasmuch as in this case the actual processes that lead to cell death are likely to be very complex. Instead, the so-called *formal analysis* is applied to arrive at a theory of RBE. Briefly, the rationale is as follows.

One starts with the single-lesion-type equation (1.37) and considers a radiation of sufficiently low LET (at best ^{60}Co γ-rays) so that U_k becomes nearly a Poisson distribution with the mean value given by Eq. (1.39):

$$S(D) = \sum_{k=0}^{\infty} H(k) \frac{(sYD)^k}{k!} e^{-sYD}. \qquad (1.46)$$

The corresponding experimental $S(D)$ serves to obtain $H(k)$ by curve fitting on the basis of the general expression (1.38). The ambiguity due to the unknown parameter sY is unimportant, since for the following the choice of Y plays no role because of the relation of equivalence (see the preceding subsection). The prediction of survival curves for other radiations depends on the lesion type described by λ_j. A suitable set of λ_j can be found, again, by formal analysis of an entire family of different-LET experimental curves. Since this analysis could be based on the RBE for one definite survival level only, it is not trivial when the theory yields the complete survival curves (the shape) correctly for each radiation. Genuine predictions resting on nothing but the experimental $S(D)$ for a low-LET radiation will become possible for mammalian cells owing to the finding (chapter 7) that, in this case, there exists a *universal* λ_j-set which is also representative for DNA double-strand breaks.

activation with radiations in a broad range of LET (GÜNTHER and SCHULZ, 1972). In addition, the model explains the variation of sensitivity and survival curve shape with cell age (SCHULZ and GÜNTHER, 1973). It is important to remark that in either case alone one could succeed with a simpler model.

2) With the exception of the first division for which $P = 1$ is assumed.

1.6. Some descriptive approaches to survival curves and RBE

In addition to what has been discussed in the previous sections, there are various attempts to promote an understanding of cell survival curves and their change with radiation quality. The problem has mostly been attacked in a phenomenological way. Rather than starting from rigorous theoretical conceptions, one has introduced models that describe what is observed, and the adjustable parameters determined mainly by curve fitting are the target of theoretical considerations. With one exception, this has not led to a complete predictive algorithm regarding the RBE of different radiations.

The three models selected to be reviewed and critically discussed in the following are considered in some detail, because these models have become more widely known and have been repeatedly used in discussing experimental data. It is also these approaches, above all, that have been the topic of critical examinations by other workers (Neufeld and Hamm, 1976; Kellerer, 1976), however from a somewhat different point of view focusing the attention on different aspects.

1.6.1. The two-component theory of radiation

Since a long time (Bender and Gooch, 1962) the so-called *two-component model* is commonly in use as a convenient mathematical description of survival curves. Originally it is descended from the multi-target model (1.2) extended by adding an exponential factor and giving up its original interpretation (n, the number of targets, is replaced by the generally non-integer p, the so-called extrapolation number). In Wideröe's (1967) notation the model reads:

$$S(D) = e^{-\alpha D/D_\alpha}[1 - (1 - e^{-(1-\alpha)D/D_\beta})^p]. \tag{1.47}$$

Apart from the redundant parameter α, the meaning of which will be seen below, the three independent parameters of this expression nearly always ensure a good fit for any survival curves found (Millar *et al.*, 1978).

The exponential factor at the right-hand side of Eq. (1.47) is said to describe single-hit action. Indeed, unless the initial slope $S'(0) = -\alpha/D_\alpha$ vanishes, cells can be killed by *one* energy deposition event. As $D \to 0$, this is the only mode of action, which should continue to reduce survival according to $\exp(-\alpha D/D_\alpha)$ as D increases, while only the survivors of this process undergo a cumulative type of action described by the factor in square brackets. This is the *general* rationale of the two-component model. However, it is not self-evident to think that this reflects the existence of two distinct sorts of energy deposition events. E.g., it may be a component of *cells* that constitutes the single-hit component. In general, one can only define a component of *potentially* single-hit-killing events with the associated dose αD, but the *same* events may be partly involved in the multi-hit killing mode, too.

The two-component theory of radiation (Wideröe, 1967) is an attempt to look at Eq. (1.47) as if each radiation consisted of an α-component[1]) that produces *nothing but* single-hit killing and a β-component responsible for multi-hit killing. Accordingly, α and $(1 - \alpha)$ have been introduced explicitly to specify the dose fractions due to the α- and β-component, respectively. If one takes this concept literally, α is the only 'radiation quality parameter', whereas the other parameters characterize the cell line considered. Hence, two questions have to be asked. Are survival curves for different radiations consistent with constant values of D_α, D_β, and p? How can α be estimated for each radiation?

As to the first question, it is clear from the outset that D_α cannot be constant in the case of very high LET, where $\alpha = 1$

---

[1]) This naming expresses the standpoint that the single-hit component is equivalent to high-LET α-particles which do not produce sublethal damage (no interference with low-LET components).

must hold due to the absence of any cumulative action, but the exponential curve slope D_α^{-1} is known to decrease if the LET is further increased. Thus, one must at least allow for overkill to increase D_α. As to the second question, WIDERÖE has tried using the LET concept. It is evident that α must be an increasing function of L, which is in keeping with the influence of L on the shape of survival curves. However, it is hardly possible to arrive at a predictory scheme. Radiations involving a single LET value L_∞ (ion bombardment) indicate a *gradual* behaviour of $\alpha(L_\infty)$ in view of experimental data, whereas in cases of a broad LET spectrum it has been suggested to identify α with the dose fraction above some *threshold value* L_0 dividing the dose-spectrum of LET, $d(L)$, into two parts. This procedure was recently applied to a large series of mammalian cell survival curve data for X-rays, 15-MeV neutrons, and peak pions (WIDERÖE, 1978). The analysis revealed that $L_0 = 12$ keV/µm is a good value to render D_α independent of the radiation in all cases. Thus the *initial slope* of survival curves could be well attributed to a changing 'α-component' defined by a threshold LET, but the complementary factor $(1 - \alpha)$, then, did *not* account for the influence of radiation quality on the cumulative response. In other words, the parameters D_β and p are also dependent on radiation quality, as shown by individual curve fitting.

Apart from the insufficiency of the LET concept in general (see section 1.3), WIDERÖE's analysis demonstrates the inability of the two-component model as such to promote the solution of the RBE problem, since nothing but some systematization of the numerous experimental data can be achieved by merely discussing the influence of radiation quality in terms of three, rather than one, variable parameters: α, D_β, and p. As to the intrinsic reason, note that the most general way to define two components of radiation would be distinguishing two sorts of absorption events by a criterion of single-hit and multi-hit action, which is certainly impossible as argued about above.

1.6.2. The track-structure theory of survival

This theory developed by KATZ *et al.* (1971) concerns the RBE problem direct. Without any biological modelling, the dose-survival relation $S_\gamma(D)$ for photon radiations is solely considered as a given characterization of the radiobiological properties of the object with respect to its average response to a *mean* energy D delivered by electrons. Now, by merely physical reasoning, an attempt is made to reduce the action of single heavy charged particles to this response function $S_\gamma(D)$. In final formulation the theory takes the form of the three-parameter two-component model (1.47), but this is neither an essential of the theory nor does it mean adopting the view of two components of radiation.

Let us consider a beam of monoenergetic ions having the stopping power L_∞ and define the so-called *ion-kill mode* (non-cumulative action) by[1])

$$S_i(D) = e^{-\frac{\sigma}{L_\infty} D}, \qquad (1.48)$$

which alone determines the total survival $S(D)$ as $D \to 0$. The effective cross-section, σ, associated with a cell is related to the geometrical cross-section, σ_0, of the sensitive volume and the effectivity of a single ion passage to kill the cell. The estimation of the latter is based on KATZ's crucial assumption that it would be adequate to look upon an ion track as *nothing but* a blurred shower of δ-electrons which have the *same quality* as photon-produced electrons. Thus, if an ion passes

[1]) The notation used here corresponds to a definition of dose as energy absorbed per unit volume.

in the distance t from the centre of the sensitive volume V, the mean energy deposited in V defines the dose, $\bar{D}(t)$, to which the cell will react according to the response function $S_\gamma(\bar{D})$:

$$\sigma = 2\pi \int_0^\infty \mathrm{d}t\, t\left[1 - S_\gamma\big(\bar{D}(t)\big)\right]. \tag{1.49}$$

Using a radial dose profile for the ion tracks, this expression can be evaluated and expressed by

$$\sigma = P\sigma_0, \tag{1.50}$$

where P can be interpreted as the probability of killing per traversal through the sensitive volume.

KATZ's approach can be characterized as an approximate version of the general 'energy-only concept' discussed in section 1.4.2. Indeed, if there is no *qualitative* difference between energies imparted by electrons and heavy ions, Eq. (1.24) must hold strictly, and the derived equation (1.25) referring to ion kill (note that $-S'(0) = \sigma/L_\infty$) must be closely related to Eq. (1.49). This can be seen as follows. If the absorption events described by the distribution $f_1(z)$ in Eq. (1.25) are grouped according to the distance parameter t, one has distributions $f_1{}^{(t)}(z)$ with the mean value $\bar{z}_F{}^{(t)} = \bar{D}(t)$. Now it is just the essence of the approach that the remaining fluctuations of z at constant t are considered *equivalent* to the fluctuations of the cumulative z in a γ-ray field of dose $D = \bar{D}(t)$. Therefore, instead of the z-integral of $[1 - S_0(z)]\, f_1{}^{(t)}(z)$ one has simply got $\left[1 - S_\gamma\big(\bar{D}(t)\big)\right]$, since $S_\gamma(D)$ by itself involves these fluctuations.[1]) Of course, the equivalence is questionable, but this is the argument on which Eq. (1.49) is based.

Asserting that the track-structure theory adopts the energy-only concept which fails in predicting RBE values greater than unity for small surviving fractions (see section 1.4.2) raises the question, from the outset: how can this theory be compatible with experimental data? In fact, if the cumulative action of ions had been treated by a formalism equivalent to Eq. (1.24) in compliance with the principles of track-structure theory, the discrepancies would have become apparent. Instead, a simple *ad hoc* approach has been used, namely

$$S(D) = S_i(D)\, S_\gamma\big((1 - P)\, D\big) \tag{1.51}$$

and, furthermore, an arbitrary modification (see below) makes the now largely formal theory fit for reproducing the essential experimental findings. Eq. (1.51) is based on the argument that $(1 - P)$ can be considered as the fraction of dose which did not succeed in single-hit killing (ion kill) according to Eq. (1.50).

The explicit evaluation of the formalism has been carried out by using the pure multi-target model for the γ-ray survival curve:

$$S_\gamma(D) = 1 - (1 - \mathrm{e}^{-D/D_0})^m. \tag{1.52}$$

With varied sensitivity parameters D_0 and m, cross-sections σ_0, and radiation parameters Z_{eff} (effective ion charge) and v (ion velocity), the integral (1.49) has been numerically calculated by using a radial dose profile of the track which is, however, not very realistic (underestimation of the track-core dose, cf. data by WINGATE and BAUM (1970, 1976)). It is an important feature, regarding the practical application of the theory, that the results of this calculation have turned out to be describable in good approximation (except for too small $\beta = v/c$) by the analytical expression

$$P = \left[1 - \exp\left(-\frac{Z_{eff}^2}{\varkappa\beta^2}\right)\right]^m. \tag{1.53}$$

The parameter $\varkappa$ is an abbreviation for

$$\varkappa = \frac{D_0\sigma_0}{2.8\pi\ 10^{-9}\ \mathrm{rad\ cm}^2}. \tag{1.54}$$

[1]) Referring to our general terminology, $S_\gamma(D)$ can be said to comprehend both biological stochastics and the aspects of physical stochastics associated with γ-rays.

Hence, collecting Eqs. (1.48), (1.50), (1.51), (1.52), one arrives at the definitive expression

$$S(D) = e^{-\sigma_0 \frac{P}{L_\infty} D} \left[1 - \left(1 - e^{-\frac{(1-P)}{D_0} D}\right)^m\right] \tag{1.55}$$

for cell killing by monoenergetic ions.

What has been obtained is a straightforward procedure of survival curve prediction for arbitrary ions (radiation quality parameters L_∞ and $P(Z_{eff}, \beta)$) which rests only on the γ-ray sensitivity of the particular cell line described by the constant parameters D_0 and m, while the cross-section σ_0 roughly corresponds to the cell nucleus.[1] The nature of this predictive algorithm, irrespective of the inherent approximations and simplifications, is characteristic for the energy-only concept as represented strictly by Eq. (1.24). Consequently, it does *not* account for the experimental findings in the medium LET range. However, at this point an "inductive leap" (KATZ and SHARMA, 1973) has been made such that the mere *formalism* can nevertheless be used as a *description* of experimental data: Eq. (1.54) is rejected, so that $\varkappa$ becomes an *additional* freely adjustable parameter. It is important to emphasize that this arbitrary trick, as it lacks any theoretical justification, makes the entire theory a rather formal, but useful, mathematical construction, and its success, therefore, is largely fortuitous.

A given sort of cells is now characterized by four rather than three parameters (D_0, m, σ_0, $\varkappa$). While D_0 and m are obtained from fitting the γ-ray survival curve only, by the expression (1.52), σ_0 and $\varkappa$ follow from fitting a *family* of different-LET curves.[2] This analysis has been carried out with nearly all pertinent data available for various cell lines *in vitro* bombarded with ions of varying LET. A systematic comprehensive re-evaluation was published more recently (ROTH *et al.*, 1976). The importance of the approach, emphasized by the good fits usually obtained, consists in the RBE becoming an object-specific function of well-defined object-specific radiation quality parameters.

Two restrictions must however be noted. The special expression (1.52) does not allow for a non-vanishing initial slope of a sigmoidal γ-ray survival curve, which often makes the corresponding fits poor and leads to RBE $\to \infty$ as $D \to 0$. Another more serious fact is the inability of the theory to yield an RBE > 1 if the γ-ray survival curve is exponential, i.e. $m = 1$.[3] This is just the characteristic defect of the energy-only concept, which cannot, in this case, be compensated for by any choice of $\varkappa$. There are experimental results for haploid yeast cells (SAYEG *et al.*, 1959) and bacterial spores (POWERS *et al.*, 1968) which clearly exhibit both exponential survival for any radiation and an RBE growing above unity at increasing LET. Believing in the fundamental character of his theory, KATZ (1974) tends to call in question such experimental data, and in a recent paper (KATZ, 1978) this standpoint has even been generalized by asserting that the following sentence is an irrefutable principle: "The probability of cell killing relative to absorbed dose, after gamma irradiation, must be supralinear if

---

[1]) While D_0 and m refer to γ-rays, σ_0 determines the response to very-high-LET radiations and can, thus, be estimated by curve fitting in the range of saturation ($P = 1$, exponential survival).

[2]) The values of $\varkappa$ thus obtained for various cell lines (KATZ and SHARMA, 1973) scatter around figures an order of magnitude smaller than the expression (1.54), so that P becomes *larger* (higher RBE) compared with what the intrinsic concept would yield.

[3]) The latter problem is tightly connected with the former: already a shallow negative initial slope of an otherwise sigmoidal $S_\gamma(D)$, which is typical for mammalian cells, if it would be acknowledged in theory by generalizing the ansatz (1.52), would prevent the theory to yield *RBE* > 1 for any high-LET radiation (KATZ and SHARMA, 1974c).

an RBE greater than 1 is to be observed for any high LET radiation." Actually, this false principle is only a consequence of thinking that a cell feels nothing but total energy imparted by any particles. The counter-arguments (sensitive volume with structure) are given in section 1.4.2.

After these critical comments, let us stress again that the formalism called track-structure theory yields a good description of survival after ion bombardment in case that the expression (1.52) fits well with $m > 1$ (supralinearity of the response to γ-rays). Of course, the data used to determine $\varkappa$ and σ_0 cannot be afterwards said to be predicted by Eq. (1.55). Predictions in the true sense are obtained regarding radiations *not* involved in the analyses that have yielded the parameters, i.e. particularly fast neutrons, after accordingly generalizing the formalism. For this purpose, KATZ and SHARMA (1973) evaluated the ion-kill probability by integration along the entire tracks of the charged recoil particles. This procedure implies assuming that the sensitive volume is thin enough for being always completely traversed by the particles (track-segment approximation). It results in the survival curve formula (1.55) continuing to hold with averaged parameters $\bar{P}$ and $\bar{L}$ replacing P and L_∞. These radiation quality parameters referring to various cell lines described by the previously fitted values of m and $\varkappa$ have been tabulated for a set of neutron energies and spectra as well as pion stars (KATZ and SHARMA, 1975; KATZ *et al.*, 1976).

In this way, survival curves after neutron irradiation can be predicted completely in the sense that the necessary information stems from corresponding γ-ray and ion-bombardment results only. In contrast to BEWLEY's (1968) analogous attempt (see section 1.3), good agreement with the experimental neutron curves can be obtained (KATZ and SHARMA, 1973, 1974a), but relevant comparisons (same cell line) are possible only in a few cases. For a detailed inquiry see GÜNTHER (1976), where it has been shown that agreement is to be expected only in the upper range of neutron energies (particularly 14 MeV) due to a fortuitous compensation of errors (track-segment approximation and invalidity of the formula (1.53) for small particle velocities). In fact, the analytical expression (1.53) used throughout is well known not to apply to values of β as small as occurring in a neutron field of 1 MeV or less. The consequence are too high RBE values never observed for cell survival.

In summary it can be stated that the track-structure theory provides an algorithm which is well suited, in a variety of cases, to *reproduce* the cell-survival RBE of monoenergetic ions and less well to *predict* it for fast neutrons. Nevertheless, if one allows for the range of variation due to differing cell lines, useful calculations can be made with some relevance to RBE problems encountered in high-LET radiotherapy (KATZ and SHARMA, 1974a, 1974b, 1975; KATZ *et al.*, 1976; PFISTER *et al.*, 1977). From the theoretical point of view, however, the approach is not quite satisfactory because of several inherent inconsistencies pointed out above; but — on the other hand — just the obvious need to increase the RBE at discretion by choice of $\varkappa$ is, once again, an instructive demonstration of the inadequacy of taking only the energy in a single sensitive volume for the determinant of the biological response.

1.6.3. The molecular theory of cell survival

This approach (CHADWICK and LEENHOUTS, 1973) assumes a simple relation between cell death and the induction of DNA double-strand breaks (DSB) which are postulated to arise from both single radiation events and combination of two single-strand breaks (SSB) produced independently by two events. This implies that N, the number of DSB per cell, includes a linear and a quadratic component in dose:

$$N = \alpha D + \beta D^2. \tag{1.56}$$

The second postulate is non-cumulative action of DSB in the cell:

$$S(D) = e^{-pN} = e^{-p(\alpha D + \beta D^2)}, \tag{1.57}$$

where p means the probability of a double-strand break to kill the cell, and the Poisson distribution underlying Eq. (1.57) does not allow for overkill (saturation) when a high-LET particle traverses the cell nucleus.

Of course, the mere possibility to fit survival curves for various conditions by a linear-quadratic model like Eq. (1.57) (SINCLAIR, 1966; MILLAR *et al.*, 1978) is not a sufficient criterion for the principles of this theory. It is, in addition, possible to discuss, qualitatively, the influence of a number of factors, including radiation quality, in terms of the parameters $p\alpha$ and $p\beta$ as obtained from survival curve analysis. Thus, as long as no attempt is made to give *quantitative* predictions, one may believe in the philosophy which lies behind the expression (1.57), but there are serious arguments against it.

Mammalian cell survival curves for low-LET radiation exact that the quadratic term in Eq. (1.56) becomes appreciable or even predominating for doses as low as several hundreds of rads at the most. At such low doses the mean distance of SSB[1]) amounts to millions of DNA base pairs. DSB being formed by interaction over such large distances is not only unlikely,[2]) but there is also experimental evidence enough (see chapter 6) demonstrating that no quadratic component exists up to hundreds of kilorads,[3]) which was contested only by DUGLE *et al.* (1976). The experimental facts are in line with rigorous theoretical counter-arguments (KELLERER, 1976): energy interaction would have to be within *micrometre* distances to account for a quadratic dependence at low doses (see footnote 2 at page 20).

However, LEENHOUTS and CHADWICK (1974) have only evaluated the linear term, i.e. the coefficient $p\alpha$ as obtained, in dependence on LET, from survival curve fitting, by means of a target model of double-strand break formation[4]), arriving at a target diameter of about 1 nm. Thus, this attempt to make the approach predictive refers to the initial slope of survival curves only. As to the unpredicted quadratic term, one should at least qualitatively state that $p\beta$ would have to *decrease* with increasing LET if it actually reflected nothing but cumulative action of single-strand breaks, which are produced in lower number at higher LET (see chapter 6). This conclusion again conflicts with the known LET-dependence of survival curves if analysed according to the linear-quadratic model (1.57) (KELLERER *et al.*, 1976).

---

[1]) The experimental data accumulate around the figure 2.7×10^{-12} SSB produced per rad and Dalton in the DNA under *in-vivo* conditions (cf. chapter 6, Fig. 6.5).

[2]) To refute the argument, LEENHOUTS and CHADWICK (1976) suggested that a first SSB alters the configuration of DNA such that it becomes vulnerable to the induction of a break in the second strand, at the same place, by *different* radiolytic products having a *longer* diffusion distance as compared with those able to produce 'first' breaks. If this hypothesis were true, also the yield of SSB as measured in the alkaline sucrose gradient would have to show a linear-quadratic dependence on dose.

[3]) On the understanding of *in-vivo* conditions of irradiation, while in aqueous solution a quadratic component can be observed (FREIFELDER and TRUMBO, 1969) due to long-range radical action.

[4]) The general philosophy of this model resembles that of ours (section 1.5.1), but *two* parallel cylindrical targets of 1 nm diameter are considered assuming that *any* energy depositions in *both* adjacent targets brought about simultaneously by one event produce a double-strand break. This yields α as a function of LET going through a maximum like the initial-slope RBE for cell survival.

Finally it should be emphasized that CHADWICK's and LEENHOUTS' idea of a molecular theory of cell survival has not so much in common with the DNA-lesion theory developed in the present book, apart from the general attitude to take DNA lesions for the critical ones. If confining to double-strand breaks, which is indeed a very plausible hypothesis, one has nevertheless to assume dose-linearity of the yield, so that the cumulative action must be a matter of *biological* interaction of double-strand breaks. This is described by the general survival probability function $H(k)$ in Eq. (1.37). Furthermore, by virtue of the linear yield again, only the LET-dependent *statistics* (U_k) of breaks per cell nucleus can be responsible for the changing shape of survival curves.

It is worth while noting that Eq. (1.57) masks the latter problem, which is a general one: even though the concept were otherwise correct, the Poisson distribution by which Eq. (1.57) follows from Eq. (1.56) would not be adequate for high-LET ions that kill a cell by one traversal through the nucleus, while the Poisson distribution could only be justified by assuming less than one break to be produced on the average by one traversal. This shortcoming of the molecular theory of survival is quite the same as was encountered in the theory of dual radiation action (section 1.4.3). In fact, both theories have often been considered as closely related in the respect that an only differently interpreted linear-quadratic *mean yield* of lesions determines the probability of survival by the simple exponential relation (1.57).

Chapter 2

An approximate approach based on microdosimetry

2.1. Introduction

It is the most important statement of the preceding chapter that an adequate theory of RBE, except for very simple objects, requires to envisage a *gross sensitive volume* (GSV) both as a whole and regarding its *internal structure*. Therefore, RBE will depend on complex aspects of the radiation absorption stochastics such that customary microdosimetry does not provide a sufficient basis of description. This refers also to the DNA-lesion theory of RBE which is outlined in section 1.5. However, in addition to the strict but unwieldy way of evaluation which is the topic of the following chapters, one can make a very useful compromise by which microdosimetric distributions are utilized, nevertheless, to give the theory a more lucid comprehensive form. The present chapter is devoted to this kind of an approximate formulation of the DNA-lesion theory of RBE as well as the principles of application, the general interpretation of LET effects, and related topics. Its contents are largely those of an earlier paper (GÜNTHER *et al.*, 1977a).

The specific idea of the microdosimetric version of theory is to discriminate rigorously the two aspects of radiation quality pointed out in section 1.5.2: molecular lesion formation governed by energy deposition characteristics at the nanometre level, and statistics of lesions per GSV (cell nucleus). The latter aspect is explicit attributed to nothing but the fluctuations of specific energy z imparted to the GSV, irrespective of the former aspect which enters the theory *independently* as an *additional* radiation quality parameter r related to the intensity of lesion production (mean yield per unit dose) by the radiation considered.

This approach involves two essential advantages. First, physical particulars need not be considered; one arrives straightforwardly at the final equations in terms of the single-event distribution $f_1(z)$, which either can be directly measured or, if calculated, this is no direct topic of RBE theory. Second, it is possible *but not necessary* to base the complementary radiation quality parameter, r, upon the model of DNA-lesion production introduced in section 1.5.1 or a similar nanodosimetric model. This freedom makes the approach even more general than the strict one is. E.g., in a first attempt to analyse, say, survival curves in dependence on radiation quality, r may be treated as a freely adjustable parameter to allow subsequent inferences rather than introduce *a priori* model assumptions in this respect.

However, it is just the latter point that the approximation nature of the approach consists in, because the strict theory symbolized by the fundamental equation (1.37) includes the DNA-lesion production model itself in an involved manner, as indicated by the constant model parameters λ_j and l_l, so that using such a model from the outset is, strictly speaking, inevitable. In other words, it is not possible strictly to summarize the features of radiation quality at the nanometre level into the mere mean yield of some kind of critical molecular lesions, and eventually then to make the latter the only objective of nanodosimetric modelling as exemplified by Eq. (1.35).[1]) A profound analysis of the shortcomings and the associated errors in quantitative form must be relegated to chapter 8. There we will arrive at the following results.

The microdosimetric version of theory leads to *substantial* deviations from the strict version only for neutrons above 2 MeV,[2]) and, which is particularly important, the differences can be expressed formally by nothing but a changed single-event distribution, $f_1(z)$, of specific energy in the GSV. Hence, there are *no qualitative differences,* so that the microdosimetric version dealt with in this chapter may, as it were, be looked upon as an adequate 'model' of the strict theory.

That is why *the present chapter plays a key role* for the entire book. In fact, it does not only reveal the characteristic features and capabilities of the general theory of cellular radiation effects, but it also introduces all the essential conceptions involved and practical procedures applied, such as formal analysis of survival curves, principles of survival curve prediction, concentration and saturation effects of the first and second kinds, relationships between cellular radiosensitivity and RBE, etc. Thus, it is this chapter that elucidates the intrinsic ideas of the theory, while the following chapters enter into the particulars of either the underlying physics and mathematics or the methods and results of applications.

2.2. Formalism of the microdosimetric version of theory

2.2.1. The rationale of the approach

As a possible starting-point let us quote Eq. (1.24) together with the statement (page 15) that this equation, by itself, cannot be wrong:

$$S(D) = \int_0^\infty dz S_0(z)\, f(z; D). \tag{2.1}$$

What must be rejected is only the idea that the function $S_0(z)$, which describes the biological response (probability of survival) to a given value of z, the specific energy in the GSV, could be independent of radiation quality. Hence, it should be possible to specify that function as a dependency on radiation quality, in conformability to the principles of the DNA-lesion theory of survival outlined in section 1.5.

Let s be the radiation-quality-dependent *rate of lesion production* (number of lesions per unit dose and unit molecular weight of DNA), and Y the total amount of DNA per cell nucleus in corresponding units. Then the dose D produces

$$N = sYD \tag{2.2}$$

---

[1]) This 'non-separability' should also not be ignored when attempting, as envisaged by KELLERER (1978), to generalize the theory of dual radiation action with respect to the influence of radiation quality on the formation of sublesions (presumably double-strand breaks).

[2]) This holds at least with respect to mammalian cell killing.

lesions on the average per cell nucleus. Now it is plausible, but not entirely stringent (see below), to assume that if one singles out those nuclei which have received the 'actual dose' z the mean lesion number will read

$$N = sYz, \tag{2.3}$$

and the precise number, k, of lesions in anyone of these nuclei will be a matter of chance controlled by Poisson statistics. $S_0(z)$, the probability of survival for that sub-group, now evidently turns out to be the corresponding mean of the survival probabilities $H(k)$ introduced in section 1.5.2. Thus

$$S_0(z) = \sum_{k=0}^{\infty} H(k) \frac{(s_0 Yz)^k}{k!} e^{-s_0 Yz} \tag{2.4}$$

holds for ^{60}Co γ-rays to which the lesion production rate s_0 refers by definition. In the general case, s has to be substituted or s_0, so that in terms of the ratio

$$r = \frac{s}{s_0} \tag{2.5}$$

the initial equation (2.1) takes the form

$$S(D) = \int_0^{\infty} dz S_0(rz)\, f(z; D) \tag{2.6}$$

with $S_0(z)$ given by the expression (2.4). Thus, the objective of this subsection has been achieved.

Eq. (2.6) exhibits the twofold influence of radiation quality on dose-effect relations. In addition to the distribution of specific energy imparted to the GSV (cell nucleus), there is a dimensionless factor r, the RBE for the production of critical primary lesions, briefly referred to as the *primary lesion RBE*, the dependence of which on radiation quality is, however, unknown as yet. Nevertheless, the mere *form* of Eq. (2.6) constitutes a meaningful specification of the general statement that $S_0(z)$ in Eq. (2.1) must 'somehow' depend on radiation quality, realizing that the function $S_0(z)$ according to Eq. (2.4) *is* now postulated to be *the same for any radiation*. Hence, it is well possible to test the consistency of the approach as such, which is done in section 2.4.

Let us now point out the approximation character of the approach. If the *submicroscopic* characteristics of charged particles' tracks[1]) were always the same, relation (2.3) and also the Poisson distribution for the lesion number k could be fully justified, because then a given track length in 'nuclear matter' would always correspond to the same mean number of lesions produced, and the overall length of these homogeneous tracks within a cell nucleus would be strictly proportional to z.[2]) In actuality, however, there are various particles of changing energy, in general, and Eq. (2.2) refers to the corresponding *total* average which will not be applicable to the evidently *non-random* set of tracks in the nucleus that cause some *definite* value of z. In other words, there are statistical correlations between specific energy and involved particles. Consequently, in so far as the lesion production rate depends on the latter, neither the linear relation (2.3) nor the Poisson distribution (2.4) are strictly valid. To express it more generally, even to a well-defined radiation one cannot ascribe a certain 'submicroscopic quality' once for all such that different energies imparted to a microscopic volume by that radiation would appear as nothing but different 'quantities' of the same 'quality'.

To avoid these shortcomings is not possible in the framework of the above formalism, but one can make a shift by summarizing the final equations in the form

$$S(D) = \sum_{k=0}^{\infty} H(k)\, U_k(D), \tag{2.7}$$

$$U_k(D) = \int_0^{\infty} dz (k!)^{-1} (sYz)^k e^{-sYz} f(z; D). \tag{2.8}$$

Note that this equivalent formulation corresponds exactly to that used in section 1.5.2 where $U_k(D)$ is defined generally as the lesion number distribution at given dose D. Now it is clear: what the microdosimetric version consists in, and what would therefore have to be given up in order to evaluate the DNA-lesion theory of cellular radiation effects

---

[1]) These include, e.g., incident ions, δ-electrons, recoil electrons and nuclei set in motion by photons and neutrons, respectively.

[2]) In view of straggling, one ought to speak of the expected value of track length at fixed z.

strictly, is just the attempt to connect $U_k(D)$ with $f(z; D)$ as expressed by the very simple relation (2.8). Hence, Eq. (2.7) is the *general* fundamental equation of the DNA-lesion theory, while the evaluation of the lesion number distribution $U_k(D)$ either can approximately be based on the corresponding specific-energy distribution and the overall lesion production rate (Eq. (2.8), microdosimetric version) or must be derived more profoundly from the statistics of specified particle tracks and associated efficiencies of lesion production, which would be impossible without using a model like that described in section 1.5.1.[1]) The latter way is implemented in chapter 3, and chapter 8 afterwards demonstrates the possibility of a *unified* description in terms of 'virtual microdosimetry' (heavy symbols) such that

$$U_k(D) = \int_0^\infty \mathrm{d}\boldsymbol{z}(k!)^{-1} (s Y \boldsymbol{z})^k \, \mathrm{e}^{-sY\boldsymbol{z}} \boldsymbol{f}(\boldsymbol{z}; D) \qquad (2.9)$$

holds for the strict $U_k(D)$, in analogy to Eq. (2.8). Though (2.9) is only a formal representation ($\boldsymbol{z}$ is no energy in the physical sense), it proves to be very useful; a certain lesion type given,[2]) the virtual single-event spectrum $\boldsymbol{f}_1(\boldsymbol{z})$ underlying $\boldsymbol{f}(\boldsymbol{z}; D)$ can really be calculated and compared to $f_1(z)$.

Finally, it should be emphasized that all the properties and aspects of the theory which are pointed out in section 1.5 fully apply also to the microdosimetric version. Among other things, this refers to the interpretation of the effect-probability function $H(k)$, the principal limitations expressed in the footnote 1 at page 20 and, which is particularly worth noting, the general scope of the theory with respect to both inhomogeneous cell populations (biological variability) and the possibility of various kinds of lesions being involved in the cellular injury process. The latter two points and related proofs outstanding are dealt with in an appendix to this chapter (section 2.6).

---

[1]) This target model of lesion formation allows specifying the share of diverse charged particles and energies in the overall production of lesions described by s.

[2]) This does not matter, of course, with respect to $f(z; D)$ occurring in Eq. (2.8), and that is just the crucial difference.

2.2.2. Derivation of the fundamental equation

To obtain, from Eq. (2.6), an explicit expression for the dose-effect relation in dependence on radiation quality requires using an ansatz for $H(k)$ determining $S_0(z)$ according to (2.4). This ansatz, on the one hand, should be quite flexible in order to avoid any restraint to be imposed on $S_0(z)$, and it should, on the other hand, manage the analytical evaluation of the theory. The expression

$$H(k) = \sum_{\mu=1}^{n} B_\mu \, \mathrm{e}^{-Z_\mu k} \qquad (2.10)$$

with arbitrary parameters n, B_μ, Z_μ, which was already introduced in chapter 1 (cf. Eq. (1.38)), meets these requirements. Nothing else is the meaning of (2.10), i.e. strong emphasis must be given to the statement that the expression (2.10) does not in the least constitute anything like a 'model'. Cf. section 1.5.3 to realize what could be regarded as a biological model of action in terms of $H(k)$.

Eqs. (2.4) to (2.6) and (2.10) in united form read

$$S(D) = \int_0^\infty \mathrm{d}z \sum_{\mu=1}^{n} B_\mu \sum_{k=0}^{\infty} (k!)^{-1} \times (\mathrm{e}^{-Z_\mu} s Y z)^k \, \mathrm{e}^{-sYz} f(z; D),$$

$$S(D) = \sum_{\mu=1}^{n} B_\mu \int_0^\infty \mathrm{d}z \times \exp\left[-(1 - \mathrm{e}^{-Z_\mu}) \, s Y z\right] f(z; D). \qquad (2.11)$$

This can further be evaluated and expressed ultimately in terms of the single-event distribution $f_1(z)$ by utilizing fundamental algorithms of microdosimetry described in section 1.4.1. First, one has to substitute the series expansion (1.14) for $f(z; D)$,

$$S(D) = \sum_{\mu=1}^{n} B_\mu \sum_{m=0}^{\infty} (m!)^{-1} (D/\bar{z}_F)^m \times \mathrm{e}^{-D/\bar{z}_F} \int_0^\infty \mathrm{d}z \times \exp\left[-(1 - \mathrm{e}^{-Z_\mu}) \, s Y z\right] f_m(z), \qquad (2.12)$$

so that the residual problem is focused in an integral of the form

$$I_m = \int_0^\infty dz \exp(-\eta z) f_m(z)$$

in which $f_m(z)$ can be replaced by the recurrence expression (1.15). This yields

$$I_m = \int_0^\infty dz\, e^{-\eta z} \int_0^z dz' f_{m-1}(z')\, f_1(z-z')$$

$$= \int_0^\infty dz' f_{m-1}(z')\, e^{-\eta z'} \int_{z'}^\infty dz\, e^{-\eta(z-z')} f_1(z-z')$$

$$= I_{m-1} \int_0^\infty dz\, e^{-\eta z} f_1(z),$$

a recurrence formula with the evident solution

$$I_m = \int_0^\infty dz\, e^{-\eta z} f_m(z)$$

$$= \left[\int_0^\infty dz\, e^{-\eta z} f_1(z)\right]^m. \tag{2.13}$$

This relation makes the summation at the right-hand side of Eq. (2.12) feasible:

$$S(D) = \sum_{\mu=1}^{n} B_\mu \exp\left\{-(D/\bar{z}_F) \int_0^\infty dz \right.$$

$$\left. \times \left[1 - \exp\left(-s(1 - e^{-Z_\mu}) Yz\right)\right] f_1(z)\right\}. \tag{2.14}$$

2.2.3. The resultant procedure

Eq. (2.14) is the fundamental formula of the microdosimetric version of DNA-lesion theory of cellular radiation effects. It displays the dependence of the dose-effect relation $S(D)$ on radiation quality explicitly in terms of the single-event distribution, $f_1(z)$, and the rate of lesion production, s, bearing in mind that the parameters n, B_μ, Z_μ, and Y as well as the diameter of the GSV are *characteristic constants* associated with the given object and its radiobiological properties with respect to the given endpoint. These rather abstract parameters shall now be partly eliminated utilizing *experimentally accessible* information. In this way, Eq. (2.14) will be converted into a form suitable for direct application.

In the initial equation (2.6) it is only the function $S_0(z)$ that comprises all that does not change with radiation quality. This function has a particular meaning as follows. The defining equation (2.4) with (2.10) inserted leads to the compact expression

$$S_0(D) = \sum_{\mu=1}^{n} B_\mu\, e^{-\alpha_\mu^{(0)} D}, \tag{2.15}$$

where, as an abbreviation, the parameter set

$$\alpha_\mu^{(0)} = (1 - e^{-Z_\mu})\, s_0 Y \tag{2.16}$$

has been introduced. D is used for the independent variable in the relation (2.15), because $S_0(D)$ can be referred to as the *zero-LET dose-effect relation* if one defines a hypothetical 'zero-LET radiation' by $f(z; D) = \delta(z - D)$ and assigns to it the lesion production rate $s = s_0$ of ^{60}Co γ-rays. On this understanding, Eq. (2.6) yields $S(D) \equiv S_0(D)$, indeed.

By introducing the 'zero-LET coefficients' $\alpha_\mu^{(0)}$ in conformity with Eq. (2.16), the fundamental equation (2.14) writes

$$S(D) = \sum_{\mu=1}^{n} B_\mu\, e^{-\alpha_\mu D},$$

$$\alpha_\mu = \frac{1}{\bar{z}_F} \int_0^\infty dz(1 - e^{-r\alpha_\mu^{(0)} z})\, f_1(z). \tag{2.17}$$

Hence, the formalism describes a dose-effect curve always as a *sum of exponential functions* in which only the coefficients in the exponents change with radiation quality. Those for zero-LET radiation, $\alpha_\mu^{(0)}$, in addition to B_μ, i.e. the complete zero-LET dose-effect relation in the form of Eq. (2.15), must be known as the starting-point to predict $S(D)$ for *any* radiation. That is the spirit of the relation (2.17) in connection with (2.15). Henceforth, taking the most general point of view, we may forget particulars of the idea of damage by DNA-lesions, such as $H(k)$ and Y, and consider Eq. (2.17) as nothing but the evaluation of the *initial postulate* (2.6) with the *ansatz* (2.15) put to use.

Of course, the hypothetical $S_0(D)$ cannot be measured direct, but it is evident that a dose-effect curve measured with low-LET radiation (e.g., γ-rays or high-voltage X-rays, in the following generally signified by the index X) will not be very different from $S_0(D)$. Thus, if a low-LET dose-effect relation $S_X(D)$ has been determined experimentally and expressed mathematically as a sum of exponential functions,

$$S_X(D) = \sum_{\mu=1}^{n} B_\mu \, e^{-\alpha_\mu^{(X)} D}, \qquad (2.18)$$

we may roughly identify $\alpha_\mu^{(X)}$ with $\alpha_\mu^{(0)}$ for using it in Eq. (2.17) in order to calculate $S(D)$ for other radiations. However, it is not difficult in practice to avoid this approximation. One has only to reverse the relation (2.17) such that each of the 'measured' $\alpha_\mu^{(X)}$ appears as the result of the integral calculated with the corresponding right $\alpha_\mu^{(0)}$ one is searching for. While this reversion may raise difficulties in general, it is easily feasible on condition that $\alpha_\mu^{(X)}$ refers to low-LET X- or γ-rays, which produce only small z values with the spectrum $f_1^{(X)}(z)$, so that the possible series expansion of the exponential in the integrand can be broken off after the quadratic term. Taking into consideration the definitions (1.13) of mean values of specific energy and solving the residual quadratic equation to an adequate approximation yields

$$\alpha_\mu^{(0)} = \frac{1}{r_X} \alpha_\mu^{(X)} \left(1 + \frac{1}{2} \bar{z}_D^{(X)} \alpha_\mu^{(X)}\right), \qquad (2.19)$$

where an X-ray r-value different from unity is admitted, too.[1]) Thus, the dose-mean of specific energy for the reference X-radiation is the crucial quantity that marks the difference between 'low LET' and 'zero LET', and it is small indeed ($\bar{z}_D = 0.14$ to 6 rad for $d = 10$ to 2 μm in the case of ^{60}Co γ-rays (DVORAK, 1975), about threefold larger values for 250-kVp X-rays). Therefore, it is absolutely unnecessary to go beyond the approximation (2.19) as long as normal X- or γ-rays are involved, since $1/\alpha_\mu^{(X)}$ is of the order of the well-known D_0 (some tens or even hundreds of rads).

Now the algorithm for the description of the dependence of dose-effect curves on radiation quality is complete. From the mathematical analysis of a single experimental low-LET curve according to Eq. (2.18) the parameters n, B_μ and $\alpha_\mu^{(X)}$ are obtained.[2]) The latter are converted into $\alpha_\mu^{(0)}$ by means of relation (2 19). Finally, Eq. (2.17) makes accessible $S(D)$ for *any* radiation in terms of the microdosimetric distribution $f_1(z)$, depending on the diameter of the GSV (cell nucleus), and the primary lesion RBE, r. The latter will be studied in section 2.5, but even with r left to free choice, Eq. (2.17) predicts at least the *shape* of dose-effect curves. If agreement with experimental curves has thus been achieved for a series of radiations, the 'empirical' r-values found form a new kind of information which provides deeper insights into characteristics of the primary molecular lesions that underlie a given biological effect.

---

[1]) One may also set $r_X = 1$ stipulating that every r-value refers to the X-rays in question, i.e. changing the definition (2.5) into $r = s/s_X$.

[2]) This analysis, although by itself irrelevant to the theory, may become a cumbersome task, particularly as there is no unique solution (e.g., only a *lower* limit can exist for the number, n, of terms necessary to achieve a 'sufficient' fit of a given series of experimental values). However, a very convenient general procedure has been developed, which is described in section 9.3.

2.2.4. The $\bar{z}_D$-approximation

On the background of dual radiation action theory (see section 1.4.3), the dose-mean of specific energy, $\bar{z}_D$, is widely considered to be *the* parameter that specifies radiation quality. By contrast, the present approach does not only contain a second, non-microdosimetric parameter, r, which might well be looked upon as an analogon to k/k_X in non-orthodox DRA theory, but it also makes use of the *complete* energy deposition spectrum rather than a mere mean value of it. Nevertheless, it is worth while asking whether, in the latter respect, the mean value $\bar{z}_D$ alone may be adequate in special cases.

The thus envisaged $\bar{z}_D$-*approximation* has already been used in the preceding subsection with respect to low-LET radiation. It follows from the general relation (2.17) by break-off of series expansion in the integrand, ending up in

$$\alpha_\mu = r\alpha_\mu{}^{(0)} \left(1 - \frac{1}{2}\,\bar{z}_D\, r\alpha_\mu{}^{(0)}\right). \qquad (2.20)$$

The general condition of validity of this approximation, evidently, is that the expression in parentheses does not fall much below unity. This may always apply to low-LET radiation owing to $\bar{z}_D$ being small enough. Now, however, the question is to be asked from the viewpoint of whether *any* radiation quality allows the description in terms of the $\bar{z}_D$-approximation (2.20). It is obvious that this requires sufficiently small coefficients $\alpha_\mu{}^{(0)}$, which means (cf. Eq. (2.15)) that the object must be very *insensitive* with respect to the reaction considered. That is the prerequisite for the present theory to become largely equivalent, as to the final results, to the DRA theory in generalized form. This interesting question of DRA theory as a limiting case will be investigated in subsection 2.3.4.

2.3. Theoretical consequences for the relative biological effectiveness

2.3.1. Preliminary notes

It is a characteristic property of the present theory that it does *not* yield general expressions for the RBE in terms of somehow defined 'radiation quality parameters'. Instead, it exhibits the involvement of parameters that specify the particular biological system to determine the dependence of RBE on both radiation quality and dose. This influence comes about indirectly by the low-LET radiosensitivity parameters n, B_μ, $\alpha_\mu{}^{(0)}$ as well as the cell nucleus size entering the fundamental equation (2.17), which in turn leads to RBE, but, unfortunately, only by numerically solving the equation $S(D) = S_X(RBE \times D)$. Nevertheless, a number of analytical RBE relations can be derived either for special cases or as an approximation. In this way, a good survey will be obtained on what the theory predicts for the RBE as a function of dose, radiation quality, radiosensitivity, etc.

For the sake of definiteness, in the sequel, $S(D)$ is looked upon as a *dose-survival* relationship, i.e.

$$S(0) = 1, \quad \left.\frac{dS}{dD}\right|_{D=0} = S'(0) \leqq 0 \qquad (2.21)$$

holds for any radiation. Moreover, in the case of sigmoidal survival curves, having a shoulder,[1]) one can assume the second

---

[1]) The case of shoulder-less curves is trivial, because then relation (2.17) consists of one exponential only ($n = 1$). Hence, the purely exponential survival curves for any radiation imply a *single* RBE value to be ascribed to each radiation, so that the general final-slope RBE value (as $D \to \infty$), which is considered below, can stand for the RBE as such, in this case.

derivative to be always negative,

$$\left.\frac{d^2S}{dD^2}\right|_{D=0} = S''(0) < 0. \tag{2.22}$$

These conditions, of course, apply also to the limiting zero-LET relation $S_0(D)$, for which the same short-hand notation $S_0'(D)$, $S_0''(D)$ will be used for the derivatives. With respect to $S_0(D)$, we introduce, as a special parameter, the well-known inactivation dose D_0 which is defined as the inverse of the negative logarithmic derivative at *large* doses, and is obviously connected with the *smallest* $\alpha_\mu^{(0)}$ in the expression (2.15),

$$\alpha_1^{(0)} = D_0^{-1} = \lim_{D\to\infty} -\frac{S_0'(D)}{S_0(D)}, \tag{2.23}$$

taking $\mu = 1$ as a definition.

In general, the RBE of a given radiation can be considered as a function of dose. This will be expressed by the notation $RBE(D)$, but for the limiting cases $D = 0$ and $D = \infty$ we will adopt labelling the symbol RBE by subscripts that indicate the survival level (100% and 0%, respectively), which is equivalent in case that cell survival is the reaction in question. Another comment to be given in advance concerns the reference radiation. No special index will be used if RBE is meant relative to X-rays, as usual in practice, while from the viewpoint of theory it is most natural to relate RBE to the hypothetical zero-LET radiation characterized by $\bar{z}_D = 0$ and $r = 1$. If the latter is done, the superscript (0) will be added.

2.3.2. The limiting cases $D = \infty$ and $D = 0$

As $D \to \infty$, only the most slowly decreasing terms in both expressions, (2.17) and (2.18), need to be retained, so that the RBE becomes equal to the ratio of α_1 and $\alpha_1^{(X)}$. Substituting the $\bar{z}_D$-approximation (2.20) for the latter and using the notation (2.23) yields

$$RBE_{0\%} = \frac{1}{1 - \frac{1}{2}\, r_X \bar{z}_D^{(X)}/D_0}\, \frac{D_0}{r_X \bar{z}_F} \times \int_0^\infty dz(1 - e^{-rz/D_0})\, f_1(z). \tag{2.24}$$

In the opposite case, $D = 0$, similar reasoning leads to

$$RBE_{100\%} = \frac{-S'(0)}{-S_X'(0)} = \frac{\sum B_\mu \alpha_\mu}{\sum B_\mu \alpha_\mu^{(X)}} = \frac{\bar{z}_F^{-1} \int_0^\infty dz[1 - S_0(rz)]\, f_1(z)}{-S_0'(0)\, r_X - \frac{1}{2} S_0''(0)\, r_X^2 \bar{z}_D^{(X)}} \tag{2.25}$$

As will be seen in the following, these two formulas reflect some remarkable properties of RBE direct in terms of the zero-LET dose-survival relation $S_0(D)$, avoiding the representation (2.15). To make it more obvious, we may use $S_0(D)$ itself for the reference curve, setting $\bar{z}_D^{(X)} = 0$, $r_X = 1$, and write, in consideration of Eq. (1.13),

$$RBE_{0\%}^{(0)} = r\, \frac{\int_0^\infty dz(1 - e^{-rz/D_0})\, f_1(z)}{\int_0^\infty dz\, (rz/D_0)\, f_1(z)} < r, \tag{2.26}$$

$$RBE_{100\%}^{(0)} = r\, \frac{\int_0^\infty dz[1 - S_0(rz)]\, f_1(z)}{\int_0^\infty dz[-S_0'(0) rz]\, f_1(z)} \lesseqgtr r, \tag{2.27}$$

which is not very different from (2.24) and (2.25) unless $S_0'(0)$ vanishes.

Eq. (2.26) shows that the 'final-slope RBE' is always smaller than r, the RBE for the critical lesions underlying the effect considered. Only for very radiation-resistant objects ($D_0 \gg r\bar{z}_D$) the limiting

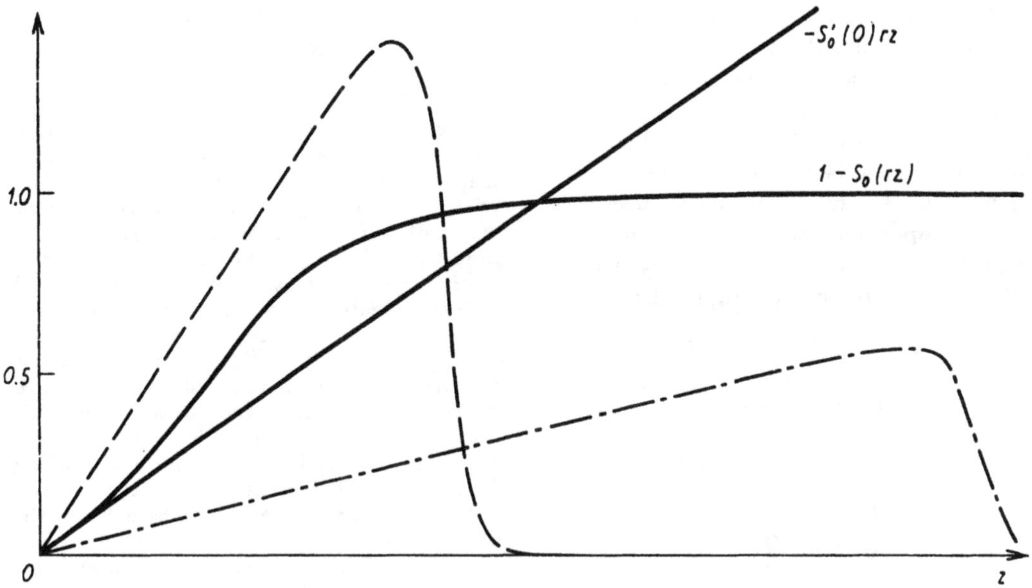

Fig. 2.1. Graphs of functions determining the 100%-survival RBE according to Eq. (2.27). Broken lines signify $f_1(z)$ for more sparsely (- - -) and more densely (-·-·-·) ionizing radiations.

value $RBE_{0\%}^{(0)} = r$ will be reached, while the general rule Eq. (2.26) displays reads: the RBE for low levels of survival is the smaller the more sensitive the cells are in terms of D_0. This rule is put to the test of experimental data in section 7.4.5 of chapter 7.

By contrast, the 'initial-slope RBE' can be both greater and smaller than r, in dependence on radiation quality. This statement is most easy to visualize by plotting the two square-bracket factors from the integrands in Eq. (2.27), on the understanding that $S_0(D)$ represents a sigmoidal curve. This plot, shown in Figure 2.1, reveals also that $RBE_{100\%}^{(0)} < r$ holds only in the range of sufficiently high LET. Low-LET radiations, on the contrary, allow series expansion of $S_0(rz)$ in Eq. (2.27), yielding the opposite result

$$RBE_{100\%}^{(0)} = r\left[1 + \frac{1}{2}\frac{S_0''(0)}{S_0'(0)}\, r\bar{z}_D - \cdots\right]. \tag{2.28}$$

This equation reflects the general experimental fact that the RBE for a high survival level increases rapidly with growing LET (approximately $\bar{z}_D \sim LET$, apart from r), particularly if the initial slope, $S_0'(0)$, of the reference curve is small. The latter implies a shallow survival curve shoulder (pronounced repair of sublethal damage), which is indeed known to cause a large RBE. Hence, with respect to high levels of survival (small doses), too, we find that radiosensitivity and RBE are negatively correlated. However, from the viewpoint of experimental evidence, this statement may appear rather trivial (reducible to the well-known change of survival curve shape with varying LET) as compared with the rule pointed out above.

These few considerations demonstrate vigorously that the RBE for biological effects at the cellular level is *not* quite simply related to the primary lesion RBE, r. Therefore, only the latter may, but the former cannot, be regarded as though directly reflecting energy deposition requirements in small targets. This question will further be studied in sections 2.4 and 2.5.

Relation (2.27) and its $\bar{z}_D$-version (2.28) become inapplicable as $S'(0)$ tends to zero. In this case, one has to go back to the full expression (2.25), i.e. to relate the RBE to a real X-radiation the non-vanishing $\bar{z}_D^{(X)}$ of which then becomes just the decisive quantity. Furthermore, it is useful to derive upper bounds to $RBE_{100\%}$ by series expansion of $S_0(rz)$, in analogy to Eq. (2.28), and reference to Figure 2.1:

$$RBE_{100\%} < \left(\frac{r}{r_X}\right)^2 \frac{\bar{z}_D + \dfrac{2}{r}\dfrac{S_0'(0)}{S_0''(0)}}{\bar{z}_D^{(X)} + \dfrac{2}{r_X}\dfrac{S_0'(0)}{S_0''(0)}} \leqq \left(\frac{r}{r_X}\right)^2 \frac{\bar{z}_D}{\bar{z}_D^{(X)}}. \tag{2.29}$$

The right-hand expression, which depends only on radiation quality[1]) provided that the critical lesion type determining r is settled, constitutes the largest RBE possible at all in every respect, i.e. regarding variations of dose, radiosensitivity, and survival curve shape. The RBE can approach this value only if there is no linear component of response $\left(S_0'(0) = 0\right)$ and the radiosensitivity is low ($\bar{z}_D$-approximation) and the dose tends to zero (100% survival).

The expressions (2.29) have much in common with results of the theory of dual radiation action (cf. Eq. (1.30) together with either (1.26) or (1.27) including the pertinent footnote 2 at page 17). Obviously r/r_X is equivalent to $\sqrt{k/k_X}$ in DRA theory. However, the point is that the present theory qualifies these formulas as referring to *limiting cases only*. On the other hand, if the associated prerequisites should be met, *complete* formal correspondence will be achieved between both theories, as demonstrated in subsection 2.3.4.

1) This implies, in particular, the fact that the *ratio* $\bar{z}_D/\bar{z}_D^{(X)}$ usually exhibits very week dependence on the size of the sensitive region.

2.3.3. Inadequacy of the energy-only concept

The investigations of the last subsection make it now very easy to supplement the definitive proof of a central theorem in the previous chapter, namely the inability of Eq. (1.24), the 'energy-only concept', to yield an RBE greater than unity for the limit of vanishing survival. Note that all the present equations are stringently derived from Eq. (2.6) which is identical with the equation (1.24) in question if one sets $r = 1$. It is true that the derivation makes use of the ansatz (2.15) for $S_0(D)$, but this does not constitute a restriction except for assuming a smallest $\alpha_{\mu_i}^{(0)}$ to exist and to be finite. This assumption means only that the survival curves do not unlimitedly continue bending downwards as D increases, but that there is a final slope to be reached somewhere. Thus, the prerequisite is the mere existence of a D_0 value occurring in the relation (2.26), and this relation then implies the proof of the theorem,[2]) since $r = 1$ and $[1 - \exp(-z/D_0)] < z/D_0$.
As an example for *non-existence* of a final slope consider the model

$$S_0(D) = e^{-\beta D^2}, \tag{2.30}$$

which cannot be expressed in the form of Eq. (2.15). Nevertheless, an approximate description in this form in the range of, say, $S_0(D) > 10^{-3}$ is possible (see section 9.3.4), and this way it is shown in section 8.4 that the model (2.30), too, bears out the inequality $RBE < 1$ for survival levels ranging from 0% up to even 50%.

2.3.4. $\bar{z}_D$-approximation as a limiting case

In this subsection, let us investigate what turns out for the RBE if Eq. (2.20) is used throughout, *not* as an approximation, but in conditions that make it strictly valid in the sense of a limiting case. As shown in subsection 2.2.4, this requires assuming $\alpha_\mu^{(0)} \to 0$. This limiting assumption of infinite radioresistance leads to two distinct alternatives depending on whether

2) One may also refer to Eq. (2.24), assuming $\bar{z}_D > \bar{z}_D^{(X)}$ as the condition for the radiation in question to be of 'higher LET' as compared with the reference X-rays.

there is no linear component of response, i.e.

$$S_0'(0) = 0, \tag{2.31}$$

or it does exist. In the latter case the result is trivial, because the prerequisite $\alpha_\mu^{(0)} \to 0$ for the formula (2.20) to become strictly valid just causes that $\bar{z}_D$ drops out altogether, so that $\alpha_\mu = r\alpha_\mu^{(0)}$ holds, which implies the RBE to be constant:

$$RBE(D) = \frac{r}{r_X}. \tag{2.32}$$

However, as shown below, this reasoning does not apply to the special case (2.31), which is, therefore, a natural implicate of the $\bar{z}_D$-limit of the theory.

Let us start with the series expansion of both $S(D)$ and $S_X(D_X)$ in the equation $S(D) = S_X(D_X)$ which relates the high-LET and low-LET doses determining the RBE by the quotient D_X/D:

$$1 + S'(0)\,D + \frac{1}{2}\,S''(0)\,D^2 + \cdots$$
$$= 1 + S_X'(0)\,D_X + \frac{1}{2}\,S_X''(0)\,D_X^2 + \cdots. \tag{2.33}$$

Substituting the expression (2.17) for both $S(D)$ and $S_X(D)$ with α_μ and $\alpha_\mu^{(X)}$, respectively, expressed according to Eq. (2.20) yields

$$\left[-r\sum B_\mu\alpha_\mu^{(0)} + \frac{1}{2}\,r^2\bar{z}_D\sum B_\mu\alpha_\mu^{(0)2}\right] D + \frac{1}{2}\,[r^2\sum B_\mu\alpha_\mu^{(0)2}]\,D^2$$
$$= \left[-r_X\sum B_\mu\alpha_\mu^{(0)} + \frac{1}{2}r_X^2\bar{z}_D^{(X)}\sum B_\mu\alpha_\mu^{(0)2}\right] \times D_X + \frac{1}{2}\,[r_X^2\sum B_\mu\alpha_\mu^{(0)2}]\,D_X^2, \tag{2.34}$$

where terms of higher than quadratic order of $\alpha_\mu^{(0)}$ are omitted due to the subsequent limiting process $\alpha_\mu^{(0)} \to 0$. It would even have been possible to neglect the quadratic terms, with the consequence (2.32), unless there were the condition (2.31) which writes

$$S_0'(0) = -\sum B_\mu\alpha_\mu^{(0)} = 0. \tag{2.35}$$

On account of Eq. (2.35), the solution of Eq. (2.34) for the ratio D_X/D reads

$$RBE(D) = \frac{2\left(\frac{r}{r_X}\right)^2\left(1 + \frac{D}{\bar{z}_D}\right)}{\frac{\bar{z}_D^{(X)}}{\bar{z}_D} + \sqrt{\left(\frac{\bar{z}_D^{(X)}}{\bar{z}_D}\right)^2 + 4\left(\frac{r}{r_X}\right)^2\left(1 + \frac{D}{\bar{z}_D}\right)\frac{D}{\bar{z}_D}}}. \tag{2.36}$$

Thus, we have found the well-known RBE-dose relationship of the theory of dual radiation action (cf. Eq. (1.29) together with Eq. (1.26)), in the following referred to as the 'DRA-limit of the theory'.- The radiation-quality-dependent ratio r/r_X, the relative effectiveness as to primary lesion production, stands for $\sqrt{k/k_X}$ in DRA theory. The pure dual nature of damage, which just makes $\bar{z}_D$ the second relevant radiation quality parameter, is due to the special condition (2.31) *and* the assumption of low sensitivity (neglect of $S_0'''(0)$ etc.).[1] However, it is important to note that $\bar{z}_D$ now refers to the gross sensitive volume as a whole instead of smaller sites in it.

---

[1]) Realize that the limiting process $\alpha_\mu^{(0)} \to 0$ as the justification for the series expansion break-off in Eq. (2.34) must not be taken literally. Actually, terms of higher order must be negligible only up to doses D large enough to permit neglect of the dose-linear terms, i.e. $D \gg \bar{z}_D$. This suffices to reach the plateau $RBE \approx r/r_X$, which then continues to hold at further increasing dose owing to the relation (2.24) for $D \to \infty$ yielding the same value on the conditions specified.
Hence, the genuine criterion for the validity of the DRA-limit of the theory reads $S(D) = 1 + S'(0)\,D + \frac{1}{2}\,S''(0)\,D^2$ up to at least $D = \bar{z}_D$. If this is fulfilled, the RBE-dose relation does not appreciably depart from (2.36), in case that $|S_0'(0)| \ll \frac{1}{2}\,r_X\bar{z}_D^{(X)}|S_0''(0)|$, and from (2.32) in the opposite case.

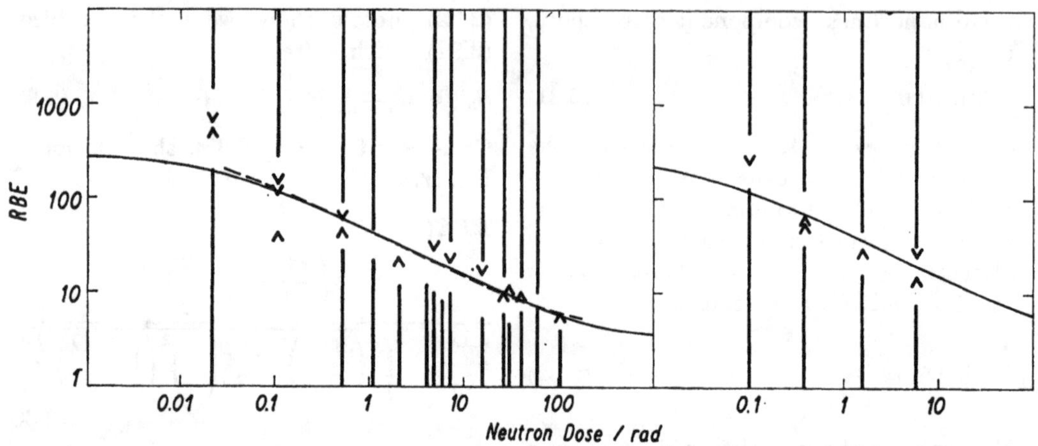

Fig. 2.2. RBE as a function of dose of 0.43-MeV neutrons relative to 250-kVp X-rays for the induction of eye lens opacification in mice (BATEMAN *et al.*, 1972; left-hand panel) and mammary neoplasms in rats (SHELLABARGER *et al.*, 1974; right-hand panel).
Bars indicate RBE ranges excluded with a certainty of 99% or 95% (level of significance); arrows mark non-significant limits.
Unbroken lines are calculated according to theory assuming $r/r_X = 3.3$ and $\sigma_0 = 5\ \mu m^2$ (sensitive volume cross-section). The latter value implies $\bar{z}_D = 228$ rad and $\bar{z}_D^{(X)} = 8.78$ rad (see Tab. II and III in the appendix).
For the dashed curve see text.

Because it constitutes the absolutely upper theoretical limit of RBE, the *low-dose plateau* (2.29) of the relationship (2.36) is most interesting, whilst in the range of medium and high doses (cf. Eqs. (1.31), (1.32)) experimental evidence for the DRA limit of the theory to be valid is never unambiguous. Only for two mammalian systems, very high RBE values (of fast neutrons) at very low doses have been reliably measured, suggesting Eq. (2.36) to hold with r/r_X clearly in excess of unity, since the data already exceed $\bar{z}_D/\bar{z}_D^{(X)}$ substantially though the plateau itself is not quite reached. These results are shown in Figure 2.2. The identical unbroken curves are a fit according to Eq. (2.36); only the size of the sensitive region to which $\bar{z}_D$ refers has been varied,[1]) while the ratio, r/r_X, of primary lesion production RBE has been taken to be 3.3 by anticipation of the so-called universal r-values for mammalian cell killing established in chapter 7 (see Tab. VIII in the appendix). A similar fit based on DRA theory with $\bar{z}_D = 150$ rad, $\bar{z}_D^{(X)} = 5$ rad, $k/k_X = 16$, was presented by the authors of one of the experiments (dashed line). Now, which is important, the DRA-limit of the DNA-lesion theory of RBE does not only bear out this earlier fit, but the latter theory also predicts the definite value of r/r_X agreeing fairly with the parameter $\sqrt{k/k_X} = 4$.

For the examples discussed, cellular dose-response curves are not available, but it was possible to infer indirectly, from measured RBE values, that these systems just obey the DRA-limit of the theory. In other words, which is very remarkable, the distinctly highest RBE factors ever found have the same magnitude for the two very different biological systems they refer to *and* coincide with the theoretical upper limit. Hence, the maximum RBE (2.29), for which the theory can predict unambiguous values after r being settled

[1]) Thereby, $\bar{z}_D^{(X)}$ follows automatically with the ratio $\bar{z}_D/\bar{z}_D^{(X)}$ being nearly independent of the region size.

in dependence on radiation quality,[1]) *has* significance, in addition to individual predictions of RBE that will be made routinely on the basis of known low-LET dose-response relations. We will revert to the question of limiting RBE in section 7.4.5 of chapter 7.

2.3.5. An approximate general RBE formula

Let us start with the DRA equation (1.29) again, but now considered from a purely formal point of view, taking

$$RBE(D) = \frac{2\frac{k}{k_X}\left(1+\frac{D}{\lambda}\right)}{\frac{\lambda_X}{\lambda}+\sqrt{\left(\frac{\lambda_X}{\lambda}\right)^2+4\frac{k}{k_X}\left(1+\frac{D}{\lambda}\right)\frac{D}{\lambda}}} \qquad (2.37)$$

merely as a three-parameter mathematical expression to describe RBE-dose relationships. As pointed out in section 1.4.3, the weakness of DRA theory does not consist in an inability of this expression to fit experimental data, but in the lack of correct predictions, or of any predictions if the most general interpretation is adopted, concerning the parameters as functions of radiation quality. This suggests asking whether the DNA-lesion theory of RBE in general can be brought into the form (2.37), what the parameters then look like, and how they depend on radiation quality.

If one adopts the expression (2.37) as an approximation and insists that the limiting values $RBE(\infty)$ and $RBE(0)$ be obtained correctly corresponding to the theory, i.e. in keeping with the formulas (2.24) and (2.25), there remains little latitude in the choice of the parameters. k/k_X and λ_X/λ are unequivocally settled this way, and the separation of λ and λ_X is subject to several conditions[2]), so that

$$\frac{k}{k_X} = \left[\frac{D_0\int\limits_0^\infty dz(1-e^{-rz/D_0})\,f_1(z)}{r_X\bar{z}_F\left(1-\frac{1}{2}r_X\bar{z}_D^{(X)}/D_0\right)}\right]^2, \qquad (2.38)$$

$$\lambda = \frac{2\bar{z}_F\int\limits_0^\infty dz[1-S_0(rz)]\,f_1(z)}{-S_0''(0)\left[D_0\int\limits_0^\infty dz(1-e^{-rz/D_0})\,f_1(z)\right]^2}, \qquad (2.39)$$

$$\lambda_X = \frac{\frac{2}{r_X}\frac{S_0'(0)}{S_0''(0)}+\bar{z}_D^{(X)}}{\left[1-\frac{1}{2}r_X\bar{z}_D^{(X)}/D_0\right]^2} \qquad (2.40)$$

constitutes the almost only reasonable possibility to choose parameters that might eventually make the expression (2.37) match what would be obtained by exact application of the theory, i.e. $S(D) = S_X(D_X)$ and $RBE = D_X/D$. Indeed, this seems to be the case, judging from several tests which have been carried out.

Very good numerical agreement was obtained for the model (2.30), which corresponds to a curve with a very large extrapolation number E and vanishing initial slope (for details see section 8.4). For another class of dose-effect relations, which just are of opposite character ($E \leqq 2$, $-S_0'(0) \geqq 0$), the test is demonstrated in the appendix to this chapter

---

[1]) Note that this is practically only a matter of r, since the ratio $\bar{z}_D/\bar{z}_D^{(X)}$, equal to $\bar{y}_D/\bar{y}_D^{(X)}$, shows little dependence on the size of the sensitive volume; it is nearly a pure physical parameter of the involved radiations as such.

[2]) λ_X should be independent of the radiation the RBE refers to; under the conditions of subsection 2.3.4 one should arrive at the limiting formula (2.36), i.e. $\lambda = \bar{z}_D$; in the special case of vanishing λ_X/λ, the resulting asymptotic relation $RBE = (k/k_X)^{1/2}(D/\lambda)^{-1/2}$ for small doses must be compatible with the corresponding asymptotic behaviour of the survival curves, i.e. $S(D) = 1 + S'(0)\,D$ and $S_X(D_X) = S_0(D_X) = 1 + S_0''(0)\,D_X^2/2$.

(section 2.6.3). Of course, the adequacy of the expression (2.37) will depend on the particular $S_0(D)$, and its general applicability is, therefore, hard to prove strictly, but there is considerable evidence for the errors to rarely exceed 10 per cent.

The thus-established approximate algorithm of direct calculating the RBE of radiations is interesting, primarily, from the theoretical point of view. The formulas exhibit obviously how the RBE is influenced by radiobiological features of the particular object in terms of $S_0(D)$ and the pertinent characteristic parameters $S_0'(0)$, $S_0''(0)$, and D_0, whereby the provisional representation (2.15) of dose-effect curves has become superfluous. Furthermore, it is important to realize that, as a consequence of Eq. (2.37) being generally applicable, the nature of the relation between RBE and dose does not depend appreciably on the underlying mechanisms of radiation action. Neither can specific conclusions be drawn from experimental facts concerning the character of this relation only, nor does it constitute a significant criterion of theoretical models. E.g., the general relation (2.37) always includes a more or less pronounced medium region of dose where $RBE \sim D^{-1/2}$ (cf. Eq. (1.31)), although the DNA-lesion theory of RBE does not generally imply the hypothesis of dual radiation action. Probably, any right theory of RBE can approximately be brought into the form (2.37), so that the crucial item, then, is what it predicts for the absolute values of the parameters as well as their dependence on radiation quality.

As to the latter question, one may analyse, by means of Eq. (2.37), experimental data obtained with several different radiation qualities. In the case of mammalian cell survival, this yields $k/k_X > 1$ for, e.g., fast neutrons and λ values which are *not* correlated with $\bar{z}_D$ in dependence on neutron energy (as a consequence of admitting k/k_X to vary with radiation quality; cf. section 1.4.3). While such a correlation is just the central postulate of DRA theory, even if its most general version is adopted, the present theory implies more latitude in this respect. It is true that, according to Eq. (2.38), k/k_X is largely, apart from the influence of D_0, a matter of the molecular level of lesion formation, as in the case of the DRA-limit of the theory where $k/k_X = (r/r_X)^2$ holds strictly (cf. Eq. (2.36)), but λ is a complex dependency on both $f_1(z)$ and r with $S_0(D)$ interfering, and reduces to $\bar{z}_D$ only in the limiting case dealt with in subsection 2.3.4 (DRA-limit). Hence, the present theory has good prospects to account better for the experimental findings. To prove this definitively, one would have to implement analyses like those mentioned above, however with only r being treated as a freely adjustable parameter for each radiation quality. In principle, this is the topic of the following section as well as the entire chapter 7, but treated explicitly in terms of dose-effect curves, which are, on the one hand, measured direct (in contrast to RBE) and, on the other hand, easier to derive from the theory than the quantities (2.38) to (2.40), moreover avoiding the approximation (2.37).

2.4. Analysis of experimental data

In this section the ability of the DNA-esion theory to predict the shape of survival curves in dependence on radiation quality will be demonstrated. This means that the primary lesion RBE, r, is treated as a freely adjustable parameter for each radiation, which evidently allows to achieve a fit of each survival curve in one point at least, or in the final slope. Only the shape of the entire curve can thus be considered as a genuine consequence of theory. Apart from this test, the obtained estimates of r themselves and their dependence on LET are of great interest.

The second parameter involved is the size of the gross sensitive volume, which is independent of radiation quality, of course. This parameter should correlate with the size of the cell nucleus. Its most suitable value, however, derives from the exponential survival curves for very high LET (Ne or Ar ions) which depend almost only on σ_0, the *mean projected area* of the gross sensitive volume. Results from experiments with accelerated ions bombarding cultured cells (track-segment method) are most appropriate for such an analysis, which will be exemplified in the following using the experimental data obtained by SKARSGARD *et al.* (1967) on Chinese hamster CHO cells.

We will not dwell on the question of how to obtain the microdosimetric energy deposition spectra $f_1(z)$ required for the analysis outlined. Numerous data and computational approaches can be found in the literature, and the calculation of microdosimetric parameters is also a main topic of this book. Therefore, the reader is referred to chapter 5, where the general procedure applied and lots of results obtained are described in detail. The mathematical and physical basis is dealt with in chapter 3. It is an important feature of the procedure that it yields directly the required integral (2.17), which can ultimately be expressed in simple formulas containing numerical quantities given in the appendix in tabulated form. This easily practicable formalism is derived in chapter 8 and compendiously presented, as a summary, in chapter 9.

2.4.1. Survival curve fitting with unknown r

The results of the analysis of SKARSGARD's data on Chinese hamster cells are reported in Table 2.1 and Figure 2.3. First of all, a sum of exponential functions (cf. Eq. (2.18)) that fits the experimental X-ray survival curve had to be found. The result of this fit (for the method see section 9.3) is given in the first three columns of Table 2.1 in terms of B_μ and $\alpha_\mu^{(X)}$, and the corresponding curve is shown in Figure 2.3 (left curve). According to Eq. (2.19), setting $r_X = 1$ as a definition, the $\alpha_\mu^{(X)}$ were converted into the $\alpha_\mu^{(0)}$ given in the fourth column. The further columns show the α_μ values calculated for the ion radiations used in the experiment, choosing the parameters as follows. $\sigma_0 = 50\ \mu m^2$ is found to be in accord with the data for Ne and Ar ions, and the r values that yield the best correspondence between theory and experiment (see Fig. 2.3) are listed for each radiation in the first row of the lower part of Table 2.1. The parentheses in the case of Ar indicate the uncertainty of this r value.

Figure 2.3 convincingly demonstrates the capability of the DNA-lesion theory of accounting for the changes in shape of survival curves due to varying radiation quality. The curves becoming practically straight lines for high-LET heavy ions is a well-known general experience. However, there are some exceptions, especially with yeast cells. In these cases the curvature does not disappear even in the LET range where the overall slope already decreases with growing LET (MORTIMER *et al.*, 1965; RAJU *et al.*, 1971). It is interesting to test whether the theory postulates this peculiar effect, too. This means that the special X-ray survival curve and nucleus cross-section, which are the only characteristics of a given object in the framework of this theory as long as r is left to free choice anyway, should bring about this unusual behaviour. Indeed, for radiation-resistant objects as are yeast cells (by contrast with mammalian cells), the α_μ's tend to be proportional to the then small $\alpha_\mu^{(0)}$'s according to the fundamental equation (2.17), which signifies that the shape of the X-ray curve is largely retained with other radiations.

Figures 2.4 and 2.5 show the results of a corresponding calculation for two strains of yeast, using, for simplicity, the same r values for the same radiations as above

Tab. 2.1. Upper part: Coefficients B_μ (second column) and $\alpha_\mu \times$ krad (other columns) describing CH2B2 cell survival curves (Skarsgard *et al.*, 1967) for photon and ion radiations (numbers in the top line signify energy per nucleon [MeV] of the ions indicated).
Lower part: r — primary lesion RBE fitted; 100%- and 0%-survival RBE calculated for the CH2B2 hamster cells and two different strains of yeast cells (*Saccharomyces cerevisiae*)

	250 kVp	X-rays	Zero LET	He	Li	Li	B	C	O	Ne	Ar
μ	B_μ	$\alpha_\mu^{(X)}$	$\alpha_\mu^{(0)}$	10.00	9.667	5.000	9.273	8.833	8.187	7.650	5.750
1	104.91	7.81	7.83	9.48	11.10	14.39	14.93	13.32	8.656	5.557	1.763
2	−67.69	8.60	8.62	10.41	12.11	15.51	15.79	13.81	8.762	5.601	1.765
3	−267.83	9.31	9.33	11.24	13.01	16.47	16.50	14.19	8.837	5.633	1.767
4	172.80	10.09	10.12	12.15	13.99	17.50	17.22	14.56	8.904	5.660	1.768
5	286.02	11.10	11.14	13.31	15.22	18.74	18.06	14.97	8.971	5.687	1.769
6	−184.53	11.88	11.93	14.21	16.16	19.67	18.65	15.24	9.012	5.704	1.770
7	−152.09	12.89	12.94	15.36	17.34	20.80	19.34	15.54	9.055	5.721	1.771
8	98.12	13.68	13.74	16.25	18.24	21.63	19.82	15.74	9.082	5.731	1.771
9	31.80	14.68	14.75	17.38	19.38	22.65	20.39	15.96	9.110	5.743	1.772
10	−20.52	15.47	15.54	18.25	20.25	23.40	20.79	16.11	9.129	5.750	1.772
LET$_\infty$ [keV/μm]			0	18.39	42.54	72.48	121.7	181.3	335.7	536.5	1758.
r			1	1.25	1.55	2.3	3.0	3.6	4.2	3.3	(2.5)
Hamster	RBE$_{100\%}$		1	1.254	1.563	2.342	3.110	3.683	4.339	3.135	1.210
CH2B2	RBE$_{0\%}$		1	1.211	1.417	1.838	1.907	1.701	1.105	0.710	0.225
Yeast	RBE$_{100\%}$		1	1.402	2.059	4.246	7.349	11.243	15.893	12.536	6.778
BZ34	RBE$_{0\%}$		1	1.237	1.505	2.120	2.566	2.746	2.452	1.717	0.696
Yeast	RBE$_{100\%}$		1	1.253	1.561	2.347	3.118	3.847	4.838	4.032	6.825
X841	RBE$_{0\%}$		1	1.242	1.522	2.188	2.724	3.039	2.962	2.144	0.985

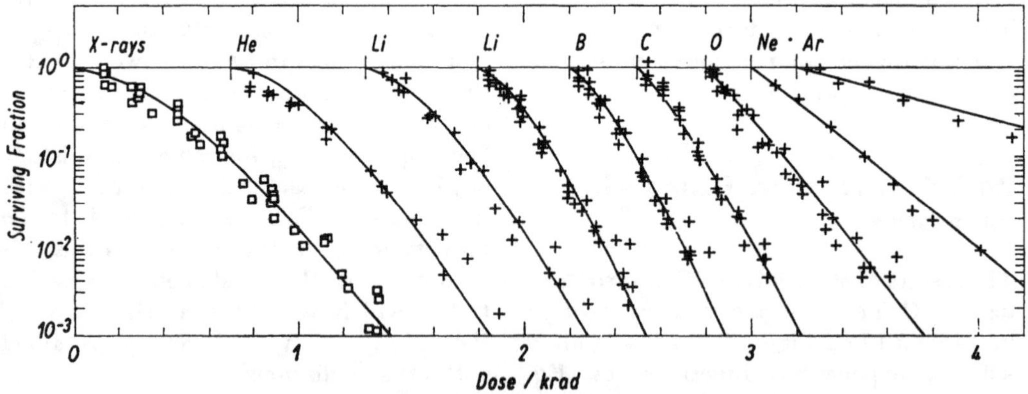

Fig. 2.3. Survival curves of hamsters cells (CH2B2) after the irradiations specified in Table 2.1. Experimental data points taken from Skarsgard *et al.* (1967) and Katz and Sharma (1973). Curves calculated according to the present theory with fitted values of the primary lesion RBE; gross sensitive volume area $\sigma_0 = 50\ \mu m^2$.

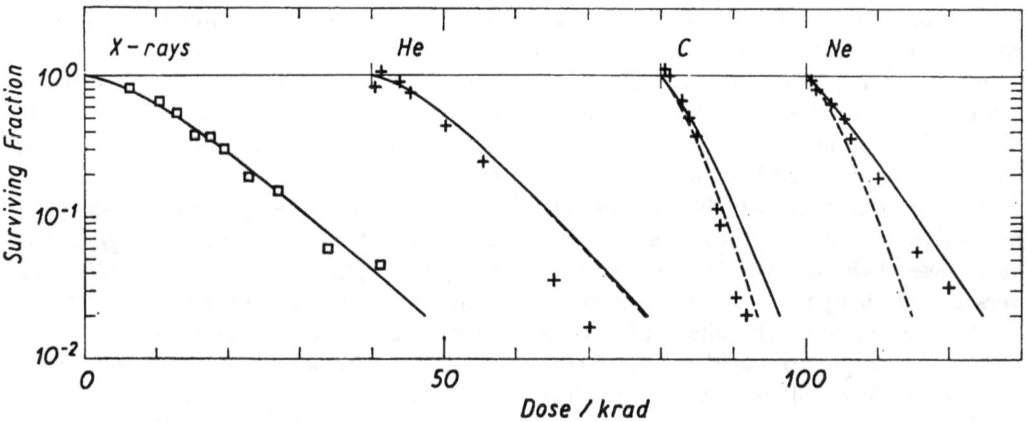

Fig. 2.4. Survival curves of yeast cells (*Saccharomyces cerevisiae* BZ34) after the irradiations specified in Table 2.1.
Experimental data points taken from RAJU *et al.* (1971).
Curves calculated according to the present theory using the primary lesion RBE for hamster cells and $\sigma_0 = 50\ \mu m^2$ (dashed lines), $\sigma_0 = 2\ \mu m^2$ (full lines).

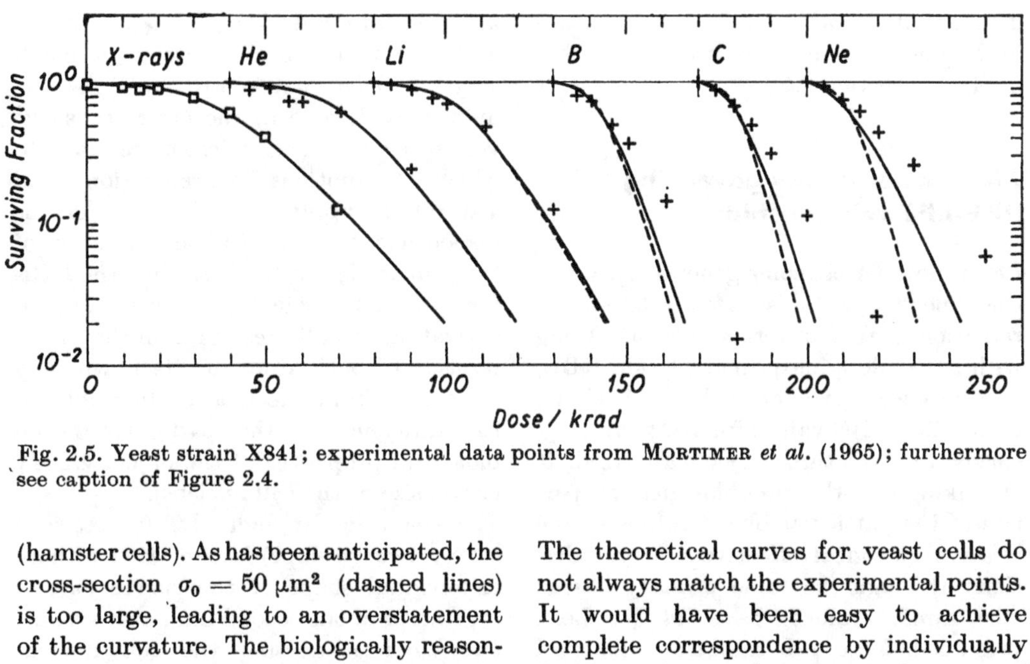

Fig. 2.5. Yeast strain X841; experimental data points from MORTIMER *et al.* (1965); furthermore see caption of Figure 2.4.

(hamster cells). As has been anticipated, the cross-section $\sigma_0 = 50\ \mu m^2$ (dashed lines) is too large, leading to an overstatement of the curvature. The biologically reasonable value $\sigma_0 = 2\ \mu m^2$, however, yields the curve shape in generally good agreement with the experimental findings. Ne ions only produce an almost exponential survival curve in the case of the less resistant strain BZ34 (Fig. 2.4), whereas the strain X841 (Fig. 2.5) has clearly sigmoidal survival curves for every radiation investigated, including Ne ions.

The theoretical curves for yeast cells do not always match the experimental points. It would have been easy to achieve complete correspondence by individually fitting both r and σ_0, as in the case of hamster cells. However, with unchanged r values the shown result more clearly exhibits the important fact that distinctive LET effects on the dose-response curves can be attributed mainly to differences in the corresponding X-ray curves, provided that the right order of magnitude of the cross-section σ_0 is observed. In view of this statement, it may further be asked whether the relative effectiveness as to primary lesion production, r, could eventually be regarded as a *universal* function of radiation quality.

The possible universality of r is a very important question which decides the predictive power and thus the practical usefulness of the DNA-lesion theory. Universality of r is by no means an implication of the theory, since the share of diverse DNA-lesion types in the damage of cells may vary from object to object. The results presented in this subsection suggest that there is no complete correspondence between the r-values for mammalian and yeast cells. Among various lines of mammalian cells, however, no significant differences seem to exist, which will be demonstrated and extensively put to use in chapter 7. The question is also discussed, briefly, in section 2.5.

2.4.2. General laws governing RBE-LET relationships

Let us now discuss some general aspects of the dependence of dose-effect relations on radiation quality in terms of RBE. Along with the underlying primary lesion RBE, r, the lower part of Table 2.1 shows theoretical RBE values for 100% and 0% survival for the three objects analysed. It is striking how the radiobiological properties of the particular object influence the relative biological effectiveness of radiations although the same primary lesions (the same r-values) have *ad hoc* been presumed. Note also that the calculated figures are in line with regularities discussed in section 2.3.2 (cf. Eqs. (2.26), (2.27) and context), but for what follows let us direct the attention to the dependence on LET.

Qualitatively, the RBE as a function of LET runs through a maximum in any case (r as well as $RBE_{100\%}$ and $RBE_{0\%}$). Quantitatively, however, this relation depends on both the biological system investigated and the level of effect considered, and it never[1]) coincides exactly with the RBE for the production of the primary lesions that are responsible for the effect studied.

Classical target theory already has introduced the terms *concentration effect* and *saturation effect* to account for the characteristic LET-dependence of RBE by supposing that an increasing concentration of energy in a decreasing number of particle tracks may enhance the effectiveness of radiation until saturation causes ineffective wasting of energy. From the standpoint of the DNA-lesion theory these ideas require specification by allowing for *two kinds* of either effect (GÜNTHER and SCHULZ, 1972). The two kinds are inferred from asking whether it is the number of ionizations in small targets (DNA, nanometre scale) or the number of lesions in the entire sensitive volume (cell nucleus, micrometre scale) to which the notions 'concentration' and 'saturation' apply. The former alone is responsible for the LET-dependence of the primary lesion RBE, r, while the RBE for typical biological endpoints (e.g., inactivation) usually results from the superposition of both kinds of effect, whereby the second-kind effects are responsible for the influence of the particular radiobiological properties of the object (X-ray curve, size of the cell nucleus).

The saturation at high LET (decrease of survival curve slope) is associated with an exponential shape of the curves *only if* saturation of the second kind prevails, i.e. one particle crossing the cell nucleus produces more lesions than required for inactivation. This behaviour is observed with hamster cells (Fig. 2.3); consequently, in this case, $RBE_{0\%}$ has its maximum at an appreciable smaller LET than has r (see Tab. 2.1). For radiation-resistant

[1]) Save for the hypothetical limiting case to which Eq. (2.32) applies.

objects (yeast), however, the critical number of lesions is large, or the number that can be produced by one event is small (little DNA in the nucleus), which moderates the influence of the saturation effect of the second kind. Thus, the survival curves remain sigmoidal (Figs. 2.4 and 2.5), i.e. the decrease of the slope at high LET has to be attributed mainly to the first kind of saturation (more ionizations than necessary, in a molecular target, to produce a lesion). As a consequence, there is less difference between r and $RBE_{0\%}$ for yeast (see Tab. 2.1).

While the $RBE_{0\%}$ is, hence, more or less affected only by the *saturation* effect of the second kind (note that always holds $RBE_{0\%} < r$), the RBE for higher levels of survival may additionally depend on the *concentration* effect of the second kind. This means that the *increase* of RBE with growing LET is enhanced relative to that of r owing to the increasing probability of exceeding the average critical number of lesions (tolerable sublethal damage) by a *single* particle passage. It is evident that this effect should be most pronounced with respect to $RBE_{100\%}$ and in the case of an X-ray survival curve with a marked shoulder (small initial slope). This is exemplified by the yeast strain BZ34 (see Tab. 2.1), whereas in the case of the strain X841 the shoulder is so large that the concentration effect of the second kind becomes strong only for Ar ions, i.e. in the range of already decreasing r-values. The opposite extreme is seen with hamster cells where the short shoulder causes second kind *saturation* even with respect to $RBE_{100\%}$ for Ne and Ar ions, while at lower LET the concentration effect of the second kind does exist, to be sure, but is of only academical importance.

As in the case of saturation effects at high LET, the two kinds of concentration effects are clearly distinguishable, conceptually: physico-chemical co-operation of several ionizations necessary to produce a relevant molecular lesion — first kind; biological co-operation of a number of lesions necessary to kill the cell — second kind. To realize the possible significance of the latter, let us assume the concentration effect of the first kind to be absent. This implies a constant rate of lesion production per unit dose ($r =$ const.) so that the average number of lesions produced per unit path length of a particle increases proportionally to LET. The thus-caused enhancement of biological effectiveness per crossing particle may be so strong that it preponderates over the decreasing particle fluence at increasing LET. This is the essence of the second-kind concentration effect.

The considerations emphasize that the LET-dependence of biological effects at the cellular level does not, on principle, reflect directly what happens at the molecular level, due to the involvement of second-kind effects. This was often disregarded; the manner in which Fowler (1974) discussed the subject is a typical example. In the case of sigmoidal survival curves, the problem at once comes to light when we ask the question of the percentage of survival for the RBE to be chosen in an attempt to correlate both levels quantitatively. For example, the so-called track-segment analysis, which relies upon $RBE_{100\%}$ and infers, therefrom, the dimensions of submicroscopic targets and an associated energy threshold, is not quite sound. For similar reasons, the influence of modifying factors must not be expected to exhibit an equal LET-dependence with respect to radiation effects at the cellular and molecular levels. Let us take the oxygen effect: the well-known decrease of the oxygen enhancement ratio (OER) with growing LET is not necessarily the result of a similar behaviour of the underlying molecular lesions. It might be that only the saturation effect of the second kind causes the decrease. By all means it has a share in it, so that the OER-LET relationship cannot strictly be a universal one.

It is important to note that the significance of the question raised and its qualitative aspects are independent of the special features and possible shortcomings of the present theoretical framework on which the arguments have been based.[1] The genuine nature of the subject can only be more, but not less, complex. From this general point of view, the DNA-lesion theory of RBE, in particular the microdosimetric version dealt with in this chapter, may be looked upon as a possible tool for quantitatively describing and thus eliminating the effects of the second kind in order to push on to the molecular level characterized by the quantity r as a function of radiation quality. This is the very essence of analysing a family of different-LET dose-effect curves as exemplified above. Hence, as long as the fundamental assumptions (see footnote 1 at page 20) are at least approximately justified, which is just corroborated by the very good survival curve fits obtainable for hamster cells (Fig. 2.3), r-values such as shown in Table 2.1 can be considered as a clue to the characterization of the primary lesions that are responsible for the biological effect under study.

It is worth while dwelling on the above-mentioned track-segment analysis which has been applied by Howard-Flanders (1958), Brustad (1962), and Barendsen (1966, 1967) to bacteria, yeast cells, and mammalian cells (human kidney T1). The effectiveness per particle as obtained from the initial slope of ion bombardment survival curves (equivalent to $RBE_{100\%}$) is interpreted as the probability to exceed a critical ionization number n within a small distance t along the ion track. Since this idea neglects the biological co-operation of several lesions produced along a track, it is not surprising that the critical length $t = 7-10$ nm following from T1 cell survival data (Barendsen, 1966) is in excess of molecular dimensions (of the DNA) due to the involvement of the other scale of interaction (diameter of the cell nucleus). A refined version of this analysis based on microdosimetric distributions (Hogeweg and Barendsen, 1972) yields still larger values for the target thickness t. Actually, it is the quantity r for which the track-segment analysis may be an adequate model (see section 2.5.3). From this point of view, the track-segment analysis refers intrinsically to the formation of primary lesions, and the error consists in the neglect of the second-kind effects when applied to $RBE_{100\%}$ instead of r. A critical remark on exactly this point was given by Kellerer (1972b).

However, if the track-segment analysis is considered as a part of a complete formal theory of radiation effectiveness by supposing the curvature of low-LET survival curves to result from an accumulation of ionizations produced by several particles in the *one* target t, as suggested by Barendsen (1967), this is nothing but a variant of the 'energy-only concept' (see sections 1.4.2 and 2.3.3) which has been shown to fail in predicting the RBE for low levels of survival. Now, that approach can be said to *neglect the first-kind effects*. This interpretation of track-segment analysis, although unacceptable anyway, is interesting because it demonstrates how thin must a formal sensitive volume be made in order to account for the LET-dependence of $RBE_{100\%}$. It degenerates into a very flat disk since an effective cross-section as large as 35 μm^2 is indispensable for T1 cells in consideration of the high-LET survival curves.

2.4.3. Conclusions

The above reflections may be generalized, pointing out that most of the previous attempts to correlate energy deposition characteristics with the biological response can be classified into two groups. Either the attention was focused on the submicroscopic level by ignoring or underestimating the role of second-kind effects, or the very opposite was done by considering the energy deposition distribution on the micrometre scale only.

Track-segment analysis and related attempts are examples of the first stand-

[1]) For example, the role of the cell nucleus may as well be played by sites of lesion interaction, so that the effects of second-kind concentration and saturation would be a matter of 'formation of elementary lesions from sublesions', to use the language of dual radiation action theory.

point which can be shorthanded in present notation by $RBE = r$ and $r \neq 1$. On the other hand, various efforts made to employ the customary microdosimetric concept exemplify the second standpoint signified by $r = 1$ and something like Eq. (1.24) or Eq. (1.26).

It is interesting to note that KELLERER (1972a), admitting the possibility "that RBE may also to some extent be determined by the submicroscopic energy concentrations", calls it a "potential second order effect" just what is named here *effects of the first kind*. However, our opposite denomination is not to suggest an opposite emphasis. The present theory rather is an attempt to unite *both* phenomena in *one* systematic theory, giving up any kind of biased emphasis. This attitude is exhibited by $RBE \neq r$ and $r \neq 1$. With the DNA-lesion production model (next section) added, this theory must be regarded as the indispensable *minimum* of energy deposition pattern complexity that has to be taken into account. This means that it cannot be excluded, particularly as there are some serious hints,[1]) that still a third, intermediate, structural level plays a role, as a consequence of, e.g., the existence of certain threshold distances of a specific DNA-lesion interaction (NEARY *et al.*, 1967), non-uniform distribution of DNA in the nucleus (COLE *et al.*, 1974; DATTA *et al.*, 1976), non-randomness of spatial arrangement of individual chromatin subunits which exhibit some kind of functional interdependence, etc. The present 'minimal theory' may thus be improved in the future by allowing, in addition, for what is just a central idea of the theory of dual radiation action, namely a possible significance of *subnuclear* sites.

[1]) Such as the too-small size of the sensitive region turning out in the analysis presented in Figure 2.2. Further discrepancies of similar nature, particularly with fast neutrons, are discussed in section 7.4.2, where an analysis of the problem is given on the whole.

2.5. Primary lesions and the relevant RBE

2.5.1. Preliminary reflections

The quantity r referred to as the *primary lesion RBE* plays a central role in the DNA-lesion theory of RBE. Along with $f_1(z)$, it is *the* parameter that depends on radiation quality. If this dependence were known in detail, the theory would be complete in the sense that dose-effect relations for any radiations could be predicted straightforwardly from the corresponding X-ray relation and the size of the gross sensitive volume (cell nucleus).

There are three possible ways of attacking the r-problem.

(i) The phenomenological way: determination of empirical r-values by survival curve fitting (section 2.4.1) as an attempt to arrive at sets of universal values applicable at least to entire classes of objects (e.g., all mammalian cells).

(ii) The experimental way: direct measurements of s, the yield of DNA-lesions (preferably single- and double-strand breaks), in dependence on radiation quality, supposing that these lesion types — eventually in combination — determine $r = s/s_0$.

(iii) The theoretical way: derivation of s from target models of lesion production based on the same principles as track-segment analysis.

Of course, the three approaches must be seen in close connection. While the advantage of the theoretical approach (iii) consists in the possibility to calculate r for any radiation of interest, the inherent parameters of such a target model must be derived previously from data obtained by (i) or (ii). The formal way (i) alone is free from additional suppositions, but it will hardly be possible to complete the DNA-lesion theory in this phenomenological manner only. Hence, (i) is mainly a prerequisite to (ii) and (iii). From the viewpoint of understanding rather than

only describing the mechanisms of radiation action, the approach (ii) is most important, because it could bridge the gap between the numerous experimental results, from the last decade, at the molecular level and the observable biological effects at the cellular level. This means that a measured breakage frequency for a given radiation (e.g., neutrons), divided by the γ-ray breakage frequency to obtain r, can be taken in order to test whether the correct survival curve does result from this molecular parameter. This kind of inference would not be possible without a theory that describes the statistics of lesions in a cell nucleus, in other words, the effects of the second kind.
Some fragmentary examples illustrating these lines of reasoning will be discussed in the following.

2.5.2. Experimental data on DNA-strand breakage

Let us compare the values of r obtained from the analysis of hamster cell survival curves (Tab. 2.1) with experimental data on the induction of primary DNA-strand breaks. Unfortunately, such data, which are numerous only for X- and γ-rays, show considerable scatter, and the relatively few results in the range of higher LET come from different authors having used diverse objects and methods. Therefore, it would be an exaggeration to speak of experimentally established quantitative break-number-LET relationships in the present state of knowledge. A noticeable exception is the extensive work of Christensen *et al.* (1972), which has, however, other drawbacks (see below). A graphical compilation of data given by Günther and Schulz in 1972 can essentially be maintained also in the light of results published since that time (see chapter 6). Therefore, with the values from Table 2.1 included, that graph is presented here again (Fig. 2.6).

The correlation of the hamster cell r-values (crosses in Fig. 2.6) with the break production rates s for *double-strand breaks* is obvious. There is in fact very good agreement[1]) with the theoretical curves which generalize the experimental breakage data (see next subsection). This interesting result suggests that mammalian cells die only, or predominantly, of DNA double-strand breaks. It might be misleading to express this important statement by saying that double-strand breaks are the 'lethal lesions' or 'lethal events'. The cell may survive a large number of double-strand breaks, in compliance with the probability $H(k)$ being at the bottom of survival curve theory (see Eq. (2.4)). In other words, sublethal damage too must consist of double-strand breaks. Thus, our finding signifies, strictly speaking, that no molecular damage *other* than DNA double-strand breaks seems to be important for the inactivation of mammalian cells. This knowledge may be valuable for any attempts to elucidate the chain of biological events which lead to the loss of colony-forming ability of cells. A possible dependence of cell survival on DNA double-strand breaks was repeatedly discussed in the past, e.g., by Neary (1970) and recently by Kellerer (1978).
The experimental results of Christensen *et al.* (1972) given in Table 2.2 confirm the conclusion drawn, although a direct quantitative comparison with the r-values in Table 2.1 is difficult because of the features of the experiment.[2]) Never-

---

[1]) With the exception of Ar ions, for which the fitted value of r is necessarily uncertain due to its little influence on the survival curve.

[2]) The ions were slowed down, partly until stopping, within the irradiated sample, so that the LET mean values given in Table 2.2 are relatively high, which means that the energy of δ-rays was smaller than it was with the irradiations to which Table 2.1 refers. Therefore, neither corresponding LET values nor identical ions are exactly comparable. Moreover, the experimental conditions were greatly different from those of DNA in a living mammalian cell.

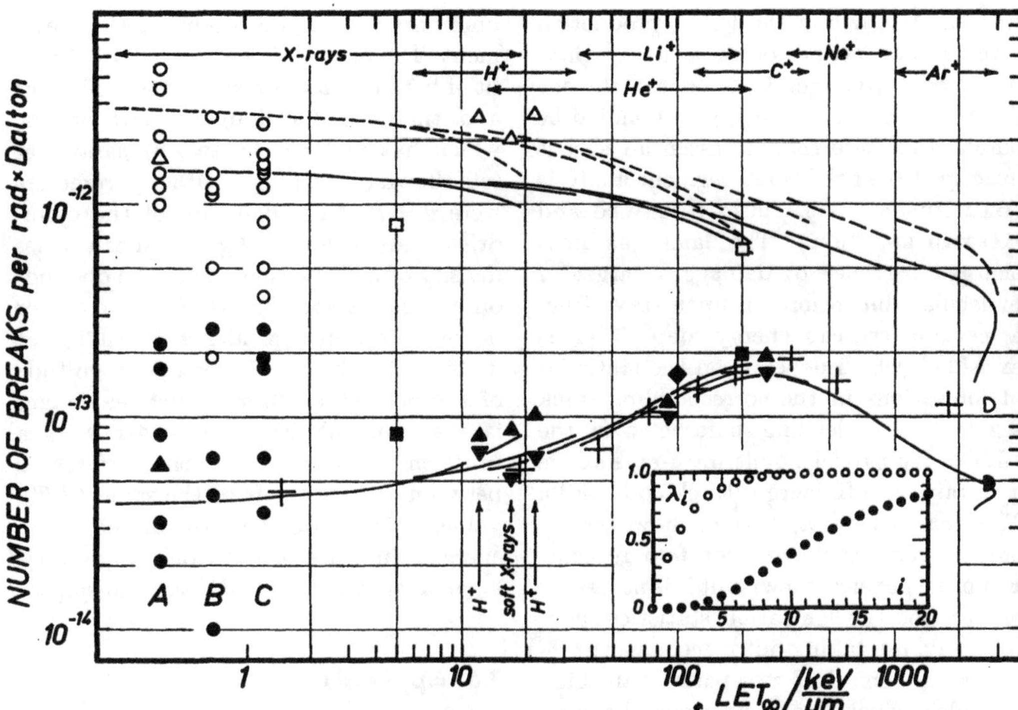

Fig. 2.6. Single-strand breaks (open symbols) and double-strand breaks (closed symbols) in dependence on radiation quality (sort of ions indicated at the top). Equal symbols belong to the same experiment except for circles which refer to X- or γ-rays separated into three groups (A, B, C) indicating phages or isolated DNA, bacteria, and mammalian cells, respectively. For the origin of data see GÜNTHER and SCHULZ (1972).
Crosses represent the values $s = rs_0$ obtained from the analysis of hamster cell survival curves (r according to Tab. 2.1, $s_0 = 4.5 \times 10^{-14}$ rad^{-1} Dalton^{-1} by discretion).
For the theoretical curves see section 2.5.3.

Tab. 2.2. r-values derived from DNA-strand breakage measurements (CHRISTENSEN *et al.*, 1972) on the phage ΦX-174 (Replicative Form DNA) irradiated with photons and various ions. In parentheses: absolute s-values in units of 10^{-14} per rad and Dalton for ^{60}Co γ-rays

	^{60}Co-γ		H	He	Li	B	C	O	Ar
LET_∞ [keV/µm]	0.3		5.8	34.3	78.5	225	332	550	2420
Single-strand breaks	(34)	1	0.74	0.71	0.68	0.59	0.47	0.32	0.37
Double-strand breaks	(1.3)	1	0.69	1.6	1.9	3.5	4.3	4.7	3.6

theless, it is remarkable that the height of the maximum of r is very similar in both cases. On the other hand, single-strand breaks are again clearly ruled out to constitute the decisive lesions in mammalian cell killing.

2.5.3. The model of DNA-lesion production

In addition to the phenomenological and experimental ways of obtaining r-values, let us briefly characterize the theoretical way, i.e. modelling the lesion production rate s in dependence on radiation quality. The model envisaged is, of course, the one described in section 1.5.1, but it should be emphasized here that its essential idea is exactly that of the track-segment analysis (Barendsen, 1966, 1967; Hogeweg and Barendsen, 1972). The latter assumes charged particles to damage a target of molecular dimensions if, upon traversing, a certain critical energy deposition is exceeded, which depends on the statistics of ionizations in the corresponding track segment. The leading parameter is the LET of the particle, omitting δ-rays above a certain cut-off energy (track-core LET). This customary approach, however, is only a very crude one for two reasons. From the physical viewpoint it is necessary to describe the ionization statistics more precisely by taking into account the δ-electrons themselves as separate damaging particles which may, however, be considerably slowed down or stopped inside the target. Moreover, variations of the ionization cluster frequencies (especially with respect to slow electrons) and the chord length spectrum of the particular target should be considered. On the other hand, the model itself ought to be generalized in regard of the sharp energy threshold like a trigger of damage, which is unlikely to exist in actuality.

All these requirements can be met by writing the model of lesion production in the general form

$$s = \frac{1}{\bar{z}_F} \sum_{j=0}^{\infty} \lambda_j f_1(j) \tag{2.41}$$

in which it is introduced and discussed in section 1.5.1, evaluated in section 3.5, and put to use in chapter 6. Further comments in the present context are therefore superfluous.

Let us only allude to Figure 2.6 again, which includes theoretical curves calculated earlier (Günther and Schulz, 1972) according to the model (2.41), using some unimportant simplifications in the treatment. The curves should be regarded as a fit of both the data for single-strand breaks and the data for double-strand breaks, which has yielded the two λ_j parameter sets shown in the inset.[1]) Other parameters were $d = 2.24$ nm diameter of the cylindrical target and 1.3 g/cm^3 mean mass density of it. Note that the model does not only account for the *relative* increase or decrease of s in dependence on LET, but it also yields the correct *absolute* magnitude of s in both cases. Another noticeable fact is the impossibility to consider s as a function of LET_∞ only: the differences between different ions at the same LET_∞ value, which are due to unlike δ-ray energies, indicate that it is only a grammalogue to speak of an r-LET relationship.

2.6. Appendix

2.6.1. More than one effective lesion type

Let us consider two different types of DNA-lesions distinguished by the indices I and II. Extension to any number of lesion types will be obvious. Now, the survival probability depends on the number of lesions of either type, k_{I} and k_{II}. Given a specific energy z, these numbers are statistically independent of each other in keeping with the Poisson distribution in Eq. (2.4). Hence, Eq. (2.4) and the representation (2.10) have to be generalized in the form

$$S_0(z) = \sum_{k_{\mathrm{I}}=0}^{\infty} \sum_{k_{\mathrm{II}}=0}^{\infty} H(k_{\mathrm{I}}, k_{\mathrm{II}}) \frac{(s_0{}^{\mathrm{I}} Yz)^{k_{\mathrm{I}}}}{k_{\mathrm{I}}!} \mathrm{e}^{-s_0{}^{\mathrm{I}} Yz} \times \frac{(s_0{}^{\mathrm{II}} Yz)^{k_{\mathrm{II}}}}{k_{\mathrm{II}}!} \mathrm{e}^{-s_0{}^{\mathrm{II}} Yz}, \tag{2.42}$$

$$H(k_{\mathrm{I}}, k_{\mathrm{II}}) = \sum_{\mu=1}^{n} B_\mu \mathrm{e}^{-Z_\mu{}^{\mathrm{I}} k_{\mathrm{I}} - Z_\mu{}^{\mathrm{II}} k_{\mathrm{II}}}. \tag{2.43}$$

The latter expression is again suited to describe any reasonable function of the two variables k_{I}, k_{II} accurately enough.

Eq. (2.42) refers to zero-LET radiation. Therefore, in the general case, $s_0{}^{\mathrm{I}}$ and $s_0{}^{\mathrm{II}}$ have

[1]) The dashed lines refer to the 'single-ionization model' $\lambda_j = 1 - \delta_{j0}$.

to be replaced by s^{I} and s^{II}, after which the expression can be inserted in Eq. (2.1). Type-specific r-values may be defined by

$$r^{\mathrm{I}} = \frac{s^{\mathrm{I}}}{s_0^{\mathrm{I}}}, \qquad r^{\mathrm{II}} = \frac{s^{\mathrm{II}}}{s_0^{\mathrm{II}}}, \tag{2.44}$$

but this does not imply the possibility to write the equation to start with in a form similar to Eq. (2.6). The evaluation, nevertheless, proceeds exactly as described in section 2.2.2. Eq. (2.11) changes into

$$S(D) = \sum_{\mu=1}^{n} B_\mu \int_0^\infty \mathrm{d}z \exp\left[-(1 - \mathrm{e}^{-Z_\mu^{\mathrm{I}}})\, s^{\mathrm{I}} Yz - (1 - \mathrm{e}^{-Z_\mu^{\mathrm{II}}})\, s^{\mathrm{II}} Yz\right] f(z; D) \tag{2.45}$$

from which it can easily be seen that carrying out all necessary steps leads to the fundamental equation (2.17) again if the quantity r is being defined by the weighted mean

$$r = \frac{(1 - \mathrm{e}^{-Z_\mu^{\mathrm{I}}})\, s_0^{\mathrm{I}} r^{\mathrm{I}} + (1 - \mathrm{e}^{-Z_\mu^{\mathrm{II}}})\, s_0^{\mathrm{II}} r^{\mathrm{II}}}{(1 - \mathrm{e}^{-Z_\mu^{\mathrm{I}}})\, s_0^{\mathrm{I}} + (1 - \mathrm{e}^{-Z_\mu^{\mathrm{II}}})\, s_0^{\mathrm{II}}}. \tag{2.46}$$

In the general case, the *effective primary lesion RBE* defined by Eq. (2.46) is not independent of μ. Hence, there is another r for each term in the sum $S(D) = \sum B_\mu \exp(-\alpha_\mu D)$, which means that the formal analysis or prediction of survival curves as described in section 2.2.3 and exemplified in section 2.4.1 may be, strictly speaking, unwarranted if more than one lesion type is involved in the injury process. Actually, however, this problem seems not to be very serious since, firstly, there are a number of quite realistic possibilities that the effective r, notwithstanding, becomes independent of μ; secondly, in the most important case of mammalian cell killing the formal analysis *was* successful and the results indicate that it is only one type of lesion, indeed, that is decisive in this case, namely double-strand breaks.

As to the former point, there are the following conditions for μ dropping out of Eq. (2.46).

(i) The two types exhibit the same dependence on radiation quality, i.e. $r^{\mathrm{I}} = r^{\mathrm{II}}$, which may occur if both types originate from the same pre-lesions which decay in a constant ratio into the two types (e.g., chemically different single-strand lesions).

(ii) The two types play the same role in the injury process, i.e. $Z_\mu^{\mathrm{I}} = Z_\mu^{\mathrm{II}}$ for every μ.

(iii) The interplay of the two types of lesions in the cumulative impairment of a cell may be such that the expression (2.46) does not depend on μ even though the two foregoing conditions are not fulfilled. A model for bacteria which has this property was considered in an earlier paper (GÜNTHER and SCHULZ, 1972): single-strand breaks, which can, in contrast to double-strand breaks, be repaired with the probability $1 - u$, were assumed to be otherwise equivalent to double-strand breaks. This ends up in $r = (us_0^{\mathrm{I}} r^{\mathrm{I}} + s_0^{\mathrm{II}} r^{\mathrm{II}})/(us_0^{\mathrm{I}} + s_0^{\mathrm{II}})$ irrespective of μ.

Hence, in the general case, it may be assumed that a common effective value of r can always be assigned to each radiation, as a good approximation at least. In this way, however, which is the most important thing to be learned from this investigation, different objects and endpoints may well give rise to a different primary lesion RBE by differently 'mixing' the eventually same lesion types.

2.6.2. Inhomogeneity of the cell population

In the framework of the DNA-lesion theory of radiation action, the properties of the cells are described by the survival probability, $H(k)$, the content of DNA per cell, Y, and the size of the gross sensitive volume, σ_0. These quantities may vary in different subpopulations, e.g., in the case of asynchronous cells.

Inhomogeneity with respect to $H(k)$ is easily seen to have no influence since $S(D)$ depends linearly on $H(k)$ (cf. Eqs. (2.4) and (2.6)). Thus, $H(k)$ must only be considered as a *weighted average* over all subpopulations.

In the case of different DNA contents in different subpopulations, let Y denote the maximum value occurring in the entire population and Y^* the value for an arbitrary subpopulation singled out. With respect to the latter we define the parameter set Z_μ^* by

$$(1 - \mathrm{e}^{-Z_\mu^*})\, Y = (1 - \mathrm{e}^{-Z_\mu})\, Y^*. \tag{2.47}$$

Now have a look at Eq. (2.14) to realize that the asterisked subpopulation has got the apparent DNA content Y if the original $H(k)$ is changed into a formal $H^*(k)$ according to

$$H^*(k) = \sum_{\mu=1}^{n} B_\mu \mathrm{e}^{-Z_\mu^* k}. \tag{2.48}$$

Since this can be done for *any* subpopulation, it is possible to describe the *entire* population by the now *common* parameter Y and a weighted average of the $H^*(k)$'s as above.

Consequently, the general formalism expressed by the equations (2.17) to (2.19) remains valid if there is biological variability with respect to $H(k)$ and Y, but not with respect to σ_0. The latter kind of variability, however, should become visible by the very-high-LET survival curves bending *upwards*, which has rarely been observed.

Tab. 2.3. Test of the approximate RBE-dose relationship; deviations in per cent

	$q = 1.9$				$q = 1.6$				$q = 1.3$				$q = 1.1$			
$p =$	0.0	0.1	0.2	0.4	0.0	0.1	0.2	0.4	0.0	0.1	0.2	0.4	0.0	0.1	0.2	0.4
D/λ																
0.001	−0.1	0.0	0.0	0.0	−0.6	0.1	0.1	0.1	−1.1	0.1	0.1	0.2	−1.4	0.1	0.1	0.2
0.003	−0.2	0.1	0.1	0.1	−1.0	0.2	0.2	0.3	−1.8	0.1	0.3	0.5	−2.3	0.0	0.4	0.6
0.01	−0.4	0.2	0.2	0.3	−1.7	0.4	0.7	1.0	−3.0	−0.1	0.8	1.5	−3.9	−0.6	0.7	1.7
0.03	−0.6	0.5	0.6	0.7	−2.5	0.5	1.4	2.5	−4.6	−0.9	1.2	3.4	−6.0	−2.1	0.5	3.6
0.1	−0.7	1.0	1.3	1.8	−3.3	0.4	2.6	5.4	−6.4	−2.2	1.1	6.4	−8.8	−4.5	−0.8	5.8
0.3	−0.5	1.5	2.2	2.8	−2.9	0.9	3.7	7.8	−6.7	−2.5	1.3	8.6	−10.2	−6.0	−1.8	6.8
1	0.3	2.0	2.5	2.8	−0.0	2.9	5.2	8.0	−3.1	0.4	3.6	9.2	−7.5	−3.9	−0.3	6.9
3	1.3	2.1	2.2	1.7	3.5	4.8	5.5	5.1	2.7	4.4	5.7	6.6	−1.3	0.8	2.8	5.5
10	1.7	1.6	1.3	0.6	3.9	3.5	2.9	1.8	3.9	3.8	3.4	2.5	2.3	2.6	2.8	2.6
30	1.3	0.8	0.5	0.2	1.6	1.3	1.0	0.6	1.6	1.4	1.2	0.8	1.5	1.4	1.3	0.9
100	0.5	0.2	0.1	0.1	0.5	0.4	0.3	0.2	0.5	0.4	0.4	0.3	0.5	0.4	0.4	0.3

2.6.3. Investigation of the RBE-dose relationship

There exists a fairly representative class of zero-LET dose-effect relations which allow a strict analytical solution for the RBE, thus enabling a simple test of considerable generality for the adequacy of the approximation (2.37):

$$S_0(D) = (2 - p)\,e^{-D/D_0} - (1 - p)\,e^{-2D/D_0}. \tag{2.49}$$

D_0, as usual, describes the final slope, while p signifies the initial to final slope ratio. For an arbitrary radiation, the dose-effect relation that results from the general procedure (2.17) is described by

$$S(D) = (2 - p)\,e^{-\alpha_1 D} - (1 - p)\,e^{-\alpha_2 D}. \tag{2.50}$$

$$RBE^{(0)} = D_0\alpha_1 \left[1 + \frac{1}{\chi D/\lambda} \ln \frac{\dfrac{1 - p/2}{1 - p} + \sqrt{\left(\dfrac{1 - p/2}{1 - p}\right)^2 - \dfrac{2 - p}{1 - p}\,e^{-\chi D/\lambda} + e^{-q\chi D/\lambda}}}{\dfrac{2 - p}{1 - p} - e^{-(q-1)\chi D/\lambda}}\right] \tag{2.54}$$

By setting

$$\alpha_2 = q\alpha_1 \tag{2.51}$$

we introduce the 'radiation quality parameter' q which ranges from 2 (zero LET) to 1 (infinite LET).

Now the evaluation of the parameters (2.38) to (2.40) leads to simple formulas if one realizes that the integrals needed are easy to express by the coefficients α_1 and α_2 according to the definition in Eq. (2.17). In addition, let us set $\bar{z}_D^{(X)} = 0$ and $r_X = 1$, i.e. consider the RBE relative to zero-LET radiation. This yields

$$\frac{k}{k_X} = D_0^2\alpha_1^2, \qquad \lambda = \frac{2 - p - (1 - p)\,q}{1 - 3p/2}\,\frac{1}{\alpha_1},$$

$$\lambda_X = \frac{pD_0}{1 - 3p/2}, \tag{2.52}$$

so that the approximate RBE-dose relation (2.37) reads

$$RBE^{(0)} = D_0\alpha_1 \times \frac{2[2 - p - (1 - p)\,q]\,(1 + D/\lambda)}{p + \sqrt{p^2 + 4[2 - p - (1 - p)\,q]^2 (1 + D/\lambda)\,D/\lambda}}. \tag{2.53}$$

The corresponding strict solution to the equation $S_0(D \times RBE^{(0)}) = S(D)$ is obtained straightforwardly using the expressions (2.49) and (2.50) and eliminating α_1 and α_2 by introducing q and λ:

[Eq. (2.54) above]

where

$$\chi = \frac{2 - p - (1 - p)\,q}{1 - 3p/2}. \tag{2.55}$$

Table 2.3 displays the numerical comparison of the strict expression (2.54) with the approximate expression (2.53) by listing the percentages of departure of the latter. It can be seen that, within reasonable ranges of the two parameters, the deviation does not exceed ±10 per cent.

Chapter 3

The mathematical formalism of the DNA-lesion theory of radiation action

3.1. Introduction

The fundamental principles of the DNA-lesion theory of radiation action were introduced in section 1.5. The present chapter deals with the mathematical treatment of the statistics of energy deposition (physical stochastics) as needed for this theory. As shown in chapter 2, customary microdosimetry (Rossi, 1959; Rossi *et al.*, 1961; Hug and Kellerer, 1966; Rossi, 1967; Kellerer, 1970; Kellerer and Chmelevsky, 1975) can only unsatisfactorily solve this problem by approximately deriving the statistical distribution of DNA-lesions per cell nucleus, $U_k(D)$, from the distribution of specific energy, $f(z; D)$, and by excluding the model description of DNA-lesion formation. The latter, to be sure, can be expressed in terms of the microdosimetric single event size distribution, but the diameter of the sensitive volume (cylinder) is of the order of one or few nanometres, which requires other methods considering single ionizations ($f_1(j)$ instead of $f_1(z)$).
The basic equations to be evaluated (see sections 1.5.1 and 1.5.2) are

$$s = \frac{1}{\bar{z}_F} \sum_{j=0}^{\infty} \lambda_j f_1(j), \quad \lambda_j = \sum_{\nu} b_\nu \, e^{-\zeta_\nu j}, \quad (3.1)$$

$$S(D) = \sum_{k=0}^{\infty} H(k) \, U_k(D; \lambda_j, l_l, Y), \qquad H(k) = \sum_{\mu} B_\mu \, e^{-Z_\mu k}. \quad (3.2)$$

Eq. (3.1) describes the number of DNA-lesions of a certain type per unit dose and unit length (or molecular weight) of a DNA molecule, while $S(D)$ is a dose-effect relation for cellular objects and effects. The physical and biological meanings of the probabilities λ_j and $H(k)$, which are discussed in chapters 1 and 2, do not matter here, because for the following mathematical treatment it is only important that λ_j and $H(k)$ can be expressed by finite sums of exponential functions as indicated by (3.1) and (3.2). Our concern in the following sections is only the physical stochastics expressed by the quantities

- $1/\bar{z}_F$ — mean number of absorption events per unit dose in a cylindrical target pertinent to a segment of the DNA molecule;
- $f_1(j)$ — probability for the production of j ionizations in the cylindrical DNA-target by one absorption event in it.

$U_k(D)$ – probability of k primary DNA-lesions of a certain type being produced in the cellular object during irradiation with the dose D.

It is an important feature of the following treatment of physical stochastics that quite general principles for the description of radiation fields and absorption events apply to both $f_1(j)$ and U_k or to other probability distributions of any random quantities associated with physical stochastics. Another unifying aspect consists in the fact that something like the characteristic functions of the probability distributions, namely $\sum e^{-\zeta j} f_1(j)$ and $\sum e^{-Zk} U_k$, are finally calculated only, which makes it possible to arrive at analytical formulas for what is actually needed according to Eqs. (3.1) and (3.2), circumventing the explicit calculation of the distributions themselves.

In the following sections, the entire subject of the stochastics of radiation absorption is dealt with comprehensively in a systematic manner. Although these considerations go beyond the concepts of present microdosimetry, they are closely related to this field, since the physical and mathematical problems involved are the same, and most ideas have been taken from the corresponding literature. Therefore, familiarity of the reader with microdosimetry, especially with its theoretical component (HUG and KELLERER, 1966; CASWELL, 1966; CORMACK, 1968; KELLERER, 1968, 1970; KELLERER and ROSSI, 1970; OLDENBURG and BOOZ, 1972b; GUBIN and SAKOVICH, 1974; BOOZ and COPPOLA, 1974; PARETZKE *et al.*, 1974; KELLERER and CHMELEVSKY, 1975; CASWELL and COYNE, 1976), will be extremely useful for reading this chapter. Besides, our approach comprises the central quantity of microdosimetry, the single event size distribution $f_1(z)$, too, for the theory of $f_1(j)$ is general enough to apply also to micrometre tissue spheres.[1]) In this way, the formalism yields

$$\int dz[1 - \exp(-\eta z)]\, f_1(z)/\bar{z}_F$$

in analytical form, which either may be used in the framework of the approximate version of the DNA-lesion theory (chapter 2, Eq. (2.17)) or from which $f_1(z)$ itself can be derived, subsequently. The latter problem is solved numerically (inverse Laplace-like transformation) in chapter 5, which is devoted to the 'microdosimetric branch' of our approach. In this way, the whole concept of description of physical stochastics developed in the present chapter is checked up in chapter 5 by means of theoretical and experimental data of the microdosimetric quantities $\bar{z}_F$, $\bar{z}_D$, and $f_1(z)$ for X-rays, fast neutrons, and directly ionizing radiations.

The problem to be solved in this chapter is not only a mathematical one. Determining are the physical characteristics of interaction of fast charged particles with matter. In this respect, by no means all questions are so far investigated exhaustively in a quantitative fashion. However, throughout this chapter, all the corresponding physical laws required are assumed to be known (from theory or experiment, as analytical functions or in tabulated form), and are described by some functions defined in the following subsection. This leads to a general formalism independent of the special form of these functions which will be considered at the present state of knowledge in chapter 4. It will there be shown that the experimental and theoretical data now available are, to a first approximation, sufficient for our purposes. This subdivision of problems has two advantages: for the sake of clearness, the complexity of the entire subject is reduced, and the

---

[1]) The problem of relating the energy z to the ionization number j is the same as in corresponding experimental investigations using proportional counters which also display j rather than z (BICHSEL, 1974).

obtained formalism is invariant with respect to further progress in radiation physics.

3.1.1. The underlying physical quantities and functions

In the following, the above-mentioned physical quantities are defined. These quantities depend on the absorbing medium, but the influence of its (inhomogeneous) chemical composition (e.g., water around the cellular objects, protoplasma, cell nucleus, DNA) will be neglected. Otherwise, $f_1(j)$ and U_k would depend on the special structure of the objects and even on the particular irradiation assay. Water will be taken as the unit absorbing medium since it represents the main constituent of most biological objects and their environment in most irradiation assays. However, in some cases it is necessary to take into account different mass densities. This will be done later on by introducing dimensionless density factors explicitly, while the following physical quantities always refer to water with the density 1 g/cm³.

$L_\infty(E)$ is the *linear energy transfer* (LET) or the *stopping power* (of water) for electrons $\left(L_\infty^{(e)}(E)\right)$ or ions $\left(L_\infty^{(i)}(E)\right)$ as a function of the energy E of the particle. Different values of the index i denote different ions ($i = 1, 2, \ldots$). $L_\infty(E)$ is defined by $\mathrm{d}E/\mathrm{d}x$, the differential mean energy loss of the particle per differential path length.

$L_\delta(E)$ is the *restricted stopping power,* in which all energy loss events (collisions) that consist in an ionization[1]) of a molecule with an energy transfer above δ are omitted.

$N(E, E')$ as a function of E' means the *initial energy spectrum of secondary electrons* (δ-electrons) produced by electrons $\left(N_e(E_e, E')\right)$ or by ions $\left(N_i(E_i, E')\right)$ with energies E_e and E_i, respectively. More exactly, $N(E, E')\,\mathrm{d}E'$ is the mean number of ionizing collisions with an energy transfer[2]) between E' and $E' + \mathrm{d}E'$ per unit path length of the charged particle at the energy E. In terms of atomic physics, $N(E, E')$ is the differential macroscopic ionization cross-section of water.

E' may have any positive value, but one must realize that

$$N(E, E') = 0 \quad \text{for} \quad E' > E_{\max}. \tag{3.3}$$

The *threshold energy* $E_{\max}$ is a function of E. In the case of electrons, the colliding and the encountered electrons are quantum-mechanically undistinguishable. Therefore, the less energetic of both is taken for the secondary one by definition. This implies for electrons

$$E_{\max} = \frac{1}{2} E_e. \tag{3.4}$$

According to the definitions given above it holds:

$$\int_\delta^\infty \mathrm{d}E' E' N(E, E') = L_\infty(E) - L_\delta(E). \tag{3.5}$$

$M(E, E')$ as a function of E' is the *cumulative energy distribution of secondary electrons* per unit path length of the ionizing particle:

$$M(E, E') = \int_{E'}^\infty \mathrm{d}E'' N(E, E''). \tag{3.6}$$

$R_e(E)$ is the *mean range of electrons* in water as a function of the initial electron energy E. Let $g(x)$ be the probability density for an electron which starts with the energy E at the origin of a co-ordinate

[1]) Other kinds of energy losses (Bremsstrahlung, elastic and inelastic nuclear collisions) are still counted in.

[2]) Equating the energy transfer E' with the energy of the produced δ-electron means that the energy spent to overcome the ionization potential is regarded as an energy loss of the δ-electron at the beginning of its track.

system into the direction of the x-axis to come to rest at a point with the abscissa x. Then, the mean range is defined by the expected value

$$R_e(E) = \int_{-\infty}^{+\infty} \mathrm{d}x \; x \; g(x). \tag{3.7}$$

$\tilde{R}_e(E, l)$, the so-called *modified range*, is the mean smoothed track length of electrons with the initial energy E in water. Smoothing means neglect of track curvature within segments of the length l. $\tilde{R}_e(E, l)$ always has a value between the mean range and the mean path length:

$$R_e(E) \leqq \tilde{R}_e(E, l) < \int_0^E \mathrm{d}E' \, \frac{1}{L_\infty^{(e)}(E')}. \tag{3.8}$$

For a quantitative definition of $\tilde{R}_e(E, l)$ see Eqs. (3.40), (3.41).

$p^{(\delta)}(E)$ is the *number of primary ionizations* per unit path length of electrons $\left(p_e^{(\delta)}(E)\right)$ or ions $\left(p_i^{(\delta)}(E)\right)$ as a function of the energy E of the particle, omitting primary ionizations with an energy transfer greater than δ. Although $p^{(\delta)}(E)$ can be expressed formally through

$$p^{(\delta)}(E) = M(E, 0) - M(E, \delta), \tag{3.9}$$

it must be considered as an independent quantity since $M(E, 0)$ cannot be obtained in the same way as $M(E, E')$ for large E' (i.e. about $E' > 200$ eV).

$\omega^{(\delta)}(E, j)$ are the *cluster frequencies* in dependence on the particle energy E, i.e. $\omega_e^{(\delta)}(E, j)$ and $\omega_i^{(\delta)}(E, j)$ for electrons and ions, respectively, denote the probability that a primary ionization (omitting those with an energy transfer greater than δ) yields an ionization cluster (short δ-track) with a total of j ionizations. If δ is large, a 'cluster' may consist of a rather long δ-track. The normalization condition reads

$$\sum_{j=1}^{\infty} \omega^{(\delta)}(E, j) = 1. \tag{3.10}$$

3.2. Prerequisites and basic principles of approximation

The physical stochastics of the absorption of radiation cannot but be treated approximately. However, certain suppositions concerning the irradiation conditions (section 3.2.1) should be distinguished from actual approximations, since the former can be fulfilled nearly exactly. The term 'principles of approximation' means general rules of how to treat the complex random patterns of primary and secondary particle tracks and the random processes of energy transfer along them. These principles may therefore be applied to the calculation of $f_1(j)$ as well as U_k or of any other probability field for a random variable that depends, generally speaking, on what 'happens' inside a micro-volume upon irradiatingit. The meaning of the latter and some questions of terminology should be considered first.

We are dealing with populations of discrete cellular or molecular objects of irradiation in an environment with similar absorption properties (e.g., an aqueous solution). The so-called *gross sensitive volume* (GSV) of each object is to be understood as the smallest convex volume great enough for local energy deposition by fast charged particles outside of it to have no share in the radiation effect under consideration. This definition corresponds to that given earlier by Rossi (1964) and Kellerer and Rossi (1972). Thus, the GSV may depend on the effect as well as on the environmental conditions; it may be larger (indirect action) or smaller than the object itself. It should be emphasized that this definition of the GSV does not imply any model of radiation action, since the mere existence of such a volume is self-evident, and nothing is said about the interior of the GSV, which may have a complex structure including sensitive sites or targets. Thus, the GSV is by no means presumed generally to be homogeneously sensitive to radiation.

Further properties of the GSV are necessarily connected with a specific theory or model of radiation action, but in this section such specifications are needless. In the framework of the DNA-lesion theory the GSV is identical with the region in which the entire[1]) DNA is located, i.e. the *cell nucleus* in the case of eukaryotic cells, while the sensitive volume with respect to a DNA molecule is called a *target*. In order to avoid confusion, we will use these terms only with this concrete meaning, although a 'target' could also be referred to as a GSV according to the above general definition if only the production of DNA-lesions is considered as the effect. Therefore, in all physical considerations that apply to either case, i.e. to arbitrary micro-volumes, we prefer the indifferent term 'sensitive volume' (SV).

3.2.1. Prerequisites concerning the conditions of irradiation

Let us consider irradiations with heavy charged particles (ions), photons (γ- or X-rays), and neutrons, which will be called the *primary incident particles*. Concerning the irradiation conditions it shall be assumed that

(i) secondary-particle equilibrium is reached at the position of the SV embeded in a sufficiently thick layer of matter;
(ii) the absorbed dose is constant within the SV, i.e. alterations of number and energy or energy spectrum of the primary incident particles are small over a distance that corresponds to the mean thickness of the SV;
(iii) number and energy or energy spectrum of the primary incident particles change only insignificantly within a distance corresponding to the maximum range of their secondaries;
(iv) SV's are oriented at random unless they are completely isotropic with respect to shape and composition.

Supposition (i) implies that the layer of matter between radiation source and SV, which may consist of parts of the object itself or of a suitable medium, is not thinner than the maximum range of secondary particles in it. In addition, the atomic composition of the medium outside the SV must not significantly differ from that inside it, which is especially important in the case of neutron and photon irradiation. Either of the requirements (ii) and (iii) may be the more severe one, depending on the size of the SV. (ii) represents a serious restriction only for irradiations with ions[2]) since the absorption lengths of neutrons and photons[3]) in tissue are of the order of centimetres or decimetres. Due to the conditions (i)—(iii), the radiation field information to start with must be only a local one (primary particle spectra at the position of the SV), i.e. macroscopic features of the radiation field are not implicated in the theory. This important simplification is, however, impossible for electron radiation because, in view of Eq. (3.4), condition (iii) is not fulfilled.[4]) Electron radiation is, therefore, excluded from the consideration. On the other hand, X-rays of a suitable energy produce almost the same biological effects and may, hence, replace electrons in the theory.

---

[1]) Excluding the so-called extrachromosomal DNA.

[2]) Irradiations with ions observing the condition (ii) are just what has been called the 'track-segment method'.

[3]) The mean free path of neutrons in tissue-equivalent matter ranges from 0.98 cm to 6.8 cm for the neutron energy interval 0.1 to 5.8 MeV (Coppola and Booz, 1975), that of photons from 4.2 cm to 9.8 cm at 0.1—1 MeV (Kunze, 1959).

[4]) A transport equation (Harder, 1966; Harder and Schulz, 1968; Schulz and Harder, 1970; Patau *et al.*, 1970; Terrissol and Patau, 1974) would have to be solved for the particular geometry, to get the electron spectrum at the position of the SV.

3.2.2. Approximations concerning the description of absorption events

For an understanding of the stochastics of radiation absorption the conception of absorption events introduced by microdosimetry is fundamental. A comprehensive discussion of the related questions may be found elsewhere (HUG and KELLERER, 1966; KELLERER, 1968a; KELLERER and CHMELEVSKY, 1975). The point is that a microscopic volume does not 'feel' irradiation as a continuous process, but only from time to time, in a random manner, something happens in the SV. These statistically independent 'happenings' (practically instantaneous energy transfers to the SV) are called *absorption events*. Several particles may be involved in one absorption event, but these must always have a common origin in order to arise simultaneously. In this way, one obtains a great variety of qualitatively and quantitatively different absorption events, of which Figure 3.1 (upper part) gives some examples. One must realize, however, that Figure 3.1 displays only schematic track patterns but not the single elementary energy transfer processes (collisions) along each track, which obey probability laws, too.

The total energy imparted (or the number of ionizations produced) in one event microdosimetry has dealt with so far is only one parameter among others by which absorption events can be characterized quantitatively.[1]) The following considerations, however, are not restricted to this or any other such parameter, but rather provide us with a general basis for the mathematical description of absorption events and their statistics.

From the physical point of view, all the particles involved in a radiation field may be classified into *primary* and *secondary* particles. However, for the description of absorption events, i.e. seen from the SV, we still need another classification in terms of the notation *incident* and *dependent* particles. The former are statistically independent of one another by definition. One absorption event is always due to exactly one incident particle which

[1]) Other parameters would be, e.g., the number of particles involved, the energy imparted by electrons only, or, as in our case, the number of DNA-lesions produced. To take the total energy for the biologically relevant quantity is a bias originating from the predominant role dose plays in radiation biology. In any case, it is not first of all physics that can tell us the proper parameter(s).

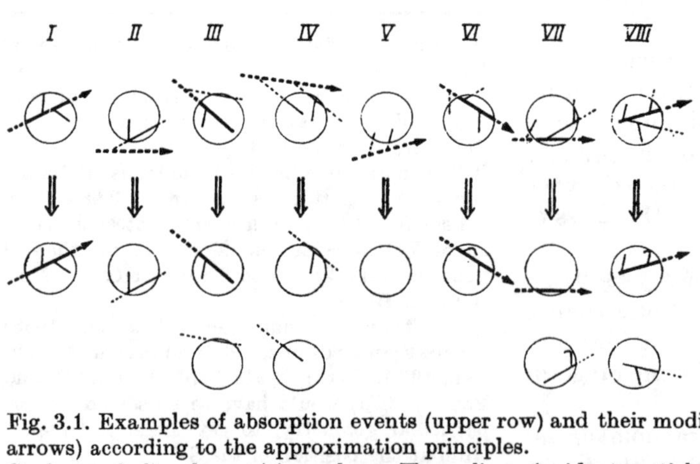

Fig. 3.1. Examples of absorption events (upper row) and their modifications (indicated by double arrows) according to the approximation principles.
Circles symbolize the sensitive volume. Heavy lines: incident particle tracks. Thin lines: dependent particle tracks (short ones for $E' < \delta$, long ones for $E' > \delta$).

itself ionizes in the SV (by passing through, or starting in, it) and/or indirectly by means of its secondaries — the dependent particles. The latter are characterized by their dependence upon incident particles by which they are produced directly or indirectly. Several of them may simultaneously participate in one event together or not with the one incident particle they are descended from. This is illustrated by Figure 3.1.

Particles coming direct from the radiation source are always incident ones called primary incident particles. Moreover, photoelectrons are also incident particles in a strict sense, whereas further charged particles produced in the medium may only approximately be considered as incident ones, as follows.

Approximation principles:

1. In addition to photoelectrons, Compton electrons[1]) as well as recoil nuclei produced by neutron impact or nuclear reaction[1]) are treated as incident particles generated independently of one another at random in the irradiated medium. In this way, all uncharged particles need not be explicitly considered further on.

2. Energetical cut-off for δ-electrons: By means of some cut-off energy δ, which depends on the size of the SV, secondary electrons ejected by charged particles (δ-electrons) are subdivided into two groups, depending on whether their initial energy E' is greater or smaller than δ.

(a) All charged particles (including δ-electrons themselves) are considered to produce δ-electrons with $E' < \delta$ only. The latter alone are the dependent particles.

(b) All δ-electrons of the radiation field with $E' > \delta$ are retaken into account by considering them to be generated at random in the medium independently of one another and of the charged particles they are intrinsically due to. Hence, these δ-electrons are added to the incident particles.

3. Reduction of the diversity of events:

(a) δ-electrons of an initial energy $E' < \delta$ (dependent particles) entering the SV from outside are completely neglected, and events that are solely due to such δ-electrons are, hence, omitted. Consequently, only two types of absorption events remain: an incident particle either goes across, or starts in, the SV. Further particles contributing to an event can only be dependent particles produced inside the SV by the incident particle or by such dependent particles themselves.

(b) All δ-electrons with $E' < \delta$ (dependent particles) produced inside the SV by virtue of the one incident particle are assumed always to lose their entire energy within the SV, i.e. not to leave it. This means that actually outgoing δ-ray tracks must be thought of as accordingly bent.

4. Simplified treatment of straggling:

(a) Continuous slowing down is presumed for all charged particles in any case, so that energy and track length of a particle are well-related quantities, while the discontinuous character of the transfer of energy to the SV is properly considered.

(b) Collisions (primary ionizations including δ-ray production) along the fixed path of a charged particle are treated as statistically independent of one another. This implies Poisson statistics for the number of any specified collisions on every path segment.

(c) Track curvature within the SV is completely disregarded for ions.

(d) For incident electrons, only the fluctuations of the path segment length inside the SV which are due to random curvature of the track are neglected, but the average effect of track curvature upon the path segment (lengthening) is taken into consideration.

---

[1]) Other elementary processes by which photons or neutrons create charged particles can be ignored for the energy range considered throughout this book.

(e) In the case of primary incident ions the decrease of energy upon traversing through the SV is neglected.

These approximation principles on which the mathematics of the subsequent sections is based may be explained and justified by the following arguments.

Principle 1 is warrantable by the large mean free path of photons and neutrons in matter (see footnote 3 at page 63). Secondary particles produced simultaneously by successive Compton effects and nuclear collisions, respectively, are very unlikely together to affect the same SV, which has microscopic dimensions.

Principles 2 and 3 must be considered in close connection since the errors introduced by them compensate for each other to a certain degree depending on the value of δ. *δ-ray cut-off* is a well-known method in radiation biophysics, first introduced by Lea (1946). But there are important differences between our procedure and most preceding work (Barendsen, 1966; Kellerer, 1968a, b; Cormack, 1968; Bewley, 1968; Oda *et al.*, 1971). As shown in Figure 3.2, δ-tracks with $E' > \delta$ neither are neglected nor are those with $E' < \delta$ assigned to the primary track core. Thus, not only the total absorbed energy remains unaltered by principle 2, but also each single particle track is kept up. It is merely the temporal and spatial coherence of tracks that is partly destroyed by principle 2. Besides, the cut-off energy δ is not fixed once for all but will be chosen in dependence on the dimensions of the SV (see below).

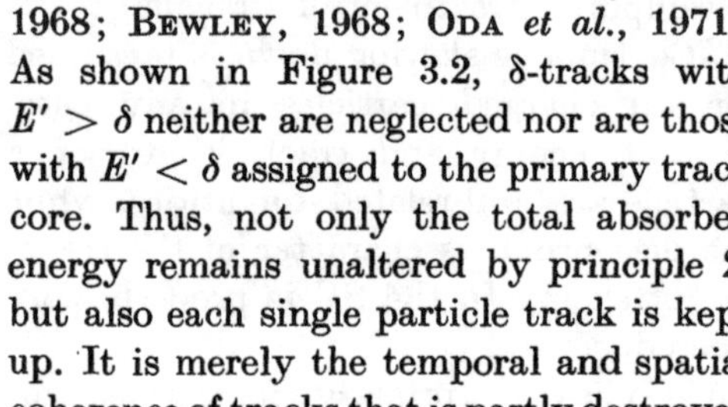

Fig. 3.2. Decomposition of a pattern of joint primary and secondary particle tracks (a) into single independent incident particle tracks (b) according to the approximation principle 2. Dashed tracks indicate δ-electrons with an initial energy $E' < \delta$.

As it were, by virtue of principle 2, the radiation field itself is replaced by a spectrally equivalent artificial one without explicit reference to the SV, whereas principle 3 forbids crossing the SV boundary by dependent particles ($E' < \delta$). Again, exact energy and particle balance evidently is maintained between 3a and 3b, i.e. only assignments of track sections and the corresponding energy depositions from one event to another take place in such a way that the same absorbed energy appears more concentrated in a smaller number of events. Principle 2, on the contrary, causes splitting of events into several, i.e. the number of events is increased and their energies are decreased. In this way, principle 3 partly compensates for the errors due to principle 2 and vice versa.

Figure 3.1 illustrates the consequences of the approximation principles 2 and 3. Provided that δ has been chosen appropriately (see below), most of events fall into the types I and II, i.e. nothing is altered and they are treated correctly. III and IV show the splitting of events caused by principle 2. Events of the type V are omitted according to 3a, while 3a and 3b together are responsible for the type VI modification. The joint action of both the approximation principles is indicated by the more complex examples VII and VIII.

An open question as yet is the choice of the cut-off energy δ. Of course, the proportions of δ-electrons subject to the principles 2 or 3 are determined by δ. Consequently, for these principles to be good approximations the former requires a large δ, the latter, however, a small one. In addition, as long as short δ-rays (compared with the dimensions of the SV) are concerned, the majority of them is in accord with the assumptions 3a and 3b, which favours a small δ. On the other hand, a large δ could guarantee that a coincidence of an incident δ-track ($E' > \delta$) with its primary track in the SV (type III event in Fig. 3.1) would be relatively rare[1]) in correspondence with principle 2.

Because of these contradicting requirements, the choice of δ can only be a compromise in such a manner that the mean δ-electron range $R_e(E')$ for $E' = \delta$ is of the order of magnitude of the SV (CORMACK, 1968). The latter may be measured by the *mean thickness* l (expected value of the *chord length distribution* (KELLERER, 1971a) of the SV).

As the δ chosen should minimize the errors caused by the approximation principles 2 and 3, it would be extremely important to have a quantitative measure for these errors. In the appendix (section 3.8.1) the calculation of such a quantity is performed by a Monte Carlo method. This quantity, η, is the percentage of 'incorrectly apportioned energy':

$$\eta = \frac{\Delta' + \Delta''}{\textit{total energy absorbed}} \times 100\%. \tag{3.11}$$

Δ' describes that part of the energy deposition by δ-electrons with $E' < \delta$ which is omitted from the corresponding events according to the approximation principle 3a. As this energy is exactly added to other events on account of 3b, it is incorrectly apportioned, apart from the possibility of compensation in the same event as shown by the type VI in Figure 3.1. Δ'' referring to principle 2 describes the energy deposition in the SV by δ-electrons with $E' > \delta$ that is inadmissibly separated to form additional artificial events.

It is clear that Δ' increases and Δ'' decreases with growing δ. Therefore, the sum η reaches a minimum for some value of δ we are interested in. The calculation of η (section 3.8.1) has been carried out for incident ions[2]) of all charge numbers and energies usually occurring in radiation biology. Spherical and cylindrical[3]) sensitive volumes with mean thicknesses in the range of 10^{-3} μm to 10 μm have been considered. In all cases, the minimum of η either is reached, or η is only insignificantly above its minimum, at a δ value determined by

$$R_e(\delta) = \varrho l. \tag{3.12}$$

l is the mean thickness of the SV and ϱ the mass density coefficient (relative to the density of water). Thus, the 'best compromise' for δ has been found. *Now the cut-off energy δ will always be chosen according to the relation* (3.12).

Of course, the minimum of η with respect to δ is one thing, but the absolute value of the percentage of incorrectly apportioned energy η for the δ value given by Eq. (3.12) is also very interesting, for it provides us with an idea about the errors introduced by the principles 2 and 3. Figures 3.3 and 3.4 show the results for spherical and cylindrical sensitive volumes, respectively. η depends on the mean thickness l of the SV (abscissa) and the energy per nucleon (ordinate) of the ion under consideration. The latter quantity is closely related to the maximum δ-electron energy $E_{\max}$ which is given at the right-hand side. The third independent variable is the charge number Z of the ion.[4]) The upper graphs in Figures 3.3 and 3.4 have been calculated for $Z \to 0$; they apply to light ions up to about α-particles. The lower graphs are for $Z = 10$ (Ne ions). The mass density coefficient ϱ of the SV is assumed to be equal to that of the environment. If this common value differs from unity, l must be replaced by ϱl. A higher mass density of the SV as compared with the exterior would only diminish η.

---

[1]) Save for ion-produced δ-electrons with an initial energy close to the maximum one, which fly almost in the direction of the ion.

[2]) Electrons are not so crucial because of a more advantageous scattering angle relation between primary and secondary electrons.

[3]) The endless cylinder was simulated by a long spheroid (axis ratio 1:1000).

[4]) Z characterizes the sort of ion only, whereas the *mean effective charge* used in the calculations depends on energy.

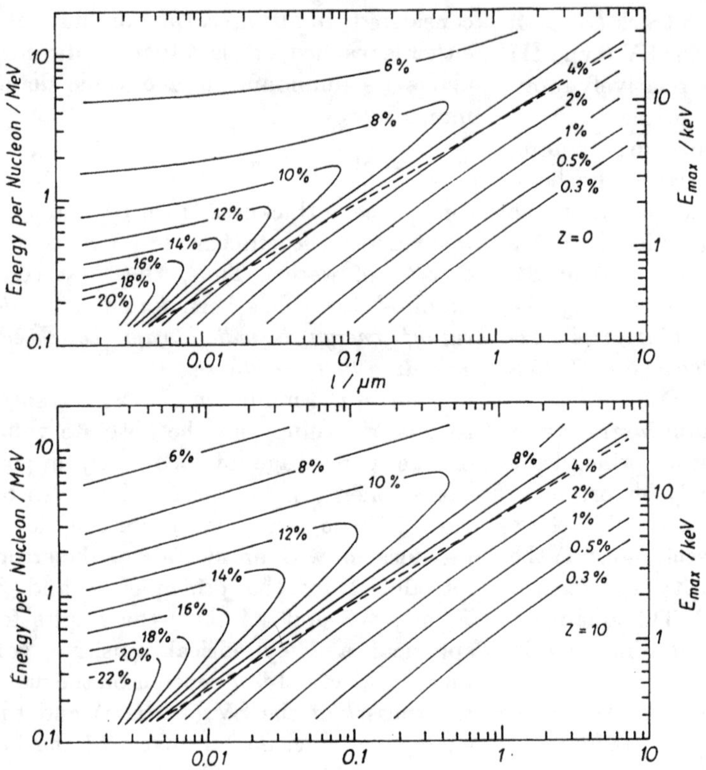

Fig. 3.3. η, the percentage of energy incorrectly apportioned on the strength of the approximation principles 2 and 3, in dependence on the mean thickness l of the sensitive volume and the energy per nucleon of the incident ions (right-hand ordinate scale: maximum δ-electron energy E_{max}.)
Upper part for light ions, lower part for Ne ions.
Shape of the sensitive volume: sphere.
The dashed lines indicate the boundary below which there are no incident δ-electrons because $E_{max} < \delta$ (the term Δ'' in Eq. (3.11) vanishes).

In discussing Figures 3.3 and 3.4 it must be born in mind that the percentage of incorrectly apportioned energy, η, can only roughly intimate the order of errors that must be expected when using the δ-ray cut-off method for the description of radiation absorption stochastics. Naturally, a sufficiently small η guarantees slightness of these errors, but the consequences of the modifications of events by the approximation principles cannot always be judged from η, since energy is not the only biologically relevant parameter (see footnote 1 at page 64). In the case of heavy ion irradiation, for example, most of events may be pure δ-electron ones, but the relatively few primary particle events with their high energy deposition make the denominator of Eq. (3.11) large, which leads to a small η in spite of possibly drastic modifications of the majority of events, including their number. If, in this case, the biological effect considered requires only small energy depositions, so that almost every event is effective, the error solely caused by an incorrect *number* of events may be considerably higher than indicated by η. Opposite cases are, of course, conceivable, too. Generally speaking, it is difficult to value any neglects as long as it is not stipulated which features of the events are the crucial ones. It

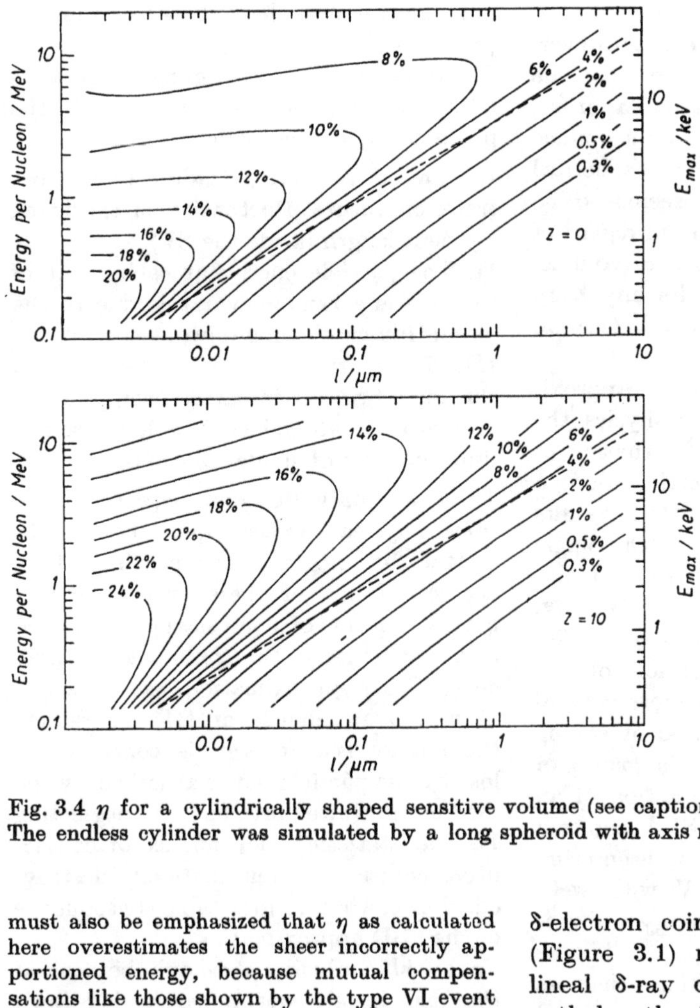

Fig. 3.4 η for a cylindrically shaped sensitive volume (see caption of Fig. 3.3). The endless cylinder was simulated by a long spheroid with axis ratio 1:1000.

must also be emphasized that η as calculated here overestimates the sheer incorrectly apportioned energy, because mutual compensations like those shown by the type VI event in Figure 3.1 and those resulting from the opposite character of the approximation principles 2 and 3 (see Fig. 3.1: the splitted types III and IV may partly compensate for the omitted type V) are not taken into account by our η.

As may be seen from Figures 3.3 and 3.4, η is higher for cylinders than for spheres and increases with growing charge number Z. This must be attributed to Δ'' only, since Δ' does not depend on Z and it differs only quite insignificantly for spheres and cylinders. As the δ-electron spectrum is unchanged by the charge number Z at constant energy per nucleon, the increase of η with growing Z is only due to the increasing number of incident δ-electron coincidences of the type IV (Figure 3.1) resulting from the higher lineal δ-ray density (number per unit path length of the ion). At this rate in the worst case, Ne ions and cylindrical SV, the high η values result from simultaneous δ-electron traversals rather distant from one another along the cylinder intrinsically standing for the DNA. Therefore, regarding the formation of DNA-lesions, it may be even more suitable to treat these only temporally but not spatially joint traversals as statistically independent ones in accord with principle 2, since they are likely to produce separate lesions.[1]) For this reason, the high η

[1]) However, whether the thus clustered lesions (let them be breaks) are also *detected* separately by means of the sucrose gradient assay is another question.

values of the lower part of Figure 3.4 need not be considered alarming. However, for charge numbers beyond 10 the procedure based upon the above approximation principles becomes dubious, since then η gets rather large for the spherical SV, too. The upper limit seems to be reached with Ar ions, at least in regard of DNA targets as the sensitive volume. The corresponding η values for any Z up to 18 are given in the appendix (section 3.8.1).

Let us now discuss the fourth approximation principle. In radiation physics, the term *straggling* means the fluctuation of the total energy loss of a particle along a definite section of its path, which results from the random character of the single elementary energy loss processes (collisions) the particle undergoes (VAVILOV, 1957; TSCHALÄR and MACCABEE, 1970). For the purpose of description of absorption events, however, a more general definition should be introduced by saying that 'straggling' designates the *totality* of factors owing to which the action of an incident particle inside the SV becomes a random process even for a particle entering into, or starting in, the SV with well-defined initial conditions (energy and geometry).

By this definition, it is noticed that we are not only interested in imparted energy and that the path segment in the SV is, itself, a random variable of the straggling process. If, however, scattering (track curvature) and often also the diminution of energy of the particle are negligible, as taken for granted in some previous calculations (KELLERER, 1968a, b; COPPOLA and BOOZ, 1973), the path segment gets fixed and the single collisions are automatically independent of one another, which makes straggling a 'simple' Poisson process. Thus, the approximation principle 4 only illuminates what has mostly been done before[1]) by tacit agreement. However, 4d and the restriction of 4e only to primary incident ions go further than these simple approximations.

Three aspects of straggling each contributing to the randomness of the process may be distinguished:

I. Random scattering leading to random track curvature (fluctuation of the track segment length inside the SV).

II. The particle energy at each point of its path is a random variable due to the discontinuous slowing down.

III. Disregarding track curvature and slowing down, the imparted energy nevertheless fluctuates since it results from a finite number of single collisions.

These are only different aspects of the *same* physical process: interaction with matter by single collisions. Therefore, neglect of one of these aspects always means violation of a physical conservation law. Particularly, the *continuous-slowing-down* (CSD) approximation, i.e. neglect of II with maintenance of III, contradicts the energy law: energy is continuously lost by the particle but is absorbed by the medium in finite portions.[2]) By neglecting II, the statistical dependence of consecutive collisions is automatically destroyed. Hence, 4b is an inevitable consequence of the CSD approximation.

To avoid confusion, it should be noticed that the term 'continuous-slowing-down approximation' has sometimes been used for total neglect of straggling (CASWELL and COYNE, 1976), i.e. for an assumed continuous *transfer* of energy to the medium. Throughout this book, however, CSD does not mean more than a mere continuous *diminution* of energy of charged particles.

Appendix 3.8.2 is devoted to the in-

---

[1]) With the exception of some Monte Carlo approaches (GUBIN and SAKOVICH, 1974; PARETZKE *et al.*, 1974).

[2]) Of course, the total imparted energy averaged over all events equals exactly the total energy loss of all particles, provided that the particle fluence spectrum is also based on the CSD approximation (see sections 3.3 and 3.4).

vestigation of the CSD approximation (principles 4a and 4b). The result is that this approach is very good for ions in any case. For incident electrons, CSD causes a marked enhancement of the mean imparted energy per event and of the variance of the latter only for electron energies $E \approx \delta$ (see Eq. (3.12)).

The path of a continuously slowed down electron has a definite total length, but it may nevertheless be randomly bent. It is possible to treat this random curvature in the framework of the CSD approximation. This approach, which will be described in detail in the following chapter (section 4.5.2), yields an estimate of the *mean value* of the energy loss $\Delta E(l')$ for an electron that would traverse the path l' through the SV if track curvature were lacking (if the path were straight). This mean 'energy segment' ΔE corresponds to a mean path length segment which is greater than l'. This approach is based on modified LET and modified range $\tilde{R}_e$ considered in section 3.4.

In this way, however, the *fluctuations* of ΔE for a given l' that are due to the randomness of track curvature are neglected (approximation principle 4d). To estimate the errors introduced by this approximation is difficult. One may only say that the CSD-caused overestimation of the variance of the energy imparted in an event (see appendix 3.8.2) may have some compensatory effect. In addition, the following argument can be put forward: KELLERER (1968a, b) has shown that the total variance of the energy imparted in an event must mainly be attributed to the dispersion of the energy deposition by the single collisions, while the geometrically determined chord length distribution $w(l')$ is of minor importance. Hence, the same should be true of the influence of the path segment fluctuations that result from track curvature, realizing that the latter might be described by a modified (broadened) chord length distribution $\tilde{w}(l')$.

For ions the situation is simple. Track curvature is indeed negligible (4c) and 4e is nothing but a consequence of the prerequisite (ii) (dose homogeneity within the SV). The errors introduced by the CSD approximation (4a and 4b) are always small as shown in appendix 2.8.2.

3.3. Description of radiation fields

If the prerequisites discussed in section 3.2.1 are met, the radiation field is completely described by the *fluence spectra* $\varphi(E)$ of the *incident particles* (see section 3.2.2) at the place of the sensitive volume (SV). The definition is as follows. For a certain sort of particles, $\varphi(E)\,\mathrm{d}E$ means the number of particles per unit projected area of a surface element (projection into the plane perpendicular to the particle path) that cross this element with an energy between E and $E + \mathrm{d}E$. An equivalent definition reads: $\varphi(E)\,\mathrm{d}E$ is the total track length per unit volume of those differential track sections in which particles traverse the energy interval $\mathrm{d}E$ at the energy E.[1])

We will be concerned with three different fluence spectra:

- $\varphi_i^{(0)}(E)$ — ions incident directly from the radiation source;
- $\varphi_i^{(n)}(E)$ — ions set in motion in the medium by fast neutrons (recoil nuclei and nuclear reaction products);
- $\varphi_e(E)$ — electrons ejected in the medium by photons or by any charged particles which produce δ-rays.

The fluence spectra $\varphi_i^{(n)}(E)$ and $\varphi_e(E)$ that describe particles generated in the medium are closely related to another kind of spectra called the *initial energy*

[1]) Therefore, the name 'track length spectrum' could also be used, but this term should be reserved only for the otherwise normalized fluence spectrum of δ-electrons, $T_i^{(\delta)}(E_i, E)$, defined below.

spectra, $q_i^{(n)}(E)$ and $q_e(E)$, of these incident particles. $q(E)\,dE$ is the number of particles ejected per unit volume with initial energies between E and $E + dE$. Initial spectra and fluence spectra[1]) are connected by the relation

$$\varphi_i^{(n)}(E) = \frac{1}{L_\infty^{(i)}(E)} \int_E^\infty dE' q_i^{(n)}(E'), \quad (3.13\text{a})$$

$$\varphi_e(E) = \frac{1}{L_\infty^{(e)}(E)} \int_E^\infty dE' q_e(E'). \quad (3.13\text{b})$$

The integral represents the number of particles per unit volume that pass through the energy E at some instant in the course of their slowing down. Realizing that $dE/L_\infty(E)$ is the length of the differential track segment corresponding to the energy interval dE, Eq. (3.13) becomes evident.

Notice that Eq. (3.13) is already a consequence of the continuous-slowing-down approximation (principle 4a in section 3.2.2), i.e. for a given initial spectrum $q(E)$ the fluence spectrum $\varphi(E)$ is in fact slightly different from (3.13). This difference is elucidated at the end of appendix 3.8.2. It proves quite unimportant for ions and also for electrons of not too-small energy. The reliability of the CSD approximation with respect to the fluence spectrum of electrons was investigated first by Spencer and Fano (1954), who solved the stochastic slowing down problem strictly for realistic physical parameters by numerical methods. In the case of a monoenergetic electron source, $q_e(E) = \delta(E - E_0)$, when the CSD solution is simply the reciprocal stopping power, the agreement is very good.[2]) For a comparison in case that secondary electrons are added to the source term see Table 3.1. More recent investigations of the problem are by Peterson (1969) and Douthat (1975).

In the case of direct ion bombardment (protons, α-particles, heavy nuclei), the fluence spectra themselves, $\varphi_i^{(0)}(E)$, are considered as the initially given quantities (mostly only one sort of monoenergetic particles, i.e. $\varphi_1^{(0)}(E) \sim \delta(E - E_0)$). In the case of fast neutrons (monoenergetic or with an energy spectrum), this primary information consists in the initial energy spectra of the nuclear secondaries, $q_i^{(n)}(E)$, from which the fluence spectra must be calculated by means of Eq. (3.13a). The electron fluence spectrum, however, cannot be obtained so easily since neither $q_e(E)$ nor $\varphi_e(E)$ is given directly, realizing that all δ-electrons above the cut-off energy are involved.

Let us first consider the δ-electrons generated by an ion of the sort i and the energy E_i within a short segment of its track. Relating the fluence spectrum of these δ-electrons to this 'parental' ion track segment rather than to unit volume, the thus defined *track length spectrum* $T_i^{(\delta)}(E_i, E)$ reads in compliance with Eq. (3.13b):

$$T_i^{(\delta)}(E_i, E) = \frac{1}{L_\infty^{(e)}(E)} \int_{\max(E,\delta)}^\infty dE' [N_i(E_i, E') + q_e^{(e)}(E')]. \quad (3.14)$$

$N_i(E_i, E')$ describes the δ-electrons ejected by the considered ion direct, while $q_e^{(e)}(E')$ is the source strength arising from the δ-electrons themselves. Note that the lower limit of integration is a consequence of approximation principle 2 (section 3.2.2). The definition of $T_i^{(\delta)}(E_i, E)$ is illustrated by Figure 3.5. The small section dx of an electron track at the energy E_e contributes $dx N_e(E_e, E')$ to the

---

[1]) The term 'initial spectrum' has been generally accepted, while $\varphi_i^{(n)}(E)$ calculated from Eq. (3.13) is often called the 'slowing-down spectrum' (Caswell and Coyne, 1976). In accord with Dennis (1974), we prefer the more general denotation 'fluence spectrum' which is also appropriate to $\varphi_i^{(0)}(E)$ and $\varphi_e(E)$ which do not (or not only) result from slowing down of the same particle.

[2]) Considerable deviations are only found if bremsstrahlung energy losses become important, which is the case only at uninterestingly high energies and for heavy materials (lead).

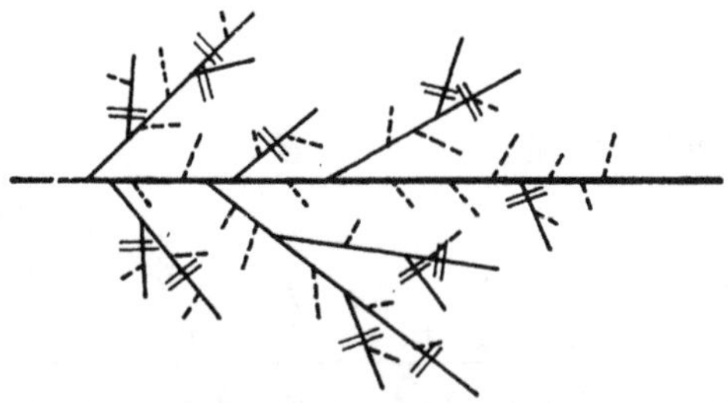

Fig. 3.5. Definition of the track length spectrum.
$T_i^{(\delta)}(E_i, E)\,\mathrm{d}E$ is the total length of all the differential track segments (indicated by the thin double bars) divided by the reference track length of the parental ion the δ-rays are descended from. Each of the differential track segments is at a point where an electron runs trough the energy E, and their lengths correspond to the energy differential $\mathrm{d}E$. If $E < \delta$, contributions from δ-electrons with an *initial* energy less than δ (dashed tracks) are omitted.

spectrum $q_e^{(e)}(E')$. All these track elements $\mathrm{d}x$, which are just given by $\mathrm{d}x = T_i^{(\delta)}(E_i, E_e)\,\mathrm{d}E_e$, together yield

$$q_e^{(e)}(E') = \int_0^\infty \mathrm{d}E_e T_i^{(\delta)}(E_i, E_e)\, N_e(E_e, E'). \tag{3.15}$$

In this way, from Eqs. (3.14), (3.15), one obtains a linear integral equation for $T_i^{(\delta)}(E_i, E)$, which can be written in the form (see Eq. (3.6)):

$$L_\infty^{(e)}(E)\, T_i^{(\delta)}(E_i, E) = \begin{cases} M_i(E_i, E) + \int_0^\infty \mathrm{d}E_e T_i^{(\delta)}(E_i, E_e)\, M_e(E_e, E) & \text{for} \quad E \geqq \delta \\ L_\infty^{(e)}(\delta)\, T_i^{(\delta)}(E_i, \delta) & \text{for} \quad E < \delta. \end{cases} \tag{3.16}$$

It is easy to solve Eq. (3.16) by numerical methods since $M_e(E_e, E)$ vanishes for $E_e < 2E$ (see Eqs. (3.3), (3.4)). Beginning at the highest energy $E = E_{\max}$, the track length spectrum $T_i^{(\delta)}(E_i, E)$ can be calculated direct from the right-hand side of Eq. (3.16) for decreasing values of E. In this calculation, the integral either vanishes (if $E > E_{\max}/2$) or depends only on the already known values of $T_i^{(\delta)}$ for $E_e > 2E$.

According to the definition of the track length spectrum $T_i^{(\delta)}(E_i, E)$, which gives the contribution of a differential track segment of an ion's track, the *total* fluence spectrum of electrons reads in the general case:

$$\varphi_e(E) = \sum_i \int_0^\infty \mathrm{d}E_i \varphi_i^{(0)}(E_i)\, T_i^{(\delta)}(E_i, E) + \varphi_n^{(\delta)}(E) + \varphi_\gamma^{(\delta)}(E),^{1)} \tag{3.17}$$

where

$$\varphi_n^{(\delta)}(E) = \sum_i \int_0^\infty \mathrm{d}E_i \varphi_i^{(n)}(E_i)\, T_i^{(\delta)}(E_i, E), \tag{3.18}$$

and $\varphi_\gamma^{(\delta)}(E)$ refers to γ- or X-radiation. This latter spectrum remains to be considered.

Let $q_e^{(\gamma)}(E)$ be the initial energy spectrum of Compton- and photoelectrons, which we will assume to be known. Since, in addition to these 'primary' electrons, also secondary δ-rays contribute to the electron fluence spectrum $\varphi_\gamma^{(\delta)}(E)$, one has to write

$$\varphi_\gamma^{(\delta)}(E) = \frac{1}{L_\infty^{(e)}(E)} \left[\int_E^\infty \mathrm{d}E' q_e^{(\gamma)}(E') + \int_{\max(E,\delta)}^\infty \mathrm{d}E' q_e^{(e)}(E') \right] \tag{3.19}$$

in analogy to Eq. (3.14), with the only difference that the first term in brackets (electrons ejected direct by photons) takes into account electrons in the range $E' < \delta$, too. As before in the case of ion-produced δ-electrons, we arrive finally

1) Note that this equation is based on the prerequisites discussed in section 3.2.1 (in particular, point (iii)).

at a similar integral equation for $\varphi_\gamma^{(\delta)}(E)$:

$$L_\infty^{(e)}(E)\,\varphi_\gamma^{(\delta)}(E) = \begin{cases} \int\limits_E^\infty \mathrm{d}E'\,q_e^{(\gamma)}(E') + \int\limits_0^\infty \mathrm{d}E_e\,\varphi_\gamma^{(\delta)}(E_e)\,M_e(E_e, E) & \text{for} \quad E \geqq \delta \\ L_\infty^{(e)}(\delta)\,\varphi_\gamma^{(\delta)}(\delta) + \int\limits_E^\delta \mathrm{d}E'\,q_e^{(\gamma)}(E') & \text{for} \quad E < \delta, \end{cases} \tag{3.20}$$

which can be solved in the same way as outlined above.

All necessary fluence spectra $\varphi(E)$ are now reduced to some initial spectra which, in the case of direct ion bombardment, must be given from the outset (fluence spectra $\varphi_i^{(0)}(E)$), whereas the initial spectra $q_i^{(n)}(E)$ and $q_e^{(\gamma)}(E)$ which describe neutron and photon radiations, respectively, must first be calculated from the energy spectrum of the uncharged primary particles and the composition of the absorbing medium. The latter question does not fall under the topic of this chapter; it is dealt with in chapter 4 (sections 4.6 and 4.7). The following scheme shows comprehensively how the fluence spectra $\varphi(E)$ are obtained:

Regarding the integral equation (3.16) or (3.20), some remarks on previous work concerning fluence spectra of secondary electrons should be added. Although the CSD approach to these spectra appears to have never been expressed in the concise form of these equations, corresponding numerical calculations have a rather long history. In literature, the subject is found mostly under the heading 'LET distributions', which means that the differential fluence (or the differential dose = LET × fluence) is expressed in terms of differential LET rather than energy (see ICRU Report 16, 1970). The first calculations for photon radiations were made by Cormack and Johns (1952), who omitted, however, the electron-produced δ-electrons (second term at the right-hand side of Eq. (3.20)). Since subsequent efforts (e.g., Bruce *et al.*, 1963) to include these δ-electrons above a cut-off energy as separate particles were usually based on *strict* spectra for *monoenergetic* initial electrons (McGinnies

$$\begin{array}{lllllll}
 & (3.16) & \rightarrow T_i^{(\delta)}(E_i, E) & \searrow & & \swarrow & \varphi_i^{(0)}(E) \\
q_i^{(n)}(E) \rightarrow & (3.13\text{a}) & \rightarrow \varphi_i^{(n)}(E) & \rightarrow & (3.18) \rightarrow & (3.17) \rightarrow & \varphi_e(E) \\
q_e^{(\gamma)}(E) \rightarrow & (3.20) & \rightarrow \varphi_\gamma^{(\delta)}(E) & & & \nearrow &
\end{array}$$

The sum spectrum $\varphi_n^{(\delta)}(E)$ defined by Eq. (3.18) was introduced because this term can also be calculated *directly* from an integral equation like (3.16). Namely, multiplying the latter equation by $\varphi_i^{(n)}(E_i)$ and performing the integration and summation according to (3.18) leads to an integral equation of the same form with the source term

$$M_n(E) = \int\limits_0^\infty \mathrm{d}E_i \sum_i \varphi_i^{(n)}(E_i)\,M_i(E_i, E). \tag{3.21}$$

This way of calculating the δ-ray contribution, $\varphi_n^{(\delta)}(E)$, from the neutron-produced secondaries is more expedient than Eq. (3.18).

1959), it became customary to identify δ-ray neglect with CSD approximation (ICRU Report 16, 1970). In contrast to this misleading terminology, an alternative way to consider the δ-electrons, just consisting in a consequent employment of the CSD approximation, is described by Eqs. (3.16), (3.20). This way was followed first by Burch (1955; 1957a, b) for both photon and ion irradiations including also the case of completely stopped ions. His 'step-wise method' is nothing but the above-mentioned numerical algorithm of solving our integral equations.

While Burch gave his results in terms to distributions in LET, similar calculations by Oda *et al.* (1971) concerned exactly the spectra $T_i^{(\delta)}(E_i, E)$, $\varphi_n^{(\delta)}(E)$, and $\varphi_\gamma^{(\delta)}(E)$. Thus, the results of these authors and our solutions to the above integral equations should agree, the more so as essentially the same physical parameters (see chapter 4) were used. Three

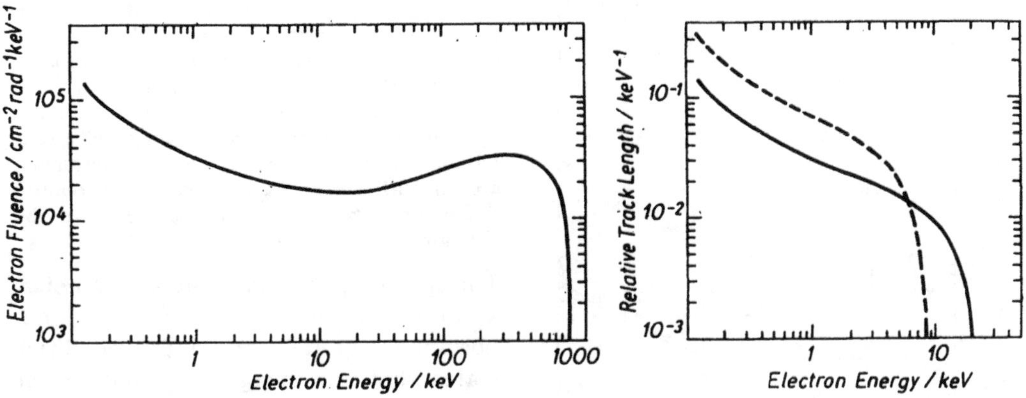

Fig. 3.6. Secondary electron fluence spectra $\varphi_\gamma^{(\delta)}(E)$ and $T_i^{(\delta)}(E_i, E)$, respectively, redrawn from ODA *et al.* (1971) as well as calculated from the integral equations (3.16) and (3.20), yielding complete coincidence (cut-off energy $\delta = 125$ eV). Left-hand panel: ^{60}Co γ-rays; right-hand panel: protons of 10 MeV (full line) and 4 MeV (broken line).

examples of these spectra, exhibiting, in fact, no appreciable difference between both calculations[1]), are shown in Figure 3.6.

Other above-mentioned spectra that are *not* based on the CSD approximation might be used for testing the goodness of the latter. These are usually derived from a set of strictly calculated electron fluence spectra, $y(E_0, E)$, for electrons starting with sharp initial energies E_0. The most prominent data on $y(E_0, E)$ are those by McGINNIES (1959), who employed the theory of SPENCER and FANO (1954). For the sake of comparison, analogous CSD spectra can be obtained, e.g., from Eq. (3.20) by setting $q_e^{(\gamma)}(E') = \delta(E' - E_0)$[2]). Table 3.1 shows this comparison for the initial electron energy $E_0 = 1.038$ MeV in water.[3]) The agreement is fairly good, confirming the results of the general model-based considerations in appendix 3.8.2 as well as the above-mentioned findings by SPENCER and FANO (1954) (see page 72), which refer only to the contribution of the *primary* electron with the initial energy E_0. The addition of the *secondary* electrons (in the range $E < E_0/2$), however, is seen from Table 3.1 to turn the CSD error to the opposite direction and to increase it markedly at smaller energies.

It is worth while to present still another comparison of the electron fluence spectrum $y(E_0, E)$ resulting from Eq. (3.20) with literature data (Fig. 3.7). While Table 3.1 is only to test for the CSD approximation from the *mathematical* viewpoint, resting on identical physical base (i.e. secondary electron distribution, $M_e(E_e, E)$, and stopping power, $L_\infty^{(e)}(E)$),

[1]) Our calculations were made, in this case, only for comparison's sake since, actually, a cut-off energy $\delta < 200$ eV is never met in our theory, according to Eq. (3.12). Really, in view of the various radiations and δ-values that may occur, we renounced extensive tabulations at all, including the equations of this section in the complex computer code into which the entire theory had to be converted, anyway.

[2]) $\delta(x)$ means Dirac's δ-function.

[3]) The physical parameters in our calculation were adapted to those used by McGINNIES.

Tab. 3.1. Comparison of the strict (McGINNIES, 1959) and the CSD-approximated (Eq. (3.20)) electron fluence spectrum, resulting from slowing down of electrons with the initial energy $E_0 = 1.038$ MeV in water, in terms of differential track length per initial electron in mass units [gcm^{-2}(mc^2)$^{-1}$]. Cut-off energy $\delta < 450$ eV

Electron energy [keV]	925.0	654.0	462.5	231.2	81.76	28.91	10.22	5.110	2.555	1.277	0.6387	0.4517
strict calculation	0.280	0.259	0.243	0.202	0.147	0.127	0.143	0.169	0.211	0.277	0.390	0.482
CSD approximation	0.271	0.261	0.244	0.201	0.144	0.123	0.136	0.159	0.196	0.253	0.347	0.414

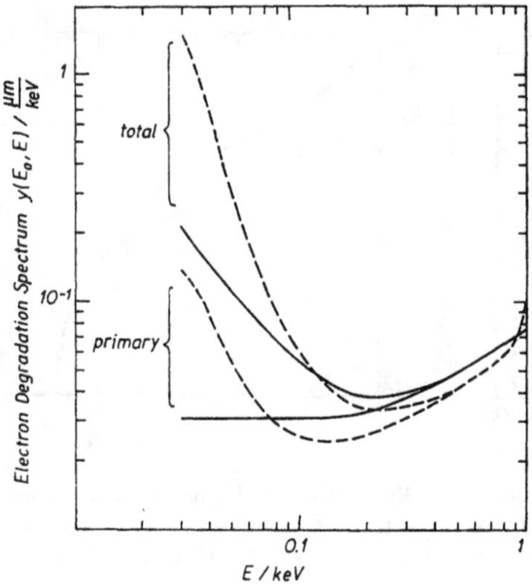

Fig. 3.7. Electron fluence spectrum (also called 'electron degradation spectrum') for a monoenergetic source of $E_0 = 1$ keV. Full lines: calculated from Eq. (3.20) with $q_e^{(\gamma)}(E') = \delta(E' - E_0)$, and $\delta = 0$ (CSD approximation). Broken lines: Monte Carlo calculation by HAMM *et al.* (1976) for liquid water.

it is just this *physical* background that Figure 3.7 is concerned with, bearing in mind that our using the MØLLER formula for $M_e(E_e, E)$ down to electron energies below 1 keV is not quite justified (see section 4.3). Therefore, the Monte Carlo simulation by HAMM *et al.* (1976) for *liquid* water, which started from 'first principles' considering practically all elementary interaction processes with their correct cross-sections, offers a good overall test for our physical approach in terms of the total electron fluence spectrum $y(E_0, E)$ for $E_0 = 1$ keV. Figure 3.7 exhibits two facts: (i) From the contribution of the *primary* electron, which is $L_\infty^{(e)}(E)^{-1}$ in CSD approximation, it can be seen that the stopping power we use is somewhat too small in the region $0.1 \text{ keV} < E < 1 \text{ keV}$, which must certainly be attributed to the peculiarities of the liquid state. For $E < 134$ eV we keep the stopping power constant, which is necessary in the framework of CSD approximation (see appendix 3.8.2). Hence, below 100 eV, we underestimate the track lengths of all electrons markedly. This is, however, irrelevant to the application of the theory since the cut-off energy δ (see Eq. (3.12)) never becomes smaller than 200 eV for the smallest volumes occurring (DNA targets). (ii) The *number* of *secondary* electrons involved is proportional to the ratio of the total and the primary $y(E_0, E)$ irrespective of stopping power. In this respect, Figure 3.7 shows an only slight underestimation by our approach, so that the total electron fluence is well assessed at $E > 100$ eV. Summing up: number and fluence of electrons generated by electrons having an energy greater than the marginal δ are in sufficient correspondence with reality. Some improvement could be achieved by changing the stopping power only.

The spectra $T_i^{(\delta)}(E_i, E)$ and $\varphi_\gamma^{(\delta)}(E)$ obey a normalization condition which is important in the following.[1]) This condition states that the energy imparted in the *track core* of all δ-rays equals the *total energy* released by ions or photons through secondary electrons:

$$\int_0^\infty \mathrm{d}E\, T_i^{(\delta)}(E_i, E)\, L_\delta^{(e)}(E) = L_\infty^{(i)}(E_i) - L_\delta^{(i)}(E_i),$$

$$\int_0^\infty \mathrm{d}E\, \varphi_\gamma^{(\delta)}(E)\, L_\delta^{(e)}(E) = \int_0^\infty \mathrm{d}E\, E q_e^{(\gamma)}(E). \tag{3.22}$$

These relations can be derived from the determining equations (3.16) and (3.20), respectively, as follows. By means of Eqs. (3.5), (3.6), express the stopping power for electrons by

$$L_\infty^{(e)}(E) = L_\delta^{(e)}(E) + \int_\delta^\infty \mathrm{d}E'\, E' N_e(E, E')$$

$$= L_\delta^{(e)}(E) + \delta M_e(E, \delta) + \int_\delta^\infty \mathrm{d}E'\, M_e(E, E')$$

and substitute this at the left-hand side of the integral equations (3.16) and (3.20), respectively. Subsequent integration of these equations over E from 0 to ∞ yields

$$\int_0^\infty \mathrm{d}E\, L_\delta^{(e)}(E)\, T_i^{(\delta)}(E_i, E) = \int_\delta^\infty \mathrm{d}E\, M_i(E_i, E) + \delta M_i(E_i, \delta),$$

$$\int_0^\infty \mathrm{d}E\, L_\delta^{(e)}(E)\, \varphi_\gamma^{(\delta)}(E) = \int_\delta^\infty \mathrm{d}E \int_E^\infty \mathrm{d}E'\, q_e^{(\gamma)}(E') + \delta \int_\delta^\infty \mathrm{d}E'\, q_e^{(\gamma)}(E') + \int_0^\delta \mathrm{d}E \int_E^\delta \mathrm{d}E'\, q_e^{(\gamma)}(E'),$$

[1]) Our computer programmes for these spectra have been checked up in this way.

where a number of terms have canceled out. Finally, integration by parts of the E-integrals gives the relations (3.22), considering Eqs. (3.5) and (3.6) in the case of ions.

Calling to mind the definition of fluence spectra ($\varphi(E)\,\mathrm{d}E$ = differential track length per unit volume), it is now easy to calculate the total energy imparted to unit volume, which is dose, D, times mass density (1 g/cm³):

$$1\,\frac{\mathrm{g}}{\mathrm{cm}^3}\,D = \sum_i \int_0^\infty \mathrm{d}E_i \varphi_i^{(0)}(E_i)\, L_\delta^{(i)}(E_i) + \sum_i \int_0^\infty \mathrm{d}E_i \varphi_i^{(n)}(E_i)\, L_\delta^{(i)}(E_i) + \int_0^\infty \mathrm{d}E \varphi_e(E)\, L_\delta^{(e)}(E). \quad (3.23)$$

Substituting $\varphi_e(E)$ according to Eq. (3.17) and also using (3.18), (3.22), and (3.13a) yields

$$1\,\frac{\mathrm{g}}{\mathrm{cm}^3}\,D = \sum_i \int_0^\infty \mathrm{d}E_i \varphi_i^{(0)}(E_i)\, L_\infty^{(i)}(E_i) + \sum_i \int_0^\infty \mathrm{d}E_i E_i q_i^{(n)}(E_i) + \int_0^\infty \mathrm{d}E E q_e^{(\gamma)}(E). \quad (3.24)$$

This relation shows that, by δ-ray cut-off and continuous-slowing-down approximation, the total energy dissipated in any radiation field is exactly distributed among the different particles if the mean 'local' energy transfer per unit track length is always taken to be $L_\delta(E)$.

3.4. Description of absorption events

At first, the geometrical properties of the sensitive volume (SV) must be appropriately described. The fundamental quantities in this respect are the so-called *chord length distribution*, $w(l')$, its expected value (*mean thickness*), l, and the *mean cross-section* (mean projected area), σ_0, of the SV, which is assumed to be a convex body randomly oriented in space (prerequisite (iv) at page 63). These things have been profoundly dealt with by KELLERER (1971a). The chord length spectrum, $w(l')$, is the normalized differential distribution of path segments inside the SV of particles that traverse the SV along straight lines. For a sphere of diameter d, $w(l')$ is a simple 'triangular' spectrum:

$$w(l') = \begin{cases} (2/d^2)\, l' & \text{for } l' \leq d \\ 0 & \text{for } l' > d. \end{cases} \quad (3.25)$$

In the case of cylinders, l'-spectra were derived by CASE *et al.* (1953) and by WILSON and EMERY (1968).
The mean thickness,

$$l = \int_0^\infty \mathrm{d}l' l' w(l'), \quad (3.26)$$

and the mean cross-section, σ_0, are related to the volume, V, of the considered body by

$$V = l\sigma_0. \quad (3.27)$$

The validity of this relation can be demonstrated as follows. Let $w(\Omega, l')$ be the differential distribution of chords that are parallel to the direction Ω (solid angle) and let $A(\Omega)$ be the projected area perpendicular to this direction. Then, $w(l')$ can be obtained as a weighted mean over Ω with the weighting factor $A(\Omega)/(4\pi\sigma_0)$:

$$w(l') = \frac{1}{4\pi\sigma_0} \int \mathrm{d}\Omega A(\Omega)\, w(\Omega, l').$$

The expected value l reads

$$l = \int \mathrm{d}l' l' w(l') = \frac{1}{4\pi\sigma_0} \int \mathrm{d}\Omega A(\Omega) \int \mathrm{d}l' l' w(\Omega, l') = \frac{1}{4\pi\sigma_0} \int \mathrm{d}\Omega A(\Omega)\, \bar{l}(\Omega).$$

$A(\Omega)\,\bar{l}(\Omega)$ is, however, independent of Ω, being equal to the volume V. Hence, it follows that $l = V/\sigma_0$.

In addition to $w(l')$, we still need another distribution, $u(l')$, which describes the path segments in the SV of particles that start randomly *inside* the SV. This case, named 'internal source randomness', was also considered by KELLERER (1971a), who showed that

$$u(l') = \frac{1}{l} \int_{l'}^{\infty} \mathrm{d}l'' w(l''). \tag{3.28}$$

The following simple proof of Eq. (3.28) may be given. For particles starting on a chord of the definite length l'', the l'-spectrum is 'rectangular':

$$u(l'', l') = \begin{cases} 1/l'' & \text{for} \quad l' \leqq l'' \\ 0 & \text{for} \quad l' > l''. \end{cases}$$

It is easy to see that the probability of a particle's start on a chord of the length l'' is given by $\mathrm{d}l'' l'' w(l'')/l$. Thus, we have

$$u(l') = \frac{1}{l} \int_{0}^{\infty} \mathrm{d}l'' w(l'')\, l'' u(l'', l')$$

which yields Eq. (3.28).

Let us now take up the general description of absorption events which is based on the approximation principles laid down in section 3.2.2. Each absorption event is brought about by one incident particle which either comes into the SV from outside or is set in motion inside the SV. Either kind of event can be characterized by three parameters: the sort of particle, i, its initial energy, E (at penetrating or starting), and its geometrically possible track length inside the SV, l' (if the path were straight). The number of the thus-characterized events is given by the differentials

$$\sigma_0 \,\mathrm{d}E \varphi(E)\, \mathrm{d}l' w(l') \quad \text{and} \quad \varrho V \,\mathrm{d}E q(E)\, \mathrm{d}l' u(l'). \tag{3.29a, b}$$

The dimensionless coefficient ϱ considers the mass density of the SV.[1]) Of course, the second expression, which describes starters in the SV, does not occur in the case of primary incident ions.

Let Ψ be an arbitrary physical quantity that assumes a definite value for each event described by i, E, and l'. For example, Ψ may be the mean energy deposition in the SV, the probability of j ionizations, or of k DNA-lesions, being produced in the SV, etc. In the following, the expected value, designated by $[\Psi]$, of such a quantity with respect to all absorption events will be investigated. The spectra describing the radiation field are assumed to be normalized in such a way that the total number of events becomes unity, i.e. $[\Psi] = 1$ for $\Psi = 1$. The expected value $[\Psi]$ can then be expressed as a sum of terms resulting from the different components of a general radiation field:

$$[\Psi] = [\Psi_i^{(0)}] + [\Psi_i^{(n)}] + [\Psi_e]. \tag{3.30}$$

The case of primary incident ions is most simple. These particles always traverse the SV completely along a straight path without being stopped (approximation principles 4c and 4e), which implies that $\Psi_i^{(0)}$ must directly depend on E and l'. This yields

$$[\Psi_i^{(0)}] = \sigma_0 \sum_i \int_0^{\infty} \mathrm{d}E \varphi_i^{(0)}(E) \int_0^{\infty} \mathrm{d}l' w(l') \times \Psi_i^{(0)}(E, l'). \tag{3.31}$$

In the case of ions generated by fast neutrons, stopping in the SV as well as the occurrence of starters (3.29b) must be taken into account. Therefore, let us consider the quantity $\Psi_i^{(n)}$ as a function of E and $E^{(f)}$, the initial and the final energies of the track segment inside the SV. $E^{(f)}$, on its part, depends on the initial energy E and the residual path

[1]) At given dose, the particle fluence $\varphi(E)$ is independent of the mass density, while the number of particles starting per unit volume is proportional to it. Therefore, relation (3.13) holds for the density 1 g/cm³ only. That is why the factor ϱ has to be added in the expression (3.29b).

length l'.[1]) If the residual range, $R_i(E)$, is smaller than l', one has $E^{(f)} = 0$ (the particle comes to rest in the SV). In general, $E^{(f)} = E_i^{(f)}(E, l')$ must be calculated from the equation

$$R_i(E_i^{(f)}) = \max\,[R_i(E) - \varrho l', 0], \tag{3.32}$$

where the range function is given by

$$R_i(E) = \int\limits_0^E \mathrm{d}E_i \frac{1}{L_\infty^{(i)}(E_i)} \tag{3.33}$$

due to the neglect of track curvature (approximation principle 4c). For the expected value of $\Psi_i^{(n)}(E, E^{(f)})$ one obtains according to (3.29):

$$\begin{aligned}[\Psi_i^{(n)}] = {} & \sigma_0 \sum_i \int\limits_0^\infty \mathrm{d}E\varphi_i^{(n)}(E) \int\limits_0^\infty \mathrm{d}l' w(l') \\ & \times \Psi_i^{(n)}\big(E, E_i^{(f)}(E, l')\big) \\ & + \varrho V \sum_i \int\limits_0^\infty \mathrm{d}E q_i^{(n)}(E) \int\limits_0^\infty \mathrm{d}l' u(l') \\ & \times \Psi_i^{(n)}\big(E, E_i^{(f)}(E, l')\big).\end{aligned} \tag{3.34}$$

An advantage of this formalism consists in the possibility to unite both terms of Eq. (3.34), which simplifies the theory considerably. This is achieved by integration by parts, using the relations

$$w(l') = -l \frac{\mathrm{d}u(l')}{\mathrm{d}l'}, \tag{3.35a}$$

$$q_i^{(n)}(E) = -\frac{\mathrm{d}\varphi_i^{(n)}(E)\, L_\infty^{(i)}(E)}{\mathrm{d}E} \tag{3.35b}$$

which follow from Eqs. (3.13a) and (3.28) by inspection. After substituting the functions (3.35) in Eq. (3.34), integration by parts of the first term with respect to l' and the second term with respect to E yields

$$\begin{aligned}[\Psi_i^{(n)}] = {} & V \sum_i \int\limits_0^\infty \mathrm{d}E\varphi_i^{(n)}(E) \int\limits_0^\infty \mathrm{d}l' u(l') \\ & \times \left[\frac{\mathrm{d}\Psi_i^{(n)}}{\mathrm{d}l'} + \varrho L_\infty^{(i)}(E) \frac{\mathrm{d}\Psi_i^{(n)}}{\mathrm{d}E}\right]\end{aligned} \tag{3.36}$$

if $\Psi_i^{(n)} = 0$ for a vanishing track segment, i.e.

$$\Psi_i^{(n)}(E, E) = 0, \tag{3.37}$$

is assumed. Regarding the total derivative $\mathrm{d}\Psi_i^{(n)}/\mathrm{d}E$, note that $\Psi_i^{(n)}$ depends on E both direct and through $E_i^{(f)}(E, l')$. Considering Eqs. (3.32) and (3.33), one can write

$$\begin{aligned}\frac{\mathrm{d}\Psi_i^{(n)}}{\mathrm{d}E} &= \frac{\partial \Psi_i^{(n)}}{\partial E} - \frac{\mathrm{d}R_i(E)}{\mathrm{d}E} \frac{1}{\varrho} \frac{\mathrm{d}\Psi_i^{(n)}}{\mathrm{d}l'} \\ &= \frac{\partial \Psi_i^{(n)}}{\partial E} - \frac{1}{\varrho L_\infty^{(i)}(E)} \frac{\mathrm{d}\Psi_i^{(n)}}{\mathrm{d}l'},\end{aligned} \tag{3.38}$$

where $\partial \Psi_i^{(n)}/\partial E$ means the derivative at constant $E_i^{(f)}$. Insertion of Eq. (3.38) in Eq. (3.36) leads to the final result

$$\begin{aligned}[\Psi_i^{(n)}] = {} & \varrho V \sum_i \int\limits_0^\infty \mathrm{d}E\varphi_i^{(n)}(E) \int\limits_0^\infty \mathrm{d}l' u(l') \\ & \times L_\infty^{(i)}(E) \frac{\partial \Psi_i^{(n)}\big(E, E_i^{(f)}(E, l')\big)}{\partial E}.\end{aligned} \tag{3.39}$$

Hence, we have gained a relatively simple expression depending, in analogy to (3.31), only on the fluence spectra $\varphi_i^{(n)}(E)$, which constitute a sufficient information about the radiation field.

For the electron term $[\Psi_e]$, an analogous formula can be derived. The only additional complication is the curvature of electron tracks inside the SV. According to approximation principle 4d, this effect is to be taken into account by a *mean* lengthening of the inside path segment. In other words, the final energy $E^{(f)}$ gets smaller than what would result from formulas like (3.32), (3.33). A corresponding correction must, however, depend on the size of the SV, which may be expressed through the mean thickness l, because track curvature does not matter if l becomes so small that the probability

---

[1]) $E^{(f)}$ being determined definitively by E and l' is an implication of the approximation principles 4a, c, d.

of a collision with an appreciable scattering angle tends to zero. Therefore, let us replace Eqs. (3.32) and (3.33) by

$$\tilde{R}_e(E_e^{(f)}, \varrho l) = \max\,[\tilde{R}_e(E, \varrho l) - \varrho l', 0],$$

$$\tilde{R}_e(E, l) = \int\limits_0^E \mathrm{d}E_e \frac{1}{\tilde{L}_\infty^{(e)}(E_e, l)}, \tag{3.40}$$

introducing, in this way, a *modified range function*, $\tilde{R}_e$, and *modified stopping power*, $\tilde{L}_\infty^{(e)}$, which depend on l as a parameter.

The theory of $\tilde{R}_e$ and $\tilde{L}_\infty^{(e)}$ is a topic of chapter 4. This theory is based on the following consideration. Take electrons that pass through a layer of matter with the thickness t, entering perpendicularly. To each incident energy, E, a *mean* exit energy $E_e(E, t)$ at leaving the layer can be assigned (assume $R_e(E) > t$). This relation is obtained by theory. Now the modified stopping power, $\tilde{L}_\infty^{(e)}(E, t)$, is defined by the requirement for the equation

$$\int\limits_{E_e(E,t)}^E \mathrm{d}E' \frac{1}{\tilde{L}_\infty^{(e)}(E', t)} = t \tag{3.41}$$

to be true of all values of E. This equation implies that the stopping power $\tilde{L}_\infty^{(e)}$ always yields the correct *mean* energy loss on a *straightened* path through the layer.

This concept of treating electron events may be illustrated by saying that the bent electron tracks are replaced by smoothed ones along which the linear energy transfer is accordingly enhanced because smoothing implies shortening. As it depends on l, this 'smoothing' does not mean total straightening of tracks; rather it is adapted to the SV under consideration in such a way that track sections within intervals of the order of the SV size can be treated as straight. Thus, as exhibited by Eqs. (3.40), complete analogy to the treatment of ion events is achieved, save for the following point: by smoothing, the track lengths are decreased, which influences the fluence spectrum $\varphi_e(E)$. too. Evidently, in Eqs. (3.13b) and (3.35), $L_\infty^{(e)}(E)$ must be replaced by $\tilde{L}_\infty^{(e)}(E, \varrho l)$. Otherwise, retaining the definition of $\varphi_e(E)$, we must then add the factor $L_\infty^{(e)}(E)/\tilde{L}_\infty^{(e)}(E, \varrho l)$ to the expression (3.29a). Now, in a way similar to that in the case of neutron produced ions, it is easy to derive the final formula for the expected value of $\Psi_e(E, E^{(f)})$:

$$\begin{aligned}
[\Psi_e] &= \sigma_0 \int\limits_0^\infty \mathrm{d}E \varphi_e(E) \frac{L_\infty^{(e)}(E)}{\tilde{L}_\infty^{(e)}(E, \varrho l)} \int\limits_0^\infty \mathrm{d}l' w(l') \\
&\quad \times \Psi_e\big(E, E_e^{(f)}(E, l')\big) \\
&\quad + \varrho V \int\limits_0^\infty \mathrm{d}E q_e(E) \int\limits_0^\infty \mathrm{d}l' u(l') \\
&\quad \times \Psi_e\big(E, E_e^{(f)}(E, l')\big) \\
&= V \int\limits_0^\infty \mathrm{d}E \varphi_e(E) \int\limits_0^\infty \mathrm{d}l' u(l') \\
&\quad \times \left[\frac{L_\infty^{(e)}(E)}{\tilde{L}_\infty^{(e)}(E, \varrho l)} \frac{\mathrm{d}\Psi_e}{\mathrm{d}l'} + \varrho L_\infty^{(e)}(E) \frac{\mathrm{d}\Psi_e}{\mathrm{d}E} \right],
\end{aligned}$$

$$\begin{aligned}
\frac{\mathrm{d}\Psi_e}{\mathrm{d}E} &= \frac{\partial \Psi_e}{\partial E} - \frac{\mathrm{d}\tilde{R}_e(E, \varrho l)}{\mathrm{d}E} \frac{1}{\varrho} \frac{\mathrm{d}\Psi_e}{\mathrm{d}l'} \\
&= \frac{\partial \Psi_e}{\partial E} - \frac{1}{\varrho \tilde{L}_\infty^{(e)}(E, \varrho l)} \frac{\mathrm{d}\Psi_e}{\mathrm{d}l'},
\end{aligned}$$

$$\begin{aligned}
[\Psi_e] &= \varrho V \int\limits_0^\infty \mathrm{d}E \varphi_e(E) \int\limits_0^\infty \mathrm{d}l' u(l') \\
&\quad \times L_\infty^{(e)}(E) \frac{\partial \Psi_e\big(E, E_e^{(f)}(E, l')\big)}{\partial E},
\end{aligned} \tag{3.42}$$

where the presumption

$$\Psi_e(E, E) = 0 \tag{3.43}$$

has again been used. There is no formal difference between the equations (3.39) and (3.42). Thus, the electron track curvature, being expressed by the modified range $\tilde{R}_e(E, l)$, enters the theory only through Eqs. (3.40) which determine the final energy $E_e^{(f)}(E, l')$. This is a further simplifying aspect of this approach.

Let us now calculate $\bar{z}_F$, the *mean dose per absorption event*. In view of the assumed normalization of the fluence spectra corresponding to *one* event in the SV, this quantity, $\bar{z}_F$, should equal the common dose D according to Eq. (3.24).
$\bar{z}_F$ is related to the expected value of the mean energy deposition of an event:

$$\bar{z}_F = \frac{1}{V\varrho \text{ g/cm}^3} [\Psi]. \quad (3.44)$$

The mean energy deposition (in the SV) per unit track length is $\varrho L_\delta(E)$, and the differential of track length reads $\mathrm{d}E/[\varrho L_\infty(E)]$.[1]) Consequently, the mean energy deposition of an event characterized by E and l', or by E and $E^{(f)}$, takes the form

$$\Psi_i^{(0)}(E, l') = \varrho l' L_\delta^{(i)}(E)$$

$$\Psi_i^{(n)}(E, E^{(f)}) = \int\limits_{E^{(f)}}^{E} \mathrm{d}E_i \frac{L_\delta^{(i)}(E_i)}{L_\infty^{(i)}(E_i)}, \quad (3.45)$$

$$\Psi_e(E, E^{(f)}) = \int\limits_{E^{(f)}}^{E} \mathrm{d}E_e \frac{L_\delta^{(e)}(E_e)}{L_\infty^{(e)}(E_e)}.\ ^{2)}$$

With these expressions inserted into Eqs. (3.31), (3.39), and (3.42), the expected value $[\Psi]$ is obtained according to Eq. (3.30). Considering Eqs. (3.26), (3.27) and $\int \mathrm{d}l' u(l') = 1$, this yields exactly the expression (3.23) for $\bar{z}_F$ defined by Eq. (3.44). As this expression leads directly to Eq. (3.24), one has

$$1 \frac{\text{g}}{\text{cm}^3} \bar{z}_F = \sum_i \int_0^\infty \mathrm{d}E \varphi_i^{(0)}(E)\, L_\infty^{(i)}(E) + \sum_i \int_0^\infty \mathrm{d}E E q_i^{(n)}(E) + \int_0^\infty \mathrm{d}E E q_e^{(\gamma)}(E). \quad (3.46)$$

[1]) The unmodified quantities referring to the true path must be used here.
[2]) Note that, in general, $\Psi(E, E^{(f)}) < E - E^{(f)}$ due to escape of δ-rays.

Thus, it has been proved that the number of different events as described by (3.29) and the energy depositions the theory assigns to them are consistent in the sense that the correct total energy deposition (dose) turns out. This holds irrespective of the special choice of the modified quantities $\tilde{R}_e$ and $\tilde{L}_\infty^{(e)}$ in the case of electrons, since $E^{(f)}$ drops out upon inserting the expressions (3.45) into Eqs. (3.39), (3.42).
In the following, the theory reduces always to the calculation of some expected values $[\Psi]$ divided by $\bar{z}_F$, the dose per event. As this quotient contains the spectra describing the radiation field in both the numerator and the denominator, the normalization of the spectra does not play any role, henceforth. Let us write down this general quantity for the most important cases.

Irradiation with ions of only one sort and energy:

$$\varphi_i^{(0)}(E) = \delta(E - E_i^{(0)}),$$
$$q_i^{(n)}(E) \equiv 0, \quad q_e^{(\gamma)}(E) \equiv 0.$$

From Eqs. (3.31), (3.42) and Eq. (3.46) under consideration of Eq. (3.17), it follows

$$\frac{[\Psi]}{\bar{z}_F} = \frac{\sigma_0 \text{ g/cm}^3}{L_\infty^{(i)}(E_i^{(0)})} \left[\int_0^\infty \mathrm{d}l' w(l')\, \Psi_i^{(0)}(E_i^{(0)}, l') + \varrho l \int_0^\infty \mathrm{d}E T_i^{(\delta)}(E_i^{(0)}, E) \times \int_0^\infty \mathrm{d}l' u(l')\, L_\infty^{(e)}(E) \frac{\partial \Psi_e}{\partial E} \right]. \quad (3.47)$$

Irradiation with fast neutrons:

$$\varphi_i^{(0)}(E) \equiv 0, \qquad q_e^{(\gamma)}(E) \equiv 0.$$

The initial energy spectra, $q_i^{(n)}(E)$, of the secondary ions are assumed to be known with correct relative normalization. From Eqs. (3.39), (3.42) and Eq. (3.46) under consideration of Eq. (3.17), one obtains

$$\frac{[\Psi]}{\bar{z}_F} = \frac{V\varrho \text{ g/cm}^3}{\sum_i \int_0^\infty \mathrm{d}E E q_i^{(n)}(E)} \times \left[\sum_i \int_0^\infty \mathrm{d}E\, \varphi_i^{(n)}(E) \times \int_0^\infty \mathrm{d}l' u(l')\, L_\infty^{(i)}(E) \frac{\partial \Psi_i^{(n)}}{\partial E} + \int_0^\infty \mathrm{d}E \varphi_n^{(\delta)}(E) \times \int_0^\infty \mathrm{d}l' u(l')\, L_\infty^{(e)}(E) \frac{\partial \Psi_e}{\partial E} \right]. \quad (3.48)$$

Irradiation with photons:

$$\varphi_i^{(0)}(E) \equiv 0, \qquad q_i^{(n)}(E) \equiv 0.$$

The initial energy spectrum, $q_e^{(\gamma)}(E)$, of Compton- and photoelectrons is assumed to be known. Eqs. (3.42), (3.46), and (3.17) yield

$$\frac{[\Psi]}{\bar{z}_F} = \frac{V\varrho \text{ g/cm}^3}{\int_0^\infty \mathrm{d}E E q_e^{(\gamma)}(E)} \int_0^\infty \mathrm{d}E \varphi_\gamma^{(\delta)}(E) \times \int_0^\infty \mathrm{d}l' u(l')\, L_\infty^{(e)}(E) \frac{\partial \Psi_e}{\partial E}. \quad (3.49)$$

Finally, a short note should be made on the case of *mixed radiations* (e.g., a γ-component usually connected with a neutron field). Corresponding formulas could, of course, be derived from Eqs. (3.30) and (3.46). The following simple consideration seems however more instructive. Let the different components of the radiation field, numbered by the index (a), be given by their *dose fractions* $\mu_D^{(a)}$, and assume the individual quantities $[\Psi]^{(a)}/\bar{z}_F^{(a)}$ to be known. By definition, the expected values themselves have to be averaged, however, by the *number* fractions (frequencies) $\mu_F^{(a)}$:

$$[\Psi] = \sum_a \mu_F^{(a)} [\Psi]^{(a)}. \quad (3.50)$$

The relation between $\mu_F^{(a)}$ and $\mu_D^{(a)}$ is obvious. The total number of events is unity for the total dose $\bar{z}_F$. Then, the dose delivered by the component a is equal to $\mu_F^{(a)}\bar{z}_F^{(a)}$. The dose *fraction* reads therefore $\mu_D^{(a)} = \mu_F^{(a)}\bar{z}_F^{(a)}/\bar{z}_F$. By substitution in Eq. (3.50) one has

$$\frac{[\Psi]}{\bar{z}_F} = \sum_a \mu_D^{(a)} \frac{[\Psi]^{(a)}}{\bar{z}_F^{(a)}}. \quad (3.51)$$

Hence, in a mixed radiation field, the general quantity $[\Psi]/\bar{z}_F$ is a weighted mean of the corresponding individual quantities with the dose fractions as the weights.

3.5. The model of DNA-lesion production

From the formal point of view, our concern is now the mathematical evaluation of the expression (3.1) for the *lesion production rate* s (number of DNA-lesions per unit dose and unit length of DNA):

$$s = \frac{1}{\bar{z}_F} \sum_{j=0}^\infty \lambda_j f_1(j). \quad (3.52)$$

The discrete single event size spectrum $f_1(j)$, being the probability of the generation of j ionizations by a general absorption event in a DNA target, can be expressed as the expected value (according to section 3.4) of the analogous conditioned probability v_j that depends on the parameters characterizing an event:

$$f_1(j) = [v_j(\textit{sort of incident particle}, E, l')]. \quad (3.53)$$

Due to the linearity of the operation $[\ldots]$, this implies

$$s = \frac{[\Psi]}{\bar{z}_F}, \qquad \Psi = \sum_{j=0}^\infty \lambda_j v_j. \quad (3.54)$$

This expression can be evaluated in closed form if the representation

$$\lambda_j = \sum_\nu b_\nu \, e^{-\zeta_\nu j}, \qquad \sum_\nu b_\nu = 0 \qquad (3.55\text{a, b})$$

is used. As λ_j constitutes the probability of a DNA-lesion (of the type considered) being formed by j simultaneous ionizations, condition (3.55b), i.e. $\lambda_0 = 0$, is self-evident. We may write

$$s = \frac{[\Psi]}{\bar{z}_F}, \qquad \Psi = \sum_\nu b_\nu \psi(\zeta_\nu), \qquad \psi(\zeta) = \sum_{j=0}^{\infty} e^{-\zeta j} v_j, \qquad (3.56)$$

where the type of lesion is characterized henceforth by the finite number of parameters b_ν and ζ_ν. The remaining task consists only in the calculation of the function $\psi(\zeta)$ for each sort of incident particles, since $[\Psi]/\bar{z}_F$ has already been evaluated in a general manner in the foregoing section. Note that condition (3.37) (or (3.43)) is fulfilled by virtue of (3.55b) and $v_j = \delta_{j0}$ for a vanishing track segment.

Let us first consider $\psi(E, E^{(f)}, \zeta)$ for ions and electrons set in motion by neutrons and photons, respectively. Along a track segment that reaches from the initial energy E to the final energy $E^{(f)}$ inside the target, *primary ionizations* are produced statistically independently of one another (approximation principle 4b). With each primary ionization, a cluster (short δ-ray) of secondary ionizations may be associated. Therefore, let us speak, generally, of *statistically independent clusters*, the number of which per unit track length is designated by $\varrho p^{(\delta)}(E)$, bearing in mind that δ-rays of an initial energy greater than the cut-off energy δ have to be disregarded (approximation principle 2a). The latter refers also to the so-called *cluster frequencies* $\omega^{(\delta)}(E, j)$ (probability for a primary ionization to lead to a total of j ionizations in the entire cluster), which are subject to the normalization condition (3.10). To assume the clusters always to lie completely inside the target is in conformity with approximation principle 3b.

For the probability, $v_j(E, E^{(f)})$, of the production of j ionizations within the track segment $(E, E^{(f)})$, an infinite system of differential equations can be derived as follows.[1]) The relation

$$v_j(E + dE, E^{(f)}) = \sum_{j'=0}^{j} v_{j-j'}(E, E^{(f)}) \, v_{j'}(E + dE, E) \qquad (3.57)$$

holds as a consequence of the assumed statistical independence of the primary ionizations. Using the normalization condition $\sum v_{j'} = 1$, Eq. (3.57) can be rewritten in the form

$$dE \frac{\partial v_j}{\partial E} = \sum_{j'=1}^{j} v_{j-j'} \, v_{j'}(E + dE, E) - v_j \sum_{j'=1}^{\infty} v_{j'}(E + dE, E), \qquad (3.58)$$

apart from terms of higher order in dE. For a differential track segment $dx = dE/[\varrho L_\infty(E)]$, no more than *one* cluster must be considered. Therefore, save for terms of higher order in dE, we have

$$v_{j'}(E + dE, E) = \frac{dE}{L_\infty(E)} \, p^{(\delta)}(E) \, \omega^{(\delta)}(E, j') \quad \text{for} \quad j' \geqq 1. \qquad (3.59)$$

By substitution in Eq. (3.58), taking into consideration the normalization condition

[1]) This is a version of the theory of the so-called compound Poisson process (KELLERER, 1968a), where v_j is equivalent to the straggling distribution. It can also be based on multiple 'folding' of the collision spectrum being equivalent to the cluster frequencies. This procedure (KELLERER, 1968a, b), however, requires energy independence of the involved physical quantities, while the differential approach used here is easier and more general.

(3.10) for $\omega^{(\delta)}(E, j')$, one obtains

$$\frac{\partial v_j}{\partial E} = \frac{p^{(\delta)}(E)}{L_\infty(E)} \left[\sum_{j'=1}^{j} v_{j-j'} \omega^{(\delta)}(E, j') - v_j \right]. \tag{3.60}$$

This infinite system of differential equations for the probabilities $v_j(E, E^{(f)})$ can in principle be solved successively, beginning with $j = 0$. In this way, realizing that

$$v_j(E, E^{(f)}) = \delta_{j0} \quad \text{at} \quad E = E^{(f)}, \tag{3.61}$$

the v_j are unequivocally determined, but this is no practicable procedure. That is why the single event size spectrum itself, according to Eq. (3.53), is difficult to obtain, while the quantity $\psi(E, E^{(f)}, \zeta)$ defined by Eq. (3.56) can easily be derived: multiply Eq. (3.60) by $e^{-\zeta j}$ and sum over $0, \infty$. Then inversion of the double sum yields

$$\frac{\partial \psi}{\partial E} = \frac{p^{(\delta)}(E)}{L_\infty(E)} \times \left[\sum_{j'=1}^{\infty} \omega^{(\delta)}(E, j') \sum_{j=j'}^{\infty} e^{-\zeta j} v_{j-j'} - \psi \right],$$

$$\frac{\partial \psi}{\partial E} = -\frac{\varkappa(E, \zeta)}{L_\infty(E)} \psi, \tag{3.62}$$

where

$$\varkappa(E, \zeta) = p^{(\delta)}(E) \left[1 - \sum_{j=1}^{j_{\max}} e^{-\zeta j} \omega^{(\delta)}(E, j) \right]. \tag{3.63}$$

The sum in Eq. (3.63) is finite because the cluster frequencies $\omega^{(\delta)}(E, j)$ expire at some maximum cluster size $j_{\max}$ depending on the cut-off energy δ or the particle energy E. With the initial condition $\psi(E, E^{(f)}, \zeta) = 1$ at $E = E^{(f)}$ following from Eqs. (3.56), (3.61), the solution to the differential equation (3.62) reads

$$\psi(E, E^{(f)}, \zeta) = \exp \left[-\int_{E^{(f)}}^{E} dE' \frac{\varkappa(E', \zeta)}{L_\infty(E')} \right]. \tag{3.64}$$

The quantity $\psi_i^{(0)}(E, l', \zeta)$ for *primary incident ions* is yet easier to calculate since the decrease of the particle's energy upon traversing the target will be neglected in this case (approximation principle 4e). The result can most simply be derived from Eq. (3.64) by taking the integrand constant and replacing the factor $(E - E^{(f)})/L_\infty(E)$ by the length of the track segment, l', times mass density factor:

$$\psi_i^{(0)}(E, l', \zeta) = e^{-\varrho l' \varkappa_i(E, \zeta)}. \tag{3.65}$$

Thus, the problem is solved for all radiations. The lesion production rate, s, is given by Eq. (3.56), using the general formulas (3.47) to (3.49) with the expressions (3.65) and (3.64), the latter for electrons (ψ_e) and secondary ions ($\psi_i^{(n)}$) as well.

It should be emphasized that the formalism as such is not restricted to the model of DNA-lesion production. It can also be used to gain information on the single event size spectrum $f_1(j)$ or $f_1(z)$. Neither the cylindrical shape nor the smallness of the target did enter the theory so far. Look only at Eqs. (3.52) and (3.55) to realize that, with $\nu = 1, 2$; $b_1 = 1, \zeta_1 = 0$; $b_2 = -1$, $\zeta_2 = \zeta$, the quantity s as a function of ζ constitutes something like the Laplace transform of $f_1(z)$, from which the mean values $\bar{z}_F$ and $\bar{z}_D$, for example, can be derived for any size and shape of the sensitive volume (e.g., for a sphere).[1] This 'microdosimetric branch' of the theory is comprehensively dealt with in chapter 5.

Explicit reference to the lesion production model will be indicated in the following by the subscript l: $w_l(l')$, $u_l(l')$, l_l, σ_{0l}, V_l, δ_l, ϱ_l. These geometrical quantities apply to an infinitely long circular cylinder with σ_{0l} and V_l to be understood per unit length of it. Since the mean

[1] It proved possible even to make the computer code general enough to cover these different aims.

thickness l_l is equal to the diameter d,[1]) one has, by the help of Eq. (3.27),

$$V_l = \frac{\pi}{4} l_l^2, \qquad \sigma_{0l} = \frac{\pi}{4} l_l. \tag{3.66}$$

Thus, the mean projected area of a long cylinder is $\pi/4$ times the maximum projected area.

Now the final formula of the DNA-lesion production model in its three versions for the different radiations can be written down by abridging the set of equations (3.47 to 3.49), (3.66), (3.56), and (3.64, 3.65):

$$s = \frac{\pi}{4} \frac{l_l \text{ g/cm}^3}{L_\infty^{(i)}(E_i^{(0)})} \left[G_i(E_i^{(0)}) + \int_0^\infty \mathrm{d}E T_i^{(\delta_i)}(E_i^{(0)}, E)\, \Gamma_e(E) \right], \tag{3.67}$$

$$s = \frac{\pi}{4} \frac{l_l \text{ g/cm}^3}{\sum_i \int_0^\infty \mathrm{d}E E q_i^{(n)}(E)} \times \left[\sum_i \int_0^\infty \mathrm{d}E \varphi_i^{(n)}(E)\, \Gamma_i(E) + \int_0^\infty \mathrm{d}E \varphi_n^{(\delta_i)}(E)\, \Gamma_e(E) \right], \tag{3.68}$$

$$s = \frac{\pi}{4} \frac{l_l \text{ g/cm}^3}{\int_0^\infty \mathrm{d}E E q_e^{(\gamma)}(E)} \int_0^\infty \mathrm{d}E \varphi_\gamma^{(\delta_i)}(E)\, \Gamma_e(E), \tag{3.69}$$

with the abbreviations

$$G_i(E) = \int_0^\infty \mathrm{d}l' w_l(l') \sum_\nu b_\nu \psi_i^{(0)}(E, l', \zeta_\nu) = \int_0^\infty \mathrm{d}l' w_l(l') \sum_\nu b_\nu \exp[-\varrho_l l' \varkappa_i(E, \zeta_\nu)], \tag{3.70}$$

$$\Gamma(E) = \varrho_l l_l \int_0^\infty \mathrm{d}l' u_l(l') \times \sum_\nu b_\nu L_\infty(E) \frac{\partial \psi(E, E^{(f)}, \zeta_\nu)}{\partial E} = -\varrho_l l_l \int_0^\infty \mathrm{d}l' u_l(l') \sum_\nu b_\nu \varkappa(E, \zeta_\nu) \times \exp\left[-\int_{E^{(f)}}^E \mathrm{d}E' \frac{\varkappa(E', \zeta_\nu)}{L_\infty(E')} \right], \tag{3.71}$$

bearing in mind that $E^{(f)}$ depends on E and l' (Eqs. (3.32) and (3.40), respectively).

The quantities (3.70) and (3.71) allow an obvious interpretation. By comparing Eqs. (3.54) and (3.56), it can be seen that the sum in the integrand of expression (3.70) is equal to $\sum \lambda_j v_j$, being the probability that an ion at given energy and path length in the target brings about a DNA-lesion. Therefore, $G_i(E)$ is the probability of the production of a DNA-lesion by an ion that crosses a target with energy E. The function $\Gamma(E)$ referring to electrons as well as ions has a similar meaning except that particles starting inside the target are additionally included in this quantity implicitly (i.e. $\Gamma(E) > 1$ is possible).

The latter point can be elucidated as follows. In Eq. (3.71), say, for ions[2]), substitute for $\partial\psi/\partial E$ by means of relation (3.38). $\Gamma_i(E)$ is thus splitted into two terms, one of which can be integrated by parts, using Eq. (3.35a). This yields

$$\Gamma_i(E) = \int_0^\infty \mathrm{d}l' w_l(l') \sum_\nu b_\nu \psi_i^{(n)}(E, E^{(f)}, \zeta_\nu) + L_\infty^{(i)}(E) \frac{\mathrm{d}}{\mathrm{d}E} \varrho_l l_l \int_0^\infty \mathrm{d}l' u_l(l') \times \sum_\nu b_\nu \psi_i^{(n)}(E, E^{(f)}, \zeta_\nu). \tag{3.72}$$

---

[1]) This follows from the general relation $Sl = 4V$, where S is the surface of the body (Kellerer, 1971a).

[2]) For electrons, the fluence correction $L_\infty^{(e)}(E)/\tilde{L}_\infty^{(e)}(E)$ due to track curvature is involved, in addition, in $\Gamma_e(E)$.

The first term of the right-hand side is equal to the expression (3.70) (save for the difference between $\psi_i^{(n)}$ and $\psi_i^{(0)}$), while the second term describes the starters inside the target. Thus, the representations (3.72), (3.70) for $\Gamma_i(E)$ and $G_i(E)$, respectively, clearly exhibit the difference: neither slowing-down nor start of particles is comprehended in $G_i(E)$. However, a representation like (3.71) is also possible for $G_i(E)$ (integration by parts of the integral (3.70)):

$$G_i(E) = -\varrho_l l_l \int_0^\infty \mathrm{d}l' u_l(l') \sum_\nu b_\nu \varkappa_i(E, \zeta_\nu) \times \exp\left[-\varrho_l l' \varkappa_i(E, \zeta_\nu)\right]. \tag{3.73}$$

In this representation, the difference between $G_i(E)$ and $\Gamma_i(E)$ according to Eq. (3.71) *seems* to consist only in the neglect of slowing-down in $G_i(E)$ (see the exponents). In other words, the starters are *automatically* taken into account by changing nothing but the exponent in a way to describe internal slowing-down. This is an important facilitating property of the theory.

For the numerical evaluation of $\Gamma(E)$, it is advantageous to introduce the new integration variable

$$x = \mathrm{e}^{-\varrho_l l' \varkappa(E, \zeta_\nu)}$$

to replace l':

$$\Gamma(E) = -l_l \sum_\nu b_\nu \int_0^1 \mathrm{d}x\, u_l\left(\frac{-\ln x}{\varrho_l \varkappa(E, \zeta_\nu)}\right) \frac{1}{x} \times \exp\left[-\int_{E^{(l)}}^{E} \mathrm{d}E' \frac{\varkappa(E', \zeta_\nu)}{L_\infty(E')}\right].$$

Neglect of internal slowing-down would mean $\exp[\ldots] = x$. It is then easy to show the equivalence $\Gamma_i(E) \equiv G_i(E)$ (integration by parts and return to the variable l'). Hence, the starters are taken into account only by the departure of $\exp[\ldots]$ from x.

3.6. The theory of cellular radiation effects

Biological radiation effects that are due to DNA-lesions do not only depend on the mean rate of lesion production of the particular radiation. One must also know the *statistics* of the discrete number of lesions per reacting unit (cell). Generally speaking, the lesions are not distributed 'at random' among the objects of a great population. This distribution, $U_k(D)$, being the probability of incidence of k DNA-lesions of a certain type in a reacting unit upon irradiation with the dose D of a particular radiation, is the topic of this section. It depends on geometric parameters (mean thickness l, mean projected area σ_0) and the DNA content, Y, of the *gross sensitive volume* (GSV) associated with the reacting unit, and on the characteristics of the lesion type considered (l_l and λ_j or b_ν, ζ_ν). However, fortunately, Eq. (3.2) does not require the distribution $U_k(D)$ in explicit form. Owing to the representation of the biological probability $H(k)$ by a finite sum of exponential functions according to (3.2), we must only know the expression

$$S(D, Z) = \sum_{k=0}^{\infty} \mathrm{e}^{-Zk} U_k(D) \tag{3.74}$$

which shall be named the 'partial dose-effect relation'. Thus, Eq. (3.2) takes the form

$$S(D) = \sum_\mu B_\mu S(D, Z_\mu). \tag{3.75}$$

Let us first define a *single-event distribution* $[V_k]$ which is related to the dose-dependent distribution $U_k(D)$ in the same manner as, in microdosimetry, the distributions $f_1(z)$ and $f(z; D)$. $[V_k]$ is the probability of the production of k DNA-lesions by *one* absorption event in the GSV. As indicated by the notation used, it is to be understood as an expected value of a corresponding conditioned probability, $V_k(E, l')$, in the sense of section 3.4. Realize, however, that all pertinent considerations and quantities now refer to the gross sensitive volume rather than the DNA-target as treated in the foregoing section.

The dose-dependent distribution $U_k(D)$ results from a compound Poisson process (see footnote 1 at page 83) with the spectrum $[V_k]$. Therefore, from the mathematical viewpoint, the theory of $U_k(D)$ and $S(D, Z)$ is very similar to the theory of $v_j(E, E^{(f)})$ and $\psi(E, E^{(f)}, \zeta)$ in section 3.5. The statistical independence of absorption events evidences

$$U_k(D + \mathrm{d}D) = \sum_{k'=0}^{k} U_{k-k'}(D)\, U_{k'}(\mathrm{d}D),$$

$$\mathrm{d}D \frac{\mathrm{d}U_k(D)}{\mathrm{d}D} = \sum_{k'=1}^{k} U_{k-k'}(D)\, U_{k'}(\mathrm{d}D) - U_k(D) \sum_{k'=1}^{\infty} U_{k'}(\mathrm{d}D). \quad (3.76)$$

The event frequency per unit dose is given by $1/\bar{z}_F$, where $\bar{z}_F$ is the mean dose per event with respect to the GSV. Thus:

$$U_{k'}(\mathrm{d}D) = \frac{\mathrm{d}D}{\bar{z}_F} [V_{k'}] \quad \text{for} \quad k' \geqq 1. \quad (3.77)$$

Eqs. (3.76) and (3.77) yield the infinite system of differential equations[1])

$$\frac{\mathrm{d}U_k}{\mathrm{d}D} = \frac{1}{\bar{z}_F} \left(\sum_{k'=0}^{k} U_{k-k'}[V_{k'}] - U_k \right) \quad (3.78)$$

which, after multiplication by e^{-Zk} and summation over $0, \infty$, changes into a simple differential equation for the partial dose-effect relation defined by Eq. (3.74):

$$\frac{\mathrm{d}S(D, Z)}{\mathrm{d}D} = \frac{1}{\bar{z}_F} \left(\sum_{k'=0}^{\infty} \mathrm{e}^{-Zk'}[V_{k'}] - 1 \right) \times S(D, Z). \quad (3.79)$$

The initial condition reads $S(0, Z) = 1$ due to $U_k(D) = \delta_{k0}$ at $D = 0$. This yields

$$S(D, Z) = \mathrm{e}^{-D\Pi(Z)} \quad \text{or} \quad S(D) = \sum_{\mu} B_\mu \mathrm{e}^{-D\Pi(Z_\mu)}, \quad (3.80)$$

where

$$\Pi(Z) = \frac{[\Psi(Z)]}{\bar{z}_F}, \qquad \Psi(Z) = 1 - \sum_{k=0}^{\infty} \mathrm{e}^{-Zk} V_{k'}, \quad (3.81)$$

having made use of the linearity of the operation [...].

Hence, the dose-effect relation $S(D)$ has been reduced to an expected value, Π, of the form evaluated in Eqs. (3.47 to 49). The quantity Ψ as a function of E and l' or E and $E^{(f)}$ remains to be determined for the three kinds of incident particles. The cases of secondary particles (electrons, ions set in motion by neutrons) can again be treated jointly. This is to be done first.

$V_k(E, E^{(f)})$ is the probability of k DNA-lesions being generated by an absorption event that is given by the incident particle track segment inside the GSV described by E and $E^{(f)}$, the initial and the final energy of the segment. These DNA-lesions are, however, not only due to the incident particle directly. δ-rays branching with an initial energy $E' < \delta$ off the considered track segment, which are the dependent particles according to approximation principle 2a, may also contribute to the formation of lesions, eventually rather distant from the incident track. In conformity with the lesion production model and the approximation principle 3b, we must take all the δ-electrons with initial energies $\delta_l < E' < \delta$ as *separate dependent particles* which belong to the event considered. As it were, these δ-tracks are simply considered to lie at random in the GSV simultaneously with the incident track segment they are descended from. In this way, the actual geometrical pictures of tracks in the GSV are disarranged (also in the topological respect). This is evidently allowed provided that the DNA molecules are distributed in the GSV uniformly and randomly in regard of their biological function. *This is the main prerequisite to the theory of cellular radiation effects.* It implies the possibility to split V_k into two independent probabilities, V_k' and V_k'', referring to the share of the incident 'track core' (including δ-rays with $E' < \delta_l$) and the dependent δ-tracks

[1]) There is a slight formal difference with Eq. (3.60): the lower limit of the sum, $k' = 0$. The reason is that $[V_0]$ does not vanish, i.e. the normalization condition reads $\sum_{k=0}^{\infty} [V_k] = 1$ in contrast to Eq. (3.10).

($\delta_l < E' < \delta$), respectively. Independence means

$$V_k(E, E^{(f)}) = \sum_{k'=0}^{k} V'_{k-k'}(E, E^{(f)})\, V''_{k'}(E, E^{(f)}). \tag{3.82}$$

Similarly, the quantity $\Psi(Z)$ defined by Eq. (3.81) can be factorized by insertion of Eq. (3.82):

$$\Psi(Z) = 1 - \Psi'(Z)\,\Psi''(Z), \tag{3.83}$$

where

$$\Psi'(Z) = \sum_{k=0}^{\infty} e^{-Zk} V_k', \qquad \Psi''(Z) = \sum_{k=0}^{\infty} e^{-Zk} V_k''. \tag{3.84}$$

Let us first take up V_k'. Suppose the geometrical density of DNA in the GSV to be fairly small so that the total volume YV_l of the hose-like targets is negligible in comparison with the gross sensitive volume. Then, V_k' must well obey a Poisson distribution:

$$V_k' = \frac{\Lambda'^k}{k!} e^{-\Lambda'}. \tag{3.85}$$

$\Lambda'(E, E^{(f)})$ is the mean number of DNA-lesions that are caused *directly* by the electron or ion track segment bounded by E, $E^{(f)}$. According to the supposition $YV_l \ll V$, a great deal of this track is outside of targets, which makes the single traversals largely independent of one another. This justifies a Poisson distribution for the number of traversals through targets and, all the more, for the number of DNA-lesions produced thereby.

In order to determine $\Lambda'(E, E^{(f)})$, imagine particle tracks with initial energy E in a medium containing randomly distributed DNA with the same geometrical density as in the GSV under discussion. Then, $\Lambda'(E, 0)$ is equal to the number of DNA-lesions per one total track, i.e. for $E^{(f)} = 0$ the problem can be reduced to Eq. (3.68) or (3.69), writing

$$\Lambda'(E, 0) = DYs, \quad D = \frac{E}{V\varrho\ \mathrm{g/cm^3}} \tag{3.86}$$

and

$$q(E') = \delta(E' - E),$$
$$\varphi(E') = \begin{cases} 1/L_\infty(E') & \text{for } E' \leqq E \\ 0 & \text{for } E' > E. \end{cases} \tag{3.87}$$

These spectra, which are related according to Eq. (3.13), simulate the situation we are considering. They have to be emplayed in Eq. (3.68) (omitting the second term in brackets) or in Eq. (3.69)[1]) to obtain the pertinent s. In either case, the result reads

$$\Lambda'(E, 0) = \frac{h}{\varrho} \int_0^E dE' \frac{\Gamma(E')}{L_\infty(E')}, \tag{3.88}$$

where the parameter

$$h = \frac{\pi}{4} \frac{Yl_l}{V} \tag{3.89}$$

has the obvious meaning of the reciprocal *mean free path* between targets, which is equivalent to the so-called macroscopic cross-section in the terminology of atomic and nuclear physics.

The mean number of DNA-lesions given by Eq. (3.88) applies exactly to an incident particle that *starts* in the GSV with the energy E and comes to rest in it, too. With a certain probability, such a particle track begins inside a target, which is automatically taken into account by the formula (3.88) (see Eq. (3.72) and context). For this reason, however, Eq. (3.88) is *not* quite correct in case the particle with the energy E *enters* the GSV from outside. Events of this kind are exactly counterbalanced by an equal number of events in which the incident particle *leaves* the GSV with the energy E. Writing $\Lambda'(E^*, E) = \Lambda'(E^*, 0) - \Lambda'(E, 0)$ for the latter events, i.e. generalizing Eq.

---

[1]) In the latter case, nothing has to be omitted since the δ-rays are already disregarded by using $\varphi(E)$ according to (3.87) rather than $\varphi_\gamma^{(\delta_l)}(E)$ as defined by Eq. (3.20).

(3.88) in the form

$$\Lambda'(E, E^{(f)}) = \frac{h}{\varrho} \int_{E^{(f)}}^{E} \mathrm{d}E' \, \frac{\Gamma(E')}{L_\infty(E')}, \qquad (3.90)$$

ensures complete mutual compensation of errors with respect to the *mean number* of lesions on an average over all events.

This question becomes clearer by substituting the expression (3.72) at the right-hand side of Eq. (3.90). In addition to an integral over $E, E^{(f)}$, two integral-free terms of the form $(hl_l\varrho_l/\varrho)\, g(E)$ and $-(hl_l\varrho_l/\varrho)\, g(E^{(f)})$ are obtained, where $g(E)$ signifies the probability of a DNA-lesion originating with a particle's start (energy E) in a target. The mutual compensation of these terms in the sum over all events is obvious. In other words, Eq. (3.90) is an approximation which, in principle, influences V_k as well as $[V_k]$, but leaves the expected value $\bar{k} = \sum kV_k$ unchanged (see section 3.7.1). By the way, it would not be possible to avoid this approximation without going beyond the general framework of section 3.4 (characterization of absorption events by $E, E^{(f)}$ only).

Summarizing Eqs. (3.84), (3.85), and (3.90), the quantity Ψ' for electrons and secondary ions is given by

$$\Psi'(E, E^{(f)}, Z) = \exp\left[-(1 - \mathrm{e}^{-Z})\, \frac{h}{\varrho} \int_{E^{(f)}}^{E} \mathrm{d}E' \, \frac{\Gamma(E')}{L_\infty(E')}\right]. \qquad (3.91)$$

Eq. (3.91) describes only the contribution of the 'track core' of the incident particle. The outstanding contribution of the dependent δ-electrons ejected inside the GSV is not so easy to obtain, because number and energy of these δ-rays are left to chance despite the fixed parameters $E, E^{(f)}$. Only a *mean value* like (3.90) for the number of lesions, which is related to the track length spectrum $T^{(\delta_l,\delta)}(E, E_e)$[1]) of the δ-electrons in question, can immediately be found by stressing the analogy to Eq. (3.90):

$$\Lambda''(E, E^{(f)}) = \frac{h}{\varrho} \int_{E^{(f)}}^{E} \frac{\mathrm{d}E'}{L_\infty(E')} \int_0^\infty \mathrm{d}E_e \times T^{(\delta_l,\delta)}(E', E_e)\, \Gamma_e(E_e). \qquad (3.92)$$

The corresponding V_k'', however, is *not* a Poisson distribution as in the case (3.85). Only the *expected value* (3.92) can be based on the *mean* δ-track length, while the *fluctuations* of this number of lesions are determined by the casual appearance or non-appearance of definite δ-rays with definite energies. In other words, these DNA-lesions are not statistically independent because several lesions can be formed *at once* by *one* δ-electron. However, for a particular radiation and in dependence on the set of parameters l, h and l_l, λ_j, it may be that the probability of more than one lesion produced by a δ-ray is small enough for the statistical dependence of lesions to be negligible, justifying a Poisson distribution for V_k'' with the mean value (3.92) as a good approximation. Then, Eqs. (3.83), (3.84), and (3.91) yield

$$\Psi(E, E^{(f)}, Z) \approx 1 - \exp\left\{-(1 - \mathrm{e}^{-Z})\, \frac{h}{\varrho} \int_{E^{(f)}}^{E} \frac{\mathrm{d}E'}{L_\infty(E')} \times \left[\Gamma(E') + \int_0^\infty \mathrm{d}E_e T^{(\delta_l,\delta)}(E', E_e)\, \Gamma_e(E_e)\right]\right\} \qquad (3.93)$$

under the condition:

mean number of lesions per δ-electron

$$\approx \frac{h}{\varrho} \int_0^{\min(E_{\max},\delta)} \mathrm{d}E_e \, \frac{\Gamma_e(E_e)}{L_\infty^{(e)}(E_e)} \ll 1. \qquad (3.94)$$

Although condition (3.94) will be fulfilled in a great number of cases, the general theory cannot be based on Eq. (3.93) only, which is to be illustrated by the following estimation.

---

[1]) The cut-off energy in the sense of section 3.3, where the spectrum was defined, is indicated by δ_l, while δ signifies an additional omission of electrons with initial energy *above* δ.

The mean thickness (2/3 times diameter) of the nuclear equivalent (nucleoid) of a bacterium *E. coli* amounts to about $l = 0.5\ \mu m$ (MUNSON *et al.*, 1967; ZUSMAN *et al.*, 1972). A 5-keV electron has roughly this range, i.e. such a δ-track would lie completely (approximation principle 3b) in the GSV. Expressed as a dose, this δ-electron is equivalent to 363 rad ($\varrho = 1$) in the GSV. Experimental data on single-strand breakage (see Fig. 6.5 at page 196) give about 2×10^{-12} breaks per rad and Dalton for low-LET radiations. Since the genome of *E. coli* represents an amount of 2.6×10^9 Dalton of DNA (CAIRNS, 1963; LAIRD, 1971), one obtains a mean value of 1.9 single-strand breaks produced by a 5-keV δ-electron in the nucleoid of *E. coli*. Hence, in this case, condition (3.94) is far from being fulfilled. Also in mammalian cell nuclei, the geometrical density of DNA is of the same order of magnitude. Of course, for double-strand breaks a much smaller number would be obtained.

In the following, a strict expression for Ψ'' will be derived. At first, due to the statistical independence of δ-rays ejected along the track segment $(E, E^{(f)})$ (approximation principle 4b), one can write

$$V_k''(E + \mathrm{d}E, E^{(f)}) = \sum_{k'=0}^{k} V''_{k-k'}(E, E^{(f)})\, V''_{k'}(E + \mathrm{d}E, E)\,. \quad (3.95)$$

This and the subsequent reasoning is entirely analogous to the formalism based on Eq. (3.57) in section 3.5. Therefore, only a short-handed treatment will be given. The 'spectrum' of the Poisson process reads

$$V''_{k'}(E + \mathrm{d}E, E) = \frac{\mathrm{d}E}{L_\infty(E)} \int_{\delta_t}^{\delta} \mathrm{d}E' N(E, E')\, V^{(e)}_{k'}(E', 0)$$

for $k' \geqq 1$. (3.96)

$V_k^{(e)}(E', 0)$ is the still unknown *total* distribution of lesions for a *complete* electron track of the initial energy E'. In this way, the higher generations of δ-rays are also included. By analogy, the differential equation

$$\frac{\partial \Psi''}{\partial E} = -\frac{1}{L_\infty(E)} \int_{\delta_t}^{\delta} \mathrm{d}E' N(E, E') \times \left[1 - \sum_{k'=0}^{\infty} \mathrm{e}^{-Zk'} V^{(e)}_{k'}(E', 0)\right] \Psi'' \quad (3.97)$$

for the quantity $\Psi''(E, E^{(f)}, Z)$ defined by Eq. (3.84) is obtained. With the initial condition $\Psi''(E, E^{(f)}, Z) = 1$ at $E = E^{(f)}$ due to $V_k''(E^{(f)}, E^{(f)}) = \delta_{k0}$, it can be solved at once. Beforehand, however, note that the expression in brackets is just the quantity Ψ (Eq. (3.81)) for electrons of the initial energy E' that come to rest in the GSV ($E^{(f)} = 0$). As it is of special importance, let us designate this quantity as

$$\xi(E, Z) \equiv \Psi_e(E, 0, Z) = 1 - \sum_{k=0}^{\infty} \mathrm{e}^{-Zk} V_k^{(e)}(E, 0)\,. \quad (3.98)$$

Now the solution to Eq. (3.97) together with (3.91) yields

$$\Psi(E, E^{(f)}, Z) = 1 - \exp\left\{-\int_{E^{(f)}}^{E} \frac{\mathrm{d}E^*}{L_\infty(E^*)} \times \left[(1 - \mathrm{e}^{-Z}) \frac{h}{\varrho} \Gamma(E^*) + \int_{\delta_t}^{\delta} \mathrm{d}E' N(E^*, E')\, \xi(E', Z)\right]\right\}. \quad (3.99)$$

Eq. (3.99) is already the final formula for the quantity Ψ which yields the survival curve $S(D)$ by means of Eqs. (3.80), (3.81) for the cases of ions set in motion by neutrons ($\Psi_i^{(n)}$) and of electrons (Ψ_e). However, the function $\xi(E, Z)$ remains to be determined. It is Eqs. (3.98), (3.99) themselves that determine $\xi(E, Z)$, since Eq. (3.99) applies also to electrons (subscript e). Thus, $\xi(E, Z)$ is the solution to the non-linear integral equation

$$\xi(E, Z) = 1 - \exp\left\{-\int_0^E \frac{dE_e}{L_\infty^{(e)}(E_e)} \times \left[(1 - e^{-Z}) \frac{h}{\varrho} \Gamma_e(E_e) + \int_{\delta_l}^{\delta} dE' N_e(E_e, E')\, \xi(E', Z)\right]\right\}. \quad (3.100)$$

This equation, being a counterpart of Eq. (3.16), can be solved numerically in an analogous way, beginning with $E = 0$ where $\xi = 0$. According to the relations (3.3), (3.4), the ξ-containing integral at the right-hand side vanishes as long as $E < 2\delta_l$. For higher values of E, it depends only on $\xi(E', Z)$ at $E' < E/2$.

Eqs. (3.99, 3.100) replace the approximate expression (3.93). There is a high degree of correspondence between both descriptions. Instead of the *linear* integral equation (3.16) for $T^{(\delta_l,\delta)}(E', E)$, we have the *non-linear* integral equation (3.100) for $\xi(E, Z)$. It is just this non-linearity of the latter that makes the entire difference between both descriptions. This will be demonstrated in section 3.7 by linearization of Eq. (3.100). Regarding the procedure for the numerical solution of either equation, no additional complications arise from the non-linearity. Therefore, the approximate version (3.93) will never be used in actual calculations. It is, however, of some importance for subsequent theoretical investigations.

There is an alternative description in terms of the derivative $\xi'(E, Z) = d\xi/dE$ rather than $\xi(E, Z)$, which employs $M(E^*, E')$ instead of $N(E^*, E')$ (see Eq. (3.6)):

$$\int_{\delta_l}^{\delta} dE' N(E^*, E')\, \xi(E', Z) = M(E^*, \delta_l)\, \xi(\delta_l, Z) - M(E^*, \delta)\, \xi(\delta, Z) + \int_{\delta_l}^{\delta} dE' M(E^*, E')\, \xi'(E', Z). \quad (3.101)$$

By means of this relation, Eq. (3.99) as well as Eq. (3.100) can be brought into another form, where the new integral equation for $\xi'(E, Z)$ is obtained by forming the derivative of $\ln[1 - \xi(E, Z)]$:

$$\frac{L_\infty^{(e)}(E)\, \xi'(E, Z)}{1 - \xi(E, Z)} = (1 - e^{-Z}) \frac{h}{\varrho} \Gamma_e(E) + M_e(E, \delta_l)\, \xi(\delta_l, Z) - M_e(E, \delta)\, \xi(\delta, Z) + \int_{\delta_l}^{\delta} dE' M_e(E, E')\, \xi'(E', Z), \quad (3.102)$$

$$\xi(E, Z) = \int_0^E dE' \xi'(E', Z).$$

This integral equation can be solved in the same way as Eq. (3.100). While the double integral at Eq. (3.100) does not occur any more, the additional integral $\xi(E, Z)$ must now be calculated simultaneously. Thus, almost nothing is gained from the practical point of view. Eq. (3.100) even has slight numerical advantages regarding the accuracy attainable. For some formal investigations, however, Eq. (3.102) is more suitable.

For the explicit evaluation of the expected value $[\Psi]/\bar{z}_F$ for *all* kinds of radiation, which yields the exponents $\Pi(Z_\mu)$ of the dose-effect relation according to Eqs. (3.80), (3.81), the corresponding quantity $\Psi_i^{(0)}(E, l', Z)$ for *primary incident ions* remains to be found. It can directly be derived from $\Psi_i^{(n)}$ for secondary ions given by Eq. (3.99). This is the same procedure as in the foregoing section (change of Eq. (3.64) into Eq. (3.65)): neglect of the decrease of energy within the GSV (approximation principle 4e) by removing the integral over $E^{(f)}$, E in the exponent of Ψ. In addition to this, the impossibility of the particles' starting or stopping inside the GSV and, hence, inside of targets must also be considered by replacing $\Gamma_i(E)$ by $G_i(E)$.[1)]

---

1) Strictly speaking, the mean number of lesions, Λ_i', the incident ion produces directly must be properly calculated along the lines similar to Eqs. (3.86, 3.90), which yields $\Lambda_i' = hl'G_i(E)$. See also the comments given on Eqs. (3.70), (3.71).

This yields

$$\Psi_i^{(0)}(E, l', Z) = 1 - \exp\left[-(1 - e^{-Z})\, hl' G_i(E) - \varrho l' \int_{\delta_t}^{\delta} dE' N_i(E, E')\, \xi(E', Z)\right]. \quad (3.103)$$

Now the explicit final formulas for $\Pi(Z)$ can be given by applying the rules (3.47 to 49) to the expected value (3.81), using the expressions (3.99) and (3.103). In the partial derivative $\partial\Psi_e/\partial E$, the exponential function may be expressed through $\xi(E, Z)$ defined by Eq. (3.98). Thus, one obtains for the three kinds of radiation:

With these equations, the mathematical formulation of the DNA-lesion theory of radiation action is complete. $\Pi(Z_\mu)$, the coefficients in the exponents of the dose-effect relation (3.80), can now be calculated for any radiation (bear in mind Eq. (3.51) for mixed radiations) and for any $Z = Z_\mu$, i.e. for arbitrary effect-probability functions $H(k) = \sum B_\mu \times \exp(-Z_\mu k)$. In addition to Eqs. (3.104 to 108), there are a number of auxiliary relations dealt with throughout this chapter. Table I in the appendix gives a survey of *all* equations and input quantities that are finally needed for the calculation of s, the rate of DNA-lesion production, and of $S(D)$, the dose-effect

$$\Pi(Z) = \frac{\sigma_0\ \text{g/cm}^3}{L_\infty^{(i)}(E_i^{(0)})}\left\{\int_0^\infty dl' w(l')\,[1 - e^{-\varrho l' K_i^{(0)}(E_i^{(0)}, Z)}] + \varrho l \int_0^\infty dE T_i^{(\delta)}(E_i^{(0)}, E) \int_0^\infty dl' u(l')\, K_e(E, Z)\, \frac{1 - \xi(E, Z)}{1 - \xi(E_e^{(f)}, Z)}\right\}, \quad (3.104)$$

$$\Pi(Z) = \frac{V\varrho\ \text{g/cm}^3}{\sum_i \int_0^\infty dE E q_i^{(n)}(E)}\left\{\sum_i \int_0^\infty dE \varphi_i^{(n)}(E) \int_0^\infty dl' u(l')\, K_i(E, Z) \exp\left[-\int_{E_i^{(f)}}^{E} dE_i \frac{K_i(E_i, Z)}{L_\infty^{(i)}(E_i)}\right] + \int_0^\infty dE \varphi_n^{(\delta)}(E) \int_0^\infty dl' u(l')\, K_e(E, Z)\, \frac{1 - \xi(E, Z)}{1 - \xi(E_e^{(f)}, Z)}\right\}, \quad (3.105)$$

$$\Pi(Z) = \frac{V\varrho\ \text{g/cm}^3}{\int_0^\infty dE E q_e^{(\gamma)}(E)} \int_0^\infty dE \varphi_\gamma^{(\delta)}(E) \int_0^\infty dl' u(l')\, K_e(E, Z)\, \frac{1 - \xi(E, Z)}{1 - \xi(E_e^{(f)}, Z)}, \quad (3.106)$$

where

$$K_i^{(0)}(E, Z) = (1 - e^{-Z})\, \frac{h}{\varrho}\, G_i(E) + \int_{\delta_t}^{\delta} dE' N_i(E, E')\, \xi(E', Z), \quad (3.107)$$

$$K(E, Z) = (1 - e^{-Z})\, \frac{h}{\varrho}\, \Gamma(E) + \int_{\delta_t}^{\delta} dE' N(E, E')\, \xi(E', Z). \quad (3.108)$$

Remember that $E_i^{(f)}$ and $E_e^{(f)}$ depend on E and l' (Eqs. (3.32) and (3.40)).

relation for cellular radiation effects. In principle, this table contains all one must know for the elaboration of a corresponding computer code.

3.7. Formal properties and consistency of the theory

In this section some relations important for the application of the theory will be derived and discussed. Among other things, this concerns the compatibility of the results of the foregoing section with the theory of DNA-lesion production (section 3.5) and the generalization of the theory to the case of several types of DNA-lesions operating simultaneously at the cellular level.

First, as a premise to the following, the relation between the final strict theory as expressed by Eq. (3.99) in terms of ξ and the approximation (3.93) in terms of $T^{(\delta_l,\delta)}$ (Poisson statistics of lesions for a definite absorption event) must be elucidated. This relation, which will be proved below, reads

$$\int_{\delta_l}^{\delta} \mathrm{d}E' N(E, E')\, \xi(E', Z) = (1 - \mathrm{e}^{-Z}) \frac{h}{\varrho} \int_0^{\infty} \mathrm{d}E_e T^{(\delta_l,\delta)}(E, E_e)\, \Gamma_e(E_e) + \textit{terms of higher order.} \qquad (3.109)$$

'Higher order' refers to the parameter $(1 - \mathrm{e}^{-Z})\, h/\varrho$, i.e. the expressions (3.93) and (3.99) are equivalent if this parameter is small, in concordance with condition (3.94). In other words, one has *exact* equivalence if restricting ξ to the *linear* portion. Thus, it is not the other *form* of description (ξ instead of $T^{(\delta_l,\delta)}$) but only the *non-linearity* of the determining integral equation (3.102) that accounts for the deviations from Poisson statistics. One must only linearize Eq. (3.102) by omitting the denominator $1 - \xi(E, Z)$, so that ξ becomes proportional to $(1 - \mathrm{e}^{-Z}) \times h/\varrho$, to obtain the approximate theory (3.93) now in terms of ξ:

$$L_{\infty}^{(e)}(E)\, \xi'(E, Z) = (1 - \mathrm{e}^{-Z}) \frac{h}{\varrho} \Gamma_e(E) + M_e(E, \delta_l)\, \xi(\delta_l, Z) - M_e(E, \delta)\, \xi(\delta, Z) + \int_{\delta_l}^{\delta} \mathrm{d}E' M_e(E, E')\, \xi'(E', Z). \qquad (3.110)$$

Hence, for the approximate theory, we have the option between two equivalent formalisms: either Eq. (3.93) requiring the solution of a linear integral equation like (3.16) for $T^{(\delta_l,\delta)}$ or Eq. (3.99) requiring the solution of the linear integral equation (3.110) for ξ. Only in the framework of the latter formalism, however, the strict theory can be practised by nothing but adding the denominator $1 - \xi(E, Z)$ at the left-hand side of Eq. (3.110), which destroys its linearity. Besides, the ξ-formalism in any case has several advantages from the practical point of view: ξ is a function of the *one* variable E only and this function is universal irrespective of the special radiation, i.e. it is determined only by the object considered (critical lesion type and the parameters h and Z). Therefore, all practical calculations have been based on the ξ-formalism expressed by the final formulas (3.104 to 108) and the strict integral equation (3.102).

Proof of Eq. (3.109):

Starting from the linearized equation (3.110) we will demonstrate the identity (3.109) with respect to the terms linear in $(1 - \mathrm{e}^{-Z})\, h/\varrho$. The track length spectrum $T^{(\delta_l,\delta)}(E, E_e)$, on the other hand, obeys an integral equation like (3.16) (cf. footnote 1 at page 89):

$$L_{\infty}^{(e)}(E_e)\, T^{(\delta_l,\delta)}(E, E_e) = \begin{cases} M(E, E_e) - M(E, \delta) + \int_0^{\delta} \mathrm{d}E_e' T^{(\delta_l,\delta)}(E, E_e')\, M_e(E_e', E_e) & \text{for } E \geqq \delta_l \\ L_{\infty}^{(e)}(\delta_l)\, T^{(\delta_l,\delta)}(E, \delta_l) & \text{for } E < \delta_l, \end{cases} \qquad (3.111)$$

with $T^{(\delta_l,\delta)}(E, E_e) \equiv 0$ for $E_e \geqq \delta$. Consider the integral

$$I = \int_0^\delta dE_e \xi'(E_e, Z)\, L_\infty^{(e)}(E_e)\, T^{(\delta_l,\delta)}(E, E_e). \quad (3.112)$$

Substitution of Eq. (3.111) yields

$$\begin{aligned} I &= \xi(\delta_l, Z)\, L_\infty^{(e)}(\delta_l)\, T^{(\delta_l,\delta)}(E, \delta_l) \\ &\quad + \int_{\delta_l}^{\delta} dE_e \xi'(E_e, Z)\, M(E, E_e) - [\xi(\delta, Z) \\ &\quad - \xi(\delta_l, Z)]\, M(E, \delta) + \int_{\delta_l}^{\delta} dE_e \xi'(E_e, Z) \\ &\quad \times \int_0^\delta dE_e' T^{(\delta_l,\delta)}(E, E_e')\, M_e(E_e', E_e) \\ &= M(E, \delta_l)\, \xi(\delta_l, Z) - M(E, \delta)\, \xi(\delta, Z) \\ &\quad + \xi(\delta_l, Z) \int_0^\delta dE_e' T^{(\delta_l,\delta)}(E, E_e')\, M_e(E_e', \delta_l) \\ &\quad + \int_{\delta_l}^{\delta} dE_e M(E, E_e)\, \xi'(E_e, Z) \\ &\quad + \int_0^\delta dE_e' T^{(\delta_l,\delta)}(E, E_e') \int_{\delta_l}^{\delta} dE_e \xi'(E_e, Z) \\ &\quad \times M_e(E_e', E_e). \end{aligned} \quad (3.113)$$

Evaluation of the expression (3.112) by means of Eq. (3.110) leads to[1])

$$\begin{aligned} I &= (1 - e^{-Z}) \frac{h}{\varrho} \int_0^\delta dE_e \Gamma_e(E_e)\, T^{(\delta_l,\delta)}(E, E_e) \\ &\quad + \xi(\delta_l, Z) \int_0^\delta dE_e T^{(\delta_l,\delta)}(E, E_e)\, M_e(E_e, \delta_l) \\ &\quad + \int_0^\delta dE_e T^{(\delta_l,\delta)}(E, E_e) \int_{\delta_l}^{\delta} dE' M_e(E_e, E') \\ &\quad \times \xi'(E', Z). \end{aligned} \quad (3.114)$$

Now, by equating the expressions (3.113) and (3.114), several terms are canceled, and the identity (3.109) is obtained in consideration of relation (3.101).

3.7.1. Mean DNA-lesion number and statistical independence of lesions as a limiting case

The properties of the theory to be elucidated in this subsection reduce to a formal mathematical relation which will be proved first. Eqs. (3.104 to 108) constitute the final result of the theory of cellular radiation effects in terms of the coefficient Π being a function of the parameter $x = (1 - e^{-Z})\, h/\varrho$. Likewise, the formulas (3.67 to 3.71) for the mean number, s, of DNA-lesions per unit dose and unit length of the DNA-molecule are the final outcome of the DNA-lesion production model. The relationship in question states a general connection of both results:

$$\lim_{x\to 0} \Pi = 0, \quad \lim_{x\to 0} \frac{\pi}{4} \frac{l_l}{\varrho V} \frac{d\Pi}{dx} = s,$$
$$x = (1 - e^{-Z}) \frac{h}{\varrho} = \frac{\pi}{4} \frac{Y l_l}{\varrho V} (1 - e^{-Z}). \quad (3.115)$$

Proof:
$\Pi = 0$ at $x = 0$ is obvious by inspection. The validity of the second relation (3.115) will be demonstrated only for ion radiations, which will make analogous straightforward calculations for the other radiations evident. For the limiting process $x \to 0$, in Eq. (3.104) the terms linear in x need to be retained only. Consequently, regarding the function ξ we can use Eq. (3.109) with higher order terms omitted. This makes $K_i^{(0)}$ and K_e proportional to x, and now the evaluation of $d\Pi/dx$ according to Eq. (3.104) is straightforward:

$$\begin{aligned} &\lim_{x\to 0} \frac{\pi}{4} \frac{l_l}{\varrho V} \frac{d\Pi}{dx} \\ &= \frac{\pi}{4} \frac{l_l\, \mathrm{g/cm^3}}{L_\infty^{(i)}(E_i^{(0)})} \Big\{ G_i(E_i^{(0)}) \\ &\quad + \int_0^\infty dE_e T_i^{(\delta_l,\delta)}(E_i^{(0)}, E_e)\, \Gamma_e(E_e) \\ &\quad + \int_0^\infty dE T_i^{(\delta)}(E_i^{(0)}, E) \Big[\Gamma_e(E) \\ &\quad + \int_0^\infty dE_e T_e^{(\delta_l,\delta)}(E, E_e)\, \Gamma_e(E_e) \Big] \Big\}, \end{aligned} \quad (3.116)$$

where the relations (3.26 to 3.28) have been used. Look at Eq. (3.67) to see that this expression equals s, *Q.E.D.*, provided that the sum

$$T_i^{(\delta)}(E_i^{(0)}, E) + T_i^{(\delta_l,\delta)}(E_i^{(0)}, E) + \int_0^\infty dE' T_i^{(\delta)}(E_i^{(0)}, E')\, T_e^{(\delta_l,\delta)}(E', E) \quad (3.117)$$

[1]) Note that $M_e(E_e, \delta)$ vanishes for $E_e < \delta$ according to Eqs. (3.3), (3.4), and (3.6).

can be shown to be the track length spectrum $T_i^{(\delta_l)}(E_i^{(0)}, E)$ that is determined by the integral equation (3.16) for the cut-off energy δ_l. Intuitively this is obvious: according to the classification of δ-electrons into incident and dependent particles, the first term of the expression (3.117) describes *all* δ-electrons with an initial energy beyond δ, while the last term stands for those which are produced by the former ones in the energy range *below* δ. The remainder (second term) are δ-electrons also below δ, but descended from the ion's track and from themselves. These comments make the formal proof following almost superfluous. Both $T_i^{(\delta_l,\delta)}$ and $T_e^{(\delta_l,\delta)}$ obey an integral equation of the form (3.111). Addition of these two equations after having multiplied the electron equation (subscript e) by $T_i^{(\delta)}$ and integrated in compliance with the last term in the expression (3.117) yields the following equation for $\tilde{T}_i^{(\delta_l,\delta)}(E_i^{(0)}, E)$, the sum of the second and third terms in (3.117):

$$L_\infty^{(e)}(E)\,\tilde{T}_i^{(\delta_l,\delta)}(E_i^{(0)}, E) = \begin{cases} M_i(E_i^{(0)}, E) - M_i(E_i, \delta) + \int\limits_0^\infty \mathrm{d}E' T_i^{(\delta)}(E_i^{(0)}, E')\, M_e(E', E) \\ -\int\limits_0^\infty \mathrm{d}E' T_i^{(\delta)}(E_i^{(0)}, E')\, M_e(E', \delta) + \int\limits_0^\delta \mathrm{d}E_e' \tilde{T}_i^{(\delta_l,\delta)}(E_i^{(0)}, E_e')\, M_e(E_e', E) & \text{for} \quad E \geqq \delta_l \\ L_\infty^{(e)}(\delta_l)\,\tilde{T}_i^{(\delta_l,\delta)}(E_i^{(0)}, \delta_l) & \text{for} \quad E < \delta_l . \end{cases} \tag{3.118}$$

This equation holds for $E < \delta$, while for $E \geqq \delta$ we have to consider only the term $T_i^{(\delta)}(E_i^{(0)}, E)$ which obeys Eq. (3.16) anyway. For the region $E < \delta$ the latter equation states that

$$L_\infty^{(e)}(E)\, T_i^{(\delta)}(E_i^{(0)}, E) = M_i(E_i, \delta) + \int\limits_0^\infty \mathrm{d}E' T_i^{(\delta)}(E_i^{(0)}, E')\, M_e(E', \delta) \tag{3.119}$$

is a constant. By addition of Eqs. (3.118) and (3.119) one obtains again the integral equation (3.16), now for the complete expression (3.117) and for the cut-off energy δ_l, *Q.E.D.*

As a question of consistency, the expected value associated with the probability distribution U_k (cf. Eq. (3.2)) for the number k of DNA-lesions per cell nucleus should exactly equal the mean number of lesions as it results from the lesion production model:

$$\sum_{k=0}^{\infty} k U_k(D; \lambda_j, l_l, Y) = sYD. \tag{3.120}$$

The theory of U_k developed in section 3.6 does not yield U_k explicitly; instead the results are expressed in terms of the 'partial dose-effect relation' $S(D, Z)$ defined by Eq. (3.74) and given by Eq. (3.80). This formalism can also be used for the calculation of the expected value (3.120) in the following way, including the relation (3.115):

$$\begin{aligned} \sum_{k=0}^{\infty} k U_k &= -\lim_{Z \to 0} \frac{\mathrm{d}}{\mathrm{d}Z} \sum_{k=0}^{\infty} \mathrm{e}^{-Zk} U_k \\ &= \lim_{Z \to 0} D \frac{\mathrm{d}\Pi}{\mathrm{d}Z}\, \mathrm{e}^{-D\Pi} \\ &= \lim_{x \to 0} D \frac{\pi}{4} \frac{Y l_l}{\varrho V} \frac{\mathrm{d}\Pi}{\mathrm{d}x} = sYD. \end{aligned} \tag{3.121}$$

Thus Eq. (3.120) proves correct. This general relation must exactly be fulfilled for arbitrary radiations and, what is important, the mean value (3.120) must not depend on the size of the gross sensitive volume, in contrast to U_k or Π. In this way, the corresponding numerical calculations have been checked by studying the behaviour of Π at small values of x, to verify relation (3.115) numerically.

Let us now discuss limiting cases in which the lesion number distribution U_k should approach a simple Poisson distribution:

$$U_k \approx \frac{(sYD)^k}{k!}\, \mathrm{e}^{-sYD}. \tag{3.122}$$

These limiting cases are characterized by the requirement that one absorption event in the gross sensitive volume does not produce more than one DNA-lesion with appreciable probability. Obviously, this situation is met if the gross sensitive volume is very large (large V) or it contains little DNA (Y small). Note that in both cases the combined parameter x of Eq. (3.115) becomes small. Another possibility is low-LET radiation, which does not influence x but makes the probabilities $G_i(E)$ and $\Gamma(E)$ (cf. the context of Eqs. (3.70), (3.71)) small. Note that the latter quantities are multiplied by x in the final formulas of the theory (Eqs. (3.104 to 108) and Eq. (3.100)). Hence, for these limiting cases, Π can always be approximated by

$$\Pi \approx sY(1 - e^{-Z}) \tag{3.123}$$

as an implication of relation (3.115).
Whether U_k is a Poisson distribution can be studied by means of the sum (3.74), which is easily evaluated with Eq. (3.122) to yield

$$S(D; Z) = \sum_{k=0}^{\infty} e^{-Zk} U_k = \exp\left[-sY(1 - e^{-Z})\, D\right]. \tag{3.124}$$

This is identical with the general result (3.80) if Eq. (3.123) holds, i.e. U_k then *is* a Poisson distribution, indeed. Hence, all that is 'non-poissonian' in U_k is reflected in Π by the terms non-linear with respect to x.

It should be mentioned that the limiting case (3.123) is equivalent to what has been called 'zero-LET' limit in chapter 2. Zero LET means $f_1(z) = \delta(z)$, leading to Eq. (2.15), i.e. the general coefficients α_μ according to Eq. (2.17) become equal to $\alpha_\mu^{(0)} = (1 - e^{-Z_\mu})\, Ys_0$. This is identical with Eq. (3.123), realizing that α_μ in the microdosimetric theory is the same quantity as $\Pi(Z_\mu)$ in the present strict theory. Thus, both approaches are equivalent in the zero-LET limit. These relationships in general will be elucidated in chapter 8.

3.7.2. The relation of equivalence

This subsection is devoted to a property of the theory which has only been touched on in the appendix to chapter 2: equivalence of parameters leading to their indistinguishability from the standpoint of formal analysis of dose-effect relations. Look at Eqs. (3.104 to 3.108) to realize that $(1 - e^{-Z})$ and h enter the theory only as a product. h is a combination of several parameters (cf. Eq. (3.89)), out of which only Y does not occur otherwhere in the theory. Hence, the actual equivalent parameters are the total amount, Y, of DNA per reacting unit and the parameter Z involved in the effect-probability' $H(k)$.
For the sake of simplicity, let us first discuss the problem in the case of pure exponential survival curves, i.e. $H(k)$ in Eq. (3.2) consists of one term only:

$$H(k) = e^{-Zk},$$
$$S(D) = \sum e^{-Zk} U_k(D; \lambda_j, l_l, Y) = e^{-D\Pi(Z)}. \tag{3.125}$$

In this case, the relation of equivalence states that biological objects differing only in Z and Y with the product $(1 - e^{-Z})\, Y$ being equal do not show any difference in their reaction to arbitrary radiations. Consequently, by formal survival curve analysis in the ignorance of both $H(k)$ and Y one can obtain only this product while either individual parameter is left open. However, this equivalence is not complete because $(1 - e^{-Z})$ cannot grow larger than unity, which leads to a lower limit, $Y_{\min}$, for the possible amount of DNA in the cell.
It may be instructive to stress the analogy to classical hit theory by taking the number, k, of DNA-lesions as the number of hits. Hit theory does not admit gradually decreasing survival probability functions $H(k)$ like Eq. (3.125), but it introduces a so-called probability of action, P, for a hit to be effective. Then, single-hit objects are assumed to be killed definitively by any *effective* hit. It is easy to adopt this philosophy with respect to Eq. (3.125): $(1 - e^{-Z})$ has to be identified with the action probability P, which implies $H(k)$

to be the probability that none of k hits is effective. According to the relation of equivalence, nothing is altered when we now take $H(k) = \delta_{k0}$ and simultaneously put the factor $(1 - e^{-Z})$, i.e. P, to the 'target size' Y. This description in terms of *effective hits* is exactly what hit theory does when it includes an unknown action probability in the target size. Hence, the ambiguity established by the relation of equivalence is nothing but the well-known assertion of hit theory that action probability and target size are interchangeable, i.e. undeterminable, parameters. What goes beyond hit theory, however, is the finding that this equivalence keeps valid for *arbitrary* radiations when U_k is no longer a Poisson distribution as presumed in hit theory.

In the general case, $H(k)$ consists of several terms with different Z_μ. There is equivalence then for different Y-values associated with different effect-probability functions $\tilde{H}(k)$ defined by

$$\tilde{H}(k) = \sum_{k'=0}^{k} \binom{k}{k'} \left(\frac{1}{P}\right)^{k'} \left(1 - \frac{1}{P}\right)^{k-k'} H(k'). \tag{3.126}$$

Substituting the sum of exponential functions (3.2) for $H(k')$ yields

$$\tilde{H}(k) = \sum_{\mu} B_\mu [1 - (1 - e^{-Z_\mu})/P]^k. \tag{3.127}$$

Hence, using $\tilde{H}(k)$ instead of $H(k)$ means replacing e^{-Z_μ} in the final formulas by $[1 - (1 - e^{-Z_\mu})/P]$. This alteration can be canceled by simultaneously taking PY rather than Y. Consequently, the *general relation of equivalence* reads

$$S(D) = \sum_{k=0}^{\infty} H(k)\, U_k(D; \lambda_j, l_l, Y)$$
$$\equiv \sum_{k=0}^{\infty} \tilde{H}(k)\, U_k(D; \lambda_j, l_l, PY). \tag{3.128}$$

The most important consequence of the general equivalence relation is the fact that the theory of cellular radiation effects applies also to the case of inhomogeneous populations of cells that differ in the DNA content Y, a special kind of *biological variability*. It is evident from the identity (3.128) that the survival curves for different subpopulations, i, can be understood as brought about by the *same* DNA content $P_i Y_i = Y = const.$ and modified $H(k)$-functions according to Eq. (3.126). The superposition of the latter is then representative for the inhomogeneous population as a whole.

It is worth while to discuss again the relatedness to classical multi-hit theory. There would be complete analogy if $\tilde{H}(k)$ were a step function: $\tilde{H}(k) = 1$ for $k < m$ and $\tilde{H}(k) = 0$ for $k \geqq m$. Then the right-hand side of Eq. (3.128) would correspond to hit theory, in which the object is definitively killed if and only if the number of *effective* hits exceeds $m - 1$, while some action probability P is additionally taken into account as a factor at Y. From this point of view, our $H(k)$ is nothing but an equivalent survival probability that by itself allows for this action probability P. This may be seen from the inversion of Eq. (3.126) which easily proves to be

$$H(k) = \sum_{k'=0}^{k} \binom{k}{k'} P^{k'} (1 - P)^{k-k'}\, \tilde{H}(k'). \tag{3.129}$$

This is self-explanatory, k' being the number of effective hits (note the binomial distribution occurring).

Actually, the present theory is more general than multi-hit theory, regarding $\tilde{H}(k)$. Starting from an arbitrary decreasing function $H(k)$, the associated $\tilde{H}(k)$'s according to Eq. (3.126) with $P < 1$ decrease more steeply but never become a step function. Instead, there exists a threshold value $P = P_{\min}$ below which $\tilde{H}(k)$ becomes unreasonable (reasonableness means $1 \geqq H(k) \geqq 0$ and $H(k+1) \leqq H(k)$). $P_{\min}$ defines the most steep reasonable $\tilde{H}(k)$, which is nevertheless

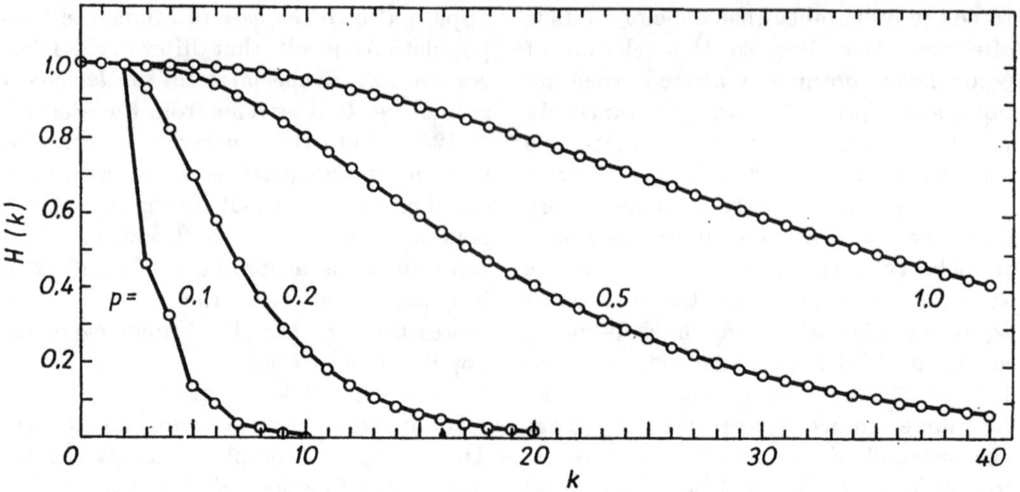

Fig. 3.8. Equivalent survival probability functions for different P, as described by Eq. (3.127): $B_1 = 3.9966$, $B_2 = -3.4264$, $B_3 = 0.4298$; $Z_1 = 0.04858$, $Z_2 = 0.07737$, $Z_3 = 0.16555$. The most steep curve corresponds to $P_{\min} = 0.1$.

gradually decreasing in general.[1]) An example of such associated $H(k)$-functions depending on the parameter P is given in Figure 3.8.

The value $P_{\min}$ is of great importance, since it determines a lower limit, $Y_{\min}$, for the possible content of DNA in the object. There is no general procedure to derive $P_{\min}$ from a given $H(k)$; this depends on the special mathematical expressions for $H(k)$, i.e. for B_μ and Z_μ, which will be dealt with in section 9.3.2. One can only infer lower bounds to $P_{\min}$ from Eq. (3.127). Upon decreasing P the terms in brackets can become negative, beginning with the term for the largest $Z_\mu = Z_n$. If the absolute value of the latter grows higher than the still positive $[1 - (1 - e^{-Z_1})/P]$, where Z_1 is the smallest Z_μ, the sign of $H(k)$ becomes alternating at large k, i.e. unreasonableness is reached. This yields

$$1 - \frac{1}{2}\left(e^{-Z_1} + e^{-Z_n}\right) \leqq P_{\min} \leqq 1\,. \quad (3.130)$$

What is the practical meaning of $P_{\min}$? Suppose one has analysed survival curves (usually a low-LET one) as demonstrated in section 2.4.1. This means that coefficients B_μ and $\alpha_\mu^{(0)}$ describing the zero-LET survival curve (2.15) have been found. We know that $\alpha_\mu^{(0)}$ is related to Z_μ and Y by Eq. (2.16):

$$\alpha_\mu^{(0)} = (1 - e^{-Z_\mu})\, Y s_0\,. \quad (3.131)$$

Assume any value $Y = Y_0$ not too small. Thus the corresponding Z_μ's, and hence $H(k)$, are fixed.[2]) Now $P_{\min}$ can be determined (in principle), and we arrive at the following statement regarding the amount of DNA compatible with the survival

---

[1]) If this limiting $\tilde{H}(k)$ happens to be an exact step function, we have exactly the situation of hit theory, and then $P_{\min}$ is an action probability in full sense. However, this case cannot be met with $H(k)$'s described by a *finite* sum according to Eq. (3.2).

[2]) The absolute rate of lesion production, s_0, for ^{60}Co γ-rays must be known in addition. Therefore, this analysis can be based on nothing but experimental survival curves only if one relies also on the lesion production model, which yields s_0, too, after being adapted so as to fit the empirical values r (primary lesion RBE).

curve we started with:

$$Y \geqq Y_{\min} = P_{\min} Y_0 \, .^{1)} \tag{3.132}$$

Apart from the general insights gained in this subsection (biological variability!), Eq. (3.132) exhibits the practical importance of the relation of equivalence. Survival curves can tell us something about the DNA in the cell, if a certain critical type of lesion is assumed. The essence of this reasoning is as follows: at a given rate of lesion production, s_0, the total amount of DNA cannot be arbitrarily small, since otherwise the mean number of lesions per cell would have to be so small that even Poisson statistics (zero LET) would yield a too large variance for the survival curve, even if additional biological stochastics were lacking (as assumed by $\tilde{H}(k)$ for $P = P_{\min}$). This kind of inference is analogous to that by HUG and KELLERER (1966), who derived a lower limit for the gross sensitive volume by attributing the variance of survival curves to physical stochastics (energy instead of lesion number) only.

3.7.3. Generalization of the theory: more than one lesion type involved in the injury process

This question was already dealt with in section 2.6.1 for the approximate microdosimetric version of the theory. The following treatment in the framework of the strict theory is completely analogous and leads to similar results. We confine ourselves again to only two types of lesion, since the generalization to an arbitrary number is obvious and straightforward. These two types will be denoted by the subscripts or superscripts I and II, respectively.

The generalized fundamental equation (3.2) reads

$$S(D) = \sum_{k_I, k_{II}=0}^{\infty} H(k_I, k_{II}) \times U_{k_I, k_{II}}(D; \lambda_j^{I}, \lambda_j^{II}, l_l^{I}, l_l^{II}, Y),$$

$$H(k_I, k_{II}) = \sum_{\mu} B_\mu \, e^{-Z_\mu^{I} k_I - Z_\mu^{II} k_{II}} . \tag{3.133}$$

The physical characteristics of lesion formation for the two types are described by the probability fields λ_j^{I} and λ_j^{II}[2]) and by the eventually differing target diameters l_l^{I} and l_l^{II}. The survival probability H depending on the corresponding two numbers of lesions, k_I and k_{II}, shall be equal to, or approximated by, the expression (3.133) with an arbitrary number of terms.

The evaluation of Eq. (3.133) proceeds in full analogy to section 3.6. The first step leading to

$$S(D) = \sum_{\mu} B_\mu \, S(D, Z_\mu^{I}, Z_\mu^{II}),$$

$$S(D, Z^{I}, Z^{II}) = e^{-D\Pi(Z^{I}, Z^{II})},$$

$$\Pi(Z^{I}, Z^{II}) = \frac{[\Psi(Z^{I}, Z^{II})]}{\bar{z}_F}, \tag{3.134}$$

$$\Psi(Z^{I}, Z^{II}) = 1 - \sum_{k_I=0}^{\infty} \sum_{k_{II}=0}^{\infty} e^{-Z^{I} k_I - Z^{II} k_{II}} V_{k_I, k_{II}}$$

consists only in replacing the lesion number sums in the formalism (3.74 to 81) by corresponding double sums. This refers also to the subsequent evaluation of $V_{k_I, k_{II}}$, i.e. Eq. (3.83) remains valid and the statistical independence of lesions produced direct along the incident particle track segment causes the Poisson distribution (3.85) to change into a product of two corresponding Poisson distributions having the mean values Λ_I' and Λ_{II}'. These mean values are given by two expressions like (3.90) for either lesion type, i.e. we

---

[1]) The inequality (3.130) implies that $Y_{\min}$ is usually smaller than what would turn out by requiring all $(1 - e^{-Z_\mu})$ to lie below unity at given $\alpha_\mu^{(0)}$ (see Eq. (3.131)), which holds in the above-discussed single-hit case only. In terms of P, this means that in Eq. (3.127) some terms in brackets may well become negative.

[2]) The condition $\lambda_j^{I} + \lambda_j^{II} \leqq 1$ is self-evident.

have now $\Gamma^{\rm I}(E)$ and $\Gamma^{\rm II}(E)$ depending on $\lambda_j^{\rm I}$, $l_l^{\rm I}$ and $\lambda_j^{\rm II}$, $l_l^{\rm II}$, respectively, according to Eqs. (3.71), (3.55). In this way, the quantity Ψ' becomes

$$\Psi' = \sum_{k_{\rm I}=0}^{\infty} \sum_{k_{\rm II}=0}^{\infty} e^{-Z^{\rm I}k_{\rm I}-Z^{\rm II}k_{\rm II}} V'_{k_{\rm I},k_{\rm II}}$$

$$= \exp\left[-\frac{h}{\varrho}\int_{E^{(I)}}^{E} dE' \frac{(1-e^{-Z^{\rm I}})\,\Gamma^{\rm I}(E') + (1-e^{-Z^{\rm II}})\,\Gamma^{\rm II}(E')}{L_\infty(E')}\right]. \tag{3.135}$$

At this point, it is worth while introducing the abbreviation

$$(1 - e^{-Z^{\rm I}})\,\Gamma^{\rm I}(E) + (1 - e^{-Z^{\rm II}})\,\Gamma^{\rm II}(E)$$
$$= (1 - e^{-Z})\,\Gamma(E), \tag{1.136}$$

which makes Eq. (3.135) formally identical with Eq. (3.91). (3.136) is not only a short-hand notation, since $\Gamma(E)$ can be considered (and calculated) as the quantity (3.71) for a formal combined lesion type characterized by

$$\lambda_j = \frac{(1 - e^{-Z^{\rm I}})\,\lambda_j^{\rm I} + (1 - e^{-Z^{\rm II}})\,\lambda_j^{\rm II}}{1 - e^{-Z}}, \tag{3.137}$$

if $l_l^{\rm I} = l_l^{\rm II}$ is assumed. The choice of Z is, of course, arbitrary. To complete the derivation, one must only realize that the final result of section 3.6, i.e. Eqs. (3.99) and (3.100), depends on nothing but the expression (3.91), apart from the replacement of the lesion number sums by double sums. This means that the theory of section 3.6 remains fully valid; it must only be applied to the formal combined lesion type described by (3.137), and the coefficients $\Pi(Z_\mu^{\rm I}, Z_\mu^{\rm II})$ needed for the dose-effect relation (3.134) can be calculated by Eqs. (3.104 to 108) according to

$$\Pi(Z^{\rm I}, Z^{\rm II}) = \Pi(Z). \tag{3.138}$$

Note that the arbitrary Z enters into the expression (3.137), too, along with $Z^{\rm I}$ and $Z^{\rm II}$.

Let us now discuss the significance of the result obtained. At first glance, any several-lesion-type model seems to be completely reducible to an equivalent formal one-lesion-type model, i.e. from the phenomenological point of view (survival curves for arbitrary radiations) no difference between both cases seems to exist if one admits quite general λ_j-sets. This is however not true. In the above derivation, we did not retain the index μ of Eq. (3.134). Actually, one has another pair $Z_\mu^{\rm I}$, $Z_\mu^{\rm II}$ for each term of the dose-effect relation, so that the formal combined λ_j defined by Eq. (3.137) depends on μ, in general. This means that each $\Pi(Z_\mu^{\rm I}, Z_\mu^{\rm II})$ is equal to a one-lesion-type $\Pi_\mu(Z_\mu)$ for another formal type, which is indicated by the subscript μ at Π. Therefore, if a family of survival curves for different radiations are actually the result of an injury mechanism that involves several lesion types, they could not, in a strict sense, be reproduced by the one-lesion-type theory, whichever λ_j would be chosen. At first glance, this statement is disencouraging in respect of formal survival curve analyses that aim at finding out the underlying lesion type, described by λ_j or $r = s/s_0$ (primary lesion RBE), just in the case of ignorance about the action mechanism; and formal analysis is possible only in the framework of one-lesion-type theory.

However, one can find a number of conditions under which several lesion typs lead nevertheless to a universal, i.e. μ-independent, λ_j. These conditions were considered in section 2.6.1 and need not be repeated here. What has been stated there applies to the strict theory also, which becomes obvious by eliminating $(1 - e^{-Z})$ from Eq. (3.137). Inserting the expression (3.137) into Eq. (3.1) yields a relation between the corresponding lesion production rates s^{I}, s^{II}, and s, which must be considered especially for ^{60}Co γ-radiation (subscript 0). After that, Eq. (3.137) can be written in the form

$$\lambda_j = s_0 \frac{(1 - e^{-Z^{\mathrm{I}}})\,\lambda_j^{\mathrm{I}} + (1 - e^{-Z^{\mathrm{II}}})\,\lambda_j^{\mathrm{II}}}{(1 - e^{-Z^{\mathrm{I}}})\,s_0^{\mathrm{I}} + (1 - e^{-Z^{\mathrm{II}}})\,s_0^{\mathrm{II}}}, \tag{3.139}$$

which is obviously equivalent to Eq. (2.46).

As to formal survival curve analysis, it was argued in section 2.6.1 why we can hope that $r = s/s_0$, i.e. λ_j here, is largely independent of μ, in sufficient approximation at least. Survival curve analysis starts from the coefficients $\alpha_\mu^{(0)}$ of the zero-LET dose-effect relation $S_0(D) = \sum B_\mu \exp(-\alpha_\mu^{(0)} D)$. Now any trial λ_j first establishes s_0 so that the parameters $(1 - e^{-Z_\mu})\,Y$ needed to calculate the Π's for other radiations are known for that trial λ_j according to Eq. (3.131). Having found a λ_j-set yielding a good fit, one should not forget that it may intrinsically describe the interplay of several lesion types according to Eq. (3.139) involving survival probability parameters, Z^{I} and Z^{II}, which have nothing to do with the physico-chemical processes of primary lesion formation. For this reason, an effective λ_j-set thus obtained may eventually depend on purely biological conditions and peculiarities of the special object, which should influence nothing but the survival probability $H(k)$ in case that actually one lesion type only was responsible for the effect.

3.8. Appendix: estimation of errors

3.8.1. Incorrectly apportioned energy

The percentage, η, of incorrectly apportioned energy is defined by Eq. (3.11) and context. The three energies in question will be chosen in relation to unit track length of the monoenergetic ions considered. Consequently, the denominator of the quotient (3.11) is the linear energy transfer given by the well-known Bethe formula (see section 4.2.1):

$$L_\infty^{(i)}(E_i) = 2\,\frac{A_0 Z_{eff}^2}{E_{\max}} \ln \frac{E_{\max}}{I}. \tag{3.140}$$

I is the effective ionization energy taken to be 70 eV for water. $E_{\max}$ is the maximum δ-electron energy following from classical collision laws:

$$E_{\max} = 4\,\frac{m_0}{M_i}\,E_i. \tag{3.141}$$

The energies in the numerator of Eq. (3.11) depend on the δ-electron spectrum $N_i(E_i, E)$ determined in sufficient approximation[1]) by the Rutherford scattering law (see section 4.3):

$$N_i(E_i, E) = \frac{A_0 Z_{eff}^2}{E_{\max}}\,\frac{1}{E^2}. \tag{3.142}$$

Hence, the quantity η to be calculated writes

$$\eta = \frac{\Delta_0' + \Delta_0''}{2 \ln (E_{\max}/I)}\,100\%,$$

$$\Delta_0' = \int_0^\delta \mathrm{d}E\,\frac{\varepsilon'(E)}{E^2}, \tag{3.143}$$

$$\Delta_0'' = \int_\delta^{E_{\max}} \mathrm{d}E\,\frac{\varepsilon''(E)}{E^2}.$$

The energies $\varepsilon'(E)$ and $\varepsilon''(E)$ (incorrectly apportioned parts of the δ-electron energy E) will be explained below.

All δ-electron tracks will be approximately considered as straight branching-free lines along which a continuous energy deposition takes place according to the approximate range-energy relation in section 4.5.1:

$$R_e(E) = \alpha E^\beta, \qquad \alpha = 0.0346\,\frac{\mu\mathrm{m}}{\mathrm{keV}^\beta}, \tag{3.144}$$
$$\beta = 1.732.$$

The cut-off energy δ, which is considered a variable in the framework of this investigation, will be expressed by a coefficient f describing the *deviation* from the rule (3.12):

$$\delta = (fl/\alpha)^{1/\beta}. \tag{3.145}$$

The mass density coefficient ϱ is omitted, because it acts only as a spatial scaling factor for the entire track patterns. It can therefore

---

[1]) Eq. (3.142) failing at small E is unimportant for what follows; it can even be used down to $E = 0$ without giving rise to divergences.

be taken into account subsequently by substituting ϱl for l in the final results, provided that the sensitive volume and its surroundings have the same density. Otherwise, only Δ_0'' is affected (see below).

Geometry:

We will consider sensitive volumes (SV) of spheroidal shape, taking the axis ratios $\lambda = 1$ for the case of a sphere and $\lambda = 1000$ for simulating a long cylinder. Let a be the length of the two minor semi-axes and ϑ the angle between the z-axis of the co-ordinate system and the major semi-axis of the spheroid. Its description in vector form writes then:

$$\boldsymbol{r}\mathbf{A}\boldsymbol{r} = a^2, \qquad \boldsymbol{r} = (x, y, z)$$

$$\mathbf{A} = (A_{mn}) = \frac{1}{\lambda^2}\begin{pmatrix} 1 + (\lambda^2 - 1)\zeta^2 & 0 & -(\lambda^2 - 1)\zeta\sqrt{1-\zeta^2} \\ 0 & 1 & 0 \\ -(\lambda^2 - 1)\zeta\sqrt{1-\zeta^2} & 0 & \lambda^2 - (\lambda^2 - 1)\zeta^2 \end{pmatrix}, \quad \zeta = \cos\vartheta. \tag{3.146}$$

Mean values with respect to all orientations of the spheroid are obtained by integration over the solid angle Ω according to

$$\frac{\mathrm{d}\Omega}{4\pi} = \frac{1}{2}\,\mathrm{d}\zeta, \qquad -1 \leqq \zeta \leqq 1. \tag{3.147}$$

The projected area A on the x,y-plane is obtained by eliminating z from the two quations $\boldsymbol{r}\mathbf{A}\boldsymbol{r} = a^2$ and $\mathrm{d}(\boldsymbol{r}\mathbf{A}\boldsymbol{r})/\mathrm{d}z = 0$:

$$A(\Omega) = \pi a^2\sqrt{\lambda^2 - (\lambda^2 - 1)\zeta^2} = \pi a^2 \lambda\sqrt{A_{33}}. \tag{3.148}$$

Averaging $A(\Omega)$ according to (3.147) leads to the mean cross-section σ_0, from which the mean thickness l follows by Eq. (3.27):

$$\frac{a}{l} = \frac{3}{8}\left[\frac{1}{\lambda} + \frac{\arcsin\sqrt{1 - 1/\lambda^2}}{\sqrt{1 - 1/\lambda^2}}\right]. \tag{3.149}$$

Calculation of Δ_0':

Consider a δ-electron that starts anywhere in the SV with initial energy E. $\varepsilon'(E)$ in Eq. (3.143) is obviously the mean value of residual energy of this δ-electron when it leaves the SV. If l' is the path length within the SV, $\varepsilon'(E)$ amounts to $(E^\beta - l'/\alpha)^{1/\beta}$ according to Eq. (3.144). The corresponding spectrum of path lengths is $u(l')$ as defined by Eq. (3.28). Thus, one has

$$\Delta_0' = \int_0^\delta \frac{\mathrm{d}E}{E^2}\int_0^{\alpha E^\beta}\mathrm{d}l'(E^\beta - l'/\alpha)^{1/\beta}\,\frac{1}{l}\int_{l'}^\infty \mathrm{d}l'' w(l'') = \frac{1}{\beta}\int_0^{fl}\frac{\mathrm{d}R}{R^{1+1/\beta}}\int_0^R \mathrm{d}l'(R - l')^{1/\beta}\,\frac{1}{l}\int_{l'}^\infty \mathrm{d}l'' w(l'')$$

$$= \frac{1}{1+\beta}\left[f - \frac{1}{l}\int_0^{fl}\mathrm{d}R\int_0^R \mathrm{d}l'\left(1 - \frac{l'}{R}\right)^{1+1/\beta} w(l')\right]. \tag{3.150}$$

The chord length spectrum $w(l')$ for a spheroid can be obtained as a weighted mean of the corresponding spectrum $w(\Omega, l')$ for chords parallel to the z-axis in a spheroid of fixed orientation Ω. A simple geometrical consideration shows that this $w(\Omega, l')$ is a triangular spectrum, like that for a sphere (see Eq. (3.25)), with the maximum chord length $2a/\sqrt{A_{33}(\zeta)}$. Randomizing the orientation Ω means

$$w(l') = \int \frac{\mathrm{d}\Omega}{4\pi}\,\frac{A(\Omega)}{\sigma_0}\,w(\Omega, l')$$

$$= \frac{3}{8}\,\frac{l}{a}\int_{-1}^{+1}\mathrm{d}\zeta\,\sqrt{A_{33}(\zeta)}\,w(\zeta, l'). \tag{3.151}$$

In the way described, Eq. (3.150) can be further evaluated analytically, leading to a final formula containing only a simple integral which must be solved by numerical methods. Δ_0' thus obtained is obviously independent of charge and energy of the ion; it depends *only* on f as long as the cut-off energy expressed by Eq. (3.145) is smaller than $E_{\max}$. Otherwise, Δ_0' becomes a constant that is obtained for $f = \alpha E_{\max}^\beta/l$.

Monte Carlo calculation of Δ_0'':

Consider a δ-electron that starts anywhere inside or outside of the SV with initial energy E and let ΔE be the energy imparted thereby to the SV. ΔE as a function of the starting point $\boldsymbol{r}_0$ and the orientation of the spheroid, ζ, is easily obtained from Eq. (3.144) and Eq. (3.146), assuming the δ-electron track to be parallel to the z-axis, which is allowed because ζ is subsequently made a random quantity. The incorrectly apportioned energy $\varepsilon''(E)$ in Eq. (3.143)[1]) is the part of ΔE that is deter-

[1]) $\varepsilon''(E)$ must finally be normalized to exactly *one* particle start with energy E in the SV.

mined by the fraction of cases in which the primary ion track or another δ-electron affects the SV simultaneously. This problem was solved by a Monte Carlo procedure which also included the E-integral of Eq. (3.143).

One has a hierarchy of random variables: ζ, E, $\boldsymbol{r}_0$, φ_0, where the latter quantity is the polar angle of the primary track in the plane perpendicular to the δ-track (x,y-plane), while the angle ϑ_0 between both tracks is taken to be determined by the classical collision law $\cos\vartheta_0 = \sqrt{E/E_{\max}}$. For given values of the random variables obtained from a random number generator, it is a matter of analytical geometry based on Eq. (3.146) whether the primary track crosses the SV. In the latter case, the corresponding $\Delta E(\zeta, E, \boldsymbol{r}_0)$ has to be fully included in $\varepsilon''(E)$. Otherwise, $\varepsilon''(E)$ is nevertheless different from zero according to the probability, q, of another δ-electron reaching the SV. This probability reads

$$q = 1 - \exp\left[-M(\zeta, \boldsymbol{r}_0, \vartheta_0, \varphi_0)\right], \tag{3.152}$$

where M is the *mean* number of δ-electrons reaching the SV in the given geometrical

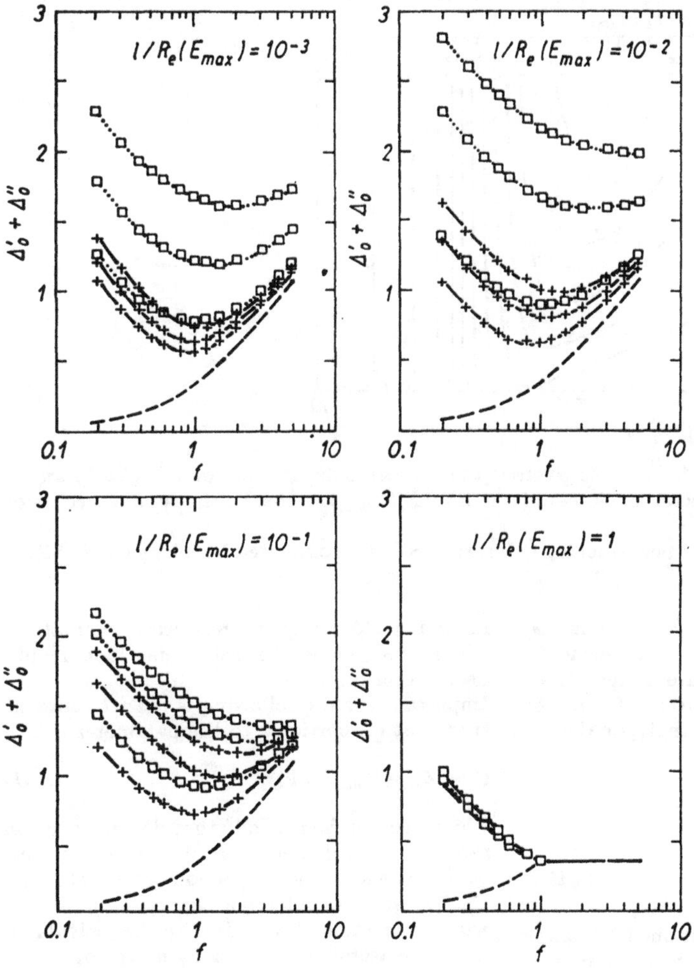

Fig. 3.9. Incorrectly apportioned energy (arbitrary units) as a function of f standing for the cut-off energy δ (see Eq. (3.145)).
Full curves: spherical SV; dotted curves: cylindrical SV approximated by a very long spheroid; dashed curves: the contribution Δ_0' only (sphere and spheroid graphically indistinguishable).
The three curves of each set are for $C = 0$, 40, 104.
Crosses and squares indicate the individual computed values, which slightly scatter due to the limited accuracy of the Monte Carlo procedure used.

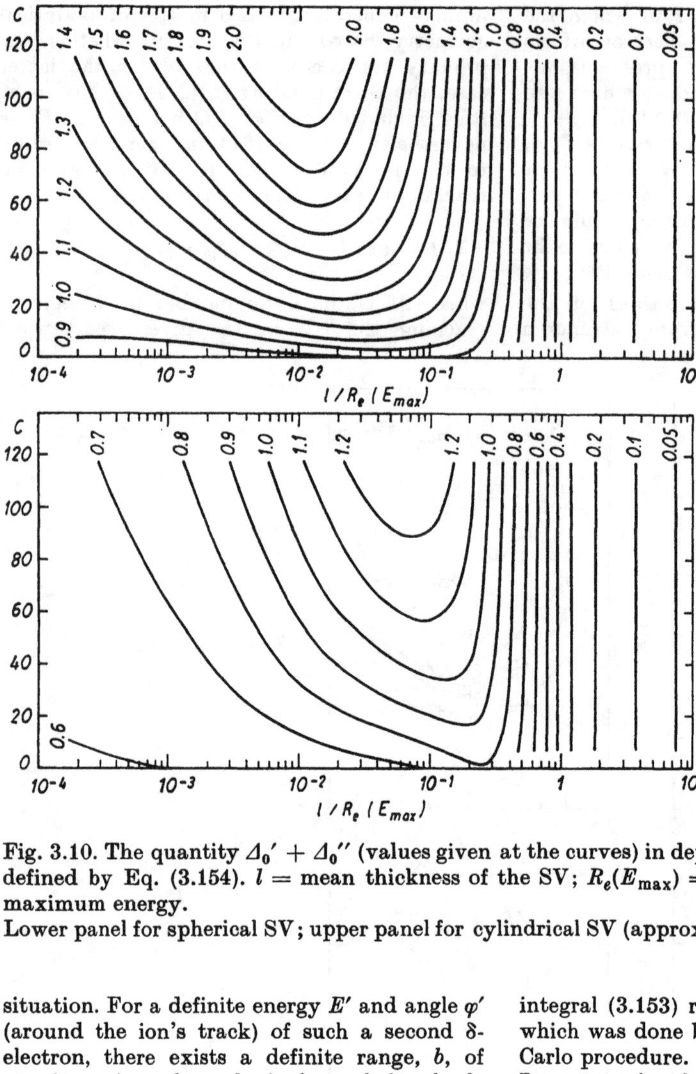

Fig. 3.10. The quantity $\Delta_0' + \Delta_0''$ (values given at the curves) in dependence on $l/R_e(E_{\max})$ and C defined by Eq. (3.154). l = mean thickness of the SV; $R_e(E_{\max})$ = mean range of δ-electrons of maximum energy.
Lower panel for spherical SV; upper panel for cylindrical SV (approximated by a long spheroid).

situation. For a definite energy E' and angle φ' (around the ion's track) of such a second δ-electron, there exists a definite range, b, of starting points along the ion's track for the δ-electron to reach the SV. Accordingly, one has

$$M = \int_{E}^{E_{\max}} \mathrm{d}E' N_i(E_i, E') \frac{1}{2\pi} \times \int_{0}^{2\pi} \mathrm{d}\varphi' b(\zeta, \boldsymbol{r}_0, \vartheta_0, \varphi_0, E', \varphi') .^{1)} \qquad (3.153)$$

The function b can still be obtained by fairly complex analytical geometry, but the double integral (3.153) requires numerical evaluation, which was done by a special inner-loop Monte Carlo procedure.

Important for the following is the fact that M turns out proportional to the parameter

$$C = Z_{eff}^2/(E_{\max}/\mathrm{keV})^{2-\beta} \qquad (3.154)$$

which comprehends all the dependence on radiation (charge and energy). It is only in this way that the charge of the ion enters into Δ_0''. The question of effective charge, Z_{eff}, is dealt with in section 4.2.1. It can be calculated from the atomic number Z by means of

$$Z_{eff} = [1 - \exp(-3.91 \sqrt{E_{\max}/\mathrm{keV}/Z^{2/3}})]\, Z . \qquad (3.155)$$

Δ_0'' depends on the parameters $l/R_e(E_{\max})$, C, and f. For each value of $l/R_e(E_{\max})$, two Monte Carlo calculations for $\lambda = 1$ and $\lambda = 1000$ were

---

[1]) The lower integration limit E ensures that not the *total* imparted energy of compound δ-electron events is taken for incorrectly apportioned. Properly speaking, the highest-initial-energy δ-electron is always considered as being treated correctly by definition.

carried out,[1]) including 10,000 throws. It was possible to obtain Δ_0'' for arbitrary values of C and f *simultaneously* from the same run.

Results:

Let us first consider the dependence of the incorrectly apportioned energy on the cut-off energy δ, i.e. on the coefficient f. In this respect, the denominator in Eq. (3.143) is unimportant. Figure 3.9 shows the sum $\Delta_0' + \Delta_0''$ as a function of f for four values of $l/R_e(E_{\max})$. These curves exhibit the most important result of this investigation: $f = 1$, i.e. the rule (3.12), is the best general way of choosing the cut-off energy δ. Most curves show a minimum almost exactly at $f = 1$, while in the other cases little would be gained by departing from the rule (3.12).
For $f = 1$, the results in terms of $\Delta_0' + \Delta_0''$ are completely mapped in Figure 3.10. For a radiation defined by the charge number Z of the ion and the energy per nucleon, E_i/M_i, one calculates $E_{\max}$ and $R_e(E_{\max})$ from Eqs. (3.141), (3.144); then C is obtained from Eqs. (3.154), (3.155). The mean thickness l of the SV, which must be multiplied by ϱ in case the mass density of the SV differs from unity, completes the set of parameters needed for reading off $\Delta_0' + \Delta_0''$ from Figure 3.10. Finally, Eq. (3.143) yields the percentage, η, of incorrectly apportioned energy. In this way, Figures 3.3 and 3.4 have been obtained.

3.8.2. Errors caused by the continuous-slowing-down approximation

The continuous-slowing-down (CSD) approximation will be studied by means of the following model which allows strict treatment of straggling for the sake of comparison. The hypothetical model particles move along straight lines in the direction of the x-axis of a coordinate system. p, the mean number of collisions per unit path length, is constant. The normalized collision spectrum $v(\varepsilon)$ (probability density for the energy loss ε by a collision)

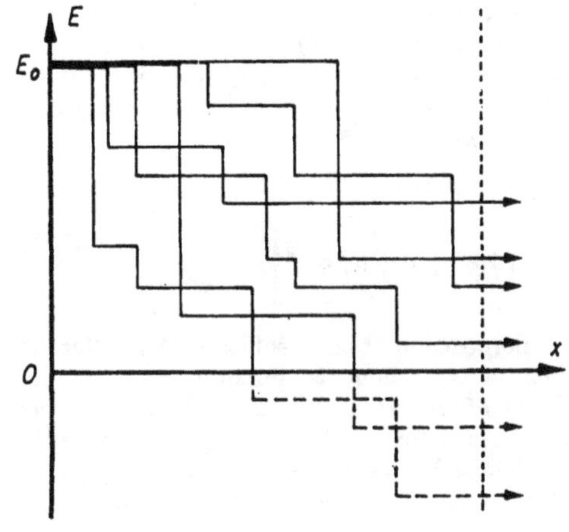

Fig. 3.11. Energy path diagram illustrating the random process of slowing down.

is also constant, i.e. independent of the particle energy E. All energy losses are assumed to be locally absorbed.
Figure 3.11 shows examples of the stochastic slowing down process for individual particles starting with initial energy E_0 at $x = 0$. Number and size of the single energy losses (collisions) within a layer of the thickness x (dotted line in Fig. 3.11) are a matter of chance. This model permits the particles to fall into the region of negative energies, but all these particles are finally counted as zero energy ones. In this way, which is mathematically advantageous, we are actually dealing with an energy-dependent collision spectrum being equal to $v(\varepsilon)$ only for $\varepsilon < E$, while to all collisions with $\varepsilon > E$ contained in $v(\varepsilon)$ the energy loss $\varepsilon = E$ is ascribed. Thus, complete stopping of particles within x is taken into account.
There is no function $E(x)$ for the discontinuous random process of slowing down. Instead, a beam of particles must be described by an x-dependent energy distribution $\Phi(E, x)$ which may also be named particle fluence spectrum. $\Phi(E, x)$ obeys the well-known transport equation

$$\frac{\partial \Phi}{\partial x} = p \left[\int_0^\infty \mathrm{d}\varepsilon \, v(\varepsilon) \, \Phi(E + \varepsilon, x) - \Phi(E, x) \right]. \tag{3.156}$$

The derivation of this equation, which was dealt with previously by several authors

[1]) Moreover, the case of a SV having a mass density higher than the environment was also considered. The result is evident: Δ_0'' becomes only smaller compared with a homogeneous mass density that corresponds to the density of the SV.

(LANDAU, 1944; VAVILOV, 1957; TSCHALÄR, 1968; KELLERER, 1968a)[1]), is obvious.
The simplest solution to Eq. (3.156) belongs to the initial energy spectrum $\Phi(E, 0) = \exp(-\alpha E)$ and reads

$$\Phi(E, x) = e^{-\alpha E - \beta x},$$
$$\beta = p\left[1 - \int_0^\infty d\varepsilon\, v(\varepsilon)\, e^{-\alpha\varepsilon}\right]. \tag{3.157}$$

By superposition of such solutions for different values of α, which is possible due to the linearity of the transport equation (3.156), one can form a solution that belongs to a sharply peaked initial energy distribution, i.e. $\Phi(E, 0) \approx \delta(E - E_0)$. This is achieved by using

$$\Phi(E, 0) = \frac{d}{dE}(1 - e^{-\alpha E})^r = \alpha \sum_{\nu=1}^{r} \binom{r}{\nu} (-1)^{\nu+1}\, \nu\, e^{-\nu\alpha E} \tag{3.158}$$

with the mean initial energy

$$E_0 = \int_0^\infty dE E\Phi(E, 0) = \frac{c_r}{\alpha},$$
$$c_r = \sum_{\nu=1}^{r} \binom{r}{\nu} (-1)^{\nu+1} \frac{1}{\nu}. \tag{3.159}$$

The corresponding solution $\Phi(E, x)$ according to Eq. (3.157) reads:

$$\Phi(E, x) = \alpha \sum_{\nu=1}^{r} \binom{r}{\nu} (-1)^{\nu+1}\, \nu\, e^{-\nu\alpha E - \beta_\nu x},$$
$$\beta_\nu = p\left[1 - \int_0^\infty d\varepsilon\, v(\varepsilon)\, e^{-\nu\alpha\varepsilon}\right]. \tag{3.160}$$

Mean and mean squared energy deposition within x:

Mean value and variance of the energy loss ΔE of the model particles will be calculated in order to compare them subsequently with the corresponding results in CSD approximation. $\overline{\Delta E}$ is easily obtained as the difference of the mean energies the particles possess before and after traversing the distance x:

$$\overline{\Delta E} = \int_0^\infty dE E\Phi(E, 0) - \int_0^\infty dE E\Phi(E, x). \tag{3.161}$$

The normalization integral of $\Phi(E, x)$ is less than unity. Consequently, the lacking particles, which have got negative energies, are treated as being stopped. Substitution of the expressions (3.159) and (3.160) yields

$$\overline{\Delta E} = \left[1 - \frac{1}{c_r} \sum_{\nu=1}^{r} \binom{r}{\nu} (-1)^{\nu+1} \frac{1}{\nu}\, e^{-\beta_\nu x}\right] E_0. \tag{3.162}$$

The variance of ΔE is obtained from the expected value $\overline{\Delta E^2}$, which can be calculated as follows. Let $\Phi_0(E', E, x)$ be the solution to the transport equation (3.156) for the sharp initial energy E', i.e.

$$\Phi_0(E', E, 0) = \delta(E - E'). \tag{3.163}$$

The corresponding mean squared energy loss for a particle of initial energy E' writes[2])

$$\Delta E^2 = \int_0^\infty dE (E' - E)^2\, \Phi_0(E', E, x) + E'^2\left[1 - \int_0^\infty dE \Phi_0(E', E, x)\right]$$
$$= E'^2 - \int_0^\infty dE(2E'E - E^2)\, \Phi_0(E', E, x). \tag{3.164}$$

This must be averaged over E' by means of $\Phi(E', 0)$ given by Eq. (3.158):

$$\overline{\Delta E^2} = \alpha \sum_{\nu=1}^{r} \binom{r}{\nu} (-1)^{\nu+1}\, \nu \left[\int_0^\infty dE' E'^2\, e^{-\nu\alpha E'} - 2\int_0^\infty dE E \int_0^\infty dE' E'\, e^{-\nu\alpha E'} \Phi_0(E', E, x) + \int_0^\infty dE E^2 \int_0^\infty dE'\, e^{-\nu\alpha E'} \Phi_0(E', E, x)\right]. \tag{3.165}$$

After substituting the derivative $-d/d(\nu\alpha) \times \exp(-\nu\alpha E')$ for $E' \exp(-\nu\alpha E')$, only the integral $\int dE' \exp(-\nu\alpha E')\, \Phi_0(E', E, x)$ remains to be evaluated. Since, by definition, $\Phi_0(E', E, x)$ is a solution to the linear equation (3.156) for arbitrary values of E', this integral must also be a solution, namely that one for the

[1]) It should be mentioned that the transport equation used by KELLERER (1968a) is not quite equivalent to Eq. (3.156). KELLERER's description is in terms of the total energy loss ΔE in the layer x ($\Phi(\Delta E, x)$ rather than $\Phi(E, x)$) so that $\Delta E > E_0$ is *not* excluded. p and $v(\varepsilon)$ being constant like ours, this is nothing but the *CSD approximation* for particles that cross the layer completely.

[2]) Note that negative-energy particles have lost the energy E'.

initial energy spectrum $\exp(-\nu\alpha E)$, in view of Eq. (3.163). This solution is just given by the expression (3.157). Now the evaluation of Eq. (3.165) is straightforward:

$$\overline{\Delta E^2} = \frac{2}{c_r^2} \sum_{\nu=1}^{r} \binom{r}{\nu} (-1)^{\nu+1} \frac{1}{\nu^2} \times \left[1 - \left(1 + \nu p x \frac{c_r}{E_0} \int_0^\infty \mathrm{d}\varepsilon\, v(\varepsilon)\, \varepsilon\, \mathrm{e}^{-\nu\alpha\varepsilon} \right) \times \mathrm{e}^{-\beta_\nu x} \right] E_0^2. \quad (3.166)$$

CSD approximation:

The picture in CSD approximation in the sense of section 3.2.2 is as follows: the energy E of the particle is continuously diminished along its path until stopping beyond or within the distance x. Energy *deposition*, however, is determined by independent collisions which must in general be considered for each differential path segment within x, since mean number, p, and spectrum, $v(\varepsilon)$, of collisions may depend on E. The function $E(x)$ derives from the continuous linear energy transfer L, which is related to p and $v(\varepsilon)$ for reasons of energy conservation *on the average:*

$$L = p\bar{\varepsilon}, \qquad \bar{\varepsilon} = \int_0^\infty \mathrm{d}\varepsilon \varepsilon v(\varepsilon). \quad (3.167)$$

These general principles must now be adequately applied to the hypothetical particles considered in this appendix. The crucial point in this respect is the collision spectrum $v(\varepsilon)$; in the framework of our model it does depend on E due to the replacement $\varepsilon = E$ for the region $\varepsilon > E$ in $v(\varepsilon)$. Such kind of dependence, however, is never considered in CSD approximation, apart from other E-dependence of both $v(\varepsilon)$ and p in the general case. Otherwise, L according to Eq. (3.167) would lead to an infinite range of CSD particles, which is easy to see.[1]) Hence, CSD approximation implies that allowance is made for collisions with an energy deposition ε greater than the energy E of the particle. Consequently, what we have to do for the sake of analogy is using the full E-independent $v(\varepsilon)$ as it enters into the transport equation (3.156), while p is constant as before. Another choice is impossible without destroying the correspondence of both approaches for large E.

Any dependence on energy being canceled, the CSD theory becomes very simple. The range reads $R = E/L = E/(p\bar{\varepsilon})$. Particles with initial energy $E < xp\bar{\varepsilon}$ are stopped within x. The mean number, $\bar{n}$, of collisions can be written in the form

$$\bar{n} = \begin{cases} E/\bar{\varepsilon} & \text{for } E < \delta \\ \delta/\bar{\varepsilon} & \text{for } E \geqq \delta, \end{cases} \qquad \delta = xL = xp\bar{\varepsilon}.^{2)} \quad (3.168)$$

The mean values of ΔE and ΔE^2 are calculated as follows. In the case of exactly n collisions with energy losses ε_i ($i = 1, \ldots, n$), it holds

$$\overline{\Delta E_n} = \overline{\sum_{i=1}^{n} \varepsilon_i} = n\bar{\varepsilon}$$

$$\overline{\Delta E_n^2} = \overline{\left(\sum_{i=1}^{n} \varepsilon_i \right)^2} = \overline{\sum_{i=1}^{n} \varepsilon_i^2} + 2 \overline{\sum_{i=1}^{n} \sum_{j>i}^{n} \varepsilon_i \varepsilon_j} = n\overline{\varepsilon^2} + n(n-1)\,\bar{\varepsilon}^2, \quad (3.169)$$

where

$$\bar{\varepsilon} = \int_0^\infty \mathrm{d}\varepsilon \varepsilon v(\varepsilon), \qquad \overline{\varepsilon^2} = \int_0^\infty \mathrm{d}\varepsilon \varepsilon^2 v(\varepsilon). \quad (3.170)$$

The expressions (3.169) have to be further averaged with respect to n, which obeys a Poisson distribution with the mean value $\bar{n}$ given by Eq. (3.168). Finally, in order to have results comparable with the strict solution (3.162) and (3.166), one must calculate corresponding mean values with respect to the initial energy E which is distributed according to the function (3.158). This yields

$$\overline{\Delta E}_{CSD} = \left[1 - \frac{1}{c_r} \sum_{\nu=1}^{r} \binom{r}{\nu} (-1)^{\nu+1} \frac{1}{\nu} \mathrm{e}^{-\nu c_r \delta/E_0} \right] E_0$$

$$\overline{\Delta E^2}_{CSD} = \frac{\overline{\varepsilon^2}}{\bar{\varepsilon}} \overline{\Delta E}_{CSD} + \frac{2}{c_r} \sum_{\nu=1}^{r} \binom{r}{\nu} (-1)^{\nu+1} \frac{1}{\nu^2} \times \left[1 - (1 + \nu c_r \delta/E_0)\, \mathrm{e}^{-\nu c_r \delta/E_0} \right] E_0^2. \quad (3.171)$$

---

[1]) This problem is usually masked due to *starting with* a linear energy transfer $L(E)$ that naturally implies a finite range. Then $v(\varepsilon)$ is chosen in a way compatible with Eq. (3.167), by which an always finite $p(E)$ is determined finally. This is also the essence of our pertinent theory in terms of discrete ionizations, which is described in section 4.4.

[2]) Note the analogy of this δ to the cut-off energy determined by Eq. (3.12) (range equal to thickness).

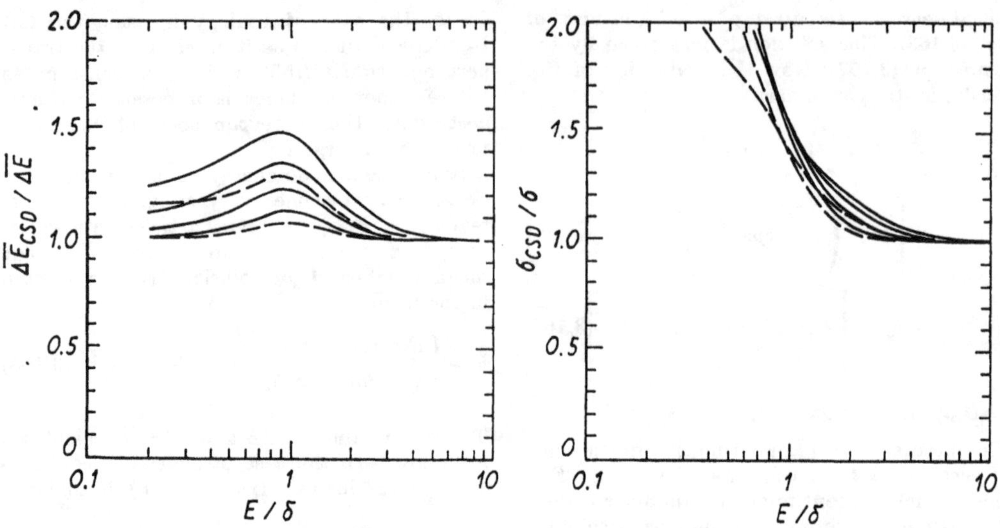

Fig. 3.12. Ratio of strict and CSD-approximated quantities: mean energy loss (left-hand panel), relative standard deviation of energy loss (right-hand panel), for particles of energy E entering a sensitive layer characterized by δ, the continuous energy decrease of crossing particles in the layer.
Collision spectra: (3.172a) — full curves; (3.172b) — broken curves.
The curve parameter is $\delta/\bar{\varepsilon}$, the mean number of collisions upon crossing the layer. Curves from up to down: $\delta/\bar{\varepsilon} = 2, 3, 5, 10$ for the full lines; $\delta/\bar{\varepsilon} = 2, 10$ for the broken lines.

Results:

To obtain definite numerical results, the two convenient expressions

$$v(\varepsilon) = \frac{1}{\bar{\varepsilon}} e^{-\varepsilon/\bar{\varepsilon}} \quad \text{and} \quad v(\varepsilon) = \delta(\varepsilon - \bar{\varepsilon}) \tag{3.172a, b}$$

are used for the collision spectrum.[1]) As to r, the value $r = 30$ used ensures a sufficiently narrow initial energy spectrum. In this way, the curves in Figure 3.12 have been calculated according to Eqs. (3.162), (3.166) and (3.171), the relative standard deviation, σ, of energy loss being defined by $\sigma = \sqrt{\overline{\Delta E^2}/\overline{\Delta E}^2 - 1}$.
To draw conclusions, from Figure 3.12, concerning the actual CSD errors regarding real charged particles involved in radiation fields is not quite easy. Since LET is not constant in reality, the parameter δ defined in Eq. (3.168) has to be equated to the mean energy loss of a given particle within the mean effective thickness ϱl of the SV considered; for particles being stopped in the SV (called 'stoppers' below), δ is the energy to which the residual range $R = \varrho l$ belongs. In this way, δ becomes dependent on the incident energy E of the particle, while $\bar{\varepsilon}$ is roughly a universal constant of about 60 eV (RAUTH and SIMPSON, 1964; KELLERER, 1968) for all kinds of charged particles. The parameter δ depends mainly on ϱl with the lower limit $\varrho l \approx 3$ nm (DNA-targets). Let us consider the different cases occurring:
(i) Primary incident ions: The requirement $E \gg \delta$ is just a prerequisite to the theory (section 3.2.1). For large E/δ, however, the CSD approximation is always correct according to Figure 3.12.
(ii) Ions set in motion by neutrons: $E \gg \delta$ is again uninteresting. The case $E \lessgtr \delta$ (stoppers or almost stoppers) must be discussed in dependence on the actual value of δ which is determined by l. If δ is large (large SV), say $\delta = 10$ keV, the CSD error of $\overline{\Delta E}$ becomes negligible due to $\delta/\bar{\varepsilon} \gg 1$, while the nevertheless greatly overestimated σ_{CSD} is unimportant as its absolute value is small anyway, i.e. straggling plays no role in this

---

[1]) A more realistic collision spectrum would have to be of the form $v(\varepsilon) \sim 1/\varepsilon^2$, but the exponential spectrum (3.172a) is a sufficient approximation for the present purpose, while (3.172b) is only to demonstrate the influence of the width of the collision spectrum.

case. For the opposite case, small δ,[1]) Figure 3.12 displays large errors, but the *fraction of total dose* going into such absorption events with, say, $E \lessapprox \delta \ll 10$ keV is negligible. Note that the initial energy of fast neutron recoils is in the range of MeV. Hence, CSD is a good approximation for fast neutron radiations.

(iii) Incident electrons: This case can be discussed in a way similar to (ii), but it is no longer true that the case $E \lessapprox \delta$ with δ being small has only a negligible share in the total dose, since the initial energy of most incident δ-electrons is not much above the cut-off energy δ. Therefore, the most unfavourable case, $\varrho l = 3$ nm, must be studied more closely. The corresponding δ amounts to about 240 eV, i.e. $\delta/\bar{\varepsilon} \approx 4$, leading to a maximum error of $\overline{\Delta E}_{CSD}$ of 25 per cent according to Figure 3.12 (full curves). σ_{CSD}, however, may exceed the true σ many times in the region $E < \delta$, because *the CSD approximation becomes unreasonable for stoppers:* the energy imparted is actually E without any fluctuations in the limit $E/\delta \to 0$, while the Poisson statistics involved in CSD makes σ_{CSD} grow at infinity due to the decreasing mean collision number $\bar{n} = E/\bar{\varepsilon}$. This finding appears disencouraging, but one should bear in mind the limitations of the underlying model. Namely, the actual theory does not concern imparted energy but ionizations in the case of DNA-targets. The number of ionizations, however, *is* fluctuating also for stoppers. This question is worth studying in more detail.

number of ionizations produced. The problem shall be solved only for complete stoppers, i.e. for the critical region $E \ll \delta$ in Figure 3.12.

We point at the fluctuation of the energy fraction spent on ionizing collisions, whereas the total energy loss being equal to the initial energy E_0 of the particle in any case is unable to fluctuate. Consequently, the energy in question is $\Delta E^* = E_0 - \lambda x'$, where the random quantity x' is the path length of the particle until stopping. The corresponding cumulative probability distribution $P(x' > x)$ following from the λ-modified particle fluence spectrum $\Phi(E + \lambda x, x)$ reads

$$P(x' > x) = \int_0^\infty \mathrm{d}E\Phi(E + \lambda x, x), \tag{3.173}$$

from which, by means of $\overline{x'}$ and $\overline{x'^2}$, the relative standard deviation, σ^*, of ΔE^* can be calculated using the expression (3.160). As to $v(\varepsilon)$, the exponential form (3.172a) will be used, which enables the limiting process $r \to \infty$ providing a sharp initial energy E_0.

The corresponding CSD treatment of the λ-modified model is easy. In the well-known Poisson expression $\sigma^{*2}_{CSD} = (\overline{\varepsilon^2}/\bar{\varepsilon}^2)/\bar{n}$, which follows from Eqs. (3.169), the mean number $\bar{n}$ of ionizing collisions is given by $pR(E_0) = pE_0/(\lambda + p\bar{\varepsilon})$.

The results obtained in the way described can be summarized as follows, mentioning only the limiting cases for the ratio $\bar{\varepsilon}/E_0$:

$$\sigma^{*2} = \begin{cases} \dfrac{4}{3}\dfrac{1}{\tau}\dfrac{\bar{\varepsilon}}{E_0} & \text{for } \dfrac{\bar{\varepsilon}}{E_0} \gg 1 \\[2ex] \dfrac{2}{\tau(1+\tau)}\dfrac{\bar{\varepsilon}}{E_0} & \text{for } \dfrac{\bar{\varepsilon}}{E_0} \ll 1, \end{cases} \qquad \sigma^{*2}_{CSD} = 2\,\frac{1+\tau}{\tau}\frac{\bar{\varepsilon}}{E_0}, \qquad \tau = \frac{p\bar{\varepsilon}}{\lambda}.$$

$$\sqrt{\frac{3}{2}(1+\tau)} < \frac{\sigma^*_{CSD}}{\sigma^*} < 1 + \tau. \tag{3.174}$$

Generalization of the model:

The actual statistics of ionization number in a track end (stopper) can be estimated on generalizing the model by an additional *continuous* energy loss process described by λ (energy transfer per unit path length) which simulates low energy collisions causing *electronic excitations*, while the parameters p and $v(\varepsilon)$ used so far shall now refer to *ionizing collisions only*. The energy going into the latter ones is considered a measure of the total

τ, the ratio of energies spent on ionizations and excitations, can be roughly estimated as follows. On an average, absorption of a total

[1]) Due to the pronounced decrease of LET in the track ends of recoil ions (Oldenburg and Booz, 1972a) δ may become very small, especially for recoil protons. For example, the residual range $R(\delta)$ assumes the value $l = 0.01\ \mu$m for the very small proton energy $\delta = 130$ eV.

of 66 eV produces two ionizations corresponding to about one ionization cluster (see section 4.4.5), i.e. one ionizing collision. The ionization potential of a molecule has the order of 10 eV. Hence, the one ionizing collision (primary ionization) must imply a mean energy loss of about $\bar{\varepsilon} = 10\text{ eV} + 33\text{ eV}$ to ensure the *average* production of the one additional, secondary ionization by the δ-electron ejected. The residual 23 eV go into electronic excitations by the primary particle, thus $\tau = 43/23 \approx 2$.[1]) This means that σ^*_{CSD}/σ^* is *never* greater than 3. Note that this maximum error applies only to extreme stoppers, for which the original model restored by $\tau = \infty$ yields infinity. For crossers, however, the error is much smaller even in the original model (Fig. 3.12).

CSD error of the particle fluence spectrum:

For real particles (electrons), corresponding comparisons with strictly calculated spectra are given in section 3.3. It is nevertheless instructive to deal with this question in the framework of the above original model, since we have seen that $\overline{\Delta E}_{CSD} > \overline{\Delta E}$, which would violate energy conservation unless the *number* of absorption events were changed correspondingly. It is the particle fluence spectrum $\varphi(E)$ that determines this number. The solution (3.160) describes a particle that starts from $x = 0$ with initial energy E_0. Consequently,

$$\varphi(E) = \int_0^\infty \mathrm{d}x\, \Phi(E, x) \tag{3.175}$$

is the fluence spectrum per started particle for a homogeneous spatial distribution of sources. Expression (3.160) with the exponential collision spectrum (3.172a) allows Eq. (3.175) to be evaluated in closed form:

$$\varphi(E) = \frac{1}{p\bar{\varepsilon}}\,[1 - (1 - \mathrm{e}^{-aE})^r] + \frac{1}{p}\frac{\mathrm{d}}{\mathrm{d}E}(1 - \mathrm{e}^{-aE})^r. \tag{3.176}$$

For $r \to \infty$ the first term becomes a constant value reaching from $E = 0$ to $E = E_0$, the second a δ-function centered at $E = E_0$.
The CSD approximation for $\varphi(E)$ (cf. section 3.3) simply consists in

$$\varphi(E) = \frac{1}{L(E)} \int_E^\infty \mathrm{d}E'\Phi(E', 0). \tag{3.177}$$

Evaluation with (3.158) and the constant $L = p\bar{\varepsilon}$ yields the first term of the expression (3.176) only. Thus, the CSD approximation proves to diminish the fluence of particles. What is the contribution of the missing term to the total fluence? The answer is obtained by integrating Eq. (3.176) for the limit $r \to \infty$:

$$\int_0^\infty \mathrm{d}E\varphi(E) = \frac{1}{p}\left(\frac{E_0}{\bar{\varepsilon}} + 1\right). \tag{3.178}$$

Hence, the term missing in CSD approximation is smaller than the residuary one by a factor $E_0/\bar{\varepsilon}$. Once more this result indicates that CSD is a very good approximation, save eventually for low-energy δ-electrons with $E_0/\bar{\varepsilon}$ becoming smaller than, e.g., ten. Just for this case the actual deviations can be seen in Table 3.1 (section 3.3).
Finally it should be noted that to use strict fluence spectra would not be compatible with any CSD treatment of straggling for reasons of energy balance.

---

[1]) The reliability of this estimation of τ as well as the entire approach may be confirmed by quoting FANO's (1947) general result that the fluctuations of the *total* number of ionizations associated with a *fixed* energy E_0 imparted on stopping correspond to a variance of about *one half* the value the Poisson distribution would yield. The latter value is $\sigma_p{}^{*2} = 33\text{ eV}/E_0$ (relative variance), while Eq. (3.174) with $\bar{\varepsilon} = 43$ eV and $\tau = 43/23$ gives $\sigma^{*2} = (23/33)^2 \times 33\text{ eV}/E_0$. Thus $\sigma^{*2} \approx 0.5\sigma_p{}^{*2}$ is obtained, indeed.

Chapter 4

The underlying physical relationships

4.1. Introduction

The mathematical formalism of theory, i.e. the lesion production model as well as the theory of cellular radiation effects, is developed in the preceding chapter starting from a number of physical relations assumed to be known. The mere definitions of the quantities in question can be found there, in subsection 3.1.1 (see also Tab. I in the appendix). Now the present chapter is to specify these quantities or relations explicitly.

The subject may also be called physics of radiation absorption dealing with the differential interaction cross-sections of radiations with atoms and molecules, from which all the quantities needed can be derived in principle.[1] However, which should be emphasized from the outset, it is neither the purpose of this chapter to enter into particulars of that field of research nor to review the latest results obtained experimentally and theoretically. The only objective is to manifest the special relations that have actually been chosen to evaluate the final equations summarized in Table I (appendix). Hence, this chapter elucidates the physical background behind all the numerical results presented in this book, including the data tabulated in the appendix. It may be pointed out that the final equations (Tab. I) contain these relations (stopping power, secondary particle spectra, cluster frequencies, etc.) in *general* notation, making it easy to replace them by amended ones if wanted. Indeed, a number of modern data are somewhat different from what we have used (e.g. W-values), but this is rather unimportant regarding the final results, as also supported by the comparisons with measurements and refined calculations presented in chapter 3 and chapter 5.

The physical relations in question rest upon essentially three types of sources. As far as possible, classical concepts (free-electron model, Born approximation) in connection with empirical parameters (effective excitation potential, effective charge) are used to obtain the fundamental quantities (stopping power, secondary electron spectrum). Initial energy spectra

---

[1]) In recent time, this purely theoretical approach has become more and more feasible in practice (Miller and Green, 1973; Paretzke *et al.*, 1974; Paretzke, 1974; Zeiss *et al.*, 1975; Hamm *et al.*, 1978).

of heavy particles set in motion by fast neutrons are taken from current compilations of data on the neutron kerma and differential scattering cross-sections of the most abundant atomic nuclei composing the standard ICRU tissue. A third group of relations is only indirectly based on available data, which then form the input of novel semi-theoretical approaches (primary ionization number and ionization cluster frequencies, mean range and modified mean range of electrons).

The fundamental quantities are usually considered for water as absorber, since this is a sufficient approximation for soft tissue or cellular objects irradiated in culture medium. It should also be noted that the theoretical background of the expressions used and most of the experimental checks refer to the gaseous state, but no substantial differences have proved to exist for condensed matter, as shown experimentally and theoretically (Cole, 1969; Kutcher and Green, 1976; Hamm *et al.*, 1978; Combecher *et al.*, 1978; Wilson *et al.*, 1978).

4.2. Linear energy transfer

A comprehensive survey of the subject can be found in the ICRU Report 16 (1970), and it is this report that all equations given in the sequel without indicating the origin are taken from. Apart from low-energy heavy charged particles (as produced by fast neutrons) and electrons below 134 eV, we use the well-known Bethe formula with the shell correction and the density correction omitted. Furthermore, due to the restricted range of energies needed, relativistic effects can be disregarded for heavy charged particles, but the latter require applying an effective charge in the Bethe formula. This simple approach is confirmed by direct numerical comparison with recommended figures of stopping power.

4.2.1. Heavy charged particles (ions)

A most compact expression for the *stopping power* $L_\infty^{(i)}$ as a function of the ion energy E_i is obtained in terms of the variable E_{max}, the maximum energy of δ-electrons produced, which is proportional to the energy per nucleon, E_i/M:

$$E_{\max} = \frac{4mm_0}{(m+m_0)^2}\,E_i \approx 4\,\frac{m_0}{m}\,E_i = 2.18\times10^{-3}\,\frac{E_i}{M}, \quad (4.1)$$

where m_0 = electron mass, m = mass of the incident particle (ion), M = relative atomic mass of the ion considered ($M = 1$ for protons). The non-relativistic Bethe equation then reads

$$L_\infty^{(i)}(E_i) = \frac{A_0 Z_{eff}^2}{E_{\max}}\ln\left(\frac{E_{\max}}{I}\right)^2, \quad (4.2)$$

which is used in the range of 1 to 20 MeV of energy per nucleon (E_i/M). The parameters for water are

$$A_0 = 8.72\,\frac{\text{keV}^2}{\mu\text{m}}, \quad I = 70\text{ eV}. \quad (4.3)$$

While A_0 is delivered by theory, the *mean excitation potential* I is an empirical parameter for which various figures have been suggested. 70 eV is the value given by Boag (1954), confirmed later on by Zeiss *et al.* (1975). It makes the stopping power for protons ($Z_{eff} = Z = 1$) very well agree with the numerical ICRU-Report-16 data (see Tab. 4.1), although that report recommends $I = 65.1$ eV for water. Certainly the difference just compensates for the shell and density corrections neglected by Eq. (4.2).

The charge correction ($Z_{eff} < Z$) is made for $Z > 1$ in the entire energy range according to the empirical expression given by Barkas (1963) which reads in terms of E_{max}:

$$Z_{eff} = Z\left[1 - \exp\left(-125\sqrt{E_{\max}/1022\text{ keV}}\;Z^{-2/3}\right)\right] \quad \text{for } Z > 1. \quad (4.4)$$

A list of stopping power values thus calculated for nine sorts of ions in the relevant range of energy per nucleon is contained in Table VI of the appendix.

Due to its limited range of validity, the algorithm presented cannot be applied to the neutron-produced heavy charged secondaries (protons; α-particles; carbon, oxygen, and nitrogen nuclei), which slow down to zero energy. Therefore, in all calculations concerning fast neutrons the stopping power has been based completely upon the pertinent formulas and tables given by OLDENBURG and BOOZ (1972a) as follows.

Three energy regions have to be distinguished:

Protons	α-particles	Higher nuclei
Low energy region:		
0 to 5	0 to 50	0 to 40
Medium energy region:		
5 to 250	50 to 250	40 to 250
High energy region:		
above 250	above 250	above 250

These figures are energy per nucleon, E_i/M, in keV.

The stopping power is represented as a sum over the chemical elements the absorber medium consists of:

$$L_\infty^{(i)}(E_i) = \sum_k N_k S_k(E_i), \tag{4.5}$$

where N_k is the number of atoms of the k-th element per gramme of absorber, and $S_k(E_i)$ is the stopping cross-section per atom of the k-th element. In the present calculations the values of N_k have been chosen according to the atomic composition of the simplified ICRU tissue as specified in a later section (see Tab. 4.7). In the following the k-th absorber element is described by its charge number Z_k and relative mass M_k, while Z and M denote the charge and mass, respectively, of the stopped particle considered.

Low energy region. The stopping cross-section is described by LINDHARD's *et al.* (1963) theoretical expression involving interaction with electrons (superscript (e)) and nuclei (superscript (n)):

$$\left.\begin{aligned} \mathrm{f}S_k(E_i) &= S_k^{(e)}(E_i) + S_k^{(n)}(E_i), \\ S_k^{(e)}(E_i) &= 3.845 \times 10^{-18}\,\mathrm{keV\ cm^2}\,\frac{Z^{7/6} Z_k}{(Z^{2/3} + Z_k^{2/3})^{3/2}} \sqrt{\frac{E_i/\mathrm{keV}}{M}}, \\ S_k^{(n)}(E_i) &= 8.477 \times 10^{-18}\,\mathrm{keV\ cm^2}\,\frac{M}{M + M_k}\,\frac{Z Z_k}{(Z^{2/3} + Z_k^{2/3})^{1/2}}\, f(w), \\ w &= 32.53\,\frac{M M_k}{M + M_k}\,\frac{(E_i/\mathrm{keV})/M}{Z Z_k (Z^{2/3} + Z_k^{2/3})^{1/2}}, \end{aligned}\right\} \tag{4.6}$$

The function $f(w)$ given as a graph can be found in the work by OLDENBURG and BOOZ (1972a).

Medium energy region. There exists no theoretical expression. Tables of $S_k(E_i)$ based on measurements can be found in the work by OLDENBURG and BOOZ (1972a) for protons and α-particles[1]), while for heavier recoils nothing but a rough interpolation can be made.

High energy region. The Bethe formula with charge correction (also for protons)

[1]) In effect, one needs only the medium-energy proton values, some of which are contained in Table 4.1, since the usual transformation by Z_{eff}^2 on the basis of equal E_i/M happens to yield the alpha particle values very well in the medium energy region, too, i.e. above 50 keV/nucl.

Tab. 4.1. Stopping power [keV/μm] for protons in water (last column for tissue)

Energy [MeV]	ICRU Report 16	Bichsel (1968)	Bethe formula Eqs. (4.1) to (4.3)	Oldenburg and Booz (1972a) Eqs. (4.5) to (4.8)	
10^{-4}	14.8			15.0	14.8
10^{-3}	22.3			22.4	22.2
10^{-2}	58.7			55.6	54.8
10^{-1}	91.5		(90.9)	96.1	95.8
0.2	75.0		(73.2)	74.0	73.5
0.5	43.7		(43.9)	44.1	43.6
1	26.8	27.1	27.5	27.6	27.3
2	17.0	16.4	16.5	16.6	16.4
4	10.0	9.67	9.65	9.66	9.57
6	7.14	7.04	6.97	6.98	6.91
10	4.68	4.68	4.59	4.59	4.55
30	1.92	1.91	1.82	1.82	1.81

Tab. 4.2. The charge correction for heavy ions

	Experimental ratio of normalized stopping power for ions and protons in aluminium				Z_{eff}^2/Z^2 according to Eq. (4.4)				Recommendation of the ICRU Report 16: no charge correction above
Energy per nucl. MeV]	1	2	5	10	1	2	5	10	
Protons	1	1	1	1	1 (0.994)	1 (0.999)	1 (1.000)	1 (1.000)	1 MeV/nucl.
α-particles	~1	1	1	1	0.948	0.988	0.999	1.000	1 MeV/nucl.
Carbon ions	0.70	0.86			0.682	0.838	0.960	0.992	10 MeV/nucl.
Neon ions	0.53	0.70	0.88	0.95	0.506	0.685	0.880	0.961	10 MeV/nucl.

is applied:

$$S_k(E_i) = 2.375 \times 10^{-16}\,\mathrm{keV^2\,cm^2} \times \frac{Z_{eff}^2 Z_k}{E_i/M} \ln\left(0.002194\,\frac{E_i/M}{I_k}\right), \tag{4.7}$$

$$Z_{eff} = Z\,\mathrm{tgh}\left(0.2516\sqrt{(E_i/\mathrm{keV})/M}\big/Z\right). \tag{4.8}$$

The values of I_k are: 0.0182 keV (hydrogen), 0.0785 keV (carbon), 0.0905 keV (nitrogen), 0.096 keV (oxygen). Although these expressions and parameters differ somewhat from the above ones (Eqs. (4.2) to (4.4)), especially in the charge correction, they have been used up to any high energies in all calculations concerning fast neutrons, for reasons of consistency.

The algorithms presented should be checked with experimental data. For protons, the ICRU Report 16 (1970) recommends a list of stopping power values for water which are based on both various measurements and theory. In Table 4.1, these figures and also those given by Bichsel (1968) are contrasted with the results of Eq. (4.2) as well as the alternative approach for neutron-produced recoil protons according to Oldenburg and Booz (1972a).[1]) The correspondence is close enough for the present calculations.

As to heavier ions, the ICRU Report 16 states that the proton values are also relevant in normalized form, i.e. as a function of energy per nucleon and multiplied by the squared ion's charge corrected for electron capture at

[1]) This work in itself confirms the approach, for heavier recoils also, by reference to experimental results.

smaller energies. Eqs. (4.1) to (4.4) just reflect this recommendation, so that only the special charge correction formula (4.4) needs to be tested. This can be done irrespective of the absorber medium by comparing merely the normalized *ratio* of measured stopping powers for higher ions and protons with Z_{eff}^2/Z^2 according to Eq. (4.4), which allows referring to the experimental results of NORTHCLIFFE (1964) for aluminium also given in the ICRU Report 16. This comparison is presented in Table 4.2 (the figures in parentheses indicate what would turn out if the charge correction were also made for protons). The agreement is seen to be fairly good.

The general concept of LET, in contrast to the *total* stopping power L_∞, involves cut-off of δ-rays leading to the so-called *restricted stopping power* which is designated as L_δ, indicating that energy transfers above some cut-off energy δ are omitted. In the framework of Bethe theory, the corresponding formula analogous to Eq. (4.2) reads:

$$L_\delta^{(i)}(E_i) = \frac{A_0 Z_{eff}^2}{E_{max}} \ln \frac{E_{max} \min(\delta, E_{max})}{I^2}. \tag{4.9}$$

This equation, properly speaking, the difference $L_\infty^{(i)} \dot{-} L_\delta^{(i)}$ (cf. Eq. (3.5)) follows from Rutherford's scattering law (energy distribution of δ-electrons given by $1/E'^2$). By experimental evidence, this law has proved valid also beyond the validity range of Bethe theory, at least for $E' > 200$ eV (ICRU Report 16). In the light of modern data, this point is discussed in section 4.3. Considering that δ does not fall below 200 eV in the framework of the present theory calculations, it seemed to be justified, therefore, to base $L_\delta^{(i)}$ upon that law throughout, i.e. also in the case of slowing-down neutron-produced recoil ions.[1]) This means writing

$$L_\delta^{(i)}(E_i) = L_\infty^{(i)}(E_i) - \frac{A_0 Z_{eff}^2}{E_{max}} \times \ln \frac{E_{max}}{\min(\delta, E_{max})}. \tag{4.10}$$

Hence, this equation constitutes the $L_\delta^{(i)}$ formula applied generally, i.e. irrespective of the algorithm by which $L_\infty^{(i)}$ is obtained.

4.2.2. Electrons

The theoretical background of electron LET is much the same as in the case of heavy charged particles, apart from the necessity to take account of relativity. The energy range needed is 0 to 2 MeV, so that only the energy losses by electron-electron impact play a role. Due to the quantum-mechanical indistinguishability of incident and ejected electrons, the maximum energy of δ-electrons produced is half the incident energy, by definition:

$$E_{max} = \frac{1}{2} E_e. \tag{4.11}$$

The relativistic Bethe formula for the stopping power reads

$$L_\infty^{(e)}(E_e) = \frac{A_0}{2m_0c^2\beta^2} \left[\ln \frac{m_0c^2\beta^2 E_e}{I^2(1-\beta^2)} - \beta^2 \right], \tag{4.12}$$

where

$$\beta^2 = \frac{(E_e + 2m_0c^2) E_e}{(E_e + m_0c^2)^2}, \qquad m_0c^2 = 511 \text{ keV}, \tag{4.13}$$

and the parameters A_0 and I are given again by Eqs. (4.3). Eq. (4.12) is a simplified version of the expression formulated by ROHRLICH and CARLSON (1954). It yields good agreement with experimental data (e.g., COLE, 1969), although the underlying theory is justified only above 10 keV. Table 4.3 presents a numerical comparison with the data recommended

---

[1]) Note that the effect of cut-off in this case usually vanishes anyway or is very small even for $\delta = 200$ eV, since $E_{max} = 200$ eV is reached at 91.9 keV/nucl. of the ions in question (cf. Eq. (4.1)).

in the ICRU Report 16 (1970). Below 10 keV, these data are purely experimental ones. The approximate coincidence breaks down below 0.5 keV. Considering that the experimental results by themselves are uncertain for these very small energies and no theory is available, we have set $L_\infty^{(e)} = \text{const.}$ beyond the maximum of the expression (4.12)[1]:

$$L_\infty^{(e)}(E_e) = 32.415 \frac{\text{keV}}{\mu\text{m}} \tag{4.14}$$

for $E_e < 0.134$ keV.

The *restricted stopping power* obtained by δ-ray cut-off depends on the relativistic secondary electron spectrum. For the latter, the expression derived by Møller (1931) using the free-electron model well applies also beyond its theoretical range of validity (i.e. down to $\delta = 200$ eV and $E_e = 400$ eV), as investigated in section 4.3. By integration (cf. Eq. (3.5)) one obtains

Tab. 4.3. Stopping power [keV/μm] for electrons in water

Energy [keV]	ICRU Report 16	Modified Bethe formula Eqs. (4.12), (4.13), (4.3)
0.02	13.3	
0.03	13.9	
0.05	23.2	
0.1	30.3	
0.2	22.0	30.5
0.3	21.5	26.2
0.5	19.5	20.2
1	13.0	13.1
2	7.75	8.10
3	5.78	6.01
5	3.92	4.08
10	2.32	2.38
20	1.35	1.40
50	0.675	0.687
100	0.420	0.426
200	0.285	0.288
500	0.207	0.207
1000	0.189	0.189
2000	0.189	0.190

$$L_\delta^{(e)}(E_e) = L_\infty^{(e)}(E_e) - \frac{A_0}{2m_0c^2\beta^2}\left[\ln \frac{E_e^2}{4(E_e-\delta)\,\delta} + 2 - \frac{E_e}{E_e-\delta} + \frac{\frac{1}{2}\left(\frac{E_e^2}{4} - \delta^2\right) - (2E_e + m_0c^2)\, m_0c^2 \ln\left(2 - 2\frac{\delta}{E_e}\right)}{(E_e + m_0c^2)^2}\right] \quad \text{for } \delta \leq \frac{E_e}{2} \tag{4.15}$$

in complete agreement, as to the *difference* $L_\infty^{(e)} - L_\delta^{(e)}$, with the formulas by Rohrlich and Carlson (1954) quoted in the ICRU Report 16 (1970).

4.3. Energy spectrum of δ-rays

Scattering theory in the Born approximation yields analytical expressions for the δ-ray spectrum known as the Rutherford scattering law for non-relativistic heavy charged particles and the Møller formula for electrons including exchange and relativity effects. Expressed as differential number of ejected electrons per unit path length of the incident particle (subscripts i and e for ions and electrons, respectively) in water (see the definition of $N(E, E')$ in section 3.1.1), the formulas read

$$N_i(E_i, E') = \frac{A_0 Z_{eff}^2}{E_{\max}} \frac{1}{E'^2}, \tag{4.16}$$

$$N_e(E_e, E') = \frac{A_0}{2m_0c^2\beta^2}\left[\frac{1}{E'^2} + \frac{1}{(E_e - E')^2} - \frac{1}{E'(E_e - E')} + \left(\frac{E_e}{E_e + m_0c^2}\right)^2 \left(\frac{1}{E_e^2} + \frac{1}{E'(E_e - E')}\right)\right]. \tag{4.17}$$

[1]) In fact, for the ultimate calculations this choice has no effect at all. This can be realized from sections 4.4 and 4.5 in connection with the final equations of theory (Tab. I, appendix), from which $L_\infty^{(e)}$ could be completely eliminated for $E_e < 0.15$ keV.

Parameters and abbreviated quantities are defined in the preceding section (Eqs. (4.1), (4.3), (4.4), (4.13)). According to Eq. (3.3), the spectra extend up to $E' = E_{max}$ which is given by Eq. (4.1) and Eq. (4.11) for ions and electrons, respectively.

From the theoretical point of view, these expressions should be valid only for sufficiently high energies of both the incident particle and the ejected electrons.[1]) However, in correspondence with the observations concerning stopping power, the validity range proves to extend to considerably smaller energies. Furthermore, in the present treatise secondary electrons are considered as separate particles only when starting with an energy E' exceeding the particular cut-off energy δ. Since the latter is never lower than $\delta = 200$ eV, the above expressions need to be checked only above, and in the vicinity of, certain lower bounds of energy, namely $E' > 200$ eV, E_i per nucleon > 0.092 MeV, and $E_e > 400$ eV.

To illustrate the justification for using Eq. (4.16) and Eq. (4.17), Figure 4.1 shows a comparison between some experimental data and calculated curves in the crucial range of energies.[2]) The $1/E'^2$ scattering law is seen to be fulfilled very well, even for $E' < 200$ eV. Apart from remaining differences due to the chemical composition of absorbers, there is also fairly good agreement with respect to the *absolute* number of ejected electrons in dependence on the incident particle energy.[3]) Increasing deviations for low-energy protons are visualized in the inlay panel. For stopping protons they will partly cancel out and, in addition, the total δ-ray contribution to the absorbed energy is very small in this case.

A special point is worth discussing, namely the so-called *resonance collisions* (WALSKE, 1952, 1956; RAUTH and SIMPSON, 1964) suggested to cause considerable upward deviations of $d\sigma/dE'$ from the $1/E'^2$ law at $E' < 200$ eV. Nothing like that is seen with the data in Figure 4.1. This depends on the interpretation of E' in Eqs. (4.16), (4.17). In compliance with the utilization of these formulas in the present treatise, we consider E' as the *kinetic energy* of the electron, produced whereas in the case of incident electrons one measures the *energy loss*, which is larger than E' by the ionization potential. The curves in the right-hand panel of Figure 4.1, according to GREEN and SAWADA (1972), do include this correction, i.e. E' means measured energy loss minus first ionization energy of O_2 and N_2, respectively. Hence, the apparent resonance effect seems to be largely a consequence of a tentative interpretation of the $1/E'^2$ law as a description of the frequency of energy losses E'. This does not matter for large E', where the free-electron theory is justified from first principles; but

---

[1]) The underlying free-electron model is justified only for knock-on collisions with E' large as compared with the binding energy of atomic electrons, and the Born approximation used requires that the incident particle velocity be large as compared with the classical orbit velocity of outer-shell electrons.

[2]) The problem involved in such kind of comparison is that the formulas are to be true of water as absorber, whereas different chemical compounds, mostly in gaseous form, have been used in experimental determinations. As a common basis of comparison, adopted in Figure 4.1, it seems to be most suitable to use the differential ionization cross-section per *effective electron*, since there is experimental evidence (WILSON and PARETZKE, 1974; TOBUREN and WILSON, 1977, 1979) that data for different compounds largely coincide when normalized in this way. Effective electrons are assumed to be all outer-shell atomic electrons. Since water of unit density contains 3.35×10^{22} molecules per cm³, each containing 8 effective electrons, the theoretical cross-section $d\sigma/dE'$ to compare with is obtained by dividing the quantity $N(E, E')$ by 2.68 $\times 10^{23}$cm^{-3}.

[3]) Further evidence in both respects, visualized by suitable plots of $d\sigma/dE'$, can be found elsewhere (GLASS *et al.*, 1972; TOBUREN and WILSON, 1977, 1979). However, the complete proton-impact secondary electron spectra (cf. Fig. 4.1) include sharp peaks of Auger electrons ejected after K-shell ionization (peak energy 240 eV, 360 eV, and 450 eV for carbon, nitrogen, and oxygen, respectively). By using Eq. (4.16), Auger electrons are disregarded, which seems to be allowed since at most a few per cent of the total secondary electron energy is due to this effect.

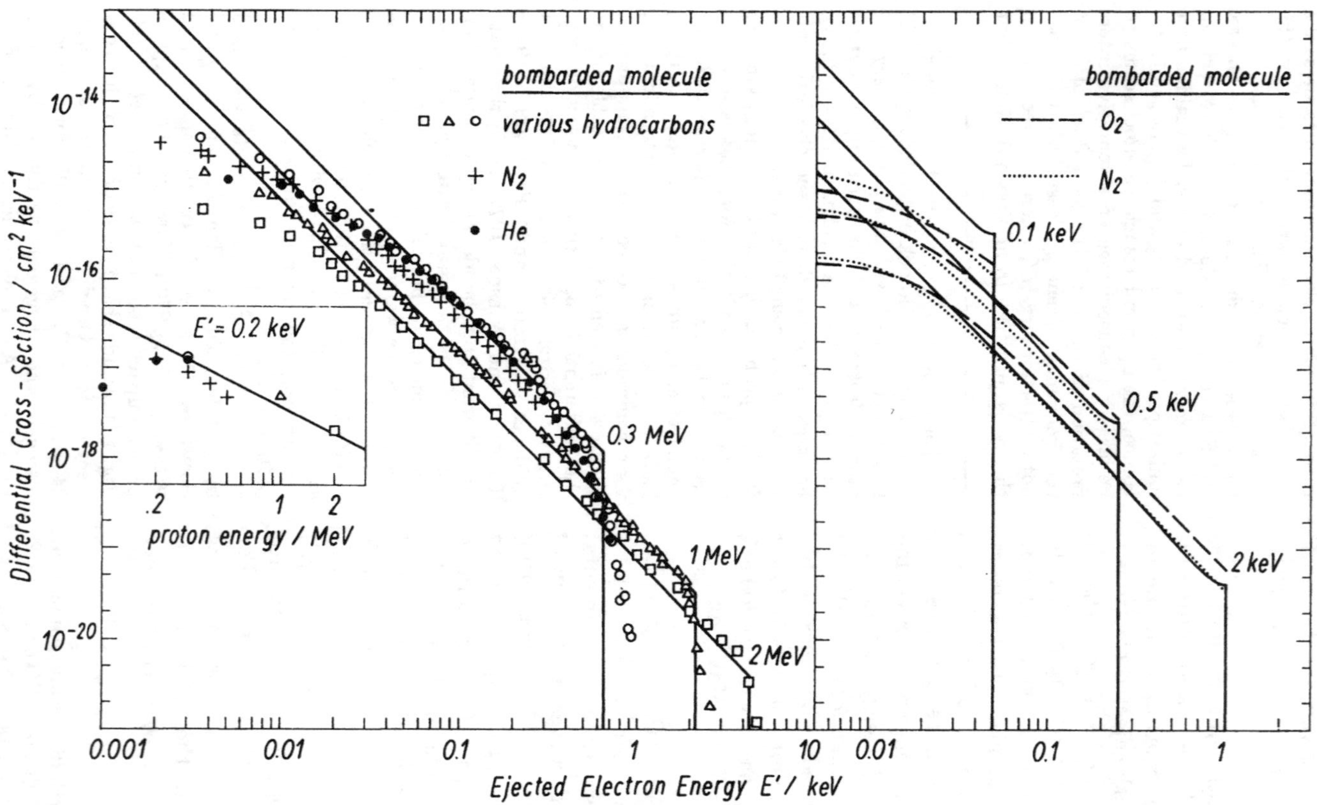

Fig. 4.1. Secondary electron energy distributions (differential cross-section per effective absorber electron) for incident protons (left) and electrons (right). The inset shows the cross-section for $E' = 200$ eV as a function of proton energy. Unbroken lines according to Eq. (4.16) and Eq. (4.17), respectively.
Origin of experimental data: open symbols — Wilson and Paretzke (1974); closed circles — Garcia (1969); crosses — Stolterfoht and Jacobi (1972); broken lines — fit of data measured by Opal *et al.* according to Green and Sawada (1972).

Tab. 4.4. Energy loss cross-section (measurement by RAUTH and SIMPSON, 1964) for 20-keV electrons

Energy loss interval $\Delta E_l'$ [eV]	Cross-section [10^{-20} cm^2] per effective electron		
	measured	calculated $E' = E_l'$	calculated $E' = E_l' - 16$ eV
0–10	2.3 ± 0.8		
10–20	17.2 ± 1.6	26.5	
20–30	21.1 ± 0.8	8.83	94.6
30–40	10.2 ± 0.8	4.41	15.8
40–50	7.0 ± 0.8	2.65	6.49
50–70	9.4 ± 1.6	3.02	5.76
70–100	4.7 ± 1.6	2.26	3.50
100–180	2.5 ± 0.8	2.34	3.06
180–455	1.8 ± 0.2	1.76	2.00
455–590	0.47 ± 0.16	0.260	0.277
590–1700	1.1 ± 0.5	0.559	0.581
1700–10000	0.39	0.220	0.223

without theoretical warranty the theory appears to remain valid down to about $E' = 30$ eV, if it is considered a description of the kinetic energy of the ejected electron rather than the energy loss of the incident one.
This conclusion can be corroborated by the energy loss spectrum for 20-keV electrons measured by RAUTH and SIMPSON (1964) in very thin Formvar foils. Table 4.4 shows this data in comparison with the theoretical values derived from Eq. (4.17), taking E' equal to energy loss (third column) and equal to energy loss minus 16 eV (fourth column). In the former case the resonance effect is clearly visible, while in the latter case the correspondence (in the limits of experimental error) reaches down to smaller energies and is then lost by the theoretical values becoming too large, in complete analogy to Figure 4.1

4.4. Primary ionizations and ionization cluster frequencies

The total number of primary ionizations produced per unit path length of a charged particle cannot be calculated by integration of the spectra (4.16) and (4.17), since the most frequent low-energy-transfer collisions are not covered by these expressions. On the other hand, there are experimental data and sophisticated theoretical approaches, but for the sake of consistency between primary ionizations with allowance for δ-ray cut-off, cluster frequency, $N(E, E')$, LET, and W-values employed in dosimetry we have preferred to develop a conceptually simple semi-theoretical approach, which brings together all these quantities. Its principal empirical component are measured W-values in dependence on the sort and energy of the ionizing particle. This will be dealt with first.

4.4.1. The W-value

An ionizing particle entering an absorber medium with the energy E and being completely stopped in the medium produces some average total number, n, of ionizations. The corresponding *total W-value*, $\overline{W}(E)$, is defined as the energy absorbed per ionization:

$$\overline{W} = \frac{E}{n}. \tag{4.18}$$

It is $\overline{W}$ that can be most easily measured, while the *differential W-value*, $W(E)$, which refers to a small track segment at energy E, follows by reason of $\mathrm{d}n = \mathrm{d}E/W$ to be

$$W = \frac{\overline{W}^2}{\overline{W} - E\dfrac{\mathrm{d}\overline{W}}{\mathrm{d}E}}. \tag{4.19}$$

From an immense number of experimental studies in gases it is well known that $\overline{W}$ increases with decreasing particle energy and depends on the kind of particle as well as the absorber medium. Starting from the existing, often contradictory, data, it is a difficult task to select reliable empirical relations appropriate for the present purposes, for both heavy charged particles and electrons. We have used the following expressions.

Heavy charged particles:

$$\overline{W}_i(E_i) = \overline{W}_i(\infty) + \frac{2.5\ \text{eV}}{\sqrt{E_{\max}/\text{keV}}}, \qquad (4.20)$$

$$\overline{W}_i(\infty) = 35\ \text{eV},$$

where $E_{\max}$ is given by Eq. (4.1).

Electrons:

$$\frac{1}{\overline{W}_e(E_e)} = \begin{cases} 0 & \text{for } E_e < 12\ \text{eV} \\ \dfrac{1}{\overline{W}_e(\infty)} - \left(\dfrac{0.002797\ \text{eV}}{E_e}\right)^{0.4207} \text{eV}^{-1} & \text{for } E_e \geqq 12\ \text{eV}, \end{cases} \qquad (4.21)$$

$$\overline{W}_e(\infty) = 33.7\ \text{eV}.$$

The mathematical form of Eq. (4.20), first suggested by Ishiwari *et al.* (1956), is in common use; the problem are the parameters. Merely for reasons of tradition, we intended to fit the parameters for air. Accordingly, the values of both $\overline{W}_i(\infty)$ and $\overline{W}_e(\infty)$ have been taken from a review by Myers (1968), where also parameters for the energy-dependent term in Eq. (4.20) are given for some heavy ions in argon gas. We found that these parameters and also Jesse's (1961) results for α-particles in nitrogen can be approximately uniformed into the value 2.5 eV after introducing the uniform energy variable $E_{\max}$, which is related to the particle velocity.

An extensive review by Dennis (1972) is in basic agreement with Eq. (4.20); it suggests a somewhat sharper increase of $\overline{W}_i$ at small energies, save for protons, which show almost no variation with energy. Recent studies and reviews of modern data for tissue-equivalent gas by Goodman (1978), Rohrig and Colvett (1978), Varma and Baum (1978), Edwards (1978) confirm the exceptional behaviour of protons, while all heavier ions fit well enough into the formula (4.20) if $\overline{W}_i(\infty)$ is taken to be 29.2 eV, the asymptotic value turning out more or less in common for all ions in tissue-equivalent gas.

Eq. (4.21) for electrons is a rough fit to Cole's (1969) measurement in air. Modern reviews (Smith and Booz, 1978; Combecher, 1978; Waibel and Grosswendt, 1978) show a general tendency to lower $\overline{W}$-values. Apart from $\overline{W}_e(\infty)$ being smaller for tissue-equivalent gas (28.2 eV) and also for water vapour (30 eV), the energy-dependent term in Eq. (4.21) overestimates the increase of $\overline{W}_e$ at decreasing energy, regardless of the absorber medium. Most of modern data show no marked increase down to electron energies far below 1 keV, in contrast to the earlier data of Cole (1969) reflected by Eq. (4.21). As liquid water or a tissue-equivalent material would be most relevant for the present purposes, it is important to note that theoretical calculations (Kutcher and Green, 1976) yield no substantial difference of $\overline{W}_e(E)$ between liquid water and water vapour. Moreover, these theoretical results fall well in line with pertinent experimental data, and a particularly close correspondence seems to exist between liquid water (calculated) and tissue-equivalent gas (measured).

In view of the mentioned shortcomings of the above $\overline{W}$-formulas, which underlie all quantitative results presented in this book, the virtual consequences of accordingly modified formulas should be discussed in brief. First, as to calculated microdosimetric parameters and distributions expressed ultimately in terms of absorbed energy rather than ionization number (chapter 5), the $\overline{W}$-values are of minor importance anyway since, as it were, they only govern the transient cutting of energy into finite pieces (ionizations) for the treatment of straggling. Second, a general modification should consist in relating all $\overline{W}$-values to tissue-equivalent material instead of air. This would be achieved by the common factor 0.835 applied to both $\overline{W}_i(\infty)$ and $\overline{W}_e(\infty)$ (see the figures given above in the text). Such a common factor cannot greatly change the *relative* quality of radiations in theory. Only the additionally necessary lowering of the energy-dependent term particularly for protons, and also for electrons, might eventually lead to non-negligible changes in the results of the lesion production model for these particles in the case of small energy. However, by virtue of the approach to primary ionization number and cluster frequencies dealt with in the sequel, the influence of $\overline{W}(E)$ is very complex. E.g., a too-large W for protons causes a corresponding too-small total number of ionizations per unit path length, but the *primary* number becomes even over-proportionally underestimated due to a simultaneous overestimation of the mean cluster size.

4.4.2. Experimental distribution of ions in clusters

For the present purposes, an ionization cluster is defined as a primary ionization plus all secondary ionizations produced by the electron ejected. This definition includes long δ-tracks, the frequency of which can be reduced to the collision cross-section $N(E, E')$. Small clusters, however, result from small-energy-transfer collisions, which cannot be described by the expressions (4.16), (4.17). Therefore, experimental information (cloud chamber studies) will be used for the frequency of small clusters up to the size of $j = 4$ ionizations.

There are two principal sets of available data, those of Wilson (1923) and the higher-resolution data of Beekman (1949). Although the former are likely to be biased by unresolved cluster overlap (Öre and Larsen, 1964), both sets of data along with their averages are listed in Table 4.5, since below we will find additional theoretical arguments to select the most reliable one. In the following, the respective figures given in Table 4.5 are denoted by $\omega_0(j)$ with $j = 1, 2, 3, 4$. The experimental data refer to electrons of a certain range of energies. In the lack of more information, we have to apply these data to any energy and also to heavy particles, however allowing for a variable normalization factor to be added to $\omega_0(j)$, so that actually the three ratios $\omega_0(2)/\omega_0(1)$, $\omega_0(3)/\omega_0(1)$, $\omega_0(4)/\omega_0(1)$ only are taken for constant. In fact, one can at best assume these ratios to be reliably measured, while the small but important proportion of larger clusters, i.e. $1 - \omega_0(1) - \omega_0(2) - \omega_0(3) - \omega_0(4)$, should depend on the maximum possible energy transfer $E_{\max}$.

Tab. 4.5. Relative frequencies of clusters

Number of positive ions, j, per cluster	$\omega_0(j)$ in per cent		
	Wilson (1923)	Beekman (1949)	Mean value
1	43	61	52
2	22	22	22
3	12	9	10.5
4	10	4	7

4.4.3. The algorithm for heavy particles

The energy-dependent relative cluster frequencies and number of primary ionizations per unit path length of the particle considered (subscript i) are denoted by $\omega_i^{(\infty)}(E_i, j)$ and $p_i^{(\infty)}(E_i)$, respectively. The superscript (∞) indicates that no δ-tracks are omitted.

The first relation between both quantities is obtained by equating the *total* number of ionizations per unit path length to the *primary* number times *mean cluster size:*

$$\frac{L_\infty^{(i)}(E_i)}{W_i(E_i)} = p_i^{(\infty)}(E_i) \sum_{j=1}^{j_{\max}} j\omega_i^{(\infty)}(E_i, j). \quad (4.22)$$

The cluster frequencies are normalized by definition:

$$\sum_{j=1}^{j_{\max}} \omega_i^{(\infty)}(E_i, j) = 1, \quad (4.23)$$

where the uppermost cluster size $j_{\max}$ is simply related to the $E_{\max}$ given by Eq. (4.1), setting

$$j_{\max} = \text{int}\,[1 + E_{\max}/\overline{W}_e(E_{\max})]. \quad (4.24)$$

For the cluster frequencies an ansatz containing adjustable parameters is used as follows. The frequency of small clusters is proportional to the empirical $\omega_0(j)$, while large clusters should follow the δ-electron distribution $N_i(E_i, E')$ with $E' = j\overline{W}_e(E_{max})$ (see Eq. (4.16)). The ansatz

$$\omega_i^{(\infty)}(E_i, j) = \begin{cases} K_1(E_i)\,\omega_0(j) & \text{for } j \leq 4 \\ K_2(E_i)\left[\dfrac{1}{(j+n-1)(j+n)} + \dfrac{a}{j(j+m)(j+2m)}\right] & \text{for } j > 4 \end{cases} \tag{4.25}$$

is meant as an interpolation between the two limiting cases. By virtue of the still unknown parameters n, m, and a, the relation

$$p_i^{(\infty)}(E_i)\,\omega_i^{(\infty)}(E_i, j) = \overline{W}_e(E_{max})\,N_i\big(E_i, j\overline{W}_e(E_{max})\big) = \frac{A_0 Z_{eff}^2}{E_{max}\overline{W}_e(E_{max})}\,\frac{1}{j^2}, \tag{4.26}$$

expected to hold as $j \to j_{max}$, is modified in the medial range of j. Regarding values of j large compared with n and m, the corresponding asymptotic expression (4.25) can be inserted at the left-hand side of Eq. (4.26), to find the condition

$$p_i^{(\infty)}(E_i)\,K_2(E_i) = \frac{A_0 Z_{eff}^2}{E_{max}\overline{W}_e(E_{max})}. \tag{4.27}$$

Another condition follows from the requirement of 'smoothness' at $j = 4$:

$$K_1(E_i)\,\omega_0(4) = K_2(E_i)\left[\frac{1}{(n+3)(n+4)} + \frac{a}{4(m+4)(2m+4)}\right]. \tag{4.28}$$

The two conditions become a linear system of equations for a and the ratio $x = K_1/K_2$ after eliminating $p_i^{(\infty)}(E_i)$ by means of Eq. (4.22) with (4.25) inserted. The solution reads

$$x = \frac{K_1}{K_2} = \frac{\dfrac{E_{max}\overline{W}_e(E_{max})}{A_0 Z_{eff}^2}\,\dfrac{L_\infty^{(i)}(E_i)}{W_i(E_i)} - S_1 + \dfrac{4(m+4)(2m+4)}{(n+3)(n+4)}\,S_2}{S_0 + 4(m+4)(2m+4)\,\omega_0(4)\,S_2}, \tag{4.29}$$

$$a = 4(m+4)(2m+4)\left[x\omega_0(4) - \frac{1}{(n+3)(n+4)}\right], \tag{4.30}$$

where

$$S_0 = \sum_{j=1}^{4} j\omega_0(j),\qquad S_1 = \sum_{j=5}^{j_{max}} \frac{j}{(j+n-1)(j+n)},\qquad S_2 = \sum_{j=5}^{j_{max}} \frac{1}{(j+m)(j+2m)}. \tag{4.31}$$

Now having the ratio K_1/K_2, we can definitively determine the relative cluster frequencies by considering the normalization condition (4.23), yielding

$$K_2 = \left[x\sum_{j=1}^{4}\omega_0(j) + \sum_{j=5}^{j_{max}}\left(\frac{1}{(j+n-1)(j+n)} + \frac{a}{j(j+m)(j+2m)}\right)\right]^{-1},$$

$$K_1 = xK_2. \tag{4.32}$$

Finally, $p_i^{(\infty)}(E_i)$ can be calculated from the total ionization number per unit path length according to Eq. (4.22):

$$p_i^{(\infty)}(E_i) = \frac{L_\infty^{(i)}(E_i)/W_i(E_i)}{K_1 S_0 + K_2(S_1 + aS_2)}. \tag{4.33}$$

Let us consider the corresponding restricted quantities, $\omega_i^{(\delta)}(E_i, j)$ and $p_i^{(\delta)}(E_i)$, obtained by omitting δ-rays with E'

above the cut-off energy δ. The highest cluster size is now given by

$$j_{\max}^{(\delta)} = \text{int}\,[1 + \delta/\overline{W}_e(\delta)], \tag{4.34}$$

and the relative cluster frequencies are obtained simply by re-normalization:

$$\omega_i^{(\delta)}(E_i, j) = \left[\sum_{j'=1}^{j_{\max}^{(\delta)}} \omega_i^{(\infty)}(E_i, j')\right]^{-1} \times \omega_i^{(\infty)}(E_i, j). \tag{4.35}$$

To derive the restricted number of primary ionizations in analogy to Eq. (4.22), we calculate the restricted total number of ionizations by subtracting those ionizations which are produced by the δ-rays omitted. By this rationale, one has

$$p_i^{(\delta)}(E_i) = \frac{L_\infty^{(i)}(E_i)/W_i(E_i) - \int\limits_\delta^{E_{\max}} \mathrm{d}E' \dfrac{E'}{\overline{W}_e(E')} N_i(E_i, E')}{\sum\limits_{j=1}^{j_{\max}^{(\delta)}} j\omega_i^{(\delta)}(E_i, j)}. \tag{4.36}$$

So we have found straightforward algorithms for the quantities in question, but the parameters n and m have as yet been left open. This means that, despite the three conditions to be fulfilled by $\omega_i^{(\infty)}(E_i, j)$, there remains considerable freedom in the medial range of j, i.e. $4 < j \ll j_{\max}$. To make the best choice of n and m, and also $\omega_0(j)$, the following criterion of consistency for the entire procedure can be applied. By adding the number of cut-off δ-rays as given by $N_i(E_i, E')$ to the restricted number of primary ionizations calculated according to Eq. (4.36), one should always obtain $p_i^{(\infty)}(E_i)$ regardless of the value of δ, provided that $\delta > 200$ eV:

$$p_i^{(\delta)}(E_i) + \int\limits_\delta^{E_{\max}} \mathrm{d}E' N_i(E_i, E') \stackrel{!}{=} p_i^{(\infty)}(E_i). \tag{4.37}$$

The results of evaluation of this criterion with respect to both parameter fixation and degree of consistency reached are given in subsection 4.4.5.

4.4.4. The algorithm for electrons

There is no difference in principle with the heavy particle procedure. In compliance with the more complex expression (4.17) for electrons, the ansatz for $\omega^{(\infty)}(E, j)$ has to be modified. It reads

$$\omega_e^{(\infty)}(E_e, j) = \begin{cases} K_1(E_e)\,\omega_0(j) & \text{for } j \leqq 4 \\ K_2(E_e)\left[\dfrac{1}{(j+n-1)(j+n)} + \dfrac{1}{(2j_{\max}-j)^2} - \dfrac{1}{j(2j_{\max}-j)} \right. \\ \left. + \left(\dfrac{E_e}{E_e + m_0c^2}\right)^2 \left(\dfrac{1}{4j_{\max}^2} + \dfrac{1}{j(2j_{\max}-j)}\right) + \dfrac{a}{j(j+m)(j+2m)}\right] & \text{for } j > 4. \end{cases} \tag{4.38}$$

$j_{\max}$ continues to be given by Eq. (4.24) with $E_{\max} = E_e/2$. Thus, the sum S_1 has to be defined by

$$S_1 = \sum_{j=5}^{j_{\max}} \left[\frac{j}{(j+n-1)(j+n)} + \frac{j}{(2j_{\max}-j)^2} - \frac{1}{2j_{\max}-j} + \left(\frac{E_e}{E_e+m_0c^2}\right)^2 \times \left(\frac{j}{4j_{\max}^2} + \frac{1}{2j_{\max}-j}\right)\right], \tag{4.39}$$

in contrast to Eq. (4.31), and in Eq. (4.32) the sum has to be expanded likewise.

By comparison of Eq. (4.17) and Eq. (4.38), the asymptotic condition (4.27) is seen to change into

$$p_e^{(\infty)}(E_e)\, K_2(E_e) = \frac{A_0}{2m_0c^2\beta^2\overline{W}_e(E_{\max})}. \tag{4.40}$$

Considering, in addition, the necessary replacements concerning LET and differential W-value, the set of final equations in the preceding subsection remains valid save for

$$x = \frac{K_1}{K_2} = \frac{\dfrac{2m_0c^2\beta^2\overline{W}_e(E_{\max})}{A_0}\,\dfrac{L_\infty^{(e)}(E_e)}{W_e(E_e)} - S_1 + \dfrac{4(m+4)\,(2m+4)}{(n+3)\,(n+4)}\,S_2}{S_0 + 4(m+4)\,(2m+4)\,\omega_0(4)\,S_2} \tag{4.41}$$

and

$$p_e^{(\infty)}(E_e) = \frac{L_\infty^{(e)}(E_e)/W_e(E_e)}{K_1S_0 + K_2(S_1 + aS_2)}. \tag{4.42}$$

An analogous substitution of quantities has to be made in the expression for $p_i^{(\delta)}(E_i)$.

4.4.5. Results

The check on consistency by means of the criterion (4.37) applied to both heavy particles and electrons has led to the following most appropriate values of the parameters n and m as well as the best choice of small-cluster frequencies out of the three sets given in Table 4.5.

Heavy particles:

$$n = -2, \quad m = -1,$$
$$\omega_0(j) = 0.52,\ 0.22,\ 0.105,\ 0.07. \tag{4.43}$$

Electrons:

$$n = -2, \quad m = -1,$$
$$\omega_0(j) = 0.61,\ 0.22,\ 0.09,\ 0.04. \tag{4.44}$$

Figure 4.2 shows the degree of coincidence reached with these parameters, regarding varying values of δ.[1])

It is quite reasonable that the same n,m-pair of values applies to both kinds of particles, and that the more reliable $\omega_0(j)$ data of Beekman prove to be suitable for electrons, for which they have actually been measured. For heavy particles the present analysis shows that the small-cluster relative frequencies should be different, but no significance can be ascribed to the particular set (4.43). Further investigations, e.g. using modern methods of particle track simulation, will be needed.

A survey of numerical results is given in Table 4.6. It is interesting to note that the mean cluster size $\bar{j}$ only weekly depends on the particle energy, although all δ-rays are included (cf. $j_{\max}$). Only at high energies, $\bar{j}$ comes close to 3, the value assumed by Lea (1946) to hold in general. $K_1(E)$ has been listed in order to demonstrate that its value is always near to unity. This confirms the measured cluster frequencies with respect to the total proportion of small clusters ($1 \leqq j \leqq 4$). According to Beekman's data this proportion is 0.96, and exactly $K_1(E_e) = 1$ is obtained just for the high electron energies used in Beekman's experiment.

The present semi-empirical theory should be regarded as an attempt to go the way back from experimentally well-established 'gross' quantities, such as stopping power and W-value, to the molecular ionization cross-section (primary ionization number) in mutual interconnection with the ionization cluster frequencies. Due to its involved nature, the formalism may be somewhat obscure, but it can be tested on the whole by comparing the calculated $p^{(\infty)}(E)$ with direct measurements of the total ionization cross-section of molecules. The problems involved in such a comparison are exactly those which are discussed in section 4.3. Again we use the ionization cross-section per effective electron (for details see footnote 2 at page 117).

---

[1]) The curves abruptly jump when the maximum cluster size changes by one. This is visible only for small energies corresponding to $j_{\max} = 1, 2$, etc.

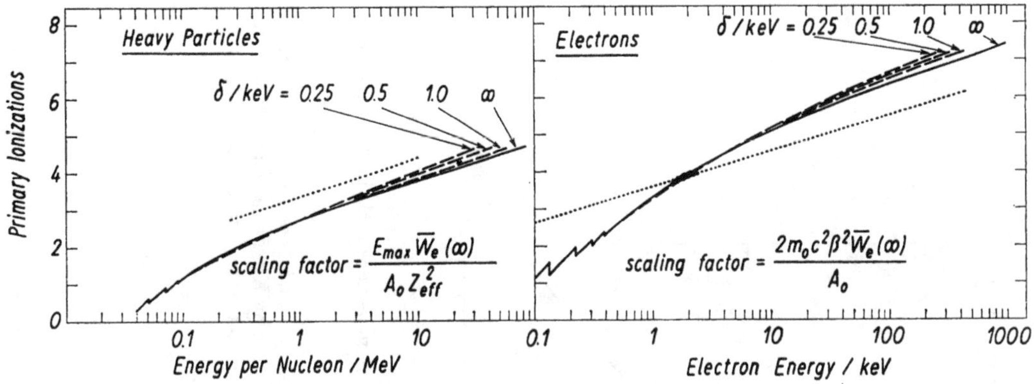

Fig. 4.2. Check of consistency between the restricted (cut-off energy δ) and unrestricted (unbroken line) primary ionization density according to the criterion (4.37). Dotted lines, for comparison, represent the values given by LEA (1946).

Tab. 4.6. Primary ionizations and ionization clusters

Heavy particles					Electrons				
Energy per nucl. [MeV]	Primary ionization density $p_i^{(\infty)}/Z_{eff}^2$ [μm⁻¹]	Mean cluster size $\bar{j}$	Maximum cluster size j_{max}	Normalization factor K_1	Electron energy [keV]	Primary ionization density $p_e^{(\infty)}$ [μm⁻¹]	Mean cluster size $\bar{j}$	Maximum cluster size j_{max}	Normalization factor K_1
0.05	1395	1.297	2	1.351	0.1	733.8	1.000	1	1.639
0.1	1386	1.733	5	1.082	0.2	587.8	1.265	2	1.205
0.2	1038	1.907	11	1.049	0.4	373.9	1.570	5	1.034
0.4	658.9	2.106	22	1.031	0.8	249.7	1.627	10	1.029
0.8	386.2	2.317	46	1.022	1.6	152.3	1.716	20	1.013
1.6	218.8	2.506	94	1.018	3.2	87.76	1.830	42	1.005
3.2	121.2	2.678	193	1.017	12.8	27.42	2.053	177	1.003
6.4	66.32	2.827	393	1.016	51.2	8.761	2.248	730	1.001
12.8	35.96	2.957	796	1.016	204.8	3.450	2.423	2972	1.000
25.6	19.36	3.070	1607	1.015	819.2	2.185	2.592	12004	1.000

Figure 4.3 shows a number of experimental data and a curve calculated for electrons in liquid water by a quasi-first-principles theory (KUTCHER and GREEN, 1976). It is interesting to note that this curve agrees best with the present results (for the leaps in the curve see footnote 1 at page 124), while other absorber materials lead to larger deviations, which could eventually be accounted for by admitting absorber-specific W-values in the present formalism. For electron energies below 100 eV, a too large cross-section turns out due to the overestimation of LET by Eq. (4.14) (see Tab. 4.3). Almost no data seems to exist for protons and heavier ions. What is shown in the left-hand panel originates from a measurement which involved large uncertainties indicated by error bars. The nevertheless significant deviation from our curve, on the one hand, is in accord with the similarly higher values for *nitrogen* bombarded with electrons (right-hand panel); on the other

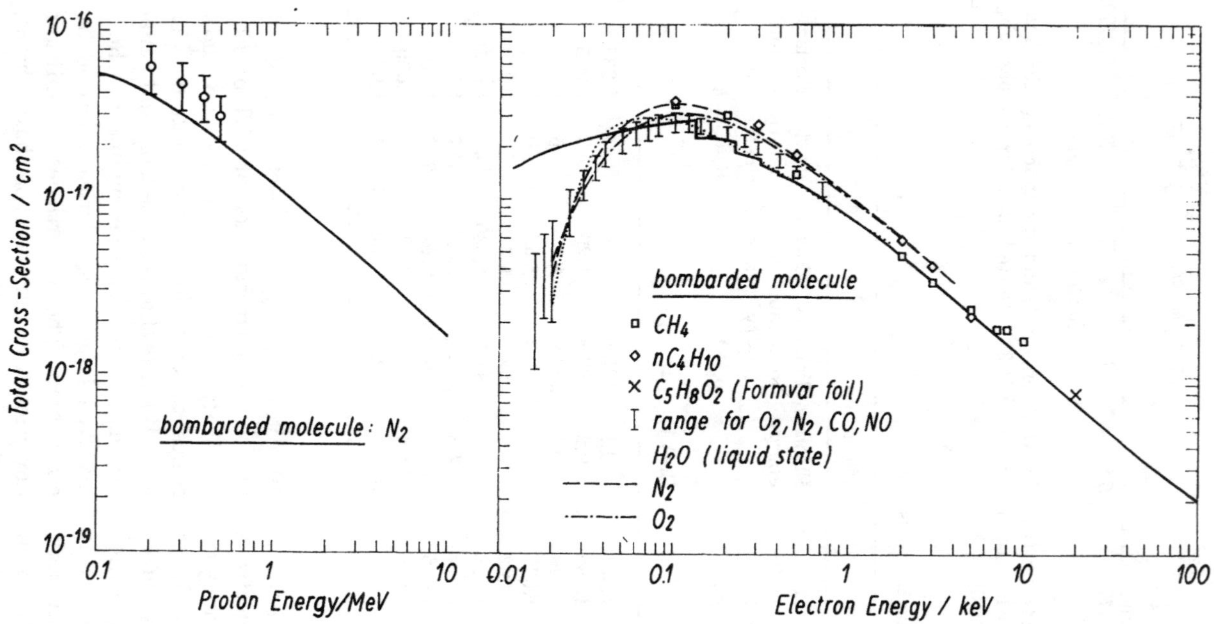

Fig. 4.3. Total ionization cross-section per effective absorber electron for bombardment of some sorts of molecules with protons (left-hand panel) and electrons (right-hand panel).
Unbroken lines according to the present theory.
Origin of experimental data: circles with error bars — STOLTERFOHT and JACOBI (1972); bars — CRAGGS and MASSEY (1959); squares — KEBARLE and GODBOLE (1962); cross — RAUTH and SIMPSON (1964); dotted line — theoretical calculation by KUTCHER and GREEN (1976); dashed and dashed-dotted lines are a fit of data measured by KIEFER and DUNN (1966) according to GREEN and SAWADA (1972).

hand, the theoretical curve may be too low as a consequence of having used overestimated W-values just for low-energy protons (see the very end of subsection 4.4.1).

4.5. Electron range

There are various definitions of range (ICRU Report 16, 1970) which are more or less appropriate from the viewpoint of either experimental determination or the intended application. Another one is added here (cf. Eq. (3.7) and context). This kind of range is correctly called *mean projected range*, described here by $R_e(E)$. It should not be confused with the 50-%-transmission range, for which the same name is in general use, though it ought to be called '*median* projected range'. However, both ranges are likely to be very similar. The main reason for the introduction of the true mean projected range is the possibility to derive it from a surprisingly simple theory, with the same theory being able to yield the so-called modified range, $\tilde{R}_e(E, l)$ (cf. Eq. (3.8) and context). This means consistency between $R_e(E)$ and $\tilde{R}_e(E, l)$, which could not be reached otherwise. Moreover, $R_e(E)$ as defined here seems to be well suited for its only purpose: to estimate cut-off energies according to the rule (3.12).

The new theory of electron range (mean projected range), presented in the following subsection, takes a simple form by introducing the *continuous-slowing-down approximation* (csda) discussed in chapter 3 (see page 70). This may also give rise to confusion, since it is the average path length traversed by the particle that is customarily referred to as the 'csda range', $R_{csda}(E)$, when calculated in continuous-slowing-down approximation (ICRU Report 16, 1970):

$$R_{csda}(E) = \int_0^E \mathrm{d}E' \frac{1}{L_\infty(E')}. \tag{4.45}$$

Clearly, to call this integral 'range' is reasonable only for heavy particles, whose trajectories are practically straight so that the integral (4.45) is a good estimate of *both* mean path length *and* range (cf. Eq. (3.33)). As to electrons, however, it is just the essential feature of the new theory that it is equivalent to Eq. (4.45) as regards path length, but does consider the random *curvature* of the csda path in order to arrive at the *range* of an electron. Hence, the *latter* might just be referred to as a csda range, because it neglects the range straggling that is solely due to path length straggling. To put it another way, csda means replacing the finite energy transfer events by an infinite number of vanishingly small ones, which obviously makes the path length a fixed non-random quantity, whereas the random bending of the path *is* maintained, because the now also infinitely small deflection angles nevertheless produce a finite cumulative effect. This is the essence of our approach.

4.5.1. A theory of mean projected range

Let $g(E, \boldsymbol{e}; \boldsymbol{r} - \boldsymbol{r}_0)$ be the normalized probability density for path endpoints in space (position vector $\boldsymbol{r}$) of particles which started with the energy E from the space point $\boldsymbol{r}_0$ in the direction of the unit vector $\boldsymbol{e}$. The 'centre of gravity' of the distribution, i.e. the expected value of $\boldsymbol{r}$, defines the mean projected range $R(E)$ according to

$$\begin{aligned}\bar{\boldsymbol{r}} &= \iiint \mathrm{d}\tau\, \boldsymbol{r} g(E, \boldsymbol{e}; \boldsymbol{r} - \boldsymbol{r}_0) \\ &= \boldsymbol{r}_0 + \boldsymbol{e}R(E).\end{aligned} \tag{4.46}$$

If $N^*(E, E')$ denotes the spectrum of energy losses E' normalized to yield, on integration, the total number of collisions per unit path length,[1]) and if the de-

[1]) The asterisk is to indicate that this spectrum is not identical with $N(E, E')$ considered in

flection angle associated with a collision is specified by the after-collision flight direction unit vector $\boldsymbol{e}'$ distributed according to $w(E, E'; \boldsymbol{e}')$, the following relation is obvious to hold:

$$g(E, \boldsymbol{e}; \boldsymbol{r}) = \left[1 - \mathrm{d}x \int \mathrm{d}E' N^*(E, E')\right] \times g(E, \boldsymbol{e}; \boldsymbol{r} - \mathrm{d}x\boldsymbol{e}) + \mathrm{d}x \int \mathrm{d}E' N^*(E, E') \times \int \mathrm{d}\boldsymbol{e}' w(E, E'; \boldsymbol{e}') \times g(E - E', \boldsymbol{e}'; \boldsymbol{r}). \tag{4.47}$$

It means that either no or at most one collision occurs along the initial *differential* path segment $\mathrm{d}x$, so that the distribution in question appears as a weighted superposition of the distributions pertinent to the two possibilities. By calculating the expected value of $\boldsymbol{r}$, relation (4.47) becomes a relation for the range, on consideration of Eq. (4.46):

$$\boldsymbol{e}R(E) = \left[1 - \mathrm{d}x \int \mathrm{d}E' N^*(E, E')\right] \times [\boldsymbol{e}\,\mathrm{d}x + \boldsymbol{e}R(E)] + \mathrm{d}x \int \mathrm{d}E'\, N^*(E, E') \times \int \mathrm{d}\boldsymbol{e}' w(E, E'; \boldsymbol{e}') \times \boldsymbol{e}' R(E - E'). \tag{4.48}$$

After multiplication by $\boldsymbol{e}$ (scalar product), the first order terms in $\mathrm{d}x$ establish the following integral equation for $R(E)$:

$$\int \mathrm{d}E' N^*(E, E') \left[R(E) - \cos \bar{\Theta}(E, E') \times R(E - E')\right] = 1, \tag{4.49}$$

where $\bar{\Theta}(E, E')$ defined by

$$\cos \bar{\Theta} = \int \mathrm{d}\boldsymbol{e}' w(E, E'; \boldsymbol{e}')\, (\boldsymbol{e}\boldsymbol{e}') \tag{4.50}$$

is the average deflection angle for given values of E and E'.

The general integral equation (4.49) describes all aspects of range.[1]) For example, if there were no scattering, i.e. $\cos \bar{\Theta} = 1$, the paths would be straight and Eq. (4.49) would account only for path length straggling. If we then want to neglect straggling too, we have to assume E' to be always small, so that $R(E - E')$ can be replaced by $R(E) - E'\,\mathrm{d}R(E)/\mathrm{d}E$, which leads exactly to the csda range (4.45) since Eq. (4.49) then changes into the differential equation

$$L_\infty(E) \frac{\mathrm{d}R}{\mathrm{d}E} = 1, \quad L_\infty(E) = \int \mathrm{d}E\, E' N^*(E, E'). \tag{4.51}$$

The same technique of introducing the continuous-slowing-down approximation, but without neglect of scattering (path bending), will now be applied to electrons. The average deflection angle at given E and E' can be sufficiently approximated by the fixed value following from classical free-electron impact theory. Taking account of relativity, it can be written in the form

$$\cos \bar{\Theta} = \frac{f_e(E - E')}{f_e(E)}, \quad f_e(E) = \sqrt{\frac{E}{2m_0c^2 + E}}, \quad m_0c^2 = 511\,\mathrm{keV}, \tag{4.52}$$

which is particularly beneficial in obtaining the csda solution to Eq. (4.49):

$$\int \mathrm{d}E' N_e^*(E, E') \left[f_e(E)\, R_e(E) - f_e(E - E')\, R_e(E - E')\right] = f_e(E),$$

---

section 4.3, since the latter refers to ionizations only, and E' there means kinetic energy of the ejected electron rather than energy loss of the incident one.

[1]) In principle, Eq. (4.49) also allows for elastic collisions characterized by $E' = 0$ and *finite* deflection angles, but we will neglect them in the sequel. As they are due to collisions with the electron-screened atomic nuclei, the following argument may be put forward. Due to the mass ratio, heavy charged particles undergo practically no scattering by the atomic electrons, but by the atomic nuclei they should be scattered to nearly the same degree as electrons, particularly if light ions (protons, α-particles) in high-atomic-number media are considered. However, as known from cloud chamber and photographic emulsion studies, appreciable deflections are rare events which, therefore, should have no large influence on the mean range.

$$L_\infty^{(e)}(E)\,\frac{\mathrm{d}}{\mathrm{d}E}\,f_e(E)\,R_e(E) = f_e(E),$$

$$R_e(E) = \frac{1}{f_e(E)}\int_0^E \mathrm{d}E'\,\frac{f_e(E')}{L_\infty^{(e)}(E')}\,. \qquad (4.53)$$

This final formula can easily be evaluated by numerical methods. Note that taking $f_e(E) = \text{const.}$ would again yield $R_{csda}(E)$ according to Eq. (4.45), while any monotonously increasing $f_e(E)$ leads to $R_e(E) < R_{csda}(E)$.

For small electron energies, say, below 0.15 keV the free-electron formula (4.52) ceases to hold and also the stopping power becomes uncertain. Therefore, the ultimate algorithm that defines the electron range for the present purposes is based on Eq. (4.53) only for $E > 0.15$ keV, substituting 10^{-3} μm for $R_e(0.15\,\text{keV})$ and using, for simplicity, a power law below 0.15 keV with the requirement of continuity fixing the parameters:

$$R_e(E) = \begin{cases} 2.643\,(E/\text{keV})^{4.15349}\ \mu\text{m} & \text{for}\quad E \leqq 0.15\ \text{keV} \\ \sqrt{1+\dfrac{2m_0c^2}{E}}\left[\displaystyle\int_{0.15\text{keV}}^{E}\frac{\sqrt{E'}\,\mathrm{d}E'}{\sqrt{2m_0c^2+E'}\,L_\infty^{(e)}(E')} + 1.2114\times10^{-5}\ \mu\text{m}\right] & \text{for}\quad E > 0.15\ \text{keV}. \end{cases} \qquad (4.54)$$

Figure 4.4 compares our theoretical $R_e(E)$ (heavy lines) with measurements of the 50 per cent and 10 per cent transmission range (closed and open circles, respectively). Good correspondence is obtained with the former.[1]) This confirms the theory at high energies where there is no influence of the arbitrary choice of 10^{-3} μm for 0.15-keV electrons, which has been just based on these data for the smallest energies covered. The so-called csda range (Eq. (4.45)) as recommended by the ICRU Report 16 (crosses) and as calculated by means of Eq. (4.12) (thin unbroken lines) is also shown. It happens to coincide greatly with the measured 10 per cent transmission range.

The lower panel of Figure 4.4 is to demonstrate that the mean projected range $R_e(E)$ can be approximated very well, over more than three orders of magnitude of E, by the following power law:

$$R_e(E) \approx 0.034625(E/\text{keV})^{1.732}\ \mu\text{m}. \qquad (4.55)$$

This formula is convenient in situations where a rough estimation of range is needed (for an example see section 3.8.1).

4.5.2. Modified stopping power and modified range

The problem to be solved concerns the mean energy loss, ΔE, of electrons on traversing a sensitive volume along a chord of the finite length l'. If l' is very small, ΔE becomes equal to $l'L_\infty^{(e)}(E)$, but this relation cannot be generalized into the implicit equation

$$l' = \int_{E_f}^{E}\mathrm{d}E'\,\frac{1}{L_\infty^{(e)}(E')} = R_{csda}(E) - R_{csda}(E_f),$$

$$\Delta E = E - E_f, \qquad (4.56)$$

since it would underestimate the true ΔE which is increased by trajectory bending. On the other hand, if l' is as large as the residual mean range or exceeds it, $\Delta E = E$ should be a good estimate, but this can also not be generalized by writing

[1]) Of course, one cannot expect exact coincidence. For small energies, in particular, Cole (1969) found nearly exponential transmission curves, so that the mean range should equal the 37 per cent, rather than the 50 per cent, transmission range. Figure 4.4 reflects this qualitatively.

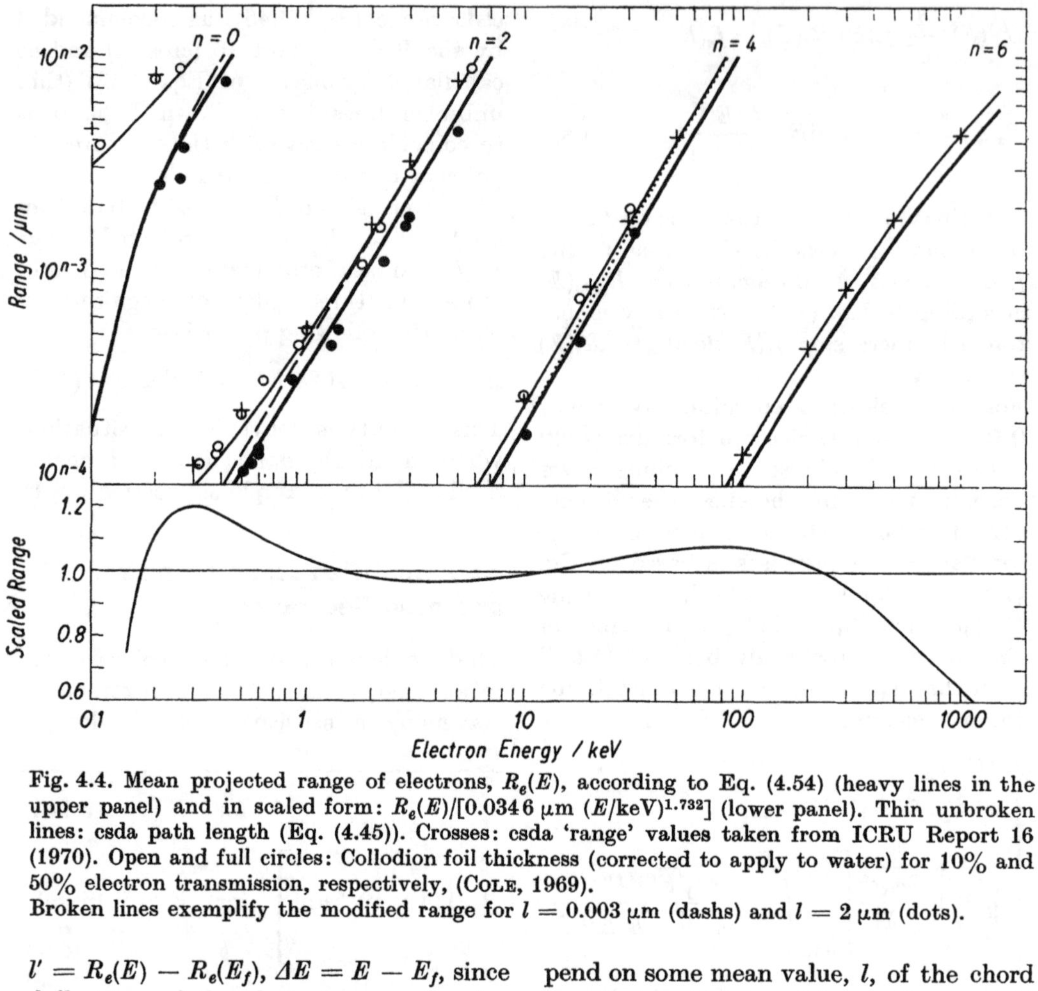

Fig. 4.4. Mean projected range of electrons, $R_e(E)$, according to Eq. (4.54) (heavy lines in the upper panel) and in scaled form: $R_e(E)/[0.0346\ \mu m\ (E/\text{keV})^{1.732}]$ (lower panel). Thin unbroken lines: csda path length (Eq. (4.45)). Crosses: csda 'range' values taken from ICRU Report 16 (1970). Open and full circles: Collodion foil thickness (corrected to apply to water) for 10% and 50% electron transmission, respectively, (COLE, 1969).
Broken lines exemplify the modified range for $l = 0.003\ \mu m$ (dashs) and $l = 2\ \mu m$ (dots).

$l' = R_e(E) - R_e(E_f)$, $\Delta E = E - E_f$, since deflections of the electron occurring between E and E_f have a larger effect on the projected range than on the projected path segment under consideration. This is evidenced by the derivative of $R_e(E)$ as follows from Eq. (4.53):

$$\frac{dR_e}{dE} = \frac{1}{L_\infty^{(e)}(E)} - R_e(E)\,\frac{d\ln f_e(E)}{dE}\,. \qquad (4.57)$$

$(dR_e/dE)^{-1}$ is always larger than $L_\infty^{(e)}(E)$, so that never can $\Delta E = l' L_\infty^{(e)}(E)$, as $l' \to 0$, result from equating the range difference to l'.
Obviously, what we need is a *modified stopping power* which is, in general, intermediate between $L_\infty^{(e)}$ and $(dR_e/dE)^{-1}$. It is denoted by $\tilde{L}_\infty^{(e)}(E, l)$ since it must depend on some mean value, l, of the chord lengths to which it will be applied to yield $\Delta E = E - E_f$ according to

$$l' = \int_{E_f}^{E} dE'\, \frac{1}{\tilde{L}_\infty^{(e)}(E', l)}$$

$$= \tilde{R}_e(E, l)\, E - \tilde{R}_e(E_f, l)\,. \qquad (4.58)$$

The *modified range*, $\tilde{R}_e(E, l)$, is defined only in order to simplify both the notation and the numerical application of the algorithm. While the underlying philosophy is in more detail discussed in chapter 3 (see the context of Eq. (3.40) to (3.42)), the task is now to find a feasable approach to $\tilde{L}_\infty^{(e)}(E, l)$.
To begin with, we must *know* the mean

energy loss that belongs to l. If we accept an approximation expressed by $l = \bar{R}_e(E, E_f)$, where $\bar{R}_e(E, E_f)$ means the mean projected 'range' for the point at which the electron has been slowed down to the energy E_f, the theory in the preceding subsection can be used. In fact, this theory does not imply any actual reference to the path endpoint except for the lower limit of integration in the final equation (4.53). Consequently, taking E_f rather than 0, $\bar{R}_e(E, E_f)$ is obtained. Hence, the relation between l, E and E_f is given by

$$l = \frac{1}{f_e(E)} \int_{E_f}^{E} \mathrm{d}E' \frac{f_e(E')}{L_\infty^{(e)}(E')}. \tag{4.59}$$

Now it is merely a mathematical problem to give the relation (4.59) the form of Eq. (4.58), i.e. to derive $\tilde{L}_\infty^{(e)}(E, l)$ from the requirement that

$$l = \int_{E_f}^{E} \mathrm{d}E' \frac{1}{\tilde{L}_\infty^{(e)}(E', l)} = \tilde{R}_e(E, l) - \tilde{R}_e(E_f, l) \tag{4.60}$$

be equivalent to Eq. (4.59). The following expression, which is derived below, solves the problem in good approximation:

$$\frac{1}{\tilde{L}_\infty^{(e)}(E, l)} = \begin{cases} \dfrac{\mathrm{d}R_e(E)}{\mathrm{d}E} & \text{for} \quad R_e(E) \leq \dfrac{l}{2} \\[2ex] \dfrac{1}{L_\infty^{(e)}(E)} - \dfrac{l}{2} \dfrac{\mathrm{d} \ln f_e(E)}{\mathrm{d}E} & \text{for} \quad R_e(E) > \dfrac{l}{2}. \end{cases} \tag{4.61}$$

Considering Eq. (4.57), it can be seen that this expression is in accord with all the conditions discussed above. Figure 4.4 illustrates what follows from Eq. (4.61) for the modified range $\tilde{R}_e(E, l)$: it equals $R_e(E)$ for small E and approaches $R_{csda}(E)$ for large E, with the transition region depending on l. Thus, Eq. (4.56) is of asymptotic validity for either small l or large E, while otherwise ΔE turns out to be larger owing to $\tilde{L}_\infty^{(e)}(E, l) > L_\infty^{(e)}(E)$.

The derivation of $\tilde{L}_\infty^{(e)}(E, l)$ is based on the easily verifiable statement that a solution does exist. This still unknown solution, in terms of $\tilde{R}_e(E, l)$, is used to transform the integral (4.59) by introducing $R = \tilde{R}_e(E', l)$ as the integration variable in place of E'. Taking account of Eq. (4.60), one obtains

$$l f_e(E) = \int_{\tilde{R}_e(E,l)-l}^{\tilde{R}_e(E,l)} \mathrm{d}R \chi(R) \tag{4.62}$$

with the integrand defined by

$$\chi(\tilde{R}_e(E, l)) = \frac{\tilde{L}_\infty^{(e)}(E, l)}{L_\infty^{(e)}(E)} f_e(E). \tag{4.63}$$

Now we will use an approximation which is warranted by the weak dependence of χ on R. In fact, the quotient $\tilde{L}_\infty^{(e)}/L_\infty^{(e)}$ must be an insignificantly *decreasing* function of E, while $f_e(E)$ is proportional to $E^{0.5}$ which is roughly proportional to $R^{0.289}$ in consideration of Eq. (4.55) and $\tilde{R}_e(E, l) \approx R_e(E)$. This may justify setting

$$l f_e(E) \approx \frac{l}{2} [\chi(\tilde{R}_e(E, l)) + \chi(\tilde{R}_e(E, l) - l)]. \tag{4.64}$$

Another equation is obtained by differentiating Eq. (4.62):

$$l \frac{\mathrm{d}f_e(E)}{\mathrm{d}E} = \frac{1}{\tilde{L}_\infty^{(e)}(E, l)} [\chi(\tilde{R}_e(E, l)) - \chi(\tilde{R}_e(E, l) - l)]. \tag{4.65}$$

The latter two equations manage to eliminate $\chi(\tilde{R}_e(E, l) - l)$, to re-substitute the expression (4.63), and so to arrive at

$$\frac{1}{\tilde{L}_\infty^{(e)}(E, l)} = \frac{1}{L_\infty^{(e)}(E)} - \frac{l}{2} \frac{\mathrm{d} \ln f_e(E)}{\mathrm{d}E}. \tag{4.66}$$

The approximation used becomes poor for small E and, on the other hand, we know from the outset that $\tilde{R}_e(E, l)$ must change into $R_e(E)$ at small E. This is the reason for the settlement (4.61) with the boundary between the two regions chosen to be $l/2$ for the sake of continuity in compliance with Eq. (4.57).[1]) As the

[1]) By contrast, Eq. (4.59) requires that boundary to be l rather than $l/2$ because of $l = R_e(E)$

approach described is no exact solution, its reliability has been examined by numerical tests concerning the required equivalence of the equations (4.59) and (4.60). Quite satisfactory results have been obtained.

4.6. Secondary electron spectrum of photon radiation

The production of energetic electrons by photons (γ-rays, X-rays) is governed by three distinct interaction processes: photoelectric effect, Compton effect, and pair production. The latter process can be neglected for the present purposes, since it requires a photon energy larger than twice the electron rest mass energy, $m_0c^2 = 0.511$ MeV, and its contribution remains smaller than one per cent for photon energies up to 2 MeV in water (EVANS, 1958). Hence, the *initial energy spectrum*, $q_e^{(\gamma)}(E)$ (see section 3.3), of secondary electrons consists of two terms,

$$q_e^{(\gamma)}(E) = q_{e,C}^{(\gamma)}(E) + q_{e,ph}^{(\gamma)}(E), \tag{4.67}$$

where the subscripts C and ph refer to the Compton and photoelectric effects, respectively.

As there is no need for normalizing the spectrum $q_e^{(\gamma)}(E)$ in a definite way (see the final equations in Tab. I, appendix), the following expressions represent the differential interaction cross-section per absorber electron, which is least cumbersome. The incident photon energy is denoted by E_γ. It is expressed also by the dimensionless parameter

$$\alpha = E_\gamma/(m_0c^2), \quad m_0c^2 = 0.511 \text{ MeV}. \tag{4.68}$$

The *Compton effect* (scattering of photons by electrons) can be described very well by the fundamental Klein-Nishina formula (EVANS, 1958):

$$q_{e,C}^{(\gamma)}(E) = \begin{cases} \dfrac{\pi r_0^2}{\alpha E_\gamma}\left[2 + \left(\dfrac{E}{E_\gamma - E}\right)^2 \left(\dfrac{1}{\alpha^2} + \dfrac{E_\gamma - E}{E_\gamma} - \dfrac{2}{\alpha}\,\dfrac{E_\gamma - E}{E}\right)\right] & \text{for} \quad E \leqq E_{\max} \\ 0 & \text{for} \quad E > E_{\max}. \end{cases} \tag{4.69}$$

The so-called Compton edge of the spectrum is given by

$$E_{\max} = \frac{2\alpha}{1 + 2\alpha}\, E_\gamma. \tag{4.70}$$

The *photoelectric effect,* which becomes predominating for small photon energies (below 0.06 MeV in the case of water), produces almost monoenergetic electrons, because the incident photon is completely annihilated and its energy becomes the energy of the recoil electron, save for the ionization potential to be overcome. In principle there is no unique value for the latter, but 0.5 keV is a good general estimate for low-atomic-number media (water), where the K-shell electrons practically alone act as the target. Therefore, and because of $E_\gamma \gg 0.5$ keV, it is felt to be sufficient taking the δ-function $\delta(E - E_\gamma + 0.5 \text{ keV})$ for the description of the photoelectron spectrum. The total collision cross-section per electron, to be added as a E_γ-dependent factor, cannot be obtained accurately enough by theory (EVANS, 1968). Therefore, an empirical expression given by VICTOREEN (1943) has been used. With the parameters given for

for $E_f = 0$, whereby a discontinuity in $\tilde{L}_\infty^{(e)}(E,l)$ would occur, which must be attributed to the approximation involved in the expression (4.66). However, the present approach is preferred primarily because it would be unreasonable to assume that the total energy E of an electron will be always lost within the distance l when its residual *mean* projected range is just equal to l. Thus, this principal shortcoming of the initial ansatz $l = \overline{R}_e(E, E_f)$ happens to be compensated for by the present approach, though in an arbitrary manner.

water it can be written in the form

$$q_{e,ph}^{(\gamma)}(E) = \pi r_0^2 \left(\frac{4.79 \times 10^{-4}}{\alpha^3} - \frac{2.85 \times 10^{-7}}{\alpha^4}\right) \times \delta(E - E_\gamma + 0.5\,\text{keV}). \quad (4.71)$$

4.7. Secondary charged particle spectra of fast neutrons

4.7.1. Principles

Neutrons produce energetic charged particles by interaction with atomic nuclei. In biological tissue, the following types of interaction account for more than 98 per cent of absorbed dose for neutron energies $E_n < 20$ MeV (AUXIER *et al.*, 1968): (i) elastic scattering by hydrogen, carbon, and oxygen; (ii) inelastic scattering by carbon and oxygen; (iii) nuclear reactions that produce α-particles from struck carbon, nitrogen, and oxygen nuclei. Only these processes are considered in the following.

$q_i^{(n)}(E)$, the *initial energy spectra* in question (see section 3.3), refer to the recoil nuclei as well as the α-particles from (n, α) reactions. For the ease of notation, let us stipulate that the subscript i be an index of both the kind of the charged particle and the particular target and reaction type it stems from. $q_i^{(n)}(E)$ is related to the differential cross-section, $\mathrm{d}\sigma_i/\mathrm{d}E$, expressed in terms of E, the energy of the secondary considered. Since the overall normalization of the spectra is unimportant (see the final equations in Tab. I, appendix), one can write

$$q_i^{(n)}(E) = N_i \frac{\mathrm{d}\sigma_i}{\mathrm{d}E}, \quad (4.72)$$

where N_i means the number of atoms of sort i per gramme of absorber (tissue).

If $N_i^{(0)}$ signifies the number of atoms per gramme of the particular element, and if 'tissue' is defined by the *weight fractions* w_i, one has

$$N_i = w_i N_i^{(0)}. \quad (4.73)$$

Tab. 4.7. Composition of wet tissue by weight fractions

	Hydrogen	Carbon	Nitrogen	Oxygen
ICRU-tissue	0.102	0.123	0.035	0.740
SHONKA plastic	0.103	0.761	0.035	0.101

With few exceptions, the present calculations have been performed for the so-called ICRU-tissue (ICRU Report 10b, 1964) with the simplified composition shown in Table 4.7 (oxygen stands for higher elements too). The exceptions are only some results in chapter 5 labelled to be for Shonka plastic with a composition also given in Table 4.7 (according to CASWELL and COYNE, 1974).

The definition (4.72) provides a simple relation between $q_i^{(n)}(E)$ and the energy transfer

$$K = \sum_i N_i \int_{E_{\min}^{(i)}}^{E_{\max}^{(i)}} \mathrm{d}E\, E \frac{\mathrm{d}\sigma_i}{\mathrm{d}E} \quad (4.74)$$

referred to as *kerma:* locally absorbed dose per unit neutron fluence (neutrons per centimetre squared). If K_i denotes the kerma for the sheer element i, the following relations are obvious to hold:

$$K_i = N_i^{(0)} \int_{E_{\min}^{(i)}}^{E_{\max}^{(i)}} \mathrm{d}E\, E \frac{\mathrm{d}\sigma_i}{\mathrm{d}E}, \quad (4.75)$$

$$w_i K_i = \int_{E_{\min}^{(i)}}^{E_{\max}^{(i)}} \mathrm{d}E\, E q_i^{(n)}(E),$$

$$K = \sum_i w_i K_i = \sum_i \int_{E_{\min}^{(i)}}^{E_{\max}^{(i)}} \mathrm{d}E\, E q_i^{(n)}(E). \quad (4.76)$$

Kerma is a well-known quantity in neutron dosimetry, for which extensive numerical sets of data calculated in a uniform way from experimental cross-sections exist (BACH and ÇASWELL, 1968), whereas various sources had to be made use of to

Tab. 4.8. Kerma of divers interactions with four elements, in units of rads per fluence of 10^9 neutrons per cm^2

Neutron energy [MeV]	H elastic scattering	C elastic scattering	C inelastic scattering	C (n, α) reaction	C (n, n3α) reaction	C total	N (n, α) reaction	O elastic scattering	O inelastic scattering	O (n, α) reaction	O total	ICRU-tissue
.25	10.46	.112	0	0	0	.112	0	.062	0	0	.062	1.13
.5	14.74	.188	0	0	0	.188	0	.109	0	0	.109	1.61
1.0	20.38	.278	0	0	0	.278	0	.418	0	0	.418	2.42
1.5	24.50	.330	0	0	0	.330	.025	.192	0	0	.192	2.68
2	27.76	.379	0	0	0	.379	.080	.219	0	0	.219	3.04
3	32.69	.734	0	0	0	.734	.420	.252	0	0	.252	3.63
4	36.33	.858	0	0	0	.858	.970	.389	0	.002	.391	4.13
6	40.90	.773	.095	0	0	.868	.470	.400	0	.044	.444	4.62
8	43.53	.748	.340	.075	0	1.16	.520	.377	.111	.162	.650	5.08
10	45.39	.411	.370	.425	.065	1.27	.675	.356	.152	.508	1.02	5.56
12	46.49	.565	.214	.742	.400	1.92	.665	.302	.269	.670	1.24	5.92
15	47.39	.433	.255	.710	1.24	2.64	.380	.232	.360	1.47	2.06	6.70
20[1]	47.92	.494	.363	.630	1.35	2.83	.217	.331	.445	1.73	2.51	7.10

[1] Extrapolated values

obtain the complete differential cross-sections needed to specify the energy distributions of the charged secondaries. Therefore, in order to avoid inhomogeneities, these cross-sections have been re-normalized so as to yield the 'exact' kerma for each particular element and reaction, i.e. the K_i values by Bach and Caswell (1968) have been required always to turn out in compliance with Eq. (4.75). Hence, what is retained from the data reports on $d\sigma_i/dE$ (see the following subsections) is nothing but the *relative* differential frequency of energies E.

The kerma values utilized for cross-section normalization are listed in Table 4.8 which contains the neutron energies used as standard in the present calculations. These figures are based on graphs of kerma pecentages due to the different interactions occurring with an element, and the listed total kermas for each element, both taken from the work of Bach and Caswell (1968). Except for hydrogen, resonances in the cross-sections cause rapid variations of energy transfer with neutron energy. However, since in practice the energy will hardly be defined so sharp that kerma values for exactly monoenergetic neutrons would be relevant, we applied an additional smoothing procedure, such that the values given in Table 4.8 must be regarded as averaged over about ±7% of neutron energy, save for hydrogen.

4.7.2. Elastic scattering

Differential cross-sections are usually given in terms of $d\sigma/d\Omega$ as a function of the scattering angle Θ in the centre-of-mass system. According to classical impact theory, it is related to the energy E of the recoil nucleus in the laboratory system by

$$E = \frac{1}{2} (1 - \cos \Theta) E_{\max}, \tag{4.77}$$

$$E_{\max} = \frac{4M}{(M+1)^2} E_n, \tag{4.78}$$

where E_n means the neutron energy and M is the mass of the struck nucleus relative to neutron mass. Since the solid-angle differential is given by $d\Omega = 2\pi\, d(\cos\Theta)$, and Eq. (4.77) is a linear relation between E and $\cos\Theta$, one has merely to substitute the expression

$$\cos\Theta = 1 - \frac{2E}{E_{\max}} \tag{4.79}$$

for $\cos\Theta$ in $d\sigma/d\Omega$ to arrive at the energy distribution, being proportional to $d\sigma/dE$. For *hydrogen* and *carbon*, experimental data on $d\sigma/d\Omega$ expressed as a sum of Legendre polynomials (HOPKINS, 1971; KALOS and GOLDSTEIN, 1956) are used in the form of relative coefficients c_L (Tab. 4.9) in the expression

$$\frac{d\sigma}{d\Omega} \sim 1 + \sum_L c_L P_L(\cos\Theta), \tag{4.80}$$

where

$$P_1(x) = x$$

$$P_2(x) = \frac{1}{2}(3x^2 - 1)$$

$$P_3(x) = \frac{1}{2}(5x^3 - 3x)$$

$$P_4(x) = \frac{1}{8}(35x^4 - 30x^2 + 3) \tag{4.81}$$

$$P_5(x) = \frac{1}{8}(63x^5 - 70x^3 + 15x).$$

Note that the kerma according to Eq. (4.75) is proportional to $(1 - c_1/3)$, which makes it easy to find the normalization constant for each neutron energy in the way described in the preceding subsection.

To estimate the values of c_L for the special neutron energies listed in Table 4.9, interpolation has been avoided, preferring to take the coefficients for a proximate energy found in the original data set. This is justified since the character of the recoil energy distribution does not abruptly change with neutron energy,[1] as do the integral cross-section and kerma due to resonances. Anway, for hydrogen there is no substantial departure from isotropic scattering (described by $c_L = 0$) up to $E_n = 20$ MeV.

For *oxygen*, only a data set (BAZAZIANC *et al.*, 1972) in terms of the scattering angle, Θ^*, in the *laboratory system* was available to us. The corresponding coefficients in the analogous expression

$$\frac{d\sigma}{d\Omega^*} \sim 1 + \sum_L c_L^* P_L(\cos\Theta^*) \tag{4.82}$$

are quoted in Table 4.9, too. However, an unwieldy relation now holds for the transformation into $d\sigma/dE$. Simple scattering theory yields

$$\cos\Theta^* = \frac{2 - (M + 1)\, E/E_n}{2(1 - E/E_n)^{1/2}}, \tag{4.83}$$

so that

$$\frac{d\sigma}{dE} = \frac{d\Omega^*}{dE}\frac{d\sigma}{d\Omega^*}$$

$$\sim \frac{2M - (M + 1)\, E/E_n}{(1 - E/E_n)^{3/2}}\frac{d\sigma}{d\Omega^*}. \tag{4.84}$$

As a consequence, the kerma depends on all terms in Eq. (4.82) and its calculation is lengthy, though it can still be expressed in analytical form.

4.7.3. Inelastic scattering

In this case the recoil nucleus is left in an excited state. According to elementary collision theory, its kinetic energy E is limited within both upper and lower bounds given by

$$E_{\max} = \frac{2M}{(M + 1)^2} E_n \times \left(1 + \sqrt{1 - \frac{M+1}{M}\frac{E^*}{E_n}}\right) - \frac{E^*}{M + 1},$$

$$E_{\min} = \frac{2M}{(M + 1)^2} E_n \times \left(1 - \sqrt{1 - \frac{M+1}{M}\frac{E^*}{E_n}}\right) - \frac{E^*}{M + 1}, \tag{4.85}$$

[1] As an argument it may be mentioned that BACH and CASWELL (1968) have shown that the mean energy transferred per collision is a rather smooth function of neutron energy.

Tab. 4.9. Relative coefficients of Legendre polynomials to describe the angular distribution of elastic neutron scattering

Neutron energy	Hydrogen				Carbon					Oxygen			
[MeV]	c_1	c_2	c_3	c_4	c_1	c_2	c_3	c_4	c_5	c_1^*	c_2^*	c_3^*	c_4^*
0.25	−.001	.000	.000	−.000	.090	.002	0	0	0	−.121	−.090	0	0
0.5	−.002	−.000	.000	.000	.131	.005	0	0	0	.991	.331	.020	−.049
1.0	−.004	−.000	−.000	.000	.176	.025	−.001	−.001	0	.230	1.40	.209	−.020
1.5	−.004	−.000	−.000	.000	.142	.072	−.012	−.004	0	.180	.180	.110	.117
2	−.008	−.000	−.000	.000	−.044	.215	−.067	−.019	0	.434	.690	.190	.041
3	−.011	−.000	−.000	.000	.135	1.31	.070	−.018	−.100	.446	.479	.193	.079
4	−.014	−.000	−.000	.000	.049	1.61	.285	.294	.055	1.34	.849	0	0
6	−.019	.000	−.001	.000	.483	.690	.256	.104	.031	.867	.771	1.20	.554
8	−.023	.001	−.001	.000	1.05	.995	.700	.281	.100	.479	.461	.985	.901
10	−.028	.004	−.002	.000	1.49	1.73	1.47	.768	.350	.869	.850	.760	1.40
12	−.032	.008	−.003	.001	1.96	2.40	2.11	1.34	0	1.50	1.68	1.40	2.00
15	−.036	.013	−.005	.002	2.28	2.81	2.49	1.78	0	1.86	2.09	1.99	2.39
20	−.042	.035	−.009	.007	2.28	2.81	2.49	1.78	0	1.86	2.09	1.99	2.39

where E^* is the excitation energy. Since differential cross-section data are scarce, it seems to be the best compromise supposing the energy spectrum to be constant between $E_{\min}$ and $E_{\max}$[1]):

$$q_i^{(n)}(E) = \begin{cases} \text{const.} & \text{for } E_{\min} < E < E_{\max} \\ 0 & \text{elsewhere.} \end{cases} \tag{4.86}$$

The constant follows from the condition of kerma normalization (Eq. (4.75) and Tab. 4.8).

As to the excitation energy E^* for carbon and oxygen, there are usually a number of values associated with different excitation levels, which ought to be treated separately, according to the pertinent partial cross-sections. Because of lacking data, however, we follow the approach by Bach and Caswell (1968) to replace E^* by a *mean excitation energy* $\overline{E^*}$ which will depend on the incident neutron energy. These empirical relations read as follows.

Carbon:

$$\overline{E^*} = \begin{cases} 4.43\ \text{MeV} & \text{for } E_n \leqq 10\ \text{MeV} \\ 0.6125 E_n - 1.695\ \text{MeV} & \text{for } 10\ \text{MeV} < E_n < 20\ \text{MeV}. \end{cases} \tag{4.87}$$

Oxygen:

$$\overline{E^*} = \begin{cases} 6.307\ \text{MeV} & \text{for } E_n \leqq 9.09\ \text{MeV} \\ 0.344 E_n - 3.18\ \text{MeV} & \text{for } 9.09\ \text{MeV} < E_n < 20\ \text{MeV}. \end{cases} \tag{4.88}$$

For the reaction to take place at all, the neutron energy must be larger than a value slightly in excess of E^* for the lowest excited state (4.43 MeV and

[1]) This assumption implies isotropic scattering in the centre-of-mass system, because E is linearly related to the cosine of the scattering angle Θ. To judge by results available for 14 MeV (Caswell and Coyne, 1972), isotropy seems to be at least a better assumption than taking, e.g., the angular distribution of the corresponding elastic scattering. Thus, moreover, our approach is consistent with the kerma as calculated by Bach and Caswell (1968), who assumed forward-backward symmetry of scattering in the centre-of-mass co-ordinates (note that the kerma does not depend on complete isotropy).

6.14 MeV[1]) for carbon and oxygen, respectively). This threshold energy, as is evident from Eqs. (4.85), reads exactly

$$E_{n,\min} = \frac{M+1}{M} E^*, \tag{4.89}$$

leading to the thresholds 4.80 MeV and 6.52 MeV of energy for a neutron to be scattered inelastically by carbon and oxygen nuclei, respectively (cf. Tab. 4.8).

4.7.4. Nuclear reactions

Of the various reactions occurring in tissue the (n, p) reaction of oxygen can be neglected due to its low probability, while in the case of nitrogen it is the low abundance in tissue that allows neglect of energy transfer by scattering as well as (n, p) and (n, d) reactions, thus retaining only the most important kind of energy transfer to nitrogen, namely the (n, α) reaction. Hence, only the nuclear reactions of carbon, nitrogen, and oxygen that produce α-particles are considered in the following.

The total kinetic energy of both the α-particle (E_α) and the recoil nucleus (E_r) arising from an (n, α) reaction is equal to the incident neutron energy E_n plus the reaction energy Q minus the excitation energy E^* to the recoil nucleus. The solution to the corresponding scattering problem reads

$$\left.\begin{aligned} E_\alpha &= \frac{4}{(M+1)^2} E_n + X + \frac{4}{M+1} \sqrt{E_n X} \cos \Theta, \\ E_r &= E_0 - E_\alpha, \\ X &= \frac{M-3}{M+1} \left(\frac{M}{M+1} E_n + Q - E^* \right), \quad E_0 = E_n + Q - E^*, \end{aligned}\right\} \tag{4.90}$$

where Θ is the 'scattering' angle of the α-particle in the centre-of-mass system.

For theoretical reasons (intermediate formation of a compound nucleus which is likely to lose the 'memory' of the neutron incidence direction), the distribution of Θ should not be far from isotropic. Isotropy implies that the differential distribution of both energies, E_α and E_r, is constant within the respective limits. However, for reasons explained below, only three discrete pairs of energy E_α, E_r belonging to the angles $\Theta_1 = 30°$, $\Theta_2 = 90°$, $\Theta_3 = 150°$ are used to form approximate energy spectra consisting of three weighted[2]) δ-functions:

$$q_\alpha^{(n)}(E) \sim \frac{1}{4} \delta(E - E_{\alpha,1}) + \frac{1}{2} \delta(E - E_{\alpha,2}) + \frac{1}{4} \delta(E - E_{\alpha,3}), \tag{4.91}$$

$$q_r^{(n)}(E) = q_\alpha^{(n)}(E_0 - E). \tag{4.92}$$

The normalization by the hand of kerma according to Eq. (4.75) is obvious: $(w_i K_i)/(E_n + Q - E^*)$ is the factor that must be added to the expression (4.91).

The necessary data for the three (n, α) reactions in question are given in Table 4.10. As in the case of inelastic scattering, E^* is replaced by an average $\overline{E^*}$ to take account of energy level multiplicity. Table 4.10 follows suggestions made by Bach and Caswell (1968) and, in addition, uses data given by Randolph (1957), Gabbard *et al.* (1959), Kunz and Schintlmeister (1965), and Caswell and Coyne (1972). To simplify the further calculations, the sparse boron nuclei arising from reacting nitrogen have been replaced in the actual calculations by ^{9}Be and ^{13}C in fifty-fifty ratio.

---

[1]) 6.307 MeV in Eq. (4.88) is an approximation valid *on an average* in the neutron energy range below 9 MeV.

[2]) The weights follow from solid-angle integration (differential $d\Omega = 2\pi \sin \Theta \, d\Theta$) over the three Θ intervals whose medians are Θ_1, Θ_2, Θ_3.

Tab. 4.10. Parameters of the (n, α) reaction and the (n, n3α) reaction

Target nucleus	Reaction products	Reaction energy Q	Excitation energy of the recoil nucleus	
^{12}C	^4He, ^9Be	−5.71 MeV	$E^* = 0$	for $E_n \leqq 8.1$ MeV
			$\overline{E^*} = 0.285E_n - 2.308$ MeV	for $E_n > 8.1$ MeV
^{14}N	^4He, ^{11}B	−0.15 MeV	$E^* = 0$	for $E_n \leqq 4.23$ MeV
			$\overline{E^*} = 0.634E_n - 2.682$ MeV	for 4.23 MeV $< E_n \leqq 7.70$ MeV
			$\overline{E^*} = 0.706E_n - 3.236$ MeV	for $E_n > 7.70$ MeV
^{16}O	^4He, ^{13}C	−2.214 MeV	$E^* = 0$	for $E_n \leqq 5$ MeV
			$\overline{E^*} = 0.522E_n - 2.610$ MeV	for $E_n > 5$ MeV
^{12}C	$3 \times {}^4He$, n	−7.275 MeV	$\overline{E_n'} = 0.267 \sqrt{E_n}$ MeV$^{1/2}$ — kinetic energy of n	

It is a peculiarity of interaction by nuclear reactions that the charged particles produced are *correlated*, which is not exhibited by the initial energy spectra $q_\alpha^{(n)}(E)$ and $q_r^{(n)}(E)$. The general treatment of absorption events in chapter 3 does not include the case of correlated particles, but a generalization in this respect is possible and has been implemented, since the correlation effect seems not to be quite unimportant just for volumes of cell nucleus dimensions, which are comparable to, or even larger than, the recoil particle ranges. A sketchy description of the approach, which requires going back to the fundamentals of chapter 3, section 3.4, may be given as follows.

In the laboratory system, too, the flight directions of the two particles are opposite for the most part. Therefore, as an approximation, the two tracks are supposed to form a common straight line. Now the general physical quantity $\Psi_i^{(n)}(E, E^{(f)})$ for $i = (\alpha, r)$ becomes a function of E_α, $E_\alpha^{(f)}$ as well as E_r, $E_r^{(f)}$ if parts of both tracks lie in the SV. It is easy to see from the form of the actual quantities in question, (3.99) and (3.64)[1]), and their derivation that

$$\Psi_{\alpha,r}(E_\alpha, E_\alpha^{(f)}; E_r, E_r^{(f)}) = \Psi_\alpha + \Psi_r - \Psi_\alpha \Psi_r. \tag{4.93}$$

The starting Eq. (3.34) for the expected value changes into

$$[\Psi_{\alpha,r}] = \sigma_0 \int_0^\infty dE \varphi_\alpha^{(n)}(E) \int_0^\infty dl' w(l') \times \Psi_\alpha(E, E_\alpha^{(f)}(E, l')) + \sigma_0 \int_0^\infty dE \varphi_r^{(n)}(E) \int_0^\infty dl' w(l') \times \Psi_r(E, E_r^{(f)}(E, l'))$$
$$+ \varrho V \int_0^\infty dE q_\alpha^{(n)}(E) \int_0^\infty dl' w(l') \times \frac{1}{l} \int_0^{l'} dl'' \Psi_{\alpha,r}(E, E_\alpha^{(f)}(E, l''); E_0 - E, E_r^{(f)}(E_0 - E, l' - l'')), \tag{4.94}$$

where reference is made to the considerations underlying Eq. (3.28). Substitution according to (4.93) and evaluation along the lines of section 3.4, in particular using the relations (3.35a) and (3.32), reveals that $[\Psi_{\alpha,r}]$ becomes the sum of the usual expressions, $[\Psi_\alpha] + [\Psi_r]$, as given by Eq. (3.34) and, *in addition*, a negative term $[\Psi_{\alpha,r}]_{corr}$ which quantifies the effect of α-r-correlation. This term contains the l''-integral in the form

$$\int_0^{l'} dl'' \Psi_\alpha(E, E_\alpha^{(f)}(E, l'')) \times \frac{d}{dl'} \Psi_r(E_0 - E, E_r^{(f)}(E_0 - E, l' - l'')),$$

which can be transformed by $d/dl' = -d/dl''$, partial integration, and introducing the integration variable $E' = E_\alpha^{(f)}(E, l'')$. This ends up in

$$[\Psi_{\alpha,r}]_{corr} = \varrho V \int_0^\infty dE q_\alpha^{(n)}(E) \int_0^\infty dl' u(l') \int_{E_\alpha^{(f)}(E,l')}^{E} dE' \times \frac{\partial \Psi_\alpha(E, E')}{\partial E'} \Psi_r(E_0 - E, E_r^{(f)}(E_0 - E, l' - R_\alpha(E) + R_\alpha(E'))). \tag{4.95}$$

[1]) Identifying $(1 - \psi)$ with the general Ψ.

Now it is easy to give this correction term the explicit form in which it has to be included in the final expressions (3.105) and (3.71), respectively.
Because (4.95) involves one integration more than usual (cf. Eq. (3.39)), it has been decided, for the sake of simplicity, to reduce the initial energy spectrum to the three δ-functions (4.91).

Another reaction of carbon nuclei is the break-up into three α-particles.[1]) The energy transferred to the α-particles reads $E_n + Q - E_n'$, where the residual energy, E_n', of the secondary neutron can be estimated as an average (see Tab. 4.10). The *distribution* of the alpha energy, however, is very difficult to obtain. There are various reaction channels (modes of interaction) which would have to be analysed and weighted by the corresponding partial cross-sections. An example was given by CASWELL and COYNE (1972), and the shape of the distribution they found for 14-MeV neutrons suggests assuming tentatively, as a rough approximation,

$$q_{3\alpha}^{(n)}(E) \sim e^{-(\ln 20)E/E_{\max}} \quad \text{for} \quad E \leqq E_{\max}. \tag{4.96}$$

This ansatz implies the relation $\bar{E} = 0.281 E_{\max}$ for the mean energy of an α-particle, and according to the above-mentioned estimation of $\bar{E}$ (BACH and CASWELL, 1968) we thus arrive at

$$E_{\max} = 1.186(E_n + Q - 0.267\sqrt{E_n}\,\text{MeV}^{1/2}). \tag{4.97}$$

For 14 MeV, this edge of the spectrum *is* in accord with the 'experimental' one, as well as the slope $(\ln 20)/E_{\max}$. The crucial test, however, is the comparison with the *theoretical* upper limit of alpha energy, which is known for *any* incident neutron energy and can be obtained, evidently, by Eq. (4.90) with $E^* = 0$ and $\cos\Theta = 1$. As easily verified, the correspondence of both expressions for $E_{\max}$ is indeed close enough to take the exponential distribution (4.96) with the *constant* parameter ln 20 for a good description of the alpha energy spectrum at *any* incident neutron energy above the threshold $-Q(M+1)/M = 7.88$ MeV.

---

[1]) The correlation of the three particles will be neglected because of mathematical complication, and can be neglected because of the relatively long average particle range.

Chapter 5

Implications concerning microdosimetry

5.1. Introduction

The mathematical formalism developed in chapter 3 is a rather general tool for the calculation of mean values as well as distributions of various random quantities associated with the stochastics of radiation absorption. This general approach is there evaluated in detail only with respect to the model of DNA-lesion production and the theory of cellular radiation effects (statistics of DNA-lesions). In this chapter, the same formalism will be used to obtain the customary microdosimetric parameters, i.e. the single event size spectrum $f_1(z)$ and its mean values $\bar{z}_F$, $\bar{z}_D$, and others surveyed in section 1.4.1. The data thus obtained, which are extensively tabulated in the appendix, are important for three reasons: to test the approximations involved in the general formalism (see section 3.2) by comparison with other computational and direct experimental results, to extend the availability of numerical data in microdosimetry, and to evaluate the approximate (microdosimetric) version of the radiation action theory as established in chapter 2.

The general basis for the following can be given at once without a lengthy derivation, since the model of DNA-lesion production dealt with in section 3.5 rests solely on $f_1(j)$, the single event size spectrum in terms of the number, j, of ionizations in the sensitive volume. Look only at the initial equation (3.1) from a formal point of view to realize that, taking[1])

$$b_1 = 1,\ b_2 = -1,\quad \zeta_1 = 0,\ \zeta_2 = \zeta, \tag{5.1}$$

one can define a function

$$\Phi(\zeta) = \frac{1}{\bar{z}_F} \sum_{j=0}^{\infty} (1 - \mathrm{e}^{-\zeta j})\, f_1(j) \tag{5.2}$$

which is identical with the quantity s for the parameters (5.1). Hence, the *fundamental function* $\Phi(\zeta)$ can be calculated by means of the explicit final formulas (3.67 to 71) for s, substituting the parameters (5.1) and re-replacing $\frac{\pi}{4}\, l_l$ by the cross-section, σ_0, of the now arbitrary sensitive volume (see Eq. (3.66) and context). This yields for the three radiations i, n, and γ:

$$\Phi_i(\zeta) = \Phi_i^{(i)}(\zeta) + \Phi_i^{(\delta)}(\zeta), \tag{5.3}$$

---

[1]) Note condition (3.55b).

$$\Phi_i^{(i)}(\zeta) = \frac{\sigma_0 \text{ g/cm}^3}{L_\infty^{(i)}(E_i^{(0)})} \int_0^\infty \mathrm{d}l' w(l') \{1 - \exp[-\varrho l' \varkappa_i(E_i^{(0)}, \zeta)]\}, \quad (5.3\text{a})$$

$$\Phi_i^{(\delta)}(\zeta) = \frac{V\varrho \text{ g/cm}^3}{L_\infty^{(i)}(E_i^{(0)})} \times \int_0^\infty \mathrm{d}E T_i^{(\delta)}(E_i^{(0)}, E) \hat{\Gamma}_e(E), \quad (5.3\text{b})$$

$$\Phi_n(\zeta) = \frac{V\varrho \text{ g/cm}^3}{\sum_i \int_0^\infty \mathrm{d}E E q_i^{(n)}(E)} \times \left[\sum_i \int_0^\infty \mathrm{d}E \varphi_i^{(n)}(E) \hat{\Gamma}_i(E) + \int_0^\infty \mathrm{d}E \varphi_n^{(\delta)}(E) \hat{\Gamma}_e(E)\right], \quad (5.4)$$

$$\Phi_\gamma(\zeta) = \frac{V\varrho \text{ g/cm}^3}{\int_0^\infty \mathrm{d}E E q_e^{(\gamma)}(E)} \times \int_0^\infty \mathrm{d}E \varphi_\gamma^{(\delta)}(E) \hat{\Gamma}_e(E), \quad (5.5)$$

where

$$\hat{\Gamma}(E) = \int_0^\infty \mathrm{d}l' u(l') \varkappa(E, \zeta) \times \exp\left[-\int_{E^{(f)}}^{E} \mathrm{d}E' \frac{\varkappa(E', \zeta)}{L_\infty(E')}\right], \quad (5.6)$$

$$\varkappa(E, \zeta) = p^{(\delta)}(E) \left[1 - \sum_{j=1}^{j_{\max}} \mathrm{e}^{-\zeta j} \omega^{(\delta)}(E, j)\right]. \quad (5.7)$$

Numerical values of Φ are obtained by means of the same computer programmes that are destined for the model of DNA-lesion production,[1]) and the microdosimetric quantities in question are then derived from $\Phi(\zeta)$ by numerical methods.

The relationships between $\Phi(\zeta)$ and the microdosimetric parameters are the topic of the following section. Let us consider here only a general aspect of this question, namely the relation between ionizations and energy. This is a conceptual problem of microdosimetry. Nominally, microdosimetry is concerned with *energy* but measures *ionization numbers* when a proportional gas counter is used. The transformation is performed using the well-known W-value, which is, however, *not* independent of kind and energy of the radiation. This leads to (usually small) errors, because even an appropriate *mean* W can never assign the correct energy to *every* absorption event, realizing that the absorption events may result from *different* ionizing particles. This question has been discussed by BICHSEL (1974). Another problem arises when the number of ionizations per event becomes small (for very small sensitive volumes). Then, the *excitations* ought to be considered separately. This has sometimes been done in theoretical computations, e.g., by BOOZ *et al.* (1972b) and PARETZKE *et al.* (1974). As an alternative, it may be more useful to create a conceptual framework that is rigorously based upon nothing but ionizations produced rather than energy imparted. This possibility was envisaged by KELLERER and CHMELEVSKY (1975) (cf. also footnote 1 at page 21).

In this monograph, the latter conception is adopted throughout, which is reflected by the discrete distribution $f_1(j)$ occurring in Eq. (5.2) instead of the continuous distribution $f_1(z)$. In fact, the present theory unambiguously aims at the *actual* number of ionizations by using *diverse* W-values for deriving ionizations from energy parameters (see section 4.4). In other words, the theory describes exactly what the experimentalist measures when scoring ionizations with a proportional counter. Therefore, we have to do the

[1]) Only some imput parameters change greatly, since the sensitive volume is now of the order of micrometres and it is a sphere in most cases, which is taken into consideration by σ_0, V, and the pertinent chord length distributions $w(l')$ and $u(l')$.

same as he does if a representation in terms of the continuous variable 'imparted energy' is wanted: transforming the number j into the specific energy z by means of a *common mean* W-value and smoothing the actual discrete distribution $f_1(j)$ in some arbitrary way. Nothing else is done when we now change Eq. (5.2) into

$$\Phi(\zeta) = \frac{1}{\bar{z}_F} \int_0^\infty dz \left(1 - e^{-\zeta \frac{m}{W} z}\right) f_1(z), \quad (5.8)$$

where $m = V\varrho$ g/cm³ is the mass of the sensitive volume. Eq. (5.8), which exhibits $\Phi(\zeta)$ to be something like the Laplace transform of $f_1(z)$, is the starting-point of all the investigations following in this chapter, considering $\Phi(\zeta)$ to be known.

$f_1(z)$ is given implicitly by the directly calculable function $\Phi(\zeta)$, i.e. relation (5.8) necessitates a procedure of inversion to arrive at $f_1(z)$. However, for the purpose of the approximate version of the radiation action theory (chapter 2), which is based on the microdosimetric concept, it is just $\Phi(\zeta)$ that is needed. Compare Eq. (2.17) with Eq. (5.8) to disclose that the formula for α_μ reads now simply:

$$\alpha_\mu = \Phi(r\alpha_\mu^{(0)} W/m). \quad (5.9)$$

Thus, this version of the theory, which will be further studied in chapter 8, is completely comprehended in Eqs. (5.3 to 7), which are only a special case of Eqs. (3.67 to 71) which determine the quantity $r = s/s_0$. In this way, the same formalism describes the stochastics at the level of cell nuclei (Eq. (5.9)) as well as the nanometre level (quantity r, i.e. DNA-lesion production).

Section 5.2 deals with the evaluation of Eq. (5.8) regarding the quantities $\bar{z}_F$, $\bar{z}_D$, and $f_1(z)$ for the three kinds of radiation, the numerical results being given in the appendix (Tab. II, III, IV, VI). Section 5.3 presents a rather large number of comparisons with other computational and experimental data. It should be emphasized that it was by no means intended to attain completeness with respect to such data available, which would have been a tremendous task. However, it is felt that the set of examples chosen is representative enough for the different radiations, volume sizes, and measurable quantities occurring. Besides, the appendix contains lots of data for the reader himself to draw further comparisons.

The intention of section 5.3 is testing the principles of our approach to radiation absorption stochastics. Of course, the scope of this test depends on the variety of experimental data available. As the latter are almost completely lacking for nanometre regions,[1]) it should be emphasized that the results of section 5.3 are relevant mainly to the description of the stochastics at the level of cell nuclei (i.e. lesion number statistics), in spite of the Eqs. (5.3 to 7) being directly related to the formalism (3.67 to 71) of the DNA-target model. Indirectly, the goodness of our principles of approximation in the case of very small regions may also be judged from the suitability of the DNA-lesion production model to describe pertinent experimental data (single- and double-strand breaks) correctly, considering the approximations involved as a constitutive part of the model. This question is the topic of chapter 6.

5.2. The mathematical formalism

5.2.1. Mean values $\bar{z}_F$, $\bar{z}_D$, and related quantities

The *frequency-mean*, $\bar{z}_F$, and the *dose-mean*, $\bar{z}_D$, of the *specific energy* imparted to a micro-volume in one event are defined as

[1]) Apart from some older computational results considered in section 5.3.3, modern calculations of $f_1(j)$ and related quantities for nanometre regions, based on strict particle track simulations, are most suited to draw comparisons. This has been relegated, however, to chapter 6 (see section 6.3.1) due to formal reasons.

the expected values belonging to the differential single event size distributions $f_1(z)$ and $zf_1(z)$, respectively (see section 1.4.1):

$$\bar{z}_F = \int_0^\infty dz z f_1(z), \qquad \bar{z}_D = \frac{1}{\bar{z}_F} \int_0^\infty dz z^2 f_1(z). \tag{5.10}$$

Along with this description in terms of specific energy, z, measured in rads, we will often use the entirely equivalent description in terms of the so-called *lineal energy*, y, which is imparted energy divided by the mean thickness l of the sensitive volume (see Eq. (3.26)). This quantity is usually given in units of keV/µm, indicating its relatedness to LET. The transformation law, which applies to the argument of the function f_1 as well as the mean values (5.10), reads

$$y = \sigma_0 z \frac{\text{g}}{\text{cm}^3} \quad \text{or} \quad y \Big/ \frac{\text{keV}}{\mu\text{m}} = 0.06242 \left(\frac{\sigma_0}{\mu\text{m}^2}\right)\left(\frac{z}{\text{rad}}\right), \tag{5.11}$$

where σ_0 is the mean cross-section of the sensitive volume.[1]) y is more suited for tabulation, because it does not so much depend on the size of the volume considered.

In the following, we will exclusively refer to the expression (5.8) which is considered a known function of ζ, given by Eqs. (5.3 to 7). By forming the first and the second derivatives of $\Phi(\zeta)$, marked by dashs, and by considering the limit $\zeta \to \infty$, one obtains at once:

$$\frac{m}{W} = \Phi'(0), \quad \text{i.e.} \quad W = \frac{V\,\text{g/cm}^3}{\Phi'(0)}, \tag{5.12}$$

$$\bar{z}_F = \frac{1}{\Phi(\infty)}, \quad \bar{z}_D = -\frac{\Phi''(0)}{[\Phi'(0)]^2}, \tag{5.13}$$

using the definitions (5.10) and the normalization condition $\int dz f_1(z) = 1$.

According to the relations (5.12) and (5.13), complete sets of these quantities have been calculated numerically for different radiations, energies and sensitive volume sizes. The derivatives $\Phi'(\zeta)$ and $\Phi''(\zeta)$ as well as the limits as $\zeta \to 0$ and $\zeta \to \infty$ have been obtained by means of simple numerical methods, starting from the corresponding computer codes of Eqs. (5.3 to 7).[2]) The results of these calculations (along with other data dealt with in chapter 8) are given in the appendix in terms of $\bar{y}_F$ and $\bar{y}_D$: Tables II, III, and VI referring to X-rays, fast neutrons, and ion radiations, respectively.

It may be illustrative for the reader himself to derive explicit formulas for W, $\bar{z}_F$, and $\bar{z}_D$ by inserting the expressions (5.3 to 7) into Eqs. (5.12), (5.13) and carrying out the corresponding operations analytically. For ion radiations, this will be partly done below, where more specified quantities are introduced in this case (Tab. IV). However, the actual calculations have been performed in the numerical way indicated above, i.e. it proved needless to elaborate special computer codes based on such explicit expressions.

Let us discuss the quantity W, being the total energy absorbed per ionization produced. W is known to depend slightly on kind and energy of the irradiation considered. The theory takes this into account by basing the inherent quantities $p^{(\delta)}(E)$ (number of primary ionizations) and $\omega^{(\delta)}(E, j)$ (ionization cluster frequencies) upon experimentally established energy-dependent differential W-values (see section 4.4.1), so that W by Eq. (5.12) should be considered as an appropriate average for each particular radiation field which may consist of quite different ionizing

---

[1]) We take $\varrho = 1$ throughout this chapter.

[2]) In this respect, there are a number of subtle numerical difficulties in finding out suitable procedures of numerical integration that are properly adapted to the problem and, at the same time, flexible enough in view of the widely varying parameters, in general, and of the limiting cases $\zeta \to 0$ and $\zeta \to \infty$, in particular.

particles, say, for a neutron radiation. Thus, these average W's given in the second columns of Tables II, III, IV constitute a significant *result* of the theory, too.[1]) On the other hand, as stressed in the introduction to this chapter, the mean values $\bar{z}_F$ and $\bar{z}_D$ intrinsically represent ionization numbers which have only been expressed in terms of energy by means of the average W given. Therefore, regarding comparisons with experimental data that are primarily based also on ionization counting, the latter ought to be rescaled whenever the author of the experiment has used a W-value different than ours.

In the remainder of this section we shall more closely study monoenergetic ion radiations by dividing the totality of absorption events into so-called direct and indirect ones (KELLERER, 1971b). Direct events are those in which a primary ion itself crosses the sensitive volume. We shall call them *ion events* (superscript (i)). Indirect or *δ-electron events* (superscript (δ)) are those caused by a by-passing ion that only injects δ-electrons into the sensitive volume. This can be described as if there were two independent radiation fields superimposed, which is indeed the case in the framework of our fundamental approximation principle 2 (see page 65).[2]) The general microdosimetric treatment of such mixed radiations was already given in section 1.4.1. Analogously, one can now define the two *partial* single event size distributions $f_1^{(i)}(z)$ and $f_1^{(\delta)}(z)$ with the mean values $\bar{z}_F^{(i)}$, $\bar{z}_D^{(i)}$ and $\bar{z}_F^{(\delta)}$, $\bar{z}_D^{(\delta)}$ defined correspondingly to Eqs. (5.10). Denoting the number fractions (frequencies) of either sort of absorption events by $\mu_F^{(i)}$ and $\mu_F^{(\delta)}$, respectively, the fundamental relation

$$f_1(z) = \mu_F^{(i)} f_1^{(i)}(z) + \mu_F^{(\delta)} f_1^{(\delta)}(z), \qquad \mu_F^{(i)} + \mu_F^{(\delta)} = 1 \tag{5.14}$$

is self-evident. By evaluation of Eqs. (5.10) with (5.14) one obtains

$$\frac{1}{\bar{z}_F} = \mu_D^{(i)} \frac{1}{\bar{z}_F^{(i)}} + \mu_D^{(\delta)} \frac{1}{\bar{z}_F^{(\delta)}}, \tag{5.15}$$

$$\bar{z}_D = \mu_D^{(i)} \bar{z}_D^{(i)} + \mu_D^{(\delta)} \bar{z}_D^{(\delta)}, \tag{5.16}$$

where the coefficients

$$\mu_D^{(i)} = \frac{\bar{z}_F^{(i)}}{\bar{z}_F} \mu_F^{(i)}, \quad \mu_D^{(\delta)} = \frac{\bar{z}_F^{(\delta)}}{\bar{z}_F} \mu_F^{(\delta)}, \qquad \mu_D^{(i)} + \mu_D^{(\delta)} = 1 \tag{5.17}$$

are obviously the dose fractions, i.e. the corresponding proportions of energy imparted to the sensitive volume by ion events and δ-electron events, respectively. In the following, we shall only use the dose fractions μ_D instead of the number fractions μ_F.

To derive these partial parameters for ion radiations from the formalism of our theory, consider Eq. (5.8) with the ex-

---

[1]) In contrast to $\bar{z}_F$ and $\bar{z}_D$, the W-value given by Eq. (5.12) must be independent of the sensitive volume, although the corresponding quantity δ (cut-off energy) does not drop out of Eq. (5.12). Consequently, the numerical results actually exhibiting this independence (within more than four digits) constitutes a significant test for the consistency of the theory as well as the reliability of the numerical procedures used. In the case of monoenergetic ions, even the absolute magnitude of W is known from the outset. It must equal the differential W-value for the particular ion energy, as given in section 4.4.1 by a simple analytical expression depending only on the energy per nucleon. That is why these charge-independent values are listed in Table IV (second column). This case exemplifies what is meant by consistency: different partial W-values are ascribed to the ion track core and the δ-electrons (corresponding to the two terms in Eq. (5.3)) in such a way that the overall average (5.12) becomes independent of the cut-off energy δ (i.e. of the volume size), on which both separate terms do depend.

[2]) Note that the *formalism* following does not depend on this approximation. It is valid in general.

pression (5.14) substituted for $f_1(z)$:

$$\Phi_i(\zeta) = \Phi_i^{(i)}(\zeta) + \Phi_i^{(\delta)}(\zeta),$$

$$\Phi_i^{(i)}(\zeta) = \frac{\mu_D^{(i)}}{\bar{z}_F^{(i)}} \int_0^\infty dz \left(1 - e^{-\zeta \frac{m}{W} z}\right) f_1^{(i)}(z),$$

$$\Phi_i^{(\delta)}(\zeta) = \frac{\mu_D^{(\delta)}}{\bar{z}_F^{(\delta)}} \int_0^\infty dz \left(1 - e^{-\zeta \frac{m}{W} z}\right) f_1^{(\delta)}(z). \quad (5.18)$$

The explicit formula (5.3) shows the same splitting of $\Phi(\zeta)$ into an ion track core term and a δ-electron term. Hence, the quantities in question are readily obtained in the same way as where Eqs. (5.12), (5.13):

$$\mu_D^{(i)} = \frac{\Phi_i^{(i)\prime}(0)}{\Phi_i'(0)}, \quad \mu_D^{(\delta)} = \frac{\Phi_i^{(\delta)\prime}(0)}{\Phi_i'(0)}, \quad (5.19\text{a, b})$$

$$\bar{z}_F^{(i)} = \frac{\mu_D^{(i)}}{\Phi_i^{(i)}(\infty)}, \quad \bar{z}_F^{(\delta)} = \frac{\mu_D^{(\delta)}}{\Phi_i^{(\delta)}(\infty)}, \quad (5.20\text{a, b})$$

$$\bar{z}_D^{(i)} = -\mu_D^{(i)} \frac{\Phi_i^{(i)\prime\prime}(0)}{[\Phi_i^{(i)\prime}(0)]^2}, \quad (5.21\text{a})$$

$$\bar{z}_D^{(\delta)} = -\mu_D^{(\delta)} \frac{\Phi_i^{(\delta)\prime\prime}(0)}{[\Phi_i^{(\delta)\prime}(0)]^2}. \quad (5.21\text{b})$$

These equations are, of course, in accordance with the relations (5.15 to 17). Using the above-mentioned numerical methods, these quantities were calculated from the expressions (5.3a, b) and listed in Table IV (in terms of $\bar{y}$). In addition to δ_2 (for this quantity see below), this table contains the (δ)-labelled parameters only, since these are independent of the ion's charge. This important point will be discussed below.

Table IV applies to any sort of ions: in combination with Table VI, using the relations (5.15 to 17), it provides the charge-dependent parameters $\bar{y}_F^{(i)}$ and $\bar{y}_D^{(i)}$, too. Thus, Tables IV and VI constitute a complete set of microdosimetric data for ion radiations. As shown in the following, there is even some redundancy since, as to the dependence on charge, $\bar{y}_F^{(i)}$ and $\bar{y}_D^{(i)}$ reduce essentially to the stopping power, $L_\infty^{(i)}(E_i^{(0)})$, which is independent of the sensitive volume. Going now into these details elucidates some important general relationships of microdosimetry as well as the particular properties of our approach.

At first, let us consider the dose fraction $\mu_D^{(\delta)}$ of pure δ-electron events. It is clear that this proportion must be strictly the same for ions that differ only in charge but have the same velocity (energy per nucleon). The velocity alone determines the energy spectrum and geometry of δ-rays, while at constant velocity a growing charge, e.g., only shrinks all distances of δ-rays along the track. Apart from this rigorous statement, we can just go further in the framework of our approximate approach, realizing that the δ-ray cut-off lets the indirect events appear as if arising from the *independent* δ-electron component of the radiation field, the energy proportion of which must be equal to $\mu_D^{(\delta)}$. This energy partition, however, is governed by the *restricted stopping power*, L_δ, in relation to the full stopping power, L_∞, of the monoenergetic ions. Consequently, it should hold

$$\mu_D^{(i)} \approx \frac{L_\delta^{(i)}(E_i^{(0)})}{L_\infty^{(i)}(E_i^{(0)})} = \frac{\ln\left[\sqrt{E_{\max} \min(E_{\max}, \delta)}/I\right]}{\ln(E_{\max}/I)}, \text{[1]} \quad (5.22)$$

$$E_{\max} = \frac{4}{1838} \frac{E_i^{(0)}}{A_i}, \quad I = 70\,\text{eV}.$$

This relation is not a quite exact one only because, in the framework of this theory, energy is assessed actually in terms of ionization number[2]) so that some bias

---

[1]) For the origin of this expression see section 4.2.1.

[2]) Indeed, the explicit formula for $\mu_D^{(i)}$ easily obtainable from Eqs. (5.19a), (5.3), (5.6), (5.7) reveals $\mu_D^{(i)}$ and $\mu_D^{(\delta)}$ to be nothing but ratios of the respective ionization numbers.

results from differences in W between the ion track core and the δ-rays. These departures however are small; the reader may verify this by means of the figures for $\mu_D^{(i)} = 1 - \mu_D^{(\delta)}$ given in Table IV where the cut-off energies δ can also be found.[1]) It is important to emphasize that the size of the sensitive volume enters Eq. (5.22) only through the cut-off energy δ. Therefore, comparing the calculated $\mu_D^{(i)}$ with corresponding experimental data in the following section means nothing but asking whether the rule (3.12) provides an adequate choice of δ.

In contrast to $\mu_D^{(\delta)}$, the mean values $\bar{z}_F^{(\delta)}$ and $\bar{z}_D^{(\delta)}$ being also independent of the ion's charge is only an artefact of our approach, because, in reality, not only the energy spectrum of δ-rays but also their spatial correlations along the ion track determine the character of δ-ray events. It is just this correlation that is disregarded in consequence of approximation principle 2b (section 3.2.2). It is easy to see that the charge of the ion drops out of Eqs. (5.20b), (5.21b) with Eqs. (5.3b), (5.6), and (5.7), since the quantities $L_\infty^{(i)}, T_i^{(\delta)}, p_i^{(\delta)}$ (and only these quantities) are all proportional to charge. While in reality, especially in the case of heavy ions, most δ-electron events are compound ones with several particles sharing in, our independent-particle approach makes the lineal energy mean values $\bar{y}_F^{(\delta)}$ and $\bar{y}_D^{(\delta)}$ listed in Table IV closely related to the so-called *track-length mean*, $\bar{L}_T$, and *dose mean*, $\bar{L}_D$, of the *LET spectrum* of δ-rays: $\bar{y}_F^{(\delta)} \approx \bar{L}_T$ and $\bar{y}_D^{(\delta)} \approx (9/8)\,\bar{L}_D$, apart from the influence of straggling and non-zero sensitive volume size.[2]) Analogous relations will now be disclosed in a rigorous manner with respect to the ion track core parameters.

Eq. (5.20a) with the expression (5.3a) inserted and transformed according to (5.11) yields

$$\bar{y}_F^{(i)} = \frac{\mu_D^{(i)} L_\infty^{(i)}(E_i^{(0)})}{\int\limits_0^\infty \mathrm{d}l' w(l')\,\{1 - \exp\,[-\varrho l' p_i^{(\delta)}(E_i^{(0)})]\}}. \tag{5.23}$$

Apart from the denominator, $\bar{y}_F^{(i)}$ is equal to the restricted stopping power of the monoenergetic ions, δ-ray cut-off being expressed by the factor $\mu_D^{(i)}$. Hence, in consideration of Eq. (5.22), one has $\bar{y}_F^{(i)} \approx L_\delta^{(i)}(E_i^{(0)})$, which is self-explanatory. However, what is the meaning of the denominator in Eq. (5.23)? It constitutes the probability of generation of at least one ionization by an ion crossing the sensitive volume. In other words, traversals without energy deposition are not acknowledged as absorption events. This is correct, bearing in mind that energy deposition means ionization in the framework of our approach. Microdosimetry usually disregards this subtlety, the more so as the mentioned probability even closer approaches unity[3]) when taking into account all kinds of elementary energy transfer acts (rigorous energy concept). However, strictly speaking, *straggling* must always cause $\bar{y}_F^{(i)} > \mu_D^{(i)} L_\infty^{(i)}$, since equality would only be compatible with a continuous transfer of energy along the track.

The dose-mean $\bar{y}_D^{(i)}$ also permits a very instructive representation in terms of $L_\infty^{(i)}$. Evaluation of Eq. (5.21a) by means

---

[1]) Within this approach, it is clear that $\mu_D^{(\delta)} = 0$ for $E_{\max} < \delta$, $\bar{y}_F^{(\delta)}$, $\bar{y}_D^{(\delta)}$ being undefined then.

[2]) The chief influence of the size of the sensitive volume as exhibited by Table IV comes however from the spectrum depending on l through the cut-off energy δ.

[3]) The departure from unity becomes appreciable only if the sensitive volume is so small that the mean number of collisions (say, primary ionizations) upon a particle crossing is not large.

of Eqs. (5.3a) and (5.11) yields

$$\bar{y}_D^{(i)} = \mu_D^{(i)} \left[L_\infty^{(i)}(E_i^{(0)}) \frac{1}{l^2} \int_0^\infty dl' l'^2 w(l') + \frac{1}{l} \frac{\overline{j^2} L_\infty^{(i)}(E_i^{(0)})}{\bar{j}^2 p_i^{(\delta)}(E_i^{(0)})} \right], \quad (5.24)$$

where

$$\bar{j} = \sum_{j=1}^{j_{\max}} j \omega_i^{(\delta)}(E_i^{(0)}, j), \qquad \overline{j^2} = \sum_{j=1}^{j_{\max}} j^2 \omega_i^{(\delta)}(E_i^{(0)}, j). \quad (5.25)$$

As could be anticipated, Eq. (5.24) contains the well-known general expression for the dose-mean of the so-called *straggling distribution* for the energy loss of charged particles in a perpendicularly crossed layer of matter (i.e. $w(l') = \delta(l' - l)$), as derived in section 1.4.1 (Eq. (1.23)). While $L_\infty^{(i)}$ describes the mean energy loss[1]) in the absence of straggling, the second term at the right-hand side of Eq. (5.24), the straggling term, is usually expressed by δ_2/l, where the energy δ_2 means the quotient of the second and first moments of the collision spectrum characterizing the elementary energy transfer acts. Since, within our approach, the latter are nothing but the primary ionizations to which the 'collision spectrum' $\omega_i^{(\delta)}(E, j)$ belongs, Eq. (5.24) implying

$$\delta_2 = \frac{\overline{j^2}}{\bar{j}} W, \quad W = \frac{\mu_D^{(i)} L_\infty^{(i)}(E_i^{(0)})}{\bar{j} p_i^{(\delta)}(E_i^{(0)})} \text{ [2])} \quad (5.26)$$

is almost trivial. Hence, for a spherical region ($w(l')$ according to Eq. (3.25)), Eq. (5.24) writes

$$\bar{y}_D^{(i)} = \frac{9}{8} \mu_D^{(i)} L_\infty^{(i)} + \delta_2/l, \quad l = \frac{2}{3} d. \quad (5.27)$$

Consequently, instead of $\bar{y}_D^{(i)}$ itself, it is only necessary to tabulate δ_2 which is independent of the ion's charge. Therefore it is given together with the other charge-independent quantities in Table IV.[3]) Although the collision spectrum is conceptually thought to be characteristic of the radiation only, Table IV shows a slight dependence of δ_2 on the size of the sensitive volume, which must be attributed to the influence of δ-ray cut-off on the cluster frequencies $\omega_i^{(\delta)}(E, j)$. Unrestricted δ_2-values are found in Table IV in those panels where $\mu_D^{(\delta)} = 0$. Experimental data on δ_2, which are usually derived from measured straggling distributions,[4]) also show some influence of size and shape of the sensitive volume realized in the measuring device, which is caused by the escape of δ-rays. A number of comparisons with such data is presented in the following section, in this way testing the capability of our whole concept of reducing straggling intrinsically to

---

[1]) It is exactly $L_\infty^{(i)}$ that occurs instead of the quotient (5.23), because the problem of 'zero events' does not matter here since their weight for the *dose-mean* vanishes.

[2]) By substituting the expression (5.19a) for $\mu_D^{(i)}$ it is easy to show that this W is identical with the overall average defined by Eq. (5.12).

[3]) These figures were calculated indirectly by means of relation (5.27) from $\bar{y}_D^{(i)}$ obtained numerically according to Eq. (5.21a). Thus actually achieving independence of charge was a further test for the correctness of the numerical procedures.

[4]) A pure straggling distribution is obtained experimentally when the measuring device collects only direct ion events and when all path lengths l of the monoenergetic ions within the sensitive volume are equal. δ_2 is then obtained from the mean value, $\bar{y}_F^{(i)}$, and the variance, σ^2, of the straggling distribution by $\delta_2 = l\sigma^2/\bar{y}_F^{(i)}$. However, this formula does not strictly follow from the generally valid equation (5.27) with the factor 9/8 omitted (realize that σ^2 is nothing but $\bar{y}_F^{(i)}(\bar{y}_D^{(i)} - \bar{y}_F^{(i)})$ according to the definitions (5.10)), unless $\bar{y}_F^{(i)} = \mu_D^{(i)} L_\infty^{(i)}$ is taken, in contrast to Eq. (5.23). In other words, the rule $\delta_2 = l\sigma^2/\bar{y}_F^{(i)} = l(\bar{y}_D^{(i)} - \bar{y}_F^{(i)})$ is not quite correct in that it counts in 'events' without energy deposition (ionization). This must in some cases (very small l) be taken into consideration when comparing with 'experimental' δ_2-values.

nothing but ionization cluster frequencies, and the semi-empirical way of obtaining the latter as described in section 4.4.

Summing up the case of monoenergetic ion radiations, it can be stated that the five microdosimetric parameters given by Eqs. (5.19—21), which depend, in general, on three variables: charge, energy per nucleon, size of the sensitive volume (diameter d), have been 'almost' expressed by five quantities that are either independent of charge ($\mu_D^{(\delta)}$, $\bar{y}_F^{(\delta)}$, $\bar{y}_D^{(\delta)}$, δ_2, contained in Tab. IV) or of d ($L_\infty^{(i)}$). Thus the necessary numerical data are considerably reduced and Table VI is nearly superfluous, because $\bar{y}_F$, $\bar{y}_D$ can be calculated from the relations (5.15) and (5.16), where $\bar{y}_D^{(i)}$ follows exactly from $L_\infty^{(i)}$, $\mu_D^{(i)} = 1 - \mu_D^{(\delta)}$, and δ_2 by means of Eq. (5.27); and $\bar{y}_F^{(i)}$ equals $\mu_D^{(i)} L_\infty^{(i)}$ in good approximation. It is only the avoidance of this approximation with respect to $\bar{y}_F^{(i)}$ that requires a table depending on all three variables (Tab. VI) since the denominator in Eq. (5.23) does so. Apart from this, Table VI serves the convenience and allows us to calculate $\bar{y}_F^{(i)}$ and $\bar{y}_D^{(i)}$ exactly from Eqs. (5.15), (5.16).

5.2.2. Event size spectrum $f_1(z)$ by inverse Laplace transformation

In the foregoing subsection the function $\Phi(\zeta)$ defined by Eq. (5.8) was evaluated only with respect to the mean values of $f_1(z)$. For obtaining $f_1(z)$ itself we need a mathematical procedure of inverting Eq. (5.8) which constitutes essentially a Laplace transformation. However, the rigorous inverse Laplace transform as obtained by functional analysis is of no practical use for our purpose. Satisfactory numerical methods starting with a finite set of values $\Phi(\zeta_n)$ are difficult to devise, because the slightest inaccuracy of $\Phi(\zeta)$ has large effects on $f_1(z)$ (e.g., arbitrary oscillations with $f_1(z) < 0$ being *not* excluded), apart from the fact that $f_1(z)$ is not, in principle, unambiguously determined by any finite number of Φ-values.[1]) Therefore, any numerical procedure implies a decision as to which of the ambiguous spectra $f_1(z)$ compatible with the Φ's given it is aiming at. In view of this, the only reasonable procedure is one which always finds out the most smooth $f_1(z)$ possible under the given conditions. Such a method is described in the following.

Let us define the gradually increasing steplike function

$$\Theta(x/x_0) = (1 - N^{-x/x_0})^N = \sum_{n=0}^{N} \binom{N}{n} (-1)^n N^{-nx/x_0} \tag{5.28}$$

and its derivative

$$\vartheta(x_0, x) \equiv \frac{d\Theta}{dx} = \frac{N \ln N}{x_0} N^{-x/x_0}(1 - N^{-x/x_0})^{N-1} = \frac{\ln N}{x_0} \sum_{n=1}^{N} \binom{N}{n} \times (-1)^n n (1 - N^{-nx/x_0}). \tag{5.29}$$

The latter is a normalized peak function with the peak at $x = x_0$, resembling a Gaussian distribution with the relative width[2])

$$\sigma_{rel} = \frac{2 \ln \left[\left(\frac{3}{2} - \frac{1}{2N} \right) + \sqrt{\left(\frac{3}{2} - \frac{1}{2N} \right)^2 - 1} \right]}{\ln N} \approx \frac{1.925}{\ln N}. \tag{5.30}$$

---

[1]) Unfortunately, increasing the number of the input Φ-values does not automatically reduce the errors of the f_1-values obtained. Only the 'wave lengths' of artificial fluctuations of $f_1(z)$, but not their amplitudes, become smaller.

[2]) Defined for convenience as the distance of the two points of inflection ($d^2\vartheta/dx^2 = 0$) divided by x_0.

Hence, the peak becomes the narrower and steeper the larger N is.
Consider the integral

$$\varphi_0(z_0) = \int_0^\infty dz\vartheta(z_0, z)\, f_1(z)/\bar{z}_F. \qquad (5.31)$$

Using the expansion (5.29), this quantity can be expressed by $\Phi(\zeta)$ and thus calculated:

$$\varphi_0(z_0) = \int_0^\infty dz \frac{\ln N}{z_0} \sum_{n=1}^{N} \binom{N}{n} (-1)^n n \times (1 - e^{-nz\ln N/z_0})\, f_1(z)/\bar{z}_F,$$

$$\varphi_0(z_0) = \frac{\ln N}{z_0} \sum_{n=1}^{N} \binom{N}{n} (-1)^n n \times \Phi\left(n \frac{W}{mz_0} \ln N\right). \qquad (5.32)$$

Owing to the properties of $\vartheta(z_0, z)$, φ_0 as a function of z_0 is already a zeroth approximation to $f_1(z_0)/\bar{z}_F$ since Eq. (5.31), considered as a transformation of $f_1(z)$ into $\varphi_0(z)$, means only a smoothing of $f_1(z)$ according to the finite width of the peak function ϑ. If the latter were the δ-function, i.e. $\vartheta(z_0, z) = \delta(z - z_0)$, we would have exactly $\varphi_0(z) = f_1(z)/\bar{z}_F$. Trying to sufficiently approach this limit by making N large is however no practicable way, because the decrease of σ_{rel} on growing N is too slow (see Eq. (5.30)).
The problem was solved in the following manner. After having calculated a series of φ_0-values for the points

$$z_0 = z_m = z_1, z_2, \ldots, z_M \qquad (5.33)$$

spread over the significant range of z, one must search for those values of $f_1(z_m)$ that reproduce all $\varphi_0(z_m)$ by computing the integral (5.31) numerically. This search is carried out by an iterative procedure, starting with the zeroth approximation $f_1(z_m)/\bar{z}_F = \varphi_0(z_m)$ and estimating the iterative improvements in a way to maintain the smoothness of the zeroth approximation as much as possible. This algorithm of iteration reads

$$\varphi_{k+1}(z_m) = \frac{\varphi_0(z_m)}{\int_0^\infty dz\vartheta(z_m, z)\, \varphi_k(z)}\, \varphi_k(z_m). \qquad (5.34)$$

After a sufficient number of iteration steps, $k = 1, 2, \ldots$, one has

$$f_1(z_m)/\bar{z}_F = \varphi_k(z_m). \qquad (5.35)$$

The integral in Eq. (5.34) becomes a linear combination of the discrete values $\varphi_k(z_m)$, $m = 1, \ldots, M$, according to the rules of numerical integration. Thus, the iteration (5.34) is simple algebra. Complete convergence, i.e. $\varphi_{k+1}(z_m) = \varphi_k(z_m)$, is therefore equivalent to the solution of a linear system of equations of the form $\sum_{m'} A_{m,m'}\varphi_k(z_{m'}) = \varphi_0(z_m)$, which could be obtained at once. However, it is the most important feature of this method *not* to attempt this 'exact' solution. Rather than do this, the iteration has to be broken off when the differences between $\int dz\vartheta(z_m, z)\, \varphi_k(z)$ and $\varphi_0(z_m)$ go beyond the magnitude of the inevitable errors of numerical integration.[1]) Otherwise, one would only compensate for these errors by large but meaningless changes of the values $\varphi_k(z_m)$ leading to increasing artificial oscillations.

The iteration procedure (5.34) has the required property of keeping the successive changes small in order to avoid a rapid destruction of the initial smoothness, which otherwise could lead to undamped oscillations. Therefore, on the other hand, the convergence is rather slow, so that the number of iteration steps would run into thousands, if one were trying to attain complete self-consistency. However, the break-off rule appointed above settles the due number of iteration steps to the order of hundred. It is not advantageous to make N large. $N = 10$ was used in all cases, while the accuracy can be readily enhanced by a large number of points, M. In most cases, $M = 25$ was taken

[1]) These intrinsic error limits for each m can be estimated by simultaneously calculating the corresponding integrals by means of the doubled step width.

with a logarithmic scale for z_m. This requires calculating $NM = 250$ values of Φ (see Eq. (5.32)), which is the most time-consuming part of the whole procedure. The integral in Eq. (5.34) was made definite by assuming $\varphi_k(z) \equiv \varphi_k(z_1)$ for $z < z_1$ and $\varphi_k(z) \equiv \varphi_k(z_M)$ for $z > z_M$. Denoting the coefficients for the numerical integration (Simpson rule) by $c_{m'}$ and using the logarithmic scale

$$z_m = z_1 \left(\frac{z_M}{z_1}\right)^{\frac{m-1}{M-1}}, \qquad m = 1, 2, \ldots, M,$$

the algebraic expression for the integral in the denominator of Eq. (5.34) reads

$$\alpha(1 - m)\, \varphi_k(z_1) + [1 - \alpha(M - m)]\, \varphi_k(z_M)$$
$$+ \sum_{m'=1}^{M} c_{m'} \beta(m' - m)\, \varphi_k(z_{m'})$$

with the constant fields

$$\alpha(j) = [1 - e^{-\gamma(j)}]^N,$$

$$\beta(j) = \frac{N \ln (z_M/z_1)}{M - 1}\, \gamma(j)\, e^{-\gamma(j)} [1 - e^{-\gamma(j)}]^{N-1},$$

where

$$\gamma(j) = \left(\frac{z_M}{z_1}\right)^{j/(M-1)} \ln N,$$

$$j = -(M - 1), \ldots, 0, \ldots, (M - 1).$$

The initial values $\varphi_0(z_m)$ the entire procedure is based upon can include large errors unless $\Phi(\zeta)$ is calculated very accurately, for the alternating sum (5.32) tends to amplify any scatter in $\Phi(\zeta)$. As the latter is obtained from multiple numerical integrations (see Eqs. (5.3 to 7)), the corresponding algorithms must be such that Φ becomes an analytical function of ζ without any discontinuities.

As the last step, the resulting spectrum $\varphi_k(z_m)$ may be normalized by numerical integration. The normalization factor thus found should be equal to $\bar{z}_F$. This is a good test for the accuracy reached. The deviations actually obtained were always much smaller then one per cent.

In order to demonstrate and to test this iterative method, a simple function $f(z)$ shown in Figure 5.1 was used, from which the zeroth approximation, $\varphi_0(z)$, was obtained analytically according to Eq. (5.31). The capability of reproducing the original function by iteration from $\varphi_0(z_m)$ to $\varphi_k(z_m)$ is illustrated in Figure 5.1. Breaking off the iteration according to the abovementioned rule leads to only slight oszillations (open circles) which originate from the two discontinuities of the original function, while driving the iteration further up to the practically 'exact' solution gives rise to more pronounced artificial oscillations (full circles). The example chosen somewhat resembles $f_1(z)$ for heavy ion radiations where the initial peak comes from the δ-ray events. Considering the un-

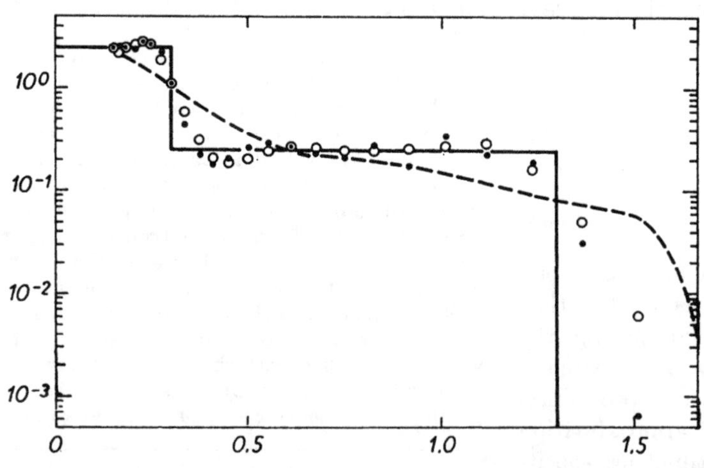

Fig. 5.1. Test of the iterative procedure for the double-rectangular function $f(z)$ (full line). Broken curve: zeroth approximation $\varphi_0(z)$; open circles: after 307 iteration steps (according to the break-off rule); full circles: after 10500 iteration steps.
Parameters: $N = 10$, $M = 25$.

realistic abrupt transitions of our test function, the method is likely to yield quite satisfactory results. This will be confirmed by the applications shown in the following section; however, it seemed appropriate to renounce displaying the single points $f_1(z_m)$, drawing continuous curves with small oscillations, if occurring, being smoothed.

The mathematical problem of obtaining the single event size spectrum $f_1(z)$ from the function $\Phi(\zeta)$ given for each particular radiation by Eqs. (5.3 to 7) is thus solved in a satisfactory manner which is quite generally applicable. In the following section, a number of examples will be shown for the three kinds of radiation. As to monoenergetic ions, not only the total spectrum $f_1(z)$ but also its components $f_1^{(i)}(z)$ and $f_1^{(\delta)}(z)$ (see Eq. (5.14) and context) can be calculated separately[1]), using either $\Phi_i^{(i)}(\zeta)$ or $\Phi_i^{(\delta)}(\zeta)$ for the input. For a spherical sensitive volume, $f_1^{(i)}(z)$ is essentially a triangular spectrum somewhat smoothed by straggling. Pure straggling distributions, however, are also obtained readily by calculating $\Phi_i^{(i)}(\zeta)$ according to Eq. (5.3a) with the cord length spectrum $w(l')$ replaced by a δ-function.

5.3. Comparison with experimental data

This section, in addition to Tables II, III, IV, and VI, presents a number of numerical results regarding microdosimetric parameters as compared with analogous data from literature. The latter are either experimental ones or have also been obtained by theory, often by means of Monte Carlo simulation of absorption stochastics. Almost all experimental results were obtained with tissue equivalent proportional gas counters[2]), which are either with a solid wall (Rossi-counter) or wall-less (i.e. grid-walled or with a sensitive region formed only by a suitably bounded electrical field in a larger gas volume).

There are of course a number of problems connected with these experimental methods: e.g., wall effects, gas amplification noise (multiplication statistics), composition of wall material and counter gas, W-value, and others. Hence, a thorough valuation of agreement or disagreement of theory and experiment in each case would require considering each particular experimental work in detail. However, this cannot be the purpose of this section, which rather is to give only a general impression about how the theory is in accord with physical 'reality'. This refers also to the W-value problem discussed in the introduction to this chapter (the W-values implicated in our theory (see Tab. II, III, IV) have not been altered to adapt them to those used by different authors). For detailed information of this sort the reader should consult the original papers. For a general insight into the pertinent problems and achievements in this field, the following papers may be recommended: X- or γ-rays (Booz, 1976; Coppola *et al.*, 1976); neutrons (Booz and Coppola, 1974; Rodgers and Gross, 1974; Caswell and Coyne, 1976); ion radiations (Glass, 1967; Gross and Rodgers, 1972; Paretzke *et al.*, 1974; Kliauga and Rossi, 1976); W-values (Bichsel, 1974). The influence of the counter wall (so-called 'wall effect') was investigated by Kellerer (1971c).

With one exception (straggling distributions of monoenergetic ions and corresponding δ_2-values), only results for *spherical* sensitive regions are considered in the following.

---

[1]) In some cases, it proved advantageous to construct the total $f_1(z)$ also by adding these two components instead of directly calculating it from the total $\Phi(\zeta)$. In this way, the problem of a steep incline in $f_1(z)$ like that shown in Figure 5.1 can be circumvented.

[2]) Other methods as, e.g., the ion recombination method or solid state detectors have not yet reached a great importance.

A general remark should be made on the ways of presenting single event size distributions, since there is some confusion in the literature regarding terminology and adequate normalization of the spectra. In the sequel, either the differential *frequency distribution*, $f_1(y)$, or the differential *dose distribution*, $d(y) = yf_1(y)/\bar{y}_F$, is shown, which are both normalized to unity with respect to a linear abscissa, y. However, if a *logarithmic* y-scale is used, both functions are multiplied by y: $yf_1(y)$ and $y^2f_1(y)/\bar{y}_F$, which means normalization with respect to equal intervals of the natural logarithm of y. Hence, one should not confound the frequency distribution $yf_1(y)$ for a logarithmic scale and the dose distribution $yf_1(y)/\bar{y}_F$ for a linear scale. In the literature, the denominator $\bar{y}_F$ has sometimes been omitted (or only not indicated) or the normalization of measured spectra is otherwise unclear. In these cases, the corresponding spectra were re-normalized either roughly by direct integration or by using $\bar{y}_F$ as given by the authors. Analogously, these comments apply if, instead of lineal energy y, the specific energy, z, or the deposited energy itself, designated as ε in the following, is used for representation.

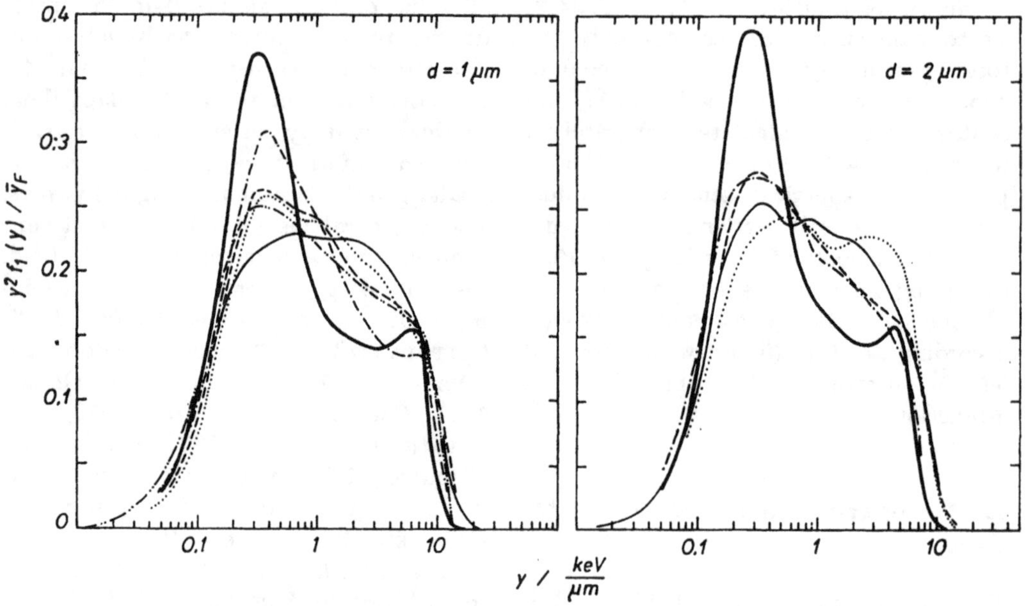

Fig. 5.2. Single event size spectra for ^{60}Co γ-rays (represented as differential fraction of energy per differential logarithmic y-interval). Heavy lines according to theory compared with several experimental results:

Left-hand panel (1 μm sphere diameter):

——— BIAVATI and BOER (1966), walled counter;
- - - - BRABY and ELLETT (1971), walled counter;
· · · · · EICKEL (1975), walled counter;
and COPPOLA *et al.* (1976), walled counter; } redrawn from Booz (1976)
—·—·— ELLETT and BRABY (1972), grid-walled counter;
—··—·· KELLERER (1968a), calculation based on multiple folding of a continuous collision spectrum.

Right-hand panel (2 μm sphere diameter):

——— BIAVATI and BOER (1966), walled counter;
- - - - MAIER, walled counter;
· · · · · EICKEL (1975), walled counter;
—·—·— DICELLO (1974), walled counter. } redrawn from Booz (1976)

5.3.1. Gamma- and X-rays

Most work was done with ^{60}Co γ-rays. Figure 5.2 shows the corresponding single event size distributions for 1 μm and 2 μm site diameter obtained by different authors in comparison with our theoretical curves. At first glance, one observes a noticeable disagreement between the bulk of experimental data and the results of theory. However, this can clearly be attributed largely to the wall effect which is known to bring about larger events at the expense of small ones (back-scattering of electrons in the wall). This is corroborated by the much better agreeing grid-walled-counter result by ELLETT and BRABY (1972) (dashed-dotted curve in the left-hand panel). As to the nevertheless remaining discrepancy, it is not unreasonable to attribute it to the limited resolution of proportional counters, realizing that the position of the main peak corresponds to only few ionizations ($y = 0.1$ keV/μm means about 2 ionizations in the 1 μm site). Hence, if a peak as sharp as shown by the theoretical curves actually occurs, it is likely to be flattened in the measurement by multiplication statistics. In view of these considerations, the good coincidence of the spectrum calculated by KELLERER (1968a) with the walled-counter results appears only as a fortuitous consequence of the somewhat arbitrary modified $1/E^2$ collision spectrum on which that calculation was based. Note that the main peak, assumed to be realistic, is a direct check for the correctness of our using ionization cluster frequencies to play the part of a collision spectrum, because the peak region corresponds to high-energy electrons that produce no more than one or two collisions (i.e. primary ionizations) in the 1 μm sphere in most cases.

Figure 5.3 must be considered with caution since the experimental X-ray results apply to a *spectrum* of γ-energies, which leads to a broadening of the y-spectra as compared with the theoretical one, while the overall displacements to the right or left are obviously due to differences in the corresponding mean γ-energies. The good fit of the dotted curve (COPPOLA *et al.*, 1976) is likely to reflect the influence of the wall effect, which may compensate for the broadening by shifting events from the region $y = 1-4$ keV/μm into the peak. In contrast to the case of ^{60}Co radiation, the 100-keV distribution does not so much depend on the cluster frequencies. It mainly depends on a correct description of fluence, range, and track curvature of the lower-energy electrons involved.

The further considerations shall be restricted to the mean values $\bar{y}_F$ and $\bar{y}_D$, which reflect the influence of the different factors discussed in the context of Figures 5.2 and 5.3 in concis eform. The right-hand panel of Figure 5.4 again suffers from uncertainties regarding the mean γ-energies and, moreover, the presence of a γ-energy spectrum must evidently increase the ratio $\bar{y}_D/\bar{y}_F$. The left-hand panel of Figure 5.4, which is free from such uncertainties, clearly exhibits how $\bar{y}_F$ and $\bar{y}_D$ are affected by the wall effect: the theoretical $\bar{y}_F$-curve practically lies in the middle of the two wall-less-counter data sets, but the walled-counter values are above it. Although the same tendency is observed with respect to $\bar{y}_D$, the theoretical $\bar{y}_D$-curve marks just the lower limit of *all* experimental values shown. This is very likely to be due to the experimental dose-means being biased additionally by multiplication statistics, which must be most pronounced just for very high energy γ-rays where most events consist of only few ionizations (note the much higher values of $\bar{y}$ in the case of 100 keV γ-energy). As to the dependence of the mean values of lineal energy on the diameter of the volume, both parts of Figure 5.4 exhibit very good agreement between theory and experimental data.

Figures 5.5 and 5.6 show the dependence

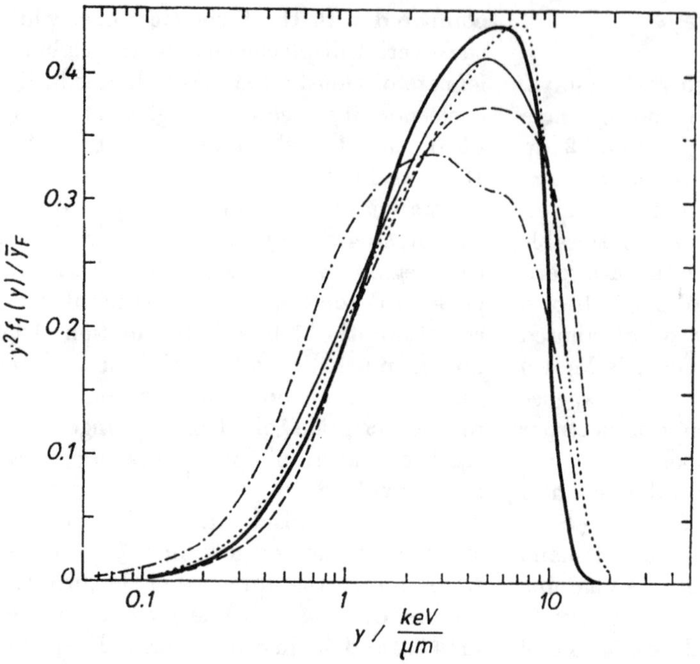

Fig. 5.3. Like Figure 5.2, but for X-rays of about 100 keV mean energy and a spherical volume of about 1 μm diameter. Theoretical curve (heavy line) for 100-keV monoenergetic photons. Origin of experimental data:

- - - - Ellett and Braby (1972), 250-kVp X-rays, HVL = 0.44 mm Cu, grid-walled counter;
-·-·- Ellett and Braby (1972), 250-kVp X-rays, HVL = 1.80 mm Cu, grid-walled counter;
——— Vidensky *et al.* (1973), 200-kVp X-rays filtered with 2 mm Cu, walled counter;
· · · · · Coppola *et al.* (1976), 200-kVp X-rays filtered with 5 mm Al + 1 mm Cu, walled counter.

of the mean values $\bar{y}_F$, $\bar{y}_D$ on the energy of photons. The good correspondence of the older results by Abel and Rossbach[1]) (Fig. 5.5) with theory is somewhat surprising in view of the Rossi-counter used. The apparently most modern and most complete grid-walled-counter data set available at present is contrasted with theory in Figure 5.6. Apart from some systematic differences which may partly be attributed to the influence of multiplication statistics (at high E_γ and small d) and the grid wall, the correspondence of theory and experiment can be called excellent, particularly with respect to the dependence on photon energy: note the undulation at about $E_\gamma \approx 50$ keV, which is caused by switching over from Compton effect to photo effect in the course of decreasing E_γ. It is also interesting to focus attention on the degree and kind of splitting of $\bar{y}_F$ for different sphere diameters. This splitting is minimal between 150 keV and 300 keV because most electrons occurring in this case are complete 'crossers' that deposite a *lineal* energy equal to their LET irrespective of the volume diameter d. At both ends of the E_γ-scale, however, the splitting becomes considerable, but it is of opposite nature: at 1000 keV *increase* of $\bar{y}_F$ for *small* d

[1]) These authors gave their results in terms of the so-called track-length mean, $\bar{L}_T$, and dose mean, $\bar{L}_D$, of LET, which were however derived from event size distributions obtained with a Rossi-counter. Actually, these parameters thus obtained are directly related to what nowadays is named 'lineal energy': $\bar{y}_F = \bar{L}_T, \bar{y}_D = (9/8)\bar{L}_D$, while the strictly defined mean values of LET cannot be measured at all with ionization counters.

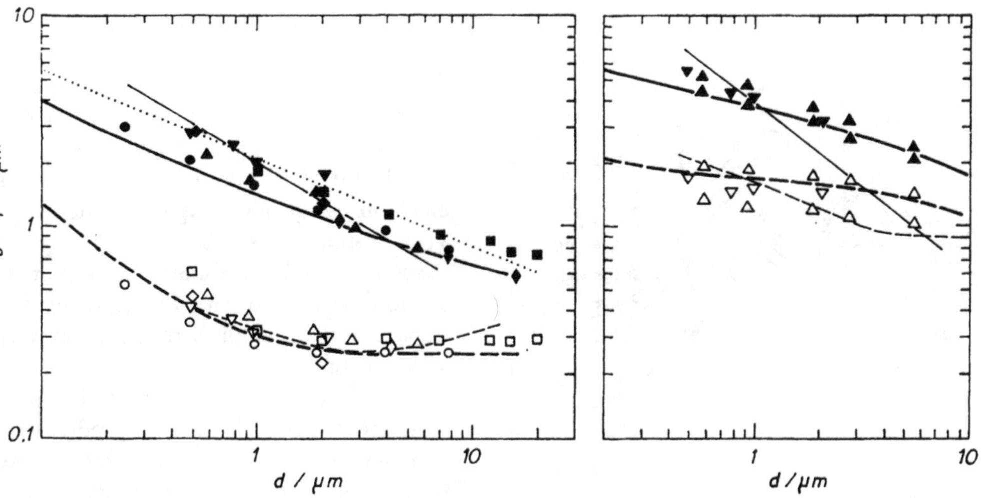

Fig. 5.4. Mean values $\bar{y}_D$ (closed symbols and unbroken curves) and $\bar{y}_F$ (open symbols and dashed curves) in dependence on the spherical volume diameter d for ^{60}Co γ-rays (left-hand panel) and X-rays of about 100 keV photon energy (right-hand panel). Heavy lines according to theory (Eqs. (5.13)).

Origin of other data:

Thin lines: Kellerer and Rossi (1971), calculated;
Dotted line: $\bar{y}_D$, Bengtsson and Lindborg (1974), variance method;
⧫ ◇ : Abel *et al.* (1966), Rossbach (1967), walled counter;
■ □: Biavati and Boer (1966), walled counter, $\bar{y}_F$ corrected here by the factor 0.85 for the wall effect;
▲ △: Braby and Ellett (1971), grid-walled counter, right-hand panel: 250-kVp X-rays with HVL = 0.44 mm Cu (upper values) and HVL = 1.8 mm Cu (lower values);
◆ ◇: Dicello (1974), walled counter;
● ○: Dvorak (1975), grid-walled counter;
▼ ▽: Coppola *et al.* (1976), walled counter.
(Biavati and Boer to Dvorak: taken from Booz (1976))

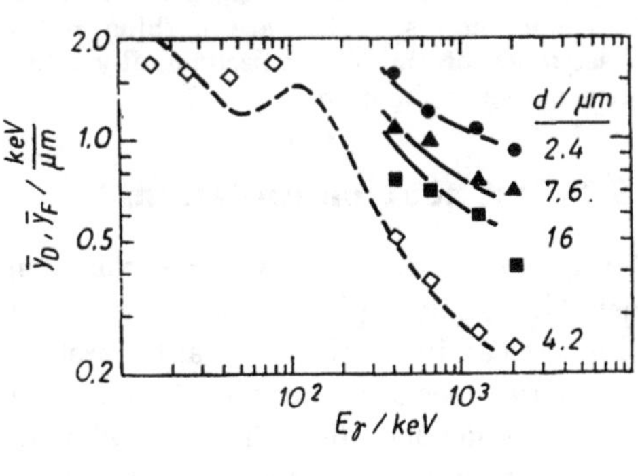

Fig. 5.5. Mean values $\bar{y}_D$ (closed symbols, full curves) and $\bar{y}_F$ (open symbols, dashed curves) in dependence on the photon energy E_γ for different spherical volume diameters d.
Theoretical curves calculated from Eqs. (5.13). Experimental data (walled counter) according to Abel *et al.* (1966) and Rossbach (1967); monoenergetic γ-ray sources for E_γ above 100 keV, X-rays for E_γ (mean energy) below 100 keV.

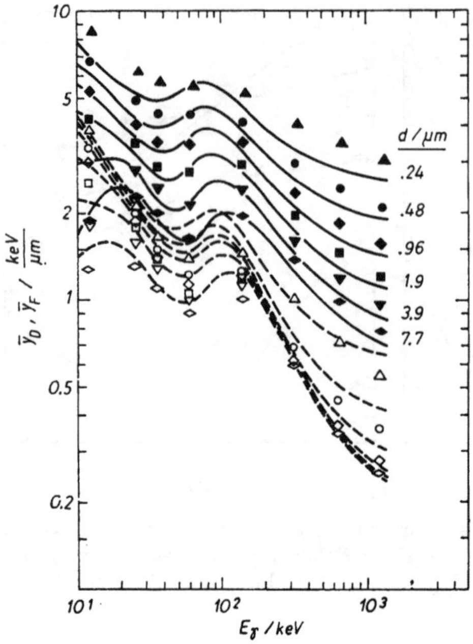

Fig. 5.6. See the caption of Figure 5.5; experimental data (grid-walled counter) according to Dvorak (1975) (taken from Booz, 1976) obtained by using different monoenergetic γ-ray sources.

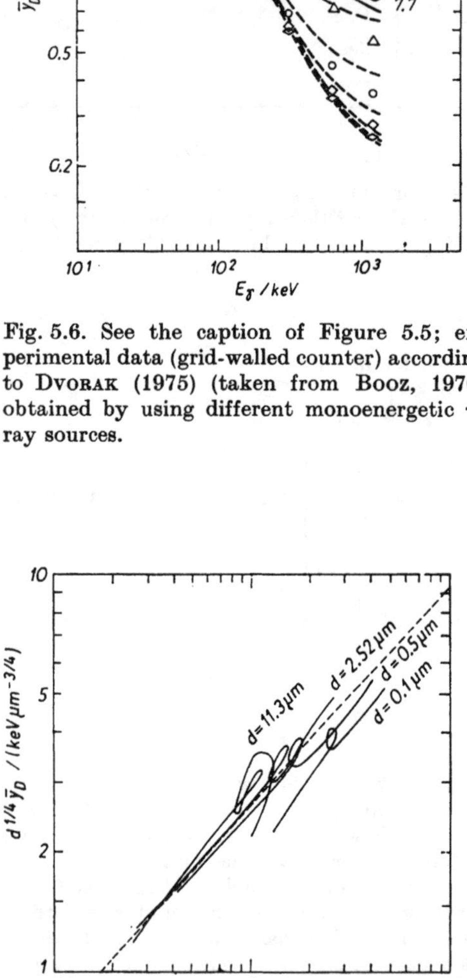

Fig. 5.7. $d^{1/4}\bar{y}_D$ versus $\bar{y}_F$ plot of theoretical results for photon radiations in the energy range 0.01 MeV $< E_\gamma <$ 1.25 MeV. Different curves for the spherical volume diameters indicated. Broken line: $d^{1/4}\bar{y}_D/\text{keV}\ \mu\text{m}^{-3/4} = 2.6\ (\bar{y}_F/\text{keV}\ \mu\text{m}^{-1})^{0.55}$.

only, at 10 keV *decrease* of $\bar{y}_F$ for *large* d only. The former is a consequence of disregarding (by theory as well as experiment) 'zero' events (passage of electrons without producing any ionization), while the latter is caused by the too-small electron range as compared with d. It is very remarkable that, regarding even these peculiarities, theory and experiment are almost in quantitative agreement (the departures are seldom much greater than 10 per cent).

Summing up, it can be stated that our analytical theory is able to predict microdosimetric parameters for photon radiations correctly for arbitrary sensitive volume diameters and photon energies. The accuracy of the results of theory obviously exceeds that of experimental data obtained with walled counters. Moreover, with some caution, the usually small departures from wall-less-counter results can be interpreted so as to find the theory likely to be even closer to reality than the wall-less counteris.

Booz (1976) has proposed an empirical relation between $\bar{y}_F$, $\bar{y}_D$ and d by showing that a log-log-plot of $d^{1/4}\bar{y}_D$ versus $\bar{y}_F$ results in a straight line of the slope 0.55 within experimental error. In Figure 5.7, the results of theory are plotted in this manner. Actually, the curves for different diameters d match such a straight line to some degree, in particular for $\bar{y}_F < 1$ keV/µm. However, in general, Booz's rule seems not to be very good in the light of our theory. The disagreements come mainly from the loops in the single curves, which are nothing but a consequence of the also experimentally established waves in Figure 5.6.

5.3.2. Fast neutron radiations

Energy deposition by fast neutrons has been the main topic of microdosimetry. A lot of experimental as well as theoretical data have been accumulated therefore. We will consider them in the following summarily in terms of the mean values $\bar{z}_F$ and $\bar{z}_D$, while the discussion of the event size spectra themselves are reduced to a few instructive examples only.

Some preliminary remarks should be made. In contrast to other radiations where straggling and δ-ray production are the most important factors, theoretical microdosimetry of neutrons depends mainly on using correct interaction cross-sections of neutrons with atomic nuclei (elastic and inelastic collisions, nuclear reactions), correct stopping power for the slowing down recoils, and a realistic atomic composition of the absorbing medium. This physical background and the choice of data made for our calculation (Tab. III in the appendix and the curves following in this section) are described in detail in chapter 4. Concisely: these data used correspond largely to those upon which the calculations by CASWELL and COYNE (1974, 1976) were based. For the atomic composition the so-called ICRU-tissue, omitting elements other than H, C, N, O, was assumed throughout (unless otherwise indicated). A particularity is our averaging all cross-sections over ± 7 per cent of the neutron energy, to smooth out sharp resonances.

Finally, the problem of δ-ray absorption events and the γ-ray contamination of experimental neutron fields should be mentioned. δ-ray events have usually been neglected in previous theoretical approaches, while the present theory includes them appropriately (see section 3.3). As to comparisons with experimental data, however, one must keep in mind two things: walled counters suppress δ-ray events (KELLERER, 1971b) and, on the other hand, an uncertain number of similarly small events due to the γ-component occur. In terms of dose, the latter can hardly be made smaller than 5 per cent. Fortunately, these uncertainties concerning electron events in neutron fields are not very important when considering *dose distributions*, $d(y) = y f_1(y)/\bar{y}_F$, or the corresponding dose-mean, $\bar{y}_D$, since then the figure 5% gives an orientation on the order of errors. The frequency-mean, $\bar{y}_F$, however may be appreciably influenced, being overestimated by theories that disregard δ-rays and underestimated by wall-less proportional counters in the presence of a γ-component, while in walled counters one has some indefinite mutual compensation of errors.

Figure 5.8 shows a number of theoretical and experimental event size spectra in comparison with the results of the present theory for three different neutron energies. The main peaks are due to recoil protons; the other peaks, if present, correspond to α-particles, C-recoils, and O-recoils, in the sequence of increasing y. This pronounced structure is somewhat obliterated in our theoretical curves, because the numerical procedure (section 5.2.2) with the moderate number of points used ($M = 39$) is too coarse to reproduce all details.

The wall-less-counter measurements by RODGERS and GROSS (thin unbroken lines) are likely to be most reliable, and it is these results that the present theory agrees with best. The deviations are comparable to those resulting from the other two theoretical approaches, except for 14-MeV neutrons where the neglect of straggling in the analytic calculation by CASWELL and COYNE (dashed line) proves perceptible in the proton peak. The various experimental spectra show considerable scatter. Reasons for this may be manyfold. One of them are distorsions caused by the counter wall, which was particularly investigated by OLDENBURG and BOOZ (1970) and COPPOLA and BOOZ (1975). Another factor are differences in the neutron energy; in particular the medial panel displays a number of curves obtained with broad neutron energy spectra, which must of course broaden the event size distributions, too, resulting in a lower main peak. The atomic composition of the counter wall and gas may also be a source of discrepancies, since our calculation is for ICRU-tissue (except for $E_n = 5.75$ MeV) while most counters correspond to the SHONKA plastic composition (see Table 4.7 at page 133). However, the changes in $f_1(y)$ found by CASWELL and COYNE (1974) when taking SHONKA plastic instead of ICRU-tissue in their calculations were not in the least an improvement in respect of experimental results. Finally, the unavoidable γ-ray component in all experiments should be kept in mind. Fortunately, this has some influence only on the very left tails of the spectra shown, i.e. at $y < 10$ keV/μm. This is also the region where pure δ-electron events may have a share. As these are suppressed in walled counters, it is interesting to note the rela-

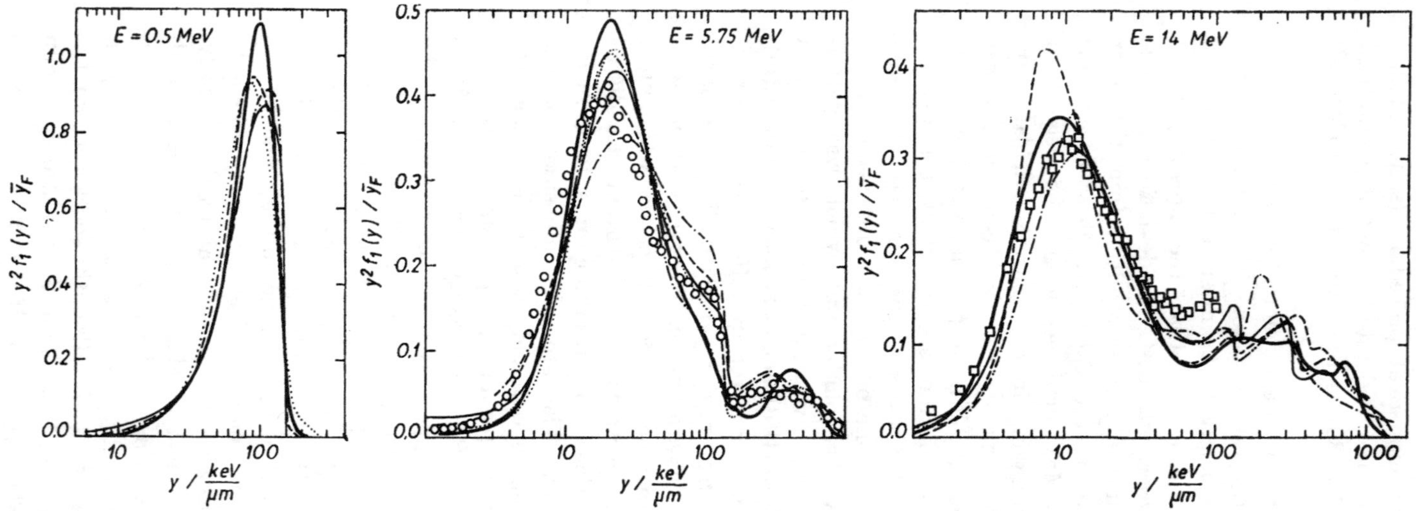

Fig. 5.8. Fast neutron single event size distributions for spherical sensitive regions of 1 μm diameter (represented as differential fraction of energy per differential logarithmic y-interval).
Heavy lines according to the present theory for the three neutron energies indicated; medial panel for SHONKA plastic, other panels for ICRU-tissue.

Other theoretical data:

· · · · · COPPOLA *et al.* (1974), BOOZ and COPPOLA (1974); Monte Carlo calculation allowing for straggling, SHONKA plastic (neutron energies: 0.62 MeV, 5.85 MeV).

– – – – CASWELL and COYNE (1974, 1976) — except for the medial panel; analytical calculation disregarding straggling, ICRU-tissue (neutron energies: 0.55 MeV, 14 MeV).

Experimental data:

——— RODGERS and GROSS (1974); wall-less counter (neutron energies: 0.55 MeV, 5.8 MeV, 15 MeV).

–·–·– BIAVATI and ROSSI (1965)*) — except for the medial panel; walled counter (neutron energies: 0.43 MeV, 14.7 MeV; $d = 1.2$ μm).

–··–·· KRÜGER and GRAUL (1970) — except for the medial panel; walled counter (neutron energy: 14 MeV).

○ WILSON and FIELD (1970); walled counter (neutron energy spectrum, 7 MeV mean energy; $d = 1.09$ μm).

□ VIDENSKY *et al.* (1973); walled counter (neutron energy: 14 MeV).

Medial panel only:

– – – – MENZEL *et al.* (1976); walled counter (neutron energy spectrum, about 7 MeV mean energy).

–·–·– CANADA *et al.* (1974); walled counter (neutron energy spectrum, 7 MeV mean energy).

–··–·· BOOZ and COPPOLA (1974); walled counter (neutron energy: 5.85 MeV).

*) Taken from ROSSI (1966)

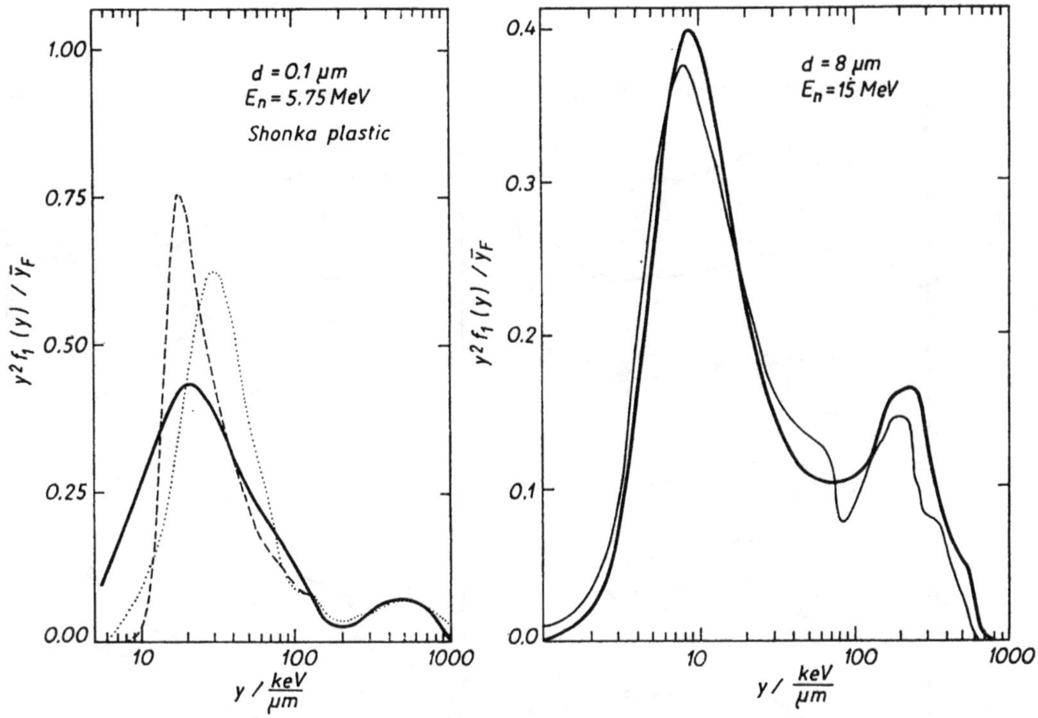

Fig. 5.9. Fast neutron single event size distributions for very small (left-hand panel) and very large (right-hand panel) spherical sensitive regions.
Heavy lines according to the present theory for the neutron energies indicated.
Origin of other data: COPPOLA *et al.* (1974), Monte Carlo calculation (5.85 MeV) including straggling (dotted line) and not including straggling (dashed line). RODGERS and GROSS (1974), wall-less counter (thin unbroken line).

tively high values of the spectra from the wall-less counter at small y and the similar tendency in our theoretical curves, which do include the δ-ray events in contrast to the other theoretical ones.

Further examples of event size distributions for small and large sensitive sites are shown in Figure 5.9. Especially the left-hand panel is to demonstrate the effect of straggling, most pronounced for small d. A pertinent theoretical investigation was made by COPPOLA *et al.* (1974), from which the example in Figure 5.9 has been taken. A marked disagreement with the present theory can be seen. Namely, our treatment of straggling does not displace the proton peak, as does the procedure of COPPOLA *et al.*, but it only broadens the peak considerably. This may be partly due to δ-ray events in the vicinity of $y = 10$ keV/μm, which are omitted in the Monte Carlo calculation. Besides, in the latter procedure straggling was included only in a rather crude manner. Thus, our results may be more reliable, which is also indicated by the good correspondence with experimental results in Figure 5.8 for 14-MeV neutrons, inasmuch as straggling plays a role in this case too. The right-hand panel of Figure 5.9 gives an example for a large sensitive site for which not the straggling, but the relatively frequent occurrence of starters, stoppers, and insiders, is characteristic. Again the noticeable agreement with the wall-less-counter measurement is exhibited, apart from the smoothing of the heavy recoil edges discussed above.

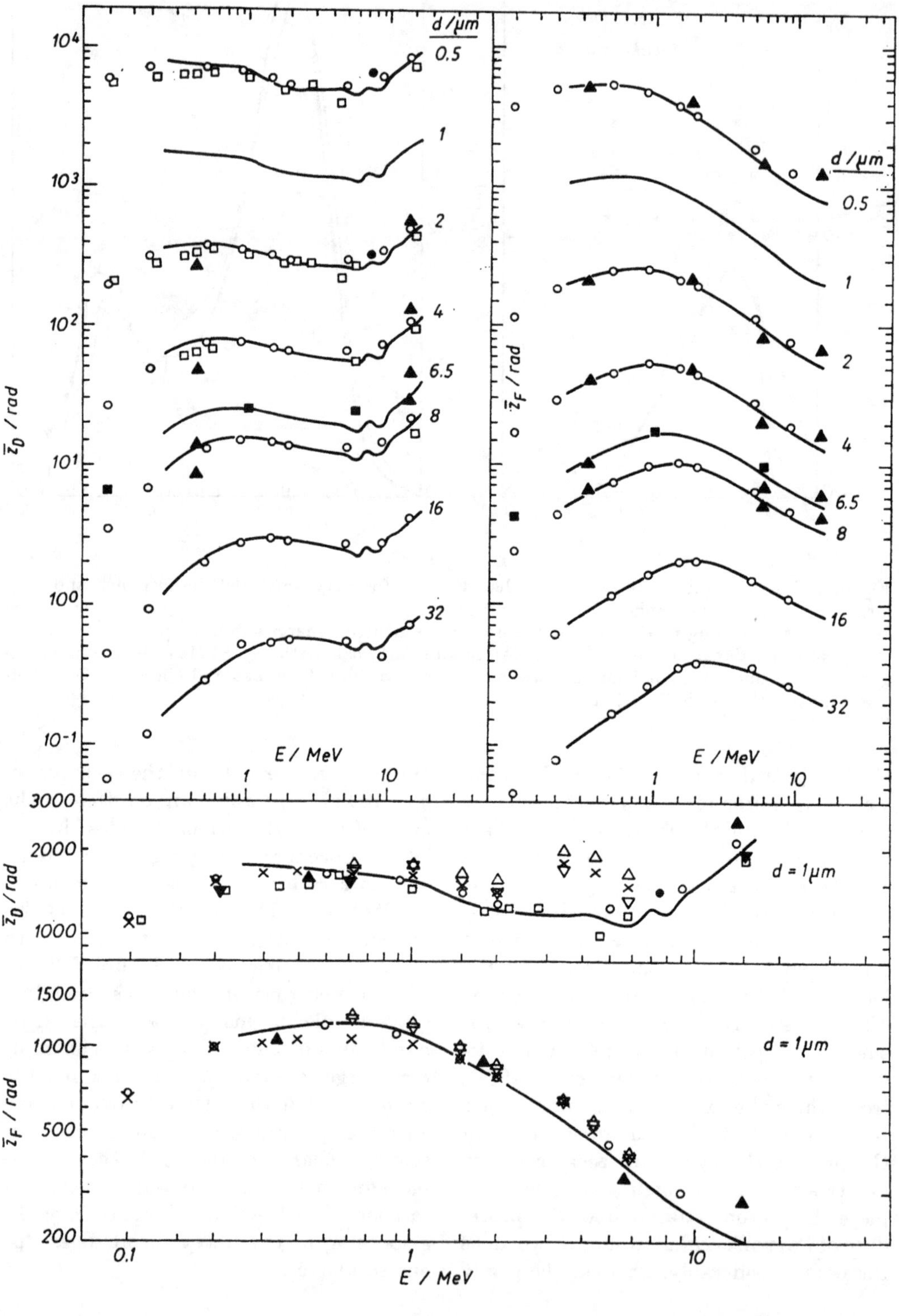
d/μm
0.5
1
2
4
6.5
8
16
32
10^4
10^3
10^2
10^1
10^0
10^{-1}
$\bar{z}_D$ / rad
$\bar{z}_F$ / rad
E / MeV
0.1
1
10
3000
2000
1000
1500
500
200
d = 1μm

After the insight into some peculiarities of fast neutron microdosimetry gained by the hand of a few examples of event size spectra, a summary of data is now given in Figure 5.10 in terms of the familiar mean values of specific energy. The sensitive site diameters have been chosen correspondingly to the calculations by CASWELL and COYNE (1974, 1976), which constitute the most extensive set of data so far. A great deal of other available data for these diameters have also been included. Moreover, $d = 6.5$ μm has been added for displaying the 1975 Monte Carlo results of COPPOLA and BOOZ. It is important to note that the dependence on neutron energy does not resemble an analytical function owing to the abruptly changing interaction cross-sections of neutrons with matter. The present theory smoothes these so-called resonances by averaging over ± 7 per cent of neutron energy, so that no sharp edges occur in the theoretical curves. This may, on the other hand, give rise to discrepancies with other data if these refer to sharp neutron energies.[1]) Figure 5.10 shows a rather good general agreement, especially with the theoretical values of CASWELL and COYNE and the wall-less-counter results of RODGERS and GROSS, while some older data show larger deviations. Systematic differences can be expected with respect to $\bar{z}_F$ owing to the γ-ray contamination and the influence of δ-ray events, which must be most pronounced for large neutron energies and small site diameters. In fact, Figure 5.10 indicates such kind of deviations in accord with the comments given above on this point. Another source of scatter, namely the varying atomic composition of the absorbing medium (in theoretical approaches as well as experiments), seems to be of minor importance, as demonstrated by the fairly small divergences of $\bar{z}_D$ only, for the two compositions used in the pertinent calculation by BOOZ *et al.* (lower panels of Fig. 5.10). However, just these theoretical results for $\bar{z}_D$ at $d = 1$ μm as well as further ones coming from the same research group (BOOZ and COPPOLA, 1974) disagree considerably with both the experimental results of GROSS and RODGERS (1974) and the theoretical curve according to the present theory.

In conclusion it can be stated that, by virtue of Figure 5.10, the results of the present theory concerning the important parameters $\bar{z}_F$, $\bar{z}_D$ for neutrons as listed in Table III of the appendix can be declared a good comprehensive representation of the hitherto existing relevant information in this field.

[1]) To this point, special attention was paid by CASWELL and COYNE (1976), who demonstrated the values for 1 MeV neutron energy to contrast markedly due to such a resonance. Therefore, these 1-MeV results have been omitted in Figure 5.10.

Fig. 5.10. Mean values, $\bar{z}_D$ and $\bar{z}_F$, of specific energy per absorption event in dependence on neutron energy for the spherical region diameters indicated.
Curves calculated from the present theory (ICRU-tissue).

Origin of other data:

○: CASWELL and COYNE (1976), analytical calculation (ICRU-tissue).
△ and ▽: BOOZ *et al.* (1972a), Monte Carlo calculation for SHONKA plastic + ROSSI-FAILLA-gas (△) and ICRU-tissue (▽).
×: BOOZ and COPPOLA (1974), Monte Carlo calculation (SHONKA plastic).
■: COPPOLA and BOOZ (1975), Monte Carlo calculation (SHONKA plastic without wall effects).
□: RODGERS and GROSS (1974), wall-less proportional counter.
▲: BIAVATI and ROSSI (1965), walled counter (taken from ROSSI, 1966; and KELLERER and ROSSI, 1972).
▼: RODGERS *et al.* (1972), walled counter (taken from CASWELL and COYNE, 1976).
●: MENZEL *et al.* (1976), walled counter (neutron energy spectrum).

5.3.3. Ion radiations

Microdosimetric description of ion radiations has some peculiarities due to the distinction of direct and indirect events (see section 5.2.1). Along with the usual overall parameters $\bar{y}_F$, $\bar{y}_D$, and $f_1(y)$, which are difficult to measure, we have to consider the analogous partial quantities for either kind of events and, in addition, the dose fraction $\mu_D^{(\delta)}$ of indirect events. $f_1^{(i)}(y)$ and its mean values for direct events in spherical sites are rather uninteresting, because the event sizes follow a triangular spectrum which largely masks the influence of straggling. Therefore, apart from sparse data on indirect events in spherical sites, most experimental work is concerned with pure straggling distributions produced by monoenergetic ions in thin gas layers. The most extensive Monte Carlo calculations just yielding all the other quantities were carried out by Gubin (1974) and Gubin *et al.* (1976). In the following we will repeatedly refer to this work while approaching the subject in the sequence: overall parameters, parameters referring to δ-rays (indirect events), straggling in thin layers.

Some examples of complete event size spectra for ion radiations are shown in Figure 5.11. One can clearly distinguish the two components in these spectra, resulting from direct (large y) and indirect (small y) events, except for the low-LET protons which produce events as small as δ-electrons do. Only in the full curve of the upper right-hand panel δ-ray events are lacking as a consequence of our fundamental approximation: δ-ray cut-off

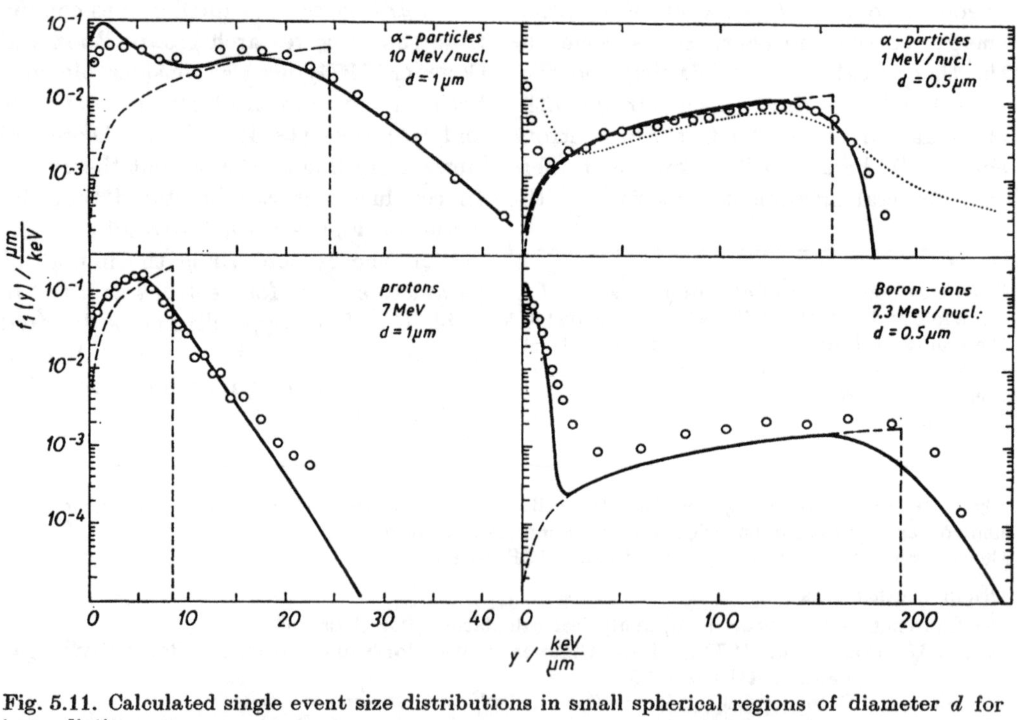

Fig. 5.11. Calculated single event size distributions in small spherical regions of diameter d for ion radiations.
Circles: Monte Carlo results of Gubin (1974) and Gubin and Sakovich (1974).
Dotted line: experimental result (grid-walled counter, ^{241}Am α-particles) by Gross *et al.* (1970). The peak at small y is due to indirect events.
The dashed curves display the triangular spectrum representative for the direct events (i.e. $\mu_F^{(i)} f_1^{(i)}(y)$ with $\bar{y}_F^{(i)} = \mu_D^{(i)} L_\infty$).

according to the rule (3.12) which yields $\delta > E_{\max}$ in this case. In the region of mere direct events, the spectra should be largely of triangular shape,

$$f_1^{(i)}(y) = \frac{8}{9}\frac{1}{\bar{y}_F^{(i)2}}\, y \quad \text{for} \quad y \leq \frac{3}{2}\,\bar{y}_F^{(i)}, \qquad \bar{y}_F^{(i)} = \mu_D^{(i)} L_\infty, \tag{5.36}$$

visualized by the dashed lines which include the additional factor $\mu_F^{(i)}$ in compliance with Eq. (5.14). The sharp edge of the latter is however smoothed in reality as an effect of straggling, which is reflected in good correspondence by both our theoretical curves[1]) and the Monte Carlo results of Gubin (circles). In the two left-hand panels the triangular component is almost completely masked by straggling and δ-ray events. Boron ions, on the contrary, produce the triangular spectrum exactly over a wide range of y. Complete agreement with the Monte Carlo data is only achieved in the case of protons, while for higher charge the present theory yields too large a number of too small indirect events, which — on the other hand — lowers the curves in the region of direct events for the sake of normalization (i.e. $\mu_F^{(i)}$ turns out too small — cf. Eq. (5.14)). The reason for this will be discussed below. The poor agreement with the experimental spectrum of Gross *et al.* (1970) for α-particles (high values beyond the normal decline of both direct and indirect events) is likely to be due to the experimental device: no sharp boundary of the sensitive region of the grid-walled counter.

The overall mean values, $\bar{z}_F$ and $\bar{z}_D$, for three different ions in dependence on energy are compared with Gubin's Monte Carlo results in Figure 5.12. One observes

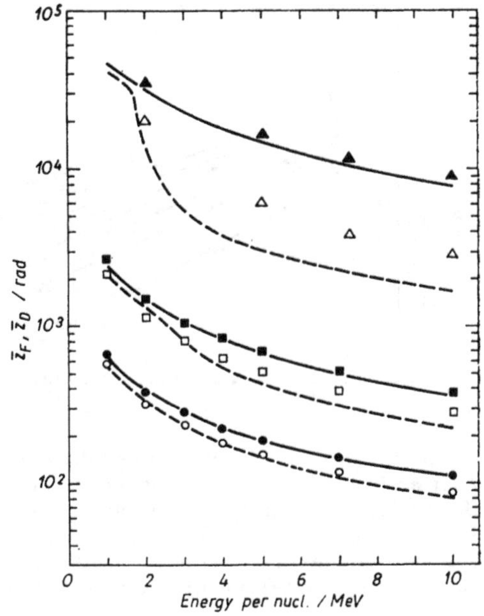

Fig. 5.12. Mean values $\bar{z}_D$ (full lines, closed symbols) and $\bar{z}_F$ (broken lines, open symbols) for three ion radiations and spherical regions of diameter d.
Curves obtained by theory (equivalent to Tab. VI).
Symbols: Monte Carlo results of Gubin (1974): ●○ protons, $d = 1\ \mu m$; ■□ α-particles, $d = 1\ \mu m$; ▲△ boron ions, $d = 0.5\ \mu m$.

a remarkable agreement, except that the present theory yields too small values of $\bar{z}_F$ at increasing charge of the ion considered. This clearly is caused by our fundamental approximation of treating the actually correlated δ-electrons in indirect events as independent particles (approximation principle 2b in section 3.2.2), which increases the number, and decreases the energy, of indirect events. It is evident that this splitting of the small, but frequent, δ-ray events into a yet larger number of smaller ones must strongly influence $\bar{z}_F$ (note that $\bar{z}_F^{-1}$ equals the total number of events per unit dose), while $\bar{z}_D$ remains largely unaffected (compare Eqs. (5.15), (5.16)).

Let us now deal with $\mu_D^{(\delta)}$, the fraction of energy imparted by indirect events. The apparently most reliable data is that

---

[1]) The numerical method described in section 5.2.2 tends to smooth sharp edges, in addition. However, what the curves show, obviously, seems to be mainly the true effect of straggling.

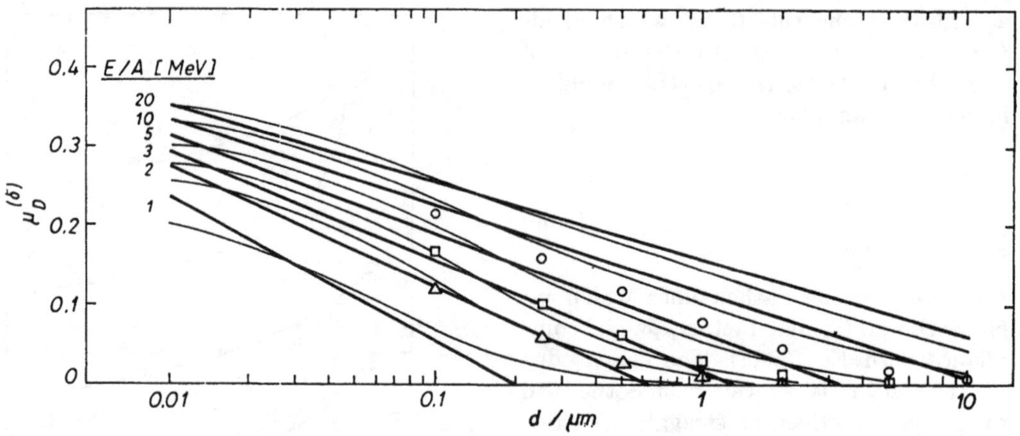

Fig. 5.13. Fraction $\mu_D^{(\delta)}$ of absorbed dose due to indirect events in spherical regions of diameter d for arbitrary ions (E/A = energy per nucleon).
Heavy lines: this theory (corresponding to Tab. IV).
Thin lines: calculated by KELLERER (1971 b).
Symbols: Monte Carlo results of GUBIN *et al.* (1976) (corresponding to Tab. VII); circles: 10 MeV/nucl.; squares: 5 MeV/nucl.; triangles: 3 MeV/nucl.

calculated by KELLERER (1971 b)[1]). Figure 5.13 shows the remarkable good agreement with the present theory. Relevant are the *absolute* deviations, which never exceed 0.02, i.e. in the worst case only 2 per cent of energy are incorrectly distributed between direct and indirect events. Note that our $\mu_D^{(\delta)}$ results from nothing but the choice of the cut-off energy δ as expressed by formula (5.22). Thus, the somewhat arbitrary rule (3.12) that determines δ proves very appropriate.[2]) GUBIN's values also shown in Figure 5.13 are somewhat too small (see also Tab. VII in the appendix). A comparison with a number of experimentally determined $\mu_D^{(\delta)}$ values is given in Tables 5.1 and 5.2, which are discussed below.

As to the microdosimetric parameters referring to *indirect events* only, i.e. $f_1^{(\delta)}(y)$, $\bar{y}_F^{(\delta)}$, and $\bar{y}_D^{(\delta)}$, the apparently most comprehensive set of data was published by GUBIN *et al.* (1976). Their normalized single-event spectra are depictured in Figure 5.14 together with the results of the present theory. The corresponding mean values, also in comparison with our results, are given in Table VII of the appendix. The most important difference is the influence of the ion's charge. The spatial correlation of δ-rays, which is neglected in our theory, leads to an increasing number of compound events (more than one δ-electron involved) at growing charge of the ion for the same energy per nucleon. The Monte Carlo spectra in Figure 5.14 as well as the corresponding mean values in Table VII clearly exhibit this effect: occurrence of large absorption events in the case of heavy ions. In the present theory, however, there is no difference at all with respect to the ion's charge, since it only admits 'single-par-

[1]) These calculations are based on radial distributions of dose in the ion's track, $D(r)$, measured by WINGATE and BAUM (1970) (see also WINGATE and BAUM, 1976). $\mu_D^{(\delta)}$ can easily be obtained by integrating $\bar{D}(r)$ over $r > d/2$, where $\bar{D}(r)$ is $D(r)$ averaged with respect to the spherical site centered at the distance r from the track.

[2]) Because the range-energy relation for electrons, which relates δ and d according to this rule, is an almost exact power law (see section 4.5.1), Eq. (5.22) yields practically straight lines in the semi-logarithmic plot of Figure 5.13.

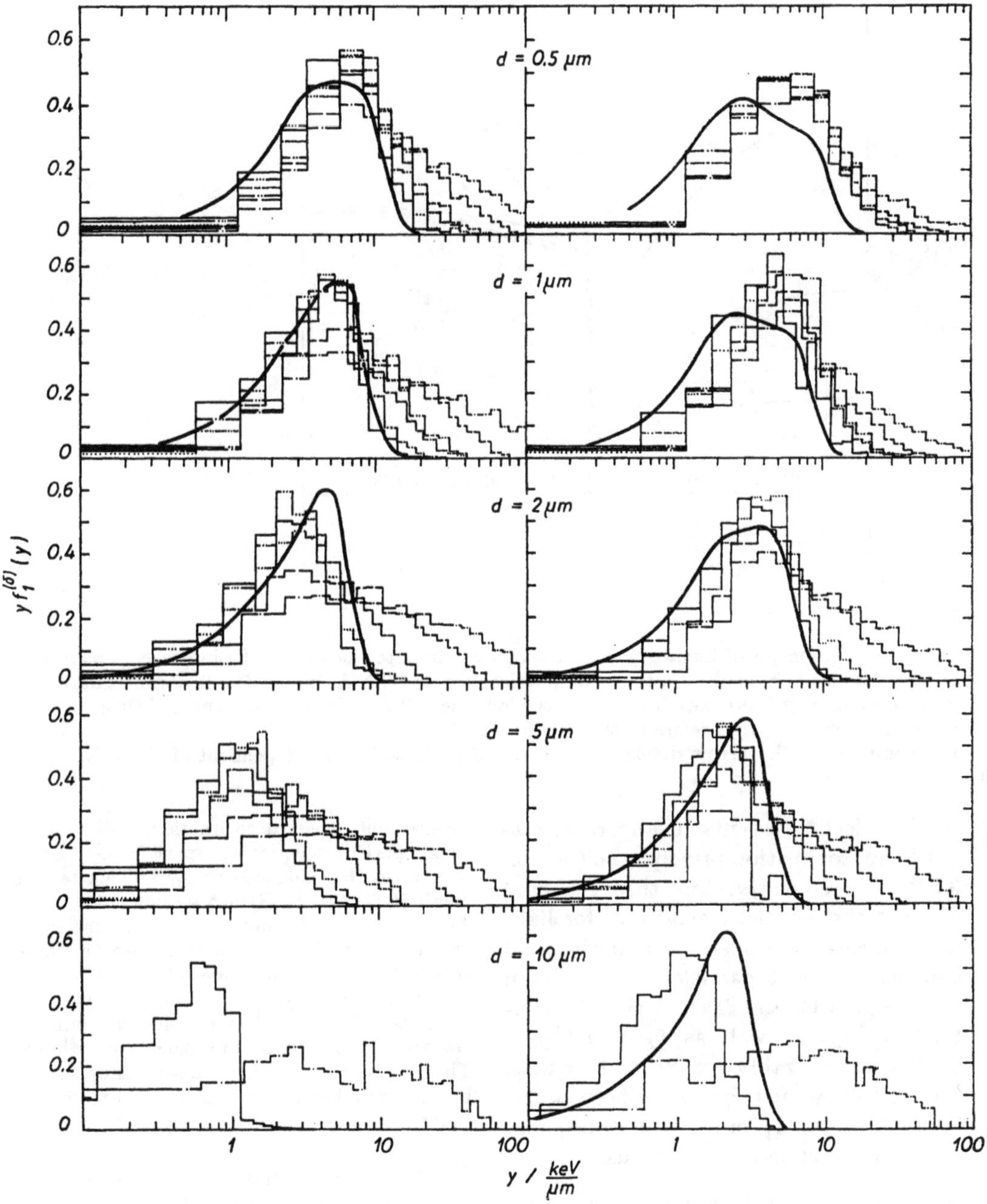

Fig. 5.14. Calculated single event size distributions in spherical regions (diameter d) of indirect events for different ion radiations: 5 MeV/nucl. — left-hand side; 10 MeV/nucl. — right-hand side.
Continuous curves according to this theory (no influence of ion charge).
Histograms: Monte Carlo results of GUBIN *et al.* (1976) for the following ions: ——— He; — ··· — Li; — ·· — B; ····· C; — — Ne; —·—·— Ar.

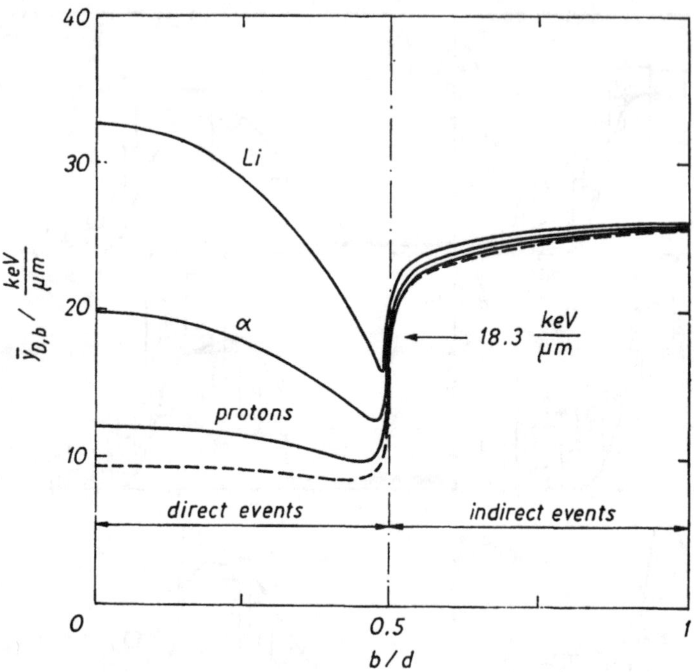

Fig. 5.15. Dose-mean of lineal energy for sharply collimated beams of the indicated ions passing in the distance b from the centre of a spherical region with diameter $d = 0.02$ μm. The curves were computed by PARETZKE from simulated ion tracks (taken from KELLERER and CHMELEVSKY, 1975). The broken curve refers to the charge zero.
For comparison, the arrow displays our value of $\bar{y}_D^{(\delta)}$, which is independent of charge.

ticle' δ-electron events in any case, due to the approximation principle 2b (section 3.2.2). Consequently, this theory should describe the indirect events best for light ions. Indeed, one observes a fairly good agreement for α-particles and lithium ions (Fig. 5.14 and Tab. VII). By contrast, $f_1^{(\delta)}(y)$ as well as $\bar{y}_F^{(\delta)}$ and $\bar{y}_D^{(\delta)}$ deviate considerably for Ne and Ar ions. This reflects what has already been stated in section 3.2.2: our theory becomes questionable at least for Ar ions.

In this context, it is important to discuss the consequences of this defect of the theory concerning the description of biological radiation effects. This will be elucidated in detail in chapter 8: the artificial further splitting of the already small δ-ray events proves to be of little influence as long as the corresponding doses (mean specific energy) are small, anyway, as compared with the inactivation dose, D_0, of the object considered, which is usually the case. This means that, regarding the indirect events, only their dose fraction, $\mu_D^{(\delta)}$, occurs in the pertinent equation, apart from a correction term that depends on $\bar{z}_D^{(\delta)}$. This term is small if $\bar{z}_D^{(\delta)} \ll D_0$. $\bar{z}_F^{(\delta)}$ does not enter at all. Besides, the re-formulation of the theory in chapter 8 makes it possible to use the more relevant Monte Carlo values of $\bar{z}_D^{(\delta)}$, while the essential quantity $\mu_D^{(\delta)}$ is more precisely estimated by the present theory than by GUBIN's Monte Carlo approach (see above). The reason for this latter discrepancy cannot be analysed here, but it is likely to bias the other result, $\bar{z}_F^{(\delta)}$ and $\bar{z}_D^{(\delta)}$, too.

Further data on indirect events are rare. Let us first consider another Monte Carlo result given in Figure 5.15. This figure is also to visualize the way in which experimental information for spherical sensitive sites can be obtained: a sharply collimated beam of ions passes in the distance b from the center of the site. Thus, all quantities become a function of the impact parameter b. With $b > d/2$

one is studying indirect events. Two characteristics of PARETZKE's results are of special interest here: in the region of indirect events the curves grow rapidly to a rather constant plateau value and there is little difference between the ions indicated. The latter confirms that the correlation effect we disregard is small up to lithium ions.[1]) The former finding makes it possible to compare our $\bar{y}_D^{(\delta)}$ roughly with that plateau value of $\bar{y}_{D,b}$. Actually, $\bar{y}_D^{(\delta)}$ is the mean value of $\bar{y}_{D,b}$ over $b > d/2$ weighted by b times the smoothed radial dose distribution $\bar{D}(b)$, which is however a steeply decreasing function of b. Accordingly, $\bar{y}_D^{(\delta)}$ should be somewhat smaller than the plateau value. Thus, the corresponding result of the present theory indicated in Figure 5.15 fits not so bad with PARETZKE's curves.

Analogous experiments with a wall-less proportional counter and collimated beams (KLIAUGA and ROSSI, 1976) also yielded this rapid flatting out of $\bar{y}_{D,b}$ for $b > d/2$. These results are listed in Table 5.1. In addition, the dose fractions due to indirect events, $\mu_D^{(\delta)}$, were also given by these authors. The noticeable disagreement of the present theory with respect to $\mu_D^{(\delta)}$ is in the same direction as in Figure 5.13 for $d = 1$ µm, i.e. KELLERER's theoretical results correspond somewhat better to, but also differ from, the experiment.

Table 5.2 shows the results of a similar experiment by GROSS and RODGERS (1972), who integrated subsequently over all values of the impact parameter b to obtain the overall frequency-mean $\bar{y}_F$ of

[1]) The dashed curve for the hypothetical case of zero charge, i.e. infinite mutual distance of δ-rays along the track of the 'ion', exactly simulates the situation that our theory applies to.

Tab. 5.1. $\bar{y}_D^{(\delta)}$ and $\mu_D^{(\delta)}$ for a spherical site of $d = 1$ µm diameter.
Experimental data by KLIAUGA and ROSSI (1976) (the experimental values for $\bar{y}_D^{(\delta)}$ are actually $\bar{y}_{D,b}$ at the impact parameter $b = 1$ µm)

	$\bar{y}_D^{(\delta)}$ [keV/µm]		$\mu_D^{(\delta)}$ [%]	
	Experiment	Present theory	Experiment	Present theory
40 MeV Li ions	7.5	5.4	6.9 ± 1.4	8.7
10 MeV protons	5.5	5.0	8.8 ± 1.8	11.7
23 MeV protons	3.6	4.4	10 ± 2	16

Tab. 5.2. $\bar{y}_F$ and $\mu_D^{(\delta)}$ for protons. Experimental values from GROSS and RODGERS (1972). Calculation according to the present theory

Sphere diameter [µm]	Proton energy [MeV]	LET [keV/µm] for water (calculated)	$\bar{y}_F$ [keV/µm]		$\mu_D^{(\delta)}$ [%]	
			measured	calculated	measured	calculated
0.18	1.7	18.7	15.5	15.7	5	7.0
0.18	4.0	9.66	7.7	7.85	10	14.3
0.25	1.7	18.7	16.0	16.3	3	4.6
0.25	4.0	9.66	9.2	7.91	10	12.4
0.50	4.0	9.66	9.6	8.14	9	8.5
1.0	2.8	12.8	12.5	12.4	1.2	0.9
1.0	4.0	9.66	10.0	8.58	4	4.5

lineal energy. They also calculated $\mu_D^{(\delta)}$ from their results. The agreement with the present theory is fairly good. However, $\bar{y}_F$ is mainly determined by the linear energy transfer, L_∞ of the protons (see Eqs. (5.15) and (5.23)), so that the remaining *difference*, which depends on the indirect events only, should be regarded as the significant quantity. Therefore, L_∞ is also given in Table 5.2. Obviously, the experimental values of $\bar{y}_F$ are not precise enough to allow a meaningful comparison with theory regarding the difference $L_\infty - \bar{y}_F$.

Dealing with direct events means dealing with straggling since in the absence of straggling one would be concerned only with LET of the particles and the trivial triangular spectra (5.36). Therefore, pertinent measurements and calculations usually refer direct to pure straggling distributions and their variance, σ^2, which means considering absorption events in thin *layers* of matter crossed perpendicularly by a beam of ions.

For the sake of comparison, our theoretical approach can easily be adapted to this situation: in Eq. (5.3a) one must only replace the chord length spectrum $w(l')$ by the δ-function $\delta(l' - l)$, where l is the layer thickness. However, problems arise from the question of δ-ray cut-off. In principle, if the layer is 'infinite' in its plane, the experiments must yield the mean value $\bar{y}_F^{(i)} \approx L_\infty$ of lineal energy (see Eq. (5.23)), i.e. no δ-ray cut-off should be made in the theory. Then however, one is dealing with energy *loss* straggling rather than energy *deposition* straggling. This means that all δ-rays are fully included, which does not allow for the escape (and equivalent injection) of δ-rays. In other words, the energy deposition straggling actually displayed by proportional gas counters rests on a collision spectrum that differs from the theoretical unrestricted one, despite of $\bar{y}_F^{(i)} = L_\infty$. The theory can roughly simulate this just by some δ-ray cut-off, with the drawback of a too-small mean $\bar{y}_F^{(i)}$ turning out. As will be seen in the following, it proved most appropriate to adopt the latter way, i.e. δ-ray cut-off according to the rule (3.12) which is intrinsically adapted to sites finite in all three dimensions (as is a sphere). To obtain reasonable agreement in this way can be anticipated because in most experiments the layer is by no means 'infinite' having a radial extension only of the order of its thickness l. Thus, there is also some radial escape of δ-electrons, which must lower the experimental values of $\bar{y}_F^{(i)}$. Of course, in case that the rule (3.12) yields no cut-off anyway (large l or small particle energy) the whole problem becomes unimportant. Then good agreement ought to be expected by all means.

As a first example let us consider KELLERER's calculation of straggling distributions (Fig. 5.16) which also refer to, strictly speaking, energy *loss* straggling. Thus, one has complete analogy to the present theory in this respect (i.e. no such problem as discussed above). The spectra in the left-hand panel are without any δ-ray cut-off, and the agreement is striking. This demonstrates, which is important to note, an almost complete equivalence of the very different ways of describing the underlying distributions of energy loss in single collisions: in our theory ionization clusters of varying size, j, described by the corresponding frequencies $\omega_i(E, j)$ (cf. section 4.4), disregarding non-ionization energy losses; in KELLERER's straggling theory a continuous spectrum of energy losses ε which was drawn from direct energy loss measurements (RAUTH and SIMPSON, 1964) at small ε and approaches $1/\varepsilon^2$ at large ε. Only for the thinnest layer of Figure 5.16[1]) a meaningful difference becomes visible: our curve somewhat reflects the single ionizations (the first sharp peak corresponds to one ionization in the layer).[2]) The right-hand panel of Figure 5.16 demonstrates the influence of δ-ray cut-off. For the curves shown, our l-dependent cut-off energy is a little smaller

---

[1]) $l = 0.003\ \mu$m, which corresponds just to the effective mean thickness of the DNA-target in the lesion production model (chapter 6).

[2]) Strictly speaking, in the framework of the present theory, any f_1-spectrum consists of a sequence of δ-functions at integer numbers of ionizations. It is only the numerical procedure described in section 5.2.2 that usually makes these spectra smooth.

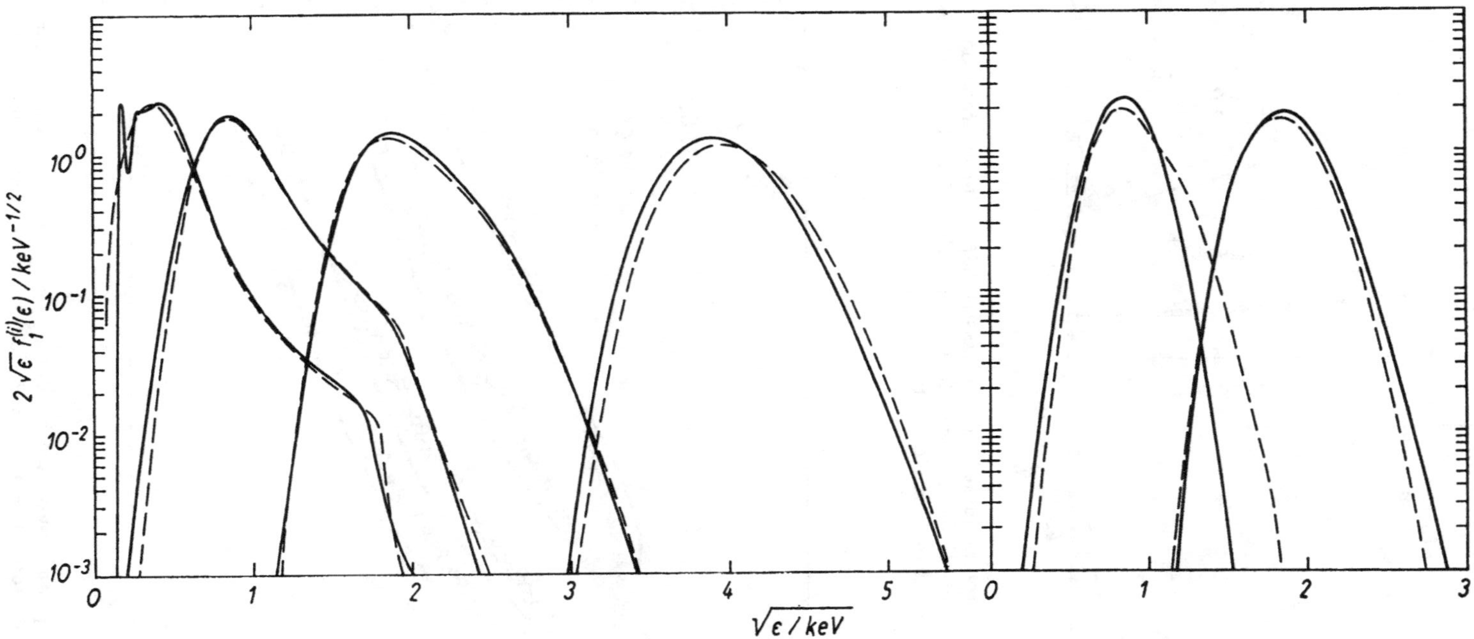

Fig. 5.16. Calculated straggling distributions of 5.75-MeV α-particles for the following layer thicknesses: left-hand panel (without δ-ray cut-off): $l/\mu m = 0.003, 0.012, 0.05, 0.19$; right-hand panel (with δ-ray cut-off): $l/\mu m = 0.012, 0.05$.
Broken curves according to KELLERER (1968): multiple folding of a modified $1/\varepsilon^2$ continuous collision spectrum (fixed cut-off energy $\delta = 1$ keV in the right-hand panel).
Full curves: from present theory (cut-off energy in the right-hand panel according to the rule (3.12)).

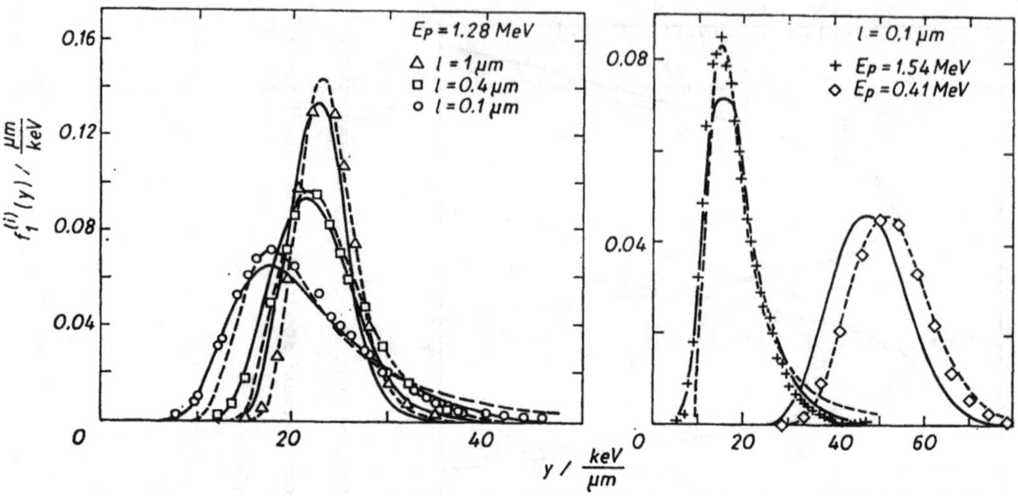

Fig. 5.17. Straggling distributions of protons measured by Glass (1968) (wall-less proportional counter). Full lines from this theory (cut-off energy according to the rule (3.12)); broken lines according to the Vavilov theory (also taken from Glass, 1968).

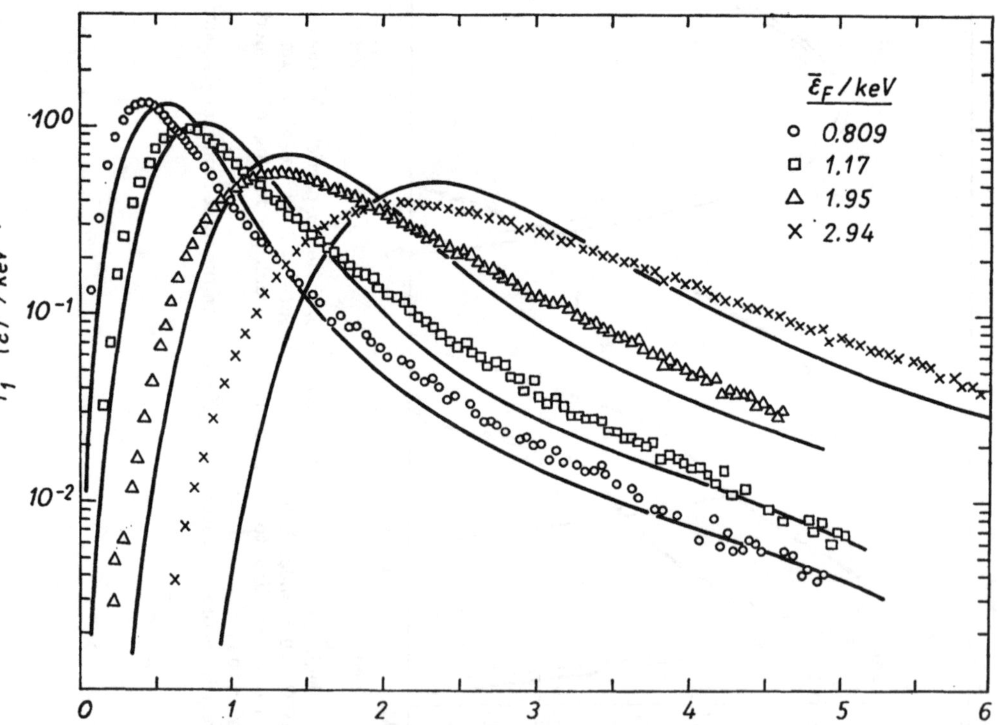

Fig. 5.18. Straggling distributions of protons measured by Baily *et al.* (1972) (wall-less proportional counter), using energy-degraded proton beams characterized by their mean energy deposition $\bar{\varepsilon}_F$ in a $l = 0.59$ μm sensitive layer. Curves from theory (cut-off energy according to the rule (3.12)) for the following proton energies [MeV]: 32.0, 22.4, 12.8, 7.6.

and a little larger, respectively, than the fixed value $\delta = 1$ keV used by KELLERER. The marked differences indicate the uncertainty (especially of the flank beyond the peak) in drawing comparisons with experimental data, in view of the discretion in the choice of δ.

Measured straggling distributions for protons are shown in Figure 5.17. The theoretical curves exhibit two facts: good correspondence for small energy depositions where the VAVILOV theory[1]) fails; discrepancies for large energy depositions in the case of the more thick layers where the VAVILOV theory, which includes no δ-ray cut-off, proves more appropriate. The displacement in the case of the 0.41-MeV protons is certainly caused by the beginning incorrectness of the Bethe formula for the linear energy transfer at small energies.

The distributions in Figure 5.18 are for higher proton energies. As the authors of the experiment did only report the mean energy depositions $\bar{\varepsilon}_F$ rather than the incident proton energies, the latter had to be 'guessed' for our calculation by means of the requirement $\bar{\varepsilon}_F = \int d\varepsilon \varepsilon f_1^{(i)}(\varepsilon)$. In this way, however, the apparent incident proton energies depend on the correctness of our δ-ray cut-off. In the experiment, different proton energies were obtained by means of an initially monoenergetic beam passing through different absorbers. This leads to an additional broadening of the experimental spectra due to the scatter in the incident energy, especially for small values of the latter. This effect which was also stressed by the authors is clearly displayed by Figure 5.18. Nevertheless, one observes remarkable agreement between theory and experiment, especially regarding the long tails of the spectra over about two orders of magnitude.

Let us finally deal with the mean values of straggling distributions, $\bar{y}_F^{(i)}$ and $\bar{y}_D^{(i)}$. Both these means would equal the linear energy transfer if there were no straggling. Therefore, the actually interesting quantity is only the *difference* which is related to the *variance*, σ^2, of the straggling distribution by $\sigma^2 = \bar{y}_F^{(i)}(\bar{y}_D^{(i)} - \bar{y}_F^{(i)})$. Corresponding results are usually given in terms of the more instructive quantity

$$l\sigma^2/\bar{y}_F^{(i)} = l(\bar{y}_D^{(i)} - \bar{y}_F^{(i)}) \approx \delta_2. \qquad (5.37)$$

For the problem expressed by writing '$\approx$' instead of '$=$' see the footnote 4 at page 147.

The energy δ_2, being equal to the quotient of the first and the second moments of the collision spectrum (see section 1.4.1), is a parameter of the given radiation only, i.e. it ought to be independent of the layer thickness l. This holds exactly with respect to energy *loss* straggling. Proportional counters, however, display energy *deposition* straggling, which makes the collision spectrum dependent on l due to escape (and injection) of δ-rays. The thus-caused dependence of δ_2 on l can be only roughly described by our theory through the δ-ray cut-off, as discussed above.

Full compatibility of theory and experiment can be expected regarding the *unrestricted* δ_2, i.e. in case that the sensitive layer is thick enough for the difference between energy loss and energy deposition straggling to become insignificant. The rule (3.12) yielding no cut-off, i.e. $R_e(E_{max}) < l$, may be regarded as a reasonable condition for the unrestrictedness. Such kind of data are summarized in Figure 5.19. The values from WILSON *et al.* (1975), resting directly on collision cross-sections rather than straggling distributions, are unrestricted ones by definition. Unfortunately, they are not quite representative for tissue or water. Nevertheless, one observes good correspondence with the wall-less-counter results of GLASS as well as with the present theory.

[1]) This 'classical' theory of energy *loss* straggling (VAVILOV, 1957) is based on the continuous collision spectrum $1/\varepsilon^2$ which reaches from a very small ε to the maximum δ-electron energy $\varepsilon = E_{max}$ without any additional cut-off.

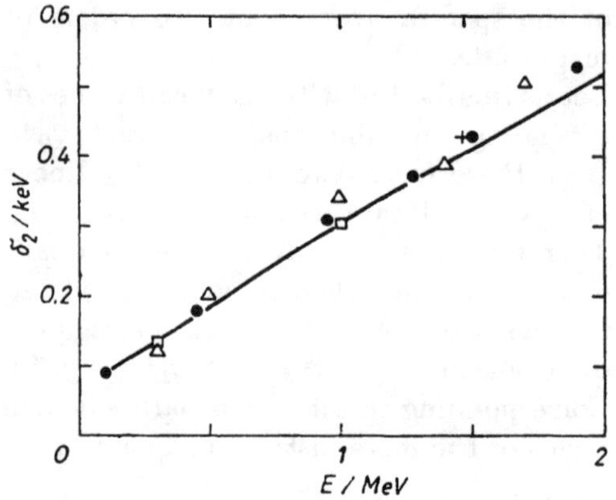

Fig. 5.19. Unrestricted δ_2 for protons of varying energy E from different sources: open symbols according to WILSON *et al.* (1975) (calculated directly from experimental collision cross-sections for N_2-gas (triangles) and CH_4-gas (squares)); closed circles according to GLASS and SAMSKY (1967) and GLASS (1968) (from the standard deviations, σ, of measured straggling distributions for $l = 1$ µm according to the relation (5.37)); cross: theoretical value calculated by KELLERER (1968) from his modified $1/\varepsilon^2$ collision spectrum (cf. Fig. 5.16).
The theoretical curve corresponds to the δ_2-values of Table IV for large sensitive regions (unrestricted values).

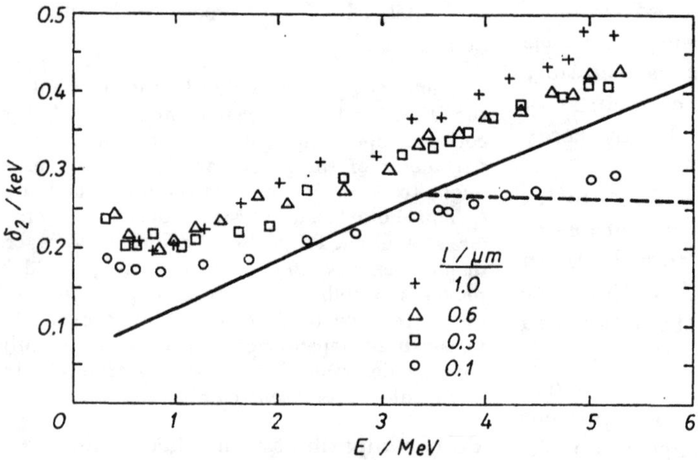

Fig. 5.20. δ_2 for α-particles of energy E for different sensitive layer thicknesses l. Experimental data (wall-less counter) from Booz *et al.* (1972b).
Theoretical curves, corresponding to the δ_2-values in Table IV, with δ-ray cut-off according to the rule (3.12), being effective only for $l = 0.1$ µm (dashed line) and resulting in no cut-off for $l = 0.3$, 0.6, 1.0 µm (full line).

The straggling of α-particles in thin sensitive gas layers (proportional counter) of various effective thicknesses was investigated for a lot of α-energies by Booz *et al.* (1972b). These results in terms of δ_2 are shown in Figure 5.20. The appreciable disagreement with the present theory is somewhat surprising considering that the foregoing figure for protons refers to the same region of energy per nucleon, i.e. to the same δ-ray spectra. Hence, the corresponding experimental δ_2-values should be equal, as the theoretical ones are. Booz' values appear as if they involved an additive constant. Apart from this, it is interesting to note that only for $l = 0.1$ µm there is a clear branching off the bulk of data at higher energies, in accordance with theory (dashed line).

The experimental work of Booz *et al.* (1972b) was completed by a parallel Monte Carlo calculation. These results are plotted in Figure 5.21 and Figure 5.22, showing

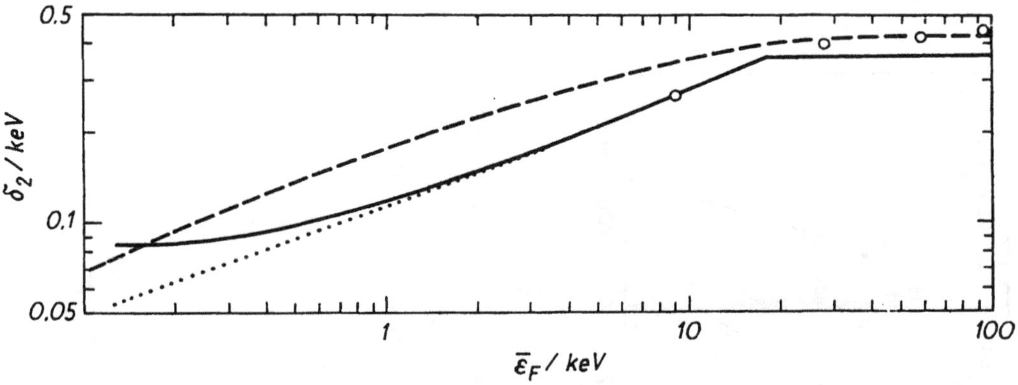

Fig. 5.21. δ_2 for 5-MeV α-particles in dependence on the mean energy deposition $\bar{\varepsilon}_F$ in layers of varying thickness.
Dashed curve and circles: Monte Carlo calculation and experimental results (wall-less counter), respectively, by Booz *et al.* (1972b).
Full and dotted curves: from our theory, calculated adequately (full line corresponding to Table IV) and in a manner equivalent to Eq. (5.37) (dotted line). δ-ray cut-off according to the rule (3.12).

good agreement with the corresponding experimental results.[1]) Moreover, these computations have been extended to very small values of l where measurements are no longer possible. This dependence, δ_2 versus $\bar{\varepsilon}_F$ standing for l, for a fixed α-particle energy is exhibited in Figure 5.21. In this case, it is merely the cut-off rule (3.12) that determines this dependence in the framework of our theory. Therefore, good agreement would only be more or less fortuitous. Actually, the disagreement is considerable but not unreasonable: roughly constant deviations of the same sign and magnitude as in Figure 5.20 at equal values of δ_2. The latter statement refers to the dotted line which was calculated according to Eq. (5.37) in analogy to the Monte Carlo treatment by Booz *et al.*, while for the correctly calculated δ_2 (full line) $\bar{y}_F^{(i)}$ in Eq. (5.37) had to be replaced by $\mu_D^{(i)}L_\infty$. This usually negligible difference, discussed in the footnote 4 at page 147, becomes appreciable here as l gets very small.

The disagreement seen in Figure 5.21 is visible again in Figure 5.22 where some examples of straggling distributions (heavy lines versus dashed lines) are compared for very thin sensitive layers. Since our spectra depend greatly on the δ-ray cut-off, for which the rule (3.12) (heavy lines) is uncertain with respect to thin layers, an attempt was made to get a better correspondence by reducing the δ-ray cut-off using the threefold value of l in formula (3.12). The thus obtained dotted curves are closer to the experimental and the Monte Carlo spectra, respectively. However, some discrepancies remain, considering that the dotted spectrum in the left-hand panel now includes no cut-off at all, while the right-hand panel demonstrates the incompatibility of our ionization cluster frequency concept with Booz' approach in this extreme case where

[1]) The Monte Carlo procedure used by Booz *et al.* (1972b) cannot be described here. It is however important to mention that this procedure refers primarily to the total energy imparted to the layer, which is subsequently 'translated' into ionizations by taking into account the so-called FANO-fluctuations (cf. footnote 1 at page 110). We quote only the latter results (given again in terms of energy by means of the common W-value 32.5 eV) in Figures 5.21 and 5.22, because it is actually the number of ionizations our theory as well as the experiment apply to.

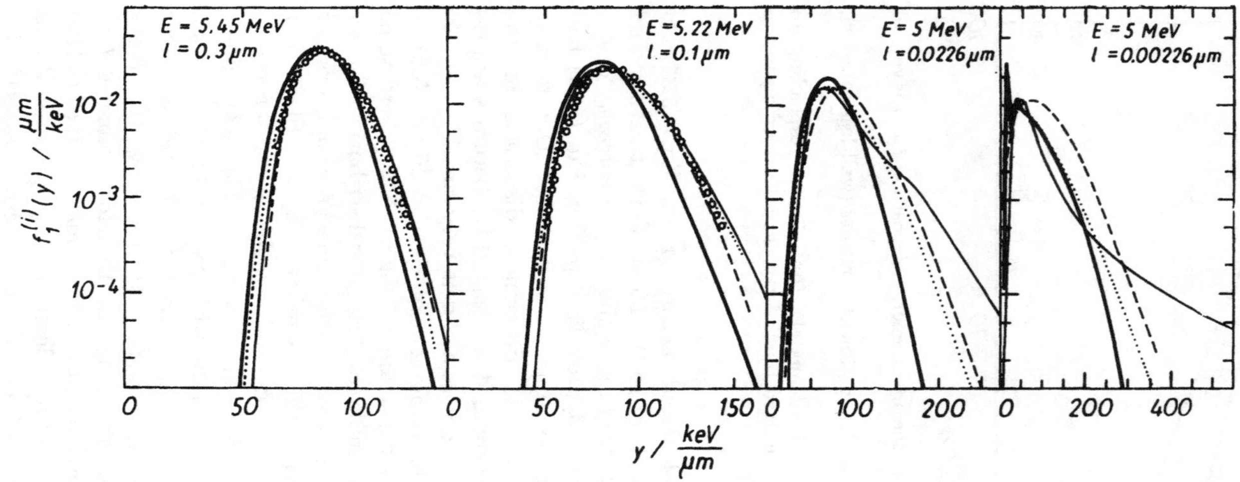

Fig. 5.22. Straggling distributions of α-particles in thin sensitive layers.
Circles and dashed curves: experimental results (wall-less counter) and Monte Carlo calculation, respectively, by Booz *et al.* (1972 b).
Thin unbroken lines: Vavilov theory (also taken from Booz *et al.*, 1972 b).
Heavy lines: present theory with 'normal' δ-ray cut-off (full heavy curves) and reduced δ-ray cut-off (pointed curves).

the single ionizations already become visible (see footnote 2 at page 168).

Some further experimental and theoretical investigations of straggling in layers of matter should only be quoted: VIDENSKY and PITKEVICH (1974); BULL *et al.* (1976); TSCHALÄR and MACCABEE (1970). The latter work, however, would not be suitable for drawing comparisons with the present theory, since it refers to thick absorbers where the primary particles lose a considerable fraction of total energy.

Chapter 6

Evaluation of the model of DNA-lesion production

6.1. Introduction

The purpose of the model of DNA-lesion production is to account for the frequency with which diverse radiation qualities produce various molecular lesions. 'Accounting for' means reducing to the frequency of energy depositions of varying density originating in single particle tracks within nanometre dimensions. If this is feasible, the model enables one (i) to obtain further *data of lesion yield* for different radiation qualities not covered by an experimental series, (ii) to make *general predictions* as to the possible dependence on radiation quality of a lesion type with a known frequency per unit dose of low-LET radiation, (iii) to derive the otherwise inaccessible *statistics of lesion number* in a larger entity of molecules (the cell nucleus) under irradiation. At bottom, the latter point constitutes the main reason for being in want of such kind of a model, as argued in section 2.2.1 and implemented in section 3.6. The objective of the present chapter, however, is to examine the model with respect to points (i) and (ii), to check it with experimental data (DNA strand breakage), and thus to give some evidence for its essential validity and relevance.

The model of lesion production is formulated in chapter 1, subsection 1.5.1, and mathematically evaluated in section 3.5. Nevertheless, to avoid any misunderstandings, it may be useful to add some comments on the principal ideas. To begin with, the model must be looked upon as a largely formal RBE theory, where 'RBE' here refers to the efficiency, measured by the reciprocal dose, of various radiations in producing a certain amount of lesions of a given kind. Thus, the complex and certainly multistep chemical and electronical processes involved in the formation of, say, a DNA double-strand break are not, and need not be, reflected by the model. Even no distinction is made between direct and indirect (i.e. by chemical radioproducts) action. The rationale of the model is much more general, as follows.

If one envisages a molecule or a particular chemical bond in it and asks about its probability to be broken or otherwise damaged by ionizing radiation, it is clear that this will ultimately depend on what initially 'happened', in terms of physics, in the molecule and *some vicinity* of it. Strictly, this vicinity region extends up to the maximum range of indirect action. Now, regardless of whether direct or indirect action prevails, it is the crucial question, as to the suitable way of modelling, whether this region is just small enough for no more than a single ionizing particle together with its correlated secondaries to be potentially involved in the formation of a lesion. By reference to the relevant dose range, the latter question can be answered by the help of microdosimetry,[1]) yielding the result that non-single-particle action can be excluded, e.g., up to 8 krads of ^{60}Co γ-rays unless the region is larger than 0.1 μm in diameter. It is very unlikely that the region of indirect action should be thus large in intracellular conditions. On the other hand, if it were so, one would expect cumulative action or any other interference of several particles resulting in a non-linear dependence on dose for the probability of lesion formation, which has not been observed, indeed.[2]) Therefore, *it is a matter of fact that the probability in question can be expressed as a sum of suitably weighted probabilities of occurrence of every kind of absorption events within a small spatial region.* It is only by the following hypothetical assumptions introduced in addition that this approach becomes a 'model': (i) absorption events are specified by nothing but the *ionizations* produced, and (ii) the mere *total number* of ionizations inside a region with a sharp boundary is thought the only relevant parameter of an absorption event. While assumption (i) is certainly not very serious (see footnote 1 at page 21), assumption (ii) clearly constitutes an idealization which may not be far from reality only in case that the region does not encompass much more than the molecule considered (very short effective range of indirect action). That is why reasonable diameters for the target region should not exceed some nanometres.

The discussion so far did not refer explicitly to DNA-lesions. In fact, the so-called 'model of DNA-lesion production' is by no means restricted to this kind of molecular damage, but its employment as a component of theory of cellular radiation effects and the availability of experimental data focus the attention on DNA single- and double-strand breaks. Nevertheless, only sections 6.3.2 and 6.4 are devoted to these types of lesion alone, while a more general view is taken in section 6.3.1, the goal of which is, first, to check the physics and mathematics underlying the calculations (all the approximations introduced in chapters 3 and 4) and, second, to present a systematic variety of results that could, for the future, be useful in any attempts to model processes in nanometre dimensions. This then demonstrates the flexibility of the lesion production model. For example, it can easily be changed into the single-ionization single-hit model which Lea has evaluated by means of his 'associated volume method' (see section 1.2.2), and on the other hand, it can be used to reproduce, and thus to make a check with, recently obtained single-event distributions and mean values calculated in terms of ionization number in spherical nanometre volumes on the basis of simulated particle tracks (Chmelevsky and Kellerer, 1977).

6.2. Methodic particulars

In chapter 3, the basic equation (cf. chapter 1, context of Eq. (1.35))

$$s = \frac{1}{\bar{z}_F} \sum_{j=0}^{\infty} \lambda_j f_1(j) \tag{6.1}$$

for the rate of lesion production per unit dose is evaluated up to explicit formulas (Eqs. (3.67) to (3.71) for every radiation

---

[1]) See section 1.4.1 and, in particular, footnote 2 at page 20.

[2]) Except for purified DNA in dilute aqueous solution where long-range radical action causes a quadratic component of dose-dependence (Freifelder and Trumbo, 1969).

considered in this book) in terms of (i) the parameters b_ν and $e^{-\zeta_\nu}$ originating from the general expression

$$\lambda_j = \sum_\nu b_\nu \, e^{-\zeta_\nu j}, \tag{6.2}$$

(ii) the mass density coefficient ϱ_l and (iii) the chord length spectrum $w_l(l')$ along with its mean value l_l.[1] This section is devoted to the choice of these parameters. Let us begin with point (iii), the geometry of the target.

As long as the model is to refer to DNA-lesions, it is reasonable to use as target an *infinite circular cylinder* of random orientation in space, and to relate both the number of absorption events, $D/\bar{z}_F$, and the number of lesions produced, Ds, to *unit length* of the cylinder, i.e. to unit length of DNA. The final formulas, (3.67) to (3.71), apply exactly on this understanding, and they should be evaluated, consequently, by inserting the $w_l(l')$ for an infinitely long cylinder. This chord length spectrum, however, is described by unwieldy elliptic integrals (Case *et al.*, 1953; Wilson and Emery, 1968). It has a narrow peak (an integrable divergence) just at $l' = l_l$, which suggests the δ-function approximation

$$w_l(l') = \delta(l' - l_l), \tag{6.3}$$

where

$$l_l = d, \tag{6.4}$$

i.e. the mean chord length for a cylinder equals the diameter of it.

The approximation (6.3) has been used throughout, not only for reasons of simplicity. In fact, for most of random particle traversals, l' comes close to the diameter l_l, while for 'grazing shots' ($l' \ll l_l$) and shots nearly parallel to the cylinder axis ($l' \gg l_l$) the model itself is likely to become inadequate, anyway. Suspecting, in the former case, the sharp target boundary and, in the latter case, the irrelevancy of the spatial arrangement of single-event ionizations inside the target (cf. the introduction to this chapter), one arrives at the conclusion that just declaring all track segments l' equal to l_l might be the *best* compromise. Therefore, Eq. (6.3) should be considered as the most adequate way of applying the model to the description of DNA-lesion formation.

For the purpose of fundamental investigations (comparison with existing results of similar models, production of nanodosimetric data), the model is applied also to *spherical targets* described by the 'triangular' spectrum (cf. Eq. (3.25))

$$w_l(l') = \begin{cases} \dfrac{8}{9}\, l_l^{-2} l' & \text{for } l' \leqq \dfrac{3}{2}\, l_l \\[2ex] 0 & \text{for } l' > \dfrac{3}{2}\, l_l \end{cases} \tag{6.5}$$

with

$$l_l = \frac{2}{3}\, d, \tag{6.6}$$

d being the sphere diameter. Furthermore, in this case of *discrete* targets, the definition of s must be changed into: number of lesions per *one* target, i.e. fraction of damaged targets. This is done by adding the factor $9l_l/4$ to the expressions (3.67) to (3.69).

The *density coefficient* ϱ_l for the target cannot be assumed to equal unity in general, particularly as DNA has a mass density considerably larger than that of water. However, in section 6.3 which is not limited to DNA-lesions, we use $\varrho_l = 1$ throughout, for reasons of simplicity and clearness. Thereby generality is not lost because of the special way in which ϱ_l enters the final equations. If s is considered a function of both ϱ_l and l_l designated as $s(\varrho_l, l_l)$, it holds for every radiation

$$s(\varrho_l, l_l) = \frac{1}{\varrho_l}\, s(1, \varrho_l l_l) \tag{6.7a}$$

---

[1]) For the related spectrum $u_l(l')$ see Eq. (3.28).

or

$$s(\varrho_l, l_l) = \frac{1}{\varrho_l^2} s(1, \varrho_l l_l) \qquad (6.7\text{b})$$

corresponding to the two above-mentioned definitions of s: Eq. (6.7a) applies to the unbounded cylindrical target with s being the number of lesions per unit length of target (DNA length); Eq. (6.7b) applies to the isolated spherical target with s being the probability for a target to receive a lesion.

For the cylindrical target to describe the production of DNA-lesions, ϱ_l can be assessed as follows. Let the target core consisting of DNA with the relative density ϱ_{DNA} occupy the fraction f of the total target volume, and assume that the residual volume contains water of unit density.[1]) This yields

$$\varrho_l = \varrho_{DNA} f + (1 - f) \qquad (6.8)$$

as the *mean* density coefficient for this actually *inhomogeneous* target. Strictly speaking, f refers to the volume of water displaced by DNA. This quantity referred to as *partial specific volume*, v_{DNA}, was measured by COHEN and EISENBERG (1968) to be 0.503 cm³ per gramme DNA (sodium salt in 0.1 mole per litre solution). Relative to one average Na-nucleotide pair with the molar mass 662 g occupying a target segment of 0.336 nm length (DNA in B-configuration), the volume fraction f can be calculated according to

$$f = \frac{v_{DNA}\, 662\text{ g } N^{-1}}{0.336\text{ nm } \pi d^2/4} = 2.09\,(d/\text{nm})^{-2}, \qquad (6.9)$$

where $N = 6.025 \times 10^{23}$ signifies Avogadro's number. The relative density ϱ_{DNA} is equal to the reciprocal ratio of specific volumes, v_{water}/v_{DNA}, where $v_{water} = 1$ cm³/g, but there is a subtlety which should be taken into account. The actually relevant parameter is not mass density, but electron density, so that the factor 0.903 by which the number of electrons per unit mass is smaller for DNA compared to water has to be added: $\varrho_{DNA} = 0.903/0.503$. Thus one arrives at the relation

$$\varrho_l = 1 + 1.66\,(d/\text{nm})^{-2}. \qquad (6.10)$$

For the evaluation of DNA single- and double-strand breaks, we have ultimately chosen the target diameter $d = 2.27$ nm corresponding to what may be called the *effective diameter* $\varrho_l d = 3$ nm. This is close to the limit value $(\varrho_l d)_{\min} = 2.6$ nm resulting from $d_{\min} = 1.45$ nm, which constitutes the minimum diameter for a target able to contain the complete DNA molecule ($f \leqq 1$). Thus, which is worth noting, most of the target mass is DNA. This is highly in accord with a general requirement pointed out in the introduction to this chapter (see page 177).

Let us now consider the *main parameters of the model:* λ_j. In principle, an infinite number of parameters are thus involved, and at first sight it seems, therefore, that one could achieve anything by choice of λ_j. However, this is not the case owing to some obvious restrictions, as follows. By definition, λ_j is a probability, which implies the condition

$$0 \leqq \lambda_j \leqq 1 \qquad (6.11)$$

to apply in general. More specific conditions can be found in the case of λ_j referring to *DNA double-strand breaks*, recalling the definition that λ_j signifies the probability of j simultaneous ionizations in the target causing the lesion. In this case, the requirements

$$\lambda_0 = 0, \quad \lambda_{j+1} \geqq \lambda_j, \quad \lim_{j\to\infty} \lambda_j = 1 \qquad (6.12\text{a, b, c})$$

are nearly self-evident, since (a) nothing will be produced by nothing, (b) each additional ionization will make it more

---

[1]) An estimation made assuming the DNA to be surrounded solely by protein yielded a 15 per cent smaller target diameter d if $\varrho_l d$ (effective diameter, see below) is assumed to be 3 nm, as an example.

likely that the result of the event as a whole is a double-strand break, (c) the latter should happen with certainty if the number of ionizations was very large.[1])
The conditions (6.12) define a step-like function λ of j which increases more or less gradually from 0 up to 1. At least two parameters are needed to describe such a function roughly (position and steepness of the 'step'). We will use the following expression which depends on three parameters (m, p, q):

$$\lambda_j = \sum_{\nu=0}^{m} \binom{m}{\nu} (-1)^\nu \times \left[1 - \frac{1-q^\nu}{1-(q+q^m)/2} p\right]^j, \qquad (6.13)$$

$$0 < q < 1, \quad 0 < p \leqq 1, \quad m = 1, 2, \dots \qquad (6.14)$$

The ansatz (6.13) meets every necessary requirement. Firstly, it has the form of expression (6.2) with b_ν equal to $(-1)^\nu \binom{m}{\nu}$ and $\exp(-\zeta_\nu)$ equal to the term in square brackets[2]). Secondly, it is compatible with the conditions (6.12), which is obvious save for (6.12b)[3]). Thirdly, it possesses any desirable flexibility yielding, e.g., a most rapid increase of λ_j if $p = 1$, while the region of j where the increase is seen depends on both q and m. The limiting case $q \to 1$ admits obtaining $\lambda_j = 0$ for $j < m$, regardless of how steep or flat the increase following at $j \geqq m$ is made by choice of p. On the other hand, anything like a step reminding of a threshold ionization number disappears by choosing $m = 1$. In this case one may set $p = 1 - q$ ending up in

$$\lambda_j = 1 - q^j. \qquad (6.15)$$

This is the generalized single-ionization model by LEA (1946), i.e. each ionization acts independently with the action probability $1 - q$. Table 6.1 lists a variety of λ-functions with which the model has been evaluated (see the succeeding sections).
It has been made plausible so far that the ansatz (6.13) is general enough to describe the formation of double-strand breaks. Yet, if we envisage different kinds of lesion, we ought to use another sort of λ-functions with λ_j approaching *zero* for large j because condition (6.12c) leaves over no chance for a very large number of ionizations to cause anything else but a double-strand break. However, we will use the expression (6.13) in every case by stipulating that the lesion kind described by any λ_j has to be always understood to *include* double-strand breaks.[4]) Due to the linearity between s and λ_j in the fundamental equation (6.1), this means that the lesion number s derived from a λ_j according to Eq. (6.13) applies either to double-strand breaks alone or to the *sum* of double-strand breaks and a further lesion type considered.
This corresponds to the actual situation on the experimental side. Due to the experimental technique (sucrose gradient

[1]) Obviously, the DNA will be left in broken state after a sufficiently large number of locally concentrated ionizations in it have formed a 'hot spot'.
[2]) Note that this term may become *negative*, i.e. the expression (6.13) does not exclude negative values of $\exp(-\zeta_\nu)$. However, this is well admissible, as may be seen from inspection of the final equations (3.67) to (3.71) and their derivation in section 3.5. ζ_ν nowhere occurs in a form other than $\exp(-\zeta_\nu)$.
[3]) It is not too difficult to demonstrate that, if condition (6.12b) is fulfilled with some set of parameters m, p, q, it is also fulfilled with smaller values of p (GÜNTHER, 1968a). Hence, the proof required can be restricted to $p = 1$. This proof, however, could till now be strictly afforded only for $m \leqq 3$.
[4]) Strictly speaking, this stipulation guarantees condition (6.12c) only, i.e. (6.13) might be insufficient if condition (6.12b) should be violated (secondary maximum at a relatively small j due to the added lesion type). However, it is unlikely that such sophisticated questions play a role.

Tab. 6.1. Examples of λ-functions according to the expression (6.13)

m	1	10	10	10	10	10	10	10	10	10	10	10	10	4	4	3
p	.168	1	1	1	.6	.6	.6	.3	.3	.3	.05	.05	.05	.314	.324	.151
q	.832	.367	.807	.997	.191	.731	.975	.067	.710	.913	.027	.176	.621	.829	.950	.920
j																
1	.168	.013	.000	.000	.079	.000	.000	.156	.000	.000	.039	.008	.000	.001	.000	.001
2	.307	.347	.000	.000	.357	.000	.000	.307	.000	.000	.076	.017	.000	.008	.001	.004
3	.423	.870	.003	.000	.605	.007	.000	.443	.002	.000	.112	.028	.000	.027	.005	.011
4	.520	.985	.036	.000	.776	.032	.000	.559	.006	.000	.147	.039	.000	.061	.020	.023
5	.601	.991	.157	.000	.879	.090	.000	.655	.014	.000	.180	.052	.001	.108	.047	.040
6	.668	.999	.378	.000	.937	.179	.000	.733	.029	.000	.212	.065	.001	.167	.088	.061
7	.723	1.00	.589	.000	.968	.289	.000	.795	.050	.000	.243	.079	.001	.232	.139	.086
8	.770	1.00	.724	.000	.984	.406	.001	.844	.078	.001	.273	.094	.002	.301	.198	.115
9	.808	1.00	.819	.012	.992	.519	.003	.881	.112	.002	.302	.109	.003	.370	.261	.147
10	.841	1.00	.885	.153	.996	.619	.009	.911	.153	.003	.330	.125	.004	.438	.327	.181
11	.867	1.00	.924	.163	.998	.703	.021	.933	.198	.005	.356	.141	.005	.503	.391	.216
12	.890	1.00	.952	.353	.999	.772	.041	.950	.246	.009	.382	.157	.006	.563	.454	.253
13	.908	1.00	.969	.360	1.00	.827	.069	.963	.297	.014	.407	.174	.007	.618	.513	.290
14	.923	1.00	.980	.535	1.00	.870	.105	.972	.350	.021	.431	.191	.009	.668	.568	.327
15	.936	1.00	.987	.539	1.00	.902	.148	.979	.402	.030	.453	.208	.011	.712	.618	.364
16	.947	1.00	.992	.676	1.00	.927	.197	.985	.453	.041	.476	.226	.013	.751	.664	.400
17	.956	1.00	.995	.679	1.00	.946	.249	.989	.503	.054	.497	.243	.015	.786	.705	.435
18	.963	1.00	.997	.779	1.00	.960	.304	.992	.551	.069	.517	.260	.018	.816	.742	.469
19	.969	1.00	.998	.780	1.00	.970	.358	.994	.595	.087	.537	.278	.020	.843	.775	.502
20	.975	1.00	.999	.851	1.00	.978	.413	.996	.637	.107	.556	.295	.023	.865	.804	.534
21	.979	1.00	.999	.851	1.00	.984	.465	.997	.676	.129	.574	.312	.026	.885	.829	.564
22	.982	1.00	.999	.900	1.00	.988	.515	.998	.712	.153	.591	.330	.030	.902	.852	.593
23	.985	1.00	1.00	.900	1.00	.991	.562	.998	.745	.179	.608	.347	.033	.917	.871	.621
24	.988	1.00	1.00	.933	1.00	.993	.606	.999	.774	.206	.624	.363	.037	.929	.888	.647
25	.990	1.00	1.00	.933	1.00	.995	.647	.999	.801	.234	.639	.380	.041	.940	.903	.671
26	.992	1.00	1.00	.955	1.00	.996	.684	.999	.825	.263	.654	.400	.046	.949	.916	.695
27	.993	1.00	1.00	.955	1.00	.997	.718	1.00	.846	.293	.669	.413	.050	.956	.928	.716
28	.994	1.00	1.00	.970	1.00	.998	.749	1.00	.865	.323	.682	.429	.055	.963	.937	.737
29	.995	1.00	1.00	.970	1.00	.999	.777	1.00	.882	.354	.695	.445	.060	.969	.946	.756
30	.996	1.00	1.00	.980	1.00	.999	.802	1.00	.897	.384	.708	.461	.065	.973	.953	.774

centrifugation), double-strand breaks are automatically included when single-strand breaks are measured,[1]) while quantitative determinations in dependence on radiation quality for other lesions are missing till now. Furthermore, when the expression (6.13) is substituted for the formal λ_j in the framework of formal survival curve analysis (chapter 7), this solely implies the assumption that double-strand breaks have at least a share in cell killing.[2])

For the purpose of fundamental investigations, yet another set of λ-functions, namely

$$\lambda_j^{(n)} = \begin{cases} 0 & \text{for } j < n \\ 1 & \text{for } j \geqq n, \end{cases} \tag{6.16}$$

[1]) Even twice are double-strand breaks counted in by this technique. Therefore, what is usually called the measured number of single-strand breaks must be diminished by the number of double-strand breaks, to compare with what (6.13) yields.

[2]) Unless double-strand breaks alone should be relevant, anyway, λ_j has to be looked upon as the weighted sum (3.137) which is likely to increase monotonously with j, provided that the share of the double-break λ_j^{II} is not too small. Since, in addition, its normalization is arbitrary, $\lambda_\infty = 1$ can be assumed. This justifies using the ansatz (6.13).

will be used in addition to (6.13). No doubt, it can be taken for granted that, in actuality, the probability of lesion formation will *gradually* increase with the number of ionizations, as adequately described just by the flexible expression (6.13). However, the 'yes-or-no approach' (6.16) is advantageous so as to gain a first general insight into the dependence on radiation quality in terms of a single most obvious parameter — the threshold number, n, of ionizations. Moreover, (6.16) corresponds to the hitherto most familiar way of thinking in terms of n-ionization models. Thus, it also allows comparisons with other nanodosimetric calculations.

To evaluate the model with $\lambda_j^{(n)}$ according to Eq. (6.16) raises up the technical problem of how to express $\lambda_j^{(n)}$ in the form of a sum of exponential functions like (6.2), since it is only the latter expression to which the final formulas refer. This can indeed be implemented in the following way. The expressions

$$\lambda_j^{(1)} = 1 - \frac{1}{\beta^j}$$

$$\lambda_j^{(2)} = 1 - \frac{1 + \alpha - \alpha(-1)^j}{(2\alpha + 1)^j}$$

$$\lambda_j^{(3)} = 1 - \frac{1 + \alpha + \alpha(-1)^j}{(2\alpha + 1)^{j-1}} + \frac{4\alpha(\alpha + 1)}{(4\alpha^2\beta)^j}$$

$$\lambda_j^{(4)} = 1 - \frac{(\alpha + 1)\,[\alpha - (\alpha - 1)\,(-1)^j]}{2(2\alpha - 1)^{j-2}} + \frac{(\alpha - 1)\,[1 + \alpha - \alpha(-1)^j]}{2(2\alpha + 1)^{j-2}}$$

$$\lambda_j^{(5)} = 1 - \frac{(\alpha + 1)\,[\alpha + (\alpha - 1)\,(-1)^j]}{2(2\alpha - 1)^{j-3}} + \frac{(\alpha - 1)\,[1 + \alpha + \alpha(-1)^j]}{2(2\alpha + 1)^{j-3}} - \frac{16\alpha^2(\alpha^2 - 1)}{(16\alpha^4\beta)^j} \qquad (6.17)$$

have the form (6.2), and they are easily seen to change over into (6.16) as

$$\alpha \to \infty, \qquad \beta \to \infty. \qquad (6.18)$$

Hence, the expressions (6.17) define the parameters b_ν and $e^{-\zeta_\nu}$ in terms of α and β which must only be made large enough in order to obtain any wanted degree of approximation to the ideal $\lambda_j^{(n)}$ given by (6.16).[1]) This shows that the mathematical advantages (see section 3.5) of evaluating the model in terms of the general expression (6.2) for λ_j has no drawback; once the final formulas (3.67) to (3.71) are known and coded for the computer, one can do almost anything with respect to λ_j.

6.3. General results

6.3.1. Spherical target and threshold number of ionizations

The topic of this subsection is the n-ionization model defined by the ansatz (6.16). Accordingly, the quantity (6.1) we calculate can be written in the form

$$s^{(n)} = \bar{z}_F^{-1} F(n) \qquad (6.19)$$

where $F(n)$ denotes the *probability distribution* (cumulative distribution) for the ionization number n produced in a single absorption event:

$$F(n) = \sum_{j=n}^{\infty} f_1(j). \qquad (6.20)$$

Hence, $F(n)$ is the probability that an absorption event includes the production of *no less* than n ionizations inside the spherical region considered. The corresponding total number of events per unit dose is given by $\bar{z}_F^{-1}$.

These definitions are not unambiguous unless it is stipulated what kind of absorption events are to be counted in. It is clear that 'zero events', in which *no* ionizations are produced, are irrelevant to the yield s, by virtue of the axioms of

[1]) One might expect numerical problems, but gradually increasing α and β yielded stable plateaus of $s^{(n)}$ over a wide range of large α and β values. Thus, the results presented in the sequel *are* reliable in this respect.

the model. Therefore, it is reasonable to interpret the above quantities with respect to *ionizing events only*,[1]) so that

$$F(1) = 1 \tag{6.21}$$

by definition, while $\bar{z}_F^{-1}$ now means the number of *ionizing* events per unit dose.[2]) The latter is just the quantity of interest in the case of the single-ionization model:

$$s^{(1)} = \bar{z}_F^{-1}. \tag{6.22}$$

This quantity should be considered by contrast with the total number of ionizations per unit dose, which is equal to W^{-1} for the radiation in question times mass m of the sensitive region. Hence, particularly as the W-value is originally a matter of dosimetry, we rather calculate the ratio

$$\bar{n} = \frac{m\bar{z}_F}{W} = \frac{m}{Ws^{(1)}} \tag{6.23}$$

when dealing with the single-ionization model.[3]) $\bar{n}$, the average number of ionizations per ionizing absorption event, should be looked upon as the genuine result of the theory, since it eliminates the problem of absolute W-values.

The single-ionization model was the subject of earlier calculations by LEA (1946). His associated volume method just yields $\bar{n}$ ($\bar{n} = V/V_{ass}$; see chapter 1, Eq. (1.7) and context), but he gave his results in terms of the dose D required to produce one ionizing event on the average per target. In the same way, applying LEA's coefficients to express $\bar{n}$ as D in roentgen units, the present theory has been used to obtain the curves in Figure 6.1 (full lines) for a number of radiations. One sees generally good correspondence with the comparatively simple approach by LEA (dashed lines). This was expected for the densely ionizing α-particles, while larger and smaller values of D had to turn out for neutrons and photons, respectively, due to LEA's neglecting heavy recoils and assuming too big ionization clusters for electrons (see lower case print at page 124).

A modern statistical evaluation of simulated particle tracks by suitable sampling procedures (CHMELEVSKY and KELLERER, 1977) has yielded, among other nanodosimetric quantities, values of $\bar{n}$ for protons which have been expressed as the frequency-mean of lineal energy by using, apparently,[4]) a constant W-value according to

$$\bar{y}_F = 33.3\,\mathrm{eV}\,\frac{\bar{n}}{\frac{2}{3}d}. \tag{6.24}$$

In terms of this parameter, the left-hand panel of Figure 6.2 compares these Monte Carlo results with the present ones. There is agreement as to the general trends. For a probable explanation of the differences see below.

To give a comprehensive set of results with the single-ionization model, $\bar{n}$ has been calculated for a series of monoenergetic ion radiations and expressed by the familiar parameter $\bar{y}_F$ as it is defined by Eq. (6.24). These figures (Tab. 6.2), which supplement Table VI (appendix) for the range of

[1]) In chapter 3, due to the more general framework of treating physical stochastics there developed, the evaluation of the model started with the alternative way of reading the fundamental equation (6.1): $f_1(0) \neq 0$ (non-ionizing traversals of particles) leading to $s^{(1)} = [1 - f_1(0)]/\bar{z}_F$ with $\bar{z}_F^{-1}$ *including* the zero events. Obviously, this is compatible with the present interpretation.

[2]) This corresponds to the ionization number concept adopted throughout chapter 5 (see the introduction) in general, and to the definition of $\bar{z}_F$ by Eq. (5.13) in particular. Thus, the present relation for $\bar{z}_F$ is in keeping with the $\bar{y}_F$ values for larger sites listed in the appendix.

[3]) The W-values that $s^{(1)}$ is to be multiplied by have to be the 'theoretical' ones defined by Eq. (5.12) and tabulated in the appendix.

[4]) This constant, 33.3 eV, is obtained by thorough inspection of several graphic representations in the paper cited.

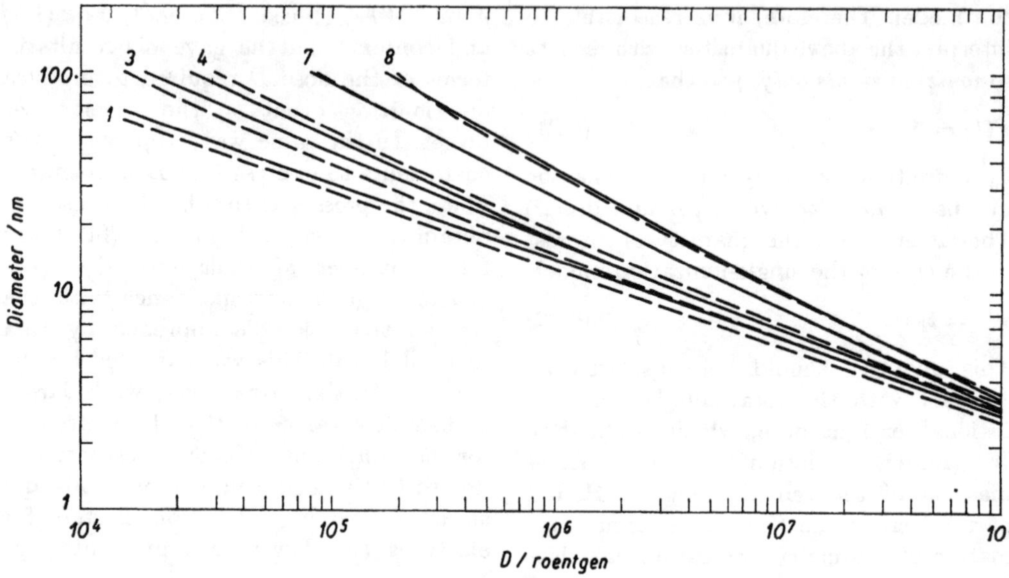

Fig. 6.1. The single-ionization model: relation between 37%-Dose D and diameter of a spherical target of unit density for the following radiations: 1 — 800-keV photons, 3 — 8.26-keV photons, 4 — d-on-Li neutrons, 7 — 6-MeV α-particles, 8 — 3-MeV α-particles.
Dashed lines according to LEA (1946), full lines according to the present calculation.

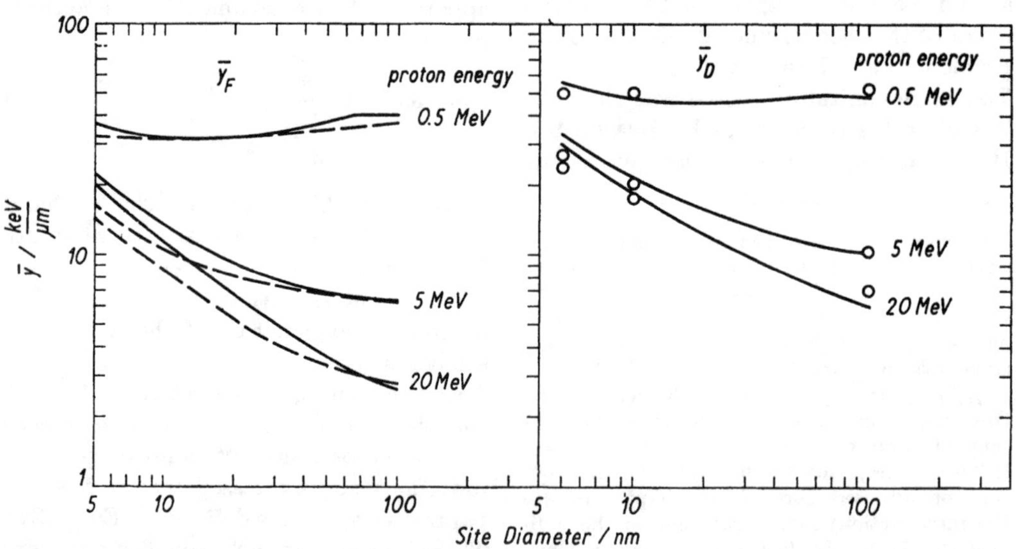

Fig. 6.2. Mean values of lineal energy for protons in spherical sites, calculated according to Eq. (6.24) (full lines) and taken from CHMELEVSKY and KELLERER (1977) (broken lines and circles).

Tab. 6.2. Results of the single-ionization model for spherical sites and monoenergetic ions in terms of the frequency-mean, $\bar{y}_F$, of lineal energy [keV/μm] as defined by Eqs. (6.24), (6.23)

Site diameter [nm]	^{60}Co γ-rays	10 MeV/nucl.								3 MeV/nucl.							
		H	He	Li	Be	B	C	O	Ne	H	He	Li	Be	B	C	O	Ne
LET_∞ $\left[\frac{keV}{\mu m}\right]$		4.60	18.4	41.4	73.5	114	164	288	442	12.1	48.3	107	187	283	395	653	947
1	74.4	78.6	81.4	86.0	92.2	99.8	108	126	141	80.9	88.8	101	116	131	144	166	179
2	40.8	45.2	48.2	53.0	59.2	66.3	73.5	86.2	95.5	47.6	56.0	68.3	81.3	92.5	101	113	119
5	17.7	21.0	24.6	30.0	36.2	42.0	46.7	53.2	57.0	23.7	33.7	45.7	55.1	61.4	65.6	70.4	73.0
10	9.41	12.1	16.3	22.1	27.6	31.8	34.8	38.5	40.4	15.2	26.8	37.6	44.4	48.5	51.2	54.1	55.6
20	5.07	7.55	12.5	18.2	22.5	25.4	27.3	29.5	30.7	11.3	24.0	33.4	38.9	42.1	44.0	46.2	47.3
50	2.27	4.74	10.4	15.4	18.5	20.4	21.7	23.1	23.8	9.52	23.1	32.0	37.0	39.9	41.7	43.6	44.6

nanometre sites,[1]) can be compared with LET_∞ according to the following general rule. Apart from large sites for which $\bar{y}_F \approx LET_\infty$ tends to turn out, one has $\bar{y}_F < LET_\infty$ for high LET_∞ values due to the importance of δ-rays (indirect events), while otherwise $\bar{y}_F > LET_\infty$ results from $\bar{n}$ becoming nearly independent of radiation quality for LET_∞ values and site diameters too small to enable more than one ionization cluster to share in an absorption event.

Let us now direct our attention to the aspect we are mainly interested in, namely the dependence of s on radiation quality for the single- as well as the multi-ionization model. Table 6.3 visualizes this 'RBE aspect' by listing the ratio $s^{(n)}/s_0^{(n)}$, where the subscript 0 stands for ^{60}Co γ-rays.[2]) This representative set of results has been calculated using the mathematical procedure signified by the relations (6.17), (6.18).

First we note that the single-ionization model never yields RBE values greater than unity; the figures always decrease with increasing LET, because for this model any concentration of ionizations means wasting of energy. This is termed *saturation effect*.[3]) If, however, more than one ionization is required to produce the lesion as described by the n-ionization models, one should expect a positive *concentration effect* at increasing LET. Table 6.3 bears out this, but it is interesting to notice that the threshold ionization number n must be rather large, $n = 2$ being too small in any case, for a distinct maximum in the RBE-LET relation to be brought about. E.g., if we envisage the primary lesion RBE values relevant to mammalian cell killing (see Tab. 2.1), which resemble those for double-strand break production, not even a crude correspondence is achieved unless n is greater than 4 in the case of the 2-nm target.

How reliable are these results in view of the numerous simplifications involved in the physical and mathematical treatment of the model (see chapters 3 and 4)? In fact, just the increase of s/s_0 with growing LET is both most important in regard to applications of the lesion production model as such and most sensitively dependent on the features of ionization patterns in particle tracks. This may be illustrated by

---

[1]) A small difference only is due to the formal normalization now, according to substituting 33.3 eV for the actual radiation-dependent W-value.

[2]) The absolute values $s^{(n)}$ are easy to obtain by the help of Table 6.2 and relations (6.24), (6.23).

[3]) 'Saturation of the first kind' according to the conceptions introduced in chapter 2.

Tab. 6.3. Results obtained from threshold-ionization-number models and spherical sites for monoenergetic ion radiations

	^{60}Co γ-rays		10 MeV/nucl.								3 MeV/nucl.							
			H	He	Li	Be	B	C	O	Ne	H	He	Li	Be	B	C	O	Ne
LET$_\infty$ $\left[\frac{\text{keV}}{\mu\text{m}}\right]$			4.60	18.4	41.4	73.5	114	164	288	442	12.1	48.3	107	187	283	395	653	947
n	$\frac{s_0^{(n)}}{s_0^{(1)}}$	$\frac{s^{(n)}}{s_0^{(n)}}$																
			Site diameter d = 1 nm,											$\jmath_{max}$ = 3				
1	1.	1.	0.91	0.88	0.83	0.78	0.72	0.66	0.57	0.51	0.88	0.80	0.71	0.62	0.55	0.49	0.43	0.40
2	.360	1.	1.03	1.05	1.06	1.07	1.05	1.02	0.92	0.81	1.04	1.06	1.06	1.00	0.91	0.82	0.68	0.60
3	.114	1.	1.14	1.24	1.39	1.54	1.67	1.75	1.75	1.59	1.21	1.45	1.69	1.80	1.77	1.65	1.35	1.12
4	.010	1.	1.59	2.55	4.05	5.90	7.85	9.60	11.7	11.7	2.29	4.80	8.03	10.7	12.2	12.3	10.5	8.47
5	.003	1.	1.56	2.56	4.52	7.64	11.8	16.5	24.8	28.4	2.27	5.58	11.9	19.6	25.9	29.1	29.3	27.8
			Site diameter d = 2 nm,											$\jmath_{max}$ = 4				
1	1.	1.	0.87	0.82	0.74	0.67	0.59	0.54	0.46	0.41	0.82	0.70	0.57	0.48	0.42	0.39	0.35	0.33
2	.398	1.	1.02	1.03	1.02	0.99	0.93	0.86	0.72	0.62	1.03	1.01	0.85	0.74	0.64	0.57	0.48	0.44
3	.162	1.	1.18	1.27	1.37	1.44	1.45	1.39	1.19	1.00	1.25	1.42	1.47	1.36	1.19	1.04	0.83	0.72
4	.058	1.	1.42	1.66	1.99	2.33	2.56	2.64	2.40	2.00	1.60	2.16	2.62	2.69	2.46	2.14	1.62	1.32
5	.009	1.	1.70	3.02	5.01	7.29	9.35	10.7	11.0	9.41	2.62	5.96	9.63	11.6	11.4	10.2	7.55	5.82
			Site diameter d = 5 nm,											$\jmath_{max}$ = 6				
1	1.	1.	0.81	0.70	0.57	0.47	0.41	0.37	0.32	0.30	0.72	0.50	0.37	0.31	0.28	0.26	0.24	0.23
2	.421	1.	1.01	0.99	0.92	0.80	0.69	0.60	0.50	0.45	1.00	0.87	0.67	0.54	0.47	0.43	0.39	0.37
3	.194	1.	1.21	1.32	1.36	1.28	1.12	0.98	0.78	0.67	1.30	1.37	1.14	0.91	0.76	0.68	0.59	0.54
4	.089	1.	1.43	1.74	2.00	2.05	1.88	1.65	1.26	1.04	1.67	2.11	1.95	1.56	1.27	1.09	0.90	0.80
5	.037	1.	1.56	2.28	3.09	3.56	3.50	3.14	2.35	1.85	2.07	3.44	3.67	3.03	2.41	2.00	1.55	1.33
			Site diameter d = 10 nm,											$\jmath_{max}$ = 8				
1	1.	1.	0.75	0.56	0.41	0.33	0.29	0.26	0.24	0.23	0.59	0.34	0.24	0.20	0.19	0.18	0.17	0.16
2	.432	1.	0.99	0.91	0.73	0.58	0.49	0.44	0.38	0.36	0.94	0.65	0.46	0.37	0.33	0.31	0.29	0.28
3	.208	1.	1.24	1.33	1.20	0.98	0.81	0.70	0.59	0.53	1.33	1.12	0.79	0.63	0.55	0.50	0.46	0.44
4	.105	1.	1.49	1.86	1.88	1.60	1.31	1.11	0.89	0.78	1.79	1.85	1.34	1.03	0.87	0.78	0.69	0.65
5	.053	1.	1.65	2.48	2.90	2.63	2.17	1.80	1.37	1.16	2.28	2.98	2.26	1.69	1.39	1.22	1.04	0.95
			Site diameter d = 20 nm,											$\jmath_{max}$ = 13				
1	1.	1.	0.65	0.39	0.27	0.22	0.19	0.18	0.17	0.16	0.43	0.20	0.15	0.12	0.12	0.11	0.11	0.10
2	.443	1.	0.94	0.72	0.51	0.40	0.34	0.31	0.28	0.27	0.78	0.41	0.29	0.24	0.22	0.21	0.20	0.19
3	.221	1.	1.25	1.18	0.87	0.67	0.56	0.50	0.44	0.41	1.25	0.75	0.51	0.42	0.38	0.35	0.33	0.32
4	.118	1.	1.57	1.78	1.41	1.07	0.88	0.77	0.65	0.60	1.82	1.27	0.84	0.68	0.60	0.55	0.51	0.49
5	.067	1.	1.81	2.51	2.18	1.66	1.33	1.14	0.94	0.85	2.46	2.02	1.33	1.04	0.90	0.82	0.75	0.71
			Site diameter d = 50 nm,											$\jmath_{max}$ = 24				
1	1.	1.	0.46	0.21	0.14	0.12	0.11	0.10	0.09	0.09	0.23	0.09	0.07	0.06	0.05	0.05	0.05	0.05
2	.464	1.	0.78	0.41	0.27	0.22	0.20	0.18	0.17	0.17	0.46	0.19	0.14	0.12	0.11	0.10	0.10	0.10
3	.241	1.	1.18	0.73	0.48	0.38	0.33	0.31	0.28	0.27	0.83	0.36	0.25	0.21	0.20	0.19	0.18	0.17
4	.134	1.	1.64	1.20	0.78	0.60	0.52	0.47	0.43	0.41	1.37	0.61	0.42	0.35	0.32	0.31	0.29	0.28
5	.081	1.	2.08	1.83	1.18	0.90	0.76	0.69	0.61	0.58	2.08	0.97	0.66	0.54	0.49	0.46	0.44	0.42

discussing an evalutation of the two-ionization model made by HOLT (1974). It should be mentioned that his work so far constitutes the almost only attempt in molecular radiation biophysics to put to use a several-ionization model as a generalization of LEA's single-ionization model.

With a view of accounting for measured rates of double-strand break production, HOLT reports results of the two-ionization model for spherical targets with diameters between 1 nm and 3.8 nm and the same ranges of LET and energy per nucleon as Table 6.3 refers to. In contrast to the present results, however, he finds the RBE (relative to ^{60}Co γ-rays) *increasing substantially* up to a maximum of 3 ($d = 3.8$ nm) and 4 ($d = 1$ nm) for He and C ions, respectively, of 10 MeV/nucl., although his calculations are based essentially on principles like the present ones (chapters 3 and 4). The main difference, to which the discrepancy must be attributed, concerns the physical parameters ascribed to low-energy electrons. The crude assumptions made in this respect cause too large a proportion of single unclustered primary ionizations along the tracks of all charged particles involved. This tends to reduce the effectiveness of radiations to the events in which two or more independent primary ionizations happen to come together within the limits of the small target. The infrequency of such events at low LET is then responsible for the increased *relative* effectiveness at higher LET.[1])

Thus, while the case of high-LET particles is not crucial, it is just the vague distribution of ionization cluster sizes that alone becomes decisive for the effectiveness of low-LET particles in the case of very small targets. The present approach also uses simplified assumptions in this respect: due to the rigid break-off at a d-dependent maximum cluster size j_{max} (see Tab. 6.3), only the experimental values (4.44)[2]) of the relative frequencies of clusters are involved if $d \leq 2$ nm. This can easily be verified since the second column of Table 6.3 just contains the event size distribution $F(n)$ for ^{60}Co γ-rays according to Eqs. (6.19), (6.22),

$$F(n) = \frac{s^{(n)}}{s^{(1)}}, \tag{6.25}$$

and these figures should be only slightly in excess of the normalized sum,

$$\sum_{j=n}^{j_{max}} \omega_0(j) \Big/ \sum_{j=1}^{j_{max}} \omega_0(j), \tag{6.26}$$

of cluster frequencies if superpositions of several clusters in an event are rare. Indeed, the values (6.26) according to the data (4.44) read: 1, 0.34, 0.10 for $j_{max} = 3$; 1, 0.36, 0.14, 0.04 for $j_{max} = 4$. Hence, in consequence of the present approach, about 40 per cent of ionizing events, at any rate, include two or more ionizations, so that the two-ionization model cannot yield an appreciable increase of RBE due to increasing LET.

The following estimation corroborates this conclusion. Let an increase of LET cause that a fraction, p, of ionization clusters, which are identical with separate events in the limiting case of very low LET, combine in complex events, which implies a reduction of the fractionary event number by a factor ν. The total number of events containing more than one ionization then reads $0.4(1-p) + p/\nu$ as compared to 0.4 for low LET. Thus, setting $\nu = 2$ (pairwise combinations only) yields the following upper bounds to the RBE that can be reached: $1 + 0.25p$, which cannot be larger than 1.25 (if $p = 1$) and would even be equal to 1 if 0.4 were replaced by 0.5.

---

[1]) This explanation is confirmed by comparing the *absolute* values of $s^{(2)}$, which turn out to be always lower in HOLT's calculation, to be sure, but there is at least rough correspondence in the region of HOLT's maximum, while it is the low-LET edge where the difference is largest.

[2]) These data are relevant if the charged particles of the radiation field include electrons only (e.g., γ-rays), while in the case of monoenergetic ions both data sets, (4.43) and (4.44), are involved due to the treatment of long-range δ-electrons as separate particles.

These considerations, on the one hand, explain the differences between HOLT's results and ours, on the other hand, the question is raised whether the present cluster-frequency approach constitutes a reasonably adequate idealization of the physical reality regarding correlated primary and secondary ionizations within the varying dimensions of small targets. In fact, apart from our derivation of cluster frequencies as such (section 4.4), it may be questionable laying down a maximum cluster size j_{max} and assuming that clusters up to $j = j_{max}$ lie always *completely* inside the target while separating larger clusters to be considered as *independent* δ-tracks. Therefore, it is of great value to make a check with the above-mentioned data obtained by CHMELEVSKY and KELLERER (1977) from simulated proton tracks. Taking the latter for the plain truth, this may be regarded as a crucial test.

CHMELEVSKY and KELLERER have calculated the event size distributions $F(n)$ which we can derive from $s^{(n)}$ according to Eq. (6.25). Besides, such distributions are also implied in the data of Table 6.3 if multiplied by the second column and divided by the pertinent first row:

$$F(n) = \frac{s_0{}^{(n)}}{s_0{}^{(1)}} \frac{s^{(n)}}{s_0{}^{(n)}} \Big/ \frac{s^{(1)}}{s_0{}^{(1)}}. \tag{6.27}$$

Rather than $F(n)$, the dose-distribution, $D(n)$, defined by

$$\begin{aligned} D(n) &= \frac{1}{\bar{n}} \sum_{j=n}^{\infty} j f_1(j) \\ &= 1 - \frac{1}{\bar{n}} \left[\sum_{j=1}^{n} F(j) - nF(n) \right], \end{aligned} \tag{6.28}$$

has been plotted as a continuous function, $D(y)$, of the variable 'lineal energy' (cf. Eq. (6.24)). In the same manner of presentation, except for forbearing the arbitrary smoothing of the actual step function $D(y)$, Figure 6.3 shows our results for the same particles and site diameters to agree fairly well with what turns out from simulated particle tracks. Again the crucical case are high-energy protons which have a small LET, while the very good correspondence for 0.5-MeV protons is not surprising.

To analyse the nevertheless visible differences in some detail, let us consider the mean values[1])

$$\begin{aligned} \bar{y}_F &= \frac{33.3\ \text{eV}}{(2/3)\, d} \sum_{n=1}^{\infty} F(n), \\ \bar{y}_D &= \frac{33.3\ \text{eV}}{(2/3)\, d} \sum_{n=1}^{\infty} D(n), \end{aligned} \tag{6.29}$$

which are compared in Figure 6.2. As to $\bar{y}_D$, significant departures, if any, are only seen with the high-energy protons and the smallest site, in conformability to Figure 6.3. $\bar{y}_F$ exhibits the same general trend, but the upward departure of the present results (full lines) is much more pronounced, particularly in conditions where absorption events are practically due to single clusters only (20 MeV, 5 nm). Hence, our method overestimates the mean cluster size[2]) by underestimating the proportion of very small clusters (i.e. $j = 1, 2$), or, which may also be an explanation, even small clusters[3]) are, in

[1]) Of course, the mean values cannot be calculated by these defining equations; they have been obtained, like the results in the appendix, by means of Eqs. (5.13) and subsequent elimination of radiation-dependent W-values (multiplication by 33.3 eV/W). However, which is worth mentioning, summation of $F(n)$ and $D(n)$ up to $n = 5$ yielded lower bounds to $\bar{y}_F$ and $\bar{y}_D$, which were used, as well as the sums (6.26) in comparison to $F(n)$, as a powerful test of all the numerical methods involved, including the sophisticated approach (6.17), (6.18).

[2]) This is likely to be the case particularly for protons, since Table 4.6 lists a markedly *larger* $\bar{j}$ for heavy particles than for electrons at equal j_{max} owing to the different data (4.43) used, the choice of which is in turn a consequence of using W-values which may not be quite suitable (see the lower case print comments at page 121).

[3]) In theory, large clusters ($j > j_{max}^{(\delta)}$) are completely omitted, for this reason, to form separate δ-tracks of which segments only will be inside a target.

Tab. 6.4. Relative effectiveness of protons in two-ionization target inactivation for 5 nm target diameter

	^{60}Co	Proton energy		
	γ-rays	20 MeV	5 MeV	0.5 MeV
LET$_\infty$ [keV/μm]		2.58	8.09	43.9
Present method	1	1.04	1.05	0.88
Track simulation method	—	1.04	1.2	0.8

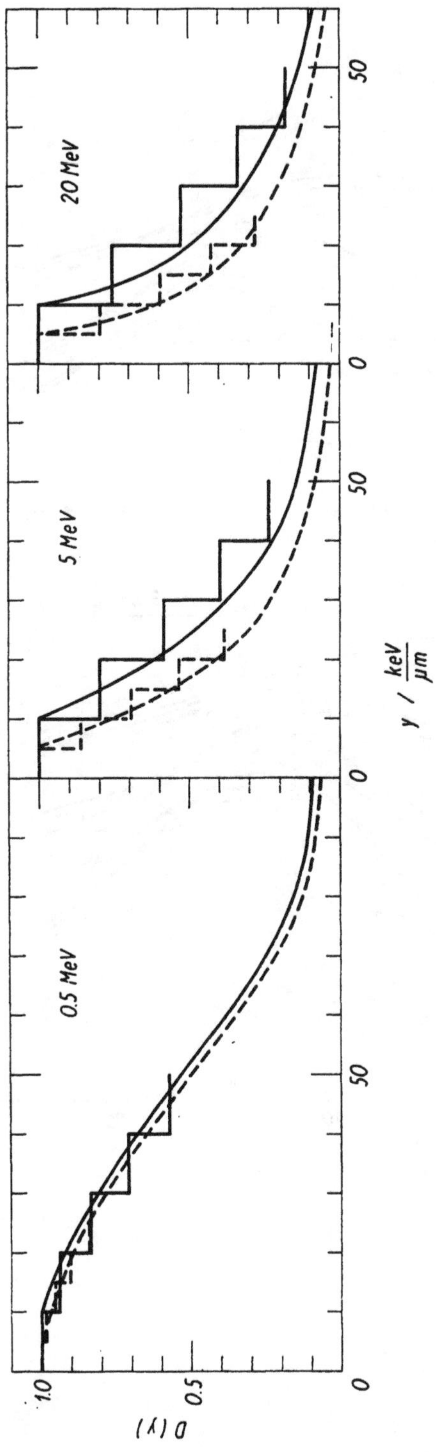

Fig. 6.3. Sum distributions, $D(y)$, of ionizations in lineal energy for protons of 0.5, 5, and 20 MeV, calculated according to Eqs. (6.28), (6.25) (step functions) and taken from CHMELEVSKY and KELLERER (1977) (smooth curves). Full lines for 5 nm and broken lines for 10 nm sphere diameter.

actuality, to some proportion segmented by the 5-nm target boundary rather than fall always completely into the target.

Having thus specified the shortcomings of our approach, let us finally return to the above-discussed crucial case of RBE with the two-ionization model, since, as argued about, it is just the clustering of ionizations, now seen to be overrated, that causes the absence of an appreciable increase of RBE. The track simulation data given by CHMELEVSKY and KELLERER (1977) allow the model to be evaluated according to

$$s^{(2)} \sim \frac{1}{\bar{n}} F(2) = \frac{1}{\bar{n}} - [1 - D(2)]. \quad (6.30)$$

Table 6.4 compares these values with ours. Due to the necessary normalization, only the figures for 5 MeV and 0.5 MeV are relevant. One has to state that our typical plateau-like RBE-LET relation is not completely borne out by track simulation, but the actual maximum, on the other hand, is by far flatter than was found by HOLT (1974).

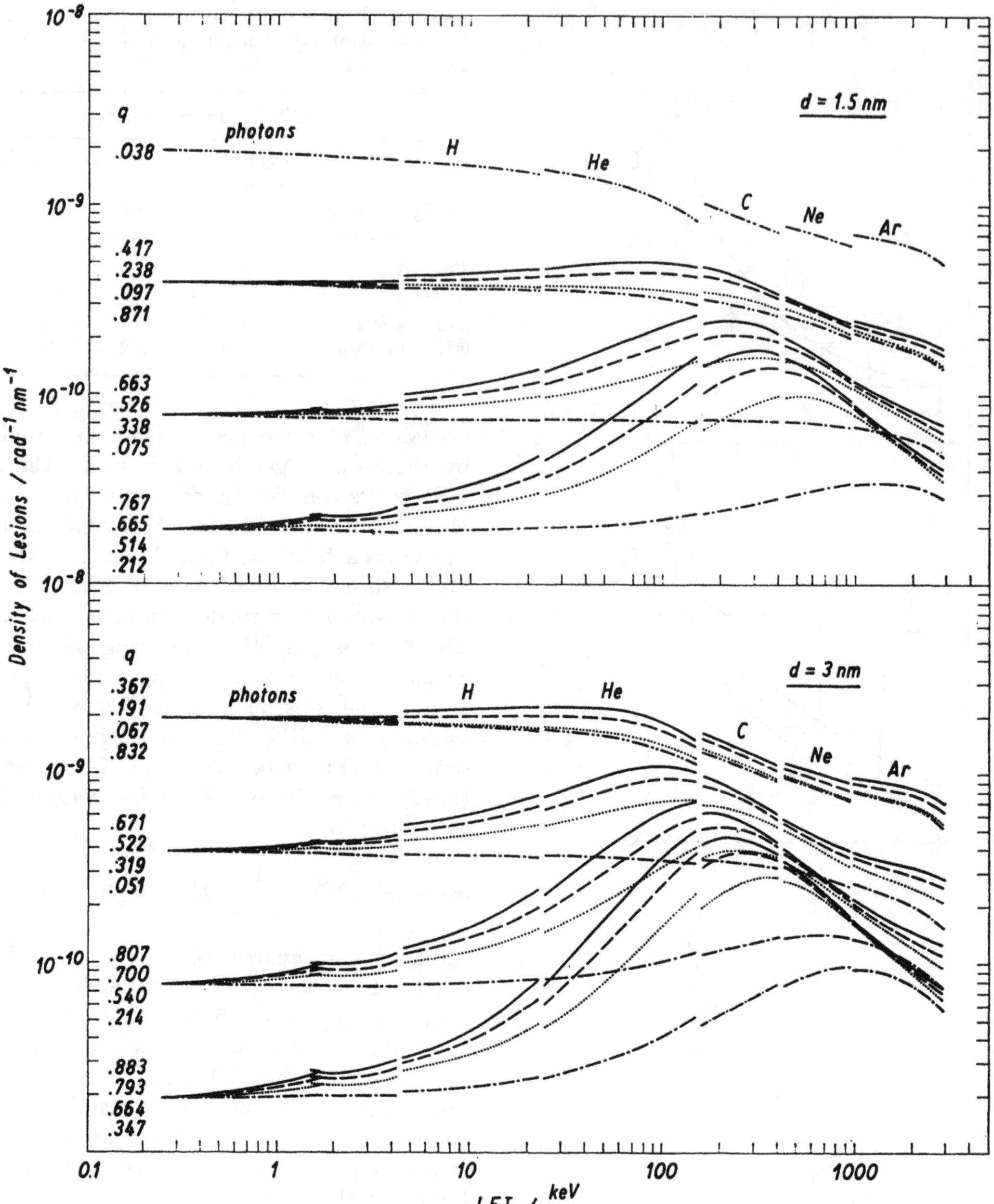
d = 1.5 nm
q
photons
.038
H
He
C
Ne
Ar
.417
.238
.097
.871
.663
.526
.338
.075
.767
.665
.514
.212
d = 3 nm
q
.367
.191
.067
.832
photons
H
He
C
Ne
Ar
.671
.522
.319
.051
.807
.700
.540
.214
.883
.793
.664
.347
10^{-8}
10^{-9}
10^{-10}
Density of Lesions / rad^{-1} nm^{-1}
0.1
1
10
100
1000
LET / $\frac{keV}{\mu m}$

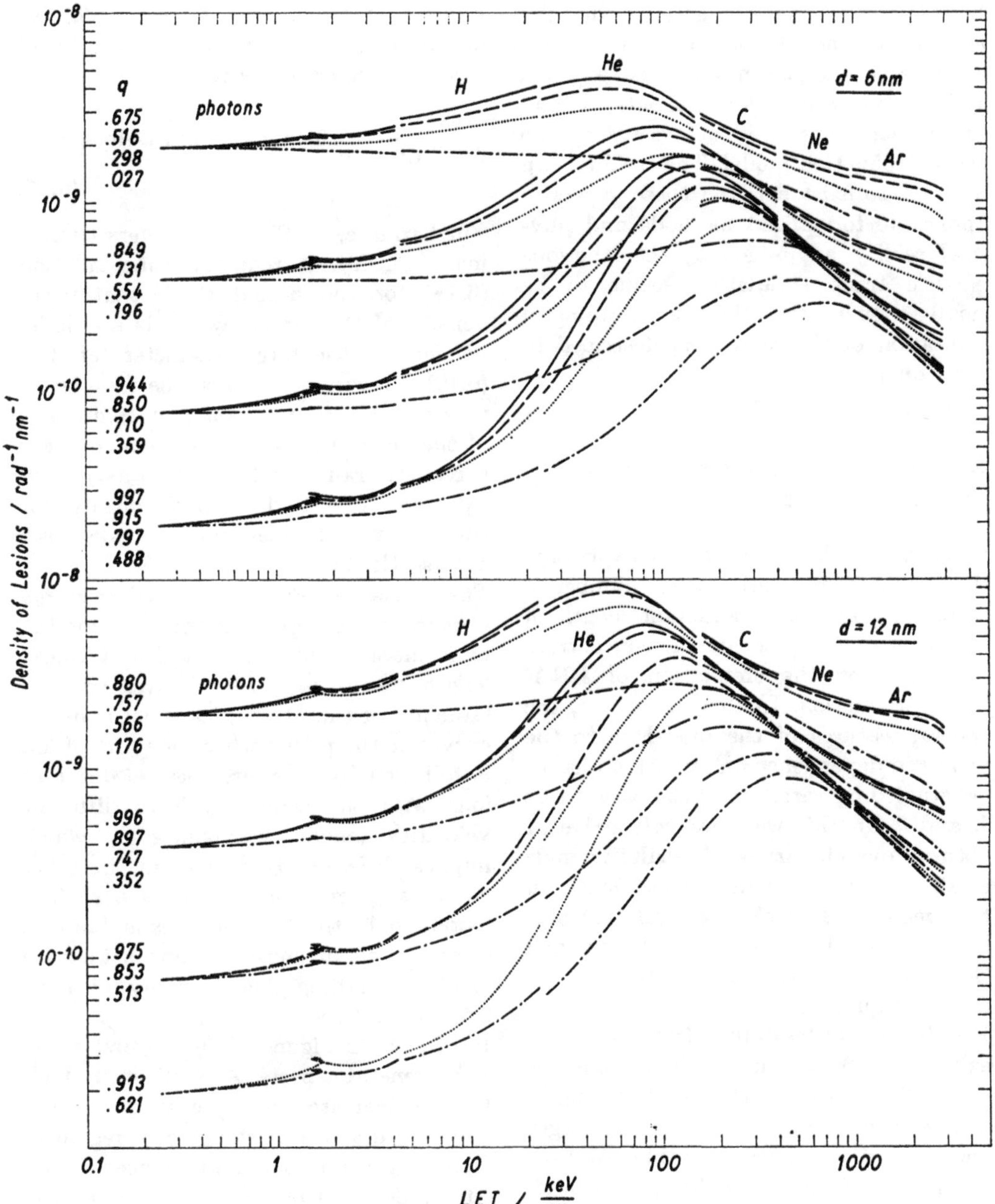

Fig. 6.4. Yield of lesions, obtained from the model of DNA-lesion production, in dependence on the LET of photons and monoenergetic ions as indicated.

Parameters: cylindrical target of unit density with the diameter d given in each panel; λ-function according to the expression (6.13) with $m = 10$, q values given at each curve, and the following p values: ——— $p = 1$, ----- $p = 0.6$, ······ $p = 0.3$, –·–·– $p = 0.05$, –··–·· $p = 1 - q$ and $m = 1$. A selection of these λ-functions is listed in Tab. 6.1.

In summary, the investigations with the n-ionization models have demonstrated a remarkable adequateness of the physical and mathematical approximations and assumptions employed in evaluating the lesion production model. The residual gap revealed to exist between these generally applicable formalized methods and physical reality as judged by the laborious particle track simulation seems to be small enough to allow a meaningful utilization of the model, as described in the sequel.

6.3.2. Lesion frequency versus radiation quality

Using a *cylindrical target* and expressing the results in terms of *absolute* number of lesions per unit dose and unit length of target, we shall, henceforth, concentrate directly upon the simulation of DNA-lesion production. In this subsection, the general features of the model as to the yield s in dependence on radiation quality are studied by variation of the parameters λ_j and d. In this way, we will arrive at both an overall survey of results possible at all and an insight into consequences of the model *as such*, i.e. relationships between the absolute yield of lesions and the characteristics of dependence on radiation quality.

For the sake of feasibility, let us use the λ_j expression (6.13) with m kept constant ($m = 10$), so that there are three parameters to be varied: p, q, and the target diameter d. A most suitable way of grouping the results is, first, by d and, second, by some reasonably chosen levels of s_0, the lesion yield for ^{60}Co γ-rays. This requires having found the variety of pairs p, q that yield a given value s_0. Accordingly, for a given target diameter, one obtains a family of curves s versus LET for each value s_0 chosen. These results are presented in Figure 6.4.

The ordinate in Figure 6.4 gives the *number of lesions per target length.* To obtain it per unit molecular weight of DNA, one must apply the factor

$$\frac{0.34\ \text{nm}}{660\,\text{Dalton}} = 5.15 \times 10^{-4}\ \text{nm Dalton}^{-1}. \tag{6.31}$$

Furthermore, as Figure 6.4 refers to unit mass density, it requires using relation (6.7a) for considering the actual mean density of the target, which is not independent of the target diameter (cf. Eq. (6.10)). According to this, the four levels of s for ^{60}Co γ-rays (common beginnings of the curves) have been chosen to cover a realistic range of lesion rate as judged by what one finds experimentally for single- and double-strand breaks (see next section).

The abscissa in Figure 6.4 is *linear energy transfer*, i.e. stopping power, L_∞, for the monoenergetic ions indicated and track-length mean, $\bar{L}_T$,[1]) for X-rays. In consequence of this choice, leaps occur in the curves at the points where one sort of ion is followed by another, because different ions with the same L_∞ have different velocities (energy per nucleon), which implies different δ-ray spectra, in other words, a different structure of tracks. The loop seen in the X-ray curves is due to a small region of non-monotonous increase of $\bar{L}_T$ with falling photon energy (cf. Figs. 5.5, 5.6, and context).

In discussing Figure 6.4, let us first stress some general aspects. At sufficiently high LET, s decreases in any case (saturation effect), while the existence of a region of increase up to a maximum (concentration effect) depends on the character of the probability set λ_j, which is given in Table 6.1 for a number of cases to be identified by the parameters p and q. In keeping

---

[1]) $\bar{L}_T$, defined by Eq. (1.10), has been calculated from $\varphi_\gamma^{(\delta)}(E)$ (Eq. (3.20)) using the usual d-dependent cut-off energy δ. The thus-caused dependence of $\bar{L}_T$ on d is quite insignificant.

with the detailed investigation in the preceding subsection, it turns out that a pronounced concentration effect (high maximum) requires that there does not exist an appreciable probability for few ionizations to produce the lesion ($\lambda_j \approx 0$ for small j). This condition cannot be fulfilled if d is small and s_0 is meant, all the same, to fall large. Therefore, one observes the general tendency that high maxima are associated with low levels of s_0 and large d values, and vice versa. Let us recall that what we are dealing with here is termed 'effects of the first kind' in chapter 2: concentration and saturation effects as a matter of insufficiency and abundance, respectively, of the single-event number of ionizations in *molecular dimensions*.

It is true that, in the ignorance of the parameters, the model admits a large variety of s-LET relations, but Figure 6.4 also demonstrates some interesting restrictions which could, on principle, conflict with experimental data, which would then falsify the model itself. For example, if the measured yield of some sort of DNA-lesions[1]) at low LET corresponds to $s_0 > 10^{-9}$ rad^{-1} nm^{-1}, there should be either no maximum at all or, at least, any maximum should not be found above 100 keV/μm. Single-strand breaks fall into that category. In this case, indeed, no increase with growing LET has been observed, which restricts the target diameter to $d < 6$ nm. A low yield, s_0, of lesions, on the contrary, implies the existence of a maximum in any case,[2]) but its position and altitude are not allowed to be arbitrary. For example, if s_0 is given to be $\sim 10^{-10}$ rad^{-1} nm^{-1}, the maximum must lie beyond 100 keV/μm and, furthermore, a relative altitude of, say, the factor 4 is correlated with the peak position not below 300 keV/μm. The latter example has been chosen to demonstrate that it is not entirely trivial that the r-values derived from survival curves in chapter 2 (Tab. 2.1) will be seen to permit a good fit by the lesion production model with $d \approx 3$ nm and $s_0 \approx 10^{-10}$ rad^{-1} $\times$ nm^{-1}, just the measured yield of double-strand breaks (see next section).

Another feature, which could be tested direct by experiments, is the tendency of leaps in s where the charge of the incident ion is increased at constant LET$_\infty$. There is a simple rule: the leap is upward in the region of saturation (falling part of the curves) and downward in the opposite case. The explanation is obvious; with a higher charge implying a larger velocity of the ion, the longer range of δ-rays diminishes the number of ionizations in the track core. Thus, direction and magnitude of a leap are nearly unambiguously predicted by the model if s has only been measured with one sort of ion in some LET range.

To sum up, it can be stated that the numerous degrees of freedom involved in the model do not make it a quite arbitrary mathematical construction. If a variety of direct and indirect[3]) experimental data on the induction of DNA-lesions can be fitted fine, it is very likely that the principal ideas of the model as stressed in the introduction to this chapter are true.

6.4. DNA-strand breaks

6.4.1. Experimental data and theoretical results

The density of radiation-induced breaks in the backbone of the DNA molecule can be measured by determining the mole-

---

[1]) With double-strand breaks always added according to the understanding of the ansatz (6.13) (see the discussion at page 180).

[2]) It is true that the maximum can artificially be avoided by making λ_j small up to very large j and non-cumulative according to $1 - \lambda_j = q^j$, but this is no realistic possibility because it implies assuming that very large numbers of ionizations would, with high probability, have no effect.

[3]) 'Indirect' means primary lesion RBE values derived from cellular dose-effect relations as exemplified in section 2.4.1.

cular weight (distribution or mean value) of the DNA fragments. With double-stranded DNA one obtains the number of double-strand breaks (DSB), while the single-strand breaks (SSB), in addition,[1]) are displayed after denaturation of the DNA. The early-used methods of viscometry and boundary sedimentation have the disadvantage that a rather large quantity of purified DNA in aqueous solution is needed, which requires DNA isolation from cells *prior to* irradiation. This does not allow studying the production of breaks under strict *in vivo* conditions in the cell nucleus. The latter has become possible by means of the sucrose gradient centrifugation technique (McGrath and Williams, 1966), which is now most widely used, apart from special methods for phage DNA (see subsection 6.4.3) and a technique (Ahnström and Edvardsson, 1974) which presently attracts growing interest (Rydberg and Johanson, 1975): determination of SSB by hydroxylapatite chromatographic separation of single-stranded fragments unwound from the DNA during a short period of alkali treatment of irradiated cells.

The major technique, that by McGrath and Williams, requires radioactive labelling of the DNA by growing the cells in, mostly, ^{3}H-thymidine medium prior to irradiation. Then, a relatively low number of cells[2]) are layered on top of a sucrose gradient in a centrifuge tube where the cells are gently lysed. Upon high speed centrifugation following, the DNA released forms a velocity sedimentation profile of radioactivity which is an image of its molecular weight distribution. In this way, one obtains, finally, the number of DSB per unit molecular weight, while SSB determination requires using an alkaline sucrose gradient, which causes denaturation of the DNA. It has to be emphasized, however, that the latter assay does not only detect primary SSB but also so-called alkali-labile sites produced by ionizing radiation, too, which are converted into single-strand breaks during exposure to high pH (Bopp and Hagen, 1970; Paterson *et al.*, 1973; Lennartz *et al.*, 1973; Lücke-Huhle, 1975).

A lot of sucrose gradient assay data have accumulated till now. Nevertheless, for a number of reasons, to be discussed in the sequel, the present situation is still far from allowing us definitively to ascribe unequivocal, precise and homogeneous values of breakage rate to a sufficiently great variety of radiation qualities. First, however, let us point out some general aspects.

In determining SSB, radiation doses of the order of 10 krads are required, while the DSB assay usually necessitates yet much higher doses. So the experiments are beyond the dose range relevant to biological effects on mammalian cells. It is customary to analyse aliquots of cells exposed to some range of doses, in addition to the unirradiated control if possible, and to plot the inverse of M_n, the so-called *number-mean of molecular weight*,[3]) versus dose. This gives a straight line whose slope is the breakage rate. Thus, which is important to note, all authors[4]) find a linear dependence of the frequency of both single- and double-strand breaks on dose within the limits of experimental error. Furthermore, no systematic differences appear to exist between different organisms (bacteria, yeast, various mammalian cells) under comparable conditions.[5])

---

[1]) It is general usage to call 'single-strand breaks' what is actually single-strand breaks plus *twice* the double-strand breaks, as an obvious consequence of the method.

[2]) 10,000 is a typical number in the case of mammalian cells.

[3]) For relationships between molecular weight distribution, mean values, and break density see Günther (1969).

[4]) With one exception: Dugle *et al.* (1976) report on a quadratic dose-dependence for DSB.

[5]) It should be mentioned, however, that four times higher rates of SSB (Young and Smith,

There are a number of difficulties which may account for the considerable scatter in the published data. A thorough evaluation of the sedimentation profiles is required to arrive at reliable estimates of M_n. Most suitable is the transformation of the profile into the number-distribution, $n(M)$, of molecular weight, which has to give a straight line in semi-logarithmic representation. Such a plot, on the one hand, indicates any irregularities of the experiment and, on the other hand, yields the best estimate of M_n by regression analysis. For details see LEHMANN and ORMEROD (1970a), PALCIC and SKARSGARD (1972), KÖRNER *et al.* (1974), GÜNTHER *et al.* (1977b). By contrast, rougher estimations have often been made since, rather than strive for a precise measurement of the primary break frequency, a good deal of investigations have other objectives (repair, influence of modifying agents, etc.). Another source of error is incomplete DNA strand separation in the alkaline gradient, and sedimentation of a complex (DNA with protein and nuclear membrane fragments) rather than free double-stranded DNA in the neutral gradient. Therefore, in the latter case at least, no unirradiated control can be analysed reliably.[1]) Moreover, anomalous sedimentation must be avoided by using a moderate speed of centrifugation (ELKIND, 1971). The consequences of this 'speed effect' in DSB determination have been investigated by HARTWIG (1977), resulting in a critical review of literature data on DSB induction by X-rays.

The data presented in Figure 6.5 have been selected on consideration of the methodical aspects mentioned, omitting older results and restricting to mammalian cells. This has considerably reduced the enormous number of low-LET results published,[2]) while homogeneous sets of data with varied LET are sparse anyway. In the latter respect, it must be mentioned that an additional problem is likely to bias high-LET measurements. Membrane-attached subunits of DNA shown to exist in mammalian cell nuclei (ORMEROD and LEHMANN, 1971), if looked upon as a superstructure of chromatin (HARTWIG, 1978), should be densely packed, so that high-LET particles may, by crossing subunits, produce clusters of adjacent breaks.[3]) The interference of a thus-caused non-random distribution of breaks along the DNA molecule is difficult to eliminate from the sucrose gradient assay results. Another feature of chromatin structure, the well-known nucleosomes, may have a marked effect: several breaks within the same nucleosome cannot be resolved.

Hence, there is enough reason, at present, to refuse the possibility of definitively fitting a selection of available data by means of the DNA-lesion production model in aiming at definite parameters λ_i and d to permit quantitative predictions of the SSB and DSB frequencies for any radiation qualities under oxic and anoxic conditions. Let us merely compare the experimental data shown in Figure 6.5 with what the model yields in a particular way as follows.

The ultimate goal of dealing with DNA-lesions here is to find out the relevant lesion type, or types, to be inserted, in terms of the model with definite parameters, into the theory of cellular radiation effects. Rather than start from the available breakage data, it is more suitable, anyway, to derive the model parameters for the 'lesion type responsible for the cellular effect considered' from analyses of the corresponding dose-re-

1976) and DSB (BURRELL *et al.*, 1971) in bacteria have been reported and valued to be relevant by HARTWIG (1977).

[1]) To overcome these difficulties, a secondary irradiation with a relatively low dose administered direct to the lysed cells on the gradient prior to centrifugation can be applied. This prevents single-strand entanglement in the alkaline gradient (KÖRNER and MALZ, 1973) as well as frees the double-stranded DNA from the complex in the neutral gradient (HANDSCHACK and MALZ, 1980).

[2]) A different compilation of data including those for bacteria, phages, and isolated DNA as well as methods other than sucrose gradient centrifugation was given earlier (GÜNTHER and SCHULZ, 1972). Figure 2.6 shows these data without comments.

[3]) This has been shown by KAMPF and REGEL (1978) and REGEL *et al.* (1983), who studied the dose-dependence of DNA release from subunits after different radiation qualities according to the method of ORMEROD and LEHMANN (1971).

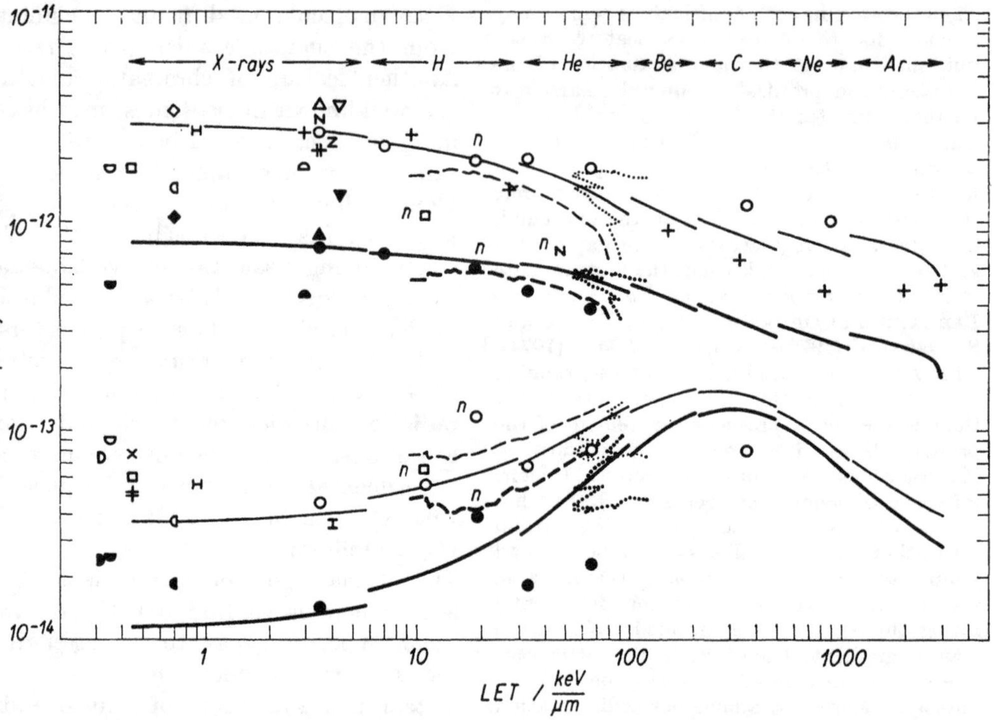

Fig. 6.5. Frequency of radiation-induced primary single-strand breaks including alkali-labile sites (upper part) and double-strand breaks (lower part) versus LET.
Curves calculated according to the model of DNA-lesion production (for details see text).
Experimental data: ○● hamster cells V79/4 (Kampf *et al.*, 1977a, b; Kampf and Regel, 1978; Kampf, 1979); N V79 (Ahnström and Edwardsson, 1974); + V79 S171 (Ritter *et al.*, 1977); □ mouse leukaemic cells L5178Y (Furuno *et al.*, 1979); ◇ ◆ mouse fibroblasts (Palcic and Skarsgard, 1972); △ ▲ hamster cells CHO (Roots and Smith, 1974); ▽ ▼ V79 (Modig *et al.*, 1974); rat thymocytes (Lennartz *et al.*, 1975); V79 (Koch and Painter, 1975); CHO (Cole *et al.*, 1975); yeast (Frankenberg-Schwager *et al.*, 1979); I L5178Y (Lehmann and Ormerod, 1970b); Z (Setlow and Setlow, 1972); ≠ HeLa (Pinto *et al.*, 1974); × human fibroblasts (Lehmann and Stevens, 1977); ⊢⊣ V79/4 (Hartwig, 1977); ╫ V79 (Koval *et al.*, 1979).
Details of presentation: *monoenergetic ions* indicated at the top: stopping power as abscissa; *photons:* dose-mean LET for 224 eV cut-off energy as abscissa; *fast neutrons:* theoretical curves (broken lines) ranging from 0.25 to 20 MeV neutron energy are plotted twice (versus dose-mean LET — dotted, versus track-length-mean LET — dashed), data points marked by "n" refer to the track-length mean. Equal symbols signify results of the same experimental series. Heavy lines and symbols indicate anoxic conditions of irradiation.

sponse relations with varied radiation quality. The rationale of the latter is introduced in chapter 2 (see sections 2.4.1 and 2.5.1), and the analyses made in chapter 7 regarding inactivation of mammalian cells end up in definite figures for d and λ_j (in terms of m, p, q) for both oxic and anoxic conditions. Within reasonable error limits, these parameters seem to apply to *any* mammalian cells. Therefore, the ratio $r = s/s_0$, correspondingly calculated for any radiation quality from the DNA-lesion production model, is referred to as the *universal* primary lesion RBE for mammalian cell killing. Let us compare the corresponding absolute values, s, with experimental frequencies of DNA strand breaks.

The parameters arrived at by survival curve analysis read[1])

$d = 2.27$ nm, $\varrho_l = 1.32$, $\varrho_l d = 3$ nm,

$m = 4$, $p = 0.314$, $q = 0.829$
for oxic conditions, (6.32)

$m = 4$, $p = 0.324$, $q = 0.950$
for anoxic conditions.

The two λ-functions are listed in Table 6.1, and the two lower sets of curves in Figure 6.5 constitute the corresponding results, which are also listed — in terms of $r = s/s_0$ — in Table VIII (appendix). In consideration of the general scatter of experimental breakage data, one can state fairly good overall agreement with the measured frequencies of double-strand breaks. That is what leads us to the conclusion that double-strand breaks are *the* lesions responsible for mammalian cell inactivation. The conformity as to the increase of RBE with growing LET suggests this anyway, but the employment of the lesion production model in the survival curve analysis has enabled us, in addition, to gather the *absolute* frequencies therefrom. The value for ^{60}Co γ-rays ($s_0 = 3.66 \times 10^{-14}$ rad^{-1} Dalton^{-1}) falling near the bulk of data for oxic conditions is very remarkable, since, as made plausible in section 6.3.2, it is the model by itself that brings about the relationship between s_0 and the relative dependence on radiation quality, regardless of whether the parameters (6.32) may be subject to considerable uncertainty. For anoxic conditions, evidence is not so clear due to few DSB data, on the one hand, and a lower significance of the survival curve analyses (see section 7.2.3), on the other hand. Yet, the low-LET oxygen enhancement ratio 3.2 turning out seems to be borne out by the DSB measurements.

As to the curves plotted along with the SSB data in Figure 6.5, the rationale is as follows. If unrealistically large target diameters are excluded, the model yields falling curves, anyway, for break frequencies as high as are those for SSB. Now, if the DSB target size $d = 2.27$ nm corresponding to $\varrho_l d = 3$ nm is retained, it is just the single-ionization model

$$\lambda_j = \begin{cases} 0 & \text{for} \quad j = 0 \\ 1 & \text{for} \quad j \geqq 1 \end{cases} \tag{6.33}$$

that accounts for the empirical SSB breakage rate for oxic conditions. These are the upper curves shown in Figure 6.5. Hence, the 2.27-nm target is the smallest one possible for oxic SSB. Note that this limiting choice of parameters engenders the most rapid decrease of s at increasing LET if $s_0 = 2.92 \times 10^{-12}$ rad^{-1} Dalton^{-1} is given. These curves fall just between the two series of data points that range to very high LET.[2]) These results imply that, if one takes the model literally, each ionization in the DNA molecule including a thin coat of surroundings causes either a primary strand break or an alkali-labile bond. Certainly, this 100 per cent efficiency will be an idealization standing for an actually somewhat larger target with a blurred boundary and a reduced probability of action per ionization. According to this interpretation, moreover, a partial probability is left over for other kind of lesions (e.g., damage to DNA bases), which would not be the case with (6.33).

The few SSB data for anoxic conditions suggest a strong decrease with growing LET, despite of the smaller value of s_0 under anoxia. It is interesting to note that this rapid decrease is already somewhat

---

[1]) In the analyses, it proved unnecessary to vary m, and $\varrho d = 3$ nm, too, has not been varied, because a greatly different value would have failed for the reason discussed in section 6.3.2.

[2]) All experimental values must be regarded as including *double* the number of DSB (see footnote 1 at page 194). Therefore, once the number of DSB, if it were well known in each case, ought to be subtracted in order to obtain data comparable exactly with the calculated curves.

beyond the range of possibilities permitted by the model. To go to the limit in this respect, the minimal target consisting of the DNA molecule only ($d = 1.45$ nm, $\varrho_l d = 2.6$ nm, cf. Eq. (6.10) and context) has been used, which brings λ_j as close as possible to the single-ionization case (6.33): $m = 1$, $p = 1 - q$, $q = 0.648$ (cf. Eq. (6.15)). These are the parameters belonging to the heavy upper lines in Figure 6.5. Thus, the calculated SSB curves for both oxic and anoxic conditions represent the limiting case of strongest saturation effect for the given respective s_0 values.

To finish the discussion of Figure 6.5, let us stress the inadequacy of any kind of LET values for specifying radiation quality. In addition to the curve leaps typical for monoenergetic ions described by L_∞, fast neutrons can by no means be brought in line with the other curves. To demonstrate this, the theoretical neutron results have been depicted twice: versus track-length-mean LET and versus dose-mean LET. In both cases, there is poor correspondence with the ion curves in the same LET range. Some experimental evidence for these differences is given by Kampf's DSB results (circles) for protons and α-particles, on the one hand, and neutrons, on the other hand.

6.4.2. Non-rejoining single-strand breaks

If irradiated mammalian cells are incubated at 37°C, the SSB induced are quickly rejoined. After incubation for several hours, at any rate, the alkaline sucrose gradient assay hardly detects any residual breaks, if low-LET radiation had been administered to the cells. Break rejoining following high-LET radiation, however, is incomplete markedly. This has been shown first by Ahnström and Edvardsson (1974).

A systematic investigation into the dependence of the frequency (relative and absolute) of non-rejoining SSB on radiation quality was made by Ritter *et al.* (1977).[1]) It is most remarkable that, according to this experiment, the number of *non-rejoined* SSB per unit dose exhibits a pattern of LET-dependence which is quite similar to our universal r-values. The corresponding curves from Figure 6.5 are shown again in Figure 6.6 to demonstrate this parallelism.[2])

The simplest explanation is to think that what is seen to be left over after repair is nothing but primary double-strand breaks. However, this has been ruled out by an investigation (Körner *et al.*, 1978) where details of the rejoining process were studied with X-rays and fast neutrons using a quite similar Chinese hamster cell line. According to these results, there is a proportion of *true* single-strand breaks induced by neutrons which does not continue lowering on extending incubation over more than half an hour. Hence, the absolute frequency of non-rejoining breaks should be *more* than twice as large as the primary DSB frequency per rad and Dalton, in contrast to the findings of Ritter *et al.* (1977), which imply that 'non-rejoining SSB' are less frequent than primary DSB (compare Fig. 6.6 with Fig. 6.5). Consequently, partial rejoining of DSB, too, must be assumed in the latter case,[3]) which might be due to the longer period of incubation (8.5 h).

---

[1]) The primary rates of SSB obtained in this experiment are shown in Figure 6.5 (crosses).

[2]) In discussing their results, Ritter *et al.* (1977) also stress the similarity to cell killing by comparing with relative single-hit efficiencies (initial slope of survival curves) of various LETs. For theoretical reasons, such kind of comparison is not quite correct (see section 2.4.2), but this does not qualify the conclusion the authors arrive at: it might be these unrepairable lesions that kill mammalian cells.

[3]) The existence of a DSB rejoining mechanism in mammalian cells was first shown by Corry and Cole (1973).

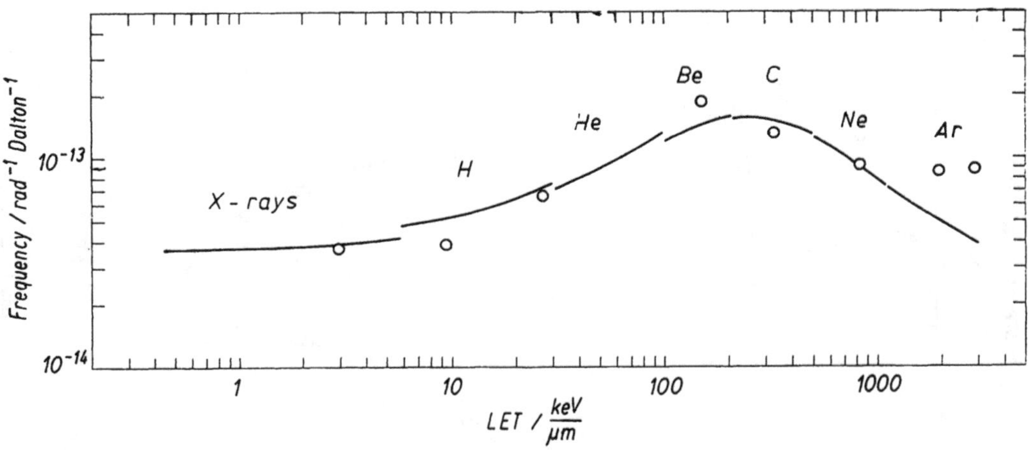

Fig. 6.6. Frequency of non-rejoining single-strand breaks produced by 300-kVp X-rays and mono energetic ions in Chinese hamster cells (V79 S171) (RITTER *et al.*, 1977). Curves redrawn from Fig. 6.5.

At any rate, it is unlikely that the lesions referred to as 'non-rejoining single-strand breaks' are simply identical with the primary double-strand breaks. Hence, whatever their nature may be,[1]) these kind of lesions could, with the same right as DSB, be considered as those which underlie inactivation of mammalian cells. They do not worse agree with the r-value curves derived from survival curve analyses (see Fig. 6.6) which yield, on principle, *relative* frequencies only. For absolute ones to be obtained, one must moreover trust in the lesion production model. Besides, from the latter point of view, too, there is no noticeable discrepancy.[2])

Thus, the DSB hypothesis established in subsection 6.4.1 is the most likely but *not the only* possibility of interpretation of the 'universal critical primary lesion RBE' based on analysing the dependence of survival curves on radiation quality. That is what should be learned from this subsection.

6.4.3. A fit for a special experimental data set

For thoroughly testing the ability of the lesion production model to account for measured frequencies of SSB and DSB, it is desirable to draw upon a homogeneous set of most accurate data obtained in the same laboratory with a variety of radiations which extend over a wide range of LET up to at least 1000 keV/µm.

An outstanding work that meets these criteria is the experiment by CHRISTENSEN *et al.* (1972) with phage ΦX 174 DNA irradiated in a radioprotective medium. While this system, unfortunately, is not comparable with mammalian DNA *in vivo*, it allows analysis of strand breaks by an advantageous special method as follows.

---

[1]) If they differ from repairable breaks chemically by specific break ends, our DNA-lesion production model should be applicable as well. Another possibility, however, is to attribute reduced repair to the breaks being partly clustered along the DNA molecule if high-LET particles interact with specific chromatin structures (RITTER *et al.*, 1977; KÖRNER *et al.*, 1978).

[2]) The coincidence of the data points with the curves of *absolute* frequency in Figure 6.6 is, to be sure, only an ostensible one, because the certainly existing DSB component in the number of non-rejoining breaks would have to be bisected before comparing with theoretical curves; but much the same agreement could easily be restituted by a slight shift of the model parameters without appreciably changing the curves of *relative* effectiveness r.

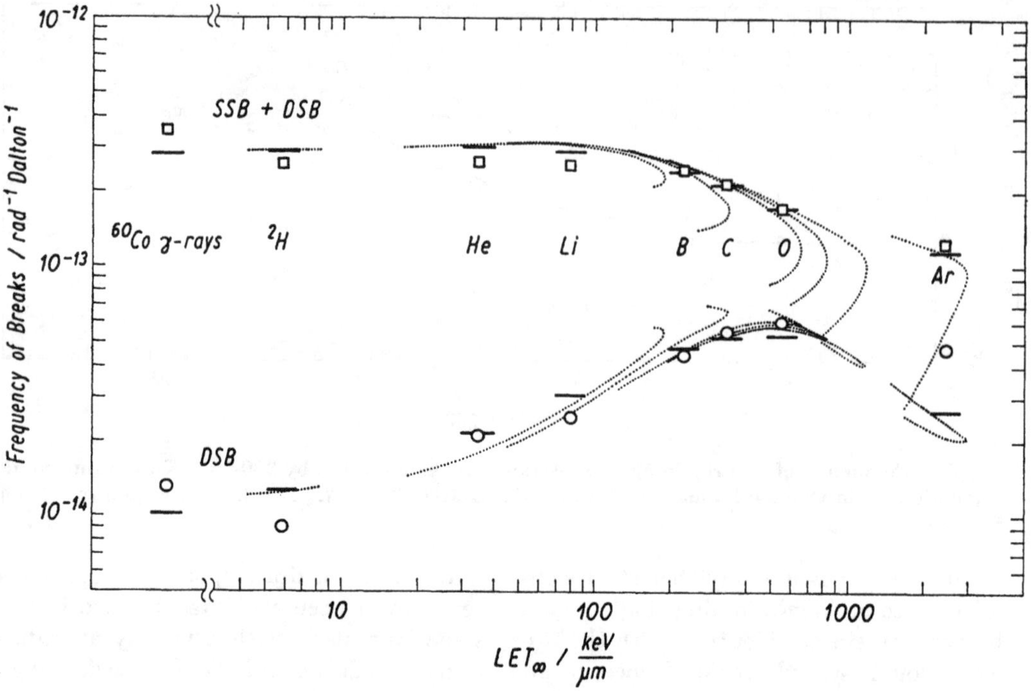

Fig. 6.7. Breakage frequency in ΦX174 replicative form DNA irradiated in a radioprotective solution with γ-rays and heavy particles. Measured values (squares and circles) from CHRISTENSEN *et al.* (1972); theoretical curves and horizontal bars according to the model of DNA-lesion production. For details see text.

The replicative form (RF) of the circular phage DNA is double-stranded and can exist in three distinct configurations depending on whether it contains no break (RF I), single-strand breaks only (RF II), or double-strand breaks (RF III). These configurations (supercoil, open coil, and linear, respectively) exhibit different velocities of sedimentation in the surcrose gradient, which thus displays the *proportions* of RF I and RF II as a function of dose. Therefrom, the induction rate of SSB and DSB per unit dose per phage genom derives in an obvious manner.

The results of CHRISTENSEN *et al.* (1972) are shown in Figure 6.7. It should be emphasized that this assay detects the genuine primary single-strand breaks without involvement of alkali-labile sites and double-strand breaks. In Figure 6.7, however, the SSB frequency is given with the DSB frequency added, which must be done by reason of compatibility with the theoretical curves based on the ansatz (6.13) (see the discussion at page 180). The latter (dotted lines) are shown as a function of L_∞ for the ions used in the experiment, yet the positions of experimental data points indicate *mean values* of LET corresponding to the irradiation technique characterized by partial or complete stopping of the ion beam within the liquid sample.[1]) Correspondingly, the theoretical break frequency had to be averaged over the entire track segment effective in each case. The horizontal bars represent these calculated mean values to be compared with the symbols.

What Figure 6.7 shows is a fit attained by varying the model parameters d and λ_j in terms of m, p, q and valueing the degree

[1]) Uniform irradiation was achieved by vigorously mixing the liquid during exposure.

of agreement by eye, ending up in: $d = 1.7$ nm; $m = 2$, $p = 0.324$, $q = 0.766$ (SSB); $m=3$, $p=0.151$, $q=0.920$ (DSB). λ_j for DSB is included in Table 6.1. The fit is good enough to demonstrate that our DNA-lesion production model is well appropriate to describe quantitatively the induction of DNA breaks by ionizing radiations, which is not trivial, as illustrated in section 6.3.2. The discrepancies with the data for γ-rays and protons cannot be explained.

Remarkable is the small target diameter, which is close to the minimal value allowed (1.45 nm, see relation (6.10) and context). With the larger diameter 2.27 nm suitable with respect to the oxic data in Figure 6.5, a significantly worse correspondence was obtained.[1]) This is likely to be due to the highly radioprotective broth in which the DNA was irradiated. This environment simulates anoxic conditions by the scavenging presence of free SH groups. In the model, therefore, the target should be small, which is not only plausible but has also been borne out by the anoxic SSB data in Figure 6.5. In the latter case, as shown in section 6.4.1, it was even necessary to take the smallest diameter possible, i.e. a target consisting of the DNA molecule only.

---

[1]) Note the shifted position of the peak of DSB.

Chapter 7

Mammalian cell killing

7.1. Introduction

The investigation of radiation effects on mammalian cells usually grown in cell culture plays an important part in modern radiation biology. This sort of objects, which became widely available after developing suitable methods of cell cultivation *in vitro* (Puck *et al.*, 1956), can probably bridge the gap between the fields of pure and applied radiation biology, between biophysics and radiation practice.

Some decades ago, relevant attempts of biophysical analysis and understanding of fundamental mechanisms were restricted to microorganisms, which could yield little reliable scientific knowledge of direct use regarding the concerns of radiation therapy and radiation protection, i.e. tissues, organs, tumours, whole animals, and men. A single *mammalian cell*, however, is certainly the smallest reacting unit that constitutes a relevant and powerful *basis model* for the processes in larger entities, without being too complex for fundamental biophysical approaches. In Elkind's (1974) words:

"The field of *radiation research*, from outward appearances, must seem to be a conglomeration of unrelated topics. Still, if there is one area I would identify as being a focus or crossroads of most of the others, it would be *cellular* and *subcellular biology*. ... It is from here, with cellular effects in mind, that the physiologist and physician seek to understand tissue and organ responses. Thus, our knowledge of the biochemistry and biology of cell effects *on the one hand* draws upon the physics of absorption processes and the chemistry following therefrom, and *on the other hand* is essential for the understanding of ecological and environmental effects of radiation, and for the improved use of radiation in medicine."

The most frequently studied *effect* is cell inactivation, i.e. the loss of the ability to form a macro-colony upon post-irradiation incubation in cell culture medium (for a review see Trott, 1972). This is briefly termed *cell killing*, described by a *survival curve*. In addition, investigations at the molecular level mainly employing the sucrose gradient centrifugation technique (McGrath and Williams, 1966) are also possible with the same cells. Furthermore, in view of the great present-day interest in high-LET radiations for cancer therapy and radiation protection in space research, it is very important that mammalian cells, in contrast to whole animals, can be bombarded readily with monoenergetic ions to study the pertinent basic aspects. This growing modern field of research was early opened by BARENDSEN *et al.* (1960) who irradiated human kidney T1 cells with α-particles. His fundamental results on RBE and OER in dependence on LET (BARENDSEN *et al.*, 1963) have remained valid up to now, however without being fully understood from the standpoint of biophysics. To achieve this understanding of these and related results in a quantitative fashion is just the central goal of the theory described in this monograph.

The present chapter is devoted to this main purpose of the theory: to understand and subsequently also to predict the *dependence* of cell survival curves *on radiation*. This includes such important subjects as relative biological effectiveness (RBE) and oxygen enhancement ratio (OER) in dependence on radiation quality as well as dose and survival level, respectively. Differences resulting from different objects or culture conditions will also be elucidated by the theory. Emphasis will be laid on fast neutrons due to their importance for radiation therapy.

From the outset, it is necessary to stress the essential point indicated by the two terms in italics some lines above. It is not an *individual* survival curve, say, for X-rays, but only the *relation between* survival curves for different radiation qualities that the theory applies to. To directly derive survival curves would require a *concrete biological model of action* expressed by a function $H(k)$ (more generally $H(k_{\text{I}}, k_{\text{II}}, \ldots)$) that is defined as the probability of survival in case of k primary DNA-lesions of the critical type(s) being generated in the cell. This would be the intrinsic theory of radiation action on cells. $H(k)$ may depend on various biological parameters, but not on radiation. It is important to understand that the mathematical formalism developed in chapter 3 is merely a general frame, a procedure of calculating survival curves $S(D)$ from a *given* model $H(k)$ or $H(k_{\text{I}}, k_{\text{II}}, \ldots)$, but not the model itself. This procedure starts with the fundamental equation

$$S(D) = \sum_{k=0}^{\infty} H(k)\, U_k(D; \lambda_j, l_l, Y) \qquad (7.1)$$

in which radiation quality enters the lesion number distribution U_k. In this state of affairs, there are only two possibilities for a practical employment of the theory: either establishing concrete biological models for the special cases of interest or eliminating $H(k)$ from the theory by introducing other, equivalent information in terms of a *single empirical survival curve* $S_X(D)$ for a low-LET radiation. The former was demonstrated for bacteria (see footnote 1 at page 24), while the present chapter concerns only the latter possibility: survival curves for arbitrary radiations are obtained from the empirical $S_X(D)$. Hence, what is committed to theory is nothing but RBE.

The topic of this chapter thus characterized may also be called *formal employment of the theory* in view of the ignorance of a biological action model $H(k)$. On the other hand, the type of DNA-lesion responsible for cell killing is also unknown *a priori*. Therefore, our first task is *formal survival curve analysis* in order to elucidate this crucial type in terms of the parameters λ_j and l_l occurring in Eq. (7.1). Only after that, intrinsic formal *survival curve prediction* for arbitrary radiations becomes possible. This means putting into practice what has already been envisaged in chapter 2, restricting oneself to mammalian cell killing.

It is very important to have a clear understanding on what is meant by formal survival curve analysis and its results. In this respect, the reader should refer to section 4 of chapter 2, where the same question has been discussed on the basis of the more comprehensible microdosimetric version of the theory. Conceptually this version is less complicated, because the model of DNA-lesion production is not involved in it from the outset. Accordingly, in place of the model parameters λ_j and l_l, one has the adjustable parameter r, the RBE for primary production of the critical molecular lesions. Analysing a family of experimental survival curves for different radiations (in particular monoenergetic ion bombardment) means simply to assign an appropriate r-value to each radiation by means of curve fitting. Thus *theory* accounts only for the *shape* of the corresponding survival curves, while the *optional* subsequent interpretation of the fitted r-values by means of the DNA-lesion production model as outlined in section 2.5,

$$r = s/s_0, \quad s = \frac{1}{\bar{z}_F} \sum_{j=0}^{\infty} \lambda_j f_1(j), \tag{7.2}$$

s_0 for ^{60}Co γ-rays,

should be regarded merely as a procedure of 'interpolation' and 'extrapolation' of the 'empirical' r for arbitrary radiations.

This understanding has to be retained also with respect to the 'strict' theory employed in this chapter, although in the *formalism* of the latter the DNA-lesion production model in terms of λ_j and l_l appears like a constitutive component, while the intrinsically relevant quantity r, which is the actual determinant for the survival curves at different radiations, looks like a secondary implication according to the model (7.2). Consequently, which is worth noting, if one doubts the validity of the lesion formation model or even the DNA-lesion hypothesis at all, the formal prediction of survival curves based on fitted parameters λ_j and l_l does not become meaningless.[1])

With these thoughts in mind, it is worth while to summarize the general results of this chapter in advance. The best fit of ion bombardment survival curves of different mammalian cell lines is obtained with different parameter sets λ_j, i.e. diverse r. This had to be anticipated not only because of the natural scatter of data, but also for theoretical reasons: even if the primary molecular lesions in question are actually DNA-lesions, of which we are thoroughly convinced, several lesion types may be involved in the then formal *effective* λ_j. In this case, λ_j is no longer independent of the sheer biological steps in the injury process; it may depend on the special object. This important conceptual question was comprehensively dealt with in section 3.7.3. However, these differences regarding r are felt to be not significant enough to be worth considering in subsequent survival curve predictions. Therefore a *universal* set λ_j implying universal r-values will be established and declared representative for mammalian cell killing in any case. This is an important hypothesis, for it opens the way for survival curve prediction based only upon the individual X-ray curve and the size of the gross sensitive volume (cell nucleus) without having need of a family of experimental ion bombardment survival curves in each particular case.[2]) The considerable success of this approach is demonstrated in sections 7.3 and 7.4.

Another question is the *interpretation* of this universal λ_j. Although this point is of great importance for radiation biology, it cannot be emphasized enough that the capability of survival curve prediction gained by the universal λ_j does in no way

---

[1]) These statements may be difficult to understand in view of the essential role the DNA-lesion model plays in the formalism of the strict version of the theory as developed in section 3.6. However, the re-formulation of the strict theory in chapter 8, where the analogy to the microdosimetric version is stressed by explicitly introducing the quantity r, clearly exhibits that only the mostly small *differences* in the final results of both versions depend on the peculiarities of this model.

[2]) This fact, seen from the formal point of view, is the essential *methodical* advantage in survival curve prediction as compared to the theory of KATZ and SHARMA (1973).

depend on the following interpretation. Speaking with caution, the universal λ_j found seems to describe nothing but *double-strand breaks*. If so, one can state that *mammalian cells* are killed by DNA double-strand breaks, i.e. by nothing else, which makes the universality by itself trivial and strict. This hypothesis is evidenced by the finding that the universal r-values fairly match the so-far-known experimental data on double-strand break production with different radiations. Furthermore, also the corresponding *absolute* values of the yield, s, according to the DNA-lesion production model (7.2) are in remarkable agreement with the experimental data (see chapter 6). Nevertheless, this remains a hypothesis. It cannot be excluded that other lesions or combinations of them follow nearly the same pattern of radiation quality dependence required by r,[1]) and on the other hand, the absolute values, s, lose their significance if what we call the DNA-lesion production model is taken for nothing but a formal description of r in dependence on radiation quality.

All these remarks should only stress that the different components, assertions, and results of the theory are of different definiteness in the light of a very critical inspection. The reader should not be confused by these hesitations. At the present state of knowledge, the authors well believe that the double-strand-break hypothesis is true. Therefore and for the sake of clear and definite terminology, we will speak of these things in the following as of facts. Just because of this, however, it was necessary to give a fair characterization of this terminology in advance.

The investigations following in the present chapter are of course based on the mathematical theory given in chapter 3, but familiarity of the reader with chapter 3 is not absolutely necessary for understanding what follows. Only some fundamental relations are needed explicitly. For this reason, it may be useful to review the general prerequisites implied in the theory:

(i) The molecular targets (DNA) are assumed to be homogeneously and randomly distributed throughout the sensitive volume (cell nucleus), regarding geometry as well as biological function.[2])

(ii) Formal employment of the theory is valid also for biologically inhomogeneous populations of cells (biological variability) including asynchrony (variable DNA-content Y), but excluding the type of critical lesions and the cell nucleus size. This was elucidated in section 3.7.2.

(iii) Despite of the formal employment being based only on the one-lesion-type theory (7.1), this approach is valid (at least approximately) also in case that several lesion types should actually be involved. This was elucidated in section 3.7.3.

---

[1]) As an example for possible candidates we quote a recent experimental result on single-strand break *rejoining* (Ritter *et al.*, 1977): radiation induced single-strand breaks include a component that is unable to be repaired even on long-term incubation of the cells, and this component was found to depend on radiation quality (ranging from low-LET protons up to argon ions) in a manner rather similar to our values of the universal r (for details see section 6.4.2). In fact, the authors of this paper suggest that it might be these lesions that are responsible for mammalian cell killing.

[2]) It is only this fundamental assumption that is likely to influence the scope of the theory and its correctness seriously, in particular for neutron radiations. Some cases of systematic disagreement between theoretically predicted and measured survival curves, demonstrated in section 7.4.2, could be explained by assuming a particular biological interaction of lesions generated in sufficient mutual proximity within the cell nucleus. This would contradict the requirement of functional randomness.

7.2. Survival curve analysis for monoenergetic ion radiations

7.2.1. Description of the method

Survival curve analysis refers to families of curves for different monoenergetic ions on the same cells, i.e. *track-segment experiments.* It starts from the experimental curve for the lowest LET, usually X-rays, and aims at a parameter set λ_j, l_l (considered as a discrete function λ of j) that yields the correct curves for all radiations. A further parameter to be fitted is σ_0, the mean cross-section of the gross sensitive volume (cell nucleus), which is however almost independently determined by the very-high-LET exponential survival curves. It should be related to the size of the cell nucleus.

The general aspects of survival curve analyses are completely analogous to what was demonstrated in section 4 of chapter 2. On the other hand, as to the additional questions connected with the DNA-lesion production model and its parameters, we can draw upon chapter 6. Therefore, we give only a concise formal account in the following, omitting background discussions.

The first task is expressing the experimental reference curve for X-rays by

$$S_X(D) = \sum_{\mu=1}^{n} B_\mu \, e^{-\alpha_\mu^{(X)} D} . \tag{7.3}$$

It is irrelevant to the theory how this mere mathematical problem is solved. E.g., one could use the only-two-parameter expression $1 - [1 - \exp(-D/D_0)]^m$ (KATZ *et al.*, 1971), which is easy to fit to a given X-ray curve. A much more general approach, based on six parameters: n, D_0, k_1, β, γ, $\lg E_1$, which can be obtained from a suitable geometrical characterization of the curve, is described in chapter 9. The coefficients B_μ and $\alpha_\mu^{(X)}$ are analytical functions of these parameters (see formulas (9.29), (9.30)). Throughout the present chapter, we will use this method and report the six parameters for each individual case.

For every radiation, the theory in chapter 3 yields the expression

$$S(D) = \sum_{\mu=1}^{n} B_\mu \, e^{-\Pi(Z_\mu) D} , \tag{7.4}$$

which is in accordance with Eq. (7.3). While the coefficients B_μ are the same for all radiations, the coefficients $\Pi(Z_\mu)$ are the actual subject of theory as given by Eqs. (3.104 to 108), where Z_μ means the corresponding coefficients of the survival probability function $H(k)$ according to Eq. (3.2). These coefficients are eliminated by expressing them through the $\alpha_\mu^{(X)}$'s of the reference curve (formal employment of the theory). In principle, this means searching after such values of Z_μ that yield $\Pi(Z_\mu) = \alpha_\mu^{(X)}$ for the X-rays that Eq. (7.3) applies to. In practice, we use the simple relation (3.123) which is true of low-LET radiations[1]):

$$Y(1 - e^{-Z_\mu}) = \alpha_\mu^{(X)}/s_X . \tag{7.5}$$

It is just this expression, $Y(1 - e^{-Z_\mu})$, rather than Z_μ alone, that is needed for the computation of Π for arbitrary radiations. This means that the DNA content Y plays no role (see section 3.7.2). The DNA-lesion production rate s_X for X-rays follows from the model (7.2) (see section 3.5), i.e. from λ_j and l_l. Thus, Eq. (7.4) can be evaluated for different radiations, starting from the parameters $\alpha_\mu^{(X)}$, σ_0 and some *trial set* λ_j, l_l.

From the formal point of view, λ_j is nothing but a function of the integer variable j, and this function can in principle be chosen arbitrarily to obtain the best fit for a family of survival curves. There are only few self-evident restric-

---

[1]) This is equivalent to the approach described in section 2.2.3 with the small correction term $\alpha_\mu^{(X)} \bar{z}_D^{(X)}/2$ in Eq. (2.19) omitted. Then, Eqs. (2.16) and (2.19) establish the same relation between $\alpha_\mu^{(X)}$ and Z_μ.

tions, namely $\lambda_0 = 0$, $\lambda_j \geqq 0$,[1]) and the obligatory mathematical form

$$\lambda_j = \sum_\nu b_\nu \, e^{-\zeta_\nu j}. \tag{7.6}$$

In chapter 6, such an expression containing only three parameters, m, p, q, was used and discussed in detail (page 180):

$$\lambda_j = \sum_{\nu=0}^{m} \binom{m}{\nu} (-1)^\nu \left[1 - \frac{1-q^\nu}{1-(q+q^m)/2}\, p\right]^j,$$
$$0 < (p, q) < 1. \tag{7.7}$$

This is a monotonously increasing function of j with a shape controlled by m, p, q, i.e. the limiting value $\lambda_\infty = 1$ is reached either gradually or after a more or less steep step, the position of which in the j-scale is also variable. For the present purpose the value $m = 4$ was used only, and it proved appropriate.

Of course, if one remembers the meaning of the parameter set λ_j (probability of a lesion being produced by j simultaneous ionizations in the DNA-target), one should ask whether the ansatz (7.7) is sufficiently general and which types of lesion (or combinations of them) may or may not be described by it. This sort of reasoning, however, is the topic of chapter 6, while for the present purpose it is only important that the expression (7.7) is able to provide reasonably good fits in all cases. This must be seen in relation to the natural scatter in the experimental survival data. In other words, we do not ask about whether the λ_j thus found are unambiguously determined by survival curve analysis. Therefore, in the following the results will be presented in terms of the parameter r defined by Eq. (7.2) rather than λ_j itself, since r in dependence on radiation quality is the intrinsic determinant of survival curves. These r-values *are* reliable if the fit was good.

As to the further parameters of the DNA-lesion production model, i.e. mean target diameter l_l, mass density coefficient ϱ_l, and chord length spectrum $w_l(l')$, it can be renounced to vary them, stressing the same argument as above: every restriction is subsequently warranted by a good fit. For these parameters we will therefore take what proved appropriate in chapter 6 for single- and double-strand breaks: $\varrho_l l_l = 3$ nm, $w_l(l') = \delta(l' - l_l)$.

Finally, for computing the Π's of Eq. (7.4), one needs the geometry of the gross sensitive volume which we assume to be the cell nucleus. It is most appropriate to take spherical shape, i.e. the triangular chord length distribution (3.25) for $w(l')$ with $d = 2\sqrt{\sigma_0/\pi}$, $l = (2/3)\, d$. This assumption is completely justified for cells irradiated in suspension, as usually done in neutron irradiations. However, monolayer cells used for ion irradiations are known to have flattened nuclei. Apart from the lack of definite pertinent data, it would not be correct to use a corresponding $w(l')$ together with the *mean* projected area for σ_0, because of the perpendicular incidence of the ions. Therefore, one has to retain the triangular distribution and to consider σ_0 as the *maximum* projected area (i.e. projection in beam direction), which becomes obvious from a thorough inspection of the first term of Eq. (3.104), while the second, i.e. the δ-electron, term often vanishes or is very small.

After these preliminaries, the 'strategy' of survival curve analysis can be described in few words. With the parameters B_μ, $\alpha_\mu^{(x)}$ being known from the experimental X-ray curve, the theoretical curves (7.4) for all radiations occurring are calculated with a trial set of parameters σ_0, p, q. Now *SSD*, the sum of squared relative deviations of all individual measured survival fractions from the corresponding theoretical values, is considered a function of σ_0, p, and q. The minimum of *SSD* with respect to varying p and q at a fixed σ_0 is then found by means of the gradient method. This least-squares fit yields the 'best' λ_j, described by p and q, for each *given* value

---

[1]) The condition $\lambda_j < 1$ is unimportant here, because λ_j could be multiplied by an arbitrary factor a, which would have no influence on the final results. Such a factor enters Eq. (7.5) through $s_X \sim a$ and Eqs. (3.107), (3.108) through $G_i(E)$, $\Gamma(E) \sim a$, and it drops out therefore. This shows again that the *absolute* rate of lesion production, s, plays no role for the formal employment of the theory.

of σ_0, nominally the cross-sectional area of the cell nucleus. The absolute minimum of *SSD* defines the *optimal* σ_0 and the corresponding λ_j, but this is of little significance. In some cases the actual cell nucleus size is known; otherwise any biologically reasonable value of σ_0 in the vicinity of the optimal one may be found true. This implies, on the one hand, some uncertainty regarding the best λ_j for each cell line, and on the other hand, more latitude for establishing a universal λ_j for all cell lines.

7.2.2. Results for oxic conditions

Six track-segment experiments with different cell lines irradiated in the presence of oxygen were evaluated by formal survival curve analysis: HeLa cells (DEERING and RICE, 1962), T1/B — human kidney T1 cells (BARENDSEN *et al.*, 1963; BARENDSEN, 1964; BARENDSEN *et al.*, 1966; BARENDSEN, 1966)[1]), T1/T — human kidney T1 cells (TODD, 1966; TODD, 1967), CH2B2 — Chinese hamster ovary cells (SKARSGARD *et al.*, 1967)[2]), P-388 — murine leukaemia cells from mice assayed *in vivo* (BERRY, 1970), M3-1 — Chinese hamster bone marrow cells (TODD, 1975). These experimental data (individual survival fractions and different radiations) are shown in Figure 7.2, together with theoretical curves discussed below.

Table 7.1 gives a concise account of the six analyses (least-squares fits). The reference X-ray curves plotted in Figure 7.2 belong to the X-ray curve parameters listed[3]) (for their meaning see chapter 9). The optimal fits (p, q, and σ_0) are visualized by the dashed lines in Figure 7.2. With few exceptions, the correspondence is fairly good. It may however be *too* good in view of the numerous sources of experimental error and data scatter, which is likely to bias the parameters p, q, and σ_0 obtained. Therefore, we must carefully consider the significance of the differences in p and q for the different cell lines as well as $\sigma_{0,\text{optimal}}$ versus $\sigma_{0,\text{observed}}$.

Observed values of the cross-sectional area, σ_0, of the cell nucleus must be considered with caution. Microscopic examination has to be carried out under the same conditions as irradiation, since the monolayer top view cross-section relevant here may depend on a number of factors; it should not be expected to be constant for a given cell line. The values listed in Table 7.1 have been given by the individual experimentalists, mostly without mentioning methodical details. The widely differing σ_0 data for T1/B cells have been taken from the following papers: BARENDSEN *et al.* (1960); BARENDSEN *et al.* (1963); BARENDSEN (1968). As a further complicating fact one usually observes a widespread *distribution* of nuclear sizes, and the most suitable mean value, $\bar{\sigma}_0$, is obtained neither from the diameters, d, nor directly from the cross-sections. Strictly speaking, this sort of biological inhomogeneity cannot be allowed for at all by a mean σ_0. For very-high-LET radiations, e.g., superposition of the corresponding sub-population survival curves given by $S(D) \approx \exp(-\sigma_0 D/L_\infty)$ should result in upwardly bent curves ('negative' shoulder). Although this is usually not observed experimentally, which may be due to the compensatory effect of small residual 'positive' shoulders our theory yields even for Ne and Ar ions, the 'correct' effective mean $\bar{\sigma}_0$ should be calculated along these lines. This was done by TODD (1975), who determined $\bar{\sigma}_0$ from the observed frequencies x_i of cross-sections $\sigma_0^{(i)}$ according to

$$e^{-\bar{\sigma}_0 f} \approx \sum_i x_i \, e^{-\sigma_0^{(i)} f}, \tag{7.8}$$

where f is the particle fluence corresponding to reasonable doses of very-high LET particles.

---

[1]) It was not easy to collect the true experimental data from the different papers. As the main source we took the 1963 and 1964 reports, supplemented by data from the later publications indicated, unless these were felt to be merely mean values from previous results.

[2]) Supplementary data taken from KATZ and SHARMA (1973).

[3]) In the two cases where the experimentalist reported additional individual X-ray control curves for each ion radiation, these were taken into account by including corresponding dose factors into the average curve. Only the latter, however, is shown in Figure 7.2 (M3-1: mean values calculated by us; T1/T: 'standard' curve as given by TODD (1966)).

Tab. 7.1. Formal analysis of experimental survival curves for six different cell lines

Cell line	X-ray curve parameters					Results of individual analyses (optimal λ_j, σ_0)			σ_0 observed	σ_0 fitted	Comments
	n	D_0 [rad]	k_1	β	γ	p	q	σ_0 [μm²]	[μm²]	[μm²]	
HeLa	3	129.4	0	0.570	2.890	0.01559	0.968 24	60	100	90	1)
T1/B	5	113.3	0.264	0.264	0.970	0.02796	0.959 88	35	30–80 ~40 ~70	60	2)
T1/T	6	99.0	0.230	0.531	0.191	0.272 10	0.883 67	80	90–100	120	3)
CH2B2	6	127.3	0.246	0.237	1.310	0.031 95	0.887 94	40	~40	50	
P-388	6	111.4	0.150	0.272	1.580	0.023 87	0.942 26	40		60	4)
M3-1	6	128.6	0.205	0.466	1.750	0.313 72	0.829 00	74	74	74	5)

1) 2 survival values omitted from fit.
2) Calculations based on the experimental LET-values (large fractionary energy loss of α-particles in the cell layer).
3) Individual X-ray control curves for each ion radiation were used; Ar ions omitted from fit.
4) X-ray curve missing, parameters refer to 6-keV/μm deuterons.
5) Individual X-ray control curves for each ion radiation were used.

The value he has obtained in this way for the M3-1 cells is $\bar{\sigma}_0 = 74\ \mu m^2$ given in Table 7.1. It is very remarkable that the optimal value resulting from our fit (survival curve analysis) was exactly the same in this case.

Neither p and q themselves nor λ_j defined by Eq. (7.7) are suited for inter-cell-line comparison. It is rather the derived quantity r (Eq. (7.2)), i.e. the RBE for the primary lesions responsible for cell killing, that has to be envisaged in dependence on radiation quality. For the sake of comprehensibility, we shall consider r only for two radiations at medium LET (80-keV/μm α-particles) and high LET (280-keV/μm carbon ions). Figure 7.1 displays what turns out for r_α and r_C from the six cell lines. The circles refer to the optimal parameters in Table 7.1, and the short curve sections indicate how r_α and r_C change when σ_0 is varied in a range compatible with observerved σ_0's and with the requirement of an only insignificant increase of the sum of squared deviations as compared to the optimal value. Note that for each σ_0 the least-squares fit for p, q was carried out separately.

Non-linear scales for both r-values have been used in Figure 7.1. This is to visualize the true effect of changing r on the corresponding survival curves. The distance from either axis is proportional to the final slope of hypothetical standard survival curves[1]) for 80 keV/μm and 280 keV/

1) The standard curves assumed for constructing the axis scales of Figure 7.1 refer to an X-ray curve with $D_0 = 125$ rad and to $\sigma_0 = 60\ \mu m^2$. The corresponding slopes for the two ion radiations have been calculated from the microdosimetric approximation of the theory (chapter 2) according to $\alpha = \int dz \times [1 - \exp(-r\alpha^{(0)}z)]\, f_1(z)/\bar{z}_F$, where $\alpha^{(0)} = D_0^{-1}$. The scales are defined by $\alpha/\alpha^{(0)}$ as a function of r. The ratio of final slopes, $\alpha/\alpha^{(0)}$, being smaller than r is just what has been termed 'saturation effect of the second kind' in chapter 2.

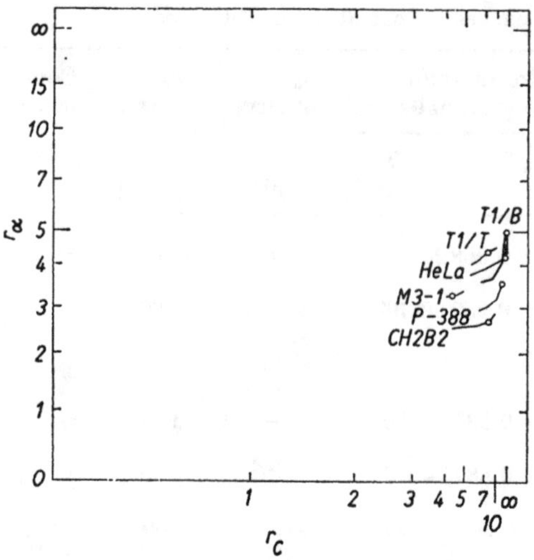

Fig. 7.1. Results of survival curve analysis from six track-segment experiments in terms of r (r_α for 80-keV/μm α-particles, r_C for 280-keV/μm carbon ions). Circles mark the optimal fits. Curves refer to the variation of σ_0 within the following limits (in μm²): HeLa: 50–80; T1/B: 30–70; T1/T: 70–100; CH2B2: 35–55; P-388: 40–100; M3-1: 65–80.

μm, respectively. In this way, Figure 7.1 gives us an idea of both how the presumed σ_0 affects the r's obtained (without appreciably making the fit poor) and which change towards worse must be expected if λ_j (parameters p, q) for one cell line would be tentatively used for another one. Especially the latter question is most interesting in regard of the possibility to fix a *universal* λ_j, i.e. universal r-values, for all cell lines. In this respect, it would be ideal if the curves in Figure 7.1 intersected in one point and if this common point referred to reasonable values of σ_0 for each cell line. Actually, there is only some proximity, which is however felt to give enough reason, nevertheless, to choose a universal λ_j (universal r-values) and to declare it representative for mammalian cell killing under oxic conditions in any case.[1]) *All survival curve predictions following in this book are based on that universal description of critical primary lesion production as a function of radiation quality.*

It has been decided to take the optimum found for M3-1 cells as the universal λ_j described by

$$m = 4, \quad p = 0.31372, \quad q = 0.82900. \tag{7.9}$$

This choice is warranted not only by the fact that the corresponding curve in Figure 7.1 lies about in the middle of the bulk, but mainly by beneficial features of this experiment. Apart from being the most recent one, it includes individual X-ray control curves[2]) and the author (Todd, 1975) has thoroughly examined the cross-sectional area of cell nuclei under his irradiation conditions. This microscopic examination yielded the mean value $\bar{\sigma}_0 = 74\ \mu m^2$ (correctly evaluated according to Eq. (7.8)), which was exactly re-found as the optimal one in the course of formal analysis (see Tab. 7.1). Thus, in this case, there was no uncertainty resulting from a poorly known σ_0. Consequently, if there does exist a universal λ_j, the M3-1 experiment is most likely to reflect it unadulteratedly.

---

[1]) It is very interesting to note that the three cell lines of human origin, i.e. T1/B, T1/T, and HeLa, form a separate group of close mutual correspondence in Figure 7.1. On the one hand, this demonstrates the ability of the theory to trace the differences in RBE versus LET of Barendsen's and Todd's results on the same cell line, T1, which were often discussed in the radiobiological literature (e.g., Bewley, 1968), back to differences in the X-ray response, nuclear area, and different ion radiations used, leaving the molecular events characterized by r nearly unchanged. On the other hand, this observation suggests establishing special common r-values for human cells as distinct from other ones for rodent cells. However, we cannot yet decide on this, all the more since these peculiarities can more likely be explained otherwise in the framework of an extension of the theory in regard of its basic principles (see the discussion of discrepancies in section 7.4.2).

[2]) This makes the formal analysis more reliable, since mammalian cells, in general, are known to change their radiation sensitivity in the course of an experiment that requires some time (Rossi, 1976).

Now applying the universal λ_j to the six experimental curve sets, σ_0 can be fitted again. The values obtained ('σ_0 fitted' in Tab. 7.1) do not match the observed σ_0's worse than the optimal ones did. The corresponding survival curves (full lines in Fig. 7.2) are of course often in poorer agreement with the experimental results. This documents the degree of accuracy that has in general to be expected for theoretical survival curve predictions based on the universal λ_j. The most dramatic deviations are found for T1/B, which is however felt to be a singular case. These experimental data were obtained successively in a long period of time and with different irradiation sources and arrangements. Individual X-ray control curves were not reported. Furthermore, the helium ions below 1.275 MeV/nucl. were energy-degraded ^{210}Po α-particles passed through different absorbers. Apart from the thus-caused scatter in the incident energy, LET and dose changed substantially over the cell nucleus in these cases of low-energy α-rays. Our calculations have been based on the mean LET-values given by BARENDSEN (1964).

From the universal λ_j described by the parameters (7.9), the corresponding universal values of r can be calculated according to the lesion production model (7.2) for arbitrary radiation qualities. These results are listed in Table VIII of the appendix for photons, fast neutrons, and ion radiations, and were discussed as a function of linear energy transfer in chapter 6 (see Fig. 6.5), where the remarkable coincidence with the so far available experimental data on the RBE for induction of DNA double-strand breaks was stressed. Not only the *relative* effectiveness, $r = s/s_0$, but also the *absolute* rate of lesion production per unit dose, i.e.

$$s_0 = 3.66 \times 10^{-14}\ \text{rad}^{-1}\ \text{Dalton}^{-1} \quad \text{for } ^{60}\text{Co radiation,} \tag{7.10}$$

is in good agreement with experimental data. This depends on the normalization $\lambda_\infty = 1$, which is appropriate just for double-strand breaks, while it is unimportant in the framework of formal survival curve analysis (see the footnote 1 at page 207). Hence, the universal type of lesion responsible for mammalian cell killing is likely to be nothing but DNA double-strand breaks, so that the universality becomes merely a matter of indifferent physico-chemical processes of primary break formation.[1])

In consequence, mammalian cells in general will henceforth be told to die from DNA double-strand breaks described by the r-values in Table VIII. The latter having been calculated by means of the DNA-lesion production model with fitted parameters (Eqs. (7.6), (7.7), and (7.9)) is the particular way of generalizing ion-irradiation data to other radiations, especially neutrons. In other words, it is the quantity r by which cell survival after qualitatively different irradiations, such as monoenergetic ions and fast neutrons, become related to each other. In addition to r, only pertinent virtual microdosimetric parameters will be needed for the prediction of a survival curve, as described comprehensively in chapter 9. Thus, the complete table of r-values, rather than the formal parameters p, q defining λ_j, is finally the substantial result of this section.

7.2.3. Results for anoxic conditions

Let us consider the well-known oxygen effect, i.e. enhanced survival in the absence of oxygen, from the viewpoint of

---

[1]) From this aspect, it would be hard to explain the inter-cell-line scatter exhibited in Figure 7.1 by different yields of double-strand break formation. On the other hand, the character of these deviations is also incompatible with the assumption that other lesions are additionally involved to different degrees (involvement of single-strand lesions diminishes the r-values). This is a further argument in favour of the 'double-strand-breaks-only' hypothesis.

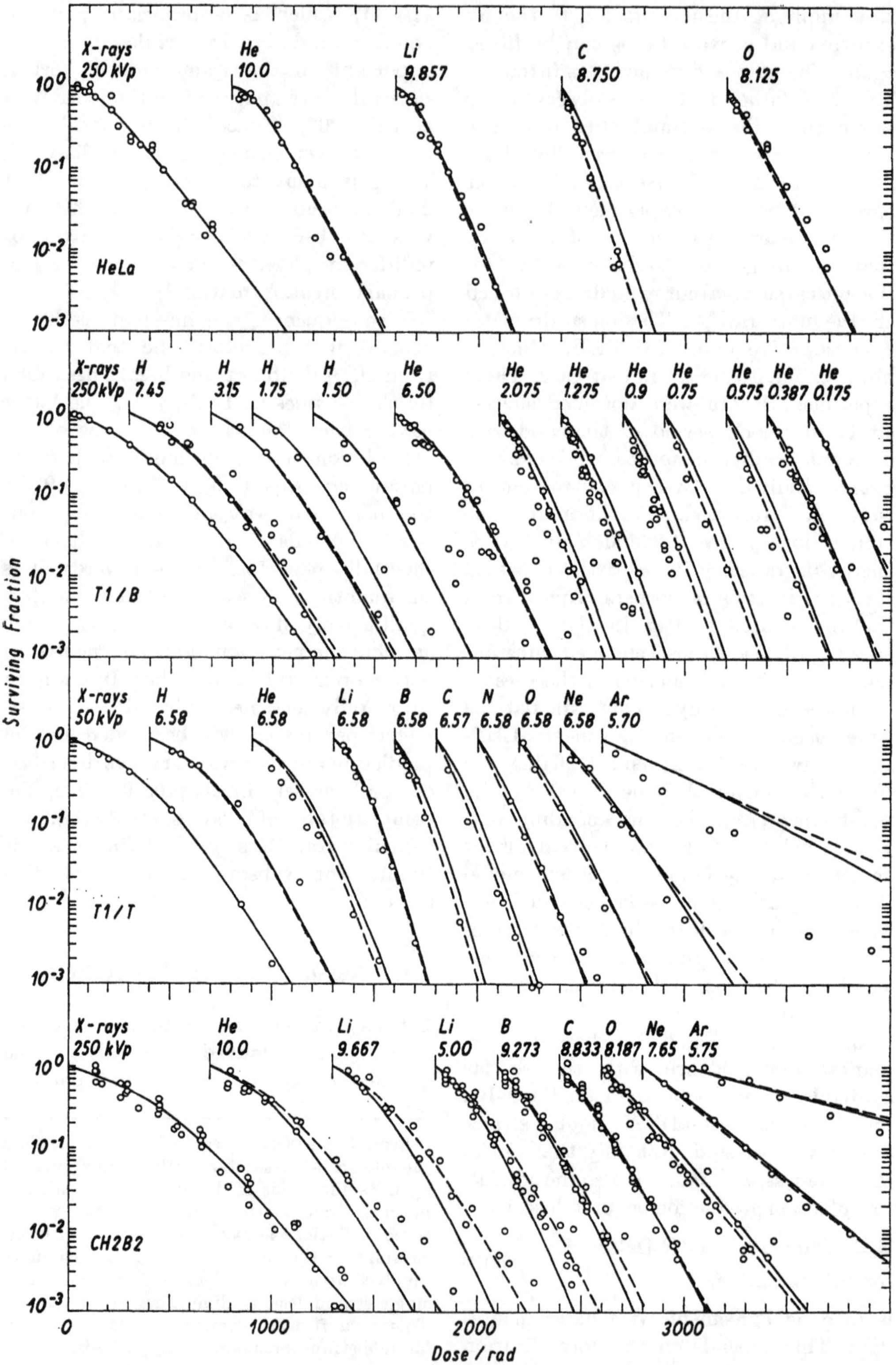
X-rays
250 kVp
He
10.0
Li
9.857
C
8.750
O
8.125
HeLa
X-rays
250 kVp
H 7.45
H 3.15
H 1.75
H 1.50
He 6.50
He 2.075
He 1.275
He 0.9
He 0.75
He 0.575
He 0.387
He 0.175
T1/B
X-rays
50 kVp
H 6.58
He 6.58
Li 6.58
B 6.58
C 6.57
N 6.58
O 6.58
Ne 6.58
Ar 5.70
T1/T
X-rays
250 kVp
He 10.0
Li 9.667
Li 5.00
B 9.273
C 8.833
O 8.187
Ne 7.65
Ar 5.75
CH2B2
Surviving Fraction
Dose / rad
0
1000
2000
3000

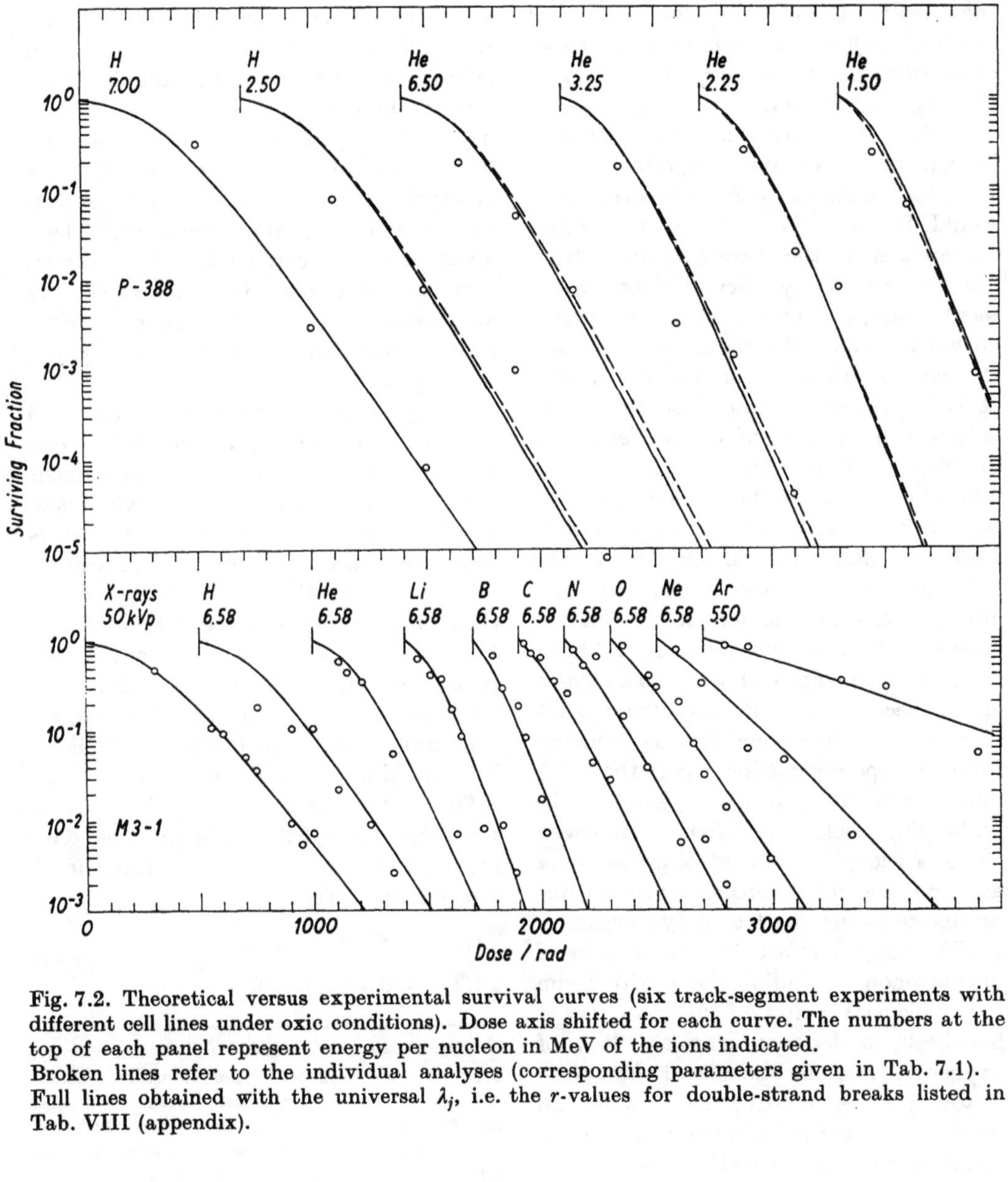

Fig. 7.2. Theoretical versus experimental survival curves (six track-segment experiments with different cell lines under oxic conditions). Dose axis shifted for each curve. The numbers at the top of each panel represent energy per nucleon in MeV of the ions indicated.
Broken lines refer to the individual analyses (corresponding parameters given in Tab. 7.1). Full lines obtained with the universal λ_j, i.e. the r-values for double-strand breaks listed in Tab. VIII (appendix).

the present theory, i.e. from a formal position. Oxygen may influence the molecular processes of *primary* lesion formation (λ_j, r, s_0) as well as the *response* of a cell $(H(k))$ to a *given* number of produced lesions, say DNA double-strand breaks. If only the latter were the case, there would be no difference in the formal employment of the theory: starting from the changed X-ray curve like an experimental fact, one could calculate survival curves for high-LET radiations by using the same r-values as for oxic conditions. It is important to note that already in this way a *decrease* of oxygen effect at growing LET is obtained due to the saturation effect of the second kind (see section 2.4.2). In particular at very high LET each passage of an ion through the cell nucleus inactivates it, so that any effect of oxygen becomes lacking if the cross-section σ_0 is left unchanged. Therefore, the mere empirical fact of the oxygen effect decreasing with increasing LET does not provide unambiguous evidence for a corresponding difference in the LET-dependence and absolute yield of the underlying molecular lesions. However, there is enough reason of other kind for assuming the primary processes of lesion formation to be *not* unaffected by oxygen.

(i) The oxygen effect is a quite general phenomenon of radiobiology. (ii) From the biological point of view, it could hardly be understood why the lack of oxygen should be beneficial for post-irradiation repair (if there were no reduction of primary lesions); the opposite seems to be true (see below). (iii) Radiochemical studies (ESR method) suggest that primary lesions are formed in part via free radical states of DNA and their interaction with oxygen. (iv) A tentative theoretical calculation of high-LET survival curves for anoxic T1 cells, using *oxic* r-values as discussed above, yields no quantitative agreement with corresponding experimental results (see below): the curves do not become steep enough, i.e. saturation of the second kind contributes, to be sure, to the diminution of oxygen effect, but does not explain it fully. Hence, there must exist also some LET-dependent oxygen effect at the molecular level, which has to be expressed by a changed RBE for primary lesion production under hypoxic conditions. These anoxic r-values can be found by formal survival curve analysis in the same way as described in the foregoing sections, starting from the anoxic X-ray curve and varying λ_j.

The oxygen effect is usually assumed to be purely dose-modifying, so that it can be expressed by the oxygen enhancement ratio, OER, of equally effective doses, valid for any level of survival.[1]) It is therefore appropriate for the analysis to use the same theoretical X-ray curve (Eq. (7.3)) as for oxic conditions, apart from a dose factor that changes the parameter D_0 in Table 7.1 only. Two track-segment experiments for anoxic conditions have been analysed: T1/B — human kidney T1 cells (Barendsen *et al.*, 1966; Barendsen, 1966) and T1/T — same cell line (Todd, 1967).[2]) The dose factors used for constructing the anoxic X-ray curves are

$$\begin{aligned} &\text{T1/B:} \quad OER^{(X)} = 3.15, \\ &\text{T1/T:} \quad OER^{(X)} = 2.53. \end{aligned} \tag{7.11}$$

[1]) This 'dose modification only' cannot with certainty be verified experimentally. There are also contradicting results (Littbrand and Révész, 1969). It should be noted that the present theory, in any case, does predict some difference in the curve shape at higher LET if strict dose modification for X-rays is presumed. This is due to the second kind effects, which suppress the shoulder somewhat more in the oxic case, but without influencing the extrapolation number. The thus-caused slight dependence of OER on the survival level could hardly be disclosed experimentally.

[2]) A third experiment of this kind, P-388 (Berry, 1970), has not been taken into account, because these results in terms of OER show too much scatter and an unusual dependence on LET.

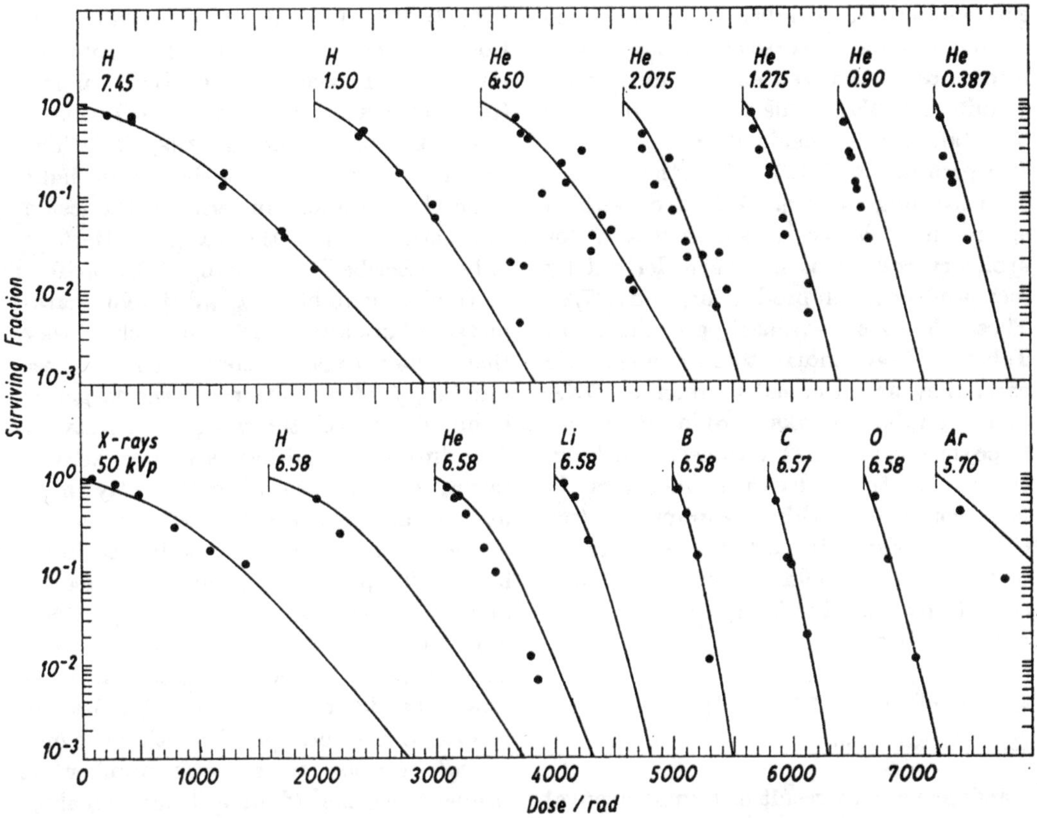

Fig. 7.3. Theoretical versus experimental survival curves for anoxic conditions (see the caption of Fig. 7.2). Theoretical curves calculated with $\sigma_0 = 60\,\mu m^2$ (T1/B, upper panel) and $\sigma_0 = 120\,\mu m^2$ (T1/T, lower panel) in accordance with the universal 'anoxic' r-values for double-strand breaks listed in Tab. VIII (appendix).

They were chosen in such a way as to ensure an optimal fit for the anoxically measured 50-kVp-X-ray curve (T1/T) and the 14.9-MeV-deuteron curve (T1/B), respectively, since in the latter case no anoxic X-ray curve was available.[1]) These fits can be seen in Figure 7.3 (left-hand curves).

The analysis for the two anoxic experiments has not been carried out in such a general manner as in section 7.2.2 for oxic conditions. First, for the cross-sectional area of the cell nucleus the values referred to as 'σ_0 fitted' in Table 7.1 have been used and left constant. Otherwise one would not obtain $OER = 1$ as LET $\to \infty$. Second, it has from the outset been searched after *common* values of r_{anoxic} for both experimental data sets by way of varying λ_j via the parameters p and q. The choice finally made reads

$$m = 4,\quad p = 0.32380,\quad q = 0.95000. \quad (7.12)$$

It may be regarded as the best compromise for both experiments, which is exhibited in Figure 7.3. A strict least squares fit would not have been suitable, because the deviations should correlate with the corresponding ones in Figure 7.2 (unbroken

[1]) OER values for X-rays given in the original papers differ somewhat from the figures (7.11). 2.53 for T1/T however is in full agreement with the measured survival curves (cf. Fig. 7.2 and Fig. 7.3), while 3.15 for T1/B is a 'theoretical' value adjusted according to the experimental curve for low-LET deuterons.

curves) for oxic conditions, in order to ensure reliable theoretical results with respect to the oxygen enhancement ratio at different LET. This point of view has been taken into consideration.

The parameters (7.12) define the 'anoxic λ_j' according to Eq. (7.7), from which again the relative effectiveness, r, for arbitrary radiations has been derived by means of the lesion production model (7.2). These 'anoxic r-values' are listed in Table VIII (appendix), too. As in the oxic case, they are assumed to refer to DNA double-strand breaks produced under hypoxic conditions. Cautiously speaking, the few pertinent experimental data do not contradict this assumption (see chapter 6, Fig. 6.5), particularly in regard of the absolute yield of hypoxic double-strand break production given by $s = s_0 r$. s_0 turns out to be

$$s_0 = 1.15 \times 10^{-14} \text{ rad}^{-1} \text{ Dalton}^{-1} \quad (7.13)$$

for ^{60}Co radiation.

This means, as a result of formal survival curve analysis and the lesion production model, a hypoxic reduction of double-strand break production by the factor 3.2 (cf. Eq. (7.10)). This is the maximum value of 'OER for primary double-strand break production', while at growing LET this $OER = s_{\text{oxic}}/s_{\text{anoxic}}$ decreases, approaching unity at very-high LET. This is seen from Table VIII: the increase with LET of r_{anoxic} is much stronger than that of r_{oxic}.

The thus-found dependence of OER on LET at the molecular level agrees qualitatively with that generally accepted hitherto by direct inference from cell inactivation and other biological radiation effects. However, it should be stressed again that the latter effects are additionally influenced by the so-called second kind effects arising from the statistics of molecular lesions within the cell nucleus.

As to the underlying molecular mechanisms, there exist different hypotheses, e.g., LET-dependent radical recombination in the tracks (Alper, 1956) or formation of oxygen in high-LET particle tracks (Neary, 1965; Alper and Bryant, 1974; Baverstock and Burns, 1976). The present theory does not shed more light upon this question: in terms of the lesion production model the oxygen effect is only 'described' by changed lesion formation probabilities λ_j (different parameters (7.9) and (7.12)) in such a way that under hypoxic conditions a larger number, j, of ionizations in the target is required tor each the same probability λ_j. This difference vanishes as $j \to \infty$, i.e. the saturation in case of sufficiently high ionization densities in a track suppresses the effect of oxygen. Roughly speaking, high-LET particles always produce a double-strand break upon crossing a DNA molecule.

The oxygen effect at the molecular level now elucidated quantitatively, let us return to the question of whether the *biological response* to a *given* number of primary lesions (double-strand breaks), i.e. $H(k)$, does also depend on the presence or absence of oxygen. If this were not the case, the OER for cell survival on low-LET irradiation would be related directly to the ratio of primary lesion numbers, i.e. it should always be equal to the above-mentioned $OER_0 = 3.2$. However, it is interesting to note that the large number of corresponding experimental values, although roughly of this magnitude, tend to *smaller* values and differ significantly[1]) from each other (for different mammalian cells the range is about 2.3–3.0 according to Trott, 1972). This suggests an inverse influence of oxygen at the cellular level, i.e. *decreased* survival of hypoxic cells,

[1]) In part, these variations may also be due to incomplete deoxygenation of the cells. It is nevertheless felt that real differences between various cell lines do exist, so that no common low-LET OER value can be set forth, irrespective of the fact that, in addition, the theoretical value $s_{0,\text{oxic}}/s_{0,\text{anoxic}} = 3.2$ may be unreliable.

which may be due to oxygen-dependent repair processes. Therefore, survival curves for anoxic conditions cannot be predicted simply by using the oxic X-ray curve and some universal value of OER_0. Instead, one has always to start from the particular X-ray curve determined under anoxic conditions, which may or may not be described as only dose-modified. The different values (7.11) used above are an example of this procedure. In this way, the formally employed theory accounts for influences of oxygen (as well as other response-modifying agents) on the *physiology* of an irradiated cell. It is this factor, together with σ_0-dependent effects of the second kind, that makes the OER for different radiations *object-specific*, in spite of the *universal* values of r_{oxic} and r_{anoxic}. The latter question is more closely studied in section 7.3.1.

7.2.4. The minimum DNA content as a result of formal analysis

Consider Eq. (7.5) which relates the coefficients $\alpha_\mu^{(X)}$ of the X-ray survival curve representation (7.3) to the coefficients Z_μ which determine the survival probability function $H(k) = \Sigma B_\mu \exp(-Z_\mu k)$ (cf. Eq. (7.1)). Since s_X, the absolute rate of lesion production (double-strand breaks), is known now (Eq. (7.10) and (7.13), respectively, and $s_X = r_X s_0$), one has an unambiguous relationship between the DNA content Y and the function $H(k)$. This means that $H(k)$ depends on what is *assumed* for Y. This 'relation of equivalence' has been studied in section 3.7.2, where it is shown that there exists a lower limit $Y_{\min}$ below which $H(k)$ becomes 'unreasonable'.

To examine the consistency of the theory, let us now consider what turns out for $Y_{\min}$ from the six experiments analysed in section 7.2.2. The corresponding mathematical procedure is outlined in section 9.3.2. It rests on the X-ray curve parameters given in Table 7.1. These parameters describe the coefficients B_μ and $\alpha_\mu^{(X)}$ in the expression (7.3). Thus, it is only the experimental survival curve for X-rays and the universal s_0-value multiplied by r_X from which $Y_{\min}$ is derived.

One obtains the following results for the theoretically possible minimum DNA content $Y_{\min}$ (in units of 10^{12} Dalton): HeLa: 0.41; T1/B: 0.30; T1/T: 0.37; CH2B2: 0.31; P-388: 0.34; M3-1: 0.48. The actual DNA contents Y should be larger than these values. Corresponding experimental data were given, e.g., by Todd (1975): 8.4 for T1 cells; 4.4 for M3-1 cells. These values are the smaller ones valid for the G_1-phase of cell cycle. Thus, the theory is consistent with reality in this respect.[1])

The survival probability function $H(k)$ given by

$$H(k) = \sum_{\mu=1}^{n} B_\mu \, e^{-Z_\mu k} = \sum_{\mu=1}^{n} B_\mu \left(1 - \frac{\alpha_\mu^{(X)}}{s_X Y}\right)^k \tag{7.14}$$

is unequivocally determined if the actual Y is known. The finding $Y \gg Y_{\min}$ means that $H(k)$ decreases with increasing k very slowly as compared with what would be obtained for $Y = Y_{\min}$. The example given in Figure 3.8 refers to the HeLa cells of Table 7.1. The steepest curve corresponds to $Y = Y_{\min} = 0.41 \times 10^{12}$ Dalton, the flattest curve marked by $p = 1$ to the tenfold value, while the actual $H(k)$ may decrease yet more slowly since the actual DNA content is certainly larger yet (human cells like T1). This means that a cell can survive a considerable number of double-strand breaks in its total DNA. This may be explained, e.g., by assuming only small parts of the DNA to be sensitive, so that most breaks are tolerated, or by a very effective repair. However, such questions are already beyond the formal employment of the theory; they are a matter of concrete biological models of action. Besides, it should not be forgotten that the formally-derived $H(k)$ describes the *average* response of an *asynchronous* cell population.

7.3. Ion radiations

Although track-segment experiments with different ions were the subject of formal analysis from which the universal r-values for survival curve prediction were *derived*, it is nevertheless worth while to examine the thus-based theoretical results in a generalized manner with respect to special questions, e.g., RBE-LET and OER-LET relationships, apart from some

---

[1]) From the experiments under hypoxic conditions one obtains almost the same values of $Y_{\min}$, because the $\alpha_\mu^{(X)}$'s and s_X only are smaller by about the same factor.

applications to experimental results that have not been included in the analysis in section 7.2.

Once the DNA-lesion type responsible for cell killing is fixed in terms of r, the whole theory of survival curves can be expressed by relatively simple analytical formulas and tabulated parameters. These procedures feasible by everyone are given in chapter 9. It is along these lines that all the following results have been obtained. Dose-survival relations have always the form

$$S(D) = \sum_{\mu=1}^{n} B_\mu \, e^{-\alpha_\mu D}. \tag{7.15}$$

The coefficients α_μ are determined by a function that depends on the radiation considered, the r-value for that radiation, the size of the cell nucleus, and the 'empirical' coefficients $\alpha_\mu^{(X)}$ describing the low-LET reference survival curve (cf. Eq. (7.3)). This scheme is valid for oxic as well as anoxic conditions, apart from the distinct r-values.

RBE and OER are ratios of doses producing the same effect. Hence, their calculation requires the numerical solution of Eq. (7.15) for D at a given survival level S.

It is customary to pay special attention to the so-called 'final slope' (high-dose slope) of survival curves, expressed by the corresponding 37%-dose D_0. In terms of the present theory, this is

$$D_0 = \frac{1}{\alpha_1}, \tag{7.16}$$

where α_1 is the smallest α_μ. It follows directly from the corresponding $\alpha_1^{(X)}$ without the need of treating complete survival curves in the form (7.15).

7.3.1. RBE and OER as a function of LET

It is important to call to mind the general assertions of chapter 2, which are fully valid also for the strict theory, in particular the terminology of *first and second kind effects*. First kind effects are the determinant of radiation-quality-dependence of primary lesion production. They are, from now on, axiomatically of universal character for mammalian cells in terms of the r-value table for double-strand breaks. Therefore, if there were no effects of the second kind, i.e. object-specific influence of DNA break number *statistics* per nucleus, we would obtain general quantitative relations of RBE, OER, and LET. Hence, it is only this statistics depending on the peculiarities of each cell type (response to X-rays, nucleus size) that has to account for the variations found in different experiments.

The dotted lines in Figure 7.4 display the universal r, i.e. the relative effectiveness in double-strand break production, as a function of LET, while the full lines show the relative *biological* effectiveness, RBE, derived from the final slopes (D_0-values,

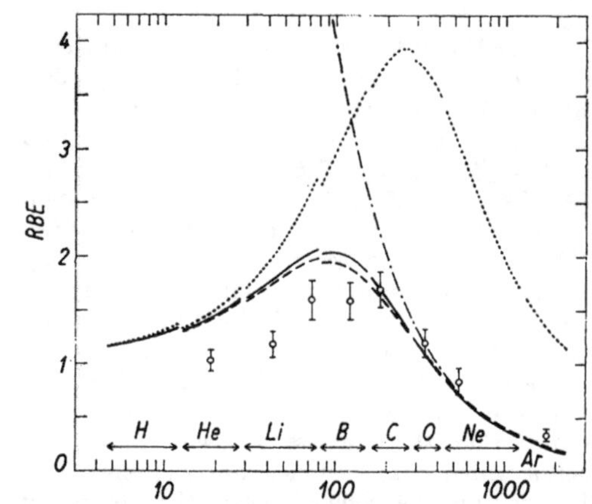

Fig. 7.4. High-dose slope RBE (defined as $D_0^{(X)}/D_0$) in dependence on linear energy transfer of different ions for CH2B2 cells (full and dashed lines, see text). $\sigma_0 = 50\ \mu m^2$; X-ray survival curve parameters see Tab. 7.1. Dashed-dotted line: according to the assumption of unit efficiency in single-ion kill. Dotted lines: universal $r/r^{(X)}$ (relative to 250-kVp X-rays). Experimental values from SKARSGARD *et al.* (1967).

Eq. (7.16)) of theoretical survival curves for CH2B2 cells. The difference between both curve sets is due to the saturation effect of the second kind, which is obviously the decisive determinant for the position and altitude of the RBE maximum. The crucial quantity in this respect is the cross-section, σ_0, of the cell nucleus, which becomes, at very high LET, the target area of a simple one-hit mechanism: survival curve slope $D_0^{-1} = \sigma_0/L_\infty$. This r-independent limit model yields the dashed-dotted curve in Figure 7.4, which becomes inadequate for smaller LET. Thus, it is just the region of medium LET values where none of the simple models, i.e. $RBE = r/r_X$ or $RBE = D_0^{(X)}\sigma_0/L_\infty$, is valid. In other words, this medium-LET region is the intrinsic scope of the present general theory.

The discrepancies between theoretical curves and experimental points in Figure 7.4 reflect, in part, the disagreements seen in the corresponding survival curves (Fig. 7.2), but they are also due to inevitable errors in deriving RBE values direct from experimental survival data. Therefore, the degree of quantitative agreement in detail between theory and experiment should always be judged from the corresponding survival curves directly. This problem is yet more serious in the shoulder region of survival curves (RBE for a high survival level). Moreover, in the present case, the 'experimental final slope' is actually the estimated slope between about 0.5 and 1 per cent survival, which does not quite correspond to the theoretical final slope (limiting value at vanishing survival given by Eq. (7.16)). Therefore, analogous theoretical values from tangents to the survival curves at 0.5 per cent survival have additionally been calculated, yielding the dashed lines in Figure 7.4, which do slightly less deviate from the experimental points.

The upper panel of Figure 7.5 shows further examples of RBE-LET relations, now for constant survival levels S, for the two T1 cell data sets (see Tab. 7.1). It is demonstrated how the saturation effect of the second kind, by altering the *shape* of survival curves, leads to a divergence of RBE curves for different constant levels of S. The twofold larger σ_0 in the case T1/T shifts the RBE maximum to a higher LET.[1]) Moreover, some differences result from the different ions used: only α-particles up to $L_\infty = 200$ keV/μm in the case T1/B (note the jumps between the r-curve segments in Fig. 7.4). The curves in Figure 7.5, especially the positions of the RBE maxima, are in general agreement with the corresponding experimental findings, which have, however, not been plotted for the reasons mentioned above. As to a quantitative comparison see the survival curves in Figure 7.2.

In the lower panel of Figure 7.5, the theoretical prediction of the oxygen enhancement ratio as a function of linear energy transfer for T1/B and T1/T cells can be seen. There are characteristic differences, in general agreement with the experimental data[2]): starting from the higher small-LET OER-value in the case T1/B — an experimental fact (see Eqs. (7.11)), the decrease at growing LET is steeper compared with T1/T for two reasons. (i) Due to different δ-ray spectra, α-particles yield a smaller ratio $r_{\text{oxic}}/r_{\text{anoxic}}$ than higher ions at the same LET (note the jumps); (ii) the saturation effect of the second kind, which shifts the OER towards unity, is more pronounced for T1/B ($\sigma_0 = 60$ μm²) than for T1/T ($\sigma_0 = 120$ μm²). The shaded stripes indicate the theoretical variation of OER in dependence on the survival level ranging from $S = 1$ to $S \to 0$. This reflects slight differences in shape of oxic and anoxic survival curves, if the same shape (i.e. dose modification only) is assumed for X-rays by definition (see footnote 1 at page 214). For every survival level,

[1]) Differences in the X-ray response are unimportant in this case, since the X-ray survival curves are very similar for T1/B and T1/T.

[2]) For T1/B taken from Barendsen *et al.* (1966), Barendsen (1966). For T1/T the values have been calculated from tabulated survival curve parameters (initial and final slopes) and their standard deviations (Todd, 1966).

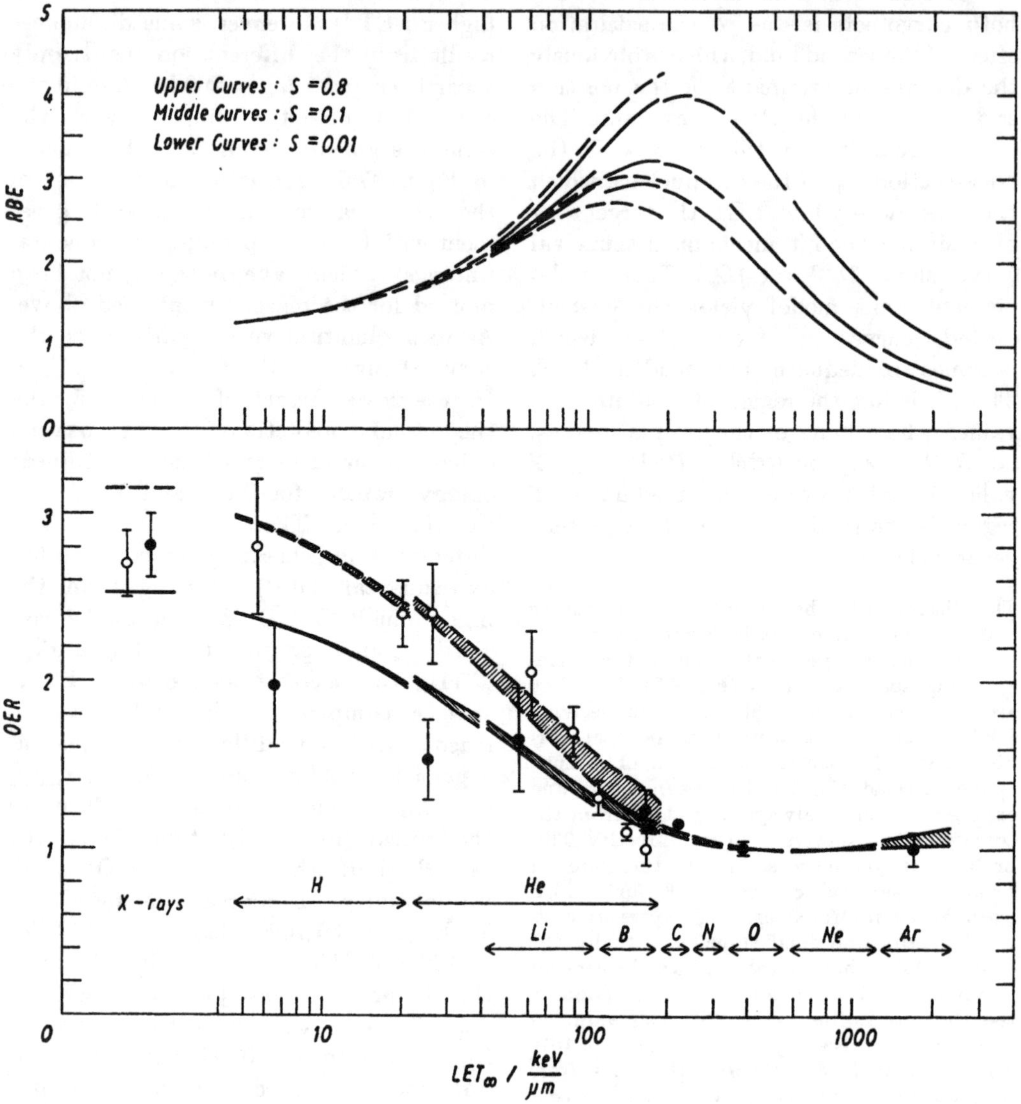

Fig. 7.5. RBE for different survival levels (upper panel) and OER (lower panel) as a function of LET for human kidney T1 cells. Open symbols (Barendsen *et al.*, 1966; Barendsen, 1966) and dashed lines for T1/B; closed symbols (Todd, 1966) and full lines for T1/T.
Theoretical curves according to the corresponding oxic X-ray curve parameters listed in Tab. 7.1. Further parameters: T1/B: $\sigma_0 = 60\ \mu m^2$, $OER^{(X)} = 3.15$; T1/T: $\sigma_0 = 120\ \mu m^2$, $OER^{(X)} = 2.53$.

OER can be expressed by

$$OER = OER^{(X)} \frac{RBE_{\text{oxic}}}{RBE_{\text{anoxic}}}, \qquad (7.17)$$

with RBE referring to the same X-rays as the constant $OER^{(X)}$.

7.3.2. Further examples of survival curve prediction

TODD (1975) reported results for human liver (Chang) cells bombarded with three different sorts of ions. These data are particularly suited for checking up the universal double-strand-break r-values as well as the theory as such, because there is no uncertainty regarding the nucleus cross-section of these cells under irradiation conditions (monolayer). Microscopic examination and averaging according to the rule of Eq. (7.8) yielded the value $\sigma_0 = 87\ \mu m^2$. Figure 7.6 demonstrates the good agreement between theory and experiment.

Figure 7.7 displays theoretical versus experimental results obtained with V79 Chinese hamster cells. Considerable scatter and partial disagreement with theory is apparent, which might be attributed to the special irradiation techniques used: a cell suspension was spouted across the beam of cyclotron-produced ions. Moreover, the higher-LET deuterons and α-particles were obtained by energy degradation in absorber foils. This causes inhomogeneities in dose and LET. The cell nucleus cross-section, $\sigma_0 = 40\ \mu m^2$, has been chosen to yield the best overall fit. It is not surprising that glass-attached V79 cells show a larger value in microscopic observations (personal communication).

This V79 experiment is of special interest for the theory since the number of DNA double-strand breaks produced by different radiations was also determined in parallel for the same cells in the same laboratory with the same irradiation techniques and dose measurement (KAMPF *et al.*, 1977). Thus, one has relevant *experimental* values of $r = s/s_0$, the relative effectiveness for the production of the primary lesions that are thought responsible for cell killing: $r = 1.43$ (7.35-keV/μm deuterons), $r = 2.09$ (32-keV/μm α-particles) — relative to ^{60}Co γ-rays. These values, which do not much differ from the corresponding universal ones (Tab. VIII), have also been used for survival curve calculation (dashed curves in Fig. 7.7). The difference is insignificant, but the somewhat better fit to the experimental survival values is nevertheless worth noting. This exemplifies how the theory, without reference to the universal double-strand-break r-values, can draw upon measured primary lesions to elucidate their relation to cell killing.

Another track-segment experiment with V79 Chinese hamster cells by BIRD and BURKI (1975) cannot be evaluated in the usual manner, because the cells were not completely singled. This constitutes a multi-target situation, while the theory refers to *single* cells. The mean cell multiplicities, which have been reported, are of little help for an appropriate modification of theoretical survival curves, since this would sensitively depend on the frequencies of cell clusters differing in size. However, all experimental survival curves show a clear exponential slope for higher doses, and this 'final' slope must evidently be independent of cell multiplicity. The corresponding D_0-values reported can therefore be compared with the theory directly according to Eq. (7.16) and context. These results, obtained with $\sigma_0 = 120\ \mu m^2$, are listed in Table 7.2. Experimental and theoretical values show a good correlation which depends, however, on the fairly large cross-section for the cell nucleus[1]) required by the steep slope (small D_0) found for argon and neon ions.

---

[1]) The threefold larger value compared with the foregoing example (also V79 cells) may not be unreasonable considering that BIRD and BURKI used monolayer cells with the medium removed by aspiration.

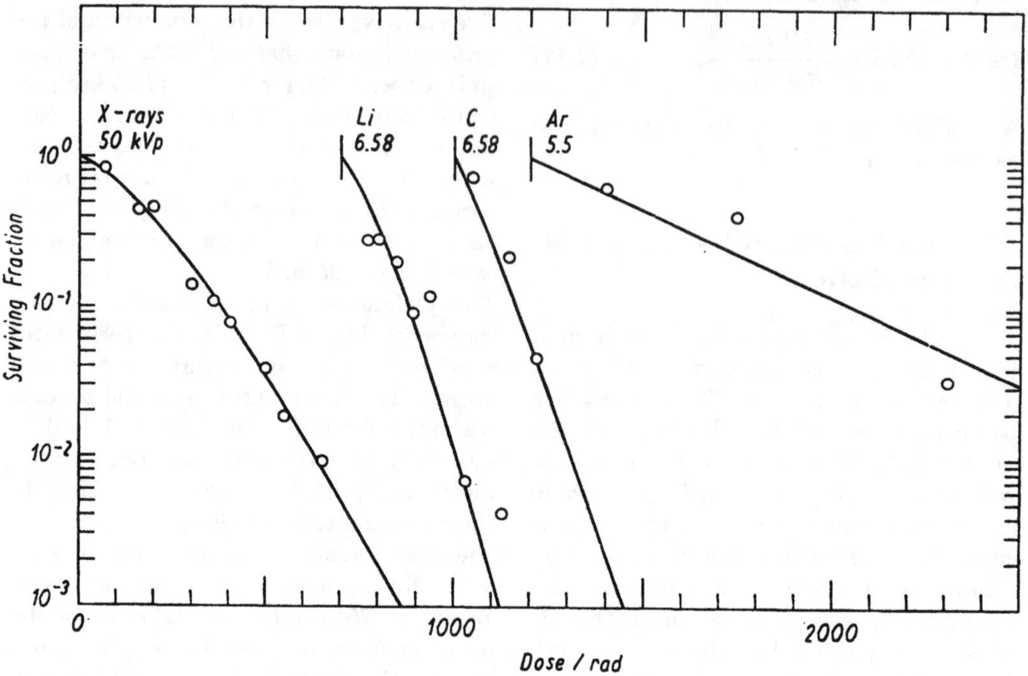

Fig. 7.6. Survival of Chang human liver cells exposed to different radiations (TODD, 1975). Numbers at curves represent energy per nucleon in MeV of the ions indicated. $\sigma_0 = 87\ \mu m^2$; X-ray curve parameters: $n = 6$, $D_0 = 92$ rad, $k_1 = 0.25$, $\beta = 0.491$, $\gamma = 2.15$, $\lg E_1 = 0.54$.

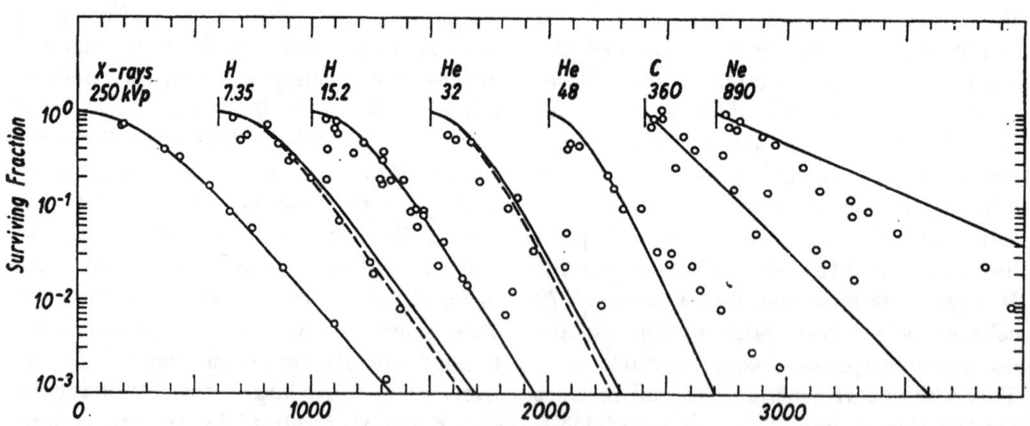

Fig. 7.7. Survival of V79 Chinese hamster cells exposed in suspension to different radiations (TOLKENDORF *et al.*, 1978). Numbers at curves represent LET_∞ in keV/μm of the ions indicated. For the broken lines see text. $\sigma_0 = 40\ \mu m^2$; X-ray curve parameters: $n = 5$, $D_0 = 152$ rad, $k_1 = 0.12$, $\beta = 0.864$, $\gamma = 0.735$, $\lg E_1 = 0$.

Tab. 7.2. D_0-values [rad] for asynchronous V79 Chinese hamster cells irradiated with different ions (BIRD and BURKI, 1975)

Ion kind Energy per nucleon [MeV]	^{60}Co γ-rays	He 9.93	Li 9.85	B 9.17	C 8.93	Ne 6.15	Ar 5.09
Experiment	—	106	67	48	63	115	258
Theory	153.85	96.7	74.3	52.6	52.9	109	291

At this place, a short comment should be given on radiation quality as a determinant of *radiosensitivity variation in cell cycle*, since this was the main subject in the work of BIRD and BURKI, who used synchronized V79 cells, too, to study these effects. From the viewpoint of the formally employed theory one can only use the different empirical X-ray survival curves for different stages of the cell cycle and tentatively use the same universal double-strand-break r-values in all cases to make predictions for higher LET. This yields, as a general rule, a reduced influence of cell cycle as LET increases, by virtue of the second kind saturation. Whether quantitative agreement with the experimental results is or is not achieved in this way depends on the validity of the 'double-strand-break-only' hypothesis for different cell cycle stages as well. E.g., if single-strand breaks have a share in the most sensitive stage[1]) (late G_1/early S), this should become apparent by a yet more pronounced reduction of the cyclic differences in sensitivity at high LET. Unfortunately, a reliable quantitative evaluation of the data by BIRD and BURKI was not possible due to the involvement of cell multiplicity. Nevertheless it can be stated that second kind saturation appears to be insufficient (large σ_0) to account for the very strong decrease and even complete vanishing of cyclic variation found when LET was increased. Some other experiments with only single radiations yielded similar results (e.g., HALL *et al.*, 1972; MASUDA, 1971; RAJU *et al.*, 1975), but contradictory findings were also reported (SKARSGARD *et al.* — cited by WHITMORE, 1972). Thus, a conclusive application of the present theory to LET effects in the cyclic survival variation will be hard before further experimental data have been accumulated.

7.3.3. Response-modifying agents and radiation quality

A considerable number of substances have been found protecting biological systems from the action of ionizing radiations. On the other hand, different compounds acting as 'sensitizers' are also known. The extent to which such an agent (or special treatment) modifies the survival curves depends on radiation quality. This *dependence* is in principle within the scope of the present theory. If the modifying action does *not* affect the molecular processes of primary lesion formation in a way to change the *relative* effectiveness of different radiations at the molecular level, one should obtain correspondence with experiment by means of the usual double-stand-break r-values, starting from either of the different (normal and modified) X-ray survival curves. Otherwise, disagreement would indicate that the agent under study interferes with the formation, or changes the involvement, of primary lesions related to the biological response of the cell. Two examples are considered in the following.

The well-known protecting SH-compound *cysteamine*, which is very effective for X-rays, was used by BARENDSEN and WALTER (1964) to demonstrate the almost complete lack of protection against high-LET α-rays (Fig. 7.8, left-hand panel). The control survival curves under normal conditions for the T1 human kidney cells used and their theoretical interpretation (broken lines) correspond to those con-

[1]) Such a contribution could be masked in asynchronous populations, so that the universal r-values may not reflect it.

sidered in section 7.2.2 (Fig. 7.2, T1/B), apart from a dose factor close to unity. The effect of cysteamine can be described by a dose-reduction factor of 3.15 for X-rays (see the corresponding curves drawn in Fig. 7.8). That the α-ray curve in the presence of cysteamine becomes almost as steep as the normal one cannot be obtained from the theory with the corresponding r-value for oxic conditions. However, the theory proves valid when using the r-value for *hypoxic conditions* (unbroken α-ray curve shown in Fig. 7.8). This interesting result suggests that cysteamine in sufficient concentration (25 mM) creates conditions equivalent to those in the absence of oxygen, which is additionally supported by the dose-reduction factor 3.15 for X-rays being equal to the pertinent $OER^{(X)}$ (see Eq. (7.11)). This mode of action of cysteamine, being a radical scavenger, is not unlikely; it has also been pointed out by other workers.

An analogous investigation with *glycerol* (10%), reported in the same paper (Barendsen and Walter, 1964), yielded a smaller effect with X-rays (factor 2.0), and the corresponding α-ray survival curve lies between our theoretical predictions with the oxic and anoxic r-values. This may indicate an incomplete anoxia simulated by that compound.

Another example of response modification investigated with different radiations is shown in the right-hand panel of Figure 7.8. Pretreatment of cells with the thymidine analog *5-iododeoxyuridine* (IUdR) makes them more sensitive to ionizing radiation by way of IUdR incorporation into the DNA. The theoretical curves for normal untreated cells, corresponding to those shown in Figure 7.2 (T1/T), are in some disagreement with experimental points only in the case of carbon ion bombardment. For IUdR cells the changed X-ray curve (smaller shoulder) leads to a very good agreement between theory and experiment for both high-LET radiations, which was achieved by using the *same* universal r-values for oxic conditions. Thus, in this case, the second kind saturation effect alone accounts for the diminution of the differences at increasing LET. This means that IUdR does apparently not change the relative

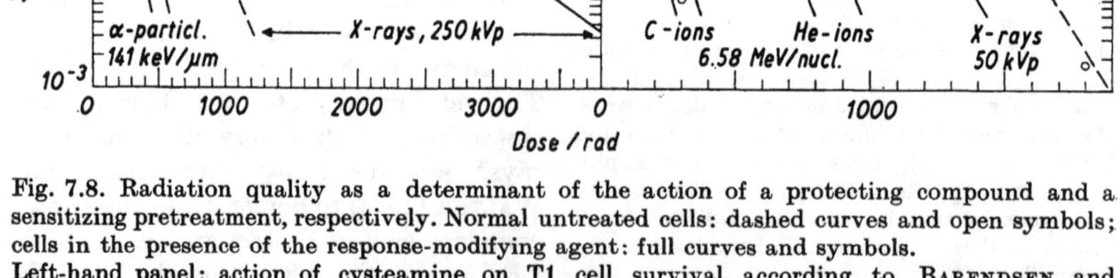

Fig. 7.8. Radiation quality as a determinant of the action of a protecting compound and a sensitizing pretreatment, respectively. Normal untreated cells: dashed curves and open symbols; cells in the presence of the response-modifying agent: full curves and symbols.
Left-hand panel: action of cysteamine on T1 cell survival according to Barendsen and Walter (1964). Theoretical curves: $\sigma_0 = 60\ \mu m^2$. X-ray curve parameters given in Tab. 7.1 except for $D_0 = 130.3$ rad (normal), 410.8 rad (with cysteamine).
Right-hand panel: effect of IUdR incorporation on T1 cell survival according to Todd (1967). Theoretical curves: $\sigma_0 = 120\ \mu m^2$. X-ray curve parameters: for normal cells given in Tab. 7.1; for IUdR cells: $n = 10$, $D_0 = 106$ rad, $k_1 = 0.3$, $\beta = 2.79$, $\gamma = 1.92$, $\lg E_1 = 0$.

effectiveness of different radiations in the production of the crucial primary lesions (double-strand breaks).

As to the mechanism of radiosensitization by 5-halogenodeoxyuridine incorporation, there exists an extensive literature which cannot be fully reviewed here. On the one hand, the yield of double-strand breaks being increased has been suggested (SZYBALSKI, 1962) but also rejected experimentally (ALEXANDER *et al.*, 1965). Moreover, it has been demonstrated that the halogenated bases act as 'energy traps' (FIELDEN *et al.*, 1971), from which one could expect alterations in the primary processes of lesion formation. On the other hand, there is clear evidence (SHIPLEY *et al.*, 1971) for the *biological* level (i.e. that which $H(k)$ describes) to be affected in any case, which shows up as a reduction in capacity for sublethal damage accumulation (reduced survival curve shoulder along with unchanged D_0 as seen in Fig. 7.8). This may be connected with effects on the repair system (LETT *et al.*, 1964). Our theoretical analysis shows that the latter aspect alone is sufficient to understand the LET-dependence of the IUdR effect in a quantitative fashion.

7.3.4. Mixed radiations

From the standpoint of theory, mixed radiations do not constitute a special problem since in chapter 3 the general theory has been elaborated for quite arbitrary radiations fields. However, experimental and practical situations (radiation protection and therapy) often bear the need to estimate the effectiveness of a mixed radiation field in case that the biological response to the single components is known. To say it in advance, there cannot exist a general rule (e.g., taking a weighted arithmetic mean of RBE) that rests on experimental information only, because damage from the *different* components acts *together* in the *same* cell, which is often termed the 'synergistic effect'. For example, if a binary radiation field is described by the dose fractions $\mu_D{}^{(a)}$ and $\mu_D{}^{(b)}$ and if $S_a(D)$ and $S_b(D)$ are the dose-survival relations for the pure components a, b, it would not, in general, be correct to assume that

$$S(D) = S_a(\mu_D{}^{(a)}D)\, S_b(\mu_D{}^{(b)}D) \tag{7.18}$$

describes the response to the mixed field. Eq. (7.18) means considering the survivors of the one component, which may be sublethally damaged, as undamaged ones with respect to the second component. This shows at once the special case in which Eq. (7.18) becomes valid: at least one of the two components must be unable to leave a cell sublethally damaged, i.e. Eq. (7.18) holds only if $S_a(D)$ or $S_b(D)$ is a single exponential.

The present theory yields a very convenient general rule for mixed radiations, provided that $S_a(D)$ and $S_b(D)$ have been derived from the theory, i.e. have the common form (7.15). Then:

$$S(D) = \sum_{\mu=1}^{n} B_\mu\, e^{-\alpha_\mu D}, \qquad \alpha_\mu = \mu_D{}^{(a)}\alpha_\mu{}^{(a)} + \mu_D{}^{(b)}\alpha_\mu{}^{(b)}. \tag{7.19}$$

This follows from the relation (3.51), realizing that the quantities α_μ are the Π's in Eqs. (3.80), (3.81). Obviously, the relation (7.19) cannot be replaced by an expression that relates $S(D)$ direct to the functions $S_a(D)$ and $S_b(D)$ as a whole. Hence, Eq. (7.19) represents the particular way in which the present theory accounts for the synergistic effect.

Special cases are now easy to find out. E.g., if $S_a(D)$ is a single exponential, i.e. $\alpha_1{}^{(a)} = \alpha_2{}^{(a)} = \ldots$, Eq. (7.18) proves trivial in accordance with the intuitive argument given above. Another special case, a rather hypothetical one, is met if both survival curves have the same shape, i.e. the $\alpha_\mu{}^{(a)}$ and $\alpha_\mu{}^{(b)}$ have constant proportions: $\alpha_\mu{}^{(a)}/\alpha_\mu{}^{(b)} = \beta$. This yields

$$S(D) = S_b(\mu_D{}^{(a)}\beta D + \mu_D{}^{(b)}D) \quad \text{with} \quad S_a(D) = S_b(\beta D). \tag{7.20}$$

Let us now discuss the problem in terms of RBE. In the field of radiation protection one usually relies on the linear relation

$$RBE = \mu_D{}^{(a)} RBE^{(a)} + \mu_D{}^{(b)} RBE^{(b)}. \quad (7.21)$$

It is important to emphasize that only Eq. (7.20) is exactly equivalent to this relation, while in general (also in cases where Eq. (7.18) is valid) RBE depends in a *non-linear* manner on the dose fractions; and the character of this dependence changes with the effect level to which the RBE refers. In respect of the latter there is, however, a limiting case in which relation (7.21) becomes always valid: for very small survival when the dose-survival relations can be assumed to be of pure exponential character. This derives obviously from Eq. (7.19) (all terms of the sum but one vanishing) as well as directly (i.e. without reference to the present theory) from the then equally applicable equations (7.18) or (7.20). The latter assertion becomes plausible by realizing that in the small-survival, i.e. high-dose, range both survival curves behave like single exponentials with the shoulder-caused displacement dose D_q being negligible compared with D. Thus, the convenient linear relation (7.21) can only be warranted for small levels of survival; it becomes increasingly wrong at higher survival, i.e. just in the range relevant to radiation protection, save for $D \to 0$.

As to the oxygen enhancement ratio, the strict solution can be obtained also only by way of the corresponding survival curves (7.19). However, in so far as OER is assumed *a priori* to be independent of S (dose modification only, see footnote 1 at page 214) it is possible to derive a general *approximate* relation that holds strictly only as $S \to 0$. This relation follows from Eq. (7.21) and the general formula (7.17), applying the latter to the mixed radiation as well as to each component, thereby eliminating $OER^{(X)}$:

$$OER = \frac{\mu_D{}^{(a)} RBE^{(a)} + \mu_D{}^{(b)} RBE^{(b)}}{\mu_D{}^{(a)} RBE^{(a)}/OER^{(a)} + \mu_D{}^{(b)} RBE^{(b)}/OER^{(b)}}. \quad (7.22)$$

This expression, which simplifies if one component, say (b), is the X-radiation to which RBE refers ($RBE^{(b)} = 1$; $OER^{(b)} = OER^{(X)}$), was earlier given by Greene and Major (1971). Note that all dose ratios (*RBE*, *OER*) to be used at the right-hand side of Eq. (7.22) have to be those for the limit of zero survival, $S \to 0$, which yields the mixed OER strictly for the same limit. This limit is sufficient if OER is assumed anyway to be independent of S. Otherwise, there is no formal justification for taking S-dependent values at the right-hand side to obtain the mixed OER S-dependently.

Let us now illustrate these theoretical considerations by experimental evidence for T1 human kidney cells irradiated with mixtures of α-particles and X-rays under both normal and anoxic conditions (Raju and Jett, 1974). The survival curves for the pure components and their theoretical counterparts obtained in the usual way are shown in the left-hand panel of Figure 7.9. The cross-section $\sigma_0 = 32\ \mu m^2$ used leads to almost straight lines for the α-rays (178 keV/μm estimated for the cell interior) in excellent agreement with the experimental results. For mixed radiations, the authors reported RBE and OER values for 10 per cent survival, which are shown in the right-hand panel. The corresponding theoretical curves (full lines) have been obtained according to Eq. (7.19).

Although the scatter in RBE data points is large, they tend to follow a bent curve in compliance with the theoretical prediction. Hence, the linear relation (7.21) would not be correct for $S = 0.1$. Instead, Eq. (7.18) is likely to apply since the pure α-ray survival curve is nearly a single exponential. Assuming the latter, Raju and Jett have calculated the dotted curve shown in Figure 7.9. It is interesting to note that this curve seems somewhat

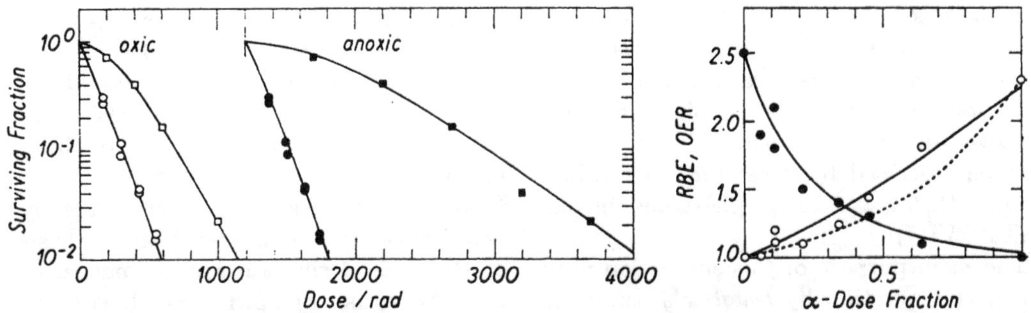

Fig. 7.9. Effect of α-rays in combination with X-rays on oxygenated and hypoxic T1 human kidney cells (experimental results of RAJU and JETT, 1974). Theoretical curves for $\sigma_0 = 32\ \mu m^2$; X-ray curve parameters: $n = 4$, $D_0 = 172$ rad (oxic), 430 rad (anoxic), $k_1 = 0.15$, $\beta = 0.264$, $\gamma = 5.48$, $\lg E_1 = 0$.
Left-hand panel: survival curves for the pure components: 250-kVp X-rays (squares), 178-keV/μm α-particles (circles).
Right-hand panel: RBE (open circles) and OER (full circles) at 10% survival as a function of the dose fraction due to α-rays.

to overstate the non-linearity, indicating that the α-ray survival curve is *not* an exact exponential one. While this is not apparent from the experimental survival points, it is suggested by the dependence of $RBE_{10\%}$ on the α-dose fraction, and the present theory does account for this subtlety.

The theoretical prediction of OER versus $\mu_D^{(\alpha)}$ also fits well to the experimental points. However, this strict curve does not substantially differ from an approximate curve based on Eq. (7.18) nor from relation (7.22) for $S \to 0$. The best correspondence, however, is obtained by applying this relation to the 10 per cent survival level. As this lacks formal justification, it must be taken for a fortuitous fact indicating the ability of relation (7.22) to account also for the small changes in OER at increasing survival level S, where the underlying relation (7.21) fails.

The pronounced curvature of the OER curve, i.e. the rapid decrease of OER at increasing α-dose fraction, is worth of emphasis. It constitutes a general law for mixtures of high-LET and low-LET radiations, as apparent from Eq. (7.22). Hence, a relatively small fraction of high-LET radiation is sufficient for a substantial reduction of OER. This fact, as pointed out by RAJU and JETT (1974), is important for the clinical radiotherapy in attacking hypoxic tumour cells by high-LET radiation: a combined treatment may be advantageous (EICHHORN and LESSEL, 1975).

7.3.5. Non-mammalian eukaryotic cells

The general theory of survival should apply to arbitrary single-cell objects, but there is no theoretical basis to suppose that the property of mammalian cells to be killed by DNA double-strand breaks only is generalizable. In particular, evolutionarily inferior objects as *yeast* and *algae*, considered below, are likely to have less efficient repair systems[1]), so that single-strand breaks may appreciably contribute to cell killing. In this case, theoretical predictions for high-LET

[1]) This does not contradict the much higher radioresistance of these biological systems because of the essential differences in size, DNA content, genetic organization, and cellular physiology.

radiations based on the 'universal' double-strand-break r-values derived from mammalian cell inactivation should yield *too-steep* survival curves, since the involvement of single-strand lesions ought to be described by lower *effective* r-values (note that $r < 1$ for single-strand breaks (Tab. VIII)).

The main purpose of this subsection is to test this question by *tentatively* using the double-strand-break r-values. To say it in advance: we will not arrive at an unambiguous answer; but there is still another question worth of attention. Does the theory correctly estimate the effects of the second kind, i.e. the *shape* of survival curves in dependence on radiation quality? The corresponding determinants are by orders of magnitude different from those for mammalian cells: the smaller σ_0 strengthens the second kind effects (suppression of survival curve shoulders), while the much higher radioresistance (large D_0) is of opposite influence (maintenance of survival curve shoulders).

Yeast cell survival curves have already been considered in chapter 2 on the basis of the microdosimetric approximation of the theory, using empirical r-values derived direct from hamster cell survival curves only and taking $\sigma_0 = 2\ \mu m^2$ without an individual fit. It is therefore worth while to re-evaluate the same two experiments along the lines of strict theory and with the universal r-values. These results are shown in Figure 7.10. In both cases, the nucleus cross-sections, $\sigma_0 = 5\ \mu m^2$ and $\sigma_0 = 0.785\ \mu m^2$, have been

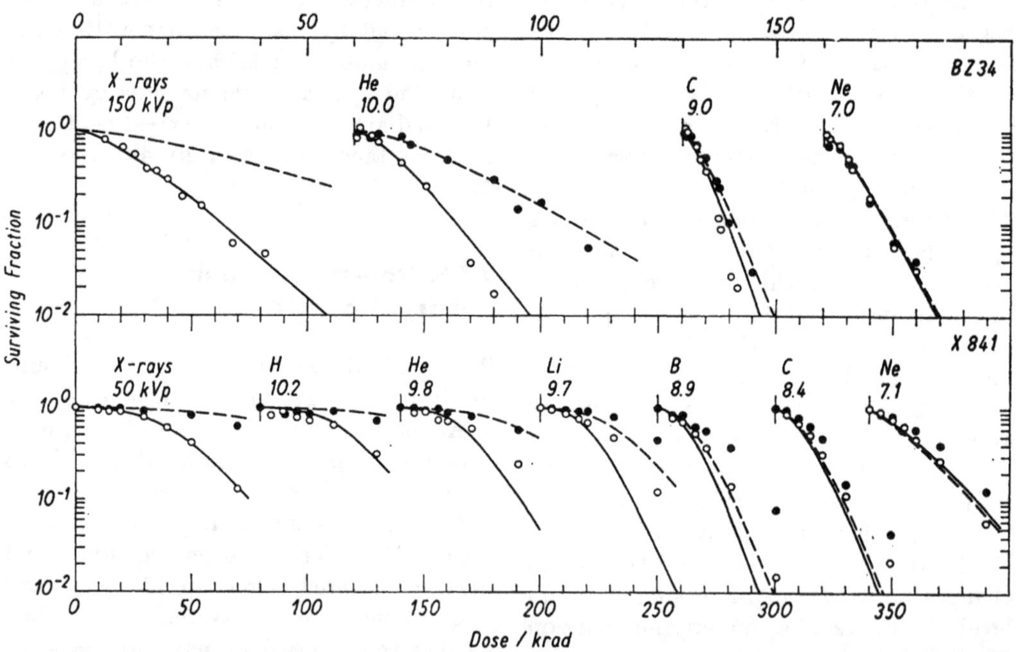

Fig. 7.10. Survival of yeast cells, *Saccharomyces cerevisiae*, strain BZ34 (upper panel, data points from Raju *et al.*, 1971) and strain X841 (lower panel, data points from Mortimer *et al.*, 1965), exposed to different radiations. Numbers at curves represent energy per nucleon in MeV of the ions indicated. Open circles and full lines for oxic conditions, full circles and dashed lines for anoxic conditions.

Theory: Upper panel: $\sigma_0 = 5\ \mu m^2$; X-ray curve parameters: $n = 2$, $D_0 = 8.42$ krad (oxic), 21.51 krad (anoxic), $k_1 = 0.077$, $\beta = 0.053$, $\gamma = 1.00$, $\lg E_1 = 0$.

Lower panel: $\sigma_0 = 0.785$; X-ray curve parameters: $n = 20$, $D_0 = 13.9$ krad (oxic), 31.8 krad (anoxic), $k_1 = 0.045$, $\beta = 0.603$, $\gamma = 1.254$, $\lg E_1 = 0.1$.

fitted to provide best correspondence for the oxic Ne-ion curves; and the dose-modifying oxygen enhancement ratios for X-rays, $OER^{(X)} = 2.56$ (BZ34) and $OER^{(X)} = 2.29$ (X841), have been chosen to produce agreement with the measured anoxic He-ion curves.

Figure 7.10 exhibits convincing evidence for the correctness of the theory and the validity of the double-strand-break r-values for the strain BZ34, whereas the strain X841 clearly conflicts with the double-strand-break-only hypothesis, yielding large deviations in the direction expected. Thus, no general conclusion for yeast cells seems to be possible. This ambiguous situation is also met with other experimental data, which we will report only briefly in relation to theory.

Survival curves for the strain BZ34 of *Saccharomyces cerevisiae* after irradiation with γ-rays and 100-keV/μm α-rays published more recently (MURTHY *et al.*, 1976) agree again very well with the theoretical prediction, thus confirming the 'mammalian-like behaviour' of that strain. This is true of both stationary growth phase cells and budding cells which were separately investigated by MURTHY *et al.*, yielding greatly different survival curves. An experiment with haploid yeast cells has also been evaluated (Tab. 7.3). As these cells have exponential survival curves for all radiations, the results of experiment and theory are fully described by the slope coefficient k listed in Table 7.3 ($k = D_0^{-1}$ or $k = \alpha_1$, cf. Eq. (7.16) and context). One finds substantial disagreement[1]) of the character to be expected: too-steep theoretical curves in the medium-LET range after fitting σ_0 for the highest LET ($\sigma_0 \approx 1\ \mu m^2$ as the area of cell nucleus was also experimentally confirmed by SAYEG *et al.*).

Returning to Figure 7.10 let us stress the tendency of sigmoidal survival curves to keep the curvature even at high LET in the region of already decreasing slope. This effect is markedly distinct from all experience with mammalian cells. The theory reduces this to nothing but X-ray sensitivity and nucleus cross-section, and it does it correctly for both strains, which show this effect to a different degree. In the case of strain X841, obviously, one would attain complete agreement by means of smaller r-values. Then, compensation by a larger σ_0 would be necessary for the Ne-ion curve, which would give it somewhat more curvature.

The second group of objects to be considered briefly in this subsection are *algae*. With vegetative cells of *Chlamydomonas reinhardii* BRYANT (1973) has made the following observation. For X-rays, deuterons, and α-particles up to 100 keV/μm the survival curves are of pronounced sigmoidal shape with a clearly exponential tail over several orders of magnitude of

[1]) However, detailed inspection shows that the correspondence overall is not so bad, especially if one looks upon the experimental X-ray value as wrong and rests the fit upon the H-ion value.

Tab. 7.3. Survival curve slope coefficients k [10^{-4} rad^{-1}] for haploid yeast cells (*Saccharomyces cerevisiae*, strain SC-7) irradiated with different ions (SAYEG *et al.*, 1959). Theoretical values calculated for $\sigma_0 = 1.1\ \mu m^2$.

Ion kind	X-rays	H	He	He	He	C	C	C	C
Energy per nucleon [MeV]	50 kVp	9.50	6.75	1.50	0.75	5.58	5.42	2.50	2.00
LET [keV/μm]		4.80	25.4	80.2	123	258	264	441	500
Experiment	2.24 ± 22	1.96 ± 11	2.40 ± 18	3.05 ± 41	4.30 ± 23 3.86 ± 23	2.51 ± 12	2.57 ± 12	1.72 ± 10	1.29 ± 08
Theory	2.25	2.59	3.45	4.44	4.12	2.61	2.56	1.50	1.33

surviving fraction. Growing LET did only reduce the shoulder width, measured by the extrapolation number E or displacement dose D_q, while the *slope* of the long exponential tail, D_0, remained nearly constant. At first glance, the present theory appears to be unable, in any case, to produce such a behaviour. This question is worth studying in general before considering BRYANT's results in particular.

Inspection of Eq. (7.15) shows that the high-dose asymptote of *any* survival curve is given by $B_1 \exp(-\alpha_1 D)$ if $\alpha_1 < \alpha_2 < \cdots < \alpha_n$. Since the coefficients B_μ are constant, this means that the *strict theoretical* extrapolation number, $E = B_1$, does not change with radiation. This strange statement is however only of academic interest in most cases, because the *practical* range of doses may be far from this asymptotic limit. High-LET radiations produce scarcely different coefficients α_μ, so that the survival curve often becomes a nearly straight line with $E \approx 1$ in the experimental range of doses (note that $\sum B_\mu = 1$ by definition). For medium values of LET a reduced residual shoulder followed by an *apparent* exponential tail which is however *not yet* the theoretical asymptote (more than one term in Eq. (7.15) contributing) can be obtained. In this way, the theory can account for the above-mentioned behaviour, and it would be inadequate to derive E and D_0 simply from the term $B_1 \exp(-\alpha_1 D)$. Instead, a 'virtual' straight line asymptote has to be constructed appropriately in order to compare with experimental data. This can be done by a numerical procedure that derives E, D_q, and D_0 from an analytic straight line going through two points, D_1, S_1, and D_2, S_2, with the doses D_1 and D_2 determined by the theoretical function $S(D)$ at two reasonably fixed survival levels S_1 and S_2.

The left-hand panel of Figure 7.11 shows the experimental survival data of *Chlamydomonas reinhardii* cells after irradiation with 8-MeV electrons[1]) (oxic conditions) and their analytical description by a twenty terms expression of the form (7.3). The also-shown theoretical curve for 80-keV/μm α-particles exhibits what was stated above: a virtual exponential tail with reduced E and D_q. The dashed lines have been constructed from the points at $S_1 = 10^{-4}$ and $S_2 = 10^{-6}$. The so-derived parameters D_q and D_0 as a function of LET for deuterons and α-particles are compared with the corresponding experimental findings[2]) in the lower rigth-hand panel of Figure 7.11, demonstrating a good general agreement. The value 2 μm² chosen for σ_0 leads to a strong saturation effect of the second kind yielding a roughly constant D_0. The flat minimum at $L_\infty = 30$ keV/μm is however in conflict with the experiment, but this is a matter of the double-strand-break r-values which may not fully apply to this object. The marked reduction of shoulder width described by D_q is rather convincingly reproduced by the theory. Thus, apart from inadequate r-values, the LET-dependent *shape* of survival curves, which is characterized by the *ratio* D_q/D_0 (distance on the logarithmic scale), is very well delivered. The oxygen enhancement ratio, based on the empirical $\text{OER}^{(X)} = 2.5$, is also in good general correspondence with the experimental data: the small but significant underestimation can again be attributed to the r-values.

As a second example for this sort of objects, algae, experimental versus theoretical survival curves of *Chlorella vulgaris* are shown in Figure 7.12 (VEKSHINA *et al.*, 1974). In this case the theory seems to fail essentially. Although the disagreements may, in part, be due again to the inapplicability of the double-strand-break concept (r-values), the magnitude and irregularity of the differences can hardly be accepted. In particular, the apparent absence of any shoulder in the experimental survival curves for helium and boron ions is difficult to explain otherwise than assuming experimental errors, which might, e.g., result from a changing inhomogeneity of the cell population with respect to

---

[1]) In the theoretical calculations, ^{60}Co γ-rays have been substituted for this radiation.

[2]) Only these parameters for different LET values were reported by BRYANT (1973).

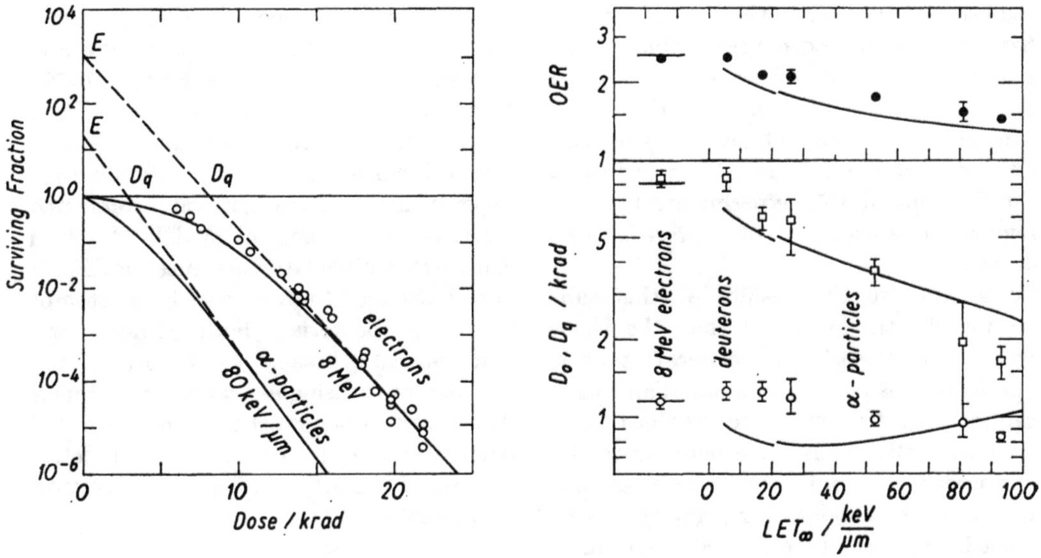

Fig. 7.11. Inactivation characteristics of the green alga *Chlamydomonas reinhardii* (BRYANT, 1973). Left-hand panel: survival curves after irradiation of single cells with γ-rays and α-particles (theory) under oxic conditions.
Right-hand panel: survival curve parameters D_0 (circles) and D_q (squares), and oxygen enhancement ratio (full circles) as a function of LET. Vertical bars represent 95% confidence intervals.
Curves according to theory: $\sigma_0 = 2\ \mu m^2$; X-ray curve parameters: $n = 20$, $D_0 = 1.08$ krad (oxic), 2.75 krad (anoxic), $k_1 = 0.051$, $\beta = 0.602$, $\gamma = 0.123$, lg $E_1 = 1$.

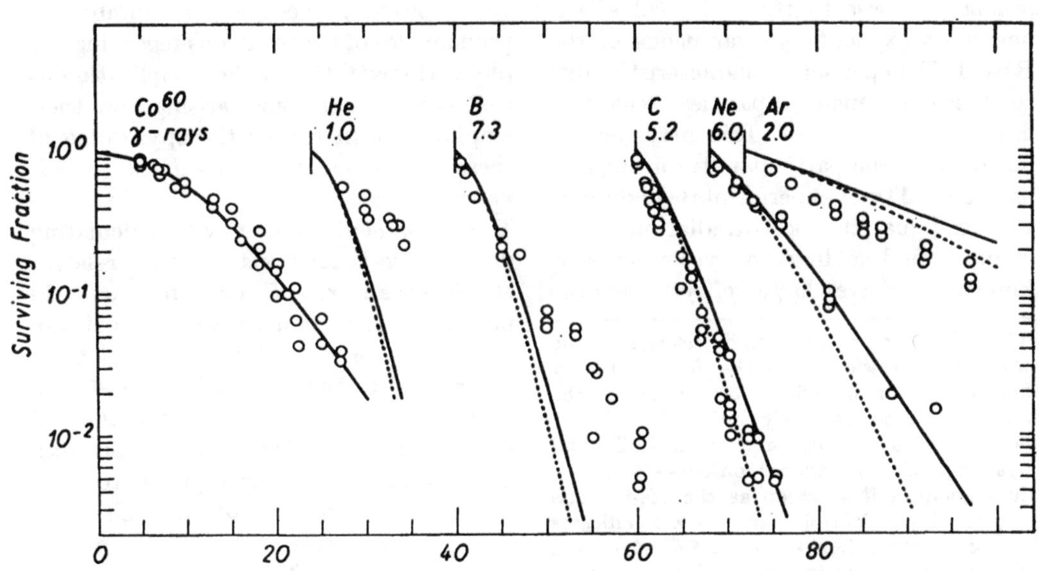

Fig. 7.12. Survival of alga cells (*Chlorella vulgaris*) exposed to different radiations (VEKSHINA *et al.*, 1974). Numbers at curves represent energy per nucleon in MeV of the ions indicated. Theoretical curves for $\sigma_0 = 3\ \mu m^2$ (full lines) and $\sigma_0 = 5\ \mu m^2$ (broken lines). X-ray curve parameters: $n = 7$, $D_0 = 4.6$ krad, $k_1 = 0.13$, $\beta = 0.43$, $\gamma = 2.27$, lg $E_1 = 0$.

radiosensitivity. The theoretical curves have been computed for two values of the parameter σ_0 (3 μm² and 5 μm²) to demonstrate that neither of them can simultaneously account for both the neon and argon curve measured.[1]) According to the authors of the experiment, the cell nucleus cross-section of *Chlorella* was 4 μm².

To summarize the results of this subsection it can be stated that the theoretical framework in general proved applicable also to non-mammalian eukaryotic cells. However, the *r*-values valid for mammalian cells corresponding to the hypothesis that only DNA double-strand breaks are the underlying primary lesions must in some cases be altered, i.e. usually *reduced* (smaller RBE). This suggests that *single-strand lesions* contribute to cell killing, but other causes for suchlike incomparabilities between qualitatively different biological systems cannot be ruled out, bearing in mind the rather formal nature of these theoretical efforts. If it should be true that *several* types of lesions give rise to the biological effect, one must expect large variations of the RBE-LET-dependence characteristics due to varying repair capacities regarding different lesion types. This may depend not only on the particular strain but also on the totality of experimental conditions. For example, the post-irradiation treatment (liquid-holding) of yeast cells is known to have large effects on the recovery from damage (Korogodin, 1966; De Langguth, 1968; Kapultsevich and Kabakova, 1971; Jain and Pohlit, 1972). Hence, one cannot hope to find generalized *r*-values to make the theory fit for formal survival curve prediction. Only detailed experimental investigations connected with concrete biological models of action can further elucidate this question. Footnote 1 at page 24 refers to such an attempt regarding bacteria: both single- and double-strand breaks are involved, and the extent of single-strand-break repair depending, among other things, on the particular mutant is an essential determinant of the LET-dependence (Munson *et al.*, 1967).

[1]) This is a good example to object against the possible impression that the formal theory working might be trivial by virtue of the several parameters freely adjustable (X-ray curve parameters, σ_0, and even individual *r*-values). E.g., that the Ar-ion curve is not less than about half as steep as the Ne-ion one despite of the fourfold larger LET indicates a far-from-saturation situation for which the theory can only yield sigmoidal curves. It is also impossible to get a single exponential for α-particles, in view of the pronounced sigmoidal shape for X-rays, unless σ_0 is chosen very small, which would hugely conflict with the high-LET findings.

7.4. Fast neutron radiations

7.4.1. Preliminary notes

The general scheme of theoretical prediction of neutron-produced survival curves is the same as for ion radiations: see the introductory remarks to section 7.3. However, there are a number of peculiarities of fast neutrons regarding the physical modalities of their application as well as special questions arising from their employment in cancer therapy. Some of these items shold be discussed in advance.

The fundamental quantity for calculating dose-survival relations is the relative effectiveness, *r*, of fast neutrons in producing the primary lesions that are responsible for cell killing. All calculations made refer to the so-called *double-strand-break r-values* listed as a function of neutron energy in Table VIII (appendix). These values can be interpreted roughly in terms of mean neutron LET: one observes an increase for small neutron energies (high-LET secondary protons) and large neutron energies (high-LET heavy recoils), with a minimum at $E_n \approx 8$ MeV. However, which is important to emphasize, it is *not* simply by LET that

Tab. 7.4. Some neutron energy spectra approximated in terms of percentage dose (kerma) ascribed to discrete energies

Label for reference	Neutron source	Neutron energy [MeV]													Literature
		.25	.5	1	1.5	2	3	4	6	8	10	12	15	20	
A	fission (^{235}U converter plate)	15	26	20	10	10	9	6	4						BEWLEY (1968)
B[1])	252-Californium	3	7	12	13	17	18	16	9	5					ANDERSON (1973)
	thick Beryllium target bombarded with:														
C	7.5 MeV deuterons		5	10	15	24	24	16	6						BURGER and BROERSE (1974)
D	15 MeV deuterons			5		14		23	21	16	9	6	6		BEWLEY (1968)
E[2])	35 MeV deuterons								7		18		32	43	BURGER and BROERSE (1974)
F	20 MeV ^{3}He-nuclei			1		3	6	13	19	17	13	12	10	6	BRUNINX and CROMBEEN (1969)

[1]) Maxwellian energy spectrum for the neutron temperature $T = 1.5$ MeV. [2]) Estimated.

these r-values are related to those for monoenergetic ion radiations (see Fig. 6.5). Hence, the DNA-lesion production model being the tool for generalizing the originally-fitted r-values[1]) becomes essential for neutron survival curve prediction. Moreover, other peculiarities of neutron radiations concern the so-called second kind effects connected with the *statistics* of lesions per cell nucleus.[2]) Consequently, just neutrons are most significant for testing the theory as a whole, considering that it has been adapted primarily to track-segment ion radiations.[3])

[1]) It is the *small average velocity* of neutron-produced charged particles that makes them incomparable, at the molecular level, with monoenergetic ion bombardment.

[2]) It is the *small residual range* of neutron-produced charged particles that causes substantial differences in the statistical distribution of energy (or lesion number) per cell nucleus as compared to track-segment ion radiations.

[3]) BEWLEY's (1968) attempt to directly relate neutron and ion bombardment survival curves in terms of equal LET failed, according to his own conclusions, just for these two reasons (see section 1.3); and this cannot be remedied without intrinsic theory.

Another point is *neutron energy spectra.* r as well as the other parameters (Tab. IX) necessary for calculating the coefficients α_μ used in expression (7.15) have been tabulated only for monoenergetic neutrons in the range of 0.25 to 20.0 MeV. Experiments are often carried out with a broad spectrum of neutron energies. In theory this constitutes a mixed-radiation situation the treatment of which has been dealt with in section 7.3.4: the spectrum has to be subdivided into intervals with mean energies E_n and corresponding dose fractions (kerma fractions) $\mu_D^{(n)}$; and α_μ reads

$$\alpha_\mu = \sum_n \mu_D^{(n)} \alpha_\mu^{(n)}, \tag{7.23}$$

where $\alpha_\mu^{(n)}$ refers to monoenergetic neutrons of energy E_n. Table 7.4 lists a number of spectra used for survival curve calculation to compare with corresponding experimental results (see the following subsections). We will refer to these spectra by the capital letters in the first column.

An often-used source of *monoenergetic neutrons* is the D-T reaction, $^3H(d, n)^4He$, yielding neutrons in the neighbourhood of

15 MeV, in some dependence on deuteron beam voltage and emission angle. In these cases, our calculations are always for 15 MeV, and corresponding experiments will be referenced simply by this value irrespective of the author's particular report.

Collimated neutron beams include some portion of lower-energy *scattered neutrons* which is seldom specified for the particular experiment. So it had to be disregarded in most theoretical calculations, although its influence on RBE may be appreciable, especially in the case of 'monoenergetic' D-T neutrons, as suggested by corresponding measurements of spectra in dependence on the collimator shape and material (HANNAN *et al.*, 1973).

Experiments with neutrons can never completely avoid some *γ-ray contamination* of the radiation field. This has been allowed for similarly to Eq. (7.23), assuming a γ-dose fraction $\mu_D^{(\gamma)}$ of 5 per cent throughout, if not otherwise stated, with $\alpha_\mu^{(\gamma)}$ referring to ^{60}Co γ-rays for convenience.

As to the *size of the gross sensitive volume* defined by its cross-sectional area σ_0, which should correlate with the cell nucleus size, definite data are usually unavailable. Fortunately, as will be seen in the following, the neutron survival curves obtained do not very sensitively depend on σ_0 within reasonable limits. Therefore, concerning comparisons between theory and experiment, the general conclusions we arrive at will not depend on the choice of σ_0 made. In some cases, σ_0 has been varied for the sake of demonstration. For those cell lines that have already been considered in the foregoing sections 7.2 and 7.3, σ_0 has been chosen not larger than the former value referring mostly to monolayer cells, considering that neutron rays have often been administered to cells in suspension or *in situ* (see the pertinent comment at page 207).

The purpose of the following subsections is twofold. *On the one hand,* the theory is put to the test by contrasting measured with calculated survival curves for a representative number of experiments with single cells. This reveals that the theory is completely or sufficiently correct in the majority of cases, but significant deviations of systematic character with respect to different groups of organisms are also found. An explanation is suggested that concerns the theory as such, i.e. these defects are not an r-value problem. *On the other hand,* the theory serves answering a number of questions pertinent to neutron radiation biology, which are of great practical importance (neutron therapy). General statements and practicable rules are derived regarding the relative biological effectiveness (RBE) of fast neutrons as a dependency on neutron energy, neutron dose, γ-dose contamination, oxygenation state, X-ray response characteristics of the particular object, and so on. These questions became an outstanding field of experimental research in the last decade, but due to its rather complex nature it deserves more theoretical foundation and generalization in order to allow definitive conclusions and quantitative estimates of direct practical use for neutron therapy and radiation protection. It is hoped that the present theory can make a weighty contribution to this, because it does *not only explain what is known,* but it is able to yield reliable (with some limitations indicated) data for so many situations that purely experimental exploration would be hard (GÜNTHER and LEISTNER, 1975). These aspects are mainly the subject of section 7.4.7, where *gross tissue reactions* are shown to be covered, too, in terms of RBE.

7.4.2. Survival curve prediction

In this subsection, a fairly representative number of comparisons between calculated and measured fast neutron survival

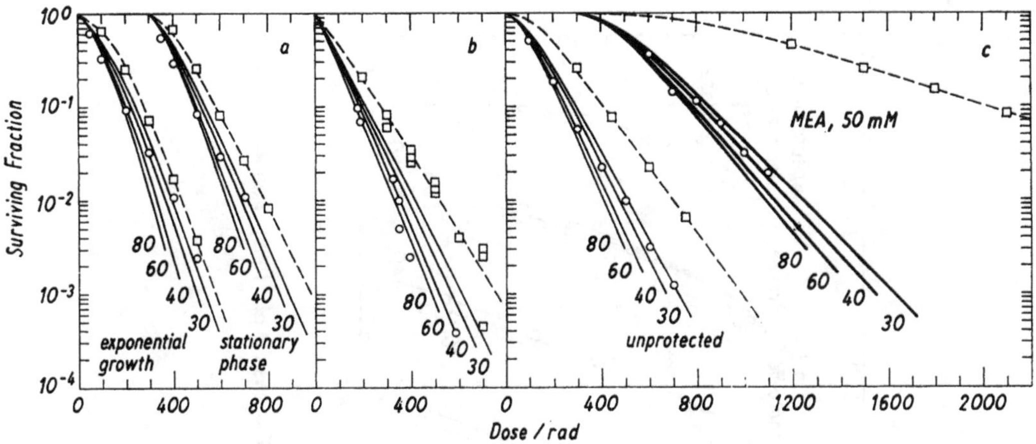

Fig. 7.13. 15-MeV neutron survival curves and low-LET controls (squares, broken lines) of cultured mouse leukaemic cells. Theoretical curves calculated for the σ_0-values indicated [μm^2], heavy lines by using the r-value for anoxic conditions.
Origin of experimental results:
panel a: TAKESHITA and SAWADA (1974), 180-kVp X-rays;
panel b: BROERSE and VAN OOSTEROM (1974), 300-kVp X-rays;
panel c: ANTOKU (1975), ^{60}Co γ-rays.

curves will be demonstrated. They are accompanied by the corresponding reference curves for X-rays, plotted without indicating the unimportant parameterization in terms of Eq. (7.3). Due to the large variety of experimental data accumulated in this field, it was out of the question to contemplate completeness. It is the goal only to impart a general impression regarding agreement or disagreement, while systematic investigations of derived questions, such as RBE and OER in dependence on the object characteristics as well as neutron energy, neutron dose, and so on, are withheld to become the subject of the following subsections.

Figure 7.13 shows results of several authors with the same cell line: cultured mouse leukaemic L5178Y cells. Different treatment of the cells brings about variations in X-ray sensitivity and curve shape, which the theory takes correctly into account, as to the derived neutron curves.[1])

[1]) This aspect, neutron RBE in dependence on X-ray sensitivity, will be specially studied in section 7.4.5, including further results of these experimental series.

Protection with mercaptoethylamine (MEA) leads to a much higher RBE of neutrons (heavy lines in the right-hand panel), which can be quantitatively accounted for in theory by using the r-value for anoxic conditions. In fact, MEA is supposed to compete with oxygen for damaged molecules (low protection under hypoxia (VERGROESEN *et al.*, 1963)).

The cell nucleus area, σ_0, has been widely varied in Figure 7.13 to demonstrate its limited influence. $\sigma_0 = 40\ \mu m^2$ seems to be the best value in all cases.

The often-quoted results on T1 cells of Barendsen's research group are contrasted with theory in Figure 7.14. There is poorer coincidence as compared with the preceding examples, especially in the intermediate energy region where the theoretical RBE turns out somewhat too small, apart from data scatter. Also by variation of σ_0, complete fitting cannot obviously be achieved, the more so as these cells, when exposed to track-segment irradiations, exhibited a nucleus area of about 60 μm^2 only (see Tab. 7.1).

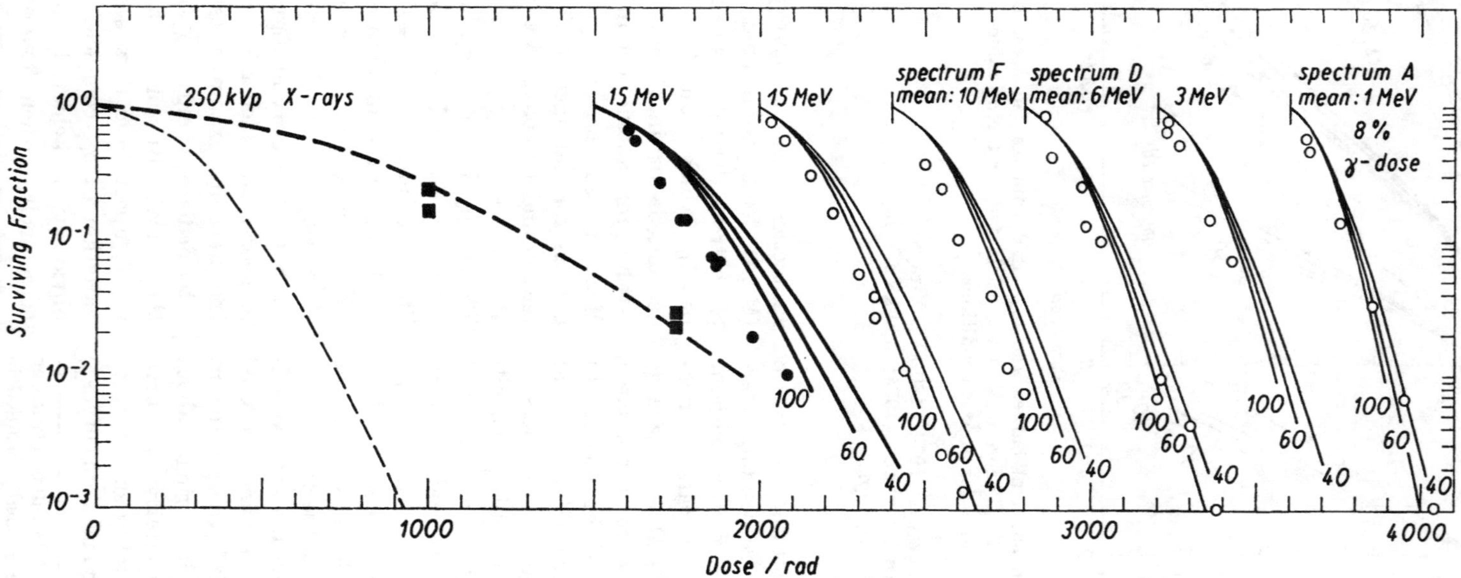

Fig. 7.14. Different-energy neutron survival curves and X-ray control of cultured human T1 cells (Broerse *et al.*, 1968). Full symbols and heavy lines for anoxic conditions. Theoretical curves calculated for the σ_0-values indicated [μm²].

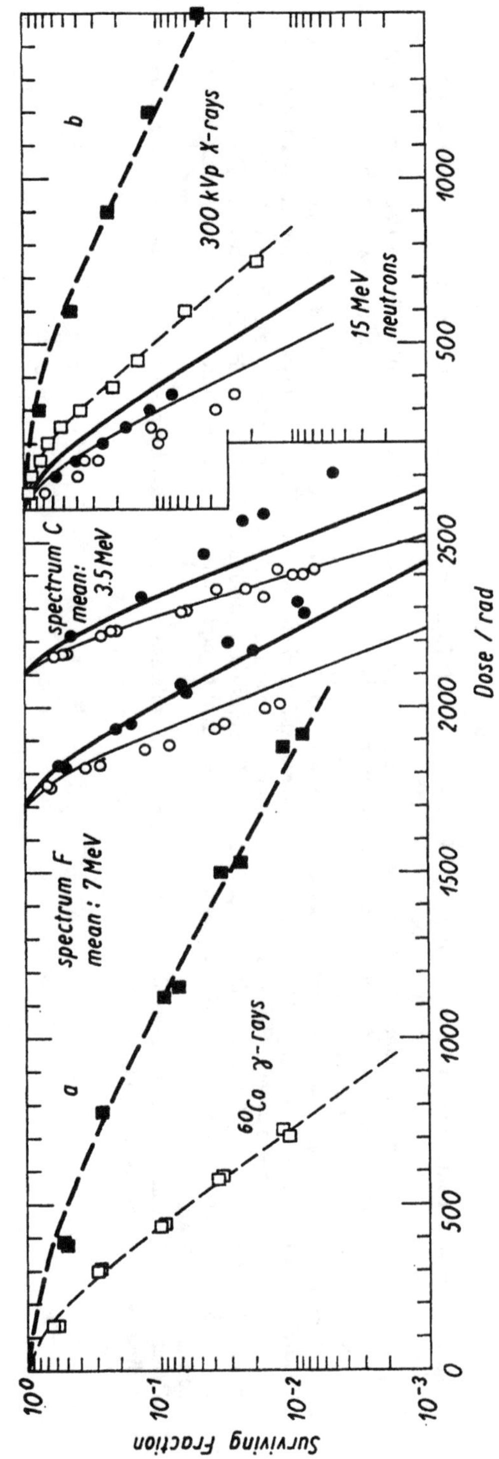

It should be mentioned, however, that TODD (1975) observed larger values for T1 cells.

Figure 7.15 presents results for HeLa cells. The degree of correspondence is about the same as for T1 cells. In particular, these examples visualize the reduced oxygen effect of fast neutrons; the theory is seen to yield this reduced oxygen enhancement ratio, too, but quantitative agreement is poor in this case.

Figure 7.16 is particularly devoted to *in vivo* cell survival. There is only a limited number of techniques that yield complete survival curves of cells under natural conditions in animals. One possibility is transplantable tumours (*a*, *b*). As to normal tissues (*c*, *d*), it is interesting and important to distinguish quickly and slowly proliferating ones, since this causes marked differences in the X-ray response characteristics, which has considerable influence on the RBE of neutrons. These relationships have been discussed by BROERSE and BARENDSEN (1973); but it is not only the usually more pronounced survival curve shoulder of slowly proliferating cells (Fig. 7.16*d*, almost vanishing initial slope for X-rays) that leads, for trivial reasons, to a large RBE (at small dose). The theory moreover gives a relationship between the out-of-shoulder radiosensitivity as measured by D_0^{-1} and the RBE, which is seen from Figure 7.16 to be in excellent quantitative agreement with the experimental findings: small D_0 and, consequently, small RBE for the very proliferative haemopoietic stem cells (*c*) and vice versa for endothelium cells (*d*) of long life span. It is important to note that cell-survival RBE data obtained on

Fig. 7.15. Different-energy neutron survival curves and low-LET controls of HeLa cells. Full symbols and heavy lines for anoxic conditions. Theoretical curves calculated for $\sigma_0 = 100\ \mu m^2$.
Origin of experimental results:
panel a: ZEITZ *et al.* (1975);
panel b: NIAS *et al.* (1967, 1971).

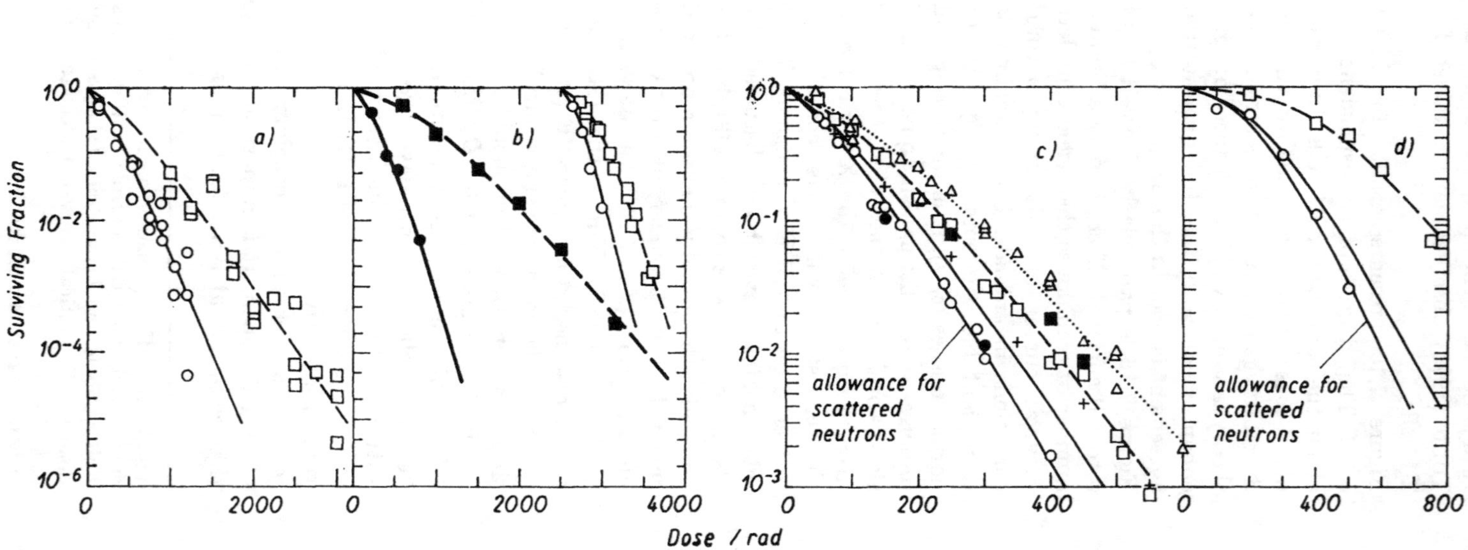

Fig. 7.16. 15-MeV neutron survival curves and 300-kVp X-ray controls (squares, dashed lines) of different tumour (a, b) and normal (c, d) cells *in vivo*. Theoretical curves calculated for $\sigma_0 = 60\ \mu m^2$.

Panel a: transplantable osteosarcoma C22LR of mice, irradiation and single cell assay (tumour origination) *in vivo* (Van Putten *et al.*, 1971).

Panel b: rat rhabdomyosarcoma R-1 irradiated *in vitro* (open symbols) or *in situ* in killed animals (hypoxic conditions, closed symbols and heavy lines) and assayed *in vitro* (Barendsen and Broerse, 1969).

Panel c: mouse bone-marrow stem cells irradiated and assayed *in vivo* (spleen-colony method of Till and McCulloch (1961)). Triangles and dotted line for γ-rays (^{60}Co and ^{137}Cs), squares and dashed line for 300-kVp X-rays, circles and crosses for collimated and open-field neutrons, respectively. Closed symbols according to Hendry *et al.* (1975); all other data from Broerse *et al.* (1971) originating, for the most part, from the following papers: △○ Duncan *et al.* (1969) and Hendry and Howard (1971); □+ Van Putten *et al.* (1970). Most symbols represent logarithmic mean values of survival at equal doses within the same experimental series. Theoretical curves corrected for dosimetry (see text).

Panel d: proliferating capacity of capillary endothelium in the subcutis of rats studied by means of fluorometry; neutron beam strongly collimated (Broerse *et al.*, 1973).

such *in vivo* test systems have been shown to apply also to solid-tumour, tissue, or whole-body reactions (e.g., mortality of mice), exhibiting corresponding differences in RBE (Broerse and Barendsen, 1973). Consequently, in so far as the theory is able to reproduce the former data, as apparent from Figure 7.16, it should be utilizable for the latter problems, too, in terms of RBE for different tumour and normal tissues. These questions, being essential for neutron therapy, are dealt with in section 7.4.7.

Comments must still be given on panels *c* and *d* of Figure 7.16. To resolve discrepancies between analogous results on haemopoietic stem cells of the Rijswijk research group (Van Putten *et al.*, 1970) and the Manchester research group (Duncan *et al.*, 1969; Hendry and Howard, 1971), a collaborative investigation was undertaken (Broerse *et al.*, 1971), which revealed that the low-energy component of scattered neutrons (about 20% in total dose) arising from the steel collimator used at Manchester is likely to be the reason for the observed higher effectiveness of this neutron field, and on the other hand, it was corroborated that X-rays produce markedly steeper survival curves thando Co or Cs γ-rays. The former has been allowed for in theory by giving 20 per cent of neutron dose the energy 1 MeV, which actually brings about good correspondence in both cases (circles and crosses[1]) in panel *c*). The same procedure applied tentatively to the endothelium cell proliferation curve (panel *d*) makes it completely match the experimental data. The second point, apart from the X-ray RBE being above unity relative to ^{60}Co γ-rays ($r = 1.075$ for 150 keV photon energy, see Tab. VIII), is moreover connected with a further correction necessary to be made: by lying adjacent to bone tissue, marrow cells *in situ* do not receive the nominal dose of both neutrons and normal X-rays owing to lacking secondary particle equilibrium.[2]) The corresponding factors of effective dose enhancement as assessed by Broerse and Barendsen (1973) are 1.09 and 0.93 for 300-kVp X-rays and 15-MeV neutrons, respectively. These corrections have been additionally included in the theoretical curves shown. Thus, only the γ-ray curve is a trivial fit to the experimental data, while the good agreement with experience of the other three curves is a significant result of theory.

A few examples for Chinese hamster cells in tissue culture are presented in Figure 7.17. Although panel *a* shows fairly good correspondence between theory and experiment, it must be stated that in most cases (including results not shown) Chinese hamster cells agree poorly with theory, especially in the shoulder region of survival curves. This applies also to the rest of monoenergetic neutron irradiations of the panel-*b* experiment, which is discussed in the following subsection (Fig. 7.18). Panel *c* will be more closely considered in section 7.4.6, where the good agreement in the case of synchronized cells will be argued to be fortuitous.

Let us now discuss the success and failure, respectively, of the theory in predicting cell survival curves for fast neutrons, in a summarizing and generalizing manner. Of course, there are numerous possible sources of error on the experimental side[3]), so that single examples are of little significance to prove or disprove the theory. Nevertheless, having examined a good deal of experimental results in addition to the shown ones, we think it possible to draw some general conclusions important with respect to both the principles underlying the theory and the possibilities of utilizing it in the field of neutron therapy.

(i) Roughly speaking, the theory works best in regard to normal and malignant *cells of murine origin* as well as almost all cell types studied *in vivo*; noticeable deviations

---

[1]) The crosses, referring to the uncollimated Rijswijk neutrons, deviating upwards from the theoretical curve may be attributed to the finding (Broerse *et al.*, 1971) that the Rijswijk dosimetry overestimated the neutron dose by 8% as compared with the Manchester one.

[2]) There exists further data on haemopoietic stem cell survival (Till and McCulloch, 1961; Schneider and Whitmore, 1963; Phillips, 1968) obtained by irradiation *in vitro* or in the recipient mice, which excludes this complication. These results have however been omitted from Figure 7.16 to avoid confusion.

[3]) Dose measurement, neutron energy spectrum and its contamination by scattered neutrons and γ-rays, secondary particle equilibrium, cell sensitivity variation in time, incomplete hypoxia, etc.

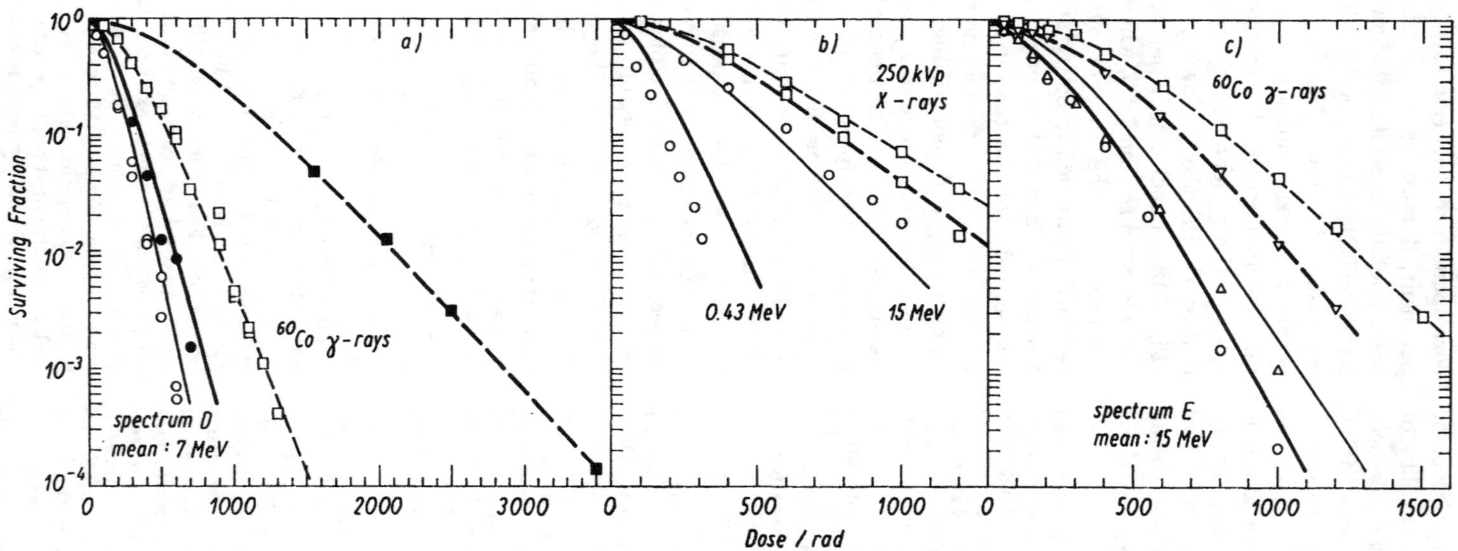

Fig. 7.17. Different-energy neutron survival curves and low-LET controls (broken lines) for Chinese hamster cells.
Panel a: CHO cells (GRAGG *et al.*, 1976); closed symbols and heavy lines for hypoxic conditions; $\sigma_0 = 60\ \mu m^2$ assumed for theory.
Panel b: V79 cells (HALL *et al.*, 1975a); differing X-ray controls distinguished by heavy and thin broken lines; $\sigma_0 = 40\ \mu m^2$ assumed for theory.
Panel c: V79 cells (HALL *et al.*, 1975b); triangles and heavy lines for synchronized G_1/S cells as distinguished from asynchronous cells (thin lines, squares and circles); $\sigma_0 = 40\ \mu m^2$ assumed for theory.

are seen when applied to cultured *cells of human origin*; and it fails substantially in the case of *Chinese hamster cells.*

(ii) With very few exceptions, the theory yields *too small RBE values* when failing.

(iii) The error of theory does not only consist in a dose factor, but it concerns the shape of survival curves above all. Erroneous theoretical curves show a *too large curvature*, so that the RBE error increases with increasing survival level. This effect of unsatisfactory shape is sometimes seen even in cases of good overall agreement (see Fig. 7.13*a* and *c*). Experimental survival curves for neutrons are often of nearly exponential nature (in spite of a pronounced sigmoidal X-ray curve), which the theory is unable to yield.

A detailed discussion of these discrepancies is extremely important because it will manifest the limitations of the theory resulting from its basic principles. At first glance, one could consider incorrect the neutron r-values, stressing their hypothetical derivation by means of the lesion production model via the λ_j parameter set fitted to track-segment irradiation results (see section 7.2.2). However, this is no likely explanation for two reasons: modified r-values would have to become cell-line-dependent in contrast to the findings with ion radiations[1]) and, what is deciding, the shortcoming (iii) cannot be remedied by r-value variation. With a larger r-value, one would only obtain a more steep survival curve on the whole without appreciably changing its shape. The same must be said regarding the second parameter: the cross-section σ_0. It has been shown to have only limited influence within a reasonable range of variation and, moreover, increasing σ_0 to make the RBE larger gives the survival curves even more curvature, i.e. the *initial* slope is not enhanced thereby.

Quite formally, in the case of disagreement, it would be possible to obtain a nearly exponential survival curve of sufficient steepness only by making σ_0 unrealistically *small* (strong saturation of the second kind) and, simultaneously, r would have to be very *large*. For example, such a formal fit of the 0.43-MeV neutron curve in Figure 7.17*b* results in $\sigma_0 \leqq 20\ \mu m^2$, $r \geqq 7.5$.

These arguments suggest that the problem concerns the theory as such, not only its parameters. This is quite another situation than that with ion radiations, where the theory always yields correctly-shaped curves at least by individually fitting r, but only r, while σ_0 has a constant, and quite reasonable, value (see Fig. 2.3 and the dashed curves in Fig. 7.2). The following hypothesis can account for all these findings: it may be an oversimplification to allow for only two levels of structure that interact with the micro-patterns of energy deposition (gross sensitive volume and DNA-targets inside). This assumption implies that a given *number* of lesions in the cell nucleus is considered to be biologically equivalent regardless of the particular *spatial distribution* of these lesions. However, it is likely that molecular lesions produced in close vicinity to each other (e.g., 1 μm or less) exhibit a special injurious interaction (other than the general cumulative action of *all* lesions a cell has received), and this intra-track effect, by being unimportant for X-rays, would additionally enhance the relative biological effectiveness of high-LET radiations. It is easy to realize that this effect should become appreciable for neutrons, in particular, since the densely ionizing but *very short* proton track ends and heavy recoil tracks will produce closely adjacent lesions of nevertheless small total number per absorption event. By contrast, with high-LET track-segment irradiations the interaction effect will be masked, because reaching the necessary interaction distance of neighbouring lesions

---

[1]) The individually-fitted optimal r-values (see Fig. 7.1) also exhibit some variation, it is true, but to a smaller extent, and there is no unequivocal correlation with the present findings for fast neutrons, as to the inter-cell-line differences.

along a track implies the total number of lesions produced by the *long* track segment inside the cell nucleus to be large, so that the cell is killed, anway, according to the response function as given by the X-ray survival curve.

Accordingly generalizing the theory means introducing a third level of structure intermediate between the molecular (3 nm) and the cell-nucleus (10 μm) levels, namely some interaction distance or site.[1]) This has been done in a preliminary approximate manner (unpublished) showing that the discrepancies in question can be resolved, indeed, in this way. It must be emphasized, however, that the additional biological parameters involved can, and have to be assumed to, be object-dependent without any possibility, at present, of their independent determination. In other words, specifying but some elements of the intrinsic biological mechanism of action, such an extended theory loses its formal character; the measured response to low-LET radiation does not any longer provide sufficient information for RBE prediction.

These statements, of course, should not be considered discouraging as to the practical usefulness of the theory in its *present* state, the more so as any improved theory must necessarily fail in reducing the RBE of neutrons for arbitrary cells so simply to nothing but the X-ray curve of survival and the nucleus cross-section. Thus, the present theory is certainly the best one can do for RBE prediction in a way like that. Just because of its conceptual simplicity it constitutes a good tool for systematizing a large body of empirical data. The following subsections will demonstrate this, regarding such determinants of RBE as neutron energy, low-LET components, sensitivity to X-rays of the particular object, and others. One must only keep in mind the possible loss of quantitative significance for small doses or high levels of survival, and even these errors are clearly pronounced only for some cell types specified above, whilst just the objects of primary interest in tumour therapy seem to be covered by the theory in good approximation.

7.4.3. RBE and OER in dependence on neutron energy

Generally speaking, the dependence of the relative biological effectiveness of fast neutrons on their energy is well established: with falling energy it increases gradually, at least down to 0.4 MeV. This is in line with the present theory as may be seen from the neutron r-values listed in Table VIII (appendix): the theoretical RBE is *mainly* determined by this relative effectiveness of primary lesion production (DNA double-strand breaks), but *in detail* it depends on the particular object (X-ray response and size of the cell nucleus) and, in any case, the RBE becomes larger for higher levels of survival. These latter determinants are summarized in the term 'effects of the second kind' (see section 2.4.2). Therefore, there is no RBE-energy relationship as such; one has to consider concrete cell lines for which energy-dependent data exist.

Figure 7.18 presents two examples to compare theory and experiment.[2]) The

---

[1]) Conceptually, by assuming a pairwise interaction, this idea has much in common with the basic principle of the "theory of dual radiation action" (Kellerer and Rossi, 1972) which neglects, however, the important role of the gross sensitive volume in which the sites are located as well as the primary lesion production efficiency (parameter r). Moreover, our present theory, also in its extended form, allows for a non-dual component of action in any case.

[2]) Apart from the theoretical interest in RBE-energy relationships, they are not well suited for the quantitative confrontation of theory and experiment because of errors possibly introduced by deriving the RBE from measured survival 'curves'. As discussed on page 219, such comparisons ought to be made in any case by the hand of what has been directly measured, namely single values of survival.

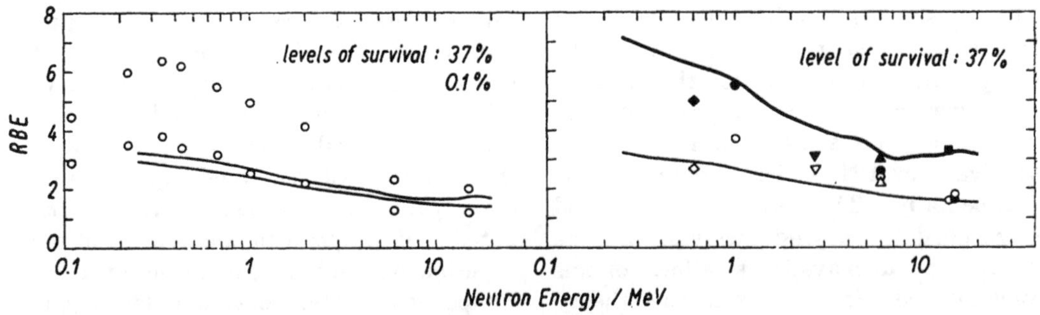

Fig. 7.18. RBE for a constant survival level in dependence on neutron energy.
Left-hand panel: V79 Chinese hamster cells (HALL *et al.*, 1975a); experimental RBE according to the fit $S(D) = \exp(-\alpha' D - \beta' D^2)$ as given by KELLERER *et al.* (1976). $\sigma_0 = 40\ \mu m^2$ assumed for theory.
Right-hand panel: P-388 murine leukaemia cells under aerobic (open symbols) and hypoxic (closed symbols) conditions, irradiated and assayed *in vivo* (BERRY, 1974); theoretical calculation based on 3-MV X-ray survival curves (BERRY and ANDREWS, 1963) and $\sigma_0 = 40\ \mu m^2$.

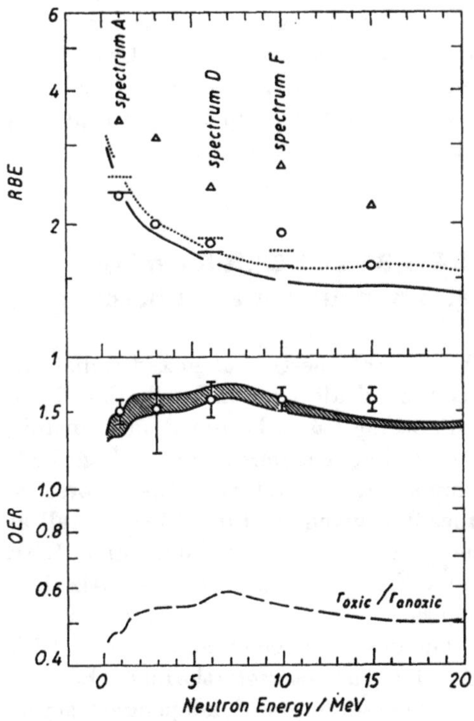

Fig. 7.19. RBE (upper panel) and OER (lower panel) for the inactivation of human kidney T1 cells (BROERSE *et al.*, 1968). Triangles and dotted line for 20% survival, circles and unbroken line for 1% survival. $\sigma_0 = 40\ \mu m^2$; horizontal bars calculated for the neutron energy spectra indicated. For the lower panel see text.

left-hand panel for Chinese hamster cells shows clearly that the theory fails in this case for the high level of survival (37%). On the other hand, for the same high survival level in the case of the murine tumour cell line to which the right-hand panel refers, one observes no systematic deviations for both aerobic and hypoxic conditions, apart from the larger scatter in BERRY's results.[1]) These statements confirm the judgements in the preceding subsection; nevertheless, the influence of neutron energy is well described by the theory.

BROERSE and co-workers (1968) were the first to measure the dependence of RBE and OER on neutron energy, using human kidney cells in tissue culture (Fig. 7.19).

---

[1]) Cell survival was assessed from recipient mice mortality after injection of known numbers of irradiated tumour cells. The experimental RBE data plotted in Figure 7.18 have been taken direct from a review paper (BERRY, 1974) in the awareness of the existence of some unexplained contradictions to the corresponding results as originally published by BERRY and co-workers (survival curves, mean neutron energies). Only one value (14 MeV, anoxic conditions) has been corrected according to the original paper (BERRY and ELLIS, 1966).

These results fall well in line with the theoretical prediction, apart from the disagreement seen again for the RBE at 20% survival. As to the oxygen enhancement ratio, no systematic influence of survival level has been detected experimentally. The error bars plotted correspond to varying survival, and the shaded area displays the smallness of this variation as derived from the theory between 80% and 0% survival. In contrast to RBE, the OER shows little dependence on neutron energy.

As exhibited by the dashed line in the lower panel of Figure 7.19, the oxygen enhancement ratio for neutrons is to good approximation proportional to the quotient $r_{\text{oxic}}/r_{\text{anoxic}}$.[1]) Hence, it is only the $OER^{(\gamma)}$ for ^{60}Co γ-rays that makes the neutron OER distinct for different objects by virtue of

$$OER^{(n)} \approx \frac{r_{\text{oxic}}}{r_{\text{anoxic}}}\, OER^{(\gamma)}. \tag{7.24}$$

In other words, the so-called *gain factor* defined by $G = OER^{(\gamma)}/OER^{(n)}$ turns out to be a universal quantity which depends but slightly on the neutron energy:

$$G \approx \frac{r_{\text{anoxic}}}{r_{\text{oxic}}}. \tag{7.25}$$

[1]) This is due to the second kind effects cancelling out, approximately, by having similar influence on the oxic and anoxic survival curves. In terms of microdosimetric theory (Eq. (2.17)):

$$\begin{aligned}(\alpha_\mu)_{\text{ox}} &= \frac{1}{\bar z_F}\int_0^\infty dz\{1 - \exp[-(\alpha_\mu{}^{(0)}r)_{\text{ox}}\, z]\}\, f_1(z)\\ &= \frac{1}{\bar z_F}\int_0^\infty dz\left\{1 - \exp\left[-OER^{(\gamma)}\frac{r_{\text{ox}}}{r_{\text{anox}}} \times (\alpha_\mu{}^{(0)}r)_{\text{anox}}\, z\right]\right\} f_1(z)\\ &\approx OER^{(\gamma)}\frac{r_{\text{ox}}}{r_{\text{anox}}}\frac{1}{\bar z_F} \times \int_0^\infty dz\{1 - \exp[-(\alpha_\mu{}^{(0)}r)_{\text{anox}}\, z]\}\, f_1(z),\\ (\alpha_\mu)_{\text{ox}} &\approx OER^{(\gamma)}\frac{r_{\text{ox}}}{r_{\text{anox}}}(\alpha_\mu)_{\text{anox}}.\end{aligned}$$

This universal G-curve is shown in Figure 7.20 together with a large number of experimental data referring, in part, to effects other than single-cell survival. There is considerable scatter between these data, with a tendency towards lower values of G, which might be attributed to incomplete anoxia in some of the experiments and the presence of a γ-component in real neutron fields. Apart from this, the nearly constant theoretical curve provides a fairly good average value, and there is even some evidence from the *vicia faba* data for the slight energy-dependence to be significant.

In summary, the RBE of fast neutrons has been shown to increase substantially as the neutron energy decreases down to about 0.3 MeV, while the OER expressed in terms of the oxygen enhancement gain factor exhibits too little neutron-energy dependence, if any, as to promise a better gain in hypoxic-tumour therapy by definite energies of neutrons.

7.4.4. RBE and OER for mixed fields of neutrons and photons

The general question of mixed radiations has been dealt with in subsection 7.3.4, and nothing has to be added, on principle, in case that neutrons form the high-LET component. We refer to the theoretical procedure given by Eq. (7.19) as well as the general statements concerning RBE and OER, which can be summarized as follows.

(i) The dose-average of component RBE's (Eq. (7.21)) *overestimates* the mixed RBE.

(ii) The assumption of independent action of components (Eq. (7.18)) *underestimates* the mixed RBE.

(iii) The mixed OER, which can to good approximation be derived from dose-average RBE's (Eq. (7.22)), exhibits considerable reduction already by a small proportion of high-LET radiation.

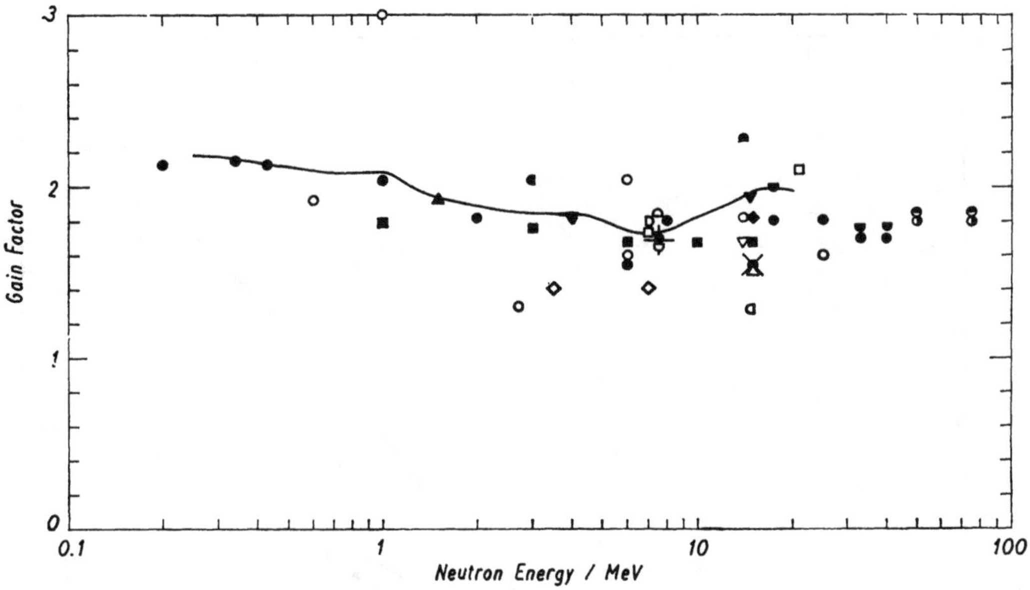

Fig. 7.20. Oxygen enhancement gain factor, $G = OER^{(\gamma)}/OER^{(n)}$, in dependence on neutron energy ($OER^{(\gamma)}$ referring to ^{60}Co γ-rays).
Theoretical curve: $G = r_{\text{anoxic}}/r_{\text{oxic}}$.
Experimental data (corrected by the factor $r_{X,\text{anoxic}}/r_{X,\text{oxic}}$ if referring to X-rays): Cell survival *in vitro*: ◖ Chinese hamster fibroblasts (SCHNEIDER and WHITMORE, 1963); ▽ HeLa (NIAS *et al.*, 1967); ■ T1 (BROERSE *et al.*, 1968); ◡ mouse ascites (MCNALLY and BEWLEY, 1969); ◆ rat rhabdomyosarcoma (BARENDSEN and BROERSE, 1969); ▼ CHO (RAILTON *et al.*, 1973); mouse haemopoietic (△) and leukaemic (×) cells (BROERSE and VAN OOSTEROM, 1974); + rat fibrosarcoma (BEWLEY *et al.*, 1974); ◇ HeLa (ZEITZ *et al.*, 1975); ▼ V79 (HALL *et al.*, 1975b; HALL *et al.*, 1977); ▲ mouse leukaemic cells (ANTOKU, 1975); □ CHO (GRAGG *et al.*, 1976); ◓ V79 (GUICHARD *et al.*, 1978). Cell survival *in vivo*: ◠ mouse ascites (BEWLEY and HORNSEY, 1964); ◖ mouse osteosarcoma (VAN PUTTEN *et al.*, 1971); ○ murine leukaemia (BERRY and ANDREWS, 1963; BERRY *et al.*, 1965; BERRY, 1971; BERRY *et al.*, 1973). Chromosomal abnormalities: ▲ Ehrlich casites cells (HORNSEY *et al.*, 1960). Gross effects: ◗ weight loss of mouse thymus (WRIGHT *et al.*, 1966); D growth of vertebrae in rat tails (DIXON, 1968); ● *vicia faba* root growth (HALL *et al.*, 1973; HALL *et al.*, 1977); ◑ the same (VAN DAM *et al.*, 1978).

Apart from the theoretical interest, the subject is of practical importance in cancer therapy with fast neutrons: statement (iii) favours mixed treatments of partially hypoxic tumours; statement (i) plays a role for treatment planning in consideration of the γ-ray contamination of neutron beams, especially in the so-called penumbra region where the relative γ-ray component is large. It is in general usage to assesss the corresponding normal tissue effects in terms of the convenient dose-averaged RBE according to

$$RBE(D) = \mu_D^{(\gamma)} + (1 - \mu_D^{(\gamma)}) \times RBE^{(n)}\big((1 - \mu_D^{(\gamma)})\,D\big), \quad (7.26)$$

where $\mu_D^{(\gamma)}$ is the γ-dose fraction and D the total dose. Eq. (7.26) yields a convex curve of RBE as a function of $\mu_D^{(\gamma)}$ due to the increase of $RBE^{(n)}$ with decreasing neutron dose. To visualize this and the degree of overestimation thus obtained, Figure 7.21 presents a theoretical example

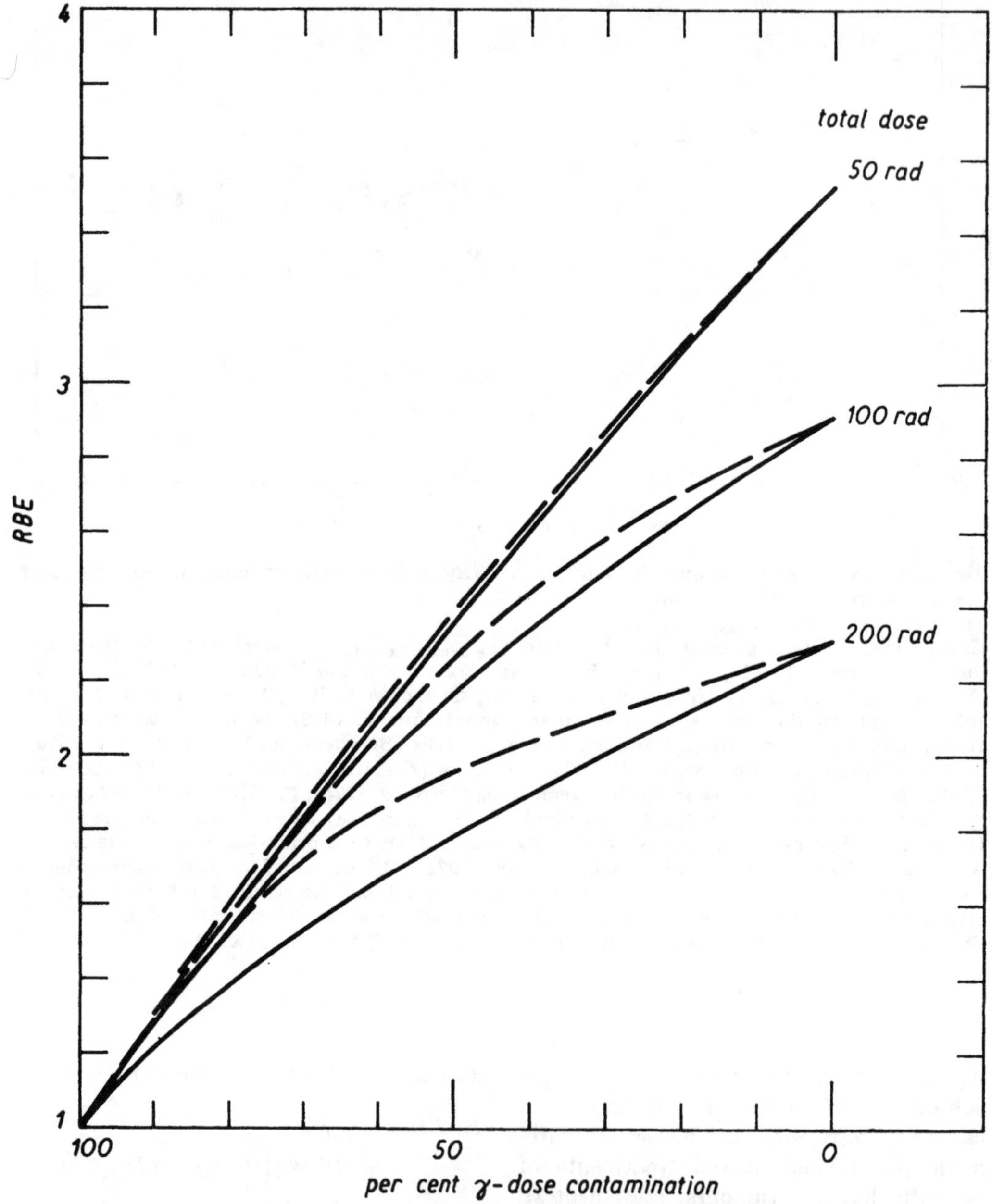

Fig. 7.21. The relative biological effectiveness of a mixed field of ^{60}Co γ-rays and fast neutrons (15 MeV and 1 MeV in the ratio 80% to 20% of dose), calculated for capillary endothelium cells according to the survival curves shown in panel d of Figure 7.16 ($\sigma_0 = 60\ \mu m^2$). Full lines: strict theory (Eq. (7.19)); broken lines: simple dose-average (Eq. (7.26)).

in which the results of Eq. (7.26) (broken lines) are contrasted with the exact theoretical solution (full lines). Of course, by referring to special cells (special X-ray survival curve and σ_0), this example can only give a general impression of the errors involved in the dose-averaged RBE.[1]) E.g., these errors become the smaller the more the X-ray survival curve is shaped exponentially.

There is only scarce experimental evidence for the theoretical results on mixed radiations. The example presented in Figure 7.9 (α-particles in combination with X-rays) confirms the theory, but only one analogous experiment with neutrons (RAILTON *et al.*, 1974) seems to have been performed. Unfortunately, this is a case in which the theory fails already for pure neutrons (Chinese hamster CHO cells). However, RAILTON's results are of great theoretical importance, and the difficulties vanish if one deals with the so-called *interaction factor*, I, which has been defined and measured in a subsequent paper (RAILTON *et al.*, 1975):

$$I(D) = \frac{S_m(D)}{S_\gamma(D_\gamma)\, S_n(D_n)}, \qquad D = D_\gamma + D_n, \tag{7.27}$$

where the three surviving fractions refer to the mixed radiation of total dose D, the pure γ-rays, and the pure neutrons, respectively.

Let us first discuss the significance of the interaction factor. $I \equiv 1$, which is equivalent to Eq. (7.18), would imply that the two components act independently of each other, i.e. by different modes of killing, so that sublethal damage from one component has no influence on the action

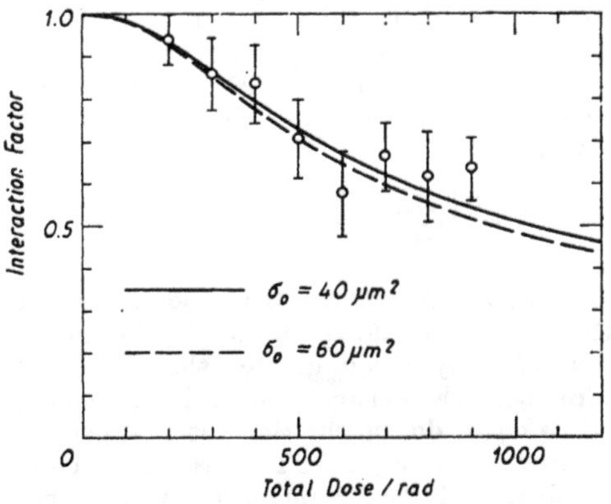

Fig. 7.22. The interaction factor defined by Eq. (7.27). Experimental values from RAILTON *et al.* (1975) for CHO cells (15-MeV neutrons and ^{60}Co γ-rays mixed in the ratio of 25% to 75%). Curves according to theory.

of the other one (each individual cell is killed either by the γ-rays or the neutrons).[2]) Any kind of cumulative or 'synergistic' action, however, gives rise to $I < 1$, and the interaction will be complete (smallest possible I) if there is not any qualitative difference between γ-ray and neutron damage, implying an equal mode of killing. Now recall that it is just *the latter idea* that constitutes a basic assumption of the present theory (only one type of molecular lesions the total number of which alone decides on the probability of survival). Consequently, measuring the interaction factor is a good test for the fundamentals of the theory.

Figure 7.22 shows the experimental results for 75 per cent γ-dose fraction and the theoretical curves for two values of the cell nucleus cross-section. The good agreement is remarkable, suggesting completeness of interaction in the sense explained above.

[1]) There is no reason to doubt the validity of these results in view of the insufficiency of the theory as such in some cases with respect to the shoulder region of neutron survival curves as discussed in section 7.4.2. On the contrary, the error of relation (7.26) would become larger in these cases.

[2]) This hypothetical possibility was disregarded in section 7.3.4, i.e. Eq. (7.18) was rejected at once, by arguments based axiomatically on the present theory (statement (ii)), taking for granted the existence of a cumulative action. The starting-point now is more general.

The phenomenon of interaction can also be studied by means of split-dose experiments. However, this method has its limitations, because for the recovery factor (ELKIND-recovery)

$$RF = \frac{S(D_1 \parallel D_2)}{S(D_1 + D_2)},$$

where $S(D_1 \parallel D_2)$ means the residual survival after the splitted doses D_1 and D_2, to indicate interaction by being greater than unity requires both the existence of sublethal damage *and its repair* during the time interval between the two doses. High-LET radiations produce only little sublethal damage (survival curves less sigmoidal than with low-LET radiation), so that the *maximum possible* recovery factor

$$RF_{\max} = \frac{S(D_1)\, S(D_2)}{S(D_1 + D_2)}$$

becomes relatively small. Hence, low values of RF obtained with high-LET radiation (e.g. TODD, 1967) must always be seen in relation to $RF_{\max}$ and, in addition, the repair may be reduced as suggested by a number of direct molecular investigations (RITTER *et al.*, 1977; KÖRNER *et al.*, 1978; HANDSCHACK, 1979). These arguments also apply if two different radiations are used in a split-dose experiment to study the effect of their interaction (BARENDSEN *et al.*, 1960; RAILTON *et al.*, 1975): $RF \approx 1$ may be due to inefficient repair during the time interval chosen. In other words, ELKIND-recovery is no good method to demonstrate interaction of damage from different radiations. Determining the coefficient $I(D)$ defined by Eq. (7.27) is a better way since repair is not involved.

7.4.5. The relationship between X-ray sensitivity and neutron RBE

It would be useful to have numerical results of the theory in a general form applicable to all situations. Unfortunately, this is impossible because the RBE for a given dose or survival level depends on the *whole* X-ray survival curve. There is only one exception: with respect to *large* doses, i.e. the 'final-slope RBE', one obtains general results in terms of nothing but the D_0-value (reciprocal final slope) for X-rays. This is evident since the 'tail' of a high-LET survival curve described by the expression (7.15) is only determined by the most slowly diminishing exponential, i.e. the smallest α_μ, which derives from the corresponding smallest $\alpha_\mu^{(X)}$ of

Tab. 7.5. High-dose RBE of fast neutrons relative to ^{60}Co γ-rays in dependence on D_0, the reciprocal final slope of the dose-effect curve for ^{60}Co γ-rays (rads, upper row)

Neutron energy [MeV]	σ_0 [μm²]	Oxic conditions 50	75	100	150	200	300	500	∞	Anoxic conditions 100	150	200	300	500	700	1000	∞
0.5	40	2.09	2.45	2.67	2.93	3.07	3.23	3.37	3.593	4.31	5.08	5.56	6.12	6.65	6.90	7.10	7.598
	60	2.55	2.84	3.00	3.18	3.27	3.37	3.46	3.593	5.31	5.92	6.28	6.67	7.02	7.18	7.30	7.598
	100	3.00	3.18	3.27	3.37	3.43	3.48	3.52	3.593	6.28	6.68	6.89	7.11	7.30	7.38	7.45	7.598
1	40	1.87	2.20	2.40	2.64	2.77	2.92	3.04	3.247	3.82	4.51	4.94	5.45	5.92	6.14	6.32	6.761
	60	2.27	2.53	2.68	2.85	2.94	3.04	3.12	3.247	4.66	5.23	5.55	5.91	6.23	6.37	6.49	6.761
	100	2.66	2.84	2.93	3.03	3.08	3.14	3.18	3.247	5.50	5.88	6.08	6.29	6.47	6.55	6.62	6.761
3	40	1.58	1.79	1.91	2.07	2.15	2.25	2.33	2.475	2.99	3.36	3.59	3.85	4.10	4.22	4.32	4.561
	60	1.82	2.01	2.09	2.20	2.26	2.33	2.38	2.475	3.42	3.72	3.90	4.09	4.26	4.35	4.41	4.561
	100	2.06	2.19	2.25	2.32	2.36	2.39	2.43	2.475	3.85	4.06	4.17	4.29	4.40	4.44	4.48	4.561
6	40	1.45	1.59	1.68	1.78	1.85	1.92	1.98	2.098	2.62	2.85	2.99	3.16	3.33	3.41	3.48	3.654
	60	1.61	1.73	1.80	1.88	1.93	1.98	2.03	2.098	2.88	3.08	3.19	3.32	3.44	3.50	3.54	3.654
	100	1.78	1.87	1.92	1.97	2.00	2.03	2.06	2.098	3.16	3.30	3.38	3.46	3.54	3.57	3.59	3.654
10	40	1.32	1.44	1.52	1.62	1.69	1.78	1.87	2.061	2.47	2.68	2.82	3.01	3.23	3.35	3.46	3.780
	60	1.45	1.57	1.64	1.74	1.80	1.88	1.94	2.061	2.71	2.92	3.05	3.22	3.40	3.49	3.57	3.780
	100	1.62	1.72	1.78	1.86	1.90	1.95	1.99	2.061	3.01	3.19	3.29	3.43	3.56	3.62	3.66	3.780
15	40	1.20	1.32	1.42	1.55	1.65	1.77	1.90	2.186	2.38	2.63	2.81	3.07	3.40	3.58	3.76	4.310
	60	1.35	1.48	1.57	1.70	1.79	1.90	1.99	2.186	2.66	2.92	3.11	3.37	3.65	3.81	3.94	4.310
	100	1.54	1.67	1.76	1.87	1.94	2.01	2.07	2.186	3.04	3.30	3.48	3.70	3.91	4.01	4.09	4.310

the X-ray curve (note Eq. (7.16) and context).

The final-slope RBE defined as the ratio of D_0-values is equal to the limiting plateau value of the RBE-dose relation as $D \to \infty$ where the RBE becomes minimal. Numerical results for different neutron energies and cell nucleus cross-sections are given in Table 7.5. These data display the following theoretical rule: *The more resistant the object as measured by the D_0 for low-LET radiations, the larger the high-dose RBE.* It approaches the value r if D_0 tends to infinity and is smaller otherwise, as a consequence of what we call the 'saturation effect of the second kind' (cf. Eq. (2.26) and section 2.4.2).

A high neutron RBE at small and medium doses for cells being resistant against X-rays by having a wide survival curve shoulder is a well-known fact and is trivial due to the always small shoulder of neutron survival curves. The above rule, however, states a non-trivial relation between radiosensitivity and RBE for large doses irrespective of the shoulder of the underlying survival curves. This deserves experimental evidence, which is not, unfortunately, easy to obtain, since the influence of the cell nucleus cross-section may disturb the correlation between RBE and D_0 if various cell lines are being compared. It would be advantageous to have results on the *same* cell line with the D_0-value made varying by special means. This is the case with the mouse leukaemic cells L5178Y: in dependence on growth phase (exponential and stationary), temperature, chemical protection, and combinations of these factors, survival curves with differing D_0-values (and extrapolation numbers) have been obtained. Table 7.6 lists these experimental results in terms of the 1-per-cent-survival RBE of 15-MeV neutrons[2]). The corresponding theoretical values have been derived from the neutron dose-survival relations calculated in the usual way for $\sigma_0 = 40\ \mu m^2$ and 5 per cent γ-dose contamination (some of them are shown in Fig. 7.13). The agreement is very good, confirming the correlation between RBE and D_0 (note that different photon reference radiations somewhat disturb the sequence, accordantly with theory).

Regarding the use of Table 7.5, it should be noted that these general theoretical data may not quite apply to practical situations because of both γ-ray contamination and reference to photon radiations other than ^{60}Co γ-rays. In these cases, a γ-dose fraction can be allowed for by forming the mean RBE according to Eq. (7.26), which is correct in this case, while to relate the RBE to lower-energy photons requires division by the corresponding r-value found in Table VIII of the appendix. A neutron energy spectrum

[2]) For the sake of accuracy, the RBE for a low constant level of survival has been used instead of the genuine final-slope RBE. Both are almost equivalent with respect to the rule in question.

Tab. 7.6. RBE of 15-MeV neutrons for 1% survival of differently-treated mouse leukaemic cells L5178Y having different D_0-values for the low-LET radiations specified (TS — TAKESHITA and SAWADA, 1974; BO — BROERSE and VAN OOSTEROM, 1974; A — ANTOKU, 1975)

Experiment	origin	TS	TS	BO	TS	A	TS	A	TS	TS
	reference radiation	X-rays 180 kV	X-rays 180 kV	X-rays 300 kV	X-rays 180 kV	^{60}Co	X-rays 180 kV	^{60}Co	X-rays 180 kV	X-rays 180 kV
	D_0 [rad]	62	85	100	116	123	141	185	238	310
	RBE	1.09	1.18	1.49	1.30	1.42	1.37	1.60	1.59	1.62
Theory	RBE	1.18	1.26	1.34	1.36	1.46	1.39	1.59	1.55	1.61

can also be taken into account by the simple arithmetic dose-mean of RBE.

Apart from these small corrections, RBE values as given in Table 7.5 can be directly compared with those for tissue and organismic reactions of experimental animals determined with large doses, provided that the tissues considered can be specified by their sensitivity to X-rays in terms of D_0 for the corresponding stem cells. These questions are dealt with in the last subsection of this chapter: one obtains further quantitative evidence therefrom for the RBE depending on D_0 as stated by the above rule, which lends theoretical support and foundation to the finding that the RBE of fast neutrons is actually different for different tissues (Broerse and Barendsen, 1973; Field and Hornsey, 1974a).

While Table 7.5 applies only to the limit $D \to \infty$, the degree of increase of the RBE at decreasing dose depends, according to the theory, on the shape of the X-ray survival curve *as a whole*, with no possibility to derive general assertions apart from considering special cases.[1]) This can be seen from Eq. (2.25) which expresses the RBE for the limit $D \to 0$ (maximum RBE) explicitly by the whole function $S_0(D)$. Besides, it is just this maximum RBE for which at least *upper bounds* can be given in definite figures shown in Table 7.7. As to the theoretical background, the reader is referred to section 2.3.2 of chapter 2, where the underlying conditions of extremity (low sensitivity and vanishing initial slope for 'zero LET') are discussed, too, ending up in Eq. (2.29). Replacing the inherent microdosimetry by 'virtual microdosimetry' (i.e. strict theory, see chapter 8), this equation reads

$$\lim_{D \to 0} RBE = r^2 \frac{(\bar{z}_D^{**})_{\text{neutrons}}}{(\bar{z}_D^{**})_{\text{photons}}}. \qquad (7.28)$$

Table 7.7 based on Eq. (7.28) exhibits the absolutely highest RBE of fast neutrons the theory can yield at all.[2]) These values are largely hypothetical since mammalian cell survival curves seem always to have a non-vanishing initial slope even for the imaginary limit of 'zero-LET' radiation. Nevertheless, the existence of critical tissues that such extremely high zero-dose RBE values pertain to cannot be entirely ruled out. An example may be eye lens opacification for which the RBE was measured in dependence on the neutron dose down to very small doses (Bateman *et al.*, 1972), and these data were very well reproduced by the present theory under the mentioned extreme conditions (see section 2.3.4). Hence, the limiting values in Table 7.7 may be of some theoretical interest for fundamental questions of radiation protection.

7.4.6. RBE and cell cycle

It is well known that the response of mammalian cells to low-LET radiation depends on the stage in cell cycle (Sinclair and Morton, 1963; Terasima and Tolmach, 1963; Whitmore *et al.*, 1965; Sinclair, 1968a; Sinclair, 1974). This concerns both the final slope and the shoulder width of survival curves of synchronized cells. Apart from substantial variations in the accumulated data on

[1]) There may be no increase at all, which the theory predicts in the case of a strictly exponential survival curve for X-rays, while a pronounced flat shoulder causes a more sharp increase.

[2]) There is little variation due to the cell nucleus cross-section. The value $\sigma_0 = 0.785\ \mu m^2$ ($d = 1\ \mu m$) has been included since Eq. (7.28) holds also in the framework of an extended theory with σ_0 referring to the 'interaction site' rather than the cell nucleus (see page 242).

Tab. 7.7. Maximum limiting zero-dose RBE of fast neutrons relative to ^{60}Co γ-rays

Neutron energy [MeV]	0.25	0.5	1	3	6	10	15
$\sigma_0 = 0.785\ \mu m^2$	568	512	478	312	263	383	540
$\sigma_0 = 40\ \mu m^2$	473	634	581	414	347	629	984
$\sigma_0 = 100\ \mu m^2$	374	545	545	406	326	627	1005

cyclic response, one can state the general rule that the border G_1/S is very sensitive and the survival increases during the S-phase, the highest resistance being reached at the end of S.

Whatever the intrinsic biological reasons of cyclic response are, the *cyclic variation of RBE* of high-LET radiations should follow from nothing but the varying X-ray survival curves, if the present theory is correct with respect to synchronized cells too. Hence, the problem constitutes a special case of what has been dealt with in the foregoing subsection, i.e. the RBE should be correlated simply with the resistance against X-rays, provided that the size of the cell nucleus remains constant.[1])

Experimental data for fast neutrons in comparison with X-rays were reported by SINCLAIR (1968b) who used synchronized V79 Chinese hamster cells. The upper panel of Figure 7.23, redrawn from his paper, shows the cyclic variation of survival after X-irradiation, while the neutron results are given in the lower panel in terms of RBE[2]). Although the latter data are of limited accuracy, in particular for the end of the cycle, there is clear evidence for the RBE to vary in the way to be expected: highest RBE in the resistant part of the cycle (late S-phase). For low dose this is mainly due to the more pronounced X-ray survival curve shoulder in middle and late S. The theory has been recognized (subsection 7.4.2) as failing in the shoulder region, particularly for hamster cells. This may explain the moderate RBE variation predicted by the theory from the X-ray survival curves given implicitly in the upper panel of Figure 7.23. However, this kind of disagreement should disappear for large

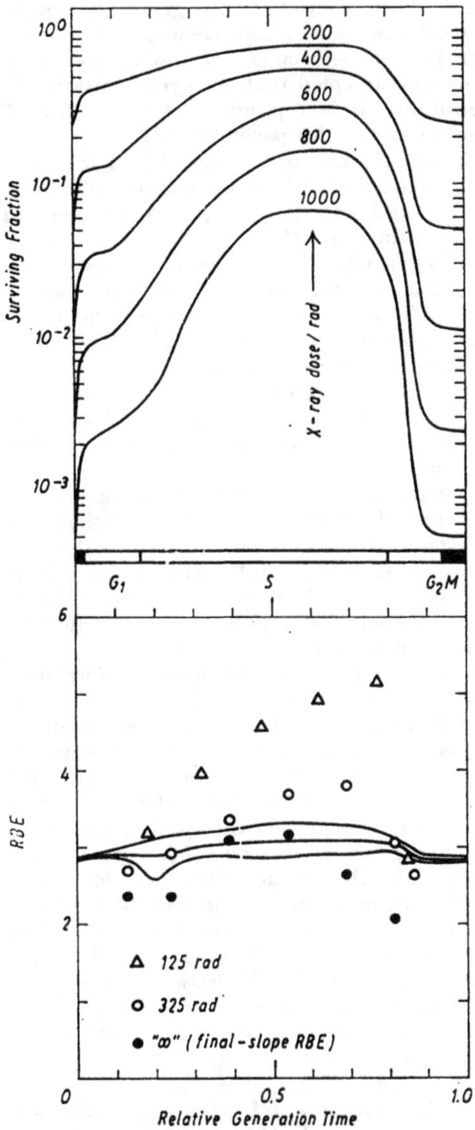

Fig. 7.23. Cell-cycle dependence of radiation response of synchronized V79 Chinese hamster cells (SINCLAIR, 1968b). Upper panel: survival after 250-kVp X-irradiation. Lower panel: RBE of fission neutrons (mean energy 0.25 MeV) for the neutron doses indicated; theoretical curves calculated for $\sigma_0 = 40\ \mu m^2$.

---

[1]) One must anyway exclude the mitosis (perhaps also the late G_2 phase) to which the theory cannot be applied, certainly. In asynchronous populations this may be masked owing to the short period of mitosis within the entire cell cycle.

[2]) SINCLAIR reported surviving fractions for three constant neutron doses (125, 225, 325 rad) and an X-ray control curve (710 rad) which has been used to calibrate the time scale. The RBE values in Figure 7.23 have been calculated from these data and the complete X-ray data.

doses, i.e. concerning the high-dose-slope RBE also shown (full circles and lowest curve). In fact, one observes better but no complete correspondence; the larger variation of the experimental high-dose-slope RBE, too, seems to be significant.

The entire subject is of great theoretical importance since it can eventually give hints for further refinements of the theory. It is generally accepted that the cyclic variation of sensitivity is less pronounced for high-LET radiations (cf. also page 223) so that the RBE varies in the manner of Figure 7.23. *Qualitatively*, in the framework of the present theory, this looks like a natural consequence of the second kind saturation effect depending on the X-ray sensitivity. By means of corresponding *quantitative* analyses, however, one can in principle decide the question of whether this is the whole matter or not. Additional factors of influence may be proposed at different conceptual levels of theory as follows:
(i) Varying size of cell nucleus in the framework of the present theory and the constant double-strand-break r-value.
(ii) Varying r-value due to changing involvement of different types of molecular lesions, but present theory otherwise.
(iii) Influence of possible changes of the inner structure, organization, and interaction of targets inside the nucleus as *not* described by the present theory.
It is clear that the experimental information available so far does not yet allow definitive conclusions concerning these possible determinants of cyclic RBE variation. In particular item (iii) must be seen in connection with what has been discussed at the end of subsection 7.4.2, i.e. an extended theory that avoids the difficulties found already for asynchronous cells can also make allowance for cyclic variations of internal structure. Such kind of changes during the cell cycle were observed by Dewey *et al.*, (1972), who suggested that alterations in the configuration of chromatin (degree of condensation and association between DNA and protein) are responsible for the variation of radiosensitivity.
For the present one can only state that the experimental data (cf. also page 223; and Baldytchev *et al.*, 1973) show a tendency of *more* cyclic RBE variation than the theory predicts without allowance for the factors (i) to (iii). It is especially the sensitive stage G_1/S that appears relatively insensitive to fast neutrons (small RBE). This is most clearly confirmed by the results of Hall *et al.* (1975b) shown in Figure 7.17 (panel *c*): synchronized G_1/S cells are even more resistant against neutrons than asynchronous cells are, despite of their higher sensitivity in the case of X-rays.[1]) This inversion of sensitivity claims to take factors like (i) to (iii) into consideration, since never can the simple theory formally employed, as in Figure 7.23, yield such a result.
It is particularly the hypothesis (ii) that is most attractive with respect to the G_1/S resistance against neutrons, if one assumes that its *sensitivity* to X-rays is caused by a contribution to cell killing of DNA *single-strand breaks* (speedy fixation by the DNA replication), which is otherwise absent due to the well-known fast rejoining of single-strand breaks.[2]) Then, the relatively enhanced resistance against neutrons follows from the *lower* effectivity of high-LET radiations in producing DNA single-strand breaks ($r < 1$, see Tab. VIII of the appendix). This would have to be approximately described in the framework of formal theory by an average r-value smaller than the commonly-used universal double-strand-break one if the border G_1/S of the cell cycle is concerned.

7.4.7. RBE for normal and malignant tissue reactions relevant to neutron therapy

7.4.7.1. Fundamentals

The first question to be answered concerns the relation between reactions of organized tissues (such as histopathological changes, tumour responses, necroses, and thus-caused death of experimental animals) and the survival of cells, because it is only the latter to which the present theory refers directly. There seems to be, indeed,

[1]) The theoretical curves in Figure 7.17, panel *c*, must therefore be considered with caution. In view of the too-flat predicted neutron curve for asynchronous cells (thin unbroken line), which means that the theory fails for this cell line, the good correspondence for G_1/S cells (heavy unbroken line and triangles) is fortuitous and should not be considered as confirming the theory with the universal r-value (note also the unsatisfactory shape).
[2]) Such a contribution could be masked in asynchronous populations, so that the universal r-value may not reflect it.

a close relation in that the tissue reaction depends mainly on the rate of cell depletion and the ability of repopulation determined by the level of induced stem cell killing. In some cases the crucial stem cells were identified (see below), and evidence was obtained for their survival *in vivo* being representative for the RBE of fast neutrons regarding gross damage to the corresponding tissues (BROERSE and BARENDSEN, 1973). It is now generally accepted that the mere *existence* of certain cells crucial in the sense that their survival curves for neutrons and X-rays imply the correct tissue RBE can be taken for granted (ALPER, 1973). Hence, the basic condition for applying the present RBE theory to tissue reactions seems to be fulfilled.

The statement made does not, however, apply to survival curves obtainable by culturing the cells *in vitro*. It must be emphasized that *in-vivo* survival curves are usually different, particularly in having larger shoulders (FIELD and HORNSEY, 1974a; DENEKAMP and FIELD, 1974). Apart from the effect of cell kinetics control in an organized tissue, the mere three-dimensional intercellular contact brings about a broad shoulder (large D_q, small initial slope), which has been demonstrated by irradiating cells in small 'spheroids' as compared with plated cells (DURAND and SUTHERLAND, 1973). While this is just a model system, various techniques are in use to determine single-cell survival curves *in situ*, i.e. mostly in mice.

(i) Single cells from tumour tissues are often transplantable from irradiated donor animals to recipient animals in which the cell survival is scored by tumour appearance. This method was applied, e.g., to ascites (BEWLEY and HORNSEY, 1964), osteosarcoma (VAN PUTTEN *et al.*, 1971), leukaemia (BERRY and ANDREWS, 1963) cells, and others (see Figs. 7.2 and 7.16).

(ii) For normal tissues, there are only some special techniques that yield the survival curve completely: spleen colony method (CFU) by TILL and MCCULLOCH (1961) for haemopoietic tissue (see Fig. 7.16); melanocyte survival assay by hair depigmentation (POTTEN and CHASE, 1970).

(iii) At outer or inner tissue surfaces, after irradiation with sufficiently high doses, one can observe separate clones of cells having repopulated from single stem cell survivors. This yields the *relative* decrease of cell survival at growing dose with the *absolute* values of the surviving fraction left open, i.e. essentially one obtains the D_0-value describing the final slope. This so-called *clonal assay* was applied to skin (WITHERS, 1967; DENEKAMP *et al.*, 1971), gut (jejunal crypt stem cells, WITHERS and ELKIND, 1969; colonic crypt stem cells, WITHERS and MASON, 1974), growth cartilage (KEMBER, 1967), and seminiferous epithelium of testis (spermatogonial stem cells; WITHERS *et al.*, 1974a; DE RUITER-BOOTSMA *et al.*, 1974). A special technique now preferred to study stem cell survival in the intestinal mucosa rests on direct scoring of the number of surviving jejunal crypts (WITHERS and ELKIND, 1970).

(iv) As suggested by WAMBERSIE and DUTREIX (1974), it is possible to construct a single-cell survival curve from the doses required in *fractionated treatments* to produce an *isoeffect*. The latter may, which is an important feature of this method, concern any, even non-quantitative, endpoints such as gross tissue damage scored by observation or mortality, tumour growth delay or cure rate, etc. In each case one obtains the underlying stem-cell survival curve with an *arbitrary scale* for log S along the following lines of reasoning: if different n-fraction schedules ($n = 1, 2, \ldots$) with the doses D_n per fraction (total dose nD_n) yield the same effect (isoeffect) associated with the unknown rate of cell survival, S, one has the sequence $S(D_1) = S$, $S(D_2) = S^{1/2}, \ldots, S(D_n) = S^{1/n}$ of survival curve points, assuming that each dose fraction reduces the survival rate in the same proportion, i.e. $S = [S(D_n)]^n$. This implies a number of assumptions (above all: complete cellular recovery but no repopulation between fractions) which are discussed elsewhere (MASUDA and WITHERS, 1976). This method, which yields only the *shape* of survival curves (particularly the initial to final slope ratio and the D_q) by referring to a *relative* scale of log S, has been applied to mouse skin, intestine, jejunum, colon, lung, and human skin (WAMBERSIE and DUTREIX, 1976 — review; FOWLER *et al.*, 1974; DOUGLAS and FOWLER, 1976; MASUDA and WITHERS, 1976).

It is important to point out that the methods (iii) and (iv) yield incomplete results in *different* respects, so that their combination can give the total survival curve unambiguously.

E.g., a survival curve obtained by the iso-effect procedure (iv) becomes definitely scaled by adding nothing but the D_0-value determined with the clonal assay, provided that the latter refers to those stem cells which are known to be responsible for the gross effect evaluated according to (iv). This nececcary correspondence is well established for intestine and skin (Withers and Elkind, 1969; Denekamp and Field, 1974). The isoeffect procedure can also be applied directly to multifraction clonal-assay data, which yields the complete survival curve composed of segments that correspond to the dose ranges in which the relative survival was measured (Hethcote *et al.*, 1976).

In summary, relevant *in-vivo* survival curves, at least the D_0-values, are available for a number of tissues to allow theoretical predictions of the RBE of fast neutrons for gross tissue reactions.

On the other hand, this constitutes an important field of experimental RBE research, which has gained fundamental significance for neutron therapy, particularly since the RBE for dose-limiting normal tissue tolerance has proved to be clearly different for various critical tissues. For reviews see Broerse and Barendsen (1973) and Field and Hornsey (1974a). It is a challenge for the theory to correlate these differences with the pertinent cellular X-ray survival curves and so to open the way for calculations of direct use in treatment planning, particularly if complex situations of local RBE variations for specified tissues in real therapeutic beams including dose gradients, γ-contamination, and scattered neutrons are concerned.

7.4.7.2. Gross reactions of dose-limiting normal tissues

Normal tissue reactions have been studied by various methods. Death of mice as the endpoint with the RBE derived from the LD_{50}'s displays *bone-marrow damage* leading to the haemopoietic syndrome (30-day mortality after whole-body irradiation, $\sim$750 rad), *gut damage* leading to the intestinal syndrome, (5-day mortality after whole-body or abdominal irradiation, $\sim$1300 rad), *lung damage* leading to late pneumonitis (180-day mortality after irradiation of the thorax, $\sim$1200 rad), and *damage to the esophagus* leading to esophagitis (30-day mortality after irradiation of the thorax, $>$2500 rad). The doses indicated are roughly those X-ray doses with which, in a single exposure, these effects can be generated (LD_{50}). Other endpoints have been used to study early and late *skin reactions* (scaled visual observations, deformities, shrinkage of the area of transplanted skin pieces), damage to growing *cartilage* (stunting of tail vertebrae), and damage to *spinal cord* (myelopathy scored by muscle atrophy and paralysis of the hind legs).

In contrast to the derivation of RBE values from cell survival curves, most gross tissue reactions permit the RBE to be determined for single-fraction irradiation in only a narrow dose range adapted to the endpoint under study. Fortunately, these doses (see above) are large enough for the measured RBE values to be theoretically interpreted as 'final-slope' RBE's (according to section 7.4.5) that should obey the rule there stated: they should correlate with the pertinent *in-vivo* stem-cell D_0-values for X-rays. Indeed, as far as the latter are known by the above-mentioned *in-situ* techniques (CFU survival for the haemopoietic syndrome, jejunal crypt survival for the intestinal syndrome, epidermal and cartilage clone-forming cell survival for skin and cartilage damage), the theoretical final-slope RBE values for different neutron energies agree fairly well with the experimental single-fraction data. This comparison is presented in Table 7.8. It should be told that there is also good correspondence, in general, between these gross-reaction RBE's and RBE values measured by the corresponding cell-survival assays for comparably large doses,

Tab. 7.8. Single-dose RBE values relative to 250—300 kVp X-rays for various gross tissue reactions compared with theoretical results derived as 'final-slope' RBE's from pertinent stem-cell D_0-values (assumptions: $\sigma_0 = 50\ \mu m^2$, 5% γ-dose contamination)

Gross tissue reaction	Stem-cell	15-MeV neutrons		Cyclotron neutrons (spectrum D)		Fission neutrons (spectrum A)	
	D_0 [rad] for X-rays	RBE calcul.	RBE measured	RBE calcul.	RBE measured	RBE calcul.	RBE measured
Lung damage	?		1.2 (a)		1.5 (i)		
Haemopoietic syndrome	73 (a)[1])		1.1 (a)		1.5 (b, a)		1.9 (h, a)
	79[3])	1.29	1.29[4])	1.58	1.63[3])	2.17	2.07[3]) (d)
Skin reactions	135 (e, f, a)	1.44	1.4 (a)	1.72	1.7 (b, a)	2.39	
Cartilage damage	165 (g)	1.50		1.77	1.7 (b)	2.46	
Intestinal syndrome	167 (a)[2])	1.50	1.4 (a)	1.77	2.3 (b, a)	2.46	3.0 (h, a)
			1.5 (j)		1.75 (c)		2.5 (d)

Origin of experimental data: (a) BROERSE and BARENDSEN (1973); (b) FIELD and HORNSEY (1974a); (c) HORNSEY (1970), jejunum clonal assay; (d) GRAHN *et al.* (1972), relative to ^{60}Co γ-rays; (e) EMERY *et al.* (1970); (f) WITHERS (1967); (g) KEMBER (1967); (h) DAVIDS (1972); (i) FIELD and HORNSEY (1974b); (j) YAMAMOTO (1967).

[1]) Quite similar D_0-values were found in (d), (h), and by PHILLIPS *et al.* (1974).
[2]) Other workers found smaller D_0-values: 157 rad (d), 151 rad (h), 136 rad (WITHERS *et al.*, 1974b), 108 rad (WITHERS and ELKIND, 1970).
[3]) Corrected for X-ray dose deviation near bone (see page 239).
[4]) Corrected for both X-ray and neutron dose deviations near bone (see page 239).

as pointed out in a review paper by BROERSE and BARENDSEN (1973) and in numerous original reports.

The RBE is known to increase with decreasing dose, and the data in Table 7.8 must be considered as the limiting values for $D \to \infty$. This dose-variation of RBE is easily derived from cellular survival curves, and also the theory predicts it in this way. For gross tissue effects, on the other hand, increased RBE values are observed after *fractionated* exposure which requires larger *total* doses of both X-rays and neutrons to produce the invariable effect, but the higher RBE must be attributed to the smaller *dose per fraction*. It is the latter that determines the repeated momentary rate of cell killing to which the RBE refers. Therefore, assuming complete cellular recovery between fractions, one may consider a plot of RBE versus neutron dose per fraction as equivalent to an RBE-single-dose plot as would be obtained from cell survival curves.[1]) This reasoning suggested by FIELD is generally accepted: "The justification for this approach is (a) that the response of many tissues is qualitatively the same after both low and high LET radiations and (b) that any cellular proliferation between fractions occurs at the same rate, regardless of radiation ... Plotted in this way RBE is no longer correlated with the number of fractions, but only with the dose per fraction" (FIELD and HORNSEY, 1974a). The latter statement is of particular importance for neutron therapy to rationalize RBE data for fractionated treatments, but it should be noted that in order to compare with the present theory we have to made the more definite assumption of full identity with a cellular single-dose RBE.

[1]) On condition that the same schedule of fractionation was used for both radiations.

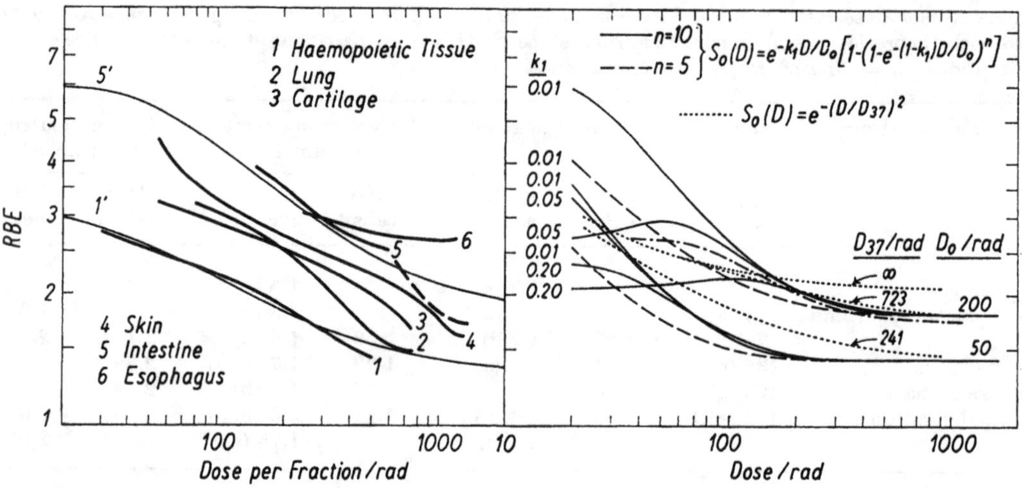

Fig. 7.24. RBE of cyclotron neutrons (16 MeV d on Be, spectrum D) relative to 250-kVp X-rays in dependence on the neutron dose.
Left-hand panel: experimental results for gross damage to various normal tissues (redrawn from Field and Hornsey, 1974a; dashed line for jejunal stem-cell survival assayed by clones (Hornsey, 1970)); for the thin lines see text.
Right-hand panel: theoretical curves corresponding to the analytical low-LET survival curve models indicated; dashed-dotted line for capillary endothelium cells (X-ray survival curve shown in Fig. 7.16, panel d). Assumptions: $\sigma_0 = 50\ \mu m^2$, 5% γ-dose contamination.

This crucial comparison is presented in Figure 7.24 for cyclotron neutrons. The theoretical RBE curves shown in the right-hand panel have been derived by the standard procedure applied to simple mathematical models of cellular X-ray survival curves[1]) rather than the partly-known tissue-specific ones, because the *general incompatibility* with the left-hand experimental curves is better demonstrated in this way. This disappointing result means that, whatever the correct cellular X-ray survival curves associated with the considered gross tissue reactions are, the theory is unable to yield the strong increase of RBE at decreasing dose found experimentally for all tissues[2]). It can be seen only a general qualitative correspondence with respect to the RBE-dose variation and possible inter-tissue variations, apart from the essentially correct theoretical values for the large single doses as listed in Table 7.8. From this momentous finding different conclusions may be drawn:

(a) The theory is erroneous substantially.
(b) The experimental RBE data are interfered with by different aspects of cell proliferation between fractions.
(c) Cellular recovery (repair of sublethal damage) is not complete between fractions.

It is true that the theory itself (assumption (a)) must be suspected regarding the tendency to predict a too-small RBE for small doses (high survival level) as pointed out in section 7.4.2. However, it is unlikely that this is the only reason, since the predictions were just very good for cell survival *in vivo* (see Figs. 7.16 and 7.18, right-hand panel), while the experimental RBE curves plotted in Figure 7.24 are incompatible with *any* curvature in the corresponding cellular survival curves for neutrons in respect of the pertinent X-ray survival curves (see below).

1) For the significance of these models see Hethcote *et al.* (1976).

2) The exceptionally high and flat curve for esophagus damage is assumed to be affected by hypoxia (Field and Hornsey, 1974a).

Assumption (b), to be sure, conflicts on principle with the above-mentioned 'dogma of dose per fraction only' which is warrantable just by suggesting that *no* differences exist in the kinetics of repopulation after neutron and X irradiation. On the other hand, such differences might still be compatible with that dogma but could make the theory fail when applied so simply to multifraction gross tissue reactions. This complicated problem cannot be in detail discussed here. E.g., it was argued (DENEKAMP and FIELD, 1974) that effects of synchronization by the first dose fractions as well as the time of 'recognition' of cell death after a fraction may be different for neutrons and X-rays, and a slower repopulation after a 14-fraction neutron exposure was found for mouse skin. Whatever the implications of these questions are for the therapy, a cellular theory of RBE can be used only if reaching the same endpoint with X-rays and neutrons implies that each of N equal fractions encounters the same kind of cell population, so that the same *rate* of cell depletion results from each fraction of either radiation. Only in this way a given dose per fraction defines an unambiguous level of cell survival from which a theoretical RBE can be derived. This prerequisite being always met can hardly be taken for granted. Otherwise, the present cellular theory will at best provide a crude approximation, which we will now show to be considerably improved by allowing for incomplete cellular repair of neutron damage.

Cellular repair (assumption (c)) of sublethal X-ray damage is usually assumed to be complete within few hours. For neutrons, however, the situation is somewhat obscure. The subject has already been discussed briefly at the end of section 7.4.4. The smaller amount of sublethal damage after neutrons makes the investigation of its repair difficult. With cultured cells, any split-dose effect was not always found (for a discussion see RAILTON *et al.*, 1975). A number of *in-vivo* data at the cellular and tissue levels indicate incomplete or no recovery from sublethal neutron damage, at least after more than 2 or 3 fractions (DENEKAMP and FIELD, 1974; WITHERS *et al.*, 1974b; WAMBERSIE *et al.*, 1977; HORNSEY *et al.*, 1977). This effect of neutron damage repair deficiency is likely to be the main key for understanding the discrepancies under discussion. Therefore, it seems worth while attempting a crude modification of the theory to make allowance for this effect.

Let us first stress the following observation. For some of the tissues occurring in Figure 7.24, the corresponding stem-cell survival curves for X-rays are available either fully (bone marrow, dashed line in Fig. 7.16, panel *c*) or in arbitrary units of logarithmic survival (intestine, skin, and lung (WAMBERSIE and DUTREIX, 1976; HETHCOTE *et al.*, 1976)). Without any theoretical reasoning, one could design the appropriate neutron survival curves that would yield the RBE-dose functions measured. This yields an interesting result: one has to take *straight lines*, i.e. sheer exponential dose-survival relations, to account for the experimental RBE. This empirical fact can easily be interpreted as the extremest consequence of lacking repair: if we have a *large* number N of dose fractions, most of them (save for the first ones) encounter predamaged cells, so that the *average* reduction of survival per fraction is no longer determined by the sigmoidal dose-survival relation, but approaches, as $N \to \infty$, a value corresponding to the final slope: $\exp(-D_{f,n}/D_{0,n})$, $D_{f,n}$ being the neutron dose per fraction.

From this rationale, it follows a very simple procedure of predicting the RBE-dose relation for gross tissue reactions after fractionated treatments. Rather than calculating the total single-dose survival curve for neutrons from the corresponding X-ray curve, one calculates only its final slope, $D_{0,n}^{-1}$, from the X-ray final slope, $D_{0,X}^{-1}$, and takes

$$S_n(D) = e^{-D/D_{0,n}} \tag{7.29}$$

as the 'virtual' neutron survival curve, from which the RBE follows in the usual way: $S_n(D_n) = S_X(D_X)$. $D_{0,n}$ is only related to $D_{0,X}$ by

$$D_{0,n} = D_{0,X}/RBE_0, \tag{7.30}$$

the 'final-slope' RBE_0 being listed in Table 7.5. Hence, the theory necessary within this approach reduces to nothing but the data in Table 7.5.

The thin lines in the left-hand panel of Figure 7.24 have been calculated in this way for bone-marrow (1') and jejunal crypt (5') stem cells[1]). They agree surprisingly well with the corresponding experimental gross-reaction RBE curves, demonstrating firstly that an exponential relation like (7.29) for neutrons is appropriate, and secondly that the value $D_{0,n}$ to be chosen is well delivered by the theory.

Of course, the general validity of this simple procedure may be suspected for a number of reasons. In view of the extreme assumptions (absence of any cellular repair of neutron damage, limit $N \to \infty$) it ought to overestimate the RBE. Hence, its success may be partly fortuitous by masking the interference with the other factors discussed above. Moreover, it cannot be excluded that the X-ray survival curves used have a too short shoulder, which can be supposed to be a consequence of the isoeffect method (see (iv) in subsection 7.4.7.1) in the case of intestine.

[1]) The X-ray survival curves used are that shown in Figure 7.16 (panel *c*, dashed line, which refers to the *nominal* X-ray dose) for haemopoietic stem cells, and a curve for jejunal crypt cell survival published by Broerse and Barendsen (1972). The latter, which is based on measured crypt survival for large doses and an extrapolation down to zero dose according to fractionation data, matches exactly the shape of the curve given by Wambersie and Dutreix (1976) using the isoeffect method (intestinal syndrome).

On the other hand, a more satisfactory theory of the RBE for normal tissue tolerance will certainly have to take into account complex processes at the tissue level, and will therefore hardly be able to give quantitative predictions in a simple way. So the algorithm (7.29), (7.30) may be recommended as a *tentative* tool for calculating normal tissue RBE data relevant to real therapy beams of neutrons (including local variations) and multi-fraction treatment schedules.

A special advantage of this approach must be emphasized: it is not necessary to possess a cellular survival curve for X-rays from the outset. Predictions for *other* neutron energy spectra can be based direct upon a known RBE-dose relation for a *given* neutron energy spectrum in the following way. Let the latter relation be denoted by $BRE^{(n)}(D)$ and realize that only the underlying virtual survival curve for neutrons changes with energy, i.e. $D_{0,n}$ in Eq. (7.29) becomes $D_{0,n'}$. At a constant survival level, this implies an alteration of both the RBE by the factor $D_{0,n}/D_{0,n'}$ and the corresponding neutron dose by the factor $D_{0,n'}/D_{0,n}$. This yields the transformation law

$$RBE^{(n')}(D) = \frac{D_{0,n}}{D_{0,n'}}\, RBE^{(n)}\left(\frac{D_{0,n}}{D_{0,n'}}\, D\right), \tag{7.31}$$

which allows us to calculate the RBE for another neutron field n'. First, however, $D_{0,X}$ must be derived from $RBE_0{}^{(n)} = RBE^{(n)}(\infty)$ by 'inverse' final-slope theory (Tab. 7.5 and context), then final-slope theory yields both $D_{0,n}$ and $D_{0,n'}$.

7.4.7.3. Tumour reactions

Finally, let us briefly discuss some main aspects of RBE for radiation responses of tumours. This is a very complex subject. Let us begin with the fact that many tumours are partly hypoxic, which was the main reason to anticipate a therapeutic gain from using neutrons in radiotherapy because of their larger RBE under anoxic conditions. This is also reflected by the present theory, but any theoretical modelling has to consider that only part of the cells are actually anoxic, so that there are at least two cell compartments (oxic and anoxic) to be distinguished.[1]) This can lead to an *increase* of RBE at increasing dose. The following example is to illustrate this feature and to demonstrate the corresponding theoretical description.

The survival curves shown in the upper panel of Figure 7.25 have been obtained by irradiating a solid tumour, rat rhabdomyosarcoma, *in situ* and subsequent cell cloning *in vitro* (Barendsen and Broerse, 1969). The most shallow curve (heavy, dashed) is for anoxic conditions by exposing dead rats.[2]) Under normal conditions, the X-ray curve becomes bi-phasic, reflecting the presence of oxic and anoxic cells.[3]) Accordingly, the corresponding smoth curve (thin, dashed) has been constructed by starting with the one fitted to the anoxic data, deriving an oxic curve of the same shape by means of the $OER = 2.4$[4]), and superposing both curves multiplied by the percentages 15% (anoxic) and 85% (oxic). This has yielded both an acceptable fit to the experimental X-ray points and the base for the theory to predict the neutron curve. The latter (unbroken line) has been obtained by an analogous superposition of curves calculated separately in the usual way for oxic and anoxic conditions (different r-values). This predicted composite neutron curve shows no bi-phasic character, in concordance with the experimental findings, and in consequence of the larger RBE for the anoxic part (i.e. small OER of neutrons).

The lower panel of Figure 7.25 shows the RBE-neutron-dose function derived from the theoretical survival curves. The increase is due to the transition from oxic to anoxic conditions.

The above example refers to the cellular level. Let us now take up implications concerning gross reactions of the whole tumour (mainly tumour cure and growth delay). This complicated problem is best discussed step by step, beginning with the simplest assumption that the reaction considered, say tumour eradication, depends only on the reduction of viable cells to a very small fraction. For the experimental rat rhabdomyosarcoma this direct relation was indeed demonstrated: 90 per cent cure rate was attained at the surviving fraction 10^{-9} (6500 rad X-rays) corresponding to 0.1 survivors per tumour (Barendsen and Broerse, 1969). Therefore, regarding *single-dose* treatments with either X-rays or neutrons, an RBE for tumour cure as high as the plateau in Figure 7.25 can be expected. The ratio of this anoxic RBE to oxic normal-tissue RBE (cf. Tab. 7.8) defines what has been called the 'therapeutic gain factor', but this constitutes only a maximum estimation which does not apply to fractionated treatments.

Concerning the RBE in the case of fractionation, let us first disregard reoxygenation and repopulation between fractions. Nevertheless, unlike normal tissues, the RBE curve in Figure 7.25 cannot be looked upon as describing RBE as a function of dose per fraction, because most of the fractions are spent on killing *anoxic* cells. Therefore, the RBE-dose-per-fraction relation approaches the one that could be derived from *purely* anoxic cell survival curves, i.e. the RBE becomes a

---

[1]) In reality, hypoxia may be gradual.

[2]) Cf. Figure 7.16, panel *b*, where this curve is also shown together with the derived theoretical neutron curve matching the corresponding experimental points.

[3]) Similar bi-phasic curves were found for cells grown as spheroids *in vitro* if the spheroid was large enough to develop hypoxia in the core (Durand and Sutherland, 1973; Lücke-Huhle *et al.*, 1977).

[4]) This value was found by *in-vitro* irradiation of rhabdomyosarcoma cells in air and nitrogen atmopheres (Barendsen and Broerse, 1969).

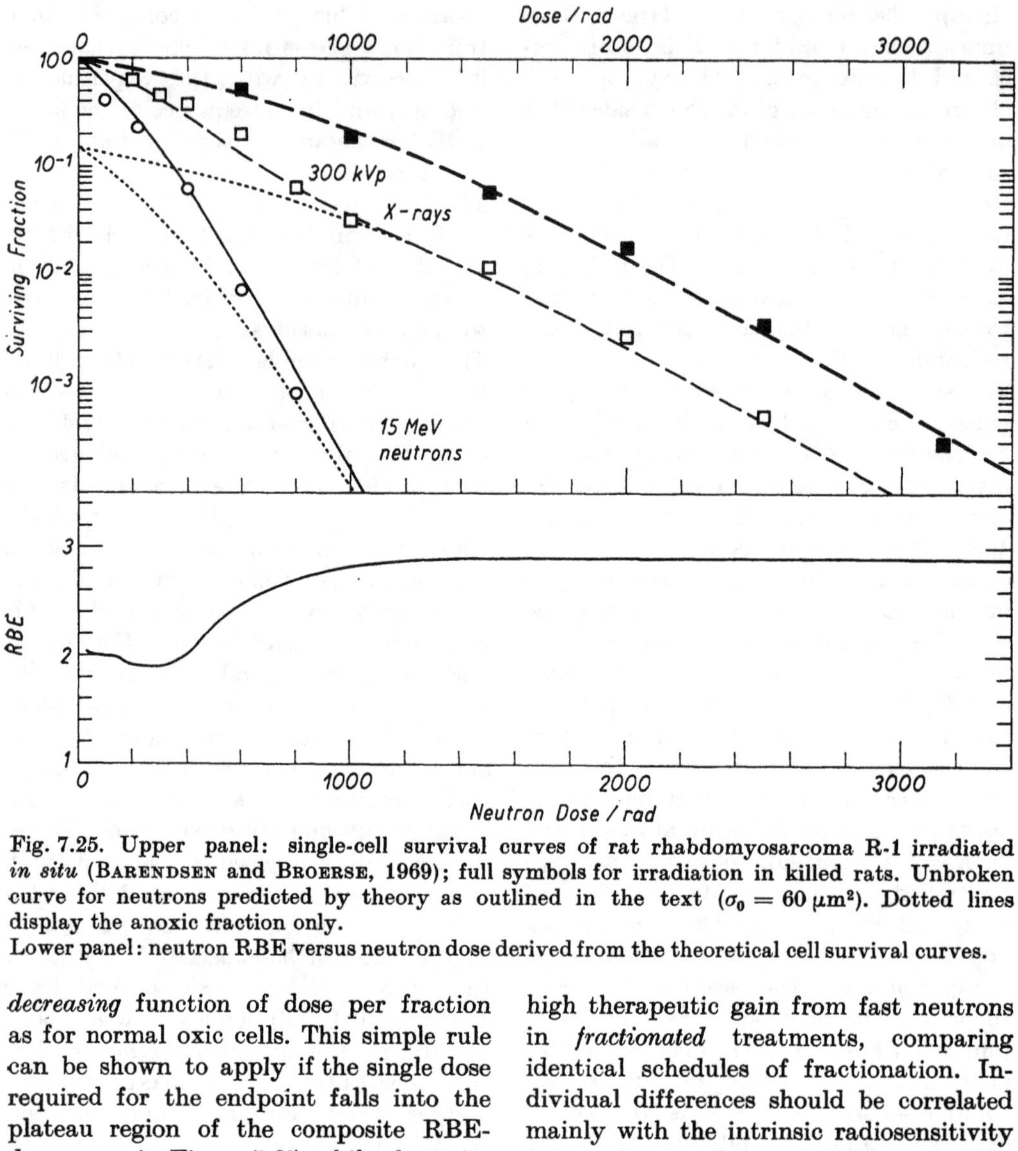

Fig. 7.25. Upper panel: single-cell survival curves of rat rhabdomyosarcoma R-1 irradiated *in situ* (BARENDSEN and BROERSE, 1969); full symbols for irradiation in killed rats. Unbroken curve for neutrons predicted by theory as outlined in the text ($\sigma_0 = 60\ \mu m^2$). Dotted lines display the anoxic fraction only.
Lower panel: neutron RBE versus neutron dose derived from the theoretical cell survival curves.

decreasing function of dose per fraction as for normal oxic cells. This simple rule can be shown to apply if the single dose required for the endpoint falls into the plateau region of the composite RBE-dose curve in Figure 7.25, while the strict formalism to be applied reads (h = proportion of anoxic cells, N = number of fractions, D_f = dose per fraction, subscripts X and n referring to X-rays and neutrons):

$$(1 - h)\,[S_n^{(\mathrm{ox})}(D_{f,n})]^N + h[S_n^{(\mathrm{anox})}(D_{f,n})]^N = (1 - h)\,[S_X^{(\mathrm{ox})}(D_{f,X})]^N + h[S_X^{(\mathrm{anox})}(D_{f,X})]^N, \tag{7.32}$$

and the $RBE = D_{f,X}/D_{f,n}$ becomes dependent on both $D_{f,n}$ *and* N.

This reasoning, under the above (unrealistic) conditions, indicates a similarly high therapeutic gain from fast neutrons in *fractionated* treatments, comparing identical schedules of fractionation. Individual differences should be correlated mainly with the intrinsic radiosensitivity (survival curve shape) of the dose-limiting normal tissue as well as the particular tumour, and in this respect the theory could eventually be useful along the lines outlined.

However, all this theorizing is far from reality because the tumour RBE (and thus the effective gain factor) is actually interfered with by too many variable factors such as reoxygenation and repopulation between fractions and presence of several sorts of cells in a tumour, e.g., proliferating and non-proliferating ones in both the oxic and anoxic compartments (BARENDSEN, 1974). They are likely, on the one hand, to differ in radiosensitivity and, on the other hand, in the repopulation kinetics of survivors, which may give rise to a disproportional participation in a possible regrowth of the tumour. The present theory is obviously unable to allow for these complex things, inasmuch as its validity for mixed cell populations regarding the *total* survival may not any longer be a sufficient prerequisite.

The latter complication does certainly not arise if the endpoint under study requires the total depletion of *all* cells as in the case of curing the rat rhabdomyosarcoma, provided that the cellular survival curve available (Fig. 7.25) is representative for the total population. Hence, the tumour-cure RBE (single dose) is equal to the low-survival RBE in this case. By contrast, the growth-delay RBE, which was also measured for that tumour as a function of single dose, reflects only the general pattern of the RBE curve in Figure 7.25 but is significantly higher. This might be attributed to differences in the post-irradiation kinetics of the constituent cells.

Reoxygenation ot tumours during protracted treatments is the main reason for the effective gain factor to be lowered due to the enhanced efficiency of X-rays. For the sake of illustration, let us again refer to the example considered in Figure 7.25. Assume that the reoxygenation in the suitably chosen intervals between fractions would always restore the initial proportion of oxic cells (85%). Then, with 250 rad neutron dose per fraction, the RBE would fall into the minimum of the curve, and the advantage of neutrons might be lost. Actually, the kinetics of reoxygenation varies from tumour to tumour, which would require individually optimalized fractionation regimens to make X-rays as effective as possible. Since there are no ways to devise such 'tailor-made' schemes for human tumours, an *average* therapeutic gain from using neutrons can be expected owing to the minor influence of fractionation. This point of view has been stressed by BARENDSEN (1974), where a more detailed discussion of these questions is given.

It is a most important problem in neutron therapy to find out ways to predict *which* tumours are better controlled by fast neutrons. Although this is essentially an RBE problem (tumour RBE versus dose-limiting normal tissue RBE), the present theory alone can little help. The intrinsic tumour RBE (apart from hypoxia) certainly varies at the cellular level[1]), but the presence of different cell compartments hinders the theory from giving the influence of this factor quantitatively on the basis of some overall cellular radiosensitivity. And on the other hand, the varying kinetics of reoxygenation is a major determinant particularly for the effectiveness of a given fractionation regimen of the X-rays to compare with. *Both* factors are obviously responsible for the fact that experimental RBE values for local tumour control and their dependence on the dose per fraction show considerable variations between types of tumours, in some cases falling below the curves for typical normal tissues (i.e. gain factor less than unity). A compilation and critical discussion of such data was given by FOWLER (1977), who stressed the importance of the intrinsic cellular radiosensitivity in regard to RBE, but more insight into the specific mechanisms at the tissue level for different tumours is required before quantitative predictions can be made.

[1]) A special experimental investigation by BROERSE *et al.* (1977) has demonstrated the existence of large variations between the survival curves (shoulder width as well as D_0) of cells from different carcinoma and sarcoma after X-rays *in vitro*, and these variations are smaller after neutrons. The present theory standard procedure, using $\sigma_0 = 100\ \mu m^2$, yields quantitative agreement with these results.

Chapter 8

Re-formulation of the theory

8.1. Introduction

The purpose of this chapter is to bring the theory of biological radiation effects at the cellular level into a convenient comprehensive form that applies to both the microdosimetric version of chapter 2 and the 'strict' version of chapter 3. Finally we will be left with simple formulas containing only parameters which are tabulated in the appendix. These parameters are quantities known from customary microdosimetry, such as $\bar{y}_F$, $\bar{y}_D$, and other ones; and the strict version of theory differs in nothing but modifications of these quantities, denoted by bold type characters and termed *virtual microdosimetric quantities*.

Furthermore, it is important that this formalution of the theory makes it possible to replace the tabulated data optionally by data of other origin, such as microdosimetric measurements or computations and measurements of DNA-lesion production rates.

In addition to this methodical side, this chapter illuminates the intrinsic skeleton of the theory and elucidates how both versions are related to one another and what the shortcomings of the microdosimetric version are.

In section 8.2 *virtual microdosimetry* is established as *a new general concept* which can be characterized as follows: the specific energy z of conventional microdosimetry is actually a measure of ionization production power, while the *virtual* specific energy $\mathbf{z}$ is a measure of lesion production power. It must be defined for a particular type of molecular lesions described by the totality of r-values (relative effectiveness of lesion production) for a sufficiently extensive variety of radiations. The strict theory of cellular radiation effects can be expressed in terms of virtual microdosimetry in total analogy to the theory in chapter 2. Thereby, our target model of DNA-lesion production is seen to play an auxiliary role only; it is not essential conceptually. Thus, this new concept is felt to be of general importance for theoretical radiation biophysics, apart from its present numerical evaluation which refers mainly to the DNA double-strand breaks set forth quantitatively in chapter 7, to make the present theory of mammalian cell killing feasible easily.

The concluding formulation of the theory of cellular radiation effects by means of

computationally simple equations, which are in part of insignificantly approximate character with general validity, is given in section 8.3 for photon, neutron, and ion radiations. Emphasis is placed on the theoretical background in tight connection with section 8.2. Therefore, readers who are interested only in practical applications are referred to chapter 9 where the ultimate formulas and corresponding instructions are laid down in compendious form.
Section 8.4 serves a special purpose; it is based on the so-called *LET-approximation* — a particularly simple version of the theory in the case of monoenergetic ion radiation. As a result of section 8.3, this approximation proves very good for mammalian cells; in it LET_∞ and r are the only radiation quality parameters that have remained. This has made it possible to give an RBE theory in explicit form (RBE-LET relation), which most clearly expresses the interplay of first and second kind effects and, on the other hand, could be useful for estimations in high-LET radiation protection.

8.2. Virtual microdosimetry

8.2.1. Formal considerations

The approximate microdosimetric version and the so-called 'strict' version of the general theory of cellular radiation effects have been dealt with in detail in chapter 2 and chapter 3, respectively. The final results of both versions can easily be brought into a simple unitary form. In any case, the dose-effect relation reads

$$S(D) = \sum_{\mu=1}^{n} B_\mu \, e^{-\alpha_\mu D}, \qquad (8.1)$$

with the coefficients B_μ coming from the effect-probability expression (2.10). The coefficients α_μ are the intrinsic subject of theory which relates them to the parameters Z_μ of the ansatz (2.10) and the DNA content Y. This relation, given by Eqs. (3.104 to 108) or Eq. (2.14), depends on the radiation considered and the size of the gross sensitive volume. Furthermore, the DNA-lesion type responsible for the effect enters this relation, but in a different way: in the microdosimetric version of theory the radiation-dependent mean rate of lesion production s (number of lesions per unit dose and unit molecular weight of DNA) occurs only, while the strict version depends explicitly on the parameters λ_j, l_l of the lesion production model. Let us now introduce a convenient common notation for this important relation concerning α_μ.
For the independent variable we take the combination

$$\alpha_\mu^{(0)} = s_0 Y(1 - e^{-Z_\mu}), \qquad (8.2)$$

which absorbs the parameter Y. s_0 refers to ^{60}Co radiation. We write

$$\alpha_\mu = A(r\alpha_\mu^{(0)}) \qquad (8.3)$$

with $r = s/s_0$ (relative effectiveness of primary lesion production), thus defining a function $A(\eta)$ that contains the whole theory in either case.[1])
In the *microdosimetric version* of the theory, the single event size distribution $f_1(z)$ alone determines the function $A(\eta)$ (cf. Eq. (2.17)):

$$A(\eta) = \frac{1}{\bar{z}_F} \int_0^\infty dz(1 - e^{-\eta z}) f_1(z). \qquad (8.4)$$

In the introduction to chapter 5, it has been shown how the expression (8.4) can be calculated for any radiation by means

---

[1]) This choice of $\alpha_\mu^{(0)}$ and the argument $\eta = r\alpha_\mu^{(0)}$ of the function A is appropriate because low-LET radiation yields $A(\eta) \approx \eta$ (see below), which means that $\alpha_\mu^{(0)}$ can be directly obtained from a representation like (8.1) of a measured low-LET dose-effect relation. In particular, for ^{60}Co γ-rays, one has $r = 1$ by definition, yielding $\alpha_\mu^{(0)} = \alpha_\mu$. Hence, the origin (8.2) of $\alpha_\mu^{(0)}$ can be forgotten for the formal employment of the theory: as it were, the $\alpha_\mu^{(0)}$ are nothing but experimental parameters.

of our general approach to energy absorption stochastics: the equivalent function $\Phi(\zeta)$ defined by Eq. (5.8) is given explicitly by Eqs. (5.3 to 7). Hence

$$A(\eta) = \Phi\left(\frac{W}{m}\,\eta\right). \tag{8.5}$$

The function $A(\eta)$ obviously has the properties

$$A(0) = 0, \quad \left.\frac{\mathrm{d}A}{\mathrm{d}\eta}\right|_{\eta=0} = A'(0) = 1, \tag{8.6}$$

and the microdosimetric mean values $\bar{z}_F$ and $\bar{z}_D$ are implied in it according to

$$\bar{z}_F = \frac{1}{A(\infty)}, \quad \bar{z}_D = -A''(0). \tag{8.7}$$

These relations, in terms of $\Phi(\zeta)$, have been considered and evaluated in section 5.2.1.

The *strict version* of the theory yields $\alpha_\mu = \Pi(Z_\mu)$ given by Eqs. (3.104 to 108). The function $A(\eta)$ is easily derived from this, realizing that Π depends actually on the variable x introduced already by Eq. (3.115):

$$x = (1 - \mathrm{e}^{-Z})\frac{h}{\varrho} = \frac{\pi}{4}\frac{Yl_l}{\varrho V}(1 - \mathrm{e}^{-Z}). \tag{8.8}$$

Henceforth Π will be considered as a function of x. In view of the definitions (8.2) and (8.3) this means

$$\boldsymbol{A}(\eta) = \Pi\left(\frac{\pi}{4}\frac{l_l}{\varrho V s}\,\eta\right). \tag{8.9}$$

The bold type character is to distinguish this $\boldsymbol{A}(\eta)$ from the microdosimetric one given by the analogous equation (8.5).

The common formulation of both versions of theory established in terms of $A(\eta)$ makes it possible to define a 'virtual microdosimetry' by means of analogy relations. This can be done in a more formal way which is outlined in the remainder of this subsection, while the following subsection is devoted to the investigation of the intrinsic meaning of this virtual microdosimetry, so elucidating the substantial differences between both versions of theory.

First we can state that the properties (8.6) hold also with respect to $\boldsymbol{A}(\eta)$:

$$\boldsymbol{A}(0) = 0, \quad \boldsymbol{A}'(0) = 1. \tag{8.10}$$

This is ascertained by the general relation (3.115) proved in section 3.7.1. Therefore, it is possible to define virtual microdosimetric mean values simply by analogy to Eqs. (8.7):

$$\bar{\boldsymbol{z}}_F = \frac{1}{\boldsymbol{A}(\infty)}, \quad \bar{\boldsymbol{z}}_D = -\boldsymbol{A}''(0). \tag{8.11}$$

According to Eq. (8.9), these quantities can be calculated numerically from $\Pi(x)$ as can the microdosimetric counterparts from $\Phi(\zeta)$ (chapter 5). The tabulation of either $\bar{z}_D$ would be sufficient for the practical application of the theory in cases where the Taylor series expansions

$$\begin{aligned} A(\eta) &= \eta - \frac{1}{2}\bar{z}_D\eta^2, \\ \boldsymbol{A}(\eta) &= \eta - \frac{1}{2}\bar{\boldsymbol{z}}_D\eta^2 \end{aligned} \tag{8.12}$$

constitute a sufficient approximation (low LET). In general, one can imagine both $A(\eta)$ and $\boldsymbol{A}(\eta)$ to be represented by an infinite Taylor series, the coefficients of which are related to the moments of the single event size distributions $f_1(z)$ and $\boldsymbol{f}_1(\boldsymbol{z})$, respectively. In other words, we can *define* a virtual single event size distribution by

$$\boldsymbol{A}(\eta) = \frac{1}{\bar{\boldsymbol{z}}_F}\int_0^\infty \mathrm{d}\boldsymbol{z}(1 - \mathrm{e}^{-\eta\boldsymbol{z}})\,\boldsymbol{f}_1(\boldsymbol{z}) \tag{8.13}$$

in analogy to Eq. (8.4). This possibility rests on nothing but the relations (8.10). This definition being more than a formal play is demonstrated in the following subsection.

An important point has not yet been mentioned: the influence of the DNA-lesion type. In the microdosimetric version, the DNA-lesion type enters the theory only through the radiation-dependent parameter r in Eq. (8.3). $A(\eta)$ does no longer depend on it. By contrast, $\boldsymbol{A}(\eta)$ given by Eq. (8.9) still depends on the parameters of the lesion production model entering $\Pi(x)$, *in addition* to r that occurs in Eq. (8.3). Thus, there does not exist a virtual microdosimetry as such. One can define it only for a particular lesion type described by definite parameters λ_j, l_l. That is why the general strict theory could not from the outset be based on the concept of virtual microdosimetry. The results of chapter 7, however, where DNA double-strand breaks described by certain parameters λ_j, l_l were acknowledged to be alone responsible for mammalian cell killing, make it now reasonable and useful to define a corresponding virtual microdosimetry and to calculate and tabulate pertinent parameters that enable $\boldsymbol{A}(\eta)$ as well as $A(\eta)$ to be practically computed in simple ways for different radiations.

8.2.2. The general concept of virtual microdosimetry

Both versions of the theory of cellular radiation effects are based on the fundamental equation

$$S(D) = \sum_{k=0}^{\infty} H(k)\, U_k(D). \tag{8.14}$$

The approximate character of the microdosimetric approach (cf. Eq. (2.8))

$$U_k(D) = \int_0^{\infty} \mathrm{d}z\, \frac{(Ys_0rz)^k}{k!}\, \mathrm{e}^{-Ys_0rz} f(z; D) \tag{8.15}$$

consists in the *undifferentiated* valuation of the specific energy z by the single factor r being defined for the *total* radiation in question, while it should actually depend on the particular incident particles that produce z. This differentiation, which will be shown to lead to the strict theory, can however not be introduced directly into the expression (8.15), since z results from *several different* absorption events (different particles). This is only possible with respect to V_k, the lesion number distribution for a single absorption event. Therefore one has first to reduce $U_k(D)$ to V_k. This was just the first step in the derivation of the strict theory in section 3.6, which we repeat here in a somewhat modified manner to stress the analogy to the formalism (2.7 to 2.17). The compound Poisson process leading to $U_k(D)$ can be described by

$$U_k(D) = \sum_{n=0}^{\infty} \frac{(D/\bar{z}_F)^n}{n!}\, \mathrm{e}^{-D/\bar{z}_F} V_k^{(n)}, \tag{8.16}$$

where $V_k^{(n)}$ denotes the probability that exactly n absorption events engender k lesions. $V_k^{(n)}$ follows from $V_k^{(1)} \equiv V_k$ by multiple folding:

$$V_k^{(n+1)} = \sum_{k'=0}^{k} V_{k-k'}^{(n)} V_{k'}. \tag{8.17}$$

From this recurrence formula it is easy to derive

$$\sum_{k=0}^{\infty} \mathrm{e}^{-Zk} V_k^{(n)} = \left(\sum_{k=0}^{\infty} \mathrm{e}^{-Zk} V_k\right)^n \tag{8.18}$$

and

$$\sum_{k=0}^{\infty} \mathrm{e}^{-Zk} U_k(D) = \exp\left[-\frac{D}{\bar{z}_F} \sum_{k=0}^{\infty} (1 - \mathrm{e}^{-Zk})\, V_k\right]. \tag{8.19}$$

Only this kind of sums with different Z are needed for the evaluation of the fundamental equation (8.14) by virtue of the representation (2.10) for $H(k)$. In this way, one arrives at the well-known expression (8.1) with

$$\alpha_\mu = \frac{1}{\bar{z}_F} \sum_{k=0}^{\infty} (1 - \mathrm{e}^{-Z_\mu k})\, V_k = \frac{1}{\bar{z}_F} \sum_{k=0}^{\infty} \left[1 - \left(1 - \frac{\alpha_\mu^{(0)}}{Ys_0}\right)^k\right] V_k, \tag{8.20}$$

introducing $\alpha_\mu^{(0)}$ according to Eq. (8.2). Eq. (8.20) is the most general form of (8.3) without any suppositions.

Now we can derive the strict theory as well as the microdosimetric approximation, both from Eq. (8.20). The ansatz

$$V_k = \int_0^\infty \mathrm{d}z \frac{(Ys_0rz)^k}{k!} \mathrm{e}^{-Ys_0rz} f_1(z) \qquad (8.21)$$

is equivalent to the ansatz (8.15); it yields, straightforwardly, the microdosimetric version of the theory (cf. Eqs. (8.3), (8.4)):

$$\alpha_\mu = \frac{1}{\bar{z}_F} \int_0^\infty \mathrm{d}z (1 - \mathrm{e}^{-r\alpha_\mu^{(0)} z}) f_1(z). \qquad (8.22)$$

Let us now modify (8.21) so as to consider the crucial point mentioned above. There may be quite different absorption events associated with different incident particles involved in the radiation field. $f_1(z)$ describes them altogether, but it can also be accordingly decomposed by writing

$$f_1(z) = \sum_a \mu_F^{(a)} f_1^{(a)}(z), \qquad \sum_a \mu_F^{(a)} = 1. \qquad (8.23)$$

The superscript a specifies the different absorption events not only with respect to the kind of the incident particle (e.g. recoil protons, α-particles, and heavy nuclei, in a fast neutron radiation field) but also with respect to its incident energy and further parameters if wanted. Thus, the imaginary decomposition (8.23) may be thought to be arbitrarily fine and the number of components may approach infinity.

To each component a one has to ascribe a particular value, r_a, of the relative effectiveness of lesion production. These *partial r-values* would be measurable quantities if it were possible to realize radiation fields that produce absorption events of one kind a only. To give an example: irradiation with high-LET ions engenders direct and indirect events. It is obvious that the r_a-values for indirect events, which are caused by δ-electrons, must be smaller than the total r, and r_a for direct events must be larger than the total r, if we are dealing with DNA double-strand breaks, while the opposite is true of single-strand breaks. This concept leads to the virtual microdosimetry which is based on a *re-valuation* of absorbed energy by virtue of relative partial r-values r_a/r. As it were, energy is replaced by relative lesion numbers.

Before completing the formalism, we must consider a subtlety which is important only for the sake of consistency. The conventional microdosimetry as dealt with in chapter 5 is based, strictly speaking, on ionization numbers rather than absorbed energy. This means that 'zero events' which produce no ionizations have been disregarded. To apply the corresponding numerical results to the microdosimetric theory of biological effects means understanding $f_1(z)$ and $\bar{z}_F$ in Eq. (8.22) in this manner. The strict theory as derived in section 3.6, however, corresponds to the energy concept without any events omitted. Therefore, the re-formulation of the strict theory in terms of virtual microdosimetry must start from the modified expression

$$\tilde{f}_1(z) = \mu_F^{(0)} \delta(z) + \sum_a (1 - \mu_F^{(0)}) \mu_F^{(a)} f_1^{(a)}(z), ^{1)} \qquad (8.24)$$

where $\mu_F^{(0)}$ denotes the total fraction of zero events, while $\mu_F^{(a)}$ and $f_1^{(a)}(z)$ continue to obey Eq. (8.23) referring to the ionization number concept, i.e.

$$\bar{z}_F = \int_0^\infty \mathrm{d}z z f_1(z) = \sum_a \mu_F^{(a)} \bar{z}_F^{(a)}. \qquad (8.25)$$

For the strict theory, however, $\bar{z}_F$ in Eq. (8.20) has to be replaced by $\bar{\bar{z}}_F$ defined by

$$\bar{\bar{z}}_F = \int_0^\infty \mathrm{d}z z \tilde{f}_1(z) = (1 - \mu_F^{(0)}) \bar{z}_F. \qquad (8.26)$$

[1]) Note that $\delta(z)$, exactly, means $\delta(z - \varepsilon)$ with a very small but non-vanishing ε. This must be kept in mind to understand what follows, especially the final result: Eq. (8.30).

$\bar{\bar{z}}_F^{-1}$ counts every event and is therefore larger than $\bar{z}_F^{-1}$.

Hence, the strict theory follows from generalizing Eq. (8.21) by means of partial r-values:

$$\begin{aligned} V_k = {} & \mu_F^{(0)}\delta_{k0} \\ & + \sum_a (1-\mu_F^{(0)})\mu_F^{(a)} \int_0^\infty \mathrm{d}z\, \frac{(Ys_0 r_a z)^k}{k!} \\ & \times \mathrm{e}^{-Ys_0 r_a z} f_1^{(a)}(z), \end{aligned} \tag{8.27}$$

which yields

$$\begin{aligned} \alpha_\mu = {} & \frac{1}{\bar{\bar{z}}_F} \sum_a (1-\mu_F^{(0)})\,\mu_F^{(a)} \\ & \times \int_0^\infty \mathrm{d}z(1-\mathrm{e}^{-r_a \alpha_\mu^{(0)} z})\, f_1^{(a)}(z). \end{aligned} \tag{8.28}$$

It is obvious how this result is related to the formal representation (8.3), (8.13): a simple transformation of the integration variable z allows us to write

$$\begin{aligned} \alpha_\mu &= \boldsymbol{A}(r\alpha_\mu^{(0)}), \\ \boldsymbol{A}(\eta) &= \frac{1}{\bar{\boldsymbol{z}}_F} \int_0^\infty \mathrm{d}\boldsymbol{z}(1-\mathrm{e}^{-\eta \boldsymbol{z}})\, \boldsymbol{f}_1(\boldsymbol{z}) \end{aligned} \tag{8.29}$$

by defining

$$\begin{aligned} \boldsymbol{f}_1(\boldsymbol{z}) = {} & \mu_F^{(0)}\delta(\boldsymbol{z}) \\ & + \sum_a (1-\mu_F^{(0)})\,\mu_F^{(a)} \frac{r}{r_a} f_1^{(a)}\left(\frac{r}{r_a}\boldsymbol{z}\right), \end{aligned} \tag{8.30}$$

$$\bar{\boldsymbol{z}}_F = \bar{\bar{z}}_F = (1-\mu_F^{(0)})\,\bar{z}_F. \tag{8.31}$$

$\boldsymbol{f}_1(\boldsymbol{z})$ is normalized and its mean value is indeed $\bar{\boldsymbol{z}}_F$, if the self-evident relation

$$r = \sum_a \mu_D^{(a)} r_a, \qquad \mu_D^{(a)} = \frac{\bar{z}_F^{(a)}}{\bar{z}_F}\,\mu_F^{(a)} \tag{8.32}$$

holds. $\mu_D^{(a)}$ denotes the *dose fraction* of the component a, so that relation (8.32) can be looked on as just the only reasonable *definition* of the total r.

Eq. (8.30) defines virtual microdosimetry. It constitutes a theoretical concept in which physical energy is replaced by a virtual energy that is a measure of lesion production power according to the factors r_a/r being different for different individual absorption events even if the physical specific energy is the same. This virtual energy is defined so as to leave the total frequency-mean $\bar{\boldsymbol{z}}_F$ unchanged ($\bar{\boldsymbol{z}}_F = \bar{\bar{z}}_F$ referring to the rigorous energy concept). The 'dose-mean' $\bar{\boldsymbol{z}}_D$, however, changes:

$$\begin{aligned} \bar{\boldsymbol{z}}_D &= \frac{1}{\bar{\boldsymbol{z}}_F} \int_0^\infty \mathrm{d}\boldsymbol{z}\boldsymbol{z}^2 \boldsymbol{f}_1(\boldsymbol{z}) \\ &= \sum_a \frac{\bar{z}_F^{(a)}}{\bar{z}_F}\,\mu_F^{(a)} \left(\frac{r_a}{r}\right)^2 \\ &\quad \times \frac{1}{\bar{z}_F^{(a)}} \int_0^\infty \mathrm{d}zz^2 f_1^{(a)}(z), \end{aligned}$$

$$\bar{\boldsymbol{z}}_D = \sum_a \left(\frac{r_a}{r}\right)^2 \mu_D^{(a)} \bar{z}_D^{(a)}, \tag{8.33}$$

while

$$\bar{z}_D = \sum_a \mu_D^{(a)} \bar{z}_D^{(a)}.^{1)} \tag{8.34}$$

Let us discuss the practical aspects of this conception. It is clear that $\boldsymbol{f}_1(\boldsymbol{z})$ is not a directly measurable physical quantity. Partial single event size distributions $f_1^{(a)}(z)$ can in some cases be measured (e.g., for direct and indirect events) and have been obtained in theoretical calculations, e.g., for the different sorts of secondary particles produced in fast neutron fields (OLDENBURG and BOOZ, 1972b). Although this kind of decomposition may not be fine enough, it would certainly constitute a useful approximation, if corresponding partial values r_a could be obtained somehow. Unfortunately, only the total r is in principle measurable or obtainable from a formal survival curve analysis. However, intuitively, it should be possible to derive r_a-values for some components of radiation fields (definite particles and energies) from known total r-values for a lot

---

[1]) This result is independent of whether the total event size distribution is defined by Eq. (8.23) or Eq. (8.24), i.e. $\bar{z}_D = \bar{\bar{z}}_D$.

of *different* radiations. In principle, this means solving a, strictly speaking infinite, system of equations like (8.32). In this way, the so-called strict theory of radiation action could be based on nothing but ordinary partial distributions $f_1^{(a)}(z)$ and *total* r-values for a *variety* of radiations, so avoiding any model of lesion production. Hence, our lesion production model, by which we have actually solved this problem, does not constitute an indispensable part of the strict theory; as it were, this model is only a convenient tool that makes r as well as r_a calculable after fitting the internal parameters for some definite lesion type or in the way of formal survival curve analysis (see chapter 7).[1])

In fact, the essence of the formalism of strict theory as developed in section 3.6 is nothing but r_a-value calculation by the help of the lesion production model, associated with a general theory of partial single event size distributions $f_1^{(a)}(z)$. This formalism does not explicitly yield the virtual distribution $\mathbf{f}_1(\mathbf{z})$ in terms of Eq. (8.30), but it yields $\Pi(x)$ that defines the function $\mathbf{A}(\eta)$ (see Eq. (8.9)) from which subsequently $\mathbf{f}_1(\mathbf{z})$ and the corresponding mean values $\bar{\mathbf{z}}_F$, $\bar{\mathbf{z}}_D$ can be derived exactly in the same manner as in chapter 5 ordinary microdosimetry has been derived from $\Phi(\zeta)$. So-obtained numerical results for virtual microdosimetric parameters are listed in the tables of the appendix. They are discussed in the following subsections.

[1]) Strictly speaking, the so-called lesion production model considered as a mere mathematical construction must only be able to fit well for arbitrary sets of total r-values. Then it automatically yields r_a-values that obey Eq. (8.32) for any radiation, i.e. the problem is solved without any additional suppositions. Therefore, it is lastly unimportant whether the underlying reality *is* some type of DNA-lesions or not. This intrinsic general character of the theory of cellular radiation effects has already been stressed without proof in the introduction to chapter 7.

8.2.3. Virtual microdosimetric parameters

At first, the lesion type for which numerical results have been obtained must be specified. For the sake of its practical importance in mammalian cell survival theory, we took the so-called universal type of critical lesions for oxic conditions, defined by the parameters (7.9) of the ansatz (7.7). We speak of this as of DNA double-strand breaks and use two asterisks to mark the corresponding parameters, e.g., $\mathbf{f}_1^{**}(\mathbf{z})$, $\bar{\mathbf{z}}_D^{**}$ and so on. In addition, a second type which can be referred to as DNA single-strand lesions defined by $\lambda_j = \delta_{j0}$ will also be considered (indication by a single asterisk) in order to have a sufficiently rich numerical basis for general applications of the theory. For both types of lesions, the pertinent r-values for a great variety of radiations are listed in Table VIII of the appendix (upper and lower values in each panel). Their dependence on radiation quality in terms of LET has been visualized and discussed in chapter 6. For a qualitative understanding of what follows, keep in mind that, roughly speaking, r^{**} increases and r^* decreases with growing LET.

The virtual parameters in question are, in the first instance, $\bar{\mathbf{z}}_F$ and $\bar{\mathbf{z}}_D$. By virtue of the fundamental definition (8.13), their numerical calculation rests on Eqs. (8.11), (8.9), (8.8), i.e.

$$\bar{\mathbf{z}}_F = \frac{1}{\Pi(\infty)}, \quad \bar{\mathbf{z}}_D = -\frac{\Pi''(0)}{[\Pi'(0)]^2}, \tag{8.35}$$

with $\Pi(x)$ computed according to Eqs. (3.104 to 108). The first derivative reproduces r:

$$r = \frac{\pi}{4} \frac{l_l}{V s_0} \Pi'(0), \tag{8.36}$$

taking $\varrho = 1$. Eq. (8.36) is nothing but the general relation (3.115), which has been used to check up the numerical procedures.

The numerical evaluation has been carried out in exactly the same way as described for the microdosimetric quantities in chapter 5. Note the complete analogy between Eqs. (8.35), (8.36) and (5.12), (5.13). The analogy between the mean values r and W^{-1} exhibits the different concepts: in the ionization number concept of conventional microdosimetry energy is ionization production power, in the concept of virtual microdosimetry it is lesion production power.

Apart from the virtual mean values (8.35), which are the subject of the following subsections, it is also possible to derive the virtual spectra $\boldsymbol{f}_1(\boldsymbol{z})$ themselves from $\Pi(x)$. Owing to the analogy between Eq. (5.8) and Eqs. (8.9), (8.13), this problem is solved by the procedure described in section 5.2.2. For the sake of illustration, three examples of $\boldsymbol{f}_1^{**}(\boldsymbol{z})$ and $\boldsymbol{f}_1^{*}(\boldsymbol{z})$ have been calculated and plotted in Figure 8.1 together with the corresponding theoretical $f_1(z)$ calculated previously (Fig. 5.3, Fig. 5.9, Fig. 5.11). The differences between normal and virtual microdosimetry are seen to be most pronounced for fast neutrons. What these plots exhibit can easily be understood by means of the defining equation (8.30).[1] Details will be clarified implicitly by the hand of the following discussions of the mean values.

It is important to emphasize that the general physical basis and the mathematical approximations used for the treatment of radiation absorption stochastics (described in the first sections of chapter 3) are exactly the same with respect to $\Phi(\zeta)$ and $\Pi(x)$. In chapter 5, the generally good agreement of the microdosimetric quantities derived from $\Phi(\zeta)$ with experimental and computational data existing in the literature has been demonstrated. Therefore, the calculated virtual parameters can be anticipated to be correct to the same degree. Nevertheless, if some shortcomings exist, they can be expected largely to drop out of the *ratios* of corresponding virtual to normal quantities. Consequently, only such dimensionless ratios have been tabulated. As it were, they constitute the residual information that cannot at present be obtained otherwise than from our approximate formalism in chapter 3. In this way, virtual microdosimetry and thus the general theory of cellular radiation effects is most tightly coupled to customary microdosimetry and its future qualitative and quantitative progress.

8.2.4. Results for X-rays and fast neutrons

By writing

$$\bar{\boldsymbol{z}}_F = \tau_F \bar{z}_F, \quad \bar{\boldsymbol{z}}_D = \tau_D \bar{z}_D \tag{8.37}$$

the τ-coefficients relating virtual parameters to normal ones (ionization number concept) are defined. According to Eqs. (8.35) and (5.13) the determining equations are

$$\tau_F = \frac{\Phi(\infty)}{\Pi(\infty)}, \quad \tau_D = \frac{\Pi''(0)}{\Phi''(0)} \left[\frac{\Phi'(0)}{\Pi'(0)}\right]^2. \tag{8.38}$$

These coefficients are given in Tables II and III of the appendix together with $\bar{y}_F$ and $\bar{y}_D$. One has $\tau_D{}^*$ and $\tau_D{}^{**}$ for single- and double-strand breaks, respectively, while τ_F does not depend on the lesion type. This is self-evident since $\bar{\boldsymbol{z}}_F$ is actually the usual frequency-mean of specific energy in the rigorous energy concept (cf. Eq. (8.31)), i.e. τ_F describes nothing but the fraction of ionizing events and is therefore always smaller than

[1]) Plotting the logarithmically scaled *frequency distribution* $y\boldsymbol{f}_1(y)$ accentuates the mutual shifts of components and the maintenance of the corresponding frequency fractions. By contrast, the *dose distribution* $y^2\boldsymbol{f}_1(y)/\bar{\boldsymbol{y}}_F$ visualizes the changed virtual dose fractions of different components.

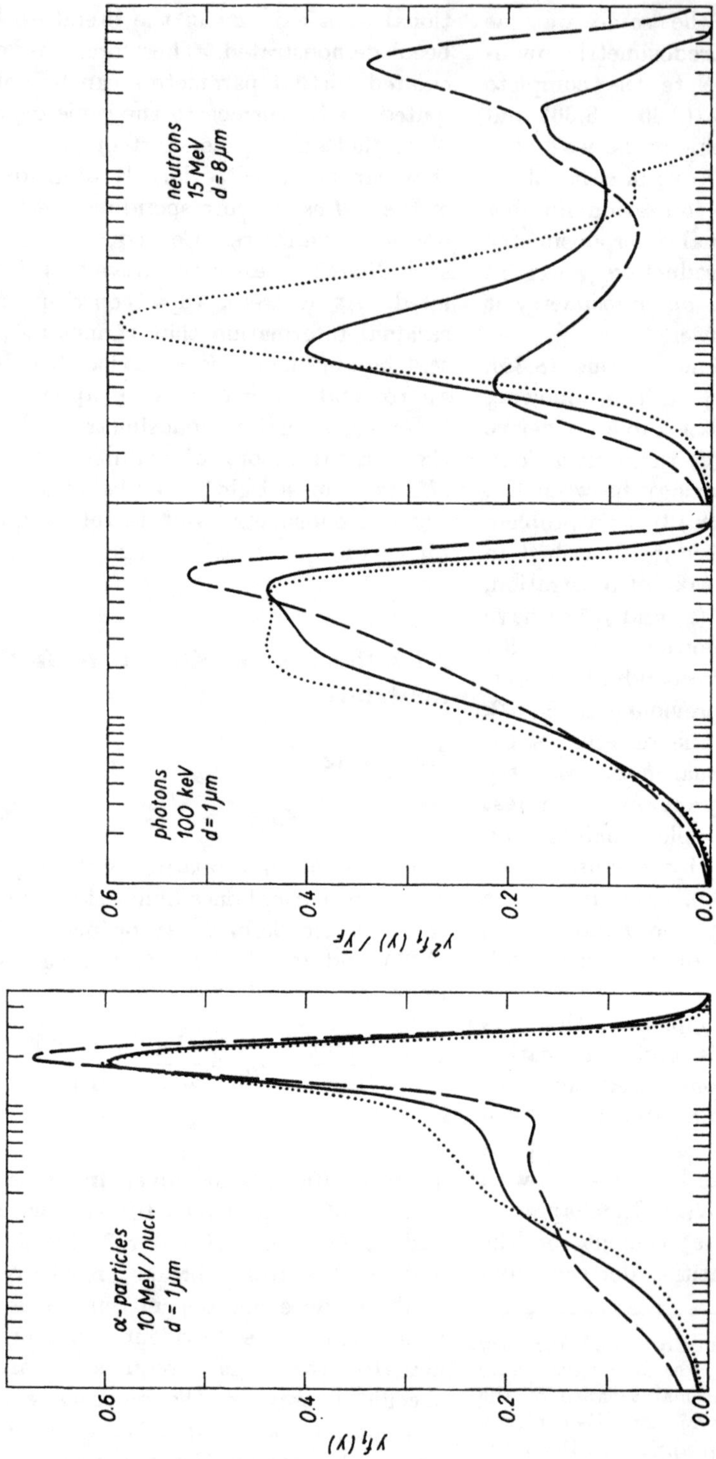

Fig. 8.1. Virtual spectra of lineal energy y, nominally referring to double-strand breaks (dashed lines), single-strand breaks (dotted lines), and ordinary microdosimetry (full lines).

unity. It is usually close to unity. This applies to neutron radiations throughout. Only in the case of high-energy X-rays and small sensitive volumes one has a considerable fraction of 'zero', i.e. non-ionizing, events, so that τ_F becomes small.

The discussion of τ_D will be restricted to τ_D^{**} since τ_D^{*} behaves just inversely. As apparent from Eqs. (8.33), (8.34), there is no general rule as to $\tau_D > 1$ or $\tau_D < 1$. If large absorption events (large z) tend to be produced by particles having a high relative effectiveness of lesion production (large r_a), one has $\tau_D > 1$. Hence $\tau_D^{**} > 1$ if large z means large LET averaged over the track segment inside the sensitive volume (SV). This is the case for small SV's where most events result from 'crossers' (the particle crosses the SV completely, i.e. $z \sim$ LET). In the opposite limiting case (very large SV) only 'insiders' occur, so that z is proportional to the initial energy of the particle, i.e. the mean LET is the *smaller* the larger z is. This must lead to $\tau_D^{**} < 1$. Consequently, in case the radiation field consists of only one sort of ionizing particles, τ_D^{**} should decrease with increasing diameter, d, of the SV. Indeed, this rule proves valid for X-rays throughout (Tab. II), while it breaks down for neutrons of higher energy owing to the occurrence of heavy recoils (Tab. III). This changes the picture: large z-values originate from higher-LET recoils (α-particles, C-nuclei, and so on), which makes τ_D^{**} always considerably larger than unity. On the other hand, the short-range protons in the case of low neutron energies give rise to $\tau_D^{**} < 1$ already for about $d > 8$ μm. Some apparent irregularities are due to the fact that the LET of very-low-energy protons *increases* with energy.

Yet, a short note on the numerical calculation of τ_F should be given. There is no problem with respect to $\Phi(\infty)$; this limiting value has been calculated directly with the general computer code for $\Phi(\zeta)$ according to Eqs. (5.4 to 7):

$$\Phi_\gamma(\infty) = \frac{V\varrho\ \mathrm{g/cm^3}}{\int_0^\infty \mathrm{d}E\,E q_e^{(\gamma)}(E)} \times \int_0^\infty \mathrm{d}E \varphi_\gamma^{(\delta)}(E) \int_0^\infty \mathrm{d}l' u(l')\, p_e^{(\delta)}(E) \times \exp\left[-\int_{E^{(f)}}^{E} \mathrm{d}E' \frac{p_e^{(\delta)}(E')}{L_\infty^{(e)}(E')}\right] \tag{8.39}$$

for X-rays, for example. $\Pi(\infty)$, however, is an intrinsic limiting value that must be derived analytically from Eqs. (3.105 to 108) to obtain a practicable formula. Remember that Π is the expected value (3.81) with Ψ given by Eq. (3.99). The final formulas for Π were obtained by applying Eq. (3.39) or Eq. (3.42). Now we must go the way (3.34 to 39) back, than the limiting process $x \to \infty$ (cf. Eq. (8.8)) becomes possible. Since $\lim_{x\to\infty} \Psi = 1$, Eq. (3.34) supplemented by the denominator (3.46) reads for X-rays

$$\Pi(\infty) = \frac{\mathrm{g/cm^3}}{\int_0^\infty \mathrm{d}E\,E q_e^{(\gamma)}(E)} \times \left[\sigma_0 \int_0^\infty \mathrm{d}E \varphi_\gamma^{(\delta)}(E) + \varrho V \int_0^\infty \mathrm{d}E q_e(E)\right]. \tag{8.40}$$

The first integral counts the number of particles entering the SV from outside, the second integral counts the number of inside starters. The latter can be expressed by $L_\infty^{(e)}(0)\varphi_e^{(\delta)}(0)$ according to the general relation (3.13) which defines the fluence spectra φ by the total initial energy spectra q. In this way, $\Pi(\infty)$ (for neutrons analogously) has been computed.

8.2.5. Results for monoenergetic ion radiations

It would be neither instructive nor useful for practical purposes to calculate the *total* mean values (8.35) only. Instead, one has to separate direct (superscript (i)) and indirect (superscript (δ)) events, as it was done in chapter 5. The theoretical basis for this is the corresponding separation of $\Pi_i(x)$ into two terms (cf. Eq. (3.104)), i.e.

$\boldsymbol{A}(\eta)$ according to Eq. (8.9) writes

$$\boldsymbol{A}(\eta) = \boldsymbol{A}^{(i)}(\eta) + \boldsymbol{A}^{(\delta)}(\eta) = \Pi_i^{(i)}\left(\frac{\pi}{4}\frac{l_l}{\varrho V s}\eta\right) + \Pi_i^{(\delta)}\left(\frac{\pi}{4}\frac{l_l}{\varrho V s}\eta\right). \tag{8.41}$$

The virtual single event size distribution can obviously be decomposed analogously,

$$\boldsymbol{f}_1(\boldsymbol{z}) = \mu_F^{(i)}\boldsymbol{f}_1^{(i)}(\boldsymbol{z}) + \mu_F^{(\delta)}\boldsymbol{f}_1^{(\delta)}(\boldsymbol{z}), \tag{8.42}$$

and the two terms resulting after insertion into Eq. (8.13) have to be identified with $\boldsymbol{A}^{(i)}(\eta)$ and $\boldsymbol{A}^{(\delta)}(\eta)$, respectively.

By virtue of these definitions, we are left with just the same formalism as developed in section 5.2.1: the partial virtual quantities are given by Eqs. (5.19 to 21) with Φ_i replaced by Π_i; the total quantities are related to the partial ones by Eqs. (5.15), (5.16), and the virtual dose fractions are defined by Eq. (5.17):

$$\mu_D^{(i)} = \frac{\bar{\boldsymbol{z}}_F^{(i)}}{\bar{\boldsymbol{z}}_F}\,\mu_F^{(i)}, \quad \mu_D^{(\delta)} = \frac{\bar{\boldsymbol{z}}_F^{(\delta)}}{\bar{\boldsymbol{z}}_F}\,\mu_F^{(\delta)}. \tag{8.43}$$

However, the so-defined partial virtual quantities are not quite appropriate for tabulation as well as application in radiation action theory. Recalling the intrinsic meaning of virtual microdosimetry as exhibited by Eq. (8.30), one sees that $\boldsymbol{f}_1^{(i)}(\boldsymbol{z})$ and $\boldsymbol{f}_1^{(\delta)}(\boldsymbol{z})$ involve the relations of 'direct' and 'indirect' r_a-values to the *total* r, while a consequent separation into direct and indirect events should relate them to some *partial mean r-values*, r_i and r_δ, which should play the role of r for the two virtual partial event size distributions. It is obvious that the latter will less differ from the usual microdosimetric distributions $f_1^{(i)}(z)$ and $f_1^{(\delta)}(z)$, respectively, since they are governed only by the 'internal' variations of r_a-values around the mean values r_i and r_δ for the separately-considered totalities of direct and indirect events.

We denote these intrinsic virtual partial distributions by $\hat{\boldsymbol{f}}_1^{(i)}(\boldsymbol{z})$ and $\hat{\boldsymbol{f}}_1^{(\delta)}(\boldsymbol{z})$ and define them by

$$\boldsymbol{f}_1(\boldsymbol{z}) = \mu_F^{(i)}\frac{r}{r_i}\,\hat{\boldsymbol{f}}_1^{(i)}\left(\frac{r}{r_i}\boldsymbol{z}\right) + \mu_F^{(\delta)}\frac{r}{r_\delta}\,\hat{\boldsymbol{f}}_1^{(\delta)}\left(\frac{r}{r_\delta}\boldsymbol{z}\right), \tag{8.44}$$

the counterpart of Eq. (8.42). The relations of the corresponding mean values are easily derived:

$$\bar{\boldsymbol{z}}_F^{(i)} = \frac{r_i}{r}\,\hat{\bar{\boldsymbol{z}}}_F^{(i)}, \quad \bar{\boldsymbol{z}}_D^{(i)} = \frac{r_i}{r}\,\hat{\bar{\boldsymbol{z}}}_D^{(i)},$$
$$\bar{\boldsymbol{z}}_F^{(\delta)} = \frac{r_\delta}{r}\,\hat{\bar{\boldsymbol{z}}}_F^{(\delta)}, \quad \bar{\boldsymbol{z}}_D^{(\delta)} = \frac{r_\delta}{r}\,\hat{\bar{\boldsymbol{z}}}_D^{(\delta)}. \tag{8.45}$$

The mean values r_i and r_δ have not yet been set forth definitively. In view of the general equation (8.36), the most natural definition reads[1])

$$r_i = \frac{1}{\mu_D^{(i)}}\frac{\pi}{4}\frac{l_l}{V s_0}\,\Pi_i^{(i)\prime}(0),$$
$$r_\delta = \frac{1}{\mu_D^{(\delta)}}\frac{\pi}{4}\frac{l_l}{V s_0}\,\Pi_i^{(\delta)\prime}(0), \tag{8.46}$$

which ensures that the total r becomes a dose-weighted mean value:

$$r = \mu_D^{(i)} r_i + \mu_D^{(\delta)} r_\delta. \tag{8.47}$$

These equations give the virtual dose fractions (cf. Eq. (5.19)) the form

$$\mu_D^{(i)} = \frac{r_i}{r}\,\mu_D^{(i)}, \quad \mu_D^{(\delta)} = \frac{r_\delta}{r}\,\mu_D^{(\delta)}. \tag{8.48}$$

Now the determining equations for $\hat{\boldsymbol{f}}_1^{(i)}(\boldsymbol{z})$ and $\hat{\boldsymbol{f}}_1^{(\delta)}(\boldsymbol{z})$ can be written down. Substituting the expression (8.44) for $\boldsymbol{f}_1(\boldsymbol{z})$ in Eq. (8.13) and considering the decompo-

---

[1]) Theoretical considerations like those in section 3.7.1 show that these partial mean r-values correspond exactly to what the lesion production model would yield for the hypothetical radiation fields consisting only of either cut-off δ-electrons or δ-electron-less ion tracks.

sition (8.41) yields

$$\begin{aligned} \boldsymbol{A}^{(i)}(\eta) &= \Pi_i^{(i)}\left(\frac{\pi}{4}\frac{l_l}{\varrho V s}\eta\right) = \frac{\mu_D^{(i)}}{\hat{\bar{z}}_F^{(i)}} \\ &\times \int_0^\infty \mathrm{d}z(1-\mathrm{e}^{-\eta(r_i/r)z})\,\hat{f}_1^{(i)}(z), \\ \boldsymbol{A}^{(\delta)}(\eta) &= \Pi_i^{(\delta)}\left(\frac{\pi}{4}\frac{l_l}{\varrho V s}\eta\right) = \frac{\mu_D^{(\delta)}}{\hat{\bar{z}}_F^{(\delta)}} \\ &\times \int_0^\infty \mathrm{d}z(1-\mathrm{e}^{-\eta(r_\delta/r)z})\,\hat{f}_1^{(\delta)}(z), \end{aligned} \tag{8.49}$$

in consideration of Eqs. (8.43), (8.45), and (8.48). From these equations, the virtual partial mean values of z can be derived in the usual way. Note the complete analogy to Eqs. (5.18). The first derivatives confirm Eqs. (8.46). Introducing τ-coefficients as before (only for the quantities referring to δ-electrons),

$$\hat{\bar{z}}_F^{(\delta)} = \tau_F^{(\delta)}\bar{z}_F^{(\delta)}, \quad \hat{\bar{z}}_D^{(\delta)} = \tau_D^{(\delta)}\bar{z}_D^{(\delta)} \tag{8.50}$$

one obtains

$$\begin{aligned} \tau_F^{(\delta)} &= \frac{\Pi_i^{(\delta)}(\infty)}{\Phi_i^{(\delta)}(\infty)}, \\ \tau_D^{(\delta)} &= \frac{\Pi_i^{(\delta)\prime\prime}(0)}{\Phi_i^{(\delta)\prime\prime}(0)}\left[\frac{\Phi_i^{(\delta)\prime}(0)}{\Pi_i^{(\delta)\prime}(0)}\right]^2 \end{aligned} \tag{8.51}$$

while the parameters referring to direct events read

$$\begin{aligned} \hat{\bar{z}}_F^{(i)} &= \mu_D^{(i)}\frac{1}{\Pi_i^{(i)}(\infty)}, \\ \hat{\bar{z}}_D^{(i)} &= -\mu_D^{(i)}\frac{\Pi_i^{(i)\prime\prime}(0)}{[\Pi_i^{(i)\prime}(0)]^2}. \end{aligned} \tag{8.52}$$

Let us briefly discuss the numerical results. The right-hand part of Table VIII contains r_δ** for oxic and anoxic conditions, and r_δ*. These values resemble those for low-LET X-rays. In fact, these r-values correspond to a hypothetical radiation field that consists only of electrons, namely the δ-electrons cut off above the energy δ. This cut-off energy depends on the diameter, d, of the sensitive volume (cf. Eq. (3.12)). Hence, the spectrum of the electrons in question, and therefore r_δ, depends on both the energy per nucleon and d, but it is independent of the ion's charge.[1]) The total r, on the contrary, by being a quantity that pertains to a given radiation as such, does not depend on d, but it depends on energy *and* charge. These properties of r_δ and r have facilitated the tabulation, while r_i depends on all the three parameters. It must be calculated from Eq. (8.47).

τ_F, τ_D**, and τ_D* are listed in Table V. They all are close to unity, i.e. $\hat{f}_1^{(\delta)}(z)$ does not much differ from $f_1^{(\delta)}(z)$ (in contrast to $\boldsymbol{f}_1^{(\delta)}(z)$, see Eqs. (8.45)). These δ-ray τ-coefficients need not be further discussed since what has been stated in section 8.2.4 holds exactly: the δ-electron field may be thought as if to result from a hypothetical X-radiation.

$\hat{\bar{z}}_F^{(i)}$ and $\hat{\bar{z}}_D^{(i)}$ given by Eqs. (8.52) can be further evaluated analytically to compare them with the microdosimetric counterparts (5.23) and (5.27). $\Pi_i^{(i)}(x)$ is the first term at the right-hand side of Eq. (3.104). Replacing specific energy by lineal energy according to Eq. (5.11), one has

$$\hat{\bar{y}}_F^{(i)} = \mu_D^{(i)} L_\infty^{(i)}. \tag{8.53}$$

As it had to be expected, the frequency-mean of virtual lineal energy for direct events is nothing but restricted stopping power. No special virtual parameter must be calculated. In comparison with Eq. (5.23), the denominator is missing, because $\hat{\bar{y}}_F^{(i)}$ involves the 'zero' events, too. For the evaluation of $\hat{\bar{y}}_D^{(i)}$ one needs only the first two terms of a power series expansion of $\Pi_i^{(i)}(x)$ (see Eqs. (8.8),

[1]) This charge-independence is, in parallel with the charge-independence of $\bar{y}_F^{(\delta)}$ and $\bar{y}_D^{(\delta)}$, only a consequence of our fundamental approximations (see page 66 and 146). It is not strict, by definition.

(3.104), (3.107), (3.100)):

$$\Pi_i^{(i)}(x) = \frac{\sigma_0 \text{ g/cm}^3}{L_\infty^{(i)}(E_i^{(0)})} \int_0^\infty dl' w(l') \times \{1 - \exp[-(ax - bx^2) l']\}$$

$$= \frac{\sigma_0 \text{ g/cm}^3}{L_\infty^{(i)}(E_i^{(0)})} \int_0^\infty dl' w(l') \times [al'x - (bl' + a^2 l'^2/2) x^2]. \tag{8.54}$$

b comes only from the non-linear portion of $\xi(E, Z)$, while a depends on both $G_i(E_i^{(0)})$ and $\xi(E, Z)$. It is important to point out the dependence of a and b on the ion's charge Z_{eff}. This dependence is obvious with respect to b; and a can be expressed through $L_\infty^{(i)} \sim Z_{eff}^2$ and r_i by means of relation (8.46):

$$b = const.\ Z_{eff}^2, \quad a = const.\ \mu_D^{(i)} L_\infty^{(i)} r_i. \tag{8.55}$$

For a spherical sensitive volume (cf. Eq. (3.25)), Eq. (8.52) yields

$$\hat{\bar{y}}_D^{(i)} = \mu_D^{(i)} L_\infty^{(i)} \left(\frac{9}{8} + \frac{2b}{la^2}\right), \tag{8.56}$$

which can be written in a form similar to $\bar{y}_D^{(i)}$ (Eq. (5.27)):

$$\hat{\bar{y}}_D^{(i)} = \frac{9}{8} \mu_D^{(i)} L_\infty^{(i)} + \sigma_2/l, \quad \sigma_2 = \frac{c}{r_i^2} \delta_2. \tag{8.57}$$

Hence, only the straggling governed by the energy δ_2 changes in comparison with normal microdosimetry. The special way of expressing this change by the factor c/r_i^2 defines a new dimensionless parameter, c, that is charge-independent on account of Eqs. (8.55).

Table V lists c^{**} and c^* as a function of d and E[1]). The coefficient c is difficult to discuss. $\sigma_2 > \delta_2$ as well as $\sigma_2 < \delta_2$ is possible. There are only two noticeable properties which derive from an inspection of the origin of b. In the range of no δ-ray events ($\delta > E_{max}$, upper right-hand corner of Tab. V) b is independent of d, and also $\mu_D^{(i)} = 1$, $r_i = r$. Therefore, σ_2 is d-independent like δ_2, and thus c is too. In the complementary range ($\delta < E_{max}$) b is proportional to Z_{eff}^2/E. This leads to the constancy of the product $c\delta_2\mu_D^{(i)} \times \ln(0.002176E/I)$ [2]) with respect to changing energy. Both statements can be verified by means of the tabulated numerical results.[3])

8.3. The ultimate general formalism of the theory of cellular radiation effects

8.3.1. Fundamentals

In effect, the remaining task is only to reap the fruits of the foregoing section. Knowing the function $A(\eta)$ for a given radiation and diameter, d, of the gross sensitive volume makes feasible the theory, completely. This important statement implies that the underlying molecular lesions have become an independent matter, i.e. the radiation-dependent parameter r enters the theory only through

$$\alpha_\mu = A(r\alpha_\mu^{(0)}). \tag{8.58}$$

Thus, one has in principle merely to tabulate $A(\eta)$ for different radiations and diameters d. This is in fact the only practicable way for neutron radiations, while X-rays and ion radiations offer to express $A(\eta)$ by approximate analytical expressions that rest upon the parameters of Tables II—VII. This question is the subject of the following subsections, but beforehand the generality of this theoretical scheme shall be asserted.

---

[1]) E signifies energy per nucleon, here.

[2]) By using the stopping power formula (4.2), where $I = 70$ eV.

[3]) These data have been computed directly by means of Eq. (8.52) and afterwards solving Eq. (8.57) for c. Thus, charge-independence and the other mentioned properties proved a true tests of the numerical procedures.

The crucial point is the unrelatedness of $A(\eta)$ and the lesion type parameter r. Exactly, this holds only in the framework of the microdosimetric approximation, while the virtual $\boldsymbol{A}(\eta)$ still depends on the lesion type characteristics (see Eqs. (8.29), (8.30)). However, this is obviously of secondary importance: the dose-effect relation (8.1) is *primarily* determined by the r in Eq. (8.58), as it is *exclusively* determined so in the microdosimetric approximation. This argument opens the way for more general applications of the 'strict' theory, too, despite $\boldsymbol{A}(\eta)$ being available only for DNA single- and double-strand breaks (see section 8.2.3). Thus, as compared to the involved formalism in chapter 3, the separation of the total r as the *main determinant* in the formulation (8.29) leaves r to *free choice*, largely, i.e. it must only *roughly* correspond to what $\boldsymbol{A}(\eta)$ has been computed for.

Let us now sum up the different possibilities of utilizing the general scheme of theory. With the microdosimetric version in mind, there are no difficulties: $A(\eta)$ is always the same. The generalization in the sense of strict theory (virtual microdosimetry) will be considered hereafter.

(i) *Formal survival curve prediction*:
Starting from the representation

$$S_X(D) = \sum_{\mu=1}^{n} B_\mu \, \mathrm{e}^{-\alpha_\mu^{(X)} D} \tag{8.59}$$

of an empirical X-ray survival curve, one derives the coefficients $\alpha_\mu^{(0)}$ from

$$\alpha_\mu^{(X)} = A(r_X \alpha_\mu^{(0)}) \tag{8.60}$$

and calculates

$$S(D) = \sum_{\mu=1}^{n} B_\mu \mathrm{e}^{-\alpha_\mu D}, \quad \alpha_\mu = A(r\alpha_\mu^{(0)}) \tag{8.61}$$

for any radiations. The necessary r-values may be taken from Table VIII (r^{**} for mammalian cells) as well as from direct measurements of absolute lesion production rates s according to $r = s/s_0$. Defining $r = s/s_X$, $r_X = 1$ is also possible.

(ii) *Formal survival curve analysis*:
The formalism is the same, but the r-values are what one is aiming at by means of curve fitting. Take $r_X = 1$; i.e. r means the relative effectiveness of lesion production with respect to the X-rays at hand.

(iii) *Theoretical calculation of dose-effect relations (one-lesion-type models)*:
One must have a concrete biological model of action that manages to calculate the so-called 'zero-LET' dose-effect relation $S_0(D)$ by using a Poisson distribution for the number of lesions per reacting unit. $S_0(D)$ either has the form of, or must be expressed approximately by, the expression

$$S_0(D) = \sum_{\mu=1}^{n} B_\mu \, \mathrm{e}^{-\alpha_\mu^{(0)} D}. \tag{8.62}$$

$S(D)$ for arbitrary radiations follows from Eq. (8.61); r-values for the lesion type in question must be known from anywhere (measurement, Tab. VIII, other theory). In deriving $S_0(D)$ one has had to use an absolute mean number, $\bar{k}D$, of primary lesions; r is related to *that* radiation for which this number was correct.

(iv) *Theoretical calculation of dose-effect relations (several-lesion-types models)*:
This is the most complex case, which has been investigated in sections 2.6.1 and 3.7.3. We restrict the discussion again to only two types of lesion, indicated by I and II. As before, assume the 'zero-LET' dose-effect relation $S_0(D)$ to have been derived directly from the axioms of the biological model by using Poisson distributions for either number of lesions with the mean values $\bar{k}_\mathrm{I} D$ and $\bar{k}_\mathrm{II} D$, which leads to the form $S_0(\bar{k}_\mathrm{I} D, \bar{k}_\mathrm{II} D)$. Its representation like (8.62) will therefore read

$$S_0(D) = \sum_{\mu=1}^{n} B_\mu \, \mathrm{e}^{-(\alpha_\mu^{(\mathrm{I})} + \alpha_\mu^{(\mathrm{II})}) D}, \tag{8.63}$$

with the coefficients $\alpha_\mu^{(\mathrm{I})}$ and $\alpha_\mu^{(\mathrm{II})}$ being *proportional* to $\bar{k}_\mathrm{I}$ and $\bar{k}_\mathrm{II}$, respectively, but the proportionality constants may be *different* and *μ-dependent.* It is obvious how this formalism is identical with what has been dealt with in section 2.6, where the two-lesion-types model has been expressed explicitly by $H(k_{\rm I}, k_{\rm II})$. Correspondingly, the theory reads now:

$$S(D) = \sum_{\mu=1}^{n} B_\mu \, e^{-\alpha_\mu D}, \quad \alpha_\mu = A(r_\mu \alpha_\mu^{(0)}), \quad \alpha_\mu^{(0)} = \alpha_\mu^{(\rm I)} + \alpha_\mu^{(\rm II)}, \tag{8.64}$$

where the so-called *formal effective r-values*, which depend on μ in general, are defined by

$$r_\mu = \frac{\alpha_\mu^{(\rm I)}}{\alpha_\mu^{(0)}} r_{\rm I} + \frac{\alpha_\mu^{(\rm II)}}{\alpha_\mu^{(0)}} r_{\rm II}. \tag{8.65}$$

$r_{\rm I}$ and $r_{\rm II}$ refer to the individual lesion types. They are assumed to be known (e.g., single- and double-strand breaks).

together with the anoxic r^{**}, which is an approximation, to be sure, but a quite insignificant one. $\boldsymbol{A}^{**}(\eta)$ can also be recommended with respect to measured or otherwise-obtained r-values that follow but roughly the LET-dependence of the present r^{**}.

Worthy of special consideration are only those cases which require r-values intermediate between r^{**} and r^*. They should, in a crude manner, be expressible by a constant linear combination

$$r \approx \vartheta^{**} r^{**} + \vartheta^* r^*, \quad \vartheta^{**} + \vartheta^* = 1, \tag{8.66}$$

for different radiation qualities. This case is met with (iv) (in the similar expression (8.65) $r_{\rm I}$ and $r_{\rm II}$ may, but have not to, be identical with r^{**} and r^*) and (ii)[2]. Having found ϑ^{**} and ϑ^*, one can use the following approximation:

$$\boldsymbol{A}(\eta) \approx \frac{(\vartheta^{**} r^{**})^2 \boldsymbol{A}^{**}(\eta) + (\vartheta^* r^*)^2 \boldsymbol{A}^*(\eta) + 2\vartheta^{**}\vartheta^* r^{**} r^* A(\eta)}{(\vartheta^{**} r^{**} + \vartheta^* r^*)^2}. \tag{8.67}$$

Implementing the strict theory means replacing $A(\eta)$ by the virtual $\boldsymbol{A}(\eta)$. However, the data in the appendix make $\boldsymbol{A}(\eta)$ available only for two particular lesion types, namely the DNA single- and double-strand breaks that the r-values in Table VIII pertain to (except the medial values in each panel: r^{**} for *anoxic* conditions). Hence, $\boldsymbol{A}(\eta) = \boldsymbol{A}^{**}(\eta)$ is *entirely* correct only in the cases (i) and (iii) if the tabulated oxic r^{**} is used.[1] For anoxic conditions, the same $\boldsymbol{A}^{**}(\eta)$ must be used

This is an interpolation with the weights chosen so as to minimize the error of the virtual dose mean $\boldsymbol{\bar{z}}_D = -\boldsymbol{A}''(0)$ (cf. Eq. (8.11) and Eq. (8.33)): note that the partial values r_a corresponding to r combine of r_a^{**} and r_a^* in keeping with Eq. (8.66), in consequence of Eq. (8.32). The relation (8.67) is easily seen to be correct with respect to $\boldsymbol{\bar{z}}_D$ as long as the quotients $(r_a^{**} r_a^*)/(r^{**} r^*)$ are close to unity, which is a reasonable assumption.

The *general* formalism is now complete. However, it is important to emphasize that the transparency of the strict theory achieved by means of the concept of virtual microdosimetry, as expressed by Eqs. (8.29), (8.30), furthers various theoretical investigations of special or general character, including the possibility of a

[1]) Of course, using $\boldsymbol{A}^*(\eta)$ together with the tabulated r^* would also be correct, but biological effects caused by single-strand lesions *alone* are not very likely.

[2]) In the entire ignorance of r from the outset, one should begin the analysis, in a first attempt, with the microdosimetric $A(\eta)$.

direct comparison to the microdosimetric version which rests, formally, on the same equations: (8.3), (8.4). For example, the *relative biological effectiveness* (RBE) and its determinants have been studied in section 2.3 on the microdosimetric basis. These results need not be repeated now: they are obvious to apply immediately to the strict theory, too, by nothing but substituting the pertinent *virtual* parameters. In this way, by the hand of the tabulated data, it is easy to elucidate the differences between both versions of theory quantitatively in terms of RBE. These differences are often small as demonstrated by some numerical examples (survival curves) below. However, regarding RBE, they may sometimes become substantial. Particularly in the case to which Eq. (2.36) applies, a factor between 2 and 3 can turn out ($\tau_D^{**} = 2.8$ occurs for neutrons in Tab. III).

For the sake of completeness, let us again note the procedure for *mixed radiations*. One may refer to the general relation (3.51) as well as to section 7.3.4: the coefficients α_μ become weighted mean values with the dose fractions $\mu_D^{(a)}$ of the individual components as the weights:

$$\alpha_\mu = \sum_a \mu_D^{(a)} \alpha_\mu^{(a)}. \tag{8.68}$$

This applies to both versions of the theory. Hence, a total $A(\eta)$ for a mixed radiation is not defined, because the different $A_a(\eta)$ are associated with different r_a in the argument $\eta = r_a \alpha_\mu^{(0)}$. Eq. (8.68) is of great practical importance, especially for neutron fields involving a spectrum of energies and a γ-ray contamination. All direct numerical data, however, are given only for monoenergetic primary radiations.

8.3.2. Photon radiations

It is sufficient to calculate $A(\eta)$ and $\boldsymbol{A}(\eta)$, respectively, in the so-called $\bar{z}_D$-*approximation* valid if $\eta \bar{z}_D = r\alpha_\mu^{(0)} \bar{y}_D / \sigma_0$ is small compared with unity. This is a matter of experience: the condition seems to be always fulfilled. Both $\alpha_\mu^{(0)}$ and σ_0 are either large (mammalian cells) or small (yeast cells, bacteria, etc.). A typical example is $\alpha_\mu^{(0)} = 10^{-2}\,\text{rad}^{-1}$, $\sigma_0 = 50\,\mu\text{m}^2$, and $r \approx 1$, $\bar{y}_D = 1\,\text{keV}/\mu\text{m}$, $\tau_D \approx 1$, which yields $\eta \bar{z}_D = 0.003$.

The approximation in question reads (cf. Eq. (8.12))

$$A_X(\eta) = \eta \left(1 - \frac{1}{2}\,\bar{z}_{D,X}\eta\right). \tag{8.69}$$

Taking $\bar{\boldsymbol{z}}_{D,X} = \tau_{D,X}\bar{z}_{D,X}$ instead of $\bar{z}_{D,X}$ provides the 'strict' theory. All necessary data are found in Table II with relation (5.11) to be used.

This approximation makes it easy to change the 'empirical' coefficients $\alpha_\mu^{(X)}$ (Eq. (8.59)) into the 'zero-LET' coefficients $\alpha_\mu^{(0)}$ (Eq. (8.60)): the expression (8.69) can be solved for η. Because of the neglect of higher order terms one can write

$$\alpha_\mu^{(0)} = \frac{\alpha_\mu^{(X)}}{r_X}\left(1 + \frac{1}{2}\,\bar{z}_{D,X}\alpha_\mu^{(X)}\right). \tag{8.70}$$

The second term in brackets constitutes but a small correction. Its dependence on σ_0 may therefore be overlooked in case that σ_0 must still be varied, subsequently.

8.3.3. Fast neutrons

In some cases only it is sufficient to use the $\bar{z}_D$-approximation like (8.69) with the pertinent data compiled in Table III. However, a generally practicable procedure cannot be established otherwise than by direct tabulation of $A_n(\eta)$, $\boldsymbol{A}_n^{**}(\eta)$, and $\boldsymbol{A}_n^{*}(\eta)$ for different neutron energies and diameters of the gross sensitive volume. This is the multi-paged Table IX, computed directly from the source equations (3.105) and (5.4) according to the relations (8.5) and (8.9).

For the sake of formal relatedness, a new parameter $\bar{z}_{D,eff}$ has been defined for tabulation, so as to have an expression for $A(\eta)$ similar to (8.69):

$$A_n(\eta) = \eta\left(1 - \frac{1}{2}\,\bar{z}_{D,eff}\eta\right), \tag{8.71}$$

with $\bar{z}_{D,eff}$ being a function of η. The first of each triple column in Table IX contains the microdosimetric effective dose mean of lineal energy $\bar{y}_{D,eff}$ defined by Eq. (8.4).[1]) It is changed into $\bar{z}_{D,eff}$ by means of relation (5.11). The second and third columns list $\tau^{**}_{D,eff}$ and $\tau^{*}_{D,eff}$ which provide $\bar{\bar{z}}_{D,eff} = \tau_{D,eff}\bar{z}_{D,eff}$ for the 'strict' virtual $A_n(\eta)$.
Obviously, for the limit $\eta \to 0$ we have $\bar{z}_{D,eff} = \bar{z}_D$ and $\tau_{D,eff} = \tau_D$ (compare the first row of Tab. IX with Tab. III). Thus, Table IX yields a direct impression about how far a given η is away from the $\bar{z}_D$-approximation.

8.3.4. Ion radiations

The separation (8.41) of $A(\eta)$ ($A(\eta)$ likewise) into the contributions of direct and indirect events leads to a generally applicable approximation which we will first derive for the 'strict' theory, starting from the expressions (8.49).
The δ-ray component is comparable with X-rays. Therefore, the $\bar{z}_D$-approximation can be applied for the same reason as discussed in section 8.3.2. Due to the analogy of the expression (8.49) of $A^{(\delta)}(\eta)$ to the expression (8.13), one can at once write down

$$A^{(\delta)}(\eta) = \mu_D{}^{(\delta)}\,\frac{r_\delta}{r}\,\eta\left(1 - \frac{1}{2}\,\hat{\bar{z}}_D{}^{(\delta)}\,\frac{r_\delta}{r}\,\eta\right). \tag{8.72}$$

As to the direct events, we make use of the fact that the corresponding single event size distribution is largely of triangular shape. This has been pointed out in chapter 5 (cf. Eq. (5.36) and context). The virtual spectrum $\hat{f}_1{}^{(i)}(z)$ does not change this

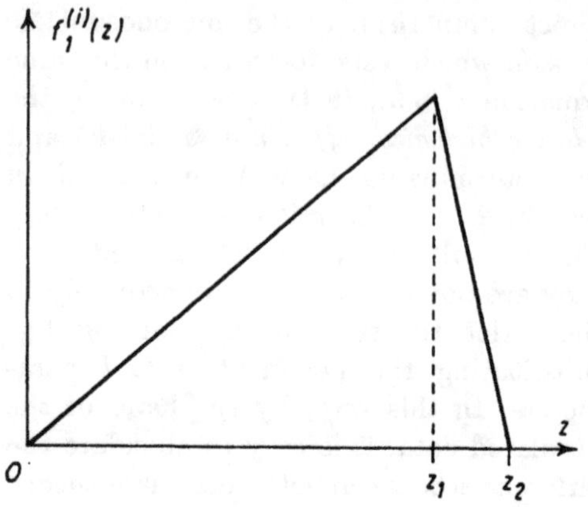

Fig. 8.2. Modified triangular spectrum of specific energy for direct ion events in a spherical volume.

situation, as indicated by the mean values (8.53), (8.57): any difference concerns only the straggling term σ_2/l which is just neglected when assuming a triangular spectrum. However, in view of the availability of σ_2 as well as δ_2, we are not forced to neglect straggling completely. It can be taken into account by means of an appropriate modification of the triangular spectrum, so that, in addition to $\bar{y}_F{}^{(i)}$, $\bar{y}_D{}^{(i)}$ becomes also correct.[2]) This modification is visualized in Figure 8.2. The parameters z_1 and z_2 of this artificial spectrum are well defined by its mean values $\bar{z}_F{}^{(i)}$ and $\bar{z}_D{}^{(i)}$ and by the requirement of normalization. This simple calculation is not worth demonstrating. $z_1 = z_2$ would result if $\bar{z}_D{}^{(i)} = (9/8)\,\bar{z}_F{}^{(i)}$ were true ($\delta_2 = 0$).
The modified triangular spectrum allows the expression (8.49) to be evaluated analytically. The result of this straight-

---

[1]) In practical use, the inevitable interpolation raises serious problems. Corresponding hints are given in chapter 9.

[2]) Therefore, this approach changes *exactly* into the $\bar{z}_D$-approximation (8.12) of the total $A(\eta)$ for *low-LET* ions, which is important because of the triangular spectrum losing its validity for low LET.

forward calculation reads as follows:

$$\begin{aligned}
A^{(i)}(\eta) &= \frac{\sigma_0\,\mathrm{g/cm^3}}{L_\infty^{(i)}}\,F(x,\varepsilon),\\
x &= \frac{3}{2}\,\frac{\mu_D{}^{(i)}L_\infty^{(i)}}{\sigma_0\,\mathrm{g/cm^3}}\,\frac{r_i}{r}\,\eta,\\
\varepsilon^2 &= \frac{8}{3}\,\frac{\delta_2/l}{\mu_D{}^{(i)}L_\infty^{(i)}},\\
F(x,\varepsilon) &= \frac{\vartheta\big((1+\varepsilon)\,x\big)-\vartheta\big((1-\varepsilon)\,x\big)}{\varepsilon x},\\
\vartheta(x) &= \frac{1-x+x^2/2-\mathrm{e}^{-x}}{x}.
\end{aligned} \tag{8.73}$$

The expressions (8.53), (8.57), along with relation (5.11), have been substituted for $\hat{\hat{z}}_F{}^{(i)}$ and $\hat{\hat{z}}_D{}^{(i)}$. It should be emphasized that ε, the contribution of straggling, is negligibly small in most cases. For low-LET protons, however, it becomes significant. Then, the underlying approximation of $\hat{f}_1{}^{(i)}(z)$ is not any longer good, to be sure, but at the same time the total $A(\eta)$ depends only on the total $\bar{z}_D$ which is just preserved exactly by this approach. The proof is straightforward by means of the relations in section 8.2.5.

Eqs. (8.72), (8.73) define a clear procedure for computing $A(\eta) = A^{(\delta)}(\eta) + A^{(i)}(\eta)$ that rests not only on the microdosimetric and virtual microdosimetric parameters listed in Tables IV and V, but also on r_δ and r_i in Table VIII. It is this partition of r that makes the *leading* difference between the strict and the microdosimetric versions of theory. Hence, by writing

$$\begin{aligned}
A^{(\delta)}(\eta) &= \mu_D{}^{(\delta)}\,\frac{r_\delta}{r}\,\eta\left(1-\frac{1}{2}\,\bar{z}_D{}^{(\delta)}\,\frac{r_\delta}{r}\,\eta\right),\\
A^{(i)}(\eta) &= \frac{\sigma_0\,\mathrm{g/cm^3}}{L_\infty^{(i)}}\,F(x,\varepsilon),\\
x &= \frac{3}{2}\,\frac{\mu_D{}^{(i)}L_\infty^{(i)}}{\sigma_0\,\mathrm{g/cm^3}}\,\frac{r_i}{r}\,\eta,\\
\varepsilon^2 &= \frac{8}{3}\,\frac{\delta_2/l}{\mu_D{}^{(i)}L_\infty^{(i)}}
\end{aligned} \tag{8.74}$$

one obtains an *intermediate approximation* which is practically equivalent to the 'strict' theory, *without* containing any virtual parameters. It can be applied almost in the same manner as the pure microdosimetric theory. The only difficulty in using model-independent r-values is to assess r_δ and r_i, connected in any case by relation (8.47). In this respect, the rough estimation $r_\delta \approx 1$ or $r_\delta \approx r_X$ is by far better than the intrinsic microdosimetric aproximation corresponding to $r_\delta = r_i = r$:

$$\begin{aligned}
A(\eta) &= \mu_D{}^{(\delta)}\eta\left(1-\frac{1}{2}\,\bar{z}_D{}^{(\delta)}\eta\right)\\
&\quad+\frac{\sigma_0\,\mathrm{g/cm^3}}{L_\infty^{(i)}}\,F(x,\varepsilon),\\
x &= \frac{3}{2}\,\frac{\mu_D{}^{(i)}L_\infty^{(i)}}{\sigma_0\,\mathrm{g/cm^3}}\,\eta.
\end{aligned} \tag{8.75}$$

That these different approximations are rather insignificant is due to one fact: the dose fraction, $\mu_D{}^{(\delta)}$, of indirect events is never large. In most cases it is even very small, particularly for mammalian cells owing to the large cross-sectional areas σ_0. In Table IV, one will often find $\mu_D{}^{(\delta)} = 0$ implying $r_i = r$ (this is a consequence of our general approach described in chapter 3). Hence, for mammalian cells at least, it is not too bad to omit $A^{(\delta)}(\eta)$ completely and to take $\mu_D{}^{(i)} = 1$, $r_i = r$. Moreover, the large l makes the straggling parameter ε negligible. This leads to the crude but often sufficient approximation

$$\begin{aligned}
A(\eta) = A(\eta) &= \frac{\sigma_0\,\mathrm{g/cm^3}}{L_\infty^{(i)}}\\
&\quad\times\left[1-2\,\frac{1-(1+x)\,\mathrm{e}^{-x}}{x^2}\right],\\
x &= \frac{3}{2}\,\frac{L_\infty^{(i)}}{\sigma_0\,\mathrm{g/cm^3}}\,\eta,
\end{aligned} \tag{8.76}$$

which we refer to as the *LET-approximation* since the only residual parameter

Tab. 8.1. Comparison of different approximations in terms of doses [rad] pertaining to constant surviving fractions: (i) — strict theory by means of $A^{**}(\eta)$; (ii) — microdosimetric theory by means of $A(\eta)$; (iii) — LET-approximation by means of Eq. (8.76). For the parameters describing the survival characteristics of the cell lines indicated see the captions of Fig. 7.7 and Fig. 7.10

surviving fraction	Yeast cells, Saccharomyces cerevisiae, strain BZ34, irradiated with heavy ions of 10 MeV/nucl.										V79 Chinese hamster cells irradiated with fast neutrons						
	X-rays	He-ions			C-ions			Ar-ions			X-rays	0.5 MeV		4 MeV		15 MeV	
		(i)	(ii)	(iii)	(i)	(ii)	(iii)	(i)	(ii)	(iii)		(i)	(ii)	(i)	(ii)	(i)	(ii)
0.8	6166	4289	4289	4289	1705	1711	1703	3613	3634	3571	180	50	50	79	81	83	88
0.2	23990	16750	16750	16760	7355	7301	7379	19050	18860	19450	511	181	181	273	261	334	290
0.05	38380	26820	26820	26830	11940	11830	11990	31970	31570	32850	744	274	275	410	388	508	431
0.01	53970	37720	37720	37740	16910	16750	16990	46030	45380	47450	996	373	374	555	523	691	582

of radiation is the unrestricted stopping power of the ions considered.[1]

8.3.5. Numerical examples

In conclusion, let us illustrate the actual significance of virtual microdosimetry by the hand of some numerical examples of survival curve calculation. X-rays are uninteresting in this respect; for ion radiations a noticeable effect can be expected only in case of small σ_0 (e.g. yeast cells); it is only fast neutrons that virtual microdosimetry is essential to, also for mammalian cells. Corresponding examples for cell lines already considered in chapter 7 are shown in Table 8.1. These results have been obtained by using the tabulated values r^{**} for oxic conditions. Despite of the small $\sigma_0 = 5\ \mu m^2$ for the yeast cells, the differences are quite unimportant, which is due to the nevertheless small $\mu_D^{(\delta)} = 0.07$. Obviously, this result warrants every approximation for ion-irradiated mammalian cells. The most crucial case, evidently, is high-energy neutron radiation: the microdosimetric approximation becomes poor, then.

8.4. Evaluation of RBE in LET-approximation for monoenergetic ions

As the LET-approximation (8.76) has proved to be quite satisfactory for the action of monoenergetic ions on mammalian cells, it is instructive and useful, regarding practical applications, to study the resultant theory of RBE, which then becomes particularly comprehensible by drawing upon the approximate RBE-dose relation (2.37). First, however, let us explain the *general* form of RBE in LET-approximation.

In addition to r, there is only one radiation quality parameter, L_∞, the stopping power for the ions considered. In place of the latter, it is advantageous to introduce the dimensionless parameter[2])

$$Q = \frac{r L_\infty}{\sigma_0 D_0}, \tag{8.77}$$

which absorbs two biological parameters that characterize the given object: σ_0, the cross-sectional area of the cell nucleus, and D_0, a characteristic dose which we shall

[1]) The fitness of this approximation for large σ_0 should not unreservedly be considered to support the widespread opinion that stopping power would be a sufficient determinant of biological radiation quality. Dose-effect relations are additionally governed by the relative effectiveness of primary lesion production which does not only depend on LET (see section 7.3.1).

[2]) In this section, dose is defined as energy per unit volume, to simplify the formulas.

identify with the usual 37-per-cent dose for the final slope of the zero-LET survival curve. Now, by (8.76), the dose-survival relation (8.1), (8.3) takes the explicit form

$$S(D) = \sum_{\mu=1}^{n} B_\mu \exp\left\{-\frac{D}{D_0} r \left[1 - 2\frac{1-(1+3QD_0\alpha_\mu^{(0)}/2)\,e^{-3QD_0\alpha_\mu^{(0)}/2}}{(3QD_0\alpha_\mu^{(0)}/2)^2}\right]\Big/Q\right\}, \tag{8.78}$$

where the coefficients B_μ and $D_0\alpha_\mu^{(0)}$ are constants that describe the shape of the zero-LET survival curve:

$$S_0(D) = \sum_{\mu=1}^{n} B_\mu \exp\left[-\frac{D}{D_0}(D_0\alpha_\mu^{(0)})\right]. \tag{8.79}$$

From these equations, it is evident that the *RBE* defined by the quotient D_X/D resulting from the equation $S(D) = S_0(D_X)$ *becomes the product of r and a function of Q.* This reflects fundamental aspects of the theory as such, which will be discussed below by the hand of a survival curve model evaluated explicitly.

Most of low-LET experimental survival curves can be fitted well by the widely used linear-quadratic mathematical model

$$S_0(D) = e^{-k_1 D/D_0 - C^2(D/D_0)^2}, \qquad C^2 = 0.0431. \tag{8.80}$$

In this special notation of the model, D_0 is something like a 37-per-cent dose for the 'final slope', and k_1 then designates the initial to 'final' slope ratio. Properly speaking, D_0 denotes the inverse negative logarithmic derivative of $S_0(D)$ at $S_0 \approx 10^{-3}$. It is true, just this model is not particularly suited to be used within the present formalism which requires expressing $S_0(D)$ in the form (8.79). However, on the one hand, the latter can be done approximately as shown in section 9.3.4; on the other hand, the expression (8.80) is advantageous in evaluating the approximate RBE formula (2.37) in LET-approximation. So we will arrive at a compact explicit RBE formula which, moreover, allows a test of the approach (2.37) on which it is based. By virtue of what is pointed out above, this formula must look as follows:

$$RBE_S = rR(S, k_1; Q), \tag{8.81}$$

since k_1 is the only curve shape parameter within the model (8.80). S denotes the reference survival level.

To derive $R(S, k_1; Q)$ from Eq. (2.37), it is necessary, firstly, to eliminate D and to introduce S by setting $D = D_X/RBE$, solving the resultant equation for *RBE*,

$$RBE = \frac{\frac{1}{2}\frac{k}{k_X}\frac{\lambda}{\lambda_X}\frac{\lambda_X}{D_X} + \sqrt{\left(\frac{1}{2}\frac{k}{k_X}\frac{\lambda}{\lambda_X}\frac{\lambda_X}{D_X}\right)^2 + \frac{k}{k_X}\left(1+\frac{\lambda_X}{D_X}\right)}}{1+\frac{\lambda_X}{D_X}}, \tag{8.82}$$

and expressing D_X by S according to the zero-LET dose-survival relation (8.80):

$$D_X = \frac{D_0}{C}\left[\sqrt{\left(\frac{k_1}{2C}\right)^2 - \ln S} - \frac{k_1}{2C}\right]. \tag{8.83}$$

The latter step implies relating RBE to zero-LET radiation, for simplicity, which also reduces the formulas for the parameters (2.38) to (2.40):

$$\frac{k}{k_X} = \left[\frac{D_0^*}{\bar{z}_F}\int_0^\infty dz(1 - e^{-rz/D_0^*})\, f_1(z)\right]^2, \tag{8.84}$$

$$\frac{k}{k_X}\frac{\lambda}{\lambda_X} = [-S_0'(0)\,\bar{z}_F]^{-1} \times \int_0^\infty dz[1 - S_0(rz)]\,f_1(z), \quad (8.85)$$

$$\lambda_X = 2\frac{S_0'(0)}{S_0''(0)}. \quad (8.86)$$

D_0^* will be settled below (see lower case print). Secondly, we have to introduce the LET-approximation in terms of the triangular spectrum

$$\bar{z}_F^{-1}f_1(z) = \begin{cases} \frac{8}{9}\left(\frac{\sigma_0}{L_\infty}\right)^3 z & \text{for } z \leqq \frac{3}{2}\frac{L_\infty}{\sigma_0} \\ 0 & \text{for } z > \frac{3}{2}\frac{L_\infty}{\sigma_0}. \end{cases} \quad (8.87)$$

Now the evaluation of Eqs. (8.84) to (8.86) yields

$$\frac{k}{k_X} = r^2 G(Q),$$

$$\frac{k}{k_X}\frac{\lambda}{\lambda_X} = \frac{C}{k_1} rF(k_1; Q), \quad (8.88)$$

$$\lambda_X = \frac{k_1 D_0}{C^2 - k_1^2/2},$$

where

$$G(Q) = \frac{(CQ)^{-4}}{9}\left[1 - \frac{8}{81} \times \frac{1 - (1 + 9C^2Q^2/2)\,e^{-9C^2Q^2/2}}{(CQ)^4}\right], \quad (8.89)$$

$$F(k_1; Q) = (CQ)^{-1} \times \left\{1 - \frac{4}{9}(CQ)^{-2}\,e^{\left(\frac{k_1}{2C}\right)^2} \times \left[e^{-\left(\frac{k_1}{2C}\right)^2} - e^{-\left(\frac{3}{2}CQ + \frac{k_1}{2C}\right)^2} - \sqrt{\pi}\frac{k_1}{2C}\left(\text{erf}\left(\frac{3}{2}CQ + \frac{k_1}{2C}\right) - \text{erf}\left(\frac{k_1}{2C}\right)\right)\right]\right\}.^{1)} \quad (8.90)$$

For this calculation, a special choice of D_0^* has been made to adapt the approach (2.37), (2.38) to the model (8.80) in which no genuine D_0 value is defined.[2]) By penetrating reflection one finds that equating D_0^* to the inverse negative logarithmic slope of $S_0(D)$ at $D = rD_1$, where D_1 is the maximum dose that can occur by a single event, should be appropriate. According to Eq. (8.87), this means $D_1 = 3/2\,\bar{z}_F = 3/2\,L_\infty/\sigma_0$ and $D_0^* = D_0/(3C^2Q)$, taking $k_1 = 0$. With this settlement involved, the numerical results of the entire approach (Fig. 8.3) have been checked with what the strict theory (8.78), (8.79) yields by means of the representation of $S_0(D)$ given in section 9.3.4. Excellent agreement was obtained, which constitutes a further test for the adequateness of the general RBE-dose relation (2.37) as such.

The equations (8.82), (8.83) and (8.88) to (8.90) constitute a *complete RBE theory in explicit form for monoenergetic ions* in LET-approximation on condition that the low-LET dose-response relation for the object and reaction considered can be described by the linear-quadratic model (8.80). Eq. (8.81) shorthands the general form of this theory. Let us detail the case $k_1 = 0$, on the one hand, and $S = 100\%$ if $k_1 \neq 0$, on the other hand[3]):

$$R(S, 0; Q) = \frac{1}{2}\frac{F(0;Q)}{\sqrt{-\ln S}} + \sqrt{\left(\frac{1}{2}\frac{F(0;Q)}{\sqrt{-\ln S}}\right)^2 + G(Q)}, \quad (8.91)$$

$$R(1, k_1; Q) = \frac{C}{k_1} F(k_1; Q). \quad (8.92)$$

---

1) erf(x) denotes the probability integral

$$\text{erf}(x) = \frac{2}{\sqrt{\pi}}\int_0^x dt\, e^{-t^2}.$$

2) Actually, D_0^* could be equated to the characteristic dose D_0 of the model (8.80), but there is no necessity to do so.

3) Only the exceptional case $k_1 = 0$ and $S = 100\%$ must be excluded since it does not allow a definition of RBE relative to zero-LET radiation. To treat this case is most easy by referring to ^{60}Co γ-rays, rendering possible to replace $S_0'(0)$ in Eqs. (8.85), (8.86) by $S_0''(0) \times \bar{z}_D^{(\gamma)}/2$, as can be seen by inspection of the starting equations (2.38) to (2.40). This ends up in

$$R(1, 0; Q) = \frac{D_0}{C\bar{z}_D^{(\gamma)}} F(0; Q).$$

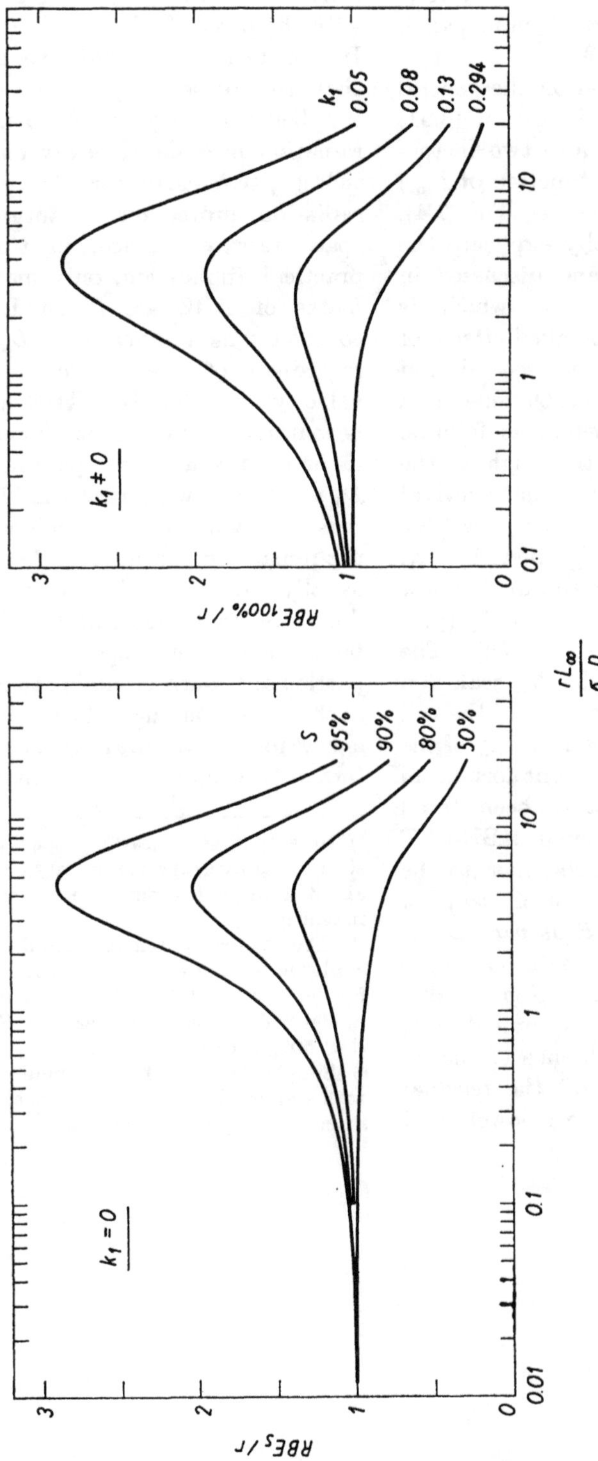

Fig. 8.3. Examples for the dependence of the second-kind-effect part of RBE on the radiation quality parameter Q. For details see text.

These functions are plotted in the left-hand panel and in the right-hand panel, respectively, of Figure 8.3.

Let us focus the attention on the general theoretical aspects of the relation (8.81) which decomposes RBE into two factors either of which is a peak function of L_∞[1]) (for the standard r curve see Fig. 7.4). These two factors formally explicate the conceptions introduced and discussed in section 2.4.2: the first, r, which is universal, contains the so-called effects of the first kind, while the second-kind effects described by $R(S, k_1; Q)$ depend on several biological parameters as follows. According to Figure 8.3, the height of the peak changes with both the survival level S considered and the low-LET survival curve shape signified by k_1. The peak may even disappear for low levels of S ($S < 50\%$) or large k_1 (pronounced initial slope of $S_0(D)$). The position of the peak in the L_∞ scale (cf. Eq. (8.77)) depends on the size of the cell nucleus (σ_0) and the radiosensitivity of the object (D_0). Hence, the present formalism bears out and reifies what has been stated in section 2.4.2: a measured RBE-LET relation exhibiting a peak has to be considered as a *superposition* of *two* peak functions one of which, R, is *variable*. If an R-peak does exist (k_1 small, S large) it may, on multiplication, heighten and displace the universal r-peak to some degree, whereas in the case of absence of the R-peak (k_1 large, or S small) the residual RBE-peak becomes always lower and shifts to a lower L_∞ value as compared with the peak of r.[2])

In conclusion, it should be emphasized that the present approach, due to its simplicity, is very beneficial in assessing radiation hazards of heavy charged particles, particularly for the purposes of radiation protection in future manned space research. Since, in this case of practical application, only marginal estimates of RBE are required, one may consider r as a function of L_∞ only (i.e. regardless of the particle's charge or velocity), so that the RBE as a whole becomes a nearly definite function of L_∞ if the usually unknown parameters σ_0 and D_0 are varied within reasonable limits to make RBE maximal in each case.[3]) As to the remaining parameters, $S = 100\%$ will usually be appropriate (prolonged low-dose-rate irradiation), while only k_1 has to be settled with respect to damage to particular tissues, assuming that the shape of the corresponding cellular dose-response curves for X-rays can always be estimated in terms of the linear-quadratic model (8.80).

---

[1]) Note that Q is nearly proportional to L_∞, since the universal r varies only between 1 and about 4 while L_∞ runs over several orders of magnitude.

[2]) The latter statement implies additional evidence for the energy-only concept (1.24) to fail (cf. section 2.3.3), since the survival curve model (8.80) with $k_1 = 0$ leads to $RBE < 1$ for $S < 50\%$ if one sets $r = 1$.

[3]) Simple folding of the function $RBE(L_\infty)$ with a given L_∞-spectrum is also allowed in the sense of marginal estimation.

Chapter 9

Concise instructions for practical calculations

9.1. Introduction

The theory of cellular radiation effects developed throughout this monograph has various aspects: its rationale is simple (chapter 1 and 2), its mathematical apparatus is complicated (chapter 3), its applications are manifold (chapter 7), but the final formulation it has been given in chapter 8 makes the calculations easy by using the tabulated parameters (appendix). The purpose of this concluding chapter is to summarize all formulas and instructions necessary to practise on the theory in compendious form, i.e. without any references to the theoretical background. In this way, the theory can be used blindly, in particular to predict *mammalian cell survival curves for high-LET radiations* from low-LET ones. This mode of application, which has been extensively exemplified in chapter 7, is the guide line of what follows. For the sake of completeness, however, the more general feasibilities of the theoretical framework are also outlined, but in lower case print to avoid confusion.

The term 'dose-effect relation' denoted by $S(D)$ will be used constantly. In the standard case this means the *rate of cell survival*, but it can also refer to other endpoints (e.g., division delay, accomplishment of the first division, aberrant mitosis), $S(D)$ always describing the rate of incidence (or failure) at the cellular level.[1] In dependence on the considered endpoint and the object, but also for mammalian cell survival, the possible variations of the standard procedure concern two aspects: (a) using other data for the parameters tabulated in the appendix, (b) starting with biological models of action of specified primary molecular lesions.

The formalism presented only aims at $S(D)$ for arbitrary radiations, since the derivation of RBE values from dose-effect relations is obvious, while the theory does not, in general, provide explicit formulas of RBE. Thus, numerical methods have to be applied, starting from the analytical expressions of $S(D)$. Nevertheless, some useful special relations concerning RBE exist, but these have not been included in

[1] Reactions which do not concern the cell as a whole and are not scored by a probability of incidence (e.g., number of chromosomal aberrations, molecular weight distribution of broken DNA) require a special adaptation of the formalism to describe the influence of radiation quality.

this chapter. The reader is referred to sections 2.3, 7.4.5, and 8.4.
The *gross sensitive volume* associated with the cellular object and the endpoint is a main determinant in the theory. It is described by the mean projected area σ_0. Regarding mammalian cell killing, the gross sensitive volume is assumed to be identical with the cell nucleus, and σ_0 has to be chosen correspondingly.
Most of the tabulated parameters occurring in the final formulas depend on σ_0. Regarding microdosimetric parameters, the formulas are written in terms of *specific energy*, z (in rads), while the tables present the figures for lineal energy, y (in keV/μm). The general rule of conversion can be given here once for all (assuming unit mass density):

$$z\,[\mathrm{rad}] = \frac{y\,[\mathrm{keV}/\mu\mathrm{m}]}{0.06242\sigma_0\,[\mu\mathrm{m}^2]}. \tag{9.1}$$

E.g., having a formula containing, say, $\bar{z}_D^{(\delta)}$ accompanied by a lapidary reference to Table IV includes the instruction to transform the value of $\bar{y}_D^{(\delta)}$ actually found in the table.
In addition to the theory as such, this chapter presents a method of expressing arbitrary experimental dose-effect curves as a *sum of exponential functions*. This is the first step in the complete procedure of formal survival curve prediction, but on principle it is irrelevant to the theory how this merely pragmatic problem is solved. The approach suggested, which is easy to practise, has therefore been separated from the intrinsic formalism of the theory. It is the topic of section 9.3.

9.2. Calculation of dose-effect relations

9.2.1. The general formalism

Dose-effect relations $S(D)$ are always expressed in the mathematical form

$$S(D) = \sum_{\mu=1}^{n} B_\mu \, \mathrm{e}^{-\alpha_\mu D}. \tag{9.2}$$

Only the coefficients α_μ depend on the radiation, and this dependence is the intrinsic subject of the theory. The radiation to start with is called 'zero-LET radiation' for which the dose-effect relation is assumed to be known:

$$S_0(D) = \sum_{\mu=1}^{n} B_\mu \, \mathrm{e}^{-\alpha_\mu^{(0)} D}. \tag{9.3}$$

For every radiation the theory provides a function $A(\eta)$ and a parameter r by which $S(D)$ is obtained according to

$$\alpha_\mu = A(r\alpha_\mu^{(0)}). \tag{9.4}$$

The zero-LET relation $S_0(D)$ in the form (9.3) derives from either a mechanistic model of action or an empirical dose-effect relation for a low-LET radiation. This will be dealt with in subsection 9.2.3, while the succeeding subsections are devoted to the function $A(\eta)$ for different radiations. This function depends on the parameter σ_0, the cross-section of the gross sensitive volume. For every radiation, $A(\eta)$ will be given in two versions referred to as the 'microdosimetric' and the 'strict' theory. The latter, being denoted by $\boldsymbol{A}(\eta)$, is recommended, particularly in the case of fast neutrons.
The parameter r describes the relative effectiveness[1]) of the given radiation in producing that type of molecular lesions that underlie the cellular effect. Concerning loss of the reproductive capacity of *mammalian cells*, the values r^{**} given in Table VIII for photons, fast neutrons, and monoenergetic ions are recommended. They differ for oxic and anoxic conditions.

This standard mode of using the theory is based on the rationale that mammalian cells are killed by DNA double-strand breaks, and that the tabulated r^{**} constitutes a good standard description for this type of lesion. However, for various purposes, it may be useful or even necessary (non-mammalian objects) to go beyond this approach.

---

[1]) By definition, $r = 1$ applies to ^{60}Co γ-radiation.

(a) The parameter r can be measured by determining the number of lesions, s, produced by different radiations per unit dose (e.g., DNA single- and double-strand breaks), yielding $r = s/s_0$, where s_0 refers to ^{60}Co γ-rays.
(b) If single-strand lesions are supposed to share in the effect, one has to try over a combination

$$r = \vartheta^* r^* + \vartheta^{**} r^{**}, \qquad \vartheta^* + \vartheta^{**} = 1, \tag{9.5}$$

with r^* and r^{**} taken either from Table VIII (nominally referring to DNA single- and double-strand breaks) or from experimental data.
(c) Formal analyses can be made in which r is nothing but an adjustable parameter for each radiation.
These and other variants regarding r do not conflict with using always the same functions $A(\eta)$ as long as the microdosimetric approximation for $A(\eta)$ is adopted. Otherwise there is a link between $A(\eta)$ and the nature of r as a function of radiation quality. Strictly speaking, only $A^*(\eta)$ and $A^{**}(\eta)$ pertaining to the tabulated r^* and r^{**}, respectively, are available.[1]) Any other r-values used have at least roughly to match either r^* or r^{**} to justify the use of $A^*(\eta)$ or $A^{**}(\eta)$. If the situation it intermediate, one should cast about for coefficients ϑ^* and ϑ^{**} such that the r-values in question can be expressed formally by the linear combination (9.5), may be in a crude manner. Then a good approximation of the strict $A(\eta)$ is given by

$$A(\eta) = \frac{(\vartheta^* r^*)^2 A^*(\eta) + (\vartheta^{**} r^{**})^2 A^{**}(\eta) + 2\vartheta^*\vartheta^{**} r^* r^{**} A(\eta)}{(\vartheta^* r^* + \vartheta^{**} r^{**})^2}. \tag{9.6}$$

9.2.2. Mixed radiations

All data defining $A(\eta)$ and r are given only for pure monoenergetic beams of primary particles (photons, neutrons, ions). Mixtures as well as energy spectra must be defined by *dose fractions*, $\mu_D^{(a)}$, of components a. E.g., a neutron energy spectrum has to be divided into intervals with those median energies that are contained in the tables, and the corresponding fractions of total kerma (not neutron fluence) have to be taken.
The dose-effect relation for a mixed field follows from

$$\alpha_\mu = \sum_a \mu_D^{(a)} \alpha_\mu^{(a)}, \tag{9.7}$$

where $\alpha_\mu^{(a)}$ is given by Eq. (9.4) for each component a separately. Thus, no total $A(\eta)$ exists for mixed radiations owing to the differing parameters r.
The photon energy spectrum of X-rays can be in sufficient approximation described by a single energy of photons, i.e. a suitably chosen mean value, avoiding the formalism (9.7).
To allow for a γ-ray contamination of neutron fields, it may be considered as a zero-LET component, i.e. $\alpha_\mu^{(a)} = \alpha_\mu^{(0)}$ for this contribution in the expression (9.7).

9.2.3. The 'zero-LET' dose-effect relation

The standard case is the so-called formal employment of the theory, i.e. the prediction of dose-effect relations based on an experimentally determined dose-effect relation for a low-LET radiation, i.e. usually X-rays. The first step is expressing the experimental data mathematically in the form

$$S_X(D) = \sum_{\mu=1}^{n} B_\mu \, e^{-\alpha_\mu^{(X)} D}. \tag{9.8}$$

This expression has no meaning in the sense of a model. The only requirement is a good fit, which can always be achieved due to the parameters n, B_μ, and $\alpha_\mu^{(X)}$ being completely arbitrary. The mathematical technique used is irrelevant, but

[1]) In the following, subsections formulas for $A(\eta)$ are given in *general* form with parameters for which the tables (appendix) provide *two* values in each case, marked by one and two asterisks, to define $A^*(\eta)$ and $A^{**}(\eta)$, respectively.

direct regression methods are often inadequate. Therefore, a convenient semigraphical procedure is suggested in section 9.3.

To arrive at the zero-LET relation (9.3), i.e. the coefficients $\alpha_\mu^{(0)}$, one has in general to use Eq. (9.4) inversely, identifying the left-hand side with the 'empirical' values $\alpha_\mu^{(X)}$ and taking both $A(\eta)$ and r corresponding to the radiation that $\alpha_\mu^{(X)}$ refers to. For X- and γ-rays the solution reads[1]):

Microdosimetric theory:

$$\alpha_\mu^{(0)} = \alpha_\mu^{(X)}\left(1 + \frac{1}{2}\,\bar{z}_D\alpha_\mu^{(X)}\right)\Big/r_X.$$

Strict theory: (9.9)

$$\alpha_\mu^{(0)} = \alpha_\mu^{(X)}\left(1 + \frac{1}{2}\,\tau_D\bar{z}_D\alpha_\mu^{(X)}\right)\Big/r_X.$$

$\bar{z}_D$ and τ_D are given in Table II. In the case of mammalian cells, take τ_D** along with r_X** (Tab. VIII). As to the photon energy spectrum of X-rays, it is sufficient to use an appropriate mean value (half the X-ray tube voltage was the rule followed throughout this book).

As it is based upon a measured dose-effect relation $S_X(D)$, the theory may be called semi-empirical, while genuine theories of radiation action which have to start with biological models of action of primary lesions allow deriving the zero-LET dose-effect relation $S_0(D)$ from 'first principles' as it were, namely by assuming Poisson statistics for the number of primary lesions per reacting unit. Thus, just the hypothetical $S_0(D)$ can most easily be obtained from a biological model, and subsequently the present theoretical apparatus serves as a tool to derive results for real radiations. Further details on this item are given in chapter 8 (Eqs. (8.62) to (8.65) and context).

9.2.4. Photon radiations

The general approximation which can always be used to a high degree of accuracy reads

$$A(\eta) = \eta\left(1 - \frac{1}{2}\,\bar{z}_D\eta\right),$$

$$A(\eta) = \eta\left(1 - \frac{1}{2}\,\tau_D\bar{z}_D\eta\right). \quad (9.10)$$

$\bar{z}_D$ and τ_D are given in Table II. In the standard case (mammalian cell killing) take τ_D**.

For the microdosimetric parameter $\bar{z}_D$ (dose-mean of specific energy) experimental data may also be used.

9.2.5. Fast neutrons

Equations similar to (9.10) with the parameters given in Table III only constitute a crude approximation which is seldom satisfactory. Therefore, the function $A(\eta)$ had to be calculated numerically for a suitable range of η. The results are expressed by the formal parameters $\bar{z}_{D,eff}$ and $\tau_{D,eff}$ which are tabulated *as a function of* η in Table IX:

$$A(\eta) = \eta\left(1 - \frac{1}{2}\,\bar{z}_{D,eff}\eta\right),$$

$$A(\eta) = \eta\left(1 - \frac{1}{2}\,\tau_{D,eff}\bar{z}_{D,eff}\eta\right). \quad (9.11)$$

In the standard case (mammalian cell killing) take $\tau^{**}_{D,eff}$.

An important note concerning numerical problems must be added. The expression (9.2) of the dose-effect relation can lead to considerable errors unless $A(\eta)$ is a completely smooth function up to a large number of decimal places. Therefore, any kind of interpolation in Table IX may often be unsatisfactory. One needs a single analytical expression of $A(\eta)$ for the *entire* range of the variable $\eta = r\alpha_\mu^{(0)}$ corresponding to $\mu = 1, \ldots, n$ in the particular case. The best way is expressing $\bar{z}_{D,eff}$ and $\tau_{D,eff}\bar{z}_{D,eff}$, respectively, as a truncated power series in η (or $\log \eta$) with the coefficients determined by a least-squares

---

[1]) In the approximation suggested for photon radiations ($\bar{z}_D$-approximation, cf. subsection 9.2.4).

fit to *all* tabulated values in the range of η required. A three-term, quadratic expression is sufficient since smoothness, not absolute precision, is the crucial point. If $\eta \bar{z}_{D,eff}$ (or $\eta \tau_{D,eff} \bar{z}_{D,eff}$) is greater than 0.5 for the median of the η-range, it is better to fit $\eta \bar{z}_{D,eff}$ (or $\eta \tau_{D,eff} \bar{z}_{D,eff}$) in the way described.

The parameter $\bar{z}_{D,eff}$ as a function of η may also be derived from a measured or calculated event size spectrum $f_1(z)$ according to

$$\bar{z}_{D,eff} = \frac{\frac{2}{\eta^2}\int\limits_0^\infty dz(e^{-\eta z} - 1 + \eta z)\, f_1(z)}{\int\limits_0^\infty dz z f_1(z)}. \tag{9.12}$$

9.2.6. Ion radiations

Beams of monoenergetic ions are described by the atomic number Z and the energy per nucleon. First, find the dimensionless parameters $\mu_D^{(\delta)}$ and r_δ as well as r in Table IV and VIII, respectively, and calculate $\mu_D^{(i)}$ and r_i/r from the relations

$$\mu_D^{(\delta)} + \mu_D^{(i)} = 1, \qquad \mu_D^{(\delta)} \frac{r_\delta}{r} + \mu_D^{(i)} \frac{r_i}{r} = 1. \tag{9.13}$$

Second, the auxiliary function

$$F(x, \varepsilon) = \frac{h(x + \varepsilon x) - h(x - \varepsilon x)}{\varepsilon x}, \qquad h(x) = \frac{1 - x + x^2/2 - e^{-x}}{x} \tag{9.14}$$

is needed. With these definitions, one has

$$A(\eta) = \mu_D^{(\delta)}\eta\left(1 - \frac{1}{2}\,\bar{z}_D^{(\delta)}\eta\right) + \frac{\sigma_0\ \mathrm{g/cm^3}}{L_\infty}\,F(x, \varepsilon),$$
$$x = \frac{3}{2}\,\frac{\mu_D^{(i)} L_\infty}{\sigma_0\ \mathrm{g/cm^3}}\,\eta, \quad \varepsilon^2 = \frac{8}{3}\,\frac{\delta_2/l}{\mu_D^{(i)} L_\infty}, \tag{9.15}$$
$$l = \frac{4}{3}\sqrt{\frac{\sigma_0}{\pi}},$$

and

$$A(\eta) = \mu_D^{(\delta)}\,\frac{r_\delta}{r}\,\eta\left(1 - \frac{1}{2}\,\tau_D^{(\delta)} \bar{z}_D^{(\delta)}\,\frac{r_\delta}{r}\,\eta\right) + \frac{\sigma_0\ \mathrm{g/cm^3}}{L_\infty}\,F(x, \varepsilon),$$
$$x = \frac{3}{2}\,\frac{\mu_D^{(i)} L_\infty}{\sigma_0\ \mathrm{g/cm^3}}\,\frac{r_i}{r}\,\eta, \tag{9.16}$$
$$\varepsilon^2 = \frac{8}{3}\,\frac{c\delta_2/l}{r_i^2 \mu_D^{(i)} L_\infty}, \qquad l = \frac{4}{3}\sqrt{\frac{\sigma_0}{\pi}}.$$

The microdosimetric parameters $\bar{z}_D^{(\delta)}$ and δ_2 are given in Table IV (together with $\mu_D^{(\delta)}$). The stopping power L_∞ can be found in Table VI (third column). For the parameters $\tau_D^{(\delta)}$ and c (Table V) take $\tau_D^{(\delta)**}$ and c^{**} (along with r^{**}, r_δ^{**}, r_i^{**}) in the standard case.

In the case of mammalian cells (small $\mu_D^{(\delta)}$), the very simple *LET-approximation* can be recommended:

$$A(\eta) = A(\eta) = \frac{\sigma_0\ \mathrm{g/cm^3}}{L_\infty}\left[1 - 2\,\frac{1 - (1 + x)\,e^{-x}}{x^2}\right],$$
$$x = \frac{3}{2}\,\frac{L_\infty \eta}{\sigma_0\ \mathrm{g/cm^3}}. \tag{9.17}$$

A special approximation which is largely equivalent to (9.16) is advantageous if one deals with non-standard applications of the theory, using r-values other than the tabulated r^{**}:

$$A(\eta) = \mu_D^{(\delta)}\,\frac{r_\delta}{r}\,\eta\left(1 - \frac{1}{2}\,\bar{z}_D^{(\delta)}\,\frac{r_\delta}{r}\,\eta\right) + \frac{\sigma_0\ \mathrm{g/cm^3}}{L_\infty}\,F(x, \varepsilon) \tag{9.18}$$
$$x = \frac{3}{2}\,\frac{\mu_D^{(i)} L_\infty}{\sigma_0\ \mathrm{g/cm^3}}\,\frac{r_i}{r}\,\eta, \quad \varepsilon^2 = \frac{8}{3}\,\frac{\delta_2/l}{\mu_D^{(i)} L_\infty},$$
$$l = \frac{4}{3}\sqrt{\frac{\sigma_0}{\pi}}.$$

In this approximation, the general influence of the r-value set on $A(\eta)$ (see page 287, lower case print) has become explicit: it is described by the ratios r_δ/r and r_i/r linked by Eq. (9.13). If r is a combination like (9.5) of the tabulated r^* and r^{**}, take the same combination for r_δ. If r comes from anywhere or is to be fitted,

even rough estimates such as $r_\delta = 1$, $r_\delta = r_\delta^{**}$, $r_\delta = r_X$ (for X-rays of relatively low energy) are much better than using the r-independent microdosimetric theory (9.15), supposing that the difference has any significance at all.
For the parameters $\mu_D^{(\delta)}$ (dose fraction due to indirect events), $\bar{z}_D^{(\delta)}$ (corresponding dose-mean of specific energy), and δ_2 (straggling parameter), data from other origin may also be used. Such an alternative is Table VII where some results of Monte Carlo calculations are given.

9.3. Expressing a dose-effect relation by a sum of exponential functions

It is always possible to describe an experimentally given survival curve (or any similar dose-response plot) by a sum of exponential functions[1]):

$$S(D) = \sum_{\mu=1}^{n} B_\mu \, e^{-\alpha_\mu D}. \qquad (9.19)$$

However, in practice it can be a troublesome task to find a smallest possible n and suitable coefficients B_μ, α_μ yielding a smooth curve $S(D)$ which well fits the given experimental values. Nonlinear regression methods tend to produce undulating curves, with regions of positive slope and even $S(D) > 1$ being not excluded. Moreover, formal handling of the experimental data can hardly allow for some kind of additional intuitive knowledge of what a survival curve should look like, i.e. even a regression curve meeting all formal requirements may appear unsatisfactory.
The *graphic procedure* described in the following avoids these difficulties. It starts with a semilogarithmic plot of $S(D)$, i.e. a smooth curve drawn by eye, which is to be characterized by *five parameters* obtained by simple geometrical operations. Smoothness means that the negative slope in semilogarithmic representation increases monotonously with growing dose, reaching a constant value as $D \to \infty$. The entire method will be presented as a 'recipe' while the mathematical background is outlined in a special appendix (section 9.3.5).
It must be emphasized that this method does not mean to propose a sophisticated five-parameter model of survival curves. The question of whether as many as five parameters can be unambiguously attributed to experimental data with reasonable significance is completely irrelevant, since this is *no model to be tested* but merely a formalism that is never to fail in providing a good fit by *any* expression of the form (9.19) which is otherwise meaningless.

[1]) With the understanding that $S(D)$ refers to a low-LET radiation, but corresponding subscripts and superscripts (see section 9.2.3) are omitted in this section.

9.3.1. Geometrical characterization of a given survival curve

Figure 9.1 shows the geometrical constructions to be made on a semilogarithmic graph of a 'measured' curve (heavy line). The straight line a which reflects the exponential asymptotic behaviour at high doses[2]) defines the *extrapolation number E* and the *final slope* described by $D_0 = D_q/\ln E$. The straight line b characterizes the *initial slope* measured by y_1, and it defines the dose D_1 by its intersection with the straight line a.[3]) So the *distance* y_2 is fixed as shown in Figure 9.1. Finally the straight line c, the tangent to the curve at the dose D_1, has to be drawn. It reflects the *slope at the dose* D_1 measured by y_3.

[2]) Note that the straight line a is an asymptote which cannot touch the curve at a finite dose. Therefore, its slope must be a little steeper than the curve end, in correlation with the gap left between the curve end and the asymptote a. If the latter is drawn wisely, the remaining ambiguity does not appreciably affect the curve finally obtained in the dose range where the starting curve was defined.
[3]) Vanishing initial slope means the line b coinciding with the dose axis, i.e. $y_1 = 0$ and $D_1 = D_q$.

Besides E and D_0, the dimensionless ratios of lengths

$$k_1 = \frac{y_1}{y}, \quad k_2 = \frac{y_2}{y}, \quad k_3 = \frac{y_3}{y} \tag{9.20}$$

are input parameters of the following mathematical procedures which yield the expression (9.19). Conversely, this expression will then describe a smooth curve exhibiting the five parameters exactly, unless the entire procedure has failed because of a parameter set incompatible with the above criterion of smoothness.

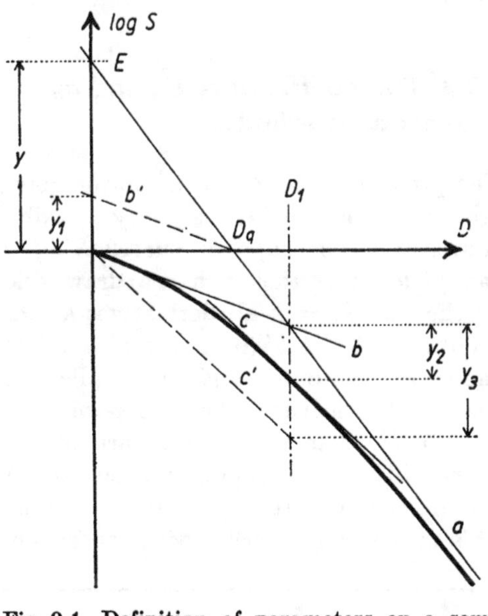

Fig. 9.1. Definition of parameters on a semilogarithmic plot of a dose-effect curve (heavy line). The straight lines b' and c' are meant to be parallel to the straight lines b and c, respectively.

9.3.2. The coefficients B_μ and α_μ (basic method)

With the auxiliary function

$$F(\Xi) = \sum_{i=2}^{n} \binom{n}{i} (-1)^i \frac{i(i-1)\,\Xi}{1+(i-2)\,\Xi} \tag{9.21}$$

the coefficients of the expression (9.19) are given by the formulas

$$\left.\begin{aligned} B_1 &= \frac{F(\gamma)}{F\left(\dfrac{\beta\gamma}{1+\beta}\right)}, \qquad \alpha_1 = \frac{1}{D_0}, \\ B_\mu &= -\frac{(-1)^\mu \dbinom{n}{\mu} \mu(\mu-1)\,\gamma}{F\left(\dfrac{\beta\gamma}{1+\beta}\right)[1+\gamma(\mu-2)]\,[1+\beta+\beta\gamma(\mu-2)]} \\ \alpha_\mu &= \frac{1+\beta(1-k_1)\,[1+\gamma(\mu-2)]}{D_0} \end{aligned}\right\} \text{ for } \mu \geqq 2. \tag{9.22}$$

Only the parameters D_0 and k_1 occur explicitly, while k_2, k_3, and E are used indirectly to fix n and the formal parameters β, γ by the help of tables (appendix) as follows.

In the framework of the formalism, k_3 is unequivocally determined by E and k_2 if n (number of terms in (9.19)) is given. This relationship is displayed numerically by Table X.[1]) Thus the first step is finding the most suitable n to ensure best matching of the actual k_3 and the value in Table X. With n having been fixed, it is easy to read off β and γ from the corresponding part of Table XII. In this way, the original parameter set k_1, k_2, k_3, E, D_0 is replaced by k_1, n, β, γ, D_0, and the coefficients B_μ, α_μ can be calculated.

It is clear that the exact value of k_3 cannot in general be reproduced and, in addition, further inaccuracies may result

[1]) The blank areas in these tables indicate unattainable pairs E, k_2 which should never occur. Otherwise change the construction exemplified in Figure 9.1, considering the hints given in footnote 2) at page 290.

from interpolation in Table XII. Therefore, before calculating the dose-response curve (9.19) to test the fit on the whole, it is useful first to check up k_2, k_3, and E by means of

$$E = B_1,$$

$$k_2 = -\frac{\log\left[1 + \sum_{\mu=2}^{n} B_\mu B_1^{-1-\beta-\beta\gamma(\mu-2)}\right]}{\log B_1},$$

$$k_3 = \frac{1 + \sum_{\mu=2}^{n} [1 + \beta + \beta\gamma(\mu - 2)] B_\mu B_1^{-1-\beta-\beta\gamma(\mu-2)}}{1 + \sum_{\mu=2}^{n} B_\mu B_1^{-1-\beta-\beta\gamma(\mu-2)}} \tag{9.23}$$

where B_1 and B_μ are the expressions (9.22).

Because the case $n = 2$ is exceptional, it has not been included in Tables X and XII. If $n = 2$, γ drops out of Eqs. (9.22), and E, k_2, k_3, being functions of β only, are unambiguously related to one another. These relations are easily derived from Eqs. (9.22), (9.23); for practical purposes we give them in the form of Table XI (appendix). If the parameters E, k_2, and k_3 from the geometrical analysis happen to form a triple contained in Table XI, take $n = 2$ and the corresponding value of β.

The present theory being interpreted as a DNA-lesion theory of cell survival, a given expression (9.19) for a low-LET survival curve implies a continuous set of *survival probability functions* $H(k)$ depending on the quantities Y (molecular weight of total DNA) and s (mean number of DNA lesions per unit dose and unit molecular weight). This problem has been dealt with in section 3.7.2 (cf. also section 7.2.4):

$$H(k) = \sum_{\mu=1}^{n} B_\mu \left(1 - \frac{\alpha_\mu}{sY}\right)^k. \tag{9.24}$$

With the coefficients (9.22) $H(k)$ reads

where $\xi = (D_0 sY - 1)/[\beta(1 - k_1)]$. *The lower bounds* $Y_{\min}$ below which $H(k)$ becomes unreasonable (i.e. violates the conditions $H(k) > 0$ and $H(k + 1) \leq H(k)$) can now be derived from Eq. (9.25). Only the *results* of this lengthy mathematical investigation will be given:

$$D_0 s Y_{\min} = 1 + \beta(1 - k_1)\, \xi_{\min} \tag{9.26}$$

accompanied by the two characteristic values

$$\xi_1 = \frac{1}{2}\,[1 + \gamma(n - 2)],$$
$$\xi_2 = \frac{1}{2}\left[1 + \gamma(n - 2) + (1 - \gamma)\left(1 - \frac{2}{n}\right)\right] \tag{9.27}$$

which determine $\xi_{\min}$ in the following way:

$\underline{\gamma \geqq 1:}$ $\quad \xi_{\min} = \xi_1;$

$\underline{\gamma < 1,\ k_1 = 0:}$ $\quad \xi_{\min} = \xi_2;$ (9.28)

$\underline{\gamma < 1,\ k_1 > 0:}$ $\quad \xi_1 < \xi_{\min} < \xi_2$ with $\xi_{\min}$ having to be calculated from the equation $H(n) = H(n + 1)$.

It should be mentioned that ξ_1 is equivalent to the lower bounds given by the inequality (3.130). This formalism (Eqs. (9.25) to (9.28)) was employed to derive the results of section 7.2.4.

9.3.3. The coefficients B_μ and α_μ (extended method)

The procedure presented in the foregoing section, although it is very flexible concerning the shape of curves $S(D)$ it can yield, nevertheless has a drawback. In the case $k_1 = 0$, all derivatives at the origin, i.e. $S'(0)$, $S''(0)$, ..., vanish up to the $(n - 1)$-th order, and this implies the second derivative to become *positive* if $k_1 > 0$, so that the initial part of the curve in *linear* representation bends upwards as does an exponential function. This may hardly conflict with experimen-

$$H(k) = \frac{\left(1 - \frac{1}{D_0 sY}\right)^k}{F\left(\frac{\beta\gamma}{1 + \beta}\right)} \left\{F(\gamma) - \sum_{\mu=2}^{n} \binom{n}{\mu} (-1)^\mu \frac{\mu(\mu - 1)\,\gamma[1 - (1 + \gamma(\mu - 2))/\xi]^k}{[1 + \gamma(\mu - 2)]\,[1 + \beta + \beta\gamma(\mu - 2)]}\right\} \tag{9.25}$$

tal data[1]), but theoretical arguments suggest that there should always exist a quadratic component of damage. i.e. a *negative* second derivative at $D = 0$. This sophistical question could be ignored unless there were the following consequences. From an X-ray curve with $S'(0) < 0$ and $S''(0) > 0$ the present theory predicts an RBE maximum at a small but *finite* dose of any high-LET radiation (some examples can be seen in the right-hand panel of Fig. 7.24). Another point concerns the feasibility of the procedure itself: in case that k_2 is relatively small so that low numbers n are excluded, one often has a k_3 value smaller than the tabulated k_3 for every n possible.

The extended algorithm including an additional formal parameter, E_1, remedies these shortcomings. It starts from the same five geometric parameters and makes use of the previous relations (9.21—23), but all the quantities in these relations *will now be asterisked* to avoid confusion. This means also Tables X to XII: what they contain must now be read as n^*, E^*, k_2^*, k_3^*, β^*, γ^*. With this notation we have

$$n = 2n^*, \tag{9.29}$$

$$B_\mu = E_1 B_\mu^* \qquad \alpha_\mu = \frac{\log B_1^*}{\log (E_1 B_1^*)}\, \alpha_\mu^* + \frac{\log E_1}{\log (E_1 B_1^*)}\, \frac{1}{D_0} \quad \text{for} \quad 1 \leqq \mu \leqq n^*,$$

$$\left.\begin{aligned} B_\mu &= -(E_1 - 1)\, B^*_{\mu - n^*} \\ \alpha_\mu &= \frac{\log B_1^*}{\log (E_1 B_1^*)}\, \alpha^*_{\mu - n^*} + \left(1 + \frac{1 - k_1}{E_1 - 1}\right) \frac{\log E_1}{\log (E_1 B_1^*)}\, \frac{1}{D_0} \end{aligned}\right\} \quad \text{for} \quad n^* + 1 \leqq \mu \leqq 2n^*, \tag{9.30}$$

and

$$E = E_1 B_1^*,$$

$$k_2 = \frac{\log B_1^*}{\log (E_1 B_1^*)}\, k_2^* - \frac{\log [1 - (E_1 - 1)\, E_1^{-E_1/(E_1 - 1)}]}{\log (E_1 B_1^*)}, \tag{9.31}$$

$$k_3 = \frac{\log B_1^*}{\log (E_1 B_1^*)}\, k_3^* + \frac{\log E_1}{\log (E_1 B_1^*)} \times \frac{1 - E_1 E_1^{-E_1/(E_1 - 1)}}{1 - (E_1 - 1)\, E_1^{-E_1/(E_1 - 1)}}. \tag{9.31}$$

Table 9.1 is to explain, by the hand of an example[2]), how E_1 as well as n^*, β^*, γ^* are obtained from the parameters k_2, k_3, and E. In brief: after transforming k_2, k_3, E into k_2^*, k_3^*, E^*, Tables X to XII are used as before. The fourth parameter, E_1, induces ambiguity which allows starting with *a series of trial values of* k_2^*, for which the figures occurring in the tables (appendix) should always be chosen as shown in the first row of Table 9.1.[3]) Calculate $(k_2^* - k_2) \lg E$ (second row) and find these values in the corresponding k_2^*-columns of Table XIII which provides $\lg E_1$ and Δk_3 from the first and last columns, respectively. Now compute $\lg E^*$ and k_3^* as indicated in Table 9.1. Finally, use the triples k_2^*, k_3^*, $\lg E^*$ for each trial separately to cast about for suitable n^*'s by the criterion of k_3^*-coincidence (Table X). The best-fitting variants for each trial have been included in Table 9.1 (last rows). Several trials usually lead to utilizable results; the trial $k_2^* = 0.225$ has been finally chosen in Table 9.1, yielding the β^* and γ^* shown according to Table XII. With these figures, the test

---

1) A finite initial region of genuine exponential character inferred from fractionation experiments has been even suggested expressly (KOGELNIK and WITHERS, 1978).

2) Adding the other two parameters $k_1 = 0.2$ and $D_0 = 141$ rad, this example corresponds to the dashed heavy line in panel *c* of Figure 7.17.

3) The last column, $k_2^* = k_2 = 0.275$, leads to $E_1 = 1$, which corresponds to the basic method dealt with in subsection 9.3.2.

Tab. 9.1. Example for the operations of parameter transformation. Starting set: $k_2 = 0.275$, $k_3 = 0.618$, $E = 20$ ($\lg E = 1.301$)

	Trial k_2^*	0.150	0.175	0.200	0.225	0.250	0.275
Calculate:	$(k_2^* - k_2) \lg E$	−0.163	−0.130	−0.098	−0.065	−0.033	0.000
From Table XIII:	$\lg E_1$	0.540	0.485	0.420	0.332	0.215	0.000
	Δk_3	0.695	0.688	0.680	0.670	0.657	0.632
Calculate:	$\lg E^* = \lg E - \lg E_1$	0.761	0.816	0.881	0.969	1.086	1.301
	$k_3^* = (k_3 \lg E - \Delta k_3 \lg E_1)/\lg E^*$	0.563	0.576	0.588	0.600	0.610	0.618
From Table X:	k_3^*	0.558	0.570	0.586	0.600	0.607	0.631
	n^*	10	8	7	6	5	5
From Table XII:	β^*				0.916		
	γ^*				0.634		

for reproducibility of the starting parameters (Eqs. (9.31)) yields: $k_2 = 0.2747$, $k_3 = 0.6175$, $E = 19.995$.

The extended method is able to give the dose-effect relation (9.19) a negative second derivative at the origin: $S''(0) < 0$. The condition for this to be the case reads

$$\sqrt{\frac{E_1}{E_1 - 1}} \lg E_1 > \frac{k_1}{1 - k_1} \lg E. \qquad (9.32)$$

To fulfil this condition may further confine the range of utilizable trials. This examination can be made after the third row of Table 9.1.

9.3.4. Special mathematical models of cell survival

Several mathematical expressions, which are partly motivated by customary target theory, are in use to smooth experimental data (HETHCOTE *et al.*, 1976; MILLAR *et al.*, 1978). The currently preferred models are the *two-component model*,

$$S(D) = e^{-k_1 D/D_0}[1 - (1 - e^{-(1-k_1)D/D_0})^n], \qquad (9.33)$$

and the *linear-quadratic model*

$$S(D) = e^{-\alpha D - \beta D^2}. \qquad (9.34)$$

If one of these models has been fitted to experimental data, it is easy to convert the result into the expression (9.19).

The conversion is trivial in the case of the two-component model:

$$S(D) = \sum_{\mu=1}^{n} (-1)^{\mu-1} \binom{n}{\mu} e^{-[k_1 + (1-k_1)\mu]D/D_0}, \qquad (9.35)$$

provided that n is an integer. This is equivalent to $\beta = \gamma = 1$ in the framework of the basic method described in subsection 9.3.2.

The linear-quadratic model can be only approximated by the expression (9.19). A useful approximation, which is based on the extended method (subsection 9.3.3), is tabulated in Table 9.2. The coefficients B_μ are constants while the α_μ's depend on the parameters α and β of Eq. (9.34):

$$\alpha_\mu = \alpha + \sqrt{\beta}\, a_\mu, \qquad (9.36)$$

with the constant auxiliary coefficients a_μ given in Table 9.2. This approach is exact with respect to the first and second derivatives $S'(0)$ and $S''(0)$, and the dose error at equal survival does not exceed ±6 per cent down to 0.001 surviving fraction. It is clear that beyond this range the approximation loses its validity since any finite sum of exponential functions

Tab. 9.2. Constant coefficients to approximate the linear-quadratic model by a n-term sum of exponential functions ($n = 10$)

μ	B_μ	a_μ	μ	B_μ	a_μ
1	11587.6904	4.8175518	6	−22838.1151	5.6902848
2	−10767.3448	4.9214144	7	−13439.9476	6.0772201
3	−25631.6053	5.0956243	8	12488.4722	6.1810827
4	23817.0266	5.1994869	9	2919.8757	6.5680180
5	24578.1121	5.5864222	10	−2713.1642	6.6718806

ends up in a straight line while the curve according to the model (9.34) continues bending downwards. This divergence is however unimportant as long as the experimental data themselves do not reach below surviving fractions of about 0.001. In these cases, it is a very convenient and often sufficient method to first fit the data with the simple linear-quadratic model and then to use Table 9.2 and Eq. (9.36), thus arriving at the expression (9.19).

9.3.5. Appendix: derivation of the mathematical formalism

Let us start with expressing $S(D)$ by a function $y(x)$ according to

$$S(D) = e^{-k_1 D/D_0} y(e^{-(1-k_1)D/D_0}). \tag{9.37}$$

$y(x)$, which is defined in the interval $0 \leq x \leq 1$, must have the properties

$$\begin{aligned} &y(1) = 1, \quad y(0) = 0, \\ &y'(1) = 0, \quad y'(0) = E, \end{aligned} \tag{9.38}$$

$$y(x) = \left\{F(\gamma)\,x - \sum_{\mu=2}^{n} (-1)^\mu \binom{n}{\mu} \frac{\mu(\mu-1)\,\gamma x^{1+\beta+\beta\gamma(\mu-2)}}{[1+\gamma(\mu-2)]\,[1+\beta+\beta\gamma(\mu-2)]}\right\} \Big/ F\left(\frac{\beta\gamma}{1+\beta}\right) \tag{9.43}$$

to ensure consistency with the parameters D_0, k_1, and E as defined by Figure 9.1 and Eqs. (9.20). For $S(D)$ to become a sum of exponential functions, $y(x)$ must be a sum of (non-integer) power functions of which the first power, i.e. the term Ex, is the smallest one.

To express the remaining parameters k_2 and k_3 by the function $y(x)$, we must start with the analytical form of definition of the dose D_1:

$\exp(-k_1 D_1/D_0) = E \exp(-D_1/D_0)$ or

$$x_1 = e^{-(1-k_1)D_1/D_0} = E^{-1}. \tag{9.39}$$

The geometrical definitions of k_2 and k_3 can be formalized as

$$k_2 = \frac{-\ln S(D_1) - k_1 D_1/D_0}{\ln E}, \qquad k_3 = \frac{-D_1 \dfrac{d \ln S(D_1)}{dD_1} - k_1 \dfrac{D_1}{D_0}}{\ln E}, \tag{9.40}$$

and the evaluation by means of Eq. (9.37) yields

$$k_2 = \frac{\ln y(E^{-1})}{\ln E^{-1}}, \quad k_3 = \frac{E^{-1} y'(E^{-1})}{y(E^{-1})}. \tag{9.41}$$

It is easy to realize that the criterion of smoothness requires that $y(x)$ be a monotonously increasing function with *monotonously decreasing slope:* $y''(x) < 0$. In compliance with this requirement the ansatz

$$y''(x) = -Cx^{\beta-1}(1 - x^{\beta\gamma})^{n-2}, \quad \beta, \gamma > 0, \tag{9.42}$$

has been used. β, γ, and n are parameters, whereas the normalization constant C as well as the integration constants rising on twice integrating the expression (9.42) have to be determined by the first three of the conditions (9.38). The result reads

with the function $F(\Xi)$ defined by Eq. (9.21). The identity of the expressions (9.37), (9.43) on the one hand and (9.19), (9.22) on the other hand is obvious. Likewise the parameter equations (9.23) follow from $E = y'(0)$ and Eqs. (9.41). The equations of E and k_2 have been considered as a (highly nonlinear) system of equations for the formal parameters β and γ which has been solved by numerical methods.

In this way Table XII has been obtained, Table X materializing by simple insertion into the k_3-equation. For the upper and lower bounds to k_2 seen in Table XII, analytical expressions can be derived by the limiting process $\gamma \to \infty$ and $\gamma \to 0$, respectively:

$$k_{2,\max} = -\frac{\ln\left[1 - (1 - E^{-1})\, E^{-1/(E-1)}\right]}{\ln E},$$

$$k_{2,\min} = -\frac{\ln\left\{1 - E^{-1/[E^{1/(n-1)}-1]} \sum_{i=0}^{n-2} \frac{1}{i!} \left[\frac{\ln E}{E^{1/(n-1)} - 1}\right]^i \left[1 - E^{i/(n-1)-1}\right]\right\}}{\ln E}. \tag{9.44}$$

In order to get the formalism manageable, it must be required that the dose D_1 given by Eq. (9.39) make the argument $x_1^* = \exp[-(1 - k_1 - \lambda)\, D_1/D_0]$ equal to E^{*-1}. This fixes λ:

$$\lambda = (1 - k_1)\, \frac{\ln E_1}{\ln (E_1 E^*)}. \tag{9.47}$$

The extended method described in subsection 9.3.3 is based on the following generalization of the ansatz (9.37):

$$S(D) = \mathrm{e}^{-k_1 D/D_0}[E_1\, \mathrm{e}^{-\lambda D/D_0} - (E_1 - 1) \times \mathrm{e}^{-\lambda (D/D_0) E_1/(E_1 - 1)}]\; y^*(\mathrm{e}^{-(1-k_1-\lambda) D/D_0}). \tag{9.45}$$

The function $y^*(x^*)$, now asterisked as well as its parameters E^*, k_2^*, k_3^*, continues to be given by the expression (9.43). The factor in square brackets added does not violate the criterion of smoothness, and the whole expression is in keeping with the definition of the parameters D_0 and k_1. The extrapolation number is seen to be

$$E = E_1 E^*. \tag{9.46}$$

Now the extended formulas (9.30) can be seen to be equivalent to the expression (9.45), and the extended parameter equations (9.31) follow from Eqs. (9.40) and Eq. (9.46). It is these parameter equations (9.31) on which the procedure described in Table 9.1 is based, which becomes obvious after rewriting them in the form

$$(k_2^* - k_2) \lg E = \lg\left[1 - (E_1 - 1)\, E_1^{-E_1/(E_1-1)}\right] + k_2^* \lg E_1,$$

$$k_3^* = \frac{k_3 \lg E - \Delta k_3 \lg E_1}{\lg E^*}, \tag{9.48}$$

$$\Delta k_3 = \frac{1 - E_1 E_1^{-E_1/(E_1-1)}}{1 - (E_1 - 1)\, E_1^{-E_1/(E_1-1)}}.$$

These expressions, $(k_2^* - k_2) \lg E$ and Δk_3, are tabulated in dependence on $\lg E_1$, k_2^* and $\lg E_1$, respectively, in Table XIII.

Appendix

Tables

The following thirteen tables constitute a compilation of data computed along the lines described throughout the book. To facilitate their use detailed legends are put together as follows.

Table I (pages 300, 301). Synopsis of final equations (chapter 3) the numerical tables are indirectly (chapter 5 and 8) based upon.

Table II (page 302). Microdosimetric parameters and virtual multipliers for photon radiations.
Head-row: size of sensitive volume (cross-section and diameter).
First column: photon energy.
Second column: mean energy per ionization.
Main area: panels containing 5 parameters:

$\bar{y}_F$	$\bar{y}_D$
τ_F	τ_D^{**}
	τ_D^{*}

$\bar{y}_F$ — frequency mean of lineal energy [keV/μm],
$\bar{y}_D$ — dose mean of lineal energy [keV/μm],
τ_F, τ_D^{**}, τ_D^{*} — dimensionless virtual multipliers.

Table III (page 303). Microdosimetric parameters and virtual multipliers for fast neutron radiations.
Head-row: size of sensitive volume (cross-section and diameter).

First column: fast neutron energy.
Second column: mean energy per ionization.
Main area: panels containing 5 parameters:

$\bar{y}_F$	$\bar{y}_D$
τ_F	τ_D^{**}
	τ_D^{*}

$\bar{y}_F$ — frequency mean of lineal energy [keV/μm],
$\bar{y}_D$ — dose mean of lineal energy [keV/μm],
τ_F, τ_D^{**}, τ_D^{*} — dimensionless virtual multipliers.

Table IV (page 304). Microdosimetric parameters for ion radiations (partial quantities).
Head-row: size of sensitive volume (cross-section and diameter) and cut-off energy of δ-electrons.
First column: energy per nucleon.
Second column: mean energy per ionization for the *total* ion track segment.
Main area: panels containing 4 parameters:

$\bar{y}_F^{(\delta)}$	$\bar{y}_D^{(\delta)}$
δ_2	$\mu_D^{(\delta)}$

$\bar{y}_F^{(\delta)}$ — frequency mean of lineal energy [keV/μm] for δ-rays only,
$\bar{y}_D^{(\delta)}$ — dose mean of lineal energy [keV/μm] for δ-rays only,
$\mu_D^{(\delta)}$ — percentage of dose due to δ-ray events [%],
δ_2 — straggling parameter [keV] for direct ion events.

Table V (page 305). Virtual multipliers for ion radiations (partial quantities).
Head-row: size of sensitive volume (cross-section and diameter).
First column: energy per nucleon.
Main area: panels containing 5 dimensionless parameters:

$\tau_F^{(\delta)}$	
c^{**}	$\tau_D^{(\delta)**}$
c^{*}	$\tau_D^{(\delta)*}$

$\tau_F^{(\delta)}$, $\tau_D^{(\delta)**}$, $\tau_D^{(\delta)*}$ pertaining to $\bar{y}_F^{(\delta)}$ and $\bar{y}_D^{(\delta)}$, respectively,
c^{**}, c^{*} pertaining to δ_2.

Table VI (pages 306—308). Microdosimetric parameters for ion radiations (total quantities).

Head-row: size of sensitive volume (cross-section and diameter).
First column: atomic number.
Second column: energy per nucleon.
Third column: stopping power.
Main area: pairs of parameters

$\bar{y}_F$	$\bar{y}_D$

$\bar{y}_F$ — total frequency mean of lineal energy [keV/μm],
$\bar{y}_D$ — total dose mean of lineal energy [keV/μm].

Table VII (page 309). Microdosimetric parameters for δ-rays of ion radiations; comparison of Monte Carlo results (GUBIN *et al.*, 1976) for specified ions with the present results.

$\mu_D{}^{(\delta)}$ — percentage of dose due to δ-ray events [%],
$\bar{y}_F{}^{(\delta)}$ — frequency mean of lineal energy [keV/μm],
$\bar{y}_D{}^{(\delta)}$ — dose mean of lineal energy [keV/μm].

Table VIII (page 310). Total r-values for photon, neutron, and ion radiations; partial r-values r_δ for δ-rays only (right-hand part of ion table).
First columns: energy or energy per nucleon.
Head-row of ion table: atomic number (left), cross-section of sensitive volume (right).
Main areas: triples of r-values for different conditions:

r^{**}
r^{**}
r^{*}

r^{**} oxic conditions, r^{**} anoxic conditions } nominally applying to DNA double-strand breaks

r^{*} nominally applying to DNA single-strand lesions.

Table IX (pages 311—323). Special parameters for neutron radiations.
Head-row: size of sensitive volume (cross-section and diameter).
First column: argument η of the function $A(\eta)$.
Main area: triples of parameters

$\bar{y}_{D,\text{eff}}$	$\tau^{**}_{D,\text{eff}}$	$\tau^{*}_{D,\text{eff}}$

$\bar{y}_{D,\text{eff}}$ — effective dose mean of lineal energy [keV/μm],
$\tau^{**}_{D,\text{eff}}$, $\tau^{*}_{D,\text{eff}}$ — dimensionless virtual multipliers.

Tables X to XIII (pages 324—339). Auxiliary tables for dose-effect curve analysis (see section 9.3).

Table I. Synopsis of final equations

Input Quantities

Primary description of radiation fields		
Photons:	$q_e^{(\gamma)}(E)$	initial energy spectrum of photo and Compton electrons
Fast neutrons:	$q_i^{(n)}(E)$	initial energy spectra of ions i set in motion by fast neutrons
Ion radiations:	i	denotes the sort of ion
	$E_i^{(0)}$	track-segment energy of ion

Physical relationships (water medium)	Electrons	Ions
Stopping power:	$L_\infty^{(e)}(E)$	$L_\infty^{(i)}(E)$
δ-electron spectrum:	$N_e(E_e, E)$	$N_i(E_i, E)$
Primary ionizations:	$p_e^{(\delta)}(E)$	$p_i^{(\delta)}(E)$
Cluster frequencies:	$\omega_e^{(\delta)}(E, j)$	$\omega_i^{(\delta)}(E, j)$
Range:	$R_e(E)$	
Modified range:	$\tilde{R}_e(E, l)$	

Derived Spectra and Functions Describing Radiation Fields

$$M_e(E_e, E) = \int_E^\infty \mathrm{d}E' N_e(E_e, E') \qquad M_i(E_i, E) = \int_E^\infty \mathrm{d}E' N_i(E_i, E')$$

Photon radiations

$$L_\infty^{(e)}(E)\, \varphi_\gamma^{(\delta)}(E) = \begin{cases} \int_E^\infty \mathrm{d}E' q_e^{(\gamma)}(E') + \int_0^\infty \mathrm{d}E_e \varphi_\gamma^{(\delta)}(E_e)\, M_e(E_e, E) & \text{for } E \geqq \delta \\ L_\infty^{(e)}(\delta)\, \varphi_\gamma^{(\delta)}(\delta) + \int_E^\delta \mathrm{d}E' q_e^{(\gamma)}(E') & \text{for } E < \delta \end{cases}$$

Fast neutron radiations

$$\varphi_i^{(n)}(E) = [L_\infty^{(i)}(E)]^{-1} \int_E^\infty \mathrm{d}E' q_i^{(n)}(E') \qquad M_n(E) = \sum_i \int_0^\infty \mathrm{d}E_i \varphi_i^{(n)}(E_i)\, M_i(E_i, E)$$

$$L_\infty^{(e)}(E)\, \varphi_n^{(\delta)}(E) = \begin{cases} M_n(E) + \int_0^\infty \mathrm{d}E_e \varphi_n^{(\delta)}(E_e)\, M_e(E_e, E) & \text{for } E \geqq \delta \\ L_\infty^{(e)}(\delta)\, \varphi_n^{(\delta)}(\delta) & \text{for } E < \delta \end{cases}$$

Ion radiations

$$L_\infty^{(e)}(E)\, T_i^{(\delta)}(E_i, E) = \begin{cases} M_i(E_i, E) + \int_0^\infty \mathrm{d}E_e T_i^{(\delta)}(E_i, E_e)\, M_e(E_e, E) & \text{for } E \geqq \delta \\ L_\infty^{(e)}(\delta)\, T_i^{(\delta)}(E_i, \delta) & \text{for } E < \delta \end{cases}$$

Auxiliary Quantities Pertaining to the Model of DNA Lesion Production

$$u_l(l') = \frac{1}{l_l} \int_{l'}^\infty \mathrm{d}l'' w_l(l'') \qquad \varkappa(E, \zeta) = p^{(\delta_l)}(E) \left[1 - \sum_{j=1}^{j_{\max}} \mathrm{e}^{-\zeta j} \omega^{(\delta_l)}(E, j)\right] \qquad R_e(\delta_l) = \varrho_l l_l$$

Electrons

$$\tilde{R}_e(E_e^{(f)}, \varrho_l l_l) = \max\,[\tilde{R}_e(E, \varrho_l l_l) - \varrho_l l', 0]$$

$$\Gamma_e(E) = -\varrho_l l_l \int_0^\infty \mathrm{d}l' u_l(l') \sum_\nu b_\nu \varkappa_e(E, \zeta_\nu) \exp\left[-\int_{E_e^{(f)}(E,l')}^{E} \mathrm{d}E' \frac{\varkappa_e(E', \zeta_\nu)}{L_\infty^{(e)}(E')}\right]$$

Ions

$$R_i(E_i^{(f)}) = \max\,[R_i(E) - \varrho_l l', 0] \qquad R_i(E) = \int_0^E \mathrm{d}E_i [L_\infty^{(i)}(E_i)]^{-1}$$

$$G_i(E) = \int_0^\infty \mathrm{d}l' w_l(l') \sum_\nu b_\nu \exp\,[-\varrho_l l' \varkappa_i(E, \zeta_\nu)]$$

$$\Gamma_i(E) = -\varrho_l l_l \int_0^\infty \mathrm{d}l' u_l(l') \sum_\nu b_\nu \varkappa_i(E, \zeta_\nu) \exp\left[-\int_{E_i^{(f)}(E,l')}^{E} \mathrm{d}E' \frac{\varkappa_i(E', \zeta_\nu)}{L_\infty^{(i)}(E')}\right]$$

Parameters describing sensitive volumes

	Gross sensitive volume	Targets related to DNA
Mean projected area:	σ_0	
Mean chord length:	l	l_ι
Chord length spectrum:	$w(l')$	$w_\iota(l')$
Mass density factor:	ϱ	ϱ_ι
Total length of DNA:	Y	

Parameters describing responses

Reactions at the cellular level (exemplified for short by cell survival):	Reactions at the molecular level (DNA lesions):
coefficients B_μ, Z_μ	coefficients b_ν, ζ_ν
in the expression for the survival probability	in the expression for the probability
$H(k) = \sum_\mu B_\mu \exp(-Z_\mu k)$	$\lambda_j = \sum_\nu b_\nu \exp(-\zeta_\nu j)$

Final Equations: Number of Lesions per Unit Dose and Unit Length of DNA

Photon radiations

$$s = \frac{\pi}{4} \frac{l_\iota \text{ g/cm}^3}{\int_0^\infty \mathrm{d}E E q_e^{(\gamma)}(E)} \int_0^\infty \mathrm{d}E \varphi_\gamma^{(\delta_\iota)}(E)\, \Gamma_e(E)$$

Ion radiations

$$s = \frac{\pi}{4} \frac{l_\iota \text{ g/cm}^3}{L_\infty^{(\iota)}(E_\iota^{(0)})} \left[G_\iota(E_\iota^{(0)}) + \int_0^\infty \mathrm{d}E T_\iota^{(\delta_\iota)}(E_\iota^{(0)}, E)\, \Gamma_e(E)\right]$$

Fast neutron radiations

$$s = \frac{\pi}{4} \frac{l_\iota \text{ g/cm}^3}{\sum_\iota \int_0^\infty \mathrm{d}E E q_\iota^{(n)}(E)} \left[\sum_\iota \int_0^\infty \mathrm{d}E \varphi_\iota^{(n)}(E)\, \Gamma_\iota(E) + \int_0^\infty \mathrm{d}E \varphi_n^{(\delta_\iota)}(E)\, \Gamma_e(E)\right]$$

Final Equations: Dose-Response Relation for Cellular Reactions

$$S(D) = \sum_\mu B_\mu \exp[-D\Pi(Z_\mu)]$$

$$u(l') = \frac{1}{l} \int_{l'}^\infty \mathrm{d}l'' w(l'') \qquad K(E, Z) = (1 - \mathrm{e}^{-Z}) \frac{h}{\varrho} \Gamma(E) + \int_{\delta_\iota}^{\delta} \mathrm{d}E' N(E, E')\, \xi(E', Z) \qquad h = \frac{\pi}{4} \frac{Y l_\iota}{V} \qquad V = \sigma_0 l$$

$$\xi(E, Z) = 1 - \exp\left\{-\int_0^E \mathrm{d}E_e [L_\infty^{(e)}(E_e)]^{-1} \left[(1 - \mathrm{e}^{-Z}) \frac{h}{\varrho} \Gamma_e(E_e) + \int_{\delta_\iota}^{\delta} \mathrm{d}E' N_e(E_e, E')\, \xi(E', Z)\right]\right\} \qquad R_e(\delta) = \varrho l$$

$$\tilde{R}_e(E_e^{(f)}, \varrho l) = \max[\tilde{R}_e(E, \varrho l) - \varrho l', 0] \qquad R_\iota(E^{(f)}) = \max[R_\iota(E) - \varrho l', 0] \qquad R_\iota(E) = \int_0^E \mathrm{d}E_\iota [L_\infty^{(\iota)}(E_\iota)]^{-1}$$

Photon radiations

$$\Pi(Z) = \frac{V\varrho \text{ g/cm}^3}{\int_0^\infty \mathrm{d}E E q_e^{(\gamma)}(E)} \int_0^\infty \mathrm{d}E \varphi_\gamma^{(\delta)}(E) \int_0^\infty \mathrm{d}l' u(l')\, K_e(E, Z) \frac{1 - \xi(E, Z)}{1 - \xi(E_e^{(f)}(E, l'), Z)}$$

Fast neutron radiations

$$\Pi(Z) = \frac{V\varrho \text{ g/cm}^3}{\sum_\iota \int_0^\infty \mathrm{d}E E q_\iota^{(n)}(E)} \left\{\sum_\iota \int_0^\infty \mathrm{d}E \varphi_\iota^{(n)}(E) \int_0^\infty \mathrm{d}l' u(l')\, K_\iota(E, Z) \exp\left[-\int_{E_\iota^{(f)}(E, l')}^{E} \mathrm{d}E_\iota \frac{K_\iota(E_\iota, Z)}{L_\infty^{(\iota)}(E_\iota)}\right]\right.$$
$$\left. + \int_0^\infty \mathrm{d}E \varphi_n^{(\delta)}(E) \int_0^\infty \mathrm{d}l' u(l')\, K_e(E, Z) \frac{1 - \xi(E, Z)}{1 - \xi(E_e^{(f)}(E, l'), Z)}\right\}$$

Ion radiations

$$K_\iota^{(0)}(E, Z) = (1 - \mathrm{e}^{-Z}) \frac{h}{\varrho} G_\iota(E) + \int_{\delta_\iota}^{\delta} \mathrm{d}E' N_\iota(E, E')\, \xi(E', Z)$$

$$\Pi(Z) = \frac{\sigma_0 \text{ g/cm}^3}{L_\infty^{(\iota)}(E_\iota^{(0)})} \left\{\int_0^\infty \mathrm{d}l' w(l') \left[1 - \mathrm{e}^{-\varrho l' K_\iota^{(0)}(E_\iota^{(0)}, Z)}\right]\right.$$
$$\left. + \varrho l \int_0^\infty \mathrm{d}E T_\iota^{(\delta)}(E_\iota^{(0)}, E) \int_0^\infty \mathrm{d}l' u(l')\, K_e(E, Z) \frac{1 - \xi(E, Z)}{1 - \xi(E_e^{(f)}(E, l'), Z)}\right\}$$

Table II. Microdosimetric parameters and virtual multipliers for photon radiations

E/MeV	W/eV	$\sigma_0 = .008\,\mu m^2$ $d = 0.1\,\mu m$	$\sigma_0 = .196\,\mu m^2$ $d = 0.5\,\mu m$	$\sigma_0 = .785\,\mu m^2$ $d = 1\,\mu m$	$\sigma_0 = 2\,\mu m^2$ $d = 1.60\,\mu m$	$\sigma_0 = 5\,\mu m^2$ $d = 2.52\,\mu m$	$\sigma_0 = 10\,\mu m^2$ $d = 3.57\,\mu m$	$\sigma_0 = 20\,\mu m^2$ $d = 5.05\,\mu m$	$\sigma_0 = 30\,\mu m^2$ $d = 6.18\,\mu m$	$\sigma_0 = 40\,\mu m^2$ $d = 7.14\,\mu m$	$\sigma_0 = 50\,\mu m^2$ $d = 7.98\,\mu m$	$\sigma_0 = 60\,\mu m^2$ $d = 8.74\,\mu m$	$\sigma_0 = 80\,\mu m^2$ $d = 10.1\,\mu m$	$\sigma_0 = 100\,\mu m^2$ $d = 11.3\,\mu m$
0.010	36.40	4.685 9.238 0.886 1.300 0.746	4.105 6.477 0.916 1.140 0.917	3.751 5.440 0.893 1.094 0.963	3.257 4.863 0.885 1.041 0.971	2.736 3.930 0.884 1.038 0.975	2.305 3.196 0.886 1.025 0.979	1.862 2.431 0.898 1.019 0.981	1.605 2.052 0.911 1.017 0.982	1.451 1.821 0.920 1.016 0.983	1.326 1.645 0.927 1.015 0.985	1.235 1.516 0.931 1.014 0.986	1.102 1.328 0.937 1.013 0.988	1.009 1.201 0.941 1.012 0.988
0.020	35.47	3.063 7.051 0.796 1.488 0.673	2.513 4.666 0.941 1.264 0.867	2.390 4.112 0.945 1.156 0.906	2.329 3.703 0.918 1.119 0.943	2.179 3.398 0.903 1.066 0.953	1.998 3.110 0.902 1.048 0.971	1.786 2.803 0.899 1.038 0.991	1.675 2.623 0.896 1.030 0.994	1.560 2.510 0.894 1.020 0.995	1.494 2.373 0.894 1.017 0.996	1.424 2.265 0.894 1.016 0.996	1.324 2.103 0.895 1.012 0.996	1.236 1.947 0.896 1.010 0.996
0.030	35.48	2.614 6.469 0.701 1.570 0.637	1.875 4.041 0.956 1.306 0.820	1.839 3.367 0.942 1.212 0.888	1.751 3.080 0.950 1.133 0.907	1.667 2.752 0.938 1.099 0.946	1.567 2.556 0.933 1.067 0.962	1.440 2.383 0.928 1.032 0.969	1.375 2.266 0.919 1.021 0.978	1.294 2.176 0.914 1.018 0.980	1.256 2.142 0.911 0.998 0.981	1.207 2.088 0.910 0.995 0.982	1.141 1.982 0.910 0.990 0.987	1.074 1.938 0.910 0.984 0.988
0.050	35.45	2.454 6.806 0.612 1.564 0.638	1.572 4.198 0.931 1.332 0.812	1.469 3.356 0.961 1.245 0.872	1.405 2.886 0.963 1.172 0.892	1.324 2.405 0.960 1.132 0.929	1.242 2.114 0.957 1.093 0.950	1.156 1.880 0.944 1.050 0.963	1.080 1.753 0.943 1.032 0.975	1.033 1.667 0.943 1.022 0.979	0.986 1.624 0.943 1.002 0.981	0.948 1.576 0.943 0.995 0.987	0.895 1.494 0.943 0.992 0.995	0.844 1.456 0.943 0.974 0.997
0.075	35.32	2.724 7.173 0.594 1.507 0.664	1.721 4.604 0.915 1.297 0.842	1.595 3.827 0.951 1.205 0.892	1.534 3.334 0.946 1.153 0.919	1.430 2.872 0.950 1.108 0.939	1.335 2.516 0.948 1.083 0.955	1.229 2.158 0.943 1.062 0.969	1.150 1.967 0.942 1.041 0.972	1.093 1.813 0.942 1.039 0.978	1.044 1.703 0.942 1.033 0.983	1.002 1.617 0.942 1.029 0.986	0.942 1.492 0.942 1.019 0.989	0.888 1.398 0.942 1.014 0.993
0.100	35.26	2.672 6.807 0.626 1.532 0.655	1.789 4.410 0.918 1.306 0.839	1.676 3.727 0.951 1.211 0.891	1.625 3.313 0.947 1.159 0.919	1.541 2.931 0.946 1.111 0.939	1.459 2.635 0.942 1.086 0.957	1.364 2.350 0.933 1.059 0.968	1.289 2.185 0.929 1.045 0.973	1.237 2.063 0.928 1.038 0.979	1.189 1.967 0.928 1.034 0.983	1.149 1.890 0.929 1.029 0.985	1.089 1.770 0.929 1.021 0.988	1.036 1.678 0.929 1.015 0.988
0.125	35.07	2.442 6.328 0.628 1.577 0.632	1.640 4.031 0.924 1.338 0.822	1.555 3.419 0.956 1.239 0.877	1.514 3.064 0.959 1.182 0.907	1.469 2.741 0.952 1.135 0.931	1.417 2.507 0.946 1.104 0.946	1.350 2.278 0.938 1.076 0.959	1.300 2.143 0.934 1.062 0.966	1.260 2.049 0.931 1.053 0.970	1.224 1.972 0.930 1.048 0.975	1.197 1.916 0.928 1.040 0.976	1.145 1.817 0.926 1.034 0.981	1.107 1.744 0.924 1.028 0.983
0.150	34.90	2.228 5.917 0.598 1.626 0.610	1.428 3.674 0.928 1.379 0.807	1.364 3.112 0.960 1.272 0.864	1.337 2.796 0.964 1.211 0.895	1.309 2.512 0.960 1.162 0.922	1.279 2.314 0.956 1.129 0.938	1.241 2.126 0.947 1.098 0.951	1.209 2.014 0.943 1.084 0.959	1.184 1.936 0.940 1 074 0.964	1.164 1.878 0.936 1.066 0.966	1.143 1.827 0.936 1.062 0.971	1.110 1.751 0.932 1.053 0.975	1.081 1.691 0.930 1.046 0.977
0.200	34.70	1.933 5.363 0.520 1.703 0.573	1.104 3.170 0.909 1.446 0.778	1.037 2.649 0.964 1.330 0.841	1.020 2.374 0.972 1.260 0.874	1.009 2.138 0.972 1.204 0.902	0.999 1.980 0.968 1.166 0.920	0.984 1.829 0.964 1.135 0.937	0.974 1.748 0.960 1.117 0.944	0.964 1.689 0.958 1.107 0.950	0.957 1.647 0.954 1.098 0.953	0.948 1.609 0.953 1.093 0.957	0.937 1.556 0.948 1.081 0.960	0.924 1.511 0.947 1.076 0.965
0.300	34.47	1.652 4.799 0.402 1.797 0.527	0.787 2.643 0.844 1.538 0.736	0.702 2.151 0.945 1.409 0.806	0.682 1.899 0.971 1.331 0.844	0.674 1.696 0.979 1.265 0.876	0.669 1.564 0.982 1.222 0.897	0.667 1.448 0.980 1.184 0.915	0.664 1.385 0.978 1.164 0.925	0.662 1.343 0.977 1.151 0.931	0.661 1.311 0.975 1.141 0.935	0.659 1.286 0.974 1.134 0.939	0.657 1.247 0.972 1.122 0.944	0.654 1.217 0.970 1.114 0.948
0.500	34.28	1.451 4.369 0.291 1.880 0.483	0.572 2.236 0.739 1.634 0.692	0.473 1.757 0.895 1.497 0.766	0.445 1.518 0.951 1.410 0.808	0.433 1.329 0.976 1.336 0.845	0.429 1.212 0.984 1.284 0.869	0.428 1.111 0.986 1.239 0.890	0.427 1.058 0.987 1.215 0.901	0.427 1.024 0.987 1.200 0.908	0.426 0.997 0.987 1.189 0.914	0.426 0.978 0.986 1.179 0.917	0.426 0.947 0.986 1.166 0.924	0.426 0.925 0.985 1.156 0.928
0.800	34.14	1.349 4.135 0.229 1.927 0.456	0.472 2.019 0.654 1.698 0.661	0.366 1.548 0.843 1.559 0.736	0.334 1.314 0.924 1.469 0.780	0.320 1.130 0.966 1.390 0.820	0.315 1.017 0.981 1.334 0.845	0.313 0.921 0.988 1.285 0.869	0.812 0.872 0.989 1.258 0.881	0.312 0.839 0.991 1.240 0.890	0.311 0.816 0.991 1.228 0.896	0.311 0.797 0.991 1.217 0.901	0.311 0.770 0.991 1.202 0.908	0.311 0.750 0.991 1.190 0.913
1.250	34.04	1.291 3.992 0.193 1.953 0.439	0.419 1.893 0.596 1.741 0.641	0.312 1.428 0.801 1.603 0.715	0.277 1.197 0.900 1.511 0.760	0.261 1.015 0.955 1.430 0.801	0.256 0.905 0.976 1.372 0.828	0.253 0.811 0.986 1.320 0.852	0.252 0.764 0.990 1.291 0.866	0.252 0.733 0.991 1.272 0.874	0.252 0.709 0.992 1.258 0.881	0.252 0.692 0.993 1.247 0.886	0.251 0.666 0.993 1.230 0.894	0.251 0.647 0.993 1.217 0.900

Table III. Microdosimetric parameters and virtual multipliers for fast neutron radiations

E/MeV	W/eV	$\sigma_0 = .008\,\mu m^2$ $d = 0.1\,\mu m$	$\sigma_0 = .196\,\mu m^2$ $d = 0.5\,\mu m$	$\sigma_0 = .785\,\mu m^2$ $d = 1\,\mu m$	$\sigma_0 = 2\,\mu m^2$ $d = 1.60\,\mu m$	$\sigma_0 = 5\,\mu m^2$ $d = 2.52\,\mu m$	$\sigma_0 = 10\,\mu m^2$ $d = 3.57\,\mu m$	$\sigma_0 = 20\,\mu m^2$ $d = 5.05\,\mu m$	$\sigma_0 = 30\,\mu m^2$ $d = 6.18\,\mu m$	$\sigma_0 = 40\,\mu m^2$ $d = 7.14\,\mu m$	$\sigma_0 = 50\,\mu m^2$ $d = 7.98\,\mu m$	$\sigma_0 = 60\,\mu m^2$ $d = 8.74\,\mu m$	$\sigma_0 = 80\,\mu m^2$ $d = 10.1\,\mu m$	$\sigma_0 = 100\,\mu m^2$ $d = 11.3\,\mu m$
0.25	41.16	76.29 103.7 0.993 1.067 0.934	64.16 95.24 0.994 1.065 0.937	53.53 86.49 0.995 1.057 0.934	44.70 76.77 0.996 1.045 0.940	35.57 63.62 0.996 1.027 0.955	28.92 52.07 0.997 1.017 0.969	22.87 40.75 0.997 1.010 0.981	19.70 34.75 0.997 1.007 0.986	17.65 30.87 0.997 1.005 0.989	16.16 28.08 0.998 1.003 0.993	15.01 25.95 0.998 1.002 0.996	13.34 22.87 0.998 1.001 0.997	12.14 20.69 0.998 0.999 0.999
0.5	39.59	72.17 93.62 0.997 1.129 0.937	65.55 87.96 0.997 1.115 0.952	58.83 83.07 0.997 1.093 0.967	52.42 78.03 0.997 1.070 0.983	44.82 71.05 0.997 1.040 1.008	38.52 64.21 0.997 1.015 1.031	32.14 56.12 0.998 0.992 1.054	28.51 50.92 0.998 0.980 1.067	26.04 47.09 0.998 0.972 1.074	24.19 44.07 0.998 0.967 1.080	22.73 41.60 0.998 0.962 1.084	20.52 37.74 0.998 0.957 1.090	18.91 34.84 0.998 0.953 1.094
1.0	39.15	59.40 89.65 0.996 1.450 0.801	57.28 82.93 0.999 1.420 0.834	54.14 76.90 0.999 1.349 0.879	50.83 71.41 0.999 1.276 0.926	46.41 65.51 0.999 1.195 0.980	42.27 60.76 0.999 1.134 1.024	37.53 55.71 0.999 1.076 1.069	34.56 52.60 0.999 1.043 1.094	32.39 50.30 0.999 1.021 1.112	30.70 48.48 0.999 1.005 1.126	29.31 46.95 0.999 0.992 1.137	27.14 44.48 0.999 0.972 1.153	25.48 42.52 0.998 0.957 1.166
1.5	37.68	43.33 70.70 0.989 1.585 0.769	44.85 67.13 0.999 1.571 0.796	43.77 64.02 0.999 1.516 0.822	42.55 60.93 0.999 1.449 0.852	40.79 57.28 0.999 1.365 0.891	38.97 54.20 0.999 1.297 0.924	36.66 50.87 0.999 1.229 0.959	35.06 48.79 0.999 1.190 0.980	33.82 47.24 0.999 1.163 0.996	32.79 46.02 0.999 1.143 1.008	31.92 44.98 0.999 1.126 1.017	30.48 43.30 0.999 1.100 1.033	29.31 41.95 0.999 1.081 1.045
2.0	37.32	35.18 65.24 0.983 1.784 0.701	38.02 62.38 0.999 1.768 0.736	37.45 59.54 1.000 1.715 0.759	36.77 56.66 1.000 1.642 0.789	35.75 53.15 1.000 1.542 0.830	34.67 50.25 1.000 1.458 0.866	33.24 47.25 1.000 1.372 0.903	32.23 45.43 1.000 1.323 0.926	31.42 44.13 1.000 1.288 0.942	30.74 43.09 1.000 1.262 0.955	30.15 42.24 1.000 1.241 0.965	29.15 40.87 1.000 1.208 0.982	28.33 39.77 1.000 1.184 0.995
3.0	36.95	26.09 61.86 0.976 2.109 0.579	29.02 59.48 0.997 2.097 0.619	29.38 56.78 1.000 2.054 0.642	29.12 53.89 1.000 1.988 0.669	28.68 50.14 1.000 1.880 0.711	28.20 46.93 1.000 1.777 0.752	27.55 43.70 1.000 1.666 0.797	27.07 41.86 1.000 1.600 0.824	26.68 40.58 1.000 1.553 0.843	26.34 39.60 1.000 1.518 0.859	26.05 38.81 1.000 1.489 0.871	25.54 37.57 1.000 1.444 0.890	25.10 36.61 1.000 1.410 0.905
4.0	36.92	21.39 63.03 0.972 2.393 0.484	23.65 60.28 0.993 2.371 0.526	24.41 57.23 0.998 2.314 0.551	24.50 53.97 1.000 2.238 0.579	24.24 49.75 1.000 2.125 0.621	23.93 46.12 1.000 2.015 0.664	23.51 42.45 1.000 1.892 0.714	23.19 40.39 1.000 1.817 0.744	22.93 38.98 1.000 1.764 0.766	22.71 37.92 1.000 1.723 0.783	22.51 37.08 1.000 1.689 0.797	22.16 35.79 1.000 1.637 0.819	21.87 34.82 1.000 1.597 0.836
6.0	36.44	15.38 57.33 0.966 2.614 0.403	16.58 55.06 0.988 2.623 0.437	17.29 52.75 0.994 2.594 0.458	17.69 50.10 0.997 2.547 0.481	17.89 46.32 0.999 2.461 0.516	17.88 42.72 1.000 2.362 0.555	17.74 38.87 1.000 2.235 0.604	17.63 36.68 1.000 2.152 0.636	17.54 35.19 1.000 2.090 0.660	17.46 34.08 1.000 2.041 0.679	17.39 33.21 1.000 2.001 0.694	17.27 31.89 1.000 1.937 0.718	17.16 30.92 1.000 1.888 0.737
8.0	36.30	12.65 61.94 0.960 2.815 0.323	13.31 59.55 0.986 2.845 0.347	13.83 57.34 0.991 2.835 0.363	14.20 54.77 0.994 2.811 0.380	14.50 50.98 0.997 2.762 0.407	14.65 47.20 0.999 2.704 0.437	14.69 42.89 1.000 2.625 0.476	14.66 40.33 1.000 2.569 0.504	14.61 38.56 1.000 2.526 0.524	14.57 37.23 1.000 2.491 0.541	14.54 36.17 1.000 2.460 0.555	14.47 34.57 1.000 2.411 0.577	14.41 33.38 1.000 2.370 0.595
10.0	36.30	11.39 76.85 0.952 2.838 0.251	11.73 74.14 0.986 2.874 0.267	12.14 71.69 0.989 2.878 0.278	12.44 68.84 0.992 2.873 0.289	12.74 64.58 0.995 2.861 0.308	12.93 60.18 0.997 2.845 0.329	13.06 54.96 0.999 2.823 0.357	13.09 51.73 0.999 2.807 0.377	13.09 49.43 1.000 2.793 0.393	13.08 47.68 1.000 2.780 0.406	13.06 46.28 1.000 2.769 0.417	13.02 44.14 1.000 2.749 0.434	12.98 42.55 1.000 2.731 0.448
12.0	36.31	10.41 87.28 0.943 2.779 0.214	10.51 84.28 0.985 2.813 0.227	10.82 81.55 0.989 2.818 0.235	11.07 78.45 0.991 2.816 0.245	11.33 73.78 0.994 2.811 0.259	11.52 68.93 0.996 2.807 0.276	11.68 63.01 0.997 2.803 0.300	11.74 59.24 0.998 2.799 0.318	11.77 56.54 0.999 2.796 0.331	11.79 54.45 0.999 2.792 0.343	11.79 52.78 1.000 2.787 0.352	11.78 50.19 1.000 2.779 0.368	11.77 48.25 1.000 2.771 0.381
15.0	36.31	9.803 108.0 0.928 2.494 0.186	9.647 104.9 0.985 2.523 0.195	9.877 102.0 0.988 2.534 0.201	10.07 98.82 0.990 2.539 0.207	10.29 93.92 0.992 2.545 0.217	10.46 88.64 0.994 2.555 0.229	10.62 81.90 0.996 2.570 0.246	10.70 77.41 0.997 2.581 0.258	10.75 74.08 0.998 2.590 0.269	10.79 71.46 0.998 2.595 0.278	10.81 69.31 0.999 2.600 0.285	10.84 65.96 0.999 2.607 0.298	10.85 63.42 1.000 2.610 0.308
20.0	36.23	8.429 125.5 0.899 2.304 0.154	7.962 121.9 0.984 2.327 0.162	8.097 118.9 0.988 2.334 0.165	8.224 115.4 0.989 2.339 0.169	8.371 110.0 0.991 2.344 0.177	8.495 104.2 0.992 2.353 0.186	8.624 96.42 0.993 2.372 0.199	8.700 91.01 0.994 2.388 0.209	8.751 86.88 0.995 2.401 0.218	8.790 83.59 0.996 2.411 0.226	8.820 80.85 0.996 2.420 0.233	8.864 76.53 0.997 2.433 0.244	8.895 73.21 0.997 2.443 0.254

Table IV. Microdosimetric parameters for ion radiations (partial quantities)

E/MeV	W/eV	$\sigma_0 = .008\,\mu m^2$ $d = 0.1\,\mu m$ $\delta = 1.46\,keV$	$\sigma_0 = .196\,\mu m^2$ $d = 0.5\,\mu m$ $\delta = 3.77\,keV$	$\sigma_0 = .785\,\mu m^2$ $d = 1\,\mu m$ $\delta = 5.60\,keV$	$\sigma_0 = 2\,\mu m^2$ $d = 1.60\,\mu m$ $\delta = 7.31\,keV$	$\sigma_0 = 5\,\mu m^2$ $d = 2.52\,\mu m$ $\delta = 9.47\,keV$	$\sigma_0 = 10\,\mu m^2$ $d = 3.57\,\mu m$ $\delta = 11.5\,keV$	$\sigma_0 = 20\,\mu m^2$ $d = 5.05\,\mu m$ $\delta = 14.0\,keV$	$\sigma_0 = 30\,\mu m^2$ $d = 6.18\,\mu m$ $\delta = 15.7\,keV$	$\sigma_0 = 40\,\mu m^2$ $d = 7.14\,\mu m$ $\delta = 17.0\,keV$	$\sigma_0 = 50\,\mu m^2$ $d = 7.98\,\mu m$ $\delta = 18.1\,keV$	$\sigma_0 = 60\,\mu m^2$ $d = 8.74\,\mu m$ $\delta = 19.0\,keV$	$\sigma_0 = 80\,\mu m^2$ $d = 10.1\,\mu m$ $\delta = 20.6\,keV$	$\sigma_0 = 100\,\mu m^2$ $d = 11.3\,\mu m$ $\delta = 21.9\,keV$
1.0	35.87	10.31 15.00												
		0.221 5.46	0.303 0.00	0.303 0.00	0.303 0.00	0.303 0.00	0.303 0.00	0.303 0.00	0.303 0.00	0.303 0.00	0.303 0.00	0.303 0.00	0.303 0.00	0.303 0.00
1.5	35.70	9.828 14.43												
		0.216 9.88	0.412 0.00	0.412 0.00	0.412 0.00	0.412 0.00	0.412 0.00	0.412 0.00	0.412 0.00	0.412 0.00	0.412 0.00	0.412 0.00	0.412 0.00	0.412 0.00
2.0	35.61	9.268 13.89	5.428 7.789											
		0.214 12.52	0.456 1.72	0.518 0.00	0.518 0.00	0.518 0.00	0.518 0.00	0.518 0.00	0.518 0.00	0.518 0.00	0.518 0.00	0.518 0.00	0.518 0.00	0.518 0.00
3.0	35.50	8.308 13.06	5.244 7.604	4.070 5.816										
		0.210 15.71	0.444 5.90	0.625 1.66	0.717 0.00	0.717 0.00	0.717 0.00	0.717 0.00	0.717 0.00	0.717 0.00	0.717 0.00	0.717 0.00	0.717 0.00	0.717 0.00
4.0	35.43	7.581 12.47	4.986 7.331	3.970 5.731	3.324 4.764									
		0.208 17.67	0.436 8.46	0.613 4.48	0.775 1.79	0.905 0.00	0.905 0.00	0.905 0.00	0.905 0.00	0.905 0.00	0.905 0.00	0.905 0.00	0.905 0.00	0.905 0.00
5.0	35.38	7.022 12.03	4.726 7.088	3.840 5.589	3.268 4.713	2.735 3.914								
		0.206 19.05	0.430 10.26	0.604 6.46	0.764 3.89	0.958 1.37	1.087 0.00	1.087 0.00	1.087 0.00	1.087 0.00	1.087 0.00	1.087 0.00	1.087 0.00	1.087 0.00
6.0	35.35	6.585 11.69	4.489 6.880	3.702 5.448	3.187 4.624	2.704 3.893	2.360 3.370							
		0.205 20.10	0.426 11.62	0.597 7.96	0.755 5.48	0.946 3.05	1.128 1.21	1.265 0.00	1.265 0.00	1.265 0.00	1.265 0.00	1.265 0.00	1.265 0.00	1.265 0.00
7.0	35.32	6.230 11.41	4.277 6.703	3.567 5.320	3.101 4.530	2.658 3.841	2.340 3.360	2.036 2.920						
		0.204 20.94	0.422 12.70	0.592 9.15	0.747 6.75	0.937 4.38	1.116 2.59	1.329 0.80	1.436 0.00	1.436 0.00	1.436 0.00	1.436 0.00	1.436 0.00	1.436 0.00
8.0	35.30	5.938 11.17	4.086 6.551	3.439 5.205	3.013 4.443	2.605 3.781	2.310 3.327	2.022 2.900	1.863 2.655	1.753 2.505				
		0.203 21.63	0.419 13.59	0.587 10.13	0.741 7.78	0.928 5.48	1.107 3.74	1.318 1.99	1.462 0.96	1.572 0.23	1.607 0.00	1.607 0.00	1.607 0.00	1.607 0.00
10.0	35.27	5.480 10.80	3.769 6.301	3.211 5.011	2.845 4.286	2.493 3.664	2.234 3.242	1.980 2.855	1.838 2.640	1.738 2.492	1.663 2.378	1.603 2.285	1.509 2.155	
		0.202 22.71	0.414 15.00	0.579 11.67	0.731 9.42	0.915 7.21	1.091 5.53	1.298 3.85	1.440 2.87	1.549 2.17	1.637 1.62	1.717 1.18	1.847 0.48	1.940 0.00
12.0	35.25	5.138 10.51	3.510 6.103	3.016 4.853	2.694 4.155	2.383 3.560	2.153 3.159	1.927 2.795	1.797 2.595	1.707 2.459	1.638 2.356	1.583 2.273	1.497 2.144	1.432 2.046
		0.201 23.55	0.410 16.07	0.573 12.85	0.723 10.67	0.905 8.53	1.078 6.90	1.283 5.28	1.423 4.32	1.531 3.64	1.618 3.12	1.696 2.69	1.825 2.01	1.930 1.48
15.0	35.22	4.757 10.18	3.203 5.873	2.776 4.664	2.501 3.995	2.235 3.427	2.037 3.048	1.841 2.706	1.728 2.520	1.650 2.394	1.589 2.300	1.540 2.224	1.464 2.109	1.406 2.021
		0.199 24.50	0.406 17.30	0.566 14.20	0.713 12.10	0.892 10.04	1.063 8.47	1.265 6.91	1.403 5.99	1.509 5.33	1.595 4.83	1.672 4.41	1.799 3.76	1.902 3.25
20.0	35.19	4.326 9.782	2.832 5.593	2.473 4.432	2.247 3.794	2.030 3.257	1.869 2.901	1.708 2.582	1.616 2.409	1.550 2.293	1.500 2.207	1.459 2.137	1.395 2.032	1.346 1.954
		0.198 25.65	0.400 18.77	0.557 15.81	0.701 13.80	0.877 11.84	1.044 10.34	1.242 8.85	1.377 7.97	1.481 7.35	1.566 6.86	1.642 6.47	1.766 5.84	1.868 5.36

Table V. Virtual multipliers for ion radiations (partial quantities)

E/MeV	$\sigma_0 = .008 \mu m^2$ $d = 0.1 \mu m$	$\sigma_0 = .196 \mu m^2$ $d = 0.5 \mu m$	$\sigma_0 = .785 \mu m^2$ $d = 1 \mu m$	$\sigma_0 = 2 \mu m^2$ $d = 1.60 \mu m$	$\sigma_0 = 5 \mu m^2$ $d = 2.52 \mu m$	$\sigma_0 = 10 \mu m^2$ $d = 3.57 \mu m$	$\sigma_0 = 20 \mu m^2$ $d = 5.05 \mu m$	$\sigma_0 = 30 \mu m^2$ $d = 6.18 \mu m$	$\sigma_0 = 40 \mu m^2$ $d = 7.14 \mu m$	$\sigma_0 = 50 \mu m^2$ $d = 7.98 \mu m$	$\sigma_0 = 60 \mu m^2$ $d = 8.74 \mu m$	$\sigma_0 = 80 \mu m^2$ $d = 10.1 \mu m$	$\sigma_0 = 100 \mu m^2$ $d = 11.3 \mu m$
1.0	0.844 2.385 0.985 0.257 0.922	2.371 0.321	2.371 0.321	2.371 0.321	2.371 0.321	2.371 0.321	2.371 0.321	2.371 0.321	2.371 0.321	2.371 0.321	2.371 0.321	2.371 0.321	2.371 0.321
1.5	0.856 2.283 1.018 0.246 0.905	2.263 0.398	2.263 0.398	2.263 0.398	2.263 0.398	2.263 0.398	2.263 0.398	2.263 0.398	2.263 0.398	2.263 0.398	2.263 0.398	2.263 0.398	2.263 0.398
2.0	0.866 2.216 1.045 0.239 0.888	0.881 2.194 1.022 0.422 0.966	2.129 0.444	2.129 0.444	2.129 0.444	2.129 0.444	2.129 0.444	2.129 0.444	2.129 0.444	2.129 0.444	2.129 0.444	2.129 0.444	2.129 0.444
3.0	0.880 2.129 1.088 0.229 0.863	0.882 2.142 1.042 0.412 0.954	0.882 2.016 1.032 0.489 0.975	1.947 0.512	1.947 0.512	1.947 0.512	1.947 0.512	1.947 0.512	1.947 0.512	1.947 0.512	1.947 0.512	1.947 0.512	1.947 0.512
4.0	0.886 2.071 1.121 0.223 0.845	0.890 2.107 1.063 0.405 0.943	0.887 1.990 1.042 0.482 0.967	0.887 1.894 1.030 0.531 0.975	1.823 0.558	1.823 0.558	1.823 0.558	1.823 0.558	1.823 0.558	1.823 0.558	1.823 0.558	1.823 0.558	1.823 0.558
5.0	0.887 2.030 1.147 0.219 0.831	0.899 2.081 1.082 0.400 0.935	0.892 1.970 1.054 0.478 0.960	0.889 1.878 1.038 0.526 0.971	0.889 1.792 1.031 0.574 0.980	1.709 0.579	1.709 0.579	1.709 0.579	1.709 0.579	1.709 0.579	1.709 0.579	1.709 0.579	1.709 0.579
6.0	0.885 1.996 1.168 0.215 0.819	0.906 2.062 1.097 0.396 0.927	0.898 1.955 1.066 0.474 0.955	0.894 1.864 1.047 0.523 0.967	0.891 1.782 1.036 0.571 0.977	0.891 1.713 1.030 0.603 0.982	1.669 0.623	1.669 0.623	1.669 0.623	1.669 0.623	1.669 0.623	1.669 0.623	1.669 0.623
7.0	0.882 1.969 1.186 0.212 0.810	0.911 2.045 1.111 0.393 0.921	0.903 1.943 1.077 0.471 0.949	0.898 1.854 1.056 0.520 0.963	0.894 1.773 1.041 0.568 0.974	0.892 1.705 1.033 0.600 0.979	0.892 1.642 1.030 0.632 0.986	1.596 0.633	1.596 0.633	1.596 0.633	1.596 0.633	1.596 0.633	1.596 0.633
8.0	0.877 1.946 1.202 0.209 0.802	0.916 2.031 1.122 0.391 0.915	0.908 1.932 1.086 0.468 0.945	0.902 1.845 1.064 0.517 0.959	0.897 1.765 1.048 0.565 0.971	0.893 1.699 1.037 0.598 0.977	0.893 1.636 1.030 0.630 0.982	0.893 1.599 1.028 0.647 0.985	0.893 1.575 1.028 0.659 0.988	1.565 0.661	1.565 0.661	1.565 0.661	1.565 0.661
10.0	0.864 1.910 1.229 0.206 0.789	0.923 2.008 1.143 0.386 0.906	0.917 1.914 1.102 0.464 0.937	0.910 1.831 1.078 0.513 0.953	0.904 1.753 1.059 0.561 0.966	0.899 1.688 1.046 0.594 0.973	0.896 1.627 1.037 0.626 0.980	0.895 1.591 1.033 0.643 0.983	0.894 1.566 1.029 0.655 0.984	0.894 1.550 1.028 0.665 0.986	0.894 1.533 1.028 0.672 0.987	0.894 1.511 1.026 0.684 0.988	1.497 0.692
12.0	0.850 1.882 1.250 0.203 0.778	0.929 1.989 1.159 0.383 0.898	0.923 1.900 1.116 0.461 0.931	0.917 1.819 1.090 0.510 0.948	0.910 1.743 1.069 0.558 0.961	0.905 1.679 1.055 0.591 0.969	0.900 1.620 1.043 0.623 0.976	0.898 1.584 1.038 0.641 0.980	0.897 1.560 1.034 0.653 0.981	0.896 1.544 1.031 0.663 0.983	0.896 1.528 1.030 0.670 0.985	0.895 1.506 1.027 0.681 0.986	0.895 1.489 1.026 0.691 0.98
15.0	0.828 1.848 1.275 0.199 0.766	0.934 1.968 1.179 0.379 0.889	0.931 1.884 1.133 0.457 0.923	0.925 1.806 1.104 0.506 0.941	0.918 1.732 1.081 0.555 0.956	0.912 1.670 1.066 0.588 0.965	0.907 1.611 1.052 0.620 0.972	0.904 1.576 1.046 0.637 0.976	0.902 1.553 1.041 0.650 0.978	0.901 1.537 1.038 0.660 0.980	0.900 1.520 1.036 0.666 0.981	0.899 1.499 1.032 0.678 0.983	0.897 1.483 1.030 0.688 0.984
20.0	0.793 1.807 1.308 0.194 0.750	0.938 1.941 1.204 0.373 0.878	0.940 1.863 1.154 0.452 0.914	0.935 1.789 1.123 0.501 0.933	0.928 1.718 1.098 0.550 0.948	0.923 1.658 1.080 0.583 0.958	0.917 1.601 1.065 0.616 0.966	0.913 1.566 1.057 0.633 0.971	0.911 1.543 1.052 0.646 0.973	0.910 1.527 1.048 0.656 0.975	0.908 1.512 1.045 0.663 0.977	0.906 1.491 1.041 0.675 0.979	0.904 1.475 1.038 0.684 0.981

Table VI. Microdosimetric parameters for ion radiations (total quantities)

Z	E/MeV	$L\big/\frac{\text{keV}}{\mu\text{m}}$	$\sigma_0 = .008\,\mu\text{m}^2$ $d = 0.1\,\mu\text{m}$	$\sigma_0 = .196\,\mu\text{m}^2$ $d = 0.5\,\mu\text{m}$	$\sigma_0 = .785\,\mu\text{m}^2$ $d = 1\,\mu\text{m}$	$\sigma_0 = 2\,\mu\text{m}^2$ $d = 1.60\,\mu\text{m}$	$\sigma_0 = 5\,\mu\text{m}^2$ $d = 2.52\,\mu\text{m}$	$\sigma_0 = 10\,\mu\text{m}^2$ $d = 3.57\,\mu\text{m}$	$\sigma_0 = 20\,\mu\text{m}^2$ $d = 5.05\,\mu\text{m}$	$\sigma_0 = 30\,\mu\text{m}^2$ $d = 6.18\,\mu\text{m}$	$\sigma_0 = 40\,\mu\text{m}^2$ $d = 7.14\,\mu\text{m}$	$\sigma_0 = 50\,\mu\text{m}^2$ $d = 7.98\,\mu\text{m}$	$\sigma_0 = 60\,\mu\text{m}^2$ $d = 8.74\,\mu\text{m}$	$\sigma_0 = 80\,\mu\text{m}^2$ $d = 10.1\,\mu\text{m}$	$\sigma_0 = 100\,\mu\text{m}^2$ $d = 11.3\,\mu\text{m}$
1	1.0	27.54	24.07 31.63	27.54 31.89	27.54 31.43	27.54 31.26	27.54 31.16	27.54 31.11	27.54 31.07	27.54 31.05	27.54 31.04	27.54 31.03	27.54 31.03	27.54 31.02	27.54 31.02
	1.5	20.52	17.06 23.10	20.53 24.32	20.52 23.71	20.52 23.47	20.52 23.33	20.52 23.26	20.52 23.21	20.52 23.19	20.52 23.17	20.52 23.17	20.52 23.16	20.52 23.15	20.52 23.14
	2.0	16.54	13.59 18.78	15.73 19.46	16.55 19.39	16.54 19.10	16.54 18.92	16.54 18.83	16.54 18.77	16.54 18.74	16.54 18.72	16.54 18.71	16.54 18.70	16.54 18.69	16.54 18.68
	3.0	12.11	9.953 14.39	10.67 13.77	11.54 14.20	12.11 14.30	12.11 14.05	12.11 13.93	12.11 13.84	12.11 13.80	12.11 13.78	12.11 13.76	12.11 13.75	12.11 13.73	12.11 13.72
	4.0	9.660	8.014 12.14	8.306 10.92	8.712 11.05	9.183 11.28	9.661 11.41	9.660 11.25	9.660 11.14	9.660 11.09	9.660 11.06	9.660 11.04	9.660 11.02	9.660 11.00	9.660 10.99
	5.0	8.086	6.792 10.76	6.885 9.212	7.120 9.168	7.376 9.275	7.772 9.464	8.086 9.553	8.086 9.419	8.086 9.360	8.086 9.325	8.086 9.301	8.086 9.283	8.086 9.258	8.086 9.241
	6.0	6.982	5.949 9.820	5.921 8.064	6.073 7.912	6.234 7.941	6.472 8.047	6.741 8.175	6.982 8.230	6.982 8.161	6.982 8.120	6.982 8.092	6.982 8.071	6.982 8.042	6.982 8.022
	7.0	6.161	5.330 9.141	5.217 7.239	5.323 7.014	5.434 6.988	5.593 7.037	5.767 7.120	6.015 7.235	6.161 7.279	6.161 7.233	6.161 7.201	6.161 7.177	6.161 7.144	6.161 7.122
	8.0	5.524	4.858 8.623	4.677 6.617	4.754 6.338	4.835 6.273	4.949 6.281	5.071 6.331	5.240 6.412	5.371 6.473	5.484 6.521	5.524 6.517	5.524 6.491	5.524 6.454	5.524 6.428
	10.0	4.598	4.183 7.882	3.900 5.739	3.943 5.389	3.992 5.270	4.059 5.223	4.128 5.229	4.221 5.263	4.291 5.296	4.349 5.324	4.401 5.348	4.448 5.370	4.532 5.407	4.598 5.431
	12.0	3.954	3.726 7.376	3.363 5.147	3.388 4.751	3.420 4.599	3.464 4.517	3.509 4.495	3.568 4.500	3.610 4.514	3.646 4.529	3.677 4.543	3.705 4.556	3.755 4.580	3.799 4.600
	15.0	3.282	3.265 6.857	2.807 4.548	2.815 4.109	2.834 3.925	2.861 3.810	2.888 3.760	2.922 3.737	2.947 3.734	2.967 3.737	2.985 3.741	3.000 3.746	3.027 3.756	3.051 3.767
	20.0	2.577	2.801 6.319	2.226 3.938	2.219 3.459	2.227 3.246	2.241 3.098	2.256 3.024	2.274 2.974	2.286 2.955	2.296 2.945	2.305 2.940	2.313 2.938	2.326 2.936	2.337 2.936
2	1.0	104.5	67.26 109.0	104.5 118.4	104.5 118.0	104.5 117.8	104.5 117.7	104.5 117.7	104.5 117.6	104.5 117.6	104.5 117.6	104.5 117.6	104.5 117.6	104.5 117.6	104.5 117.6
	1.5	80.21	44.40 77.63	80.21 91.47	80.21 90.85	80.21 90.62	80.21 90.48	80.21 90.41	80.21 90.36	80.21 90.34	80.21 90.32	80.21 90.31	80.21 90.31	80.21 90.30	80.21 90.29
	2.0	65.41	34.73 60.86	54.21 72.57	65.41 74.37	65.41 74.08	65.41 73.90	65.41 73.81	65.41 73.74	65.41 73.72	65.41 73.70	65.41 73.69	65.41 73.68	65.41 73.67	65.41 73.66
	3.0	48.27	25.25 43.29	31.29 49.79	40.33 53.54	48.27 54.98	48.27 54.73	48.27 54.61	48.27 54.52	48.27 54.48	48.27 54.46	48.27 54.44	48.27 54.43	48.27 54.41	48.27 54.40
	4.0	38.59	20.33 34.19	23.32 38.19	26.88 40.74	31.94 42.67	38.59 43.95	38.59 43.79	38.59 43.68	38.59 43.63	38.59 43.60	38.59 43.58	38.59 43.57	38.59 43.55	38.59 43.53
	5.0	32.32	17.24 28.62	19.00 31.17	20.94 33.03	23.34 34.46	27.82 35.99	32.32 36.82	32.32 36.69	32.32 36.63	32.32 36.59	32.32 36.57	32.32 36.55	32.32 36.53	32.32 36.51
	6.0	27.92	15.09 24.86	16.21 26.46	17.45 27.87	18.86 28.98	21.23 30.19	24.43 31.17	27.92 31.78	27.92 31.72	27.92 31.67	27.92 31.65	27.92 31.63	27.92 31.60	27.92 31.58
	7.0	24.64	13.50 22.14	14.23 23.08	15.10 24.17	16.04 25.07	17.52 26.04	19.35 26.84	22.46 27.69	24.64 28.07	24.64 28.02	24.64 27.99	24.64 27.97	24.64 27.93	24.64 27.91
	8.0	22.10	12.27 20.07	12.74 20.53	13.39 21.40	14.07 22.13	15.08 22.94	16.28 23.61	18.15 24.32	19.83 24.76	21.47 25.08	22.10 25.16	22.10 25.13	22.10 25.10	22.10 25.07
	10.0	18.39	10.47 17.15	10.62 16.95	11.02 17.50	11.43 18.00	12.01 18.59	12.64 19.08	13.55 19.61	14.29 19.94	14.96 20.18	15.59 20.37	16.20 20.52	17.38 20.78	18.39 20.95
	12.0	15.81	9.214 15.17	9.175 14.55	9.451 14.89	9.725 15.25	10.10 15.68	10.49 16.06	11.04 16.47	11.46 16.73	11.82 16.92	12.16 17.07	12.47 17.19	13.05 17.39	13.60 17.55
	15.0	13.13	7.890 13.17	7.683 12.12	7.855 12.26	8.029 12.48	8.259 12.77	8.491 13.04	8.796 13.34	9.024 13.52	9.216 13.66	9.385 13.77	9.540 13.87	9.820 14.02	10.07 14.14
	20.0	10.31	6.481 11.12	6.126 9.675	6.214 9.624	6.311 9.707	6.439 9.858	6.563 10.01	6.720 10.20	6.833 10.32	6.925 10.41	7.004 10.48	7.075 10.55	7.200 10.65	7.309 10.73
3	1.0	218.1	101.2 223.2	218.1 246.3	218.1 245.8	218.1 245.7	218.1 245.5	218.1 245.5	218.1 245.4	218.1 245.5	218.1 245.4	218.1 245.4	218.1 245.4	218.1 245.4	218.1 245.4
	1.5	172.7	63.10 162.1	172.7 195.5	172.7 194.9	172.7 194.6	172.7 194.5	172.7 194.4	172.7 194.4	172.7 194.3	172.7 194.3	172.7 194.3	172.7 194.3	172.7 194.3	172.7 194.3
	2.0	143.1	48.79 127.8	98.55 157.0	143.1 161.8	143.1 161.5	143.1 161.3	143.1 161.2	143.1 161.2	143.1 161.1	143.1 161.1	143.1 161.1	143.1 161.1	143.1 161.1	143.1 161.1
	3.0	107.2	35.42 90.42	48.62 108.5	74.62 117.7	107.2 121.3	107.2 121.1	107.2 121.0	107.2 120.9	107.2 120.8	107.2 120.8	107.2 120.8	107.2 120.8	107.2 120.8	107.2 120.7
	4.0	86.27	28.65 70.55	35.02 83.14	43.71 89.68	58.88 94.41	86.27 97.59	86.27 97.43	86.27 97.32	86.27 97.27	86.27 97.24	86.27 97.22	86.27 97.21	86.27 97.19	86.27 97.17
	5.0	72.48	24.43 58.22	28.17 67.55	32.66 72.55	38.90 76.19	53.18 79.93	72.48 81.99	72.48 81.86	72.48 81.80	72.48 81.77	72.48 81.74	72.48 81.72	72.48 81.70	72.48 81.68
	6.0	62.69	21.52 49.83	23.90 57.02	26.71 61.01	30.17 63.93	36.73 66.96	47.49 69.35	62.69 70.91	62.69 70.84	62.69 70.80	62.69 70.77	62.69 70.75	62.69 70.72	62.69 70.70
	7.0	55.37	19.36 43.75	20.94 49.43	22.88 52.71	25.12 55.13	28.94 57.65	34.31 59.65	45.46 61.71	55.37 62.64	55.37 62.60	55.37 62.56	55.37 62.54	55.37 62.51	55.37 62.49
	8.0	49.68	17.69 39.13	18.73 43.70	20.17 46.46	21.76 48.51	24.29 50.66	26.54 52.36	33.38 54.13	39.54 55.20	46.62 55.97	49.68 56.19	49.68 56.16	49.68 56.13	49.68 56.10
	10.0	41.37	15.25 32.59	15.63 35.63	16.53 37.67	17.46 39.21	18.84 40.84	20.43 42.15	22.92 43.51	25.14 44.33	27.29 44.92	29.47 45.38	31.72 45.77	36.59 46.38	41.37 46.80
	12.0	35.58	13.53 28.17	13.53 30.21	14.15 31.77	14.77 32.99	15.65 34.29	16.62 35.33	18.02 36.42	19.18 37.08	20.23 37.56	21.22 37.94	22.19 38.25	24.09 38.74	26.03 39.13
	15.0	29.54	11.73 23.69	11.38 24.75	11.77 25.85	12.16 26.75	12.70 27.71	13.25 28.50	14.01 29.34	14.60 29.84	15.11 30.21	15.57 30.49	16.00 30.73	16.80 31.11	17.56 31.42
	20.0	23.19	9.782 19.14	9.140 19.24	9.341 19.90	9.565 20.48	9.861 21.12	10.16 21.67	10.54 22.24	10.82 22.60	11.05 22.85	11.25 23.06	11.44 23.23	11.77 23.50	12.06 23.71
5	1.0	511.4	137.9 518.1	511.4 576.2	511.4 575.7	511.4 575.6	511.4 575.5	511.4 575.4	511.4 575.4	511.4 575.3	511.4 575.3	511.4 575.3	511.4 575.3	511.4 575.3	511.4 575.3
	1.5	426.2	80.63 393.7	426.2 480.7	426.2 480.1	426.2 479.8	426.2 479.7	426.2 479.6	426.2 479.6	426.2 479.5	426.2 479.5	426.2 479.5	426.2 479.5	426.2 479.5	426.2 479.5
	2.0	364.7	61.53 318.5	169.4 397.8	364.7 411.1	364.7 410.8	364.7 410.6	364.7 410.5	364.7 410.4	364.7 410.4	364.7 410.4	364.7 410.4	364.7 410.4	364.7 410.4	364.7 410.4
	3.0	283.4	44.57 231.2	67.70 284.0	131.5 309.3	283.4 319.5	283.4 319.2	283.4 319.1	283.4 319.0	283.4 319.0	283.4 319.0	283.4 318.9	283.4 318.9	283.4 318.9	283.4 318.9
	4.0	232.3	36.21 181.9	47.01 220.8	64.13 239.6	103.1 252.9	232.3 261.9	232.3 261.7	232.3 261.6	232.3 261.6	232.3 261.5	232.3 261.5	232.3 261.5	232.3 261.5	232.3 261.5
	5.0	197.3	31.05 150.2	37.35 180.6	45.69 195.4	58.89 205.9	99.24 216.5	197.3 222.4	197.3 222.3	197.3 222.2	197.3 222.2	197.3 222.1	197.3 222.1	197.3 222.1	197.3 222.1
	6.0	171.8	27.51 128.2	31.54 152.9	36.61 165.0	43.44 173.6	58.49 182.3	91.50 189.1	171.8 193.6	171.8 193.5	171.8 193.5	171.8 193.5	171.8 193.4	171.8 193.4	171.8 193.4

	7.0	152.3	24.89	111.9	27.57	132.6	31.05	142.8	35.31	150.0	43.37	157.4	56.66	163.2	95.20	169.1	152.3	171.7	152.3	171.7	152.3	171.7	152.3	171.6	152.3	171.6	152.3	171.6
	8.0	137.1	22.87	99.51	24.66	117.1	27.22	125.9	30.18	132.1	35.29	138.5	42.60	143.4	58.41	148.6	80.29	151.6	116.0	153.8	137.1	154.5	137.1	154.5	137.1	154.4	137.1	154.4
	10.0	114.5	19.93	81.73	20.61	95.07	22.18	101.9	23.90	106.7	26.56	111.7	29.85	115.6	35.48	119.6	41.09	121.9	47.17	123.7	54.06	125.0	62.15	126.1	84.07	127.9	114.5	129.1
	12.0	98.62	17.87	69.63	17.88	80.16	18.97	85.65	20.11	89.59	21.78	93.63	23.70	96.80	26.65	100.1	29.25	102.0	31.77	103.4	34.29	104.5	36.89	105.4	42.49	106.8	48.89	108.0
	15.0	81.97	15.70	57.31	15.10	65.08	15.79	69.28	16.51	72.32	17.51	75.45	18.59	77.92	20.12	80.45	21.35	81.97	22.45	83.06	23.49	83.92	24.48	84.63	26.41	85.75	28.33	86.63
	20.0	64.40	13.37	44.77	12.22	49.82	12.59	52.76	12.99	54.92	13.55	57.16	14.11	58.93	14.86	60.76	15.42	61.86	15.89	62.65	16.32	63.27	16.71	63.78	17.42	64.60	18.08	65.23
6	1.0	677.6	147.7	685.2	677.6	763.2	677.6	762.7	677.6	762.5	677.6	762.4	677.6	762.4	677.6	762.3	677.6	762.3	677.6	762.3	677.6	762.3	677.6	762.3	677.6	762.3	677.6	762.3
	1.5	576.1	84.80	530.7	576.1	649.4	576.1	648.8	576.1	648.5	576.1	648.4	576.1	648.3	576.1	648.3	576.1	648.2	576.1	648.2	576.1	648.2	576.1	648.2	576.1	648.2	576.1	648.2
	2.0	499.8	64.47	434.8	193.7	544.6	499.8	563.0	499.8	562.7	499.8	562.5	499.8	562.4	499.8	562.4	499.8	562.4	499.8	562.3	499.8	562.3	499.8	562.3	499.8	562.3	499.8	562.3
	3.0	395.0	46.64	320.4	72.60	395.2	151.3	430.7	395.0	445.0	395.0	444.8	395.0	444.7	395.0	444.6	395.0	444.5	395.0	444.5	395.0	444.5	395.0	444.5	395.0	444.5	395.0	444.4
	4.0	327.0	37.92	254.1	49.94	310.1	69.70	336.8	118.3	355.6	327.0	368.5	327.0	368.3	327.0	368.2	327.0	368.1	327.0	368.1	327.0	368.1	327.0	368.1	327.0	368.0	327.0	368.0
	5.0	279.5	32.56	210.8	39.55	255.1	49.03	276.3	64.56	291.3	116.5	306.5	279.5	314.9	279.5	314.8	279.5	314.7	279.5	314.7	279.5	314.6	279.5	314.6	279.5	314.6	279.5	314.6
	6.0	244.4	28.88	180.3	33.36	216.7	39.09	234.2	46.98	246.6	65.07	259.1	108.7	268.9	244.4	275.3	244.4	275.3	244.4	275.2	244.4	275.2	244.4	275.2	244.4	275.1	244.4	275.1
	7.0	217.4	26.17	157.7	29.15	188.4	33.07	203.2	37.95	213.7	47.41	224.3	63.75	232.6	117.1	241.1	217.4	244.9	217.4	244.9	217.4	244.9	217.4	244.8	217.4	244.8	217.4	244.8
	8.0	196.0	24.08	140.2	26.07	166.6	28.94	179.4	32.32	188.5	38.25	197.7	46.99	204.9	67.00	212.3	97.46	216.7	155.7	219.8	196.0	220.8	196.0	220.8	196.0	220.8	196.0	220.7
	10.0	164.2	21.04	115.1	21.79	135.5	23.57	145.5	25.52	152.6	28.57	159.8	32.41	165.5	39.15	171.2	46.10	174.7	53.88	177.2	63.08	179.1	74.37	180.7	108.1	183.2	164.2	185.0
	12.0	141.6	18.91	97.91	18.92	114.3	20.14	122.4	21.44	128.2	23.35	134.1	25.56	138.7	29.04	143.5	32.15	146.3	35.22	148.3	38.34	149.9	41.62	151.2	48.88	153.3	57.56	155.0
	15.0	117.9	16.67	80.33	16.00	92.70	16.77	99.01	17.59	103.5	18.73	108.1	19.97	111.7	21.74	115.5	23.19	117.7	24.50	119.3	25.73	120.5	26.93	121.5	29.28	123.2	31.66	124.4
	20.0	92.68	14.27	62.36	12.97	70.82	13.38	75.31	13.85	78.56	14.48	81.89	15.12	84.51	15.98	87.20	16.63	88.81	17.19	89.96	17.68	90.87	18.14	91.62	18.99	92.80	19.77	93.73
7	1.0	851.5	154.5	860.1	851.5	958.8	851.5	958.4	851.5	958.2	851.5	958.1	851.5	958.0	851.5	958.0	851.5	958.0	851.5	958.0	851.5	958.0	851.5	957.9	851.5	957.9	851.5	957.9
	1.5	736.5	87.61	677.2	736.5	829.8	736.5	829.2	736.5	829.0	736.5	828.8	736.5	828.7	736.5	828.7	736.5	828.7	736.5	828.6	736.5	828.6	736.5	828.6	736.5	828.6	736.5	828.6
	2.0	646.5	66.42	561.1	212.4	704.1	646.5	728.1	646.5	727.8	646.5	727.6	646.5	727.5	646.5	727.5	646.5	727.5	646.5	727.4	646.5	727.4	646.5	727.4	646.5	727.4	646.5	727.4
	3.0	519.0	47.99	419.5	75.93	518.8	166.5	565.7	519.0	584.5	519.0	584.3	519.0	584.2	519.0	584.1	519.0	584.0	519.0	584.0	519.0	584.0	519.0	584.0	519.0	584.0	519.0	584.0
	4.0	433.9	39.04	335.6	51.89	410.9	73.56	446.5	129.9	471.6	433.9	488.7	433.9	488.5	433.9	488.4	433.9	488.3	433.9	488.3	433.9	488.3	433.9	488.3	433.9	488.3	433.9	488.2
	5.0	373.2	33.54	279.9	41.01	340.0	51.29	368.6	68.53	388.7	130.1	409.1	373.2	420.3	373.2	420.2	373.2	420.1	373.2	420.1	373.2	420.1	373.2	420.1	373.2	420.0	373.2	420.0
	6.0	327.8	29.78	240.2	34.56	290.0	40.74	313.7	49.39	330.4	69.80	347.3	122.6	360.5	327.8	369.2	327.8	369.1	327.8	369.1	327.8	369.0	327.8	369.0	327.8	369.0	327.8	369.0
	7.0	292.6	27.01	210.5	30.19	252.8	34.41	273.0	39.73	287.2	50.23	301.6	68.94	312.8	135.9	324.3	292.6	329.5	292.6	329.4	292.6	329.4	292.6	329.4	292.6	329.4	292.6	329.3
	8.0	264.4	24.87	187.5	27.00	224.1	30.09	241.6	33.76	253.9	40.28	266.5	50.10	276.2	73.49	286.2	111.8	292.1	195.9	296.4	264.4	297.7	264.4	297.7	264.4	297.7	264.4	297.7
	10.0	222.2	21.76	154.1	22.58	182.6	24.48	196.4	26.59	206.1	29.92	216.0	34.17	223.7	41.75	231.5	49.75	236.2	58.93	239.6	70.11	242.2	84.35	244.4	130.5	247.9	222.2	250.2
	12.0	192.0	19.59	131.1	19.61	154.2	20.92	165.5	22.32	173.5	24.40	181.6	26.83	187.9	30.69	194.4	34.19	198.2	37.68	201.0	41.27	203.1	45.10	204.9	53.75	207.8	64.43	210.0
	15.0	160.1	17.31	107.4	16.59	125.2	17.43	134.0	18.31	140.2	19.55	146.6	20.90	151.5	22.85	156.6	24.46	159.6	25.92	161.8	27.30	163.5	28.66	164.9	31.34	167.1	34.07	168.9
	20.0	126.0	14.88	83.09	13.47	95.56	13.92	101.9	14.42	106.4	15.10	111.0	15.80	114.7	16.74	118.4	17.46	120.6	18.08	122.2	18.62	123.4	19.14	124.4	20.08	126.1	20.95	127.3
8	1.0	1031	159.6	1040	1031	1160	1031	1160	1031	1160	1031	1160	1031	1160	1031	1160	1031	1160	1031	1160	1031	1160	1031	1160	1031	1160	1031	1159
	1.5	904.7	89.59	830.9	904.7	1019	904.7	1018	904.7	1018	904.7	1018	904.7	1018	904.7	1018	904.7	1018	904.7	1018	904.7	1018	904.7	1018	904.7	1018	904.7	1018
	2.0	802.5	67.77	695.5	226.9	873.6	802.5	903.6	802.5	903.3	802.5	903.2	802.5	903.1	802.5	903.0	802.5	903.0	802.5	903.0	802.5	902.9	802.5	902.9	802.5	902.9	802.5	902.9
	3.0	653.4	48.92	527.0	78.29	652.7	178.3	711.9	653.4	735.8	653.4	735.5	653.4	735.4	653.4	735.3	653.4	735.3	653.4	735.2	653.4	735.2	653.4	735.2	653.4	735.2	653.4	735.2
	4.0	551.3	39.80	425.1	53.24	521.5	76.32	567.0	138.8	599.0	551.3	620.7	551.3	620.6	551.3	620.4	551.3	620.4	551.3	620.4	551.3	620.3	551.3	620.3	551.3	620.3	551.3	620.3
	5.0	477.2	34.21	356.5	42.01	434.2	52.87	470.9	71.39	496.7	140.8	522.8	477.2	537.3	477.2	537.1	477.2	537.1	477.2	537.1	477.2	537.0	477.2	537.0	477.2	537.0	477.2	537.0
	6.0	421.0	30.39	307.2	35.38	371.9	41.90	402.5	51.09	424.1	73.26	445.9	133.7	462.8	421.0	474.0	421.0	474.0	421.0	473.9	421.0	473.9	421.0	473.9	421.0	473.8	421.0	473.8
	7.0	377.0	27.58	269.9	30.91	325.2	35.34	351.4	40.97	369.8	52.23	388.5	72.79	403.0	151.7	417.8	377.0	424.5	377.0	424.4	377.0	424.4	377.0	424.4	377.0	424.4	377.0	424.3
	8.0	341.6	25.41	240.8	27.64	288.9	30.89	311.7	34.77	327.8	41.72	344.0	52.34	356.7	78.42	369.6	123.7	377.3	235.3	382.8	341.6	384.6	341.6	384.6	341.6	384.5	341.6	384.5
	10.0	288.1	22.26	198.4	23.11	236.1	25.12	254.2	27.34	266.9	30.88	279.8	35.42	289.8	43.62	300.1	52.43	306.2	62.74	310.5	75.56	314.0	92.37	316.8	150.8	321.3	288.1	324.3
	12.0	249.5	20.07	168.9	20.08	199.7	21.46	214.6	22.94	225.1	25.14	235.7	27.73	243.9	31.86	252.4	35.65	257.4	39.46	261.0	43.42	263.9	47.68	266.2	57.46	269.9	69.82	272.8
	15.0	208.4	17.76	138.4	17.00	162.4	17.88	174.0	18.81	182.2	20.12	190.6	21.56	197.1	23.63	203.7	25.36	207.7	26.93	210.5	28.43	212.8	29.90	214.6	32.83	217.5	35.84	219.8
	20.0	164.3	15.30	106.9	13.82	124.0	14.28	132.5	14.81	138.5	15.53	144.6	16.28	149.3	17.28	154.2	18.04	157.1	18.70	159.2	19.29	160.8	19.84	162.2	20.85	164.3	21.80	166.0

Table VI continued

Z	E/MeV	$L\Big/\frac{\text{keV}}{\mu\text{m}}$	$\sigma_0 = .008\,\mu\text{m}^2$ $d = 0.1\,\mu\text{m}$		$\sigma_0 = .196\,\mu\text{m}^2$ $d = 0.5\,\mu\text{m}$		$\sigma_0 = .785\,\mu\text{m}^2$ $d = 1\,\mu\text{m}$		$\sigma_0 = 2\,\mu\text{m}^2$ $d = 1.60\,\mu\text{m}$		$\sigma_0 = 5\,\mu\text{m}^2$ $d = 2.52\,\mu\text{m}$		$\sigma_0 = 10\,\mu\text{m}^2$ $d = 3.57\,\mu\text{m}$		$\sigma_0 = 20\,\mu\text{m}^2$ $d = 5.05\,\mu\text{m}$		$\sigma_0 = 30\,\mu\text{m}^2$ $d = 6.18\,\mu\text{m}$		$\sigma_0 = 40\,\mu\text{m}^2$ $d = 7.14\,\mu\text{m}$		$\sigma_0 = 50\,\mu\text{m}^2$ $d = 7.98\,\mu\text{m}$		$\sigma_0 = 60\,\mu\text{m}^2$ $d = 8.74\,\mu\text{m}$		$\sigma_0 = 80\,\mu\text{m}^2$ $d = 10.1\,\mu\text{m}$		$\sigma_0 = 100\,\mu\text{m}^2$ $d = 11.3\,\mu\text{m}$	
10	1.0	1398	166.3	1410	1398	1574	1398	1574	1398	1574	1398	1573	1398	1573	1398	1573	1398	1573	1398	1573	1398	1573	1398	1573	1398	1573	1398	1573
	1.5	1257	92.15	1153	1257	1416	1257	1415	1257	1415	1257	1415	1257	1415	1257	1415	1257	1415	1257	1415	1257	1415	1257	1415	1257	1415	1257	1415
	2.0	1135	69.49	981.8	247.4	1235	1135	1278	1135	1278	1135	1277	1135	1277	1135	1277	1135	1277	1135	1277	1135	1277	1135	1277	1135	1277	1135	1277
	3.0	947.1	50.08	761.7	81.31	945.3	194.8	1031	947.1	1066	947.1	1066	947.1	1066	947.1	1066	947.1	1066	947.1	1066	947.1	1066	947.1	1066	947.1	1066	947.1	1066
	4.0	812.2	40.74	624.1	54.95	767.5	79.87	834.8	151.0	882.1	812.2	914.3	812.2	914.1	812.2	914.0	812.2	914.0	812.2	913.9	812.2	913.9	812.2	913.9	812.2	913.9	812.2	913.8
	5.0	711.3	35.04	529.1	43.26	646.4	54.87	701.4	75.09	740.0	155.9	779.1	711.3	800.7	711.3	800.6	711.3	800.5	711.3	800.5	711.3	800.4	711.3	800.4	711.3	800.4	711.3	800.4
	6.0	633.1	31.14	459.5	36.41	558.3	43.34	604.7	53.26	637.2	77.79	670.1	149.6	695.7	633.1	712.6	633.1	712.6	633.1	712.5	633.1	712.5	633.1	712.5	633.1	712.4	633.1	712.4
	7.0	570.7	28.28	406.1	31.79	491.3	36.50	531.3	42.54	559.3	54.81	587.7	77.89	609.7	175.7	632.2	570.7	642.4	570.7	642.4	570.7	642.3	570.7	642.3	570.7	642.3	570.7	642.3
	8.0	519.8	26.08	364.0	28.43	438.6	31.88	473.7	36.02	498.3	43.54	523.1	55.24	542.5	85.12	562.2	141.2	574.0	308.0	582.4	519.8	585.1	519.8	585.1	519.8	585.0	519.8	585.0
	10.0	441.7	22.88	301.6	23.78	361.0	25.90	389.0	28.28	408.7	32.07	428.6	37.00	444.0	46.05	459.8	55.97	469.2	67.88	475.9	83.14	481.2	104.0	485.5	184.3	492.4	441.7	497.1
	12.0	384.5	20.65	257.6	20.67	306.7	22.13	329.9	23.70	346.3	26.06	362.8	28.85	375.6	33.35	388.6	37.53	396.4	41.78	402.0	46.24	406.4	51.11	410.0	62.51	415.7	77.43	420.2
	15.0	322.7	18.31	211.7	17.51	250.3	18.44	268.7	19.43	281.6	20.83	294.6	22.38	304.8	24.62	315.2	26.50	321.3	28.22	325.8	29.87	329.2	31.50	332.1	34.77	336.6	38.17	340.1
	20.0	255.6	15.82	163.7	14.24	191.7	14.74	205.2	15.30	214.7	16.08	224.3	16.88	231.8	17.95	239.4	18.78	244.0	19.49	247.3	20.13	249.8	20.73	251.9	21.84	255.3	22.88	257.9
18	1.0	2893	177.2	2913	2893	3256	2893	3256	2893	3255	2893	3255	2893	3255	2893	3255	2893	3255	2893	3255	2893	3255	2893	3255	2893	3255	2893	3255
	1.5	2755	95.97	2521	2755	3100	2755	3099	2755	3099	2755	3099	2755	3099	2755	3099	2755	3099	2755	3099	2755	3099	2755	3099	2755	3099	2755	3099
	2.0	2597	71.97	2240	282.0	2824	2597	2922	2597	2922	2597	2922	2597	2922	2597	2922	2597	2922	2597	2922	2597	2922	2597	2922	2597	2922	2597	2921
	3.0	2310	51.70	1851	85.65	2303	221.7	2514	2310	2599	2310	2599	2310	2599	2310	2598	2310	2598	2310	2598	2310	2598	2310	2598	2310	2598	2310	2598
	4.0	2073	42.03	1586	57.31	1956	84.95	2129	170.2	2251	2073	2333	2073	2333	2073	2333	2073	2333	2073	2333	2073	2333	2073	2333	2073	2333	2073	2333
	5.0	1880	36.15	1391	44.96	1705	57.63	1852	80.36	1955	180.5	2058	1880	2116	1880	2116	1880	2115	1880	2115	1880	2115	1880	2115	1880	2115	1880	2115
	6.0	1720	32.14	1240	37.78	1514	45.30	1641	56.25	1730	84.34	1820	175.8	1889	1720	1936	1720	1935	1720	1935	1720	1935	1720	1935	1720	1935	1720	1935
	7.0	1586	29.21	1120	32.97	1361	38.06	1474	44.67	1552	58.40	1632	85.35	1693	218.8	1756	1586	1784	1586	1784	1586	1784	1586	1784	1586	1784	1586	1784
	8.0	1471	26.95	1021	29.47	1237	33.19	1338	37.71	1408	46.03	1479	59.32	1534	95.20	1590	171.3	1624	499.4	1647	1471	1655	1471	1655	1471	1655	1471	1655
	10.0	1286	23.68	868.9	24.65	1047	26.94	1130	29.52	1188	33.68	1246	39.15	1292	49.43	1338	61.06	1365	75.50	1385	94.87	1400	123.0	1413	253.9	1433	1286	1447
	12.0	1143	21.41	756.4	21.43	907.8	23.01	978.1	24.71	1027	27.29	1077	30.36	1115	35.39	1154	40.13	1178	45.03	1194	50.26	1207	56.05	1218	70.07	1235	89.38	1248
	15.0	980.7	19.04	633.6	18.17	756.5	19.17	813.7	20.24	853.6	21.77	893.8	23.47	924.9	25.95	956.7	28.05	975.6	29.98	989.2	31.85	999.7	33.71	1008	37.48	1022	41.46	1033
	20.0	794.6	16.52	498.9	14.80	591.8	15.34	635.0	15.95	665.3	16.79	695.7	17.67	719.3	18.85	743.3	19.77	757.6	20.56	767.8	21.27	775.8	21.94	782.4	23.18	792.9	24.36	801.0

Table VII. Microdosimetric parameters for δ-rays of ion radiations

Parameter	Method	3 MeV/nucl.: 0.10	0.25	0.50	1.00	2.00	5.00	10.0
$\mu_D^{(\delta)}$ [%]	present theory	15.7	10.1	5.90	1.66	0	0	0.
	Monte Carlo	12.1	5.65	2.62	1.01	0.36	0.11	0.03
$\bar{y}_F^{(\delta)}$ $\left[\frac{keV}{\mu m}\right]$	present theory	8.31	6.51	5.24	4.07	–	–	–
	Monte Carlo He	–	7.38	4.77	2.75	1.58	0.77	0.42
	Monte Carlo Li	–	–	–	–	–	–	–
	Monte Carlo B	–	10.2	7.68	5.05	3.23	1.77	–
	Monte Carlo C	–	–	–	–	–	–	–
	Monte Carlo Ne	–	16.0	13.6	10.4	7.15	4.26	–
	Monte Carlo Ar	23.5	25.6	24.4	18.1	13.6	8.28	–
$\bar{y}_D^{(\delta)}$ $\left[\frac{keV}{\mu m}\right]$	present theory	13.0	9.65	7.59	5.80	–	–	–
	Monte Carlo He	–	10.7	7.32	4.38	2.59	1.32	0.75
	Monte Carlo Li	–	–	–	–	–	–	–
	Monte Carlo B	–	17.2	13.5	9.40	6.51	3.81	–
	Monte Carlo C	–	–	–	–	–	–	–
	Monte Carlo Ne	–	33.7	29.9	24.6	17.7	11.3	–
	Monte Carlo Ar	60.0	69.0	66.6	52.9	40.2	26.3	–

Parameter	Method	5 MeV/nucl.: 0.10	0.25	0.50	1.00	2.00	5.00	10.0
$\mu_D^{(\delta)}$ [%]	present theory	19.1	14.0	10.3	6.46	2.65	0	0
	Monte Carlo	17.1	10.4	6.35	3.05	1.26	0.32	0.13
$\bar{y}_F^{(\delta)}$ $\left[\frac{keV}{\mu m}\right]$	present theory	7.02	5.65	4.73	3.84	3.0	–	–
	Monte Carlo He	–	–	5.58	3.59	2.28	1.07	0.62
	Monte Carlo Li	–	8.04	6.30	4.35	2.65	1.37	–
	Monte Carlo B	–	8.94	7.59	5.84	3.93	2.19	1.38
	Monte Carlo C	–	9.66	8.34	6.64	4.47	2.68	–
	Monte Carlo Ne	–	12.0	12.0	10.2	7.65	4.92	–
	Monte Carlo Ar	16.1	17.3	18.8	17.9	14.3	9.69	6.61
$\bar{y}_D^{(\delta)}$ $\left[\frac{keV}{\mu m}\right]$	present theory	12.0	8.90	7.08	5.58	4.30	–	–
	Monte Carlo He	–	–	8.28	5.52	3.51	1.71	1.03
	Monte Carlo Li	–	12.1	9.51	6.76	4.12	2.34	–
	Monte Carlo B	–	14.1	12.0	9.91	6.97	4.23	2.62
	Monte Carlo C	–	16.4	13.7	11.5	8.70	5.28	–
	Monte Carlo Ne	–	23.4	24.5	21.8	17.6	11.9	–
	Monte Carlo Ar	42.0	44.3	48.9	49.5	40.1	27.9	20.1

Parameter	Method	7 MeV/nucl.: 0.10	0.25	0.50	1.00	2.00	5.00	10.0
$\mu_D^{(\delta)}$ [%]	present theory	20.9	16.3	12.7	9.15	5.58	0.85	0
	Monte Carlo	19.5	13.8	9.21	5.31	2.41	0.70	0.27
$\bar{y}_F^{(\delta)}$ $\left[\frac{keV}{\mu m}\right]$	present theory	6.23	5.02	4.28	3.57	2.88	2.04	–
	Monte Carlo He	–	–	–	–	–	–	–
	Monte Carlo Li	–	–	–	–	–	–	–
	Monte Carlo B	–	–	6.69	5.65	4.38	2.47	1.53
	Monte Carlo C	–	–	–	–	–	–	–
	Monte Carlo Ne	–	9.96	9.54	8.91	7.47	5.07	–
	Monte Carlo Ar	–	12.7	14.1	14.9	14.9	10.1	–
$\bar{y}_D^{(\delta)}$ $\left[\frac{keV}{\mu m}\right]$	present theory	11.4	8.41	6.69	5.31	4.17	2.90	–
	Monte Carlo He	–	–	–	–	–	–	–
	Monte Carlo Li	–	–	–	–	–	–	–
	Monte Carlo B	–	–	10.3	8.83	7.45	4.29	2.85
	Monte Carlo C	–	–	–	–	–	–	–
	Monte Carlo Ne	–	19.2	18.4	18.5	15.6	11.2	–
	Monte Carlo Ar	–	33.4	37.8	40.4	40.0	28.6	–

Parameter	Method	10 MeV/nucl.: 0.10	0.25	0.50	1.00	2.00	5.00	10.0
$\mu_D^{(\delta)}$ [%]	present theory	22.7	18.3	15.0	11.7	8.33	3.90	0.52
	Monte Carlo	21.5	16.0	11.8	7.97	4.44	1.56	0.62
$\bar{y}_F^{(\delta)}$ $\left[\frac{keV}{\mu m}\right]$	present theory	5.48	4.37	3.77	3.21	2.67	1.99	1.52
	Monte Carlo He	–	–	–	4.00	3.12	1.81	1.00
	Monte Carlo Li	–	–	–	–	–	–	–
	Monte Carlo B	–	6.66	5.91	5.14	4.14	2.71	1.78
	Monte Carlo C	–	–	6.21	5.56	4.62	3.24	–
	Monte Carlo Ne	–	7.74	7.47	6.91	6.75	4.98	3.61
	Monte Carlo Ar	–	9.24	9.75	11.0	11.5	10.2	7.81
$\bar{y}_D^{(\delta)}$ $\left[\frac{keV}{\mu m}\right]$	present theory	10.8	7.92	6.29	5.00	3.96	2.86	2.14
	Monte Carlo He	–	–	–	6.07	4.54	2.52	1.51
	Monte Carlo Li	–	–	–	–	–	–	–
	Monte Carlo B	–	11.4	9.30	8.04	6.40	4.50	3.03
	Monte Carlo C	–	–	10.7	8.71	7.35	5.79	–
	Monte Carlo Ne	–	15.2	13.8	13.1	13.4	10.8	8.60
	Monte Carlo Ar	–	22.1	24.8	28.4	31.7	28.6	21.0

Column headings give the sphere diameter in µm.

Table VIII. r-values for photon, neutron, and ion radiations

Photons		Neutrons			Ions (total)									$\sigma_0/\mu m^2$		Ions (cut-off δ-electrons only)										
E/keV		E/MeV		E/MeV	$Z=1$	$Z=2$	$Z=3$	$Z=5$	$Z=6$	$Z=7$	$Z=8$	$Z=10$	$Z=18$	0.008	0.196	0.785	2	5	10	20	30	40	50	60	80	100
10	1.161	0.25	3.771	1.0	1.990	3.733	4.553	3.436	2.818	2.373	2.059	1.661	1.083	1.583												
	1.313		8.232		2.922	7.834	11.55	9.717	7.938	6.577	5.591	4.328	2.488	2.070												
	0.845		0.235		0.606	0.357	0.264	0.213	0.203	0.198	0.194	0.188	0.181	0.710												
20	1.104	0.5	3.593	1.5	1.768	3.221	4.339	3.822	3.179	2.663	2.285	1.805	1.129	1.525												
	1.204		7.598		2.426	6.198	10.33	10.64	8.943	7.423	6.258	4.741	2.590	1.956												
	0.897		0.285		0.666	0.423	0.309	0.241	0.229	0.222	0.217	0.211	0.203	0.737												
30	1.100	1.0	3.247	2.0	1.642	2.866	4.061	4.083	3.485	2.934	2.504	1.945	1.170	1.488	1.347											
	1.188		6.761		2.162	5.167	9.151	11.16	9.754	8.207	6.914	5.166	2.701	1.883	1.625											
	0.918		0.343		0.703	0.473	0.344	0.262	0.248	0.239	0.233	0.226	0.216	0.755	0.801											
50	1.113	1.5	2.917	3.0	1.504	2.424	3.554	4.322	3.927	3.398	2.912	2.221	1.251	1.438	1.311	1.261										
	1.214		5.664		1.883	3.985	7.331	11.32	10.77	9.480	8.116	6.013	2.936	1.790	1.556	1.472										
	0.911		0.420		0.746	0.546	0.403	0.295	0.276	0.264	0.256	0.247	0.235	0.777	0.823	0.834										
75	1.121	2.0	2.720	4.0	1.428	2.164	3.168	4.329	4.169	3.745	3.265	2.487	1.330	1.406	1.285	1.242	1.219									
	1.228		5.150		1.735	3.340	6.104	10.86	11.13	10.33	9.100	6.825	3.177	1.730	1.508	1.435	1.393									
	0.901		0.459		0.772	0.596	0.450	0.322	0.297	0.283	0.273	0.262	0.247	0.792	0.836	0.849	0.860									
100	1.107	3.0	2.475	5.0	1.378	1.990	2.879	4.224	4.266	3.980	3.552	2.736	1.406	1.383	1.266	1.226	1.203	1.181								
	1.201		4.561		1.640	2.935	5.253	10.18	11.07	10.79	9.839	7.575	3.416	1.688	1.474	1.405	1.365	1.329								
	0.912		0.508		0.787	0.631	0.487	0.344	0.315	0.297	0.286	0.272	0.255	0.802	0.844	0.858	0.868	0.874								
125	1.090	4.0	2.347	6.0	1.346	1.872	2.662	4.077	4.277	4.129	3.778	2.972	1.486	1.365	1.251	1.213	1.191	1.170	1.156							
	1.168		4.316		1.578	2.664	4.645	9.469	10.79	10.98	10.35	8.252	3.661	1.655	1.449	1.382	1.344	1.309	1.285							
	0.925		0.532		0.802	0.661	0.521	0.367	0.333	0.313	0.300	0.284	0.265	0.810	0.851	0.865	0.874	0.880	0.885							
150	1.075	6.0	2.098	7.0	1.319	1.780	2.490	3.913	4.228	4.205	3.943	3.181	1.561	1.351	1.240	1.203	1.181	1.162	1.149	1.132						
	1.141		3.654		1.529	2.463	4.186	8.789	10.38	10.96	10.67	8.836	3.901	1.629	1.429	1.364	1.327	1.294	1.271	1.246						
	0.936		0.589		0.810	0.682	0.548	0.385	0.347	0.324	0.309	0.292	0.270	0.816	0.857	0.870	0.879	0.886	0.891	0.892						
200	1.054	8.0	2.020	8.0	1.301	1.711	2.354	3.754	4.150	4.232	4.060	3.369	1.639	1.339	1.231	1.194	1.173	1.154	1.142	1.129	1.122	1.114				
	1.103		3.530		1.494	2.314	3.834	8.174	9.924	10.81	10.83	9.332	4.143	1.608	1.412	1.349	1.313	1.280	1.259	1.239	1.227	1.215				
	0.952		0.609		0.819	0.701	0.572	0.404	0.363	0.337	0.320	0.301	0.276	0.821	0.861	0.874	0.883	0.889	0.894	0.898	0.900	0.900				
300	1.033	10.0	2.061	10.0	1.272	1.610	2.150	3.462	3.950	4.189	4.177	3.669	1.791	1.320	1.216	1.181	1.161	1.143	1.131	1.119	1.112	1.109	1.105	1.101	1.097	
	1.064		3.780		1.440	2.101	3.327	7.140	9.006	10.29	10.82	10.06	4.623	1.574	1.386	1.325	1.291	1.260	1.240	1.220	1.209	1.203	1.197	1.192	1.184	
	0.970		0.601		0.830	0.729	0.611	0.437	0.389	0.359	0.339	0.315	0.286	0.829	0.868	0.881	0.889	0.895	0.900	0.904	0.906	0.908	0.909	0.909	0.910	
500	1.017	12.0	2.077	12.0	1.251	1.539	2.005	3.214	3.739	4.077	4.191	3.880	1.941	1.306	1.204	1.170	1.151	1.134	1.122	1.111	1.105	1.101	1.098	1.095	1.092	1.088
	1.032		3.906		1.403	1.956	2.981	6.332	8.173	9.655	10.54	10.48	5.096	1.549	1.366	1.307	1.274	1.245	1.225	1.206	1.196	1.190	1.185	1.180	1.175	1.169
	0.985		0.597		0.838	0.750	0.641	0.465	0.413	0.378	0.355	0.328	0.294	0.836	0.873	0.886	0.893	0.900	0.904	0.908	0.910	0.913	0.913	0.914	0.916	0.916
800	1.007	15.0	2.186	15.0	1.229	1.466	1.852	2.918	3.449	3.863	4.105	4.063	2.163	1.289	1.191	1.159	1.140	1.124	1.113	1.102	1.096	1.093	1.090	1.087	1.084	1.081
	1.013		4.310		1.363	1.809	2.631	5.430	7.137	8.708	9.894	10.69	5.791	1.520	1.344	1.287	1.255	1.227	1.208	1.191	1.181	1.175	1.170	1.166	1.160	1.156
	0.994		0.576		0.847	0.772	0.675	0.502	0.445	0.405	0.377	0.344	0.303	0.843	0.879	0.891	0.899	0.905	0.909	0.913	0.915	0.917	0.918	0.919	0.921	0.921
1250	1.000	20.0	2.132	20.0	1.206	1.388	1.689	2.569	3.061	3.508	3.855	4.134	2.517	1.270	1.176	1.145	1.128	1.112	1.101	1.091	1.086	1.082	1.080	1.077	1.074	1.071
	1.000		4.218		1.321	1.659	2.276	4.440	5.884	7.376	8.725	10.40	6.876	1.485	1.317	1.264	1.234	1.207	1.189	1.172	1.163	1.158	1.153	1.149	1.144	1.140
	1.000		0.579		0.857	0.797	0.715	0.552	0.490	0.444	0.410	0.368	0.316	0.851	0.886	0.898	0.905	0.911	0.915	0.919	0.921	0.922	0.923	0.924	0.926	0.927

21*

Table IX. Special parameters for neutron radiations; neutron energy 0.25 MeV

η/rad^{-1}	lg η	$\sigma_0 = 2\ \mu m^2$ $d = 1.60\ \mu m$	$\sigma_0 = 5\ \mu m^2$ $d = 2.52\ \mu m$	$\sigma_0 = 20\ \mu m^2$ $d = 5.05\ \mu m$	$\sigma_0 = 30\ \mu m^2$ $d = 6.18\ \mu m$	$\sigma_0 = 40\ \mu m^2$ $d = 7.14\ \mu m$	$\sigma_0 = 50\ \mu m^2$ $d = 7.98\ \mu m$	$\sigma_0 = 60\ \mu m^2$ $d = 8.74\ \mu m$	$\sigma_0 = 80\ \mu m^2$ $d = 10.1\ \mu m$	$\sigma_0 = 100\ \mu m^2$ $d = 11.3\ \mu m$
zero		76.77 1.045 0.940	63.62 1.027 0.955	40.75 1.010 0.981	34.75 1.007 0.986	30.87 1.005 0.989	28.08 1.003 0.993	25.95 1.002 0.996	22.87 1.001 0.997	20.69 0.999 0.999
1.0000−04	−4.0	74.89 1.043 0.940	63.09 1.027 0.955	40.69 1.010 0.981	34.72 1.007 0.986	30.85 1.005 0.989	28.07 1.003 0.993	25.94 1.003 0.996	22.86 1.001 0.997	20.69 0.999 0.999
1.2589−04	−3.9	74.42 1.042 0.940	62.96 1.027 0.955	40.68 1.010 0.981	34.72 1.007 0.986	30.85 1.005 0.989	28.07 1.003 0.993	25.94 1.003 0.996	22.86 1.001 0.997	20.68 0.999 0.999
1.5849−04	−3.8	73.87 1.042 0.940	62.80 1.027 0.955	40.66 1.010 0.981	34.71 1.007 0.986	30.84 1.005 0.989	28.06 1.003 0.993	25.94 1.003 0.996	22.86 1.001 0.997	20.68 0.999 0.999
1.9953−04	−3.7	73.11 1.041 0.940	62.57 1.027 0.954	40.64 1.010 0.981	34.70 1.007 0.986	30.84 1.005 0.989	28.06 1.003 0.993	25.94 1.003 0.996	22.86 1.001 0.997	20.68 0.999 0.999
2.5119−04	−3.6	72.21 1.041 0.941	62.30 1.027 0.954	40.61 1.010 0.981	34.68 1.007 0.986	30.83 1.005 0.989	28.06 1.003 0.993	25.93 1.003 0.996	22.85 1.001 0.997	20.68 0.999 0.999
3.1623−04	−3.5	71.10 1.040 0.941	61.97 1.027 0.954	40.58 1.010 0.981	34.67 1.007 0.986	30.82 1.005 0.989	28.05 1.003 0.993	25.93 1.003 0.996	22.85 1.001 0.997	20.68 0.999 0.999
3.9811−04	−3.4	69.74 1.039 0.941	61.55 1.027 0.954	40.53 1.010 0.981	34.65 1.007 0.986	30.81 1.005 0.989	28.04 1.003 0.993	25.92 1.003 0.996	22.85 1.001 0.997	20.68 0.999 0.999
5.0119−04	−3.3	68.13 1.038 0.941	61.03 1.027 0.954	40.48 1.010 0.981	34.62 1.007 0.986	30.79 1.005 0.989	28.03 1.003 0.993	25.92 1.003 0.996	22.84 1.001 0.997	20.68 0.999 0.999
6.3096−04	−3.2	66.10 1.037 0.942	60.39 1.027 0.953	40.41 1.010 0.981	34.58 1.007 0.986	30.77 1.005 0.989	28.02 1.003 0.993	25.91 1.003 0.995	22.84 1.001 0.997	20.67 0.999 0.999
7.9433−04	−3.1	63.68 1.036 0.942	59.59 1.027 0.953	40.32 1.010 0.980	34.54 1.007 0.986	30.75 1.005 0.989	28.00 1.003 0.993	25.89 1.003 0.995	22.83 1.000 0.997	20.67 0.999 0.999
1.0000−03	−3.0	60.86 1.035 0.943	58.62 1.027 0.953	40.21 1.010 0.980	34.49 1.007 0.985	30.71 1.005 0.989	27.98 1.003 0.993	25.88 1.003 0.995	22.82 1.000 0.997	20.66 0.999 0.999
1.2589−03	−2.9	57.62 1.034 0.944	57.46 1.026 0.952	40.07 1.010 0.980	34.42 1.007 0.985	30.67 1.005 0.989	27.95 1.003 0.993	25.86 1.003 0.995	22.81 1.000 0.997	20.66 0.999 0.999
1.5849−03	−2.8	53.98 1.031 0.946	55.97 1.026 0.952	39.90 1.010 0.980	34.34 1.007 0.985	30.63 1.005 0.989	27.92 1.003 0.993	25.84 1.003 0.995	22.80 1.000 0.997	20.65 0.999 0.999
1.9953−03	−2.7	50.01 1.029 0.948	54.17 1.026 0.952	39.69 1.010 0.979	34.23 1.007 0.984	30.56 1.005 0.989	27.88 1.003 0.993	25.81 1.003 0.995	22.78 1.000 0.997	20.64 0.999 0.999
2.5119−03	−2.6	45.65 1.026 0.951	52.07 1.025 0.952	39.42 1.010 0.979	34.10 1.007 0.984	30.49 1.005 0.989	27.83 1.003 0.992	25.77 1.003 0.995	22.76 1.000 0.997	20.62 0.999 0.999
3.1623−03	−2.5	41.09 1.022 0.954	49.63 1.024 0.952	39.09 1.011 0.978	33.94 1.007 0.983	30.39 1.005 0.988	27.76 1.003 0.992	25.73 1.003 0.995	22.74 1.000 0.997	20.60 0.999 0.999
3.9811−03	−2.4	36.45 1.019 0.959	46.87 1.024 0.953	38.68 1.011 0.977	33.73 1.007 0.982	30.27 1.005 0.988	27.68 1.003 0.992	25.67 1.003 0.995	22.70 1.000 0.997	20.58 0.999 0.999
5.0119−03	−2.3	31.84 1.016 0.964	43.83 1.022 0.954	38.19 1.011 0.976	33.48 1.007 0.981	30.11 1.005 0.988	27.58 1.003 0.992	25.60 1.003 0.995	22.66 1.000 0.997	20.55 0.999 0.999
6.3096−03	−2.2	27.43 1.013 0.970	40.42 1.021 0.957	37.52 1.011 0.975	33.16 1.007 0.980	29.92 1.006 0.987	27.45 1.003 0.991	25.51 1.003 0.994	22.61 1.000 0.997	20.52 0.999 0.998
7.9433−03	−2.1	23.32 1.011 0.977	36.78 1.020 0.961	36.68 1.012 0.973	32.80 1.008 0.978	29.69 1.006 0.987	27.30 1.003 0.991	25.39 1.003 0.994	22.54 1.000 0.997	20.48 0.999 0.998
1.0000−02	−2.0	19.60 1.009 0.983	33.01 1.018 0.966	35.69 1.012 0.972	32.27 1.008 0.976	29.42 1.006 0.986	27.10 1.003 0.990	25.25 1.003 0.994	22.46 1.000 0.997	20.42 0.999 0.998
1.2589−02	−1.9	16.31 1.007 0.988	29.17 1.016 0.971	34.53 1.012 0.970	31.60 1.008 0.974	29.01 1.006 0.985	26.88 1.003 0.989	25.07 1.003 0.993	22.35 1.000 0.997	20.35 0.999 0.998
1.5849−02	−1.8	13.45 1.006 0.991	25.41 1.013 0.976	33.20 1.013 0.968	30.82 1.009 0.972	28.48 1.007 0.984	26.52 1.004 0.988	24.88 1.004 0.992	22.22 1.000 0.996	20.27 0.999 0.998
1.9953−02	−1.7	11.02 1.005 0.994	21.84 1.011 0.980	31.71 1.014 0.967	29.90 1.010 0.971	27.86 1.008 0.982	26.07 1.007 0.987	24.56 1.004 0.992	22.09 1.001 0.996	20.16 0.998 0.997
2.5119−02	−1.6	8.969 1.004 0.995	18.53 1.009 0.983	29.96 1.014 0.967	28.85 1.011 0.970	27.14 1.008 0.981	25.54 1.010 0.985	24.14 1.005 0.990	21.84 1.002 0.996	20.06 0.999 0.997
3.1623−02	−1.5	7.268 1.004 0.997	15.55 1.008 0.987	27.98 1.014 0.969	27.68 1.012 0.972	26.32 1.009 0.979	24.92 1.012 0.984	23.67 1.005 0.989	21.50 1.005 0.996	19.84 1.001 0.996
3.9811−02	−1.4	5.867 1.003 0.998	12.92 1.006 0.990	25.81 1.014 0.972	26.28 1.013 0.974	25.42 1.011 0.977	24.23 1.012 0.982	23.11 1.006 0.988	21.13 1.009 0.996	19.55 1.004 0.995
5.0119−02	−1.3	4.722 1.003 0.999	10.65 1.005 0.993	23.48 1.014 0.976	24.68 1.014 0.977	24.30 1.012 0.976	23.47 1.012 0.981	22.49 1.007 0.987	20.69 1.011 0.996	19.22 1.009 0.993
6.3096−02	−1.2	3.790 1.002 0.999	8.716 1.005 0.995	21.05 1.013 0.979	22.91 1.013 0.979	23.02 1.013 0.977	22.51 1.012 0.981	21.81 1.009 0.986	20.21 1.012 0.996	18.84 1.011 0.991
7.9433−02	−1.1	3.036 1.002 1.000	7.095 1.004 0.996	18.60 1.011 0.982	20.97 1.013 0.980	21.58 1.014 0.978	21.41 1.012 0.982	20.94 1.011 0.986	19.71 1.013 0.996	18.45 1.012 0.989
1.0000−01	−1.0	2.428 1.002 1.000	5.749 1.003 0.998	16.19 1.010 0.984	18.93 1.012 0.981	19.96 1.013 0.981	20.16 1.012 0.984	19.94 1.013 0.987	19.05 1.012 0.996	18.04 1.013 0.987
1.2589−01	−0.9	1.939 1.002 1.000	4.641 1.003 0.998	13.90 1.008 0.987	16.84 1.011 0.983	18.23 1.012 0.984	18.74 1.012 0.986	18.81 1.014 0.988	18.27 1.012 0.995	17.48 1.012 0.986
1.5849−01	−0.8	1.547 1.002 1.001	3.735 1.003 0.999	11.79 1.007 0.990	14.76 1.010 0.985	16.40 1.011 0.985	17.21 1.012 0.986	17.52 1.013 0.986	17.38 1.012 0.989	16.82 1.012 0.987
1.9953−01	−0.7	1.233 1.001 1.001	2.998 1.002 1.000	9.882 1.006 0.992	12.75 1.009 0.988	14.54 1.010 0.986	15.58 1.011 0.985	16.11 1.012 0.985	16.34 1.012 0.985	16.06 1.013 0.989
2.5119−01	−0.6	0.982 1.001 1.001	2.401 1.002 1.000	8.207 1.005 0.994	10.87 1.007 0.990	12.70 1.009 0.987	13.89 1.010 0.985	14.61 1.011 0.984	15.20 1.012 0.983	15.17 1.013 0.989
3.1623−01	−0.5	0.782 1.001 1.001	1.920 1.002 1.000	6.761 1.005 0.996	9.153 1.006 0.993	10.94 1.008 0.989	12.20 1.009 0.987	13.06 1.010 0.985	13.95 1.012 0.984	14.19 1.012 0.989
3.9811−01	−0.4	0.622 1.001 1.001	1.533 1.002 1.000	5.533 1.004 0.997	7.631 1.005 0.995	9.293 1.007 0.991	10.56 1.008 0.989	11.49 1.009 0.988	12.62 1.011 0.987	13.09 1.012 0.988
5.0119−01	−0.3	0.495 1.001 1.001	1.223 1.001 1.001	4.503 1.003 0.998	6.308 1.005 0.996	7.806 1.006 0.994	9.017 1.007 0.992	9.968 1.009 0.991	11.24 1.010 0.991	11.92 1.011 0.988
6.3096−01	−0.2	0.394 1.001 1.001	0.975 1.001 1.001	3.648 1.003 0.999	5.176 1.004 0.997	6.493 1.005 0.995	7.606 1.006 0.994	8.525 1.007 0.992	9.867 1.009 0.990	10.69 1.010 0.989
7.9433−01	−0.1	0.313 1.001 1.001	0.777 1.001 1.001	2.944 1.002 0.999	4.222 1.003 0.998	5.356 1.004 0.997	6.348 1.005 0.995	7.202 1.006 0.994	8.533 1.008 0.992	9.437 1.009 0.991
1.0000+00	−0.0	0.249 1.001 1.001	0.618 1.001 1.001	2.369 1.002 1.000	3.427 1.003 0.999	4.388 1.004 0.998	5.252 1.004 0.997	6.019 1.005 0.995	7.279 1.007 0.993	8.209 1.008 0.992
1.2589+00	0.1	0.198 1.001 1.001	0.492 1.001 1.001	1.901 1.002 1.000	2.770 1.002 0.999	3.574 1.003 0.998	4.313 1.004 0.998	4.985 1.004 0.997	6.134 1.006 0.995	7.041 1.007 0.994
1.5849+00	0.2	0.157 1.001 1.001	0.391 1.001 1.001	1.523 1.001 1.000	2.231 1.002 0.999	2.897 1.003 0.999	3.519 1.003 0.998	4.096 1.004 0.998	5.116 1.005 0.996	5.961 1.006 0.995
1.9953+00	0.3	0.125 1.000 1.000	0.311 1.000 1.000	1.219 1.001 1.000	1.793 1.001 0.999	2.340 1.002 0.999	2.857 1.003 0.999	3.344 1.003 0.999	4.227 1.004 0.998	4.988 1.005 0.997

Table IX continued; neutron energy 0.5 MeV

η/rad^{-1}	lg η	$\sigma_0 = 2\ \mu m^2$ $d = 1.60\ \mu m$	$\sigma_0 = 5\ \mu m^2$ $d = 2.52\ \mu m$	$\sigma_0 = 20\ \mu m^2$ $d = 5.05\ \mu m$	$\sigma_0 = 30\ \mu m^2$ $d = 6.18\ \mu m$	$\sigma_0 = 40\ \mu m^2$ $d = 7.14\ \mu m$	$\sigma_0 = 50\ \mu m^2$ $d = 7.98\ \mu m$	$\sigma_0 = 60\ \mu m^2$ $d = 8.74\ \mu m$	$\sigma_0 = 80\ \mu m^2$ $d = 10.1\ \mu m$	$\sigma_0 = 100\ \mu m^2$ $d = 11.3\ \mu m$
zero		78.03 1.070 0.983	71.05 1.040 1.008	56.12 0.992 1.054	50.92 0.980 1.067	47.09 0.972 1.074	44.07 0.967 1.080	41.60 0.962 1.084	37.74 0.957 1.090	34.84 0.953 1.094
1.0000−04	−4.0	76.17 1.067 0.983	70.41 1.039 1.007	56.02 0.992 1.054	50.86 0.980 1.066	47.05 0.972 1.074	44.04 0.967 1.080	41.58 0.962 1.084	37.73 0.957 1.090	34.83 0.953 1.094
1.2589−04	−3.9	75.70 1.066 0.983	70.25 1.039 1.007	55.99 0.992 1.054	50.85 0.980 1.066	47.04 0.972 1.074	44.04 0.967 1.080	41.57 0.962 1.084	37.73 0.957 1.090	34.83 0.953 1.094
1.5849−04	−3.8	75.15 1.065 0.983	70.06 1.039 1.007	55.96 0.992 1.054	50.83 0.980 1.066	47.03 0.972 1.074	44.03 0.967 1.080	41.57 0.962 1.084	37.72 0.957 1.090	34.82 0.953 1.094
1.9953−04	−3.7	74.40 1.064 0.983	69.79 1.038 1.007	55.92 0.992 1.054	50.81 0.980 1.066	47.01 0.972 1.074	44.02 0.967 1.080	41.56 0.962 1.084	37.72 0.957 1.090	34.82 0.953 1.094
2.5119−04	−3.6	73.50 1.063 0.983	69.47 1.038 1.007	55.87 0.992 1.054	50.78 0.980 1.066	46.99 0.972 1.074	44.00 0.967 1.080	41.55 0.962 1.084	37.71 0.957 1.090	34.82 0.953 1.094
3.1623−04	−3.5	72.40 1.061 0.984	69.08 1.038 1.006	55.80 0.992 1.054	50.74 0.980 1.066	46.97 0.972 1.074	43.99 0.967 1.080	41.54 0.962 1.084	37.70 0.957 1.090	34.81 0.953 1.094
3.9811−04	−3.4	71.06 1.059 0.984	68.58 1.037 1.006	55.72 0.992 1.054	50.70 0.980 1.066	46.94 0.972 1.074	43.97 0.967 1.080	41.52 0.962 1.084	37.69 0.957 1.090	34.81 0.953 1.094
5.0119−04	−3.3	69.42 1.057 0.984	67.96 1.036 1.005	55.61 0.992 1.054	50.64 0.980 1.066	46.90 0.972 1.074	43.94 0.967 1.080	41.50 0.962 1.084	37.68 0.957 1.090	34.80 0.953 1.094
6.3096−04	−3.2	67.40 1.054 0.984	67.20 1.036 1.005	55.48 0.992 1.054	50.56 0.980 1.066	46.86 0.972 1.073	43.91 0.967 1.079	41.48 0.962 1.084	37.67 0.957 1.090	34.79 0.953 1.094
7.9433−04	−3.1	65.01 1.051 0.984	66.26 1.035 1.004	55.32 0.992 1.053	50.47 0.980 1.066	46.80 0.972 1.073	43.87 0.967 1.079	41.45 0.962 1.084	37.65 0.957 1.090	34.78 0.953 1.094
1.0000−03	−3.0	62.20 1.048 0.984	65.11 1.034 1.004	55.11 0.992 1.053	50.36 0.980 1.065	46.72 0.972 1.073	43.81 0.967 1.079	41.41 0.963 1.084	37.63 0.957 1.090	34.76 0.953 1.094
1.2589−03	−2.9	58.97 1.043 0.984	63.70 1.032 1.003	54.85 0.992 1.053	50.21 0.980 1.065	46.63 0.972 1.073	43.75 0.967 1.079	41.36 0.963 1.084	37.60 0.957 1.090	34.74 0.953 1.094
1.5849−03	−2.8	55.33 1.038 0.984	61.96 1.031 1.003	54.53 0.992 1.052	50.03 0.980 1.064	46.51 0.972 1.072	43.67 0.967 1.079	41.30 0.963 1.083	37.56 0.957 1.090	34.71 0.953 1.094
1.9953−03	−2.7	51.27 1.033 0.985	59.89 1.030 1.002	54.13 0.992 1.051	49.81 0.980 1.064	46.37 0.972 1.072	43.56 0.967 1.078	41.22 0.963 1.083	37.51 0.957 1.090	34.68 0.953 1.094
2.5119−03	−2.6	46.84 1.028 0.985	57.45 1.029 1.002	53.64 0.992 1.050	49.53 0.980 1.063	46.18 0.973 1.071	43.43 0.967 1.078	41.13 0.963 1.083	37.45 0.957 1.089	34.64 0.953 1.094
3.1623−03	−2.5	42.18 1.022 0.986	54.64 1.027 1.002	53.02 0.992 1.049	49.18 0.981 1.062	45.95 0.973 1.070	43.27 0.967 1.078	41.00 0.963 1.082	37.37 0.957 1.089	34.59 0.953 1.094
3.9811−03	−2.4	37.40 1.017 0.987	51.45 1.024 1.002	52.28 0.993 1.047	48.74 0.981 1.060	45.67 0.973 1.069	43.07 0.967 1.077	40.85 0.963 1.082	37.28 0.957 1.089	34.53 0.953 1.093
5.0119−03	−2.3	32.66 1.013 0.989	47.88 1.021 1.002	51.34 0.993 1.045	48.20 0.981 1.058	45.31 0.973 1.067	42.82 0.968 1.076	40.66 0.964 1.081	37.16 0.957 1.089	34.44 0.953 1.093
6.3096−03	−2.2	28.11 1.009 0.990	43.95 1.018 1.002	50.17 0.994 1.042	47.55 0.981 1.056	44.87 0.974 1.065	42.50 0.968 1.075	40.43 0.964 1.081	37.02 0.958 1.088	34.34 0.954 1.093
7.9433−03	−2.1	23.86 1.006 0.992	39.78 1.015 1.003	48.77 0.995 1.038	46.72 0.982 1.053	44.34 0.974 1.063	42.11 0.968 1.074	40.13 0.964 1.080	36.83 0.958 1.088	34.22 0.954 1.093
1.0000−02	−2.0	20.02 1.004 0.994	35.46 1.012 1.003	47.10 0.996 1.034	45.69 0.983 1.049	43.65 0.975 1.060	41.64 0.969 1.073	39.77 0.964 1.079	36.60 0.958 1.087	34.06 0.954 1.092
1.2589−02	−1.9	16.62 1.003 0.996	31.14 1.010 1.003	45.16 0.997 1.030	44.45 0.985 1.046	42.79 0.976 1.057	41.03 0.970 1.071	39.33 0.965 1.077	36.31 0.959 1.086	33.86 0.954 1.091
1.5849−02	−1.8	13.67 1.002 0.997	26.95 1.007 1.002	42.94 0.998 1.026	42.97 0.987 1.043	41.75 0.978 1.055	40.25 0.972 1.069	38.75 0.966 1.075	35.97 0.959 1.085	33.61 0.955 1.090
1.9953−02	−1.7	11.17 1.002 0.998	23.00 1.005 1.001	40.41 0.998 1.023	41.25 0.988 1.040	40.51 0.980 1.053	39.32 0.974 1.066	38.02 0.968 1.073	35.52 0.960 1.084	33.32 0.955 1.089
2.5119−02	−1.6	9.077 1.002 0.998	19.38 1.004 0.999	37.57 0.999 1.021	39.28 0.990 1.038	39.06 0.982 1.051	38.20 0.976 1.063	37.15 0.970 1.071	34.94 0.962 1.082	32.93 0.956 1.088
3.1623−02	−1.5	7.341 1.002 0.999	16.16 1.003 0.998	34.48 0.999 1.018	37.02 0.991 1.035	37.39 0.984 1.049	36.90 0.978 1.059	36.10 0.973 1.067	34.23 0.964 1.080	32.42 0.958 1.086
3.9811−02	−1.4	5.917 1.002 0.999	13.35 1.002 0.997	31.20 1.000 1.016	34.49 0.992 1.031	35.46 0.985 1.045	35.40 0.980 1.055	34.88 0.975 1.064	33.38 0.967 1.077	31.80 0.961 1.083
5.0119−02	−1.3	4.754 1.001 1.000	10.95 1.002 0.998	27.81 1.000 1.013	31.73 0.994 1.028	33.26 0.987 1.041	33.66 0.981 1.050	33.48 0.977 1.059	32.39 0.969 1.073	31.06 0.963 1.080
6.3096−02	−1.2	3.812 1.001 1.000	8.919 1.002 0.999	24.43 1.000 1.010	28.78 0.995 1.023	30.83 0.989 1.036	31.66 0.983 1.045	31.84 0.979 1.054	31.25 0.971 1.068	30.20 0.966 1.076
7.9433−02	−1.1	3.050 1.001 1.000	7.231 1.002 0.999	21.15 1.000 1.007	25.72 0.997 1.019	28.20 0.991 1.030	29.44 0.986 1.039	29.96 0.981 1.048	29.92 0.973 1.062	29.21 0.968 1.071
1.0000−01	−1.0	2.437 1.001 1.001	5.839 1.002 1.000	18.07 1.001 1.005	22.66 0.998 1.014	25.42 0.993 1.024	27.02 0.988 1.034	27.88 0.984 1.042	28.36 0.976 1.056	28.04 0.970 1.065
1.2589−01	−0.9	1.945 1.001 1.001	4.700 1.001 1.000	15.25 1.001 1.003	19.67 0.998 1.011	22.59 0.995 1.019	24.46 0.991 1.028	25.61 0.986 1.035	26.60 0.978 1.049	26.66 0.973 1.059
1.5849−01	−0.8	1.551 1.001 1.001	3.773 1.001 1.000	12.73 1.001 1.002	16.84 0.999 1.008	19.77 0.996 1.014	21.81 0.993 1.022	23.19 0.989 1.029	24.65 0.982 1.042	25.10 0.975 1.052
1.9953−01	−0.7	1.235 1.001 1.001	3.022 1.001 1.000	10.53 1.001 1.002	14.25 0.999 1.006	17.07 0.997 1.011	19.17 0.995 1.017	20.69 0.991 1.024	22.53 0.985 1.036	23.35 0.979 1.046
2.5119−01	−0.6	0.984 1.001 1.001	2.417 1.001 1.000	8.648 1.001 1.001	11.92 1.000 1.004	14.54 0.998 1.008	16.60 0.996 1.013	18.19 0.993 1.019	20.31 0.988 1.030	21.45 0.982 1.040
3.1623−01	−0.5	0.783 1.001 1.001	1.930 1.001 1.001	7.057 1.001 1.001	9.878 1.000 1.003	12.24 0.999 1.006	14.19 0.997 1.010	15.77 0.995 1.015	18.04 0.990 1.024	19.42 0.985 1.034
3.9811−01	−0.4	0.623 1.001 1.001	1.540 1.001 1.001	5.730 1.001 1.000	8.123 1.001 1.002	10.20 1.000 1.004	11.98 0.998 1.008	13.48 0.997 1.011	15.78 0.993 1.020	17.33 0.988 1.028
5.0119−01	−0.3	0.495 1.001 1.001	1.227 1.001 1.001	4.633 1.001 1.000	6.638 1.001 1.001	8.427 1.000 1.003	10.01 0.999 1.006	11.39 0.998 1.008	13.61 0.994 1.015	15.23 0.991 1.023
6.3096−01	−0.2	0.394 1.001 1.001	0.978 1.001 1.001	3.732 1.001 1.000	5.395 1.001 1.001	6.912 1.000 1.002	8.284 1.000 1.004	9.517 0.999 1.006	11.59 0.996 1.011	13.19 0.993 1.018
7.9433−01	−0.1	0.313 1.001 1.001	0.778 1.001 1.001	2.999 1.001 1.000	4.366 1.001 1.001	5.636 1.001 1.002	6.807 1.000 1.003	7.882 0.999 1.005	9.753 0.997 1.009	11.28 0.995 1.013
1.0000+00	−0.0	0.249 1.001 1.001	0.619 1.001 1.001	2.404 1.001 1.001	3.520 1.001 1.001	4.572 1.001 1.001	5.559 1.000 1.002	6.479 1.000 1.003	8.125 0.998 1.006	9.523 0.996 1.010
1.2589+00	0.1	0.198 1.001 1.001	0.493 1.001 1.001	1.924 1.001 1.001	2.830 1.001 1.001	3.695 1.001 1.001	4.515 1.001 1.002	5.292 1.000 1.002	6.711 0.999 1.005	7.957 0.998 1.008
1.5849+00	0.2	0.157 1.001 1.001	0.392 1.001 1.001	1.537 1.001 1.000	2.270 1.001 1.000	2.976 1.001 1.001	3.652 1.001 1.001	4.299 1.001 1.002	5.503 1.000 1.003	6.588 0.999 1.006
1.9953+00	0.3	0.125 1.000 1.000	0.312 1.000 1.000	1.228 1.000 1.000	1.819 1.000 1.000	2.391 1.001 1.000	2.945 1.001 1.001	3.479 1.001 1.001	4.486 1.000 1.003	5.412 1.000 1.005

Table IX continued; neutron energy 1 MeV

η/rad^{-1}	lg η	$\sigma_0 = 2\ \mu m^2$ $d = 1.60\ \mu m$	$\sigma_0 = 5\ \mu m^2$ $d = 2.52\ \mu m$	$\sigma_0 = 20\ \mu m^2$ $d = 5.05\ \mu m$	$\sigma_0 = 30\ \mu m^2$ $d = 6.18\ \mu m$	$\sigma_0 = 40\ \mu m^2$ $d = 7.14\ \mu m$	$\sigma_0 = 50\ \mu m^2$ $d = 7.98\ \mu m$	$\sigma_0 = 60\ \mu m^2$ $d = 8.74\ \mu m$	$\sigma_0 = 80\ \mu m^2$ $d = 10.1\ \mu m$	$\sigma_0 = 100\ \mu m^2$ $d = 11.3\ \mu m$
zero		71.41 1.276 0.926	65.51 1.195 0.980	55.71 1.076 1.069	52.60 1.043 1.094	50.30 1.021 1.112	48.48 1.005 1.126	46.95 0.992 1.137	44.48 0.972 1.153	42.52 0.957 1.16
1.0000−04	−4.0	69.79 1.262 0.929	64.97 1.192 0.981	55.61 1.075 1.069	52.54 1.043 1.094	50.26 1.021 1.112	48.45 1.004 1.126	46.93 0.992 1.137	44.46 0.972 1.153	42.51 0.957 1.166
1.2589−04	−3.9	69.38 1.259 0.930	64.83 1.191 0.982	55.58 1.075 1.069	52.53 1.043 1.094	50.25 1.021 1.112	48.44 1.004 1.125	46.92 0.992 1.136	44.46 0.972 1.153	42.50 0.957 1.166
1.5849−04	−3.8	68.91 1.255 0.931	64.67 1.190 0.982	55.55 1.075 1.069	52.51 1.043 1.094	50.24 1.021 1.112	48.43 1.004 1.125	46.91 0.992 1.136	44.46 0.972 1.153	42.50 0.957 1.166
1.9953−04	−3.7	68.24 1.250 0.932	64.44 1.189 0.982	55.51 1.075 1.069	52.48 1.043 1.094	50.22 1.021 1.112	48.42 1.004 1.125	46.90 0.991 1.136	44.45 0.972 1.153	42.50 0.957 1.166
2.5119−04	−3.6	67.42 1.244 0.934	64.17 1.187 0.983	55.46 1.075 1.069	52.45 1.043 1.094	50.20 1.021 1.112	48.40 1.004 1.125	46.89 0.991 1.136	44.44 0.972 1.153	42.49 0.957 1.166
3.1623−04	−3.5	66.41 1.236 0.936	63.84 1.185 0.984	55.40 1.075 1.069	52.41 1.043 1.094	50.17 1.021 1.112	48.39 1.004 1.125	46.88 0.991 1.136	44.43 0.972 1.153	42.48 0.957 1.166
3.9811−04	−3.4	65.20 1.227 0.938	63.43 1.183 0.984	55.32 1.075 1.069	52.37 1.043 1.094	50.14 1.021 1.112	48.36 1.004 1.125	46.86 0.991 1.136	44.42 0.972 1.153	42.48 0.957 1.166
5.0119−04	−3.3	63.75 1.216 0.941	62.88 1.180 0.985	55.22 1.075 1.069	52.31 1.043 1.094	50.10 1.021 1.112	48.33 1.004 1.125	46.83 0.991 1.136	44.40 0.972 1.153	42.46 0.957 1.166
6.3096−04	−3.2	62.02 1.204 0.944	62.19 1.177 0.986	55.09 1.074 1.069	52.23 1.043 1.094	50.05 1.021 1.112	48.29 1.004 1.125	46.80 0.991 1.136	44.38 0.972 1.153	42.45 0.957 1.165
7.9433−04	−3.1	59.95 1.189 0.948	61.33 1.173 0.987	54.94 1.074 1.069	52.14 1.043 1.094	49.98 1.021 1.111	48.24 1.004 1.125	46.76 0.991 1.136	44.36 0.972 1.152	42.43 0.957 1.165
1.0000−03	−3.0	57.53 1.172 0.952	60.31 1.169 0.989	54.74 1.073 1.069	52.02 1.042 1.094	49.90 1.021 1.111	48.18 1.004 1.125	46.72 0.991 1.135	44.32 0.972 1.152	42.41 0.957 1.165
1.2589−03	−2.9	54.72 1.154 0.957	59.08 1.163 0.990	54.49 1.073 1.069	51.87 1.042 1.093	49.80 1.021 1.111	48.10 1.004 1.124	46.65 0.991 1.135	44.28 0.972 1.152	42.38 0.958 1.165
1.5849−03	−2.8	51.53 1.133 0.962	57.61 1.156 0.992	54.20 1.072 1.069	51.68 1.042 1.093	49.67 1.021 1.111	48.01 1.004 1.124	46.58 0.991 1.135	44.23 0.972 1.152	42.34 0.958 1.164
1.9953−03	−2.7	47.96 1.113 0.968	55.85 1.147 0.993	53.79 1.072 1.069	51.45 1.042 1.093	49.50 1.021 1.111	47.88 1.004 1.124	46.48 0.991 1.135	44.16 0.972 1.151	42.29 0.958 1.164
2.5119−03	−2.6	44.06 1.092 0.974	53.78 1.136 0.995	53.28 1.072 1.068	51.18 1.042 1.093	49.30 1.021 1.110	47.73 1.004 1.123	46.36 0.991 1.134	44.08 0.972 1.151	42.23 0.958 1.164
3.1623−03	−2.5	39.92 1.072 0.979	51.35 1.123 0.996	52.63 1.071 1.068	50.79 1.042 1.092	49.06 1.021 1.110	47.54 1.004 1.123	46.21 0.991 1.134	43.98 0.972 1.151	42.15 0.958 1.163
3.9811−03	−2.4	35.64 1.054 0.984	48.57 1.109 0.997	51.86 1.071 1.067	50.29 1.042 1.092	48.72 1.021 1.109	47.32 1.004 1.122	46.02 0.991 1.133	43.85 0.972 1.150	42.06 0.958 1.163
5.0119−03	−2.3	31.35 1.039 0.989	45.43 1.094 0.998	50.93 1.070 1.066	49.68 1.041 1.091	48.29 1.021 1.109	47.00 1.005 1.122	45.81 0.992 1.133	43.69 0.972 1.150	41.94 0.958 1.162
6.3096−03	−2.2	27.17 1.027 0.993	41.95 1.079 0.999	49.85 1.068 1.065	48.94 1.041 1.090	47.74 1.022 1.108	46.59 1.005 1.121	45.48 0.993 1.132	43.50 0.973 1.149	41.79 0.958 1.162
7.9433−03	−2.1	23.22 1.017 0.995	38.21 1.063 0.999	48.53 1.065 1.063	48.06 1.041 1.088	47.09 1.022 1.107	46.06 1.007 1.121	45.08 0.994 1.132	43.23 0.974 1.149	41.62 0.959 1.162
1.0000−02	−2.0	19.59 1.010 0.997	34.29 1.049 0.999	46.97 1.062 1.060	47.02 1.041 1.087	46.31 1.023 1.105	45.45 1.008 1.120	44.57 0.995 1.131	42.87 0.975 1.149	41.36 0.959 1.162
1.2589−02	−1.9	16.34 1.006 0.998	30.31 1.036 0.999	45.11 1.058 1.057	45.76 1.040 1.084	45.40 1.023 1.104	44.71 1.008 1.119	43.96 0.996 1.131	42.42 0.976 1.149	41.03 0.961 1.161
1.5849−02	−1.8	13.51 1.003 0.999	26.39 1.025 0.999	42.97 1.053 1.054	44.28 1.039 1.081	44.29 1.023 1.101	43.85 1.009 1.117	43.22 0.997 1.129	41.89 0.978 1.148	40.60 0.962 1.161
1.9953−02	−1.7	11.07 1.002 0.999	22.64 1.016 0.999	40.52 1.048 1.050	42.53 1.036 1.078	42.97 1.022 1.098	42.79 1.009 1.114	42.37 0.998 1.128	41.26 0.979 1.147	40.11 0.964 1.160
2.5119−02	−1.6	9.018 1.001 0.999	19.17 1.010 0.999	37.77 1.042 1.046	40.52 1.033 1.073	41.40 1.021 1.094	41.54 1.009 1.111	41.32 0.999 1.125	40.53 0.980 1.144	39.52 0.966 1.159
3.1623−02	−1.5	7.307 1.001 1.000	16.05 1.006 0.999	34.75 1.036 1.042	38.20 1.030 1.068	39.58 1.020 1.090	40.05 1.009 1.107	40.10 0.999 1.121	39.62 0.982 1.141	38.84 0.967 1.156
3.9811−02	−1.4	5.896 1.001 1.000	13.29 1.003 1.000	31.52 1.029 1.037	35.60 1.026 1.063	37.48 1.018 1.084	38.34 1.009 1.102	38.65 0.999 1.116	38.55 0.983 1.137	37.99 0.969 1.153
5.0119−02	−1.3	4.743 1.001 1.000	10.92 1.002 1.000	28.16 1.023 1.031	32.75 1.023 1.056	35.11 1.016 1.078	36.35 1.008 1.096	36.97 1.000 1.110	37.26 0.984 1.133	36.98 0.971 1.150
6.3096−02	−1.2	3.805 1.001 1.000	8.909 1.001 1.000	24.78 1.017 1.025	29.71 1.019 1.049	32.49 1.014 1.070	34.10 1.007 1.088	35.02 1.000 1.103	35.77 0.985 1.126	35.79 0.973 1.144
7.9433−02	−1.1	3.046 1.001 1.001	7.229 1.001 1.000	21.48 1.012 1.020	26.54 1.015 1.041	29.65 1.012 1.061	31.60 1.006 1.078	32.82 1.000 1.093	34.04 0.987 1.117	34.40 0.975 1.135
1.0000−01	−1.0	2.435 1.001 1.001	5.840 1.001 1.000	18.36 1.007 1.014	23.36 1.011 1.033	26.67 1.010 1.051	28.88 1.006 1.068	30.38 1.000 1.083	32.07 0.988 1.107	32.79 0.977 1.125
1.2589−01	−0.9	1.943 1.001 1.001	4.702 1.001 1.000	15.49 1.004 1.010	20.25 1.008 1.026	23.62 1.008 1.042	26.02 1.005 1.058	27.74 1.000 1.072	29.85 0.989 1.095	30.93 0.979 1.115
1.5849−01	−0.8	1.550 1.001 1.001	3.775 1.001 1.000	12.93 1.002 1.007	17.31 1.005 1.019	20.62 1.006 1.033	23.10 1.004 1.048	24.96 1.000 1.061	27.44 0.991 1.084	28.85 0.981 1.103
1.9953−01	−0.7	1.235 1.001 1.001	3.024 1.001 1.001	10.68 1.001 1.004	14.62 1.003 1.014	17.73 1.004 1.026	20.19 1.003 1.038	22.13 1.000 1.051	24.87 0.993 1.073	26.57 0.984 1.093
2.5119−01	−0.6	0.983 1.001 1.001	2.418 1.001 1.001	8.761 1.001 1.003	12.20 1.002 1.009	15.05 1.002 1.019	17.40 1.002 1.030	19.33 1.000 1.041	22.20 0.994 1.062	24.13 0.986 1.081
3.1623−01	−0.5	0.783 1.001 1.001	1.931 1.001 1.001	7.138 1.001 1.002	10.09 1.001 1.006	12.63 1.001 1.013	14.80 1.001 1.022	16.64 1.000 1.032	19.52 0.996 1.051	21.59 0.989 1.069
3.9811−01	−0.4	0.623 1.001 1.001	1.540 1.001 1.001	5.786 1.001 1.001	8.273 1.000 1.004	10.48 1.001 1.009	12.43 1.000 1.016	14.14 1.000 1.024	16.92 0.997 1.040	19.03 0.992 1.057
5.0119−01	−0.3	0.495 1.001 1.001	1.228 1.001 1.001	4.671 1.001 1.001	6.743 1.000 1.003	8.631 1.000 1.006	10.34 1.000 1.011	11.87 1.000 1.017	14.47 0.998 1.031	16.54 0.994 1.046
6.3096−01	−0.2	0.394 1.001 1.001	0.978 1.001 1.001	3.758 1.001 1.001	5.467 1.000 1.002	7.055 1.000 1.004	8.520 1.000 1.007	9.865 1.000 1.012	12.22 0.998 1.023	14.17 0.996 1.035
7.9433−01	−0.1	0.313 1.001 1.001	0.778 1.001 1.001	3.016 1.001 1.001	4.414 1.000 1.001	5.734 1.000 1.003	6.972 1.000 1.005	8.129 1.000 1.008	10.21 0.999 1.016	11.99 0.997 1.026
1.0000+00	−0.0	0.249 1.001 1.001	0.619 1.001 1.001	2.415 1.001 1.001	3.553 1.001 1.001	4.639 1.000 1.002	5.671 1.000 1.003	6.650 1.000 1.005	8.446 0.999 1.011	10.03 0.998 1.019
1.2589+00	0.1	0.198 1.001 1.001	0.493 1.001 1.001	1.931 1.001 1.001	2.852 1.001 1.001	3.739 1.000 1.001	4.592 1.000 1.002	5.409 1.000 1.003	6.935 0.999 1.008	8.318 0.999 1.013
1.5849+00	0.2	0.157 1.001 1.001	0.392 1.001 1.001	1.542 1.001 1.001	2.284 1.001 1.001	3.005 1.000 1.001	3.703 1.000 1.001	4.378 1.000 1.002	5.657 1.000 1.005	6.840 0.999 1.009
1.9953+00	0.3	0.125 1.000 1.000	0.312 1.000 1.000	1.231 1.000 1.000	1.828 1.000 1.000	2.411 1.000 1.000	2.979 1.000 1.001	3.532 1.000 1.002	4.590 1.000 1.003	5.585 1.000 1.006

Table IX continued; neutron energy 1.5 MeV

η/rad^{-1}	lg η	$\sigma_0 = 2\ \mu m^2$ $d = 1.60\ \mu m$	$\sigma_0 = 5\ \mu m^2$ $d = 2.52\ \mu m$	$\sigma_0 = 20\ \mu m^2$ $d = 5.05\ \mu m$	$\sigma_0 = 30\ \mu m^2$ $d = 6.18\ \mu m$	$\sigma_0 = 40\ \mu m^2$ $d = 7.14\ \mu m$	$\sigma_0 = 50\ \mu m^2$ $d = 7.98\ \mu m$	$\sigma_0 = 60\ \mu m^2$ $d = 8.74\ \mu m$	$\sigma_0 = 80\ \mu m^2$ $d = 10.1\ \mu m$	$\sigma_0 = 100\ \mu m^2$ $d = 11.3\ \mu m$
zero		60.93 1.449 0.852	57.28 1.365 0.891	50.87 1.229 0.959	48.79 1.190 0.980	47.24 1.163 0.996	46.02 1.143 1.008	44.98 1.126 1.017	43.30 1.100 1.033	41.95 1.081 1.045
1.0000 − 04	−4.0	59.64 1.423 0.857	56.84 1.358 0.892	50.79 1.228 0.959	48.73 1.190 0.980	47.20 1.163 0.996	45.98 1.143 1.008	44.96 1.126 1.017	43.28 1.100 1.033	41.93 1.081 1.045
1.2589 − 04	−3.9	59.29 1.417 0.859	56.72 1.357 0.893	50.77 1.228 0.959	48.72 1.190 0.981	47.20 1.163 0.996	45.98 1.142 1.008	44.95 1.126 1.017	43.28 1.100 1.033	41.93 1.081 1.045
1.5849 − 04	−3.8	58.87 1.410 0.860	56.60 1.355 0.893	50.74 1.228 0.960	48.71 1.190 0.981	47.18 1.163 0.996	45.97 1.142 1.008	44.95 1.126 1.017	43.27 1.100 1.033	41.93 1.081 1.045
1.9953 − 04	−3.7	58.34 1.401 0.862	56.41 1.352 0.894	50.71 1.227 0.960	48.68 1.189 0.981	47.17 1.163 0.996	45.96 1.142 1.008	44.94 1.126 1.017	43.27 1.100 1.033	41.92 1.081 1.045
2.5119 − 04	−3.6	57.73 1.389 0.865	56.17 1.349 0.895	50.66 1.227 0.960	48.66 1.189 0.981	47.15 1.163 0.996	45.94 1.142 1.008	44.93 1.126 1.017	43.26 1.100 1.033	41.92 1.081 1.045
3.1623 − 04	−3.5	56.99 1.375 0.868	55.87 1.345 0.896	50.61 1.226 0.960	48.63 1.189 0.981	47.13 1.163 0.996	45.93 1.142 1.008	44.91 1.126 1.017	43.25 1.100 1.033	41.91 1.081 1.045
3.9811 − 04	−3.4	56.07 1.359 0.872	55.51 1.340 0.897	50.54 1.226 0.960	48.58 1.189 0.981	47.10 1.162 0.996	45.90 1.142 1.008	44.90 1.125 1.017	43.24 1.100 1.033	41.90 1.081 1.045
5.0119 − 04	−3.3	54.93 1.340 0.876	55.07 1.333 0.898	50.46 1.225 0.960	48.53 1.188 0.981	47.06 1.162 0.996	45.88 1.142 1.008	44.87 1.125 1.017	43.23 1.100 1.033	41.89 1.081 1.045
6.3096 − 04	−3.2	53.55 1.319 0.882	54.54 1.326 0.900	50.35 1.224 0.961	48.47 1.188 0.981	47.02 1.162 0.996	45.84 1.142 1.008	44.85 1.125 1.017	43.21 1.100 1.033	41.88 1.081 1.045
7.9433 − 04	−3.1	51.90 1.295 0.888	53.90 1.317 0.902	50.22 1.223 0.961	48.38 1.187 0.981	46.96 1.161 0.996	45.80 1.141 1.008	44.81 1.125 1.017	43.18 1.100 1.033	41.86 1.081 1.045
1.0000 − 03	−3.0	49.97 1.269 0.896	53.11 1.306 0.905	50.03 1.222 0.961	48.29 1.186 0.982	46.89 1.161 0.996	45.74 1.141 1.008	44.77 1.125 1.018	43.15 1.100 1.033	41.84 1.081 1.045
1.2589 − 03	−2.9	47.74 1.240 0.905	52.13 1.292 0.909	49.80 1.220 0.962	48.14 1.185 0.982	46.80 1.160 0.996	45.68 1.140 1.008	44.71 1.125 1.018	43.11 1.100 1.033	41.81 1.081 1.045
1.5849 − 03	−2.8	45.20 1.209 0.915	50.93 1.277 0.913	49.52 1.219 0.963	47.95 1.185 0.982	46.68 1.159 0.997	45.59 1.140 1.008	44.64 1.124 1.018	43.07 1.099 1.033	41.77 1.080 1.045
1.9953 − 03	−2.7	42.35 1.178 0.925	49.49 1.259 0.917	49.18 1.216 0.963	47.72 1.184 0.983	46.52 1.159 0.997	45.48 1.139 1.008	44.57 1.124 1.018	43.01 1.099 1.033	41.73 1.080 1.045
2.5119 − 03	−2.6	39.20 1.148 0.937	47.79 1.238 0.923	48.77 1.213 0.964	47.45 1.183 0.983	46.32 1.158 0.998	45.33 1.139 1.008	44.44 1.123 1.018	42.93 1.098 1.033	41.67 1.080 1.045
3.1623 − 03	−2.5	35.84 1.119 0.948	45.79 1.216 0.929	48.28 1.209 0.965	47.14 1.181 0.984	46.07 1.158 0.998	45.13 1.138 1.009	44.27 1.123 1.018	42.84 1.098 1.033	41.60 1.080 1.045
3.9811 − 03	−2.4	32.32 1.092 0.959	43.50 1.191 0.936	47.67 1.203 0.966	46.76 1.178 0.985	45.78 1.157 0.999	44.89 1.138 1.009	44.07 1.123 1.018	42.70 1.098 1.033	41.51 1.080 1.045
5.0119 − 03	−2.3	28.74 1.069 0.969	40.90 1.165 0.944	46.91 1.197 0.968	46.28 1.174 0.985	45.43 1.155 0.999	44.61 1.137 1.009	43.84 1.123 1.019	42.52 1.098 1.033	41.38 1.080 1.045
6.3096 − 03	−2.2	25.19 1.049 0.978	38.00 1.138 0.952	45.97 1.189 0.969	45.69 1.170 0.986	45.01 1.152 1.000	44.28 1.136 1.010	43.57 1.122 1.019	42.31 1.098 1.033	41.21 1.079 1.045
7.9433 − 03	−2.1	21.78 1.033 0.985	34.87 1.112 0.961	44.84 1.181 0.971	44.95 1.165 0.987	44.49 1.149 1.000	43.88 1.133 1.010	43.26 1.120 1.019	42.06 1.098 1.033	41.01 1.079 1.045
1.0000 − 02	−2.0	18.57 1.021 0.990	31.55 1.088 0.969	43.49 1.170 0.973	44.05 1.159 0.987	43.83 1.145 1.000	43.38 1.131 1.011	42.85 1.118 1.020	41.77 1.097 1.034	40.77 1.079 1.045
1.2589 − 02	−1.9	15.65 1.013 0.994	28.14 1.066 0.976	41.90 1.158 0.975	42.96 1.152 0.989	43.02 1.140 0.999	42.75 1.127 1.012	42.35 1.115 1.020	41.42 1.096 1.034	40.50 1.079 1.045
1.5849 − 02	−1.8	13.05 1.008 0.996	24.73 1.048 0.983	40.05 1.143 0.978	41.66 1.143 0.990	42.04 1.135 1.000	41.97 1.124 1.012	41.72 1.113 1.021	40.98 1.094 1.034	40.17 1.078 1.045
1.9953 − 02	−1.7	10.77 1.004 0.998	21.43 1.033 0.989	37.90 1.127 0.981	40.12 1.133 0.992	40.86 1.128 1.001	41.03 1.119 1.013	40.95 1.109 1.021	40.42 1.092 1.035	39.75 1.077 1.046
2.5119 − 02	−1.6	8.827 1.002 0.999	18.32 1.021 0.993	35.48 1.110 0.985	38.33 1.122 0.994	39.47 1.120 1.003	39.90 1.114 1.013	40.00 1.105 1.021	39.73 1.089 1.035	39.22 1.075 1.046
3.1623 − 02	−1.5	7.186 1.001 0.999	15.46 1.013 0.995	32.81 1.093 0.988	36.26 1.109 0.996	37.83 1.110 1.005	38.56 1.107 1.014	38.88 1.101 1.022	38.89 1.087 1.035	38.57 1.073 1.046
3.9811 − 02	−1.4	5.820 1.001 1.000	12.91 1.008 0.997	29.92 1.075 0.992	33.93 1.095 0.998	35.94 1.100 1.007	36.99 1.099 1.015	37.54 1.095 1.022	37.88 1.084 1.035	37.77 1.071 1.046
5.0119 − 02	−1.3	4.694 1.001 1.000	10.67 1.004 0.998	26.90 1.059 0.995	31.35 1.080 1.001	33.78 1.088 1.008	35.17 1.089 1.016	35.98 1.087 1.023	36.68 1.079 1.035	36.81 1.069 1.045
6.3096 − 02	−1.2	3.774 1.001 1.000	8.751 1.002 0.999	23.82 1.044 0.998	28.57 1.065 1.002	31.37 1.076 1.009	33.09 1.079 1.016	34.17 1.079 1.022	35.28 1.074 1.034	35.67 1.065 1.044
7.9433 − 02	−1.1	3.027 1.001 1.000	7.129 1.001 0.999	20.78 1.031 0.999	25.67 1.051 1.003	28.75 1.063 1.009	30.76 1.069 1.015	32.11 1.070 1.021	33.64 1.068 1.032	34.34 1.061 1.042
1.0000 − 01	−1.0	2.423 1.001 1.001	5.778 1.001 1.000	17.87 1.021 1.000	22.71 1.038 1.004	25.98 1.050 1.009	28.23 1.057 1.014	29.81 1.061 1.020	31.76 1.061 1.030	32.77 1.056 1.040
1.2589 − 01	−0.9	1.936 1.001 1.001	4.663 1.001 1.000	15.17 1.013 1.000	19.79 1.027 1.004	23.12 1.038 1.008	25.54 1.046 1.013	27.31 1.051 1.018	29.64 1.053 1.028	30.97 1.051 1.037
1.5849 − 01	−0.8	1.545 1.001 1.001	3.750 1.001 1.000	12.72 1.008 1.000	17.00 1.018 1.003	20.27 1.027 1.007	22.75 1.035 1.012	24.67 1.041 1.017	27.31 1.045 1.026	28.94 1.045 1.035
1.9953 − 01	−0.7	1.232 1.001 1.001	3.009 1.001 1.000	10.55 1.004 1.000	14.42 1.011 1.003	17.50 1.019 1.006	19.97 1.026 1.011	21.94 1.031 1.015	24.81 1.037 1.024	26.70 1.039 1.033
2.5119 − 01	−0.6	0.981 1.001 1.001	2.409 1.001 1.001	8.680 1.002 1.000	12.08 1.007 1.002	14.91 1.012 1.005	17.27 1.018 1.009	19.23 1.023 1.013	22.21 1.029 1.021	24.30 1.032 1.029
3.1623 − 01	−0.5	0.781 1.001 1.001	1.925 1.001 1.001	7.090 1.001 1.000	10.01 1.004 1.001	12.55 1.007 1.004	14.73 1.012 1.007	16.60 1.015 1.010	19.58 1.022 1.018	21.78 1.025 1.026
3.9811 − 01	−0.4	0.622 1.001 1.001	1.537 1.001 1.001	5.757 1.001 1.000	8.232 1.002 1.001	10.44 1.004 1.002	12.40 1.007 1.005	14.13 1.010 1.008	17.01 1.015 1.015	19.23 1.019 1.022
5.0119 − 01	−0.3	0.495 1.001 1.001	1.225 1.001 1.001	4.654 1.001 1.000	6.721 1.001 1.000	8.611 1.002 1.001	10.33 1.004 1.003	11.89 1.006 1.006	14.56 1.010 1.012	16.73 1.014 1.018
6.3096 − 01	−0.2	0.393 1.001 1.001	0.976 1.001 1.001	3.748 1.001 1.000	5.455 1.001 1.000	7.047 1.001 1.001	8.524 1.002 1.002	9.889 1.004 1.004	12.30 1.006 1.009	14.34 1.009 1.014
7.9433 − 01	−0.1	0.313 1.001 1.001	0.777 1.001 1.001	3.010 1.001 1.000	4.409 1.001 1.000	5.732 1.001 1.001	6.980 1.001 1.001	8.153 1.002 1.002	10.28 1.004 1.006	12.14 1.006 1.010
1.0000 + 00	−0.0	0.249 1.001 1.001	0.619 1.001 1.001	2.412 1.001 1.001	3.550 1.001 1.000	4.640 1.001 1.000	5.680 1.001 1.001	6.671 1.001 1.002	8.506 1.002 1.004	10.15 1.003 1.007
1.2589 + 00	0.1	0.198 1.001 1.001	0.492 1.001 1.001	1.929 1.001 1.001	2.850 1.001 1.000	3.741 1.001 1.000	4.599 1.001 1.001	5.426 1.001 1.001	6.981 1.001 1.003	8.408 1.002 1.005
1.5849 + 00	0.2	0.157 1.001 1.001	0.392 1.001 1.001	1.541 1.001 1.000	2.284 1.001 1.000	3.007 1.001 1.000	3.710 1.001 1.000	4.391 1.001 1.001	5.691 1.001 1.002	6.906 1.001 1.003
1.9953 + 00	0.3	0.125 1.000 1.000	0.312 1.000 1.000	1.230 1.000 1.000	1.828 1.000 1.000	2.412 1.000 1.000	2.984 1.000 1.000	3.541 1.000 1.000	4.615 1.001 1.001	5.632 1.001 1.002

Table IX continued; neutron energy 2 MeV

η/rad^{-1}	lg η	$\sigma_0 = 2\ \mu m^2$ $d = 1.60\ \mu m$	$\sigma_0 = 5\ \mu m^2$ $d = 2.52\ \mu m$	$\sigma_0 = 20\ \mu m^2$ $d = 5.05\ \mu m$	$\sigma_0 = 30\ \mu m^2$ $d = 6.18\ \mu m$	$\sigma_0 = 40\ \mu m^2$ $d = 7.14\ \mu m$	$\sigma_0 = 50\ \mu m^2$ $d = 7.98\ \mu m$	$\sigma_0 = 60\ \mu m^2$ $d = 8.74\ \mu m$	$\sigma_0 = 80\ \mu m^2$ $d = 10.1\ \mu m$	$\sigma_0 = 100\ \mu m^2$ $d = 11.3\ \mu m$
zero		56.66 1.642 0.789	53.15 1.542 0.830	47.25 1.372 0.903	45.43 1.323 0.926	44.13 1.288 0.942	43.09 1.262 0.955	42.24 1.241 0.965	40.87 1.208 0.982	39.77 1.184 0.995
1.0000−04	−4.0	55.36 1.597 0.797	52.73 1.531 0.832	47.16 1.370 0.904	45.39 1.321 0.927	44.09 1.288 0.942	43.07 1.262 0.955	42.22 1.241 0.965	40.84 1.208 0.982	39.76 1.183 0.994
1.2589−04	−3.9	54.96 1.587 0.799	52.61 1.528 0.833	47.14 1.369 0.904	45.38 1.321 0.927	44.08 1.287 0.943	43.06 1.262 0.955	42.22 1.241 0.965	40.84 1.208 0.982	39.76 1.183 0.994
1.5849−04	−3.8	54.61 1.575 0.801	52.49 1.524 0.834	47.12 1.369 0.904	45.37 1.321 0.927	44.07 1.287 0.943	43.06 1.262 0.955	42.21 1.241 0.965	40.84 1.208 0.982	39.76 1.183 0.994
1.9953−04	−3.7	54.09 1.559 0.805	52.30 1.519 0.835	47.08 1.368 0.904	45.35 1.320 0.927	44.06 1.287 0.943	43.05 1.261 0.955	42.20 1.241 0.965	40.83 1.208 0.982	39.75 1.183 0.994
2.5119−04	−3.6	53.51 1.540 0.808	52.07 1.513 0.836	47.05 1.367 0.904	45.32 1.320 0.927	44.04 1.287 0.943	43.03 1.261 0.955	42.19 1.240 0.965	40.82 1.208 0.982	39.75 1.183 0.994
3.1623−04	−3.5	52.78 1.517 0.813	51.78 1.506 0.838	47.00 1.366 0.904	45.29 1.319 0.927	44.02 1.286 0.943	43.02 1.261 0.955	42.18 1.240 0.965	40.82 1.208 0.932	39.74 1.183 0.994
3.9811−04	−3.4	51.88 1.492 0.818	51.43 1.497 0.839	46.93 1.365 0.905	45.26 1.319 0.927	44.00 1.286 0.943	43.00 1.261 0.956	42.17 1.240 0.965	40.80 1.208 0.982	39.73 1.183 0.994
5.0119−04	−3.3	50.78 1.463 0.825	51.02 1.486 0.842	46.85 1.364 0.905	45.21 1.318 0.927	43.97 1.285 0.943	42.97 1.260 0.956	42.15 1.239 0.965	40.79 1.208 0.982	39.72 1.183 0.994
6.3096−04	−3.2	49.45 1.431 0.833	50.53 1.473 0.844	46.75 1.362 0.906	45.15 1.316 0.928	43.92 1.285 0.943	42.94 1.259 0.956	42.12 1.239 0.965	40.77 1.207 0.982	39.71 1.183 0.994
7.9433−04	−3.1	47.89 1.396 0.842	49.93 1.457 0.847	46.61 1.360 0.907	45.07 1.315 0.928	43.87 1.284 0.944	42.90 1.259 0.956	42.09 1.238 0.965	40.75 1.207 0.982	39.69 1.183 0.994
1.0000−03	−3.0	46.09 1.359 0.852	49.20 1.438 0.851	46.44 1.358 0.908	44.96 1.314 0.928	43.80 1.282 0.944	42.85 1.258 0.956	42.05 1.237 0.965	40.72 1.207 0.982	39.67 1.183 0.994
1.2589−03	−2.9	44.03 1.319 0.864	48.28 1.415 0.856	46.23 1.355 0.909	44.81 1.312 0.929	43.71 1.281 0.944	42.79 1.257 0.956	42.00 1.236 0.965	40.69 1.206 0.982	39.65 1.182 0.993
1.5849−03	−2.8	41.71 1.278 0.877	47.16 1.390 0.862	45.97 1.351 0.910	44.64 1.311 0.930	43.59 1.280 0.945	42.71 1.256 0.957	41.94 1.235 0.965	40.65 1.205 0.982	39.61 1.182 0.993
1.9953−03	−2.7	39.13 1.237 0.892	45.82 1.362 0.869	45.66 1.347 0.911	44.44 1.309 0.930	43.44 1.279 0.945	42.59 1.255 0.957	41.84 1.234 0.966	40.59 1.204 0.982	39.57 1.181 0.993
2.5119−03	−2.6	36.31 1.197 0.908	44.24 1.331 0.878	45.31 1.341 0.912	44.20 1.306 0.931	43.25 1.277 0.946	42.44 1.254 0.958	41.71 1.234 0.967	40.52 1.203 0.983	39.52 1.180 0.994
3.1623−03	−2.5	33.30 1.160 0.924	42.41 1.298 0.887	44.87 1.332 0.914	43.93 1.302 0.932	43.03 1.276 0.947	42.26 1.253 0.958	41.56 1.233 0.968	40.41 1.202 0.983	39.46 1.179 0.994
3.9811−03	−2.4	30.17 1.126 0.939	40.32 1.262 0.898	44.33 1.322 0.916	43.59 1.297 0.934	42.77 1.273 0.948	42.05 1.251 0.959	41.38 1.232 0.970	40.28 1.202 0.984	39.35 1.178 0.995
5.0119−03	−2.3	26.97 1.096 0.954	37.97 1.226 0.910	43.65 1.310 0.918	43.16 1.290 0.935	42.47 1.269 0.949	41.80 1.249 0.960	41.18 1.231 0.970	40.11 1.202 0.984	39.23 1.178 0.996
6.3096−03	−2.2	23.79 1.070 0.966	35.36 1.190 0.922	42.81 1.297 0.921	42.64 1.281 0.937	42.09 1.264 0.950	41.51 1.246 0.961	40.95 1.228 0.971	39.92 1.201 0.985	39.06 1.178 0.997
7.9433−03	−2.1	20.70 1.049 0.977	32.54 1.155 0.935	41.79 1.281 0.925	41.98 1.272 0.939	41.63 1.257 0.951	41.15 1.241 0.962	40.66 1.225 0.971	39.70 1.200 0.986	38.88 1.178 0.998
1.0000−02	−2.0	17.78 1.033 0.985	29.56 1.123 0.948	40.57 1.262 0.929	41.17 1.261 0.941	41.05 1.249 0.952	40.71 1.235 0.963	40.30 1.221 0.971	39.45 1.198 0.987	38.67 1.177 0.999
1.2589−02	−1.9	15.09 1.021 0.990	26.50 1.094 0.960	39.14 1.242 0.933	40.18 1.248 0.944	40.33 1.239 0.954	40.15 1.228 0.964	39.86 1.216 0.972	39.13 1.195 0.987	38.43 1.176 0.999
1.5849−02	−1.8	12.66 1.013 0.994	23.43 1.069 0.970	37.46 1.218 0.939	39.01 1.232 0.947	39.45 1.228 0.956	39.45 1.220 0.966	39.29 1.210 0.973	38.74 1.190 0.988	38.13 1.173 0.999
1.9953−02	−1.7	10.51 1.007 0.997	20.43 1.049 0.980	35.53 1.193 0.946	37.62 1.214 0.951	38.38 1.216 0.959	38.60 1.211 0.968	38.59 1.202 0.975	38.25 1.185 0.988	37.76 1.169 0.999
2.5119−02	−1.6	8.656 1.004 0.998	17.58 1.033 0.987	33.34 1.166 0.953	36.01 1.194 0.956	37.12 1.201 0.963	37.58 1.199 0.970	37.73 1.194 0.977	37.63 1.179 0.989	37.29 1.164 1.000
3.1623−02	−1.5	7.075 1.002 0.999	14.93 1.021 0.992	30.92 1.139 0.961	34.14 1.172 0.962	35.64 1.184 0.967	36.37 1.186 0.973	36.72 1.184 0.979	36.87 1.172 0.990	36.69 1.160 1.001
3.9811−02	−1.4	5.749 1.001 0.999	12.54 1.013 0.995	28.31 1.113 0.969	32.03 1.149 0.967	33.93 1.165 0.971	34.96 1.171 0.976	35.51 1.172 0.982	35.96 1.164 0.992	35.95 1.154 1.002
5.0119−02	−1.3	4.650 1.001 1.000	10.43 1.007 0.997	25.56 1.088 0.977	29.69 1.125 0.974	31.97 1.145 0.976	33.30 1.154 0.980	34.10 1.158 0.985	34.86 1.155 0.994	35.08 1.148 1.004
6.3096−02	−1.2	3.746 1.001 1.000	8.591 1.004 0.998	22.75 1.066 0.983	27.17 1.102 0.980	29.78 1.124 0.980	31.41 1.136 0.983	32.46 1.142 0.987	33.58 1.144 0.996	34.04 1.140 1.005
7.9433−02	−1.1	3.009 1.001 1.000	7.026 1.002 0.999	19.96 1.047 0.989	24.51 1.079 0.985	27.39 1.102 0.985	29.29 1.117 0.987	30.58 1.125 0.990	32.08 1.131 0.998	32.82 1.130 1.005
1.0000−01	−1.0	2.411 1.001 1.001	5.712 1.001 0.999	17.26 1.032 0.993	21.79 1.059 0.990	24.85 1.081 0.989	26.97 1.097 0.990	28.47 1.107 0.993	30.35 1.117 0.999	31.38 1.119 1.006
1.2589−01	−0.9	1.929 1.001 1.001	4.621 1.001 1.000	14.72 1.021 0.996	19.09 1.042 0.994	22.22 1.062 0.993	24.49 1.078 0.994	26.18 1.089 0.995	28.41 1.102 1.001	29.72 1.107 1.007
1.5849−01	−0.8	1.541 1.001 1.001	3.724 1.001 1.000	12.41 1.013 0.998	16.48 1.029 0.997	19.57 1.045 0.996	21.92 1.060 0.997	23.74 1.071 0.998	26.26 1.086 1.002	27.84 1.093 1.008
1.9953−01	−0.7	1.229 1.001 1.001	2.992 1.001 1.000	10.35 1.007 0.999	14.04 1.019 0.998	16.98 1.031 0.999	19.33 1.044 0.999	21.20 1.054 1.000	23.94 1.070 1.004	25.77 1.079 1.009
2.5119−01	−0.6	0.980 1.001 1.001	2.398 1.001 1.001	8.545 1.004 0.999	11.82 1.011 0.999	14.54 1.020 1.000	16.79 1.030 1.001	18.66 1.039 1.002	21.52 1.055 1.005	23.53 1.065 1.010
3.1623−01	−0.5	0.780 1.001 1.001	1.919 1.001 1.001	7.003 1.002 1.000	9.844 1.006 1.000	12.29 1.013 1.000	14.39 1.020 1.001	16.18 1.027 1.003	19.05 1.041 1.006	21.17 1.051 1.010
3.9811−01	−0.4	0.621 1.001 1.001	1.533 1.001 1.001	5.702 1.001 1.000	8.122 1.004 1.000	10.27 1.007 1.001	12.17 1.012 1.002	13.84 1.018 1.003	16.61 1.029 1.006	18.76 1.038 1.010
5.0119−01	−0.3	0.494 1.001 1.001	1.223 1.001 1.001	4.619 1.001 1.000	6.650 1.002 1.000	8.497 1.004 1.001	10.17 1.007 1.001	11.68 1.011 1.003	14.28 1.019 1.005	16.38 1.027 1.009
6.3096−01	−0.2	0.393 1.001 1.001	0.975 1.001 1.001	3.726 1.001 1.000	5.410 1.001 1.000	6.974 1.002 1.000	8.421 1.004 1.001	9.755 1.006 1.002	12.11 1.012 1.005	14.10 1.018 1.008
7.9433−01	−0.1	0.313 1.001 1.001	0.776 1.001 1.001	2.996 1.001 1.000	4.380 1.001 1.000	5.686 1.001 1.000	6.914 1.002 1.001	8.066 1.003 1.001	10.15 1.007 1.003	11.97 1.011 1.006
1.0000+00	−0.0	0.249 1.001 1.001	0.618 1.001 1.001	2.403 1.001 1.001	3.532 1.001 1.000	4.611 1.001 1.000	5.638 1.001 1.001	6.616 1.002 1.001	8.423 1.004 1.002	10.04 1.007 1.005
1.2589+00	0.1	0.198 1.001 1.001	0.492 1.001 1.001	1.924 1.001 1.001	2.839 1.001 1.000	3.723 1.001 1.000	4.573 1.001 1.000	5.391 1.001 1.001	6.928 1.002 1.002	8.338 1.004 1.003
1.5849+00	0.2	0.157 1.001 1.001	0.391 1.001 1.001	1.537 1.001 1.000	2.277 1.001 1.000	2.995 1.000 1.000	3.693 1.001 1.000	4.369 1.001 1.000	5.658 1.001 1.001	6.862 1.002 1.002
1.9953+00	0.3	0.125 1.000 1.000	0.311 1.000 1.000	1.228 1.000 1.000	1.823 1.000 1.000	2.405 1.000 1.000	2.973 1.000 1.000	3.527 1.000 1.000	4.594 1.001 1.001	5.604 1.002 1.002

Table IX continued; neutron energy 3 MeV

η/rad^{-1}	lg η	σ_0 = 2 μm², d = 1.60 μm	σ_0 = 5 μm², d = 2.52 μm	σ_0 = 20 μm², d = 5.05 μm	σ_0 = 30 μm², d = 6.18 μm	σ_0 = 40 μm², d = 7.14 μm	σ_0 = 50 μm², d = 7.98 μm	σ_0 = 60 μm², d = 8.74 μm	σ_0 = 80 μm², d = 10.1 μm	σ_0 = 100 μm², d = 11.3 μm
zero		53.89 1.988 0.669	50.14 1.880 0.711	43.70 1.666 0.797	41.86 1.600 0.824	40.58 1.553 0.843	39.60 1.518 0.859	38.81 1.489 0.871	37.57 1.444 0.890	36.61 1.410 0.905
1.0000−04	−4.0	52.18 1.899 0.684	49.61 1.851 0.716	43.62 1.662 0.798	41.81 1.598 0.825	40.55 1.552 0.844	39.58 1.516 0.859	38.78 1.488 0.871	37.55 1.444 0.890	36.60 1.409 0.905
1.2589−04	−3.9	51.78 1.878 0.687	49.47 1.844 0.717	43.60 1.661 0.798	41.80 1.597 0.825	40.54 1.551 0.844	39.57 1.516 0.859	38.78 1.488 0.871	37.55 1.444 0.890	36.59 1.409 0.905
1.5849−04	−3.8	51.35 1.853 0.690	49.30 1.836 0.719	43.57 1.660 0.798	41.79 1.597 0.825	40.53 1.551 0.844	39.56 1.516 0.859	38.77 1.487 0.871	37.54 1.444 0.890	36.59 1.409 0.905
1.9953−04	−3.7	50.73 1.823 0.695	49.07 1.825 0.721	43.54 1.658 0.798	41.77 1.596 0.825	40.52 1.551 0.844	39.55 1.516 0.859	38.77 1.487 0.871	37.54 1.443 0.890	36.59 1.409 0.905
2.5119−04	−3.6	49.99 1.790 0.702	48.79 1.813 0.723	43.50 1.656 0.798	41.74 1.595 0.825	40.50 1.550 0.844	39.54 1.515 0.859	38.76 1.487 0.871	37.53 1.443 0.890	36.58 1.409 0.905
3.1623−04	−3.5	49.09 1.751 0.710	48.45 1.798 0.726	43.45 1.652 0.799	41.71 1.593 0.826	40.48 1.549 0.845	39.53 1.515 0.859	38.75 1.487 0.871	37.52 1.443 0.890	36.58 1.409 0.905
3.9811−04	−3.4	48.03 1.709 0.719	48.05 1.779 0.729	43.38 1.648 0.800	41.68 1.590 0.826	40.45 1.548 0.845	39.51 1.514 0.860	38.73 1.486 0.872	37.51 1.442 0.891	36.57 1.409 0.905
5.0119−04	−3.3	46.77 1.661 0.730	47.59 1.756 0.733	43.28 1.644 0.800	41.63 1.587 0.826	40.42 1.545 0.845	39.48 1.513 0.860	38.71 1.485 0.872	37.50 1.442 0.891	36.56 1.408 0.905
6.3096−04	−3.2	45.32 1.610 0.743	47.03 1.729 0.738	43.16 1.639 0.802	41.56 1.582 0.827	40.38 1.542 0.846	39.45 1.510 0.860	38.69 1.484 0.872	37.48 1.441 0.891	36.55 1.408 0.905
7.9433−04	−3.1	43.66 1.556 0.758	46.35 1.697 0.744	43.00 1.634 0.803	41.47 1.579 0.828	40.32 1.538 0.846	39.41 1.507 0.861	38.66 1.481 0.872	37.46 1.440 0.891	36.53 1.408 0.905
1.0000−03	−3.0	41.80 1.500 0.775	45.50 1.661 0.751	42.81 1.628 0.805	41.35 1.575 0.829	40.24 1.534 0.847	39.35 1.503 0.861	38.62 1.478 0.873	37.44 1.438 0.891	36.51 1.407 0.906
1.2589−03	−2.9	39.75 1.443 0.794	44.48 1.621 0.760	42.59 1.621 0.807	41.19 1.571 0.830	40.14 1.531 0.848	39.28 1.500 0.862	38.56 1.474 0.873	37.41 1.435 0.892	36.49 1.404 0.906
1.5849−03	−2.8	37.51 1.387 0.814	43.27 1.577 0.770	42.33 1.612 0.809	41.02 1.567 0.831	40.00 1.528 0.849	39.18 1.497 0.862	38.48 1.471 0.874	37.35 1.431 0.892	36.46 1.401 0.906
1.9953−03	−2.7	35.12 1.331 0.835	41.85 1.529 0.783	42.02 1.601 0.812	40.81 1.561 0.832	39.84 1.524 0.850	39.05 1.494 0.863	38.38 1.468 0.875	37.28 1.428 0.893	36.42 1.397 0.906
2.5119−03	−2.6	32.57 1.278 0.858	40.23 1.480 0.797	41.65 1.587 0.814	40.57 1.554 0.834	39.65 1.520 0.851	38.89 1.491 0.864	38.24 1.465 0.876	37.19 1.425 0.893	36.35 1.393 0.907
3.1623−03	−2.5	29.92 1.229 0.880	38.41 1.428 0.813	41.21 1.568 0.817	40.30 1.544 0.837	39.44 1.514 0.853	38.72 1.487 0.866	38.10 1.463 0.877	37.07 1.423 0.894	36.26 1.391 0.907
3.9811−03	−2.4	27.20 1.183 0.903	36.40 1.376 0.831	40.68 1.545 0.821	39.96 1.531 0.839	39.19 1.506 0.855	38.51 1.483 0.867	37.92 1.460 0.878	36.93 1.421 0.895	36.14 1.390 0.908
5.0119−03	−2.3	24.44 1.143 0.923	34.21 1.324 0.850	40.02 1.518 0.826	39.55 1.515 0.842	38.90 1.496 0.857	38.28 1.476 0.869	37.72 1.456 0.879	36.77 1.419 0.896	36.00 1.388 0.909
6.3096−03	−2.2	21.71 1.109 0.942	31.84 1.273 0.870	39.20 1.488 0.832	39.06 1.495 0.846	38.54 1.483 0.859	38.00 1.467 0.871	37.49 1.451 0.881	36.59 1.416 0.898	35.84 1.387 0.911
7.9433−03	−2.1	19.06 1.080 0.958	29.34 1.226 0.890	38.20 1.456 0.839	38.43 1.473 0.850	38.11 1.467 0.862	37.67 1.456 0.873	37.23 1.442 0.883	36.38 1.413 0.900	35.68 1.384 0.913
1.0000−02	−2.0	16.52 1.056 0.971	26.74 1.182 0.910	37.02 1.421 0.848	37.66 1.447 0.855	37.57 1.448 0.865	37.26 1.441 0.876	36.90 1.430 0.885	36.15 1.407 0.901	35.48 1.381 0.915
1.2589−02	−1.9	14.16 1.038 0.981	24.09 1.142 0.928	35.64 1.383 0.859	36.72 1.419 0.861	36.91 1.426 0.869	36.76 1.423 0.879	36.50 1.416 0.887	35.87 1.399 0.903	35.26 1.376 0.917
1.5849−02	−1.8	11.99 1.025 0.988	21.44 1.108 0.946	34.06 1.342 0.870	35.61 1.388 0.869	36.08 1.402 0.875	36.12 1.404 0.882	35.99 1.399 0.890	35.53 1.386 0.905	34.99 1.368 0.917
1.9953−02	−1.7	10.05 1.015 0.993	18.85 1.080 0.961	32.28 1.300 0.883	34.30 1.354 0.878	35.08 1.375 0.881	35.33 1.382 0.887	35.34 1.381 0.894	35.09 1.371 0.906	34.68 1.357 0.918
2.5119−02	−1.6	8.345 1.009 0.996	16.37 1.056 0.973	30.30 1.257 0.897	32.81 1.317 0.889	33.90 1.345 0.889	34.38 1.357 0.893	34.56 1.360 0.898	34.53 1.355 0.909	34.26 1.344 0.919
3.1623−02	−1.5	6.871 1.005 0.998	14.04 1.038 0.983	28.14 1.216 0.912	31.11 1.278 0.900	32.54 1.312 0.898	33.27 1.330 0.900	33.62 1.337 0.904	33.82 1.338 0.913	33.73 1.330 0.922
3.9811−02	−1.4	5.617 1.003 0.999	11.90 1.025 0.989	25.84 1.176 0.928	29.21 1.239 0.913	30.98 1.277 0.908	31.97 1.299 0.908	32.52 1.311 0.911	32.98 1.319 0.918	33.07 1.315 0.926
5.0119−02	−1.3	4.564 1.001 0.999	9.982 1.015 0.994	23.43 1.139 0.942	27.13 1.200 0.926	29.23 1.241 0.919	30.47 1.267 0.917	31.23 1.283 0.918	31.99 1.297 0.924	32.26 1.298 0.930
6.3096−02	−1.2	3.692 1.001 1.000	8.291 1.009 0.996	20.97 1.107 0.956	24.90 1.163 0.939	27.28 1.204 0.930	28.78 1.233 0.927	29.75 1.252 0.926	30.84 1.272 0.929	31.32 1.279 0.935
7.9433−02	−1.1	2.975 1.001 1.000	6.829 1.005 0.998	18.53 1.079 0.968	22.57 1.129 0.952	25.16 1.169 0.942	26.89 1.198 0.937	28.07 1.219 0.935	29.50 1.245 0.936	30.22 1.257 0.940
1.0000−01	−1.0	2.390 1.001 1.000	5.584 1.003 0.999	16.15 1.056 0.978	20.18 1.099 0.964	22.92 1.135 0.953	24.83 1.164 0.947	26.21 1.187 0.944	27.95 1.216 0.943	28.93 1.232 0.946
1.2589−01	−0.9	1.915 1.001 1.001	4.539 1.001 0.999	13.90 1.038 0.986	17.80 1.073 0.974	20.60 1.104 0.964	22.65 1.132 0.958	24.17 1.154 0.954	26.22 1.186 0.951	27.45 1.206 0.952
1.5849−01	−0.8	1.532 1.001 1.001	3.672 1.001 1.000	11.82 1.025 0.991	15.50 1.051 0.983	18.26 1.078 0.974	20.38 1.102 0.968	22.01 1.124 0.964	24.31 1.156 0.959	25.78 1.178 0.959
1.9953−01	−0.7	1.224 1.001 1.001	2.959 1.001 1.000	9.937 1.015 0.995	13.32 1.035 0.989	15.97 1.056 0.983	18.08 1.077 0.977	19.77 1.096 0.973	22.26 1.127 0.968	23.93 1.150 0.966
2.5119−01	−0.6	0.976 1.001 1.001	2.377 1.001 1.000	8.269 1.009 0.997	11.30 1.022 0.993	13.78 1.038 0.989	15.82 1.055 0.985	17.51 1.071 0.981	20.10 1.100 0.976	21.94 1.123 0.973
3.1623−01	−0.5	0.778 1.001 1.001	1.905 1.001 1.001	6.821 1.005 0.998	9.485 1.014 0.996	11.74 1.025 0.994	13.66 1.038 0.991	15.30 1.051 0.988	17.90 1.075 0.983	19.84 1.097 0.980
3.9811−01	−0.4	0.620 1.001 1.001	1.524 1.001 1.001	5.584 1.003 0.999	7.881 1.008 0.998	9.889 1.015 0.996	11.64 1.025 0.995	13.18 1.034 0.993	15.72 1.054 0.989	17.69 1.073 0.987
5.0119−01	−0.3	0.493 1.001 1.001	1.218 1.001 1.001	4.544 1.001 1.000	6.492 1.004 0.999	8.241 1.009 0.998	9.809 1.015 0.997	11.21 1.022 0.996	13.61 1.038 0.994	15.55 1.053 0.992
6.3096−01	−0.2	0.393 1.001 1.001	0.971 1.001 1.001	3.678 1.001 1.000	5.309 1.002 0.999	6.806 1.005 0.999	8.176 1.009 0.998	9.431 1.014 0.998	11.63 1.025 0.997	13.47 1.037 0.996
7.9433−01	−0.1	0.312 1.001 1.001	0.774 1.001 1.001	2.966 1.001 1.000	4.315 1.001 1.000	5.577 1.003 0.999	6.753 1.005 0.999	7.848 1.008 0.999	9.816 1.015 0.999	11.52 1.024 0.998
1.0000+00	−0.0	0.248 1.001 1.001	0.617 1.001 1.001	2.384 1.001 1.000	3.491 1.001 1.000	4.541 1.001 1.000	5.534 1.003 1.000	6.473 1.004 1.000	8.196 1.009 0.999	9.727 1.015 0.999
1.2589+00	0.1	0.197 1.001 1.001	0.491 1.001 1.001	1.912 1.001 1.000	2.813 1.001 1.000	3.678 1.001 1.000	4.507 1.001 1.000	5.299 1.002 1.000	6.778 1.005 1.000	8.125 1.009 1.000
1.5849+00	0.2	0.157 1.001 1.001	0.391 1.001 1.001	1.530 1.001 1.000	2.260 1.000 1.000	2.967 1.001 1.000	3.651 1.001 1.000	4.311 1.001 1.000	5.562 1.003 1.000	6.722 1.005 1.000
1.9953+00	0.3	0.125 1.000 1.000	0.311 1.000 1.000	1.223 1.000 1.000	1.813 1.000 1.000	2.387 1.000 1.000	2.946 1.001 1.000	3.490 1.001 1.000	4.532 1.002 1.000	5.514 1.003 1.001

Table IX continued; neutron energy 4 MeV

η/rad^{-1}	lg η	$\sigma_0 = 2\ \mu m^2$ $d = 1.60\ \mu m$	$\sigma_0 = 5\ \mu m^2$ $d = 2.52\ \mu m$	$\sigma_0 = 20\ \mu m^2$ $d = 5.05\ \mu m$	$\sigma_0 = 30\ \mu m^2$ $d = 6.18\ \mu m$	$\sigma_0 = 40\ \mu m^2$ $d = 7.14\ \mu m$	$\sigma_0 = 50\ \mu m^2$ $d = 7.98\ \mu m$	$\sigma_0 = 60\ \mu m^2$ $d = 8.74\ \mu m$	$\sigma_0 = 80\ \mu m^2$ $d = 10.1\ \mu m$	$\sigma_0 = 100\ \mu m^2$ $d = 11.3\ \mu m$
zero		53.97 2.238 0.579	49.75 2.125 0.621	42.45 1.892 0.714	40.39 1.817 0.744	38.98 1.764 0.766	37.92 1.723 0.783	37.08 1.689 0.797	35.79 1.637 0.819	34.82 1.597 0.836
1.0000−04	−4.0	51.87 2.116 0.597	49.07 2.083 0.628	42.35 1.886 0.714	40.33 1.814 0.744	38.94 1.761 0.767	37.89 1.722 0.784	37.06 1.688 0.798	35.77 1.636 0.819	34.80 1.596 0.836
1.2589−04	−3.9	51.38 2.089 0.601	48.89 2.073 0.630	42.33 1.884 0.714	40.32 1.813 0.744	38.93 1.761 0.767	37.89 1.721 0.784	37.05 1.688 0.798	35.77 1.636 0.820	34.80 1.596 0.836
1.5849−04	−3.8	50.76 2.058 0.606	48.66 2.062 0.631	42.30 1.883 0.715	40.30 1.812 0.744	38.92 1.760 0.767	37.88 1.721 0.784	37.05 1.688 0.798	35.77 1.636 0.820	34.80 1.596 0.836
1.9953−04	−3.7	49.97 2.021 0.614	48.37 2.047 0.634	42.25 1.880 0.715	40.28 1.811 0.745	38.91 1.759 0.767	37.87 1.720 0.784	37.04 1.687 0.798	35.76 1.635 0.820	34.80 1.595 0.836
2.5119−04	−3.6	49.03 1.979 0.622	46.01 2.030 0.637	42.20 1.877 0.716	40.25 1.809 0.745	38.89 1.758 0.767	37.85 1.719 0.784	37.03 1.687 0.798	35.75 1.635 0.820	34.79 1.595 0.836
3.1623−04	−3.5	47.92 1.931 0.633	47.60 2.008 0.641	42.14 1.871 0.716	40.21 1.806 0.745	38.86 1.757 0.768	37.84 1.718 0.785	37.01 1.686 0.798	35.74 1.634 0.820	34.78 1.595 0.836
3.9811−04	−3.4	46.64 1.877 0.645	47.13 1.982 0.646	42.06 1.864 0.717	40.17 1.803 0.745	38.83 1.755 0.768	37.82 1.716 0.785	37.00 1.684 0.798	35.73 1.634 0.820	34.78 1.594 0.837
5.0119−04	−3.3	45.18 1.818 0.659	46.55 1.950 0.651	41.95 1.857 0.718	40.11 1.798 0.746	38.80 1.752 0.769	37.79 1.714 0.785	36.98 1.683 0.799	35.72 1.633 0.820	34.77 1.594 0.837
6.3096−04	−3.2	43.54 1.753 0.676	45.86 1.914 0.658	41.81 1.849 0.720	40.04 1.791 0.747	38.75 1.747 0.769	37.76 1.710 0.786	36.95 1.680 0.799	35.70 1.631 0.821	34.75 1.593 0.837
7.9433−04	−3.1	41.71 1.686 0.695	45.00 1.873 0.667	41.62 1.842 0.722	39.94 1.784 0.749	38.69 1.741 0.770	37.71 1.706 0.786	36.92 1.677 0.800	35.68 1.629 0.821	34.74 1.592 0.837
1.0000−03	−3.0	39.70 1.617 0.716	43.96 1.826 0.677	41.37 1.833 0.725	39.80 1.777 0.750	38.61 1.734 0.771	37.65 1.700 0.787	36.88 1.672 0.800	35.65 1.626 0.821	34.71 1.590 0.837
1.2589−03	−2.9	37.55 1.547 0.739	42.77 1.774 0.689	41.09 1.822 0.728	39.61 1.771 0.753	38.49 1.728 0.772	37.57 1.694 0.788	36.81 1.667 0.801	35.62 1.622 0.822	34.69 1.587 0.838
1.5849−03	−2.8	35.27 1.479 0.764	41.39 1.717 0.703	40.77 1.808 0.732	39.39 1.764 0.755	38.33 1.723 0.774	37.46 1.688 0.789	36.71 1.661 0.802	35.55 1.616 0.822	34.66 1.582 0.838
1.9953−03	−2.7	32.89 1.412 0.791	39.83 1.658 0.720	40.41 1.790 0.735	39.13 1.755 0.758	38.12 1.717 0.776	37.30 1.683 0.791	36.59 1.655 0.803	35.47 1.610 0.823	34.61 1.576 0.839
2.5119−03	−2.6	30.42 1.349 0.818	38.09 1.595 0.738	40.00 1.768 0.739	38.85 1.742 0.761	37.89 1.709 0.778	37.11 1.677 0.792	36.44 1.650 0.805	35.37 1.605 0.824	34.52 1.569 0.839
3.1623−03	−2.5	27.91 1.289 0.845	36.19 1.532 0.759	39.50 1.740 0.744	38.53 1.727 0.764	37.64 1.700 0.780	36.89 1.672 0.794	36.26 1.645 0.806	35.23 1.600 0.826	34.42 1.564 0.841
3.9811−03	−2.4	25.37 1.235 0.872	34.14 1.468 0.782	38.89 1.707 0.750	38.15 1.708 0.768	37.35 1.689 0.783	36.65 1.664 0.797	36.06 1.640 0.808	35.06 1.597 0.827	34.27 1.561 0.842
5.0119−03	−2.3	22.84 1.187 0.898	31.97 1.405 0.805	38.16 1.669 0.757	37.70 1.683 0.772	37.03 1.673 0.787	36.39 1.654 0.800	35.83 1.632 0.811	34.88 1.592 0.829	34.11 1.558 0.844
6.3096−03	−2.2	20.34 1.145 0.921	29.69 1.344 0.830	37.25 1.628 0.766	37.16 1.653 0.777	36.64 1.652 0.791	36.09 1.640 0.803	35.56 1.623 0.814	34.66 1.586 0.832	33.93 1.555 0.846
7.9433−03	−2.1	17.93 1.109 0.941	27.33 1.287 0.855	36.17 1.583 0.778	36.50 1.619 0.784	36.18 1.626 0.795	35.73 1.621 0.807	35.28 1.609 0.817	34.44 1.579 0.835	33.74 1.550 0.849
1.0000−02	−2.0	15.63 1.079 0.958	24.92 1.234 0.880	34.91 1.536 0.791	35.68 1.582 0.792	35.62 1.596 0.801	35.30 1.597 0.811	34.93 1.591 0.820	34.19 1.569 0.838	33.53 1.545 0.852
1.2589−02	−1.9	13.47 1.056 0.972	22.49 1.186 0.903	33.48 1.486 0.806	34.68 1.542 0.802	34.92 1.563 0.807	34.78 1.569 0.815	34.51 1.568 0.824	33.89 1.555 0.841	33.30 1.536 0.855
1.5849−02	−1.8	11.48 1.038 0.982	20.08 1.144 0.925	31.88 1.434 0.822	33.52 1.499 0.814	34.05 1.527 0.815	34.14 1.538 0.821	33.99 1.542 0.829	33.54 1.536 0.843	33.02 1.523 0.857
1.9953−02	−1.7	9.685 1.024 0.989	17.73 1.109 0.945	30.13 1.381 0.840	32.18 1.453 0.827	33.02 1.488 0.825	33.32 1.505 0.829	33.35 1.512 0.834	33.10 1.513 0.847	32.69 1.505 0.859
2.5119−02	−1.6	8.090 1.015 0.993	15.47 1.079 0.961	28.23 1.328 0.859	30.69 1.405 0.842	31.82 1.447 0.837	32.35 1.469 0.838	32.55 1.481 0.841	32.55 1.486 0.851	32.29 1.484 0.862
3.1623−02	−1.5	6.698 1.009 0.996	13.35 1.056 0.974	26.20 1.277 0.879	29.03 1.356 0.859	30.46 1.403 0.851	31.21 1.430 0.849	31.61 1.447 0.850	31.86 1.459 0.857	31.77 1.459 0.865
3.9811−02	−1.4	5.502 1.005 0.998	11.39 1.038 0.983	24.07 1.228 0.899	27.22 1.307 0.876	28.93 1.358 0.865	29.91 1.390 0.861	30.49 1.410 0.861	31.02 1.429 0.865	31.12 1.434 0.871
5.0119−02	−1.3	4.490 1.003 0.999	9.618 1.024 0.990	21.87 1.183 0.918	25.28 1.259 0.894	27.25 1.311 0.881	28.46 1.346 0.875	29.23 1.370 0.873	30.02 1.397 0.874	30.33 1.407 0.878
6.3096−02	−1.2	3.644 1.001 0.999	8.038 1.015 0.994	19.64 1.143 0.937	23.22 1.213 0.912	25.42 1.265 0.897	26.84 1.302 0.889	27.79 1.329 0.885	28.87 1.362 0.884	29.39 1.377 0.886
7.9433−02	−1.1	2.944 1.001 1.000	6.658 1.009 0.997	17.42 1.109 0.953	21.09 1.171 0.929	23.46 1.220 0.914	25.07 1.258 0.904	26.20 1.287 0.899	27.57 1.325 0.895	28.30 1.345 0.895
1.0000−01	−1.0	2.370 1.001 1.000	5.471 1.005 0.998	15.27 1.080 0.967	18.92 1.133 0.945	21.41 1.178 0.930	23.17 1.215 0.920	24.45 1.245 0.913	26.11 1.286 0.907	27.06 1.311 0.905
1.2589−01	−0.9	1.903 1.001 1.000	4.466 1.003 0.999	13.22 1.056 0.978	16.77 1.100 0.960	19.30 1.140 0.946	21.17 1.175 0.935	22.58 1.204 0.928	24.48 1.246 0.919	25.66 1.274 0.916
1.5849−01	−0.8	1.524 1.001 1.001	3.625 1.001 1.000	11.32 1.038 0.986	14.68 1.073 0.972	17.19 1.107 0.960	19.11 1.138 0.950	20.60 1.165 0.942	22.72 1.207 0.932	24.09 1.237 0.928
1.9953−01	−0.7	1.219 1.001 1.001	2.929 1.001 1.000	9.575 1.025 0.992	12.69 1.051 0.982	15.11 1.079 0.972	17.02 1.105 0.963	18.56 1.129 0.956	20.84 1.170 0.945	22.39 1.200 0.939
2.5119−01	−0.6	0.973 1.001 1.001	2.358 1.001 1.000	8.018 1.015 0.995	10.84 1.034 0.989	13.12 1.056 0.981	14.98 1.078 0.974	16.51 1.098 0.968	18.87 1.135 0.958	20.56 1.165 0.951
3.1623−01	−0.5	0.776 1.001 1.001	1.893 1.001 1.000	6.651 1.009 0.997	9.156 1.022 0.994	11.25 1.038 0.989	13.01 1.055 0.983	14.50 1.072 0.978	16.88 1.104 0.969	18.56 1.131 0.963
3.9811−01	−0.4	0.619 1.001 1.001	1.517 1.001 1.001	5.472 1.005 0.999	7.654 1.013 0.996	9.533 1.025 0.993	11.16 1.037 0.990	12.57 1.051 0.986	14.89 1.077 0.979	16.69 1.101 0.973
5.0119−01	−0.3	0.493 1.001 1.001	1.213 1.001 1.001	4.471 1.003 0.999	6.339 1.008 0.998	7.993 1.015 0.996	9.459 1.024 0.994	10.76 1.034 0.992	12.97 1.055 0.987	14.74 1.075 0.982
6.3096−01	−0.2	0.392 1.001 1.001	0.968 1.001 1.001	3.631 1.001 1.000	5.208 1.004 0.999	6.638 1.009 0.998	7.933 1.015 0.997	9.108 1.022 0.996	11.15 1.037 0.992	12.84 1.054 0.989
7.9433−01	−0.1	0.312 1.001 1.001	0.772 1.001 1.001	2.936 1.001 1.000	4.250 1.002 1.000	5.466 1.005 0.999	6.589 1.009 0.999	7.625 1.013 0.998	9.469 1.024 0.996	11.05 1.037 0.994
1.0000+00	−0.0	0.248 1.001 1.001	0.616 1.001 1.001	2.365 1.001 1.000	3.449 1.001 1.000	4.469 1.003 1.000	5.426 1.005 0.999	6.323 1.007 0.999	7.954 1.015 0.998	9.389 1.024 0.997
1.2589+00	0.1	0.197 1.001 1.001	0.490 1.001 1.001	1.900 1.001 1.000	2.787 1.001 1.000	3.632 1.001 1.000	4.436 1.002 1.000	5.200 1.004 1.000	6.615 1.009 1.000	7.889 1.015 0.999
1.5849+00	0.2	0.157 1.001 1.001	0.390 1.001 1.001	1.522 1.000 1.000	2.243 1.000 1.000	2.938 1.001 1.000	3.606 1.001 1.000	4.247 1.002 1.000	5.453 1.005 1.000	6.562 1.009 1.000
1.9953+00	0.3	0.125 1.000 1.000	0.311 1.000 1.000	1.218 1.000 1.000	1.802 1.000 1.000	2.368 1.000 1.000	2.917 1.001 1.000	3.448 1.001 1.000	4.461 1.003 1.001	5.408 1.005 1.001

Table IX continued; neutron energy 6 MeV

η/rad^{-1}	lg η	$\sigma_0 = 2\ \mu m^2$ $d = 1.60\ \mu m$	$\sigma_0 = 5\ \mu m^2$ $d = 2.52\ \mu m$	$\sigma_0 = 20\ \mu m^2$ $d = 5.05\ \mu m$	$\sigma_0 = 30\ \mu m^2$ $d = 6.18\ \mu m$	$\sigma_0 = 40\ \mu m^2$ $d = 7.14\ \mu m$	$\sigma_0 = 50\ \mu m^2$ $d = 7.98\ \mu m$	$\sigma_0 = 60\ \mu m^2$ $d = 8.74\ \mu m$	$\sigma_0 = 80\ \mu m^2$ $d = 10.1\ \mu m$	$\sigma_0 = 100\ \mu m^2$ $d = 11.3\ \mu m$
zero		50.10 2.547 0.481	46.32 2.461 0.516	38.87 2.235 0.604	36.68 2.152 0.636	35.19 2.090 0.660	34.08 2.041 0.679	33.21 2.001 0.694	31.89 1.937 0.718	30.92 1.888 0.737
1.0000 − 04	−4.0	47.86 2.354 0.499	45.46 2.397 0.524	38.75 2.224 0.606	36.62 2.146 0.637	35.15 2.087 0.661	34.05 2.039 0.679	33.19 1.999 0.694	31.87 1.936 0.719	30.90 1.887 0.737
1.2589 − 04	−3.9	47.18 2.319 0.505	45.24 2.382 0.526	38.72 2.222 0.606	36.60 2.145 0.637	35.14 2.086 0.661	34.05 2.038 0.679	33.18 1.999 0.695	31.87 1.936 0.719	30.90 1.887 0.737
1.5849 − 04	−3.8	46.37 2.279 0.513	44.96 2.363 0.528	38.69 2.218 0.606	36.58 2.143 0.638	35.13 2.085 0.661	34.04 2.038 0.679	33.18 1.998 0.695	31.87 1.936 0.719	30.90 1.886 0.737
1.9953 − 04	−3.7	45.39 2.231 0.522	44.64 2.338 0.531	38.64 2.214 0.607	36.55 2.140 0.638	35.11 2.083 0.661	34.03 2.037 0.680	33.17 1.998 0.695	31.86 1.935 0.719	30.90 1.886 0.737
2.5119 − 04	−3.6	44.27 2.176 0.533	44.30 2.308 0.535	38.58 2.209 0.607	36.52 2.137 0.638	35.09 2.081 0.662	34.01 2.035 0.680	33.16 1.997 0.695	31.85 1.935 0.719	30.89 1.886 0.737
3.1623 − 04	−3.5	43.00 2.113 0.546	43.88 2.271 0.540	38.51 2.202 0.608	36.48 2.134 0.639	35.07 2.079 0.662	33.99 2.033 0.680	33.14 1.995 0.695	31.85 1.934 0.719	30.89 1.885 0.738
3.9811 − 04	−3.4	41.57 2.044 0.562	43.35 2.230 0.546	38.42 2.195 0.609	36.43 2.129 0.640	35.04 2.076 0.662	33.97 2.031 0.680	33.13 1.993 0.695	31.83 1.933 0.719	30.88 1.884 0.738
5.0119 − 04	−3.3	39.98 1.970 0.579	42.65 2.184 0.553	38.30 2.185 0.611	36.37 2.123 0.640	35.00 2.071 0.663	33.94 2.028 0.681	33.10 1.991 0.696	31.82 1.931 0.719	30.87 1.883 0.738
6.3096 − 04	−3.2	38.24 1.891 0.600	41.75 2.133 0.562	38.15 2.173 0.613	36.29 2.116 0.641	34.95 2.066 0.664	33.91 2.024 0.681	33.08 1.988 0.696	31.80 1.929 0.720	30.86 1.882 0.738
7.9433 − 04	−3.1	36.37 1.811 0.622	40.70 2.077 0.573	37.96 2.158 0.615	36.19 2.107 0.642	34.88 2.060 0.664	33.86 2.019 0.682	33.04 1.984 0.697	31.78 1.927 0.720	30.84 1.880 0.738
1.0000 − 03	−3.0	34.40 1.731 0.648	39.50 2.014 0.587	37.71 2.141 0.618	36.06 2.096 0.644	34.80 2.051 0.665	33.81 2.013 0.683	33.00 1.979 0.697	31.75 1.923 0.720	30.82 1.878 0.738
1.2589 − 03	−2.9	32.35 1.651 0.676	38.16 1.946 0.602	37.43 2.119 0.622	35.89 2.083 0.646	34.69 2.042 0.667	33.73 2.005 0.684	32.95 1.973 0.698	31.72 1.919 0.721	30.79 1.875 0.739
1.5849 − 03	−2.8	30.26 1.574 0.705	36.67 1.873 0.621	37.10 2.094 0.625	35.68 2.068 0.648	34.55 2.032 0.668	33.63 1.997 0.685	32.87 1.965 0.699	31.68 1.913 0.721	30.76 1.870 0.739
1.9953 − 03	−2.7	28.15 1.498 0.736	35.04 1.797 0.642	36.73 2.061 0.630	35.42 2.051 0.651	34.37 2.019 0.670	33.50 1.988 0.687	32.78 1.958 0.700	31.61 1.907 0.722	30.73 1.864 0.739
2.5119 − 03	−2.6	26.03 1.427 0.767	33.29 1.720 0.665	36.29 2.024 0.635	35.13 2.029 0.655	34.16 2.006 0.673	33.34 1.977 0.688	32.65 1.949 0.702	31.53 1.900 0.723	30.67 1.858 0.740
3.1623 − 03	−2.5	23.92 1.360 0.799	31.44 1.643 0.691	35.73 1.980 0.642	34.80 2.001 0.659	33.91 1.988 0.676	33.15 1.965 0.691	32.49 1.939 0.703	31.42 1.893 0.724	30.59 1.852 0.741
3.9811 − 03	−2.4	21.82 1.298 0.830	29.53 1.568 0.718	35.03 1.930 0.650	34.42 1.968 0.664	33.62 1.965 0.679	32.92 1.950 0.693	32.30 1.928 0.706	31.28 1.885 0.726	30.49 1.846 0.742
5.0119 − 03	−2.3	19.76 1.242 0.860	27.56 1.495 0.747	34.19 1.876 0.661	33.95 1.928 0.671	33.30 1.937 0.684	32.65 1.929 0.697	32.08 1.913 0.708	31.13 1.876 0.728	30.36 1.839 0.744
6.3096 − 03	−2.2	17.74 1.193 0.888	25.57 1.424 0.777	33.21 1.818 0.674	33.36 1.882 0.679	32.91 1.903 0.689	32.36 1.904 0.701	31.83 1.895 0.712	30.95 1.864 0.731	30.21 1.832 0.746
7.9433 − 03	−2.1	15.78 1.150 0.913	23.56 1.358 0.807	32.08 1.756 0.690	32.63 1.832 0.688	32.43 1.862 0.696	32.01 1.872 0.705	31.55 1.871 0.715	30.73 1.849 0.734	30.05 1.822 0.749
1.0000 − 02	−2.0	13.90 1.114 0.935	21.55 1.297 0.837	30.81 1.690 0.707	31.76 1.778 0.699	31.84 1.818 0.703	31.58 1.835 0.711	31.22 1.841 0.720	30.49 1.830 0.737	29.85 1.809 0.752
1.2589 − 02	−1.9	12.12 1.084 0.953	19.55 1.241 0.865	29.41 1.622 0.727	30.74 1.720 0.713	31.10 1.769 0.713	31.05 1.793 0.718	30.81 1.805 0.725	30.21 1.805 0.741	29.63 1.792 0.755
1.5849 − 02	−1.8	10.46 1.060 0.967	17.58 1.192 0.892	27.90 1.554 0.749	29.60 1.659 0.729	30.22 1.716 0.724	30.38 1.748 0.727	30.29 1.765 0.732	29.86 1.775 0.745	29.37 1.771 0.758
1.9953 − 02	−1.7	8.932 1.041 0.978	15.66 1.149 0.916	26.29 1.486 0.772	28.32 1.594 0.747	29.21 1.659 0.739	29.57 1.697 0.737	29.65 1.720 0.740	29.44 1.740 0.750	29.06 1.742 0.762
2.5119 − 02	−1.6	7.551 1.027 0.986	13.81 1.113 0.938	24.59 1.420 0.797	26.91 1.529 0.767	28.06 1.599 0.755	28.62 1.644 0.750	28.87 1.672 0.751	28.91 1.701 0.757	28.68 1.710 0.766
3.1623 − 02	−1.5	6.322 1.018 0.991	12.05 1.083 0.955	22.82 1.356 0.823	25.40 1.464 0.789	26.78 1.537 0.773	27.54 1.587 0.765	27.96 1.621 0.763	28.25 1.658 0.765	28.19 1.673 0.772
3.9811 − 02	−1.4	5.246 1.011 0.995	10.41 1.059 0.969	21.00 1.297 0.849	23.79 1.400 0.812	25.39 1.474 0.793	26.34 1.527 0.782	26.91 1.566 0.777	27.46 1.612 0.776	27.59 1.634 0.780
5.0119 − 02	−1.3	4.319 1.006 0.997	8.894 1.040 0.980	19.15 1.243 0.875	22.09 1.339 0.836	23.88 1.412 0.814	25.01 1.467 0.801	25.75 1.508 0.794	26.54 1.562 0.789	26.85 1.591 0.789
6.3096 − 02	−1.2	3.532 1.004 0.998	7.521 1.026 0.987	17.30 1.194 0.899	20.34 1.282 0.861	22.27 1.352 0.836	23.56 1.406 0.821	24.45 1.449 0.812	25.49 1.509 0.803	25.99 1.545 0.801
7.9433 − 02	−1.1	2.872 1.002 0.999	6.299 1.017 0.992	15.47 1.151 0.921	18.54 1.230 0.885	20.59 1.294 0.859	22.01 1.347 0.842	23.03 1.390 0.831	24.31 1.454 0.818	25.01 1.495 0.814
1.0000 − 01	−1.0	2.324 1.001 1.000	5.228 1.010 0.996	13.68 1.115 0.941	16.74 1.183 0.908	18.85 1.241 0.882	20.38 1.291 0.864	21.51 1.333 0.851	23.01 1.398 0.836	23.90 1.443 0.828
1.2589 − 01	−0.9	1.873 1.001 1.000	4.304 1.006 0.997	11.98 1.084 0.958	14.95 1.142 0.928	17.08 1.193 0.905	18.68 1.239 0.886	19.90 1.279 0.873	21.59 1.342 0.854	22.65 1.390 0.844
1.5849 − 01	−0.8	1.505 1.001 1.000	3.520 1.003 0.999	10.37 1.060 0.971	13.21 1.107 0.947	15.32 1.151 0.925	16.94 1.191 0.908	18.22 1.228 0.894	20.07 1.289 0.874	21.29 1.336 0.862
1.9953 − 01	−0.7	1.207 1.001 1.000	2.861 1.002 0.999	8.873 1.041 0.981	11.54 1.078 0.962	13.58 1.115 0.944	15.20 1.150 0.928	16.51 1.182 0.914	18.46 1.238 0.894	19.82 1.284 0.880
2.5119 − 01	−0.6	0.966 1.001 1.001	2.315 1.001 1.000	7.516 1.027 0.988	9.969 1.055 0.975	11.91 1.085 0.960	13.49 1.114 0.946	14.79 1.142 0.933	16.80 1.192 0.913	18.26 1.235 0.899
3.1623 − 01	−0.5	0.771 1.001 1.001	1.866 1.001 1.000	6.302 1.017 0.993	8.517 1.037 0.984	10.32 1.060 0.972	11.83 1.084 0.961	13.10 1.107 0.950	15.12 1.151 0.932	16.64 1.189 0.917
3.9811 − 01	−0.4	0.616 1.001 1.001	1.499 1.001 1.000	5.236 1.010 0.996	7.200 1.024 0.990	8.850 1.041 0.982	10.26 1.060 0.973	11.47 1.078 0.965	13.45 1.115 0.948	14.99 1.149 0.935
5.0119 − 01	−0.3	0.491 1.001 1.001	1.202 1.001 1.001	4.315 1.005 0.998	6.026 1.015 0.994	7.504 1.027 0.989	8.792 1.041 0.983	9.925 1.055 0.976	11.82 1.085 0.963	13.34 1.113 0.951
6.3096 − 01	−0.2	0.391 1.001 1.001	0.961 1.001 1.001	3.530 1.003 0.999	4.997 1.009 0.997	6.297 1.017 0.993	7.455 1.027 0.989	8.491 1.037 0.985	10.27 1.060 0.975	11.73 1.084 0.965
7.9433 − 01	−0.1	0.311 1.001 1.001	0.768 1.001 1.001	2.871 1.002 0.999	4.111 1.005 0.998	5.235 1.010 0.996	6.256 1.016 0.994	7.186 1.024 0.991	8.815 1.041 0.984	10.19 1.060 0.976
1.0000 + 00	−0.0	0.248 1.001 1.001	0.613 1.001 1.001	2.323 1.001 1.000	3.359 1.002 0.999	4.316 1.005 0.998	5.201 1.010 0.996	6.019 1.015 0.995	7.484 1.027 0.990	8.753 1.041 0.984
1.2589 + 00	0.1	0.197 1.001 1.001	0.489 1.001 1.001	1.873 1.001 1.000	2.729 1.001 1.000	3.533 1.003 0.999	4.287 1.005 0.998	4.996 1.009 0.997	6.287 1.017 0.994	7.432 1.027 0.990
1.5849 + 00	0.2	0.157 1.001 1.001	0.389 1.001 1.001	1.505 1.000 1.000	2.207 1.001 1.000	2.874 1.002 0.999	3.509 1.003 0.999	4.112 1.005 0.998	5.231 1.010 0.997	6.244 1.017 0.995
1.9953 + 00	0.3	0.125 1.000 1.000	0.310 1.000 1.000	1.207 1.000 1.000	1.778 1.000 1.000	2.327 1.001 1.000	2.854 1.002 1.000	3.361 1.003 0.999	4.314 1.006 0.999	5.193 1.010 0.997

Table IX continued; neutron energy 8 MeV

η/rad^{-1}	lg η	$\sigma_0 = 2\ \mu m^2$ $d = 1.60\ \mu m$	$\sigma_0 = 5\ \mu m^2$ $d = 2.52\ \mu m$	$\sigma_0 = 20\ \mu m^2$ $d = 5.05\ \mu m$	$\sigma_0 = 30\ \mu m^2$ $d = 6.18\ \mu m$	$\sigma_0 = 40\ \mu m^2$ $d = 7.14\ \mu m$	$\sigma_0 = 50\ \mu m^2$ $d = 7.98\ \mu m$	$\sigma_0 = 60\ \mu m^2$ $d = 8.74\ \mu m$	$\sigma_0 = 80\ \mu m^2$ $d = 10.1\ \mu m$	$\sigma_0 = 100\ \mu m^2$ $d = 11.3\ \mu m$
zero		54.77 2.811 0.380	50.98 2.762 0.407	42.89 2.625 0.476	40.33 2.569 0.504	38.56 2.526 0.524	37.23 2.491 0.541	36.17 2.460 0.555	34.57 2.411 0.577	33.38 2.370 0.595
1.0000−04	−4.0	51.52 2.576 0.401	49.69 2.673 0.416	42.70 2.608 0.478	40.23 2.560 0.505	38.49 2.520 0.525	37.17 2.486 0.541	36.13 2.456 0.555	34.54 2.407 0.577	33.36 2.367 0.595
1.2589−04	−3.9	50.52 2.537 0.408	49.35 2.653 0.418	42.65 2.605 0.478	40.20 2.557 0.505	38.47 2.518 0.525	37.16 2.484 0.541	35.12 2.455 0.555	34.53 2.407 0.577	33.35 2.367 0.595
1.5849−04	−3.8	49.35 2.494 0.417	48.88 2.630 0.421	42.59 2.599 0.479	40.17 2.556 0.506	38.45 2.516 0.525	37.15 2.483 0.541	36.11 2.454 0.555	34.53 2.406 0.577	33.35 2.366 0.595
1.9953−04	−3.7	48.02 2.440 0.427	48.38 2.599 0.425	42.51 2.590 0.480	40.12 2.551 0.506	38.42 2.513 0.525	37.13 2.481 0.541	36.09 2.452 0.555	34.52 2.405 0.577	33.34 2.366 0.595
2.5119−04	−3.6	46.52 2.377 0.439	47.85 2.562 0.430	42.42 2.578 0.481	40.07 2.545 0.507	38.39 2.510 0.525	37.10 2.478 0.541	36.08 2.450 0.555	34.51 2.404 0.577	33.33 2.365 0.595
3.1623−04	−3.5	44.85 2.303 0.453	47.23 2.517 0.435	42.30 2.562 0.482	40.00 2.536 0.508	38.34 2.505 0.526	37.07 2.476 0.542	36.05 2.449 0.555	34.49 2.402 0.577	33.32 2.363 0.594
3.9811−04	−3.4	43.00 2.221 0.470	46.43 2.468 0.442	42.14 2.542 0.484	39.93 2.522 0.509	38.29 2.497 0.526	37.03 2.471 0.542	36.02 2.445 0.555	34.47 2.400 0.577	33.31 2.362 0.594
5.0119−04	−3.3	40.98 2.132 0.490	45.41 2.414 0.451	41.92 2.522 0.486	39.81 2.505 0.510	38.22 2.483 0.527	36.98 2.463 0.543	35.98 2.439 0.556	34.45 2.398 0.577	33.29 2.360 0.594
6.3096−04	−3.2	38.79 2.040 0.512	44.10 2.353 0.461	41.61 2.502 0.489	39.66 2.484 0.511	38.14 2.464 0.528	36.93 2.448 0.543	35.94 2.428 0.556	34.42 2.394 0.577	33.27 2.359 0.595
7.9433−04	−3.1	36.49 1.946 0.539	42.69 2.283 0.474	41.20 2.485 0.493	39.46 2.464 0.513	38.03 2.442 0.530	36.84 2.429 0.545	35.89 2.410 0.557	34.38 2.384 0.577	33.24 2.355 0.595
1.0000−03	−3.0	34.10 1.853 0.568	41.09 2.205 0.490	40.74 2.467 0.497	39.18 2.445 0.516	37.86 2.422 0.532	36.73 2.406 0.546	35.81 2.390 0.558	34.33 2.366 0.578	33.21 2.342 0.595
1.2589−03	−2.9	31.70 1.762 0.600	39.33 2.118 0.508	40.27 2.440 0.503	38.81 2.431 0.519	37.60 2.407 0.535	36.56 2.387 0.549	35.68 2.369 0.560	34.28 2.344 0.580	33.16 2.323 0.596
1.5849−03	−2.8	29.31 1.674 0.635	37.39 2.028 0.530	39.77 2.404 0.508	38.38 2.415 0.522	37.26 2.393 0.540	36.32 2.372 0.552	35.51 2.351 0.563	34.20 2.320 0.582	33.13 2.300 0.597
1.9953−03	−2.7	26.98 1.589 0.671	35.32 1.937 0.554	39.23 2.358 0.514	37.93 2.392 0.527	36.85 2.381 0.544	35.99 2.358 0.556	35.26 2.339 0.567	34.06 2.299 0.584	33.05 2.275 0.598
2.5119−03	−2.6	24.72 1.509 0.708	33.15 1.846 0.582	38.56 2.301 0.520	37.47 2.361 0.532	36.44 2.363 0.550	35.63 2.348 0.561	34.95 2.325 0.571	33.82 2.286 0.588	32.89 2.257 0.601
3.1623−03	−2.5	22.54 1.433 0.746	30.93 1.757 0.613	37.72 2.237 0.529	37.00 2.318 0.539	36.02 2.339 0.554	35.23 2.333 0.566	34.60 2.315 0.576	33.51 2.275 0.592	32.68 2.245 0.605
3.9811−03	−2.4	20.44 1.364 0.782	28.70 1.670 0.646	36.72 2.165 0.541	36.43 2.262 0.547	35.63 2.305 0.559	34.83 2.313 0.571	34.22 2.301 0.581	33.15 2.271 0.597	32.38 2.233 0.610
5.0119−03	−2.3	18.43 1.300 0.818	26.50 1.587 0.681	35.52 2.085 0.555	35.74 2.198 0.556	35.18 2.258 0.564	34.48 2.283 0.575	33.82 2.283 0.586	32.81 2.264 0.602	32.04 2.230 0.615
6.3096−03	−2.2	16.51 1.243 0.851	24.35 1.507 0.717	34.14 2.002 0.572	34.89 2.126 0.567	34.65 2.195 0.570	34.08 2.238 0.580	33.48 2.254 0.590	32.47 2.254 0.607	31.73 2.226 0.621
7.9433−03	−2.1	14.69 1.194 0.882	22.25 1.432 0.753	32.60 1.918 0.593	33.86 2.045 0.580	33.98 2.126 0.578	33.62 2.177 0.585	33.10 2.210 0.594	32.17 2.231 0.611	31.39 2.219 0.625
1.0000−02	−2.0	12.93 1.151 0.909	20.23 1.362 0.789	30.91 1.833 0.617	32.62 1.965 0.595	33.10 2.049 0.590	33.03 2.108 0.592	32.67 2.148 0.599	31.86 2.193 0.614	31.10 2.202 0.629
1.2589−02	−1.9	11.34 1.115 0.932	18.27 1.299 0.824	29.13 1.749 0.643	31.23 1.884 0.613	32.02 1.973 0.604	32.26 2.031 0.603	32.14 2.078 0.606	31.54 2.137 0.618	30.83 2.166 0.632
1.5849−02	−1.8	9.827 1.085 0.951	16.40 1.242 0.856	27.30 1.655 0.672	29.66 1.803 0.636	30.77 1.895 0.622	31.26 1.956 0.616	31.41 2.001 0.616	31.13 2.071 0.623	30.55 2.112 0.635
1.9953−02	−1.7	8.438 1.062 0.965	14.60 1.192 0.836	25.42 1.585 0.702	23.01 1.723 0.661	29.36 1.817 0.643	30.10 1.881 0.633	30.47 1.927 0.629	30.54 1.997 0.631	30.20 2.047 0.640
2.5119−02	−1.6	7.177 1.043 0.977	12.90 1.149 0.913	23.53 1.507 0.735	26.30 1.642 0.690	27.85 1.738 0.666	28.77 1.808 0.653	29.35 1.854 0.646	29.74 1.923 0.643	29.69 1.974 0.646
3.1623−02	−1.5	6.048 1.029 0.985	11.29 1.113 0.935	21.65 1.433 0.767	24.54 1.564 0.719	26.27 1.658 0.692	27.35 1.729 0.676	28.05 1.781 0.666	28.76 1.856 0.657	28.95 1.904 0.657
3.9811−02	−1.4	5.051 1.019 0.990	9.796 1.083 0.954	19.79 1.364 0.800	22.75 1.488 0.750	24.61 1.580 0.719	25.84 1.652 0.700	26.66 1.707 0.688	27.62 1.786 0.675	28.06 1.839 0.671
5.0119−02	−1.3	4.185 1.012 0.994	8.417 1.060 0.968	17.97 1.301 0.832	20.95 1.416 0.781	22.91 1.504 0.748	24.26 1.575 0.727	25.20 1.632 0.712	26.37 1.716 0.695	26.98 1.771 0.688
6.3096−02	−1.2	3.441 1.007 0.996	7.162 1.041 0.979	16.19 1.244 0.863	19.16 1.349 0.813	21.18 1.432 0.778	22.62 1.500 0.755	23.66 1.557 0.738	25.04 1.643 0.718	25.82 1.703 0.707
7.9433−02	−1.1	2.811 1.004 0.998	6.036 1.027 0.987	14.47 1.194 0.891	17.40 1.287 0.843	19.45 1.364 0.809	20.95 1.429 0.784	22.08 1.484 0.765	23.62 1.570 0.742	24.57 1.633 0.728
1.0000−01	−1.0	2.284 1.003 0.999	5.042 1.017 0.992	12.81 1.151 0.917	15.67 1.232 0.873	17.71 1.302 0.839	19.25 1.362 0.813	20.45 1.414 0.794	22.13 1.498 0.767	23.22 1.562 0.751
1.2589−01	−0.9	1.848 1.002 1.000	4.176 1.011 0.995	11.25 1.114 0.938	13.99 1.183 0.899	16.01 1.245 0.868	17.56 1.300 0.842	18.79 1.348 0.822	20.58 1.429 0.794	21.79 1.492 0.775
1.5849−01	−0.8	1.489 1.001 1.000	3.433 1.006 0.997	9.778 1.084 0.956	12.38 1.142 0.923	14.34 1.195 0.895	15.88 1.244 0.871	17.13 1.287 0.851	19.00 1.363 0.821	20.31 1.424 0.801
1.9953−01	−0.7	1.196 1.001 1.000	2.804 1.004 0.998	8.415 1.059 0.970	10.88 1.106 0.944	12.73 1.152 0.919	14.24 1.194 0.897	15.48 1.233 0.878	17.39 1.301 0.848	18.77 1.359 0.827
2.5119−01	−0.6	0.959 1.001 1.000	2.278 1.002 0.999	7.171 1.041 0.981	9.418 1.077 0.961	11.19 1.115 0.940	12.65 1.151 0.921	13.86 1.184 0.903	15.77 1.246 0.875	17.21 1.299 0.853
3.1623−01	−0.5	0.767 1.001 1.001	1.842 1.001 1.000	6.052 1.027 0.988	8.091 1.055 0.973	9.745 1.084 0.957	11.12 1.114 0.941	12.29 1.143 0.926	14.18 1.196 0.900	15.65 1.244 0.878
3.9811−01	−0.4	0.613 1.001 1.001	1.484 1.001 1.000	5.059 1.017 0.993	6.882 1.037 0.983	8.397 1.060 0.971	9.684 1.084 0.958	10.79 1.107 0.946	12.62 1.153 0.922	14.06 1.194 0.903
5.0119−01	−0.3	0.489 1.001 1.001	1.192 1.001 1.000	4.193 1.010 0.996	5.796 1.024 0.990	7.163 1.041 0.981	8.345 1.059 0.972	9.382 1.078 0.962	11.12 1.116 0.942	12.52 1.151 0.925
6.3096−01	−0.2	0.390 1.001 1.001	0.955 1.001 1.001	3.449 1.006 0.998	4.837 1.015 0.994	6.049 1.027 0.988	7.118 1.040 0.982	8.070 1.055 0.974	9.698 1.085 0.959	11.04 1.115 0.944
7.9433−01	−0.1	0.311 1.001 1.001	0.764 1.001 1.001	2.818 1.003 0.999	4.002 1.009 0.996	5.060 1.017 0.993	6.011 1.026 0.989	6.871 1.037 0.984	8.369 1.060 0.972	9.632 1.084 0.960
1.0000+00	−0.0	0.247 1.001 1.001	0.610 1.001 1.001	2.289 1.002 0.999	3.287 1.005 0.998	4.196 1.010 0.996	5.028 1.016 0.993	5.792 1.024 0.990	7.147 1.041 0.982	8.314 1.060 0.973
1.2589+00	0.1	0.197 1.001 1.001	0.487 1.001 1.001	1.851 1.001 1.000	2.682 1.003 0.999	3.453 1.006 0.998	4.169 1.010 0.996	4.837 1.015 0.994	6.041 1.027 0.989	7.100 1.041 0.982
1.5849+00	0.2	0.156 1.001 1.001	0.388 1.001 1.001	1.491 1.001 1.000	2.176 1.001 0.999	2.822 1.003 0.999	3.430 1.005 0.998	4.004 1.009 0.997	5.058 1.017 0.994	6.002 1.027 0.989
1.9953+00	0.3	0.125 1.000 1.000	0.309 1.000 1.000	1.198 1.000 1.000	1.758 1.001 1.000	2.293 1.002 0.999	2.802 1.003 0.999	3.289 1.005 0.998	4.195 1.010 0.997	5.023 1.017 0.994

Table IX continued; neutron energy 10 MeV

η/rad^{-1}	lg η	$\sigma_0 = 2\ \mu m^2$ $d = 1.60\ \mu m$	$\sigma_0 = 5\ \mu m^2$ $d = 2.52\ \mu m$	$\sigma_0 = 20\ \mu m^2$ $d = 5.05\ \mu m$	$\sigma_0 = 30\ \mu m^2$ $d = 6.18\ \mu m$	$\sigma_0 = 40\ \mu m^2$ $d = 7.14\ \mu m$	$\sigma_0 = 50\ \mu m^2$ $d = 7.98\ \mu m$	$\sigma_0 = 60\ \mu m^2$ $d = 8.74\ \mu m$	$\sigma_0 = 80\ \mu m^2$ $d = 10.1\ \mu m$	$\sigma_0 = 100\ \mu m^2$ $d = 11.3\ \mu m$
zero		68.84 2.873 0.289	64.58 2.861 0.308	54.96 2.823 0.357	51.73 2.807 0.377	49.43 2.793 0.393	47.68 2.780 0.406	46.28 2.769 0.417	44.14 2.749 0.434	42.55 2.731 0.448
1.0000−04	−4.0	63.60 2.651 0.311	62.49 2.773 0.317	54.61 2.806 0.359	51.53 2.795 0.378	49.30 2.784 0.393	47.59 2.773 0.406	46.21 2.763 0.417	44.09 2.744 0.434	42.51 2.727 0.448
1.2589−04	−3.9	62.00 2.620 0.318	61.95 2.752 0.320	54.53 2.801 0.359	51.48 2.792 0.378	49.27 2.782 0.394	47.57 2.772 0.406	46.19 2.762 0.417	44.08 2.743 0.434	42.50 2.726 0.448
1.5849−04	−3.8	60.27 2.582 0.326	61.25 2.727 0.323	54.42 2.796 0.360	51.42 2.790 0.379	49.23 2.779 0.394	47.54 2.770 0.406	46.17 2.760 0.417	44.07 2.742 0.434	42.49 2.726 0.448
1.9953−04	−3.7	58.29 2.532 0.336	60.42 2.696 0.327	54.28 2.785 0.361	51.34 2.784 0.379	49.18 2.776 0.394	47.50 2.767 0.406	46.14 2.758 0.417	44.05 2.741 0.434	42.48 2.725 0.448
2.5119−04	−3.6	56.11 2.472 0.348	59.50 2.659 0.331	54.11 2.772 0.362	51.24 2.779 0.380	49.11 2.772 0.394	47.45 2.764 0.407	46.10 2.755 0.417	44.02 2.739 0.434	42.46 2.723 0.448
3.1623−04	−3.5	53.67 2.400 0.362	58.39 2.619 0.337	53.91 2.752 0.363	51.12 2.768 0.381	49.03 2.768 0.395	47.39 2.760 0.407	46.06 2.752 0.417	43.99 2.736 0.434	42.44 2.721 0.448
3.9811−04	−3.4	51.01 2.319 0.378	57.05 2.574 0.344	53.60 2.727 0.365	50.97 2.752 0.382	48.93 2.758 0.396	47.32 2.757 0.407	46.00 2.750 0.417	43.96 2.733 0.434	42.41 2.719 0.448
5.0119−04	−3.3	48.13 2.231 0.398	55.41 2.522 0.353	53.21 2.701 0.368	50.79 2.728 0.383	48.80 2.743 0.397	47.22 2.746 0.408	45.93 2.743 0.418	43.91 2.732 0.435	42.38 2.716 0.448
6.3096−04	−3.2	45.04 2.135 0.421	53.44 2.465 0.364	52.68 2.679 0.371	50.52 2.702 0.385	48.67 2.717 0.398	47.11 2.730 0.409	45.84 2.731 0.418	43.85 2.724 0.435	42.33 2.715 0.448
7.9433−04	−3.1	41.81 2.038 0.448	51.29 2.395 0.377	51.95 2.654 0.375	50.13 2.676 0.388	48.44 2.689 0.400	46.99 2.702 0.410	45.73 2.708 0.420	43.77 2.712 0.435	42.28 2.708 0.448
1.0000−03	−3.0	38.52 1.941 0.479	48.94 2.315 0.393	51.15 2.634 0.380	49.61 2.654 0.392	48.08 2.660 0.403	46.79 2.672 0.412	45.61 2.675 0.421	43.68 2.695 0.436	42.21 2.694 0.449
1.2589−03	−2.9	35.24 1.846 0.513	46.37 2.227 0.412	50.30 2.610 0.386	48.90 2.630 0.396	47.61 2.640 0.406	46.47 2.641 0.415	45.44 2.645 0.423	43.59 2.665 0.438	42.12 2.670 0.450
1.5849−03	−2.8	32.05 1.754 0.551	43.59 2.132 0.434	49.34 2.574 0.393	48.10 2.613 0.402	46.97 2.620 0.411	46.01 2.620 0.419	45.11 2.620 0.427	43.42 2.630 0.440	42.05 2.634 0.451
1.9953−03	−2.7	29.00 1.666 0.592	40.63 2.034 0.460	48.28 2.530 0.400	47.29 2.591 0.408	46.20 2.602 0.417	45.38 2.602 0.424	44.53 2.603 0.431	43.15 2.596 0.443	41.91 2.597 0.454
2.5119−03	−2.6	26.13 1.581 0.635	37.58 1.938 0.490	47.11 2.476 0.408	46.38 2.558 0.415	45.44 2.583 0.423	44.60 2.587 0.430	43.82 2.592 0.437	42.74 2.569 0.447	41.62 2.566 0.457
3.1623−03	−2.5	23.44 1.502 0.679	34.48 1.843 0.524	45.64 2.404 0.418	45.50 2.515 0.421	44.63 2.557 0.429	43.83 2.572 0.437	43.03 2.577 0.444	42.14 2.555 0.452	41.12 2.548 0.462
3.9811−03	−2.4	20.95 1.427 0.722	31.44 1.753 0.561	44.04 2.323 0.431	44.54 2.458 0.429	43.78 2.524 0.436	43.06 2.553 0.443	42.32 2.562 0.451	41.40 2.542 0.459	40.47 2.540 0.468
5.0119−03	−2.3	18.64 1.359 0.765	28.53 1.665 0.601	42.24 2.232 0.447	43.30 2.382 0.438	42.97 2.477 0.442	42.28 2.516 0.449	41.62 2.535 0.456	40.66 2.538 0.467	39.73 2.527 0.476
6.3096−03	−2.2	16.51 1.297 0.805	25.76 1.580 0.643	40.16 2.131 0.466	41.88 2.298 0.450	41.97 2.408 0.451	41.56 2.466 0.456	41.03 2.501 0.461	39.91 2.525 0.474	39.06 2.525 0.483
7.9433−03	−2.1	14.55 1.241 0.843	23.17 1.501 0.686	37.83 2.034 0.489	40.24 2.205 0.465	40.81 2.323 0.460	40.68 2.395 0.463	40.38 2.451 0.467	39.22 2.508 0.479	38.42 2.512 0.490
1.0000−02	−2.0	12.75 1.192 0.877	20.75 1.426 0.729	35.34 1.939 0.516	38.31 2.103 0.484	39.46 2.229 0.473	39.66 2.309 0.472	39.57 2.380 0.473	38.68 2.473 0.485	37.81 2.495 0.495
1.2589−02	−1.9	11.09 1.150 0.906	18.49 1.357 0.771	32.69 1.846 0.547	36.19 2.005 0.506	37.81 2.126 0.490	38.43 2.218 0.484	38.60 2.293 0.483	38.03 2.409 0.490	37.31 2.462 0.500
1.5849−02	−1.8	9.588 1.115 0.930	16.40 1.295 0.811	30.06 1.756 0.582	33.89 1.910 0.533	35.86 2.029 0.510	36.94 2.120 0.499	37.38 2.199 0.495	37.32 2.315 0.497	36.74 2.398 0.505
1.9953−02	−1.7	8.222 1.086 0.950	14.47 1.239 0.848	27.48 1.669 0.619	31.40 1.920 0.564	33.73 1.934 0.536	35.12 2.020 0.519	35.83 2.094 0.511	36.40 2.217 0.507	36.15 2.304 0.510
2.5119−02	−1.6	6.995 1.062 0.965	12.69 1.190 0.881	24.99 1.585 0.658	28.91 1.733 0.598	31.38 1.843 0.565	33.10 1.927 0.544	34.10 1.997 0.531	35.13 2.109 0.520	35.35 2.205 0.519
3.1623−02	−1.5	5.902 1.044 0.976	11.06 1.148 0.910	22.61 1.505 0.699	26.46 1.648 0.635	29.02 1.756 0.598	30.85 1.837 0.572	32.15 1.904 0.556	33.54 2.015 0.539	34.20 2.097 0.532
3.9811−02	−1.4	4.940 1.030 0.985	9.563 1.112 0.934	20.35 1.430 0.739	24.10 1.565 0.674	26.68 1.670 0.633	28.57 1.751 0.605	29.98 1.816 0.585	31.77 1.926 0.562	32.72 2.004 0.550
5.0119−02	−1.3	4.103 1.020 0.990	8.206 1.083 0.953	18.22 1.360 0.779	21.82 1.487 0.713	24.38 1.587 0.670	26.28 1.667 0.639	27.79 1.731 0.617	29.75 1.838 0.589	31.04 1.917 0.572
6.3096−02	−1.2	3.383 1.012 0.994	6.984 1.060 0.968	16.23 1.297 0.818	19.66 1.413 0.753	22.16 1.507 0.708	24.07 1.584 0.676	25.58 1.648 0.651	27.73 1.753 0.619	29.11 1.831 0.599
7.9433−02	−1.1	2.772 1.008 0.997	5.894 1.042 0.979	14.36 1.241 0.854	17.61 1.345 0.792	20.03 1.431 0.747	21.91 1.504 0.713	23.43 1.566 0.687	25.66 1.669 0.651	27.18 1.748 0.628
1.0000−01	−1.0	2.258 1.005 0.998	4.933 1.028 0.987	12.63 1.191 0.886	15.68 1.284 0.829	17.99 1.362 0.785	19.83 1.429 0.751	21.32 1.488 0.724	23.59 1.586 0.686	25.20 1.665 0.660
1.2589−01	−0.9	1.830 1.003 0.999	4.097 1.018 0.992	11.03 1.148 0.914	13.88 1.229 0.863	16.07 1.299 0.822	17.83 1.360 0.789	19.30 1.415 0.762	21.56 1.508 0.721	23.22 1.583 0.693
1.5849−01	−0.8	1.478 1.002 1.000	3.377 1.011 0.995	9.560 1.112 0.937	12.20 1.180 0.894	14.26 1.242 0.857	15.94 1.297 0.825	17.35 1.347 0.798	19.59 1.433 0.757	21.28 1.505 0.727
1.9953−01	−0.7	1.189 1.001 1.000	2.766 1.007 0.997	8.218 1.082 0.956	10.64 1.139 0.921	12.56 1.192 0.888	14.15 1.241 0.859	15.50 1.286 0.834	17.68 1.364 0.793	19.35 1.431 0.762
2.5119−01	−0.6	0.954 1.001 1.000	2.252 1.004 0.998	7.004 1.058 0.971	9.209 1.104 0.943	10.99 1.149 0.915	12.48 1.191 0.890	13.75 1.230 0.867	15.84 1.301 0.828	17.49 1.362 0.797
3.1623−01	−0.5	0.764 1.001 1.001	1.825 1.002 0.999	5.918 1.040 0.981	7.904 1.076 0.961	9.535 1.113 0.939	10.92 1.148 0.917	12.11 1.182 0.897	14.10 1.244 0.861	15.70 1.299 0.831
3.9811−01	−0.4	0.611 1.001 1.001	1.473 1.002 1.000	4.957 1.026 0.988	6.725 1.053 0.974	8.205 1.083 0.957	9.476 1.112 0.940	10.59 1.140 0.923	12.46 1.194 0.891	13.98 1.242 0.864
5.0119−01	−0.3	0.488 1.001 1.001	1.185 1.001 1.000	4.119 1.017 0.993	5.673 1.036 0.984	7.000 1.059 0.971	8.155 1.082 0.958	9.178 1.105 0.944	10.92 1.150 0.918	12.36 1.193 0.893
6.3096−01	−0.2	0.389 1.001 1.001	0.951 1.001 1.000	3.397 1.010 0.996	4.744 1.023 0.990	5.919 1.040 0.982	6.957 1.058 0.972	7.887 1.077 0.962	9.492 1.114 0.940	10.84 1.149 0.919
7.9433−01	−0.1	0.310 1.001 1.001	0.761 1.001 1.001	2.782 1.006 0.998	3.935 1.015 0.994	4.962 1.026 0.989	5.883 1.039 0.982	6.717 1.054 0.975	8.180 1.083 0.958	9.428 1.113 0.941
1.0000+00	−0.0	0.247 1.001 1.001	0.609 1.001 1.001	2.266 1.003 0.999	3.240 1.009 0.997	4.125 1.016 0.993	4.931 1.026 0.989	5.671 1.036 0.984	6.986 1.059 0.972	8.126 1.083 0.959
1.2589+00	0.1	0.197 1.001 1.001	0.486 1.001 1.001	1.836 1.002 0.999	2.650 1.005 0.998	3.403 1.010 0.996	4.099 1.016 0.994	4.746 1.023 0.990	5.913 1.040 0.982	6.941 1.058 0.973
1.5849+00	0.2	0.156 1.001 1.001	0.387 1.001 1.001	1.481 1.001 1.000	2.156 1.003 0.999	2.788 1.006 0.998	3.381 1.010 0.997	3.939 1.014 0.995	4.960 1.026 0.989	5.876 1.040 0.983
1.9953+00	0.3	0.125 1.000 1.000	0.309 1.000 1.000	1.192 1.001 1.000	1.745 1.002 1.000	2.270 1.003 0.999	2.769 1.006 0.998	3.244 1.009 0.997	4.125 1.017 0.994	4.927 1.026 0.990

Table IX continued; neutron energy 12 MeV

η/rad^{-1}	lg η	$\sigma_0 = 2\ \mu m^2$ $d = 1.60\ \mu m$	$\sigma_0 = 5\ \mu m^2$ $d = 2.52\ \mu m$	$\sigma_0 = 20\ \mu m^2$ $d = 5.05\ \mu m$	$\sigma_0 = 30\ \mu m^2$ $d = 6.18\ \mu m$	$\sigma_0 = 40\ \mu m^2$ $d = 7.14\ \mu m$	$\sigma_0 = 50\ \mu m^2$ $d = 7.98\ \mu m$	$\sigma_0 = 60\ \mu m^2$ $d = 8.74\ \mu m$	$\sigma_0 = 80\ \mu m^2$ $d = 10.1\ \mu m$	$\sigma_0 = 100\ \mu m^2$ $d = 11.3\ \mu m$
zero		78.45 2.816 0.245	73.78 2.811 0.259	63.01 2.803 0.300	59.24 2.799 0.318	56.54 2.796 0.331	54.45 2.792 0.343	52.78 2.787 0.352	50.19 2.779 0.368	48.25 2.771 0.381
1.0000 − 04	−4.0	71.68 2.620 0.265	71.06 2.738 0.268	62.54 2.785 0.302	58.98 2.788 0.318	56.36 2.787 0.332	54.33 2.787 0.343	52.68 2.785 0.352	50.13 2.777 0.368	48.20 2.769 0.381
1.2589 − 04	−3.9	69.70 2.591 0.272	70.39 2.719 0.271	62.42 2.781 0.302	58.91 2.785 0.319	56.32 2.785 0.332	54.29 2.786 0.343	52.65 2.784 0.353	50.11 2.776 0.368	48.19 2.768 0.381
1.5849 − 04	−3.8	67.55 2.558 0.279	69.51 2.696 0.273	62.28 2.778 0.303	58.83 2.782 0.319	56.26 2.783 0.332	54.26 2.784 0.343	52.63 2.782 0.353	50.09 2.775 0.368	48.18 2.767 0.381
1.9953 − 04	−3.7	65.09 2.513 0.289	68.46 2.667 0.277	62.08 2.769 0.304	58.72 2.777 0.320	56.19 2.779 0.333	54.20 2.781 0.343	52.58 2.780 0.353	50.07 2.773 0.368	48.16 2.766 0.381
2.5119 − 04	−3.6	62.38 2.460 0.300	67.29 2.634 0.282	61.85 2.759 0.305	58.58 2.770 0.320	56.10 2.775 0.333	54.14 2.777 0.344	52.54 2.777 0.353	50.03 2.771 0.368	48.14 2.765 0.381
3.1623 − 04	−3.5	59.37 2.396 0.314	65.86 2.595 0.287	61.56 2.745 0.306	58.41 2.764 0.321	55.99 2.769 0.333	54.06 2.772 0.344	52.47 2.773 0.353	49.99 2.769 0.369	48.11 2.763 0.381
3.9811 − 04	−3.4	56.09 2.322 0.330	64.13 2.554 0.294	61.20 2.725 0.308	58.21 2.752 0.322	55.85 2.762 0.334	53.95 2.766 0.344	52.40 2.768 0.354	49.94 2.765 0.369	48.07 2.760 0.381
5.0119 − 04	−3.3	52.56 2.241 0.349	62.04 2.507 0.303	60.73 2.703 0.310	57.94 2.737 0.323	55.68 2.750 0.335	53.83 2.756 0.345	52.30 2.760 0.354	49.88 2.760 0.369	48.02 2.757 0.381
6.3096 − 04	−3.2	48.85 2.154 0.372	59.53 2.455 0.314	60.08 2.679 0.313	57.62 2.716 0.325	55.45 2.735 0.336	53.67 2.744 0.346	52.17 2.749 0.355	49.80 2.753 0.369	47.97 2.752 0.381
7.9433 − 04	−3.1	45.01 2.062 0.398	56.85 2.395 0.327	59.15 2.657 0.317	57.19 2.690 0.328	55.19 2.714 0.338	53.47 2.727 0.347	52.03 2.733 0.356	49.69 2.743 0.370	47.89 2.746 0.382
1.0000 − 03	−3.0	41.14 1.970 0.429	53.89 2.323 0.342	58.17 2.630 0.322	56.57 2.665 0.331	54.81 2.688 0.340	53.22 2.706 0.349	51.83 2.713 0.357	49.57 2.729 0.371	47.80 2.736 0.382
1.2589 − 03	−2.9	37.33 1.879 0.464	50.70 2.241 0.361	57.04 2.601 0.328	55.70 2.643 0.336	54.26 2.662 0.344	52.88 2.681 0.352	51.58 2.688 0.359	49.42 2.710 0.372	47.68 2.722 0.383
1.5849 − 03	−2.8	33.63 1.790 0.502	47.31 2.153 0.383	55.81 2.566 0.334	54.68 2.618 0.341	53.47 2.643 0.348	52.35 2.656 0.355	51.24 2.659 0.361	49.24 2.684 0.374	47.58 2.702 0.385
1.9953 − 03	−2.7	30.14 1.705 0.544	43.77 2.060 0.409	54.39 2.522 0.341	53.59 2.593 0.347	52.53 2.623 0.354	51.59 2.635 0.359	50.72 2.635 0.365	48.94 2.657 0.376	47.35 2.673 0.387
2.5119 − 03	−2.6	26.89 1.622 0.588	40.15 1.968 0.439	52.76 2.472 0.350	52.46 2.561 0.353	51.49 2.599 0.360	50.68 2.617 0.365	49.98 2.618 0.370	48.46 2.628 0.380	47.05 2.644 0.389
3.1623 − 03	−2.5	23.90 1.543 0.634	36.55 1.877 0.473	50.81 2.413 0.361	51.18 2.517 0.360	50.46 2.568 0.366	49.66 2.593 0.371	49.00 2.604 0.376	47.74 2.608 0.385	46.59 2.614 0.393
3.9811 − 03	−2.4	21.16 1.468 0.680	33.02 1.789 0.511	48.67 2.339 0.375	49.77 2.465 0.369	49.34 2.529 0.372	48.67 2.568 0.378	47.99 2.583 0.383	46.89 2.595 0.391	45.89 2.595 0.398
5.0119 − 03	−2.3	18.68 1.398 0.726	29.66 1.704 0.552	46.30 2.256 0.391	48.05 2.403 0.379	48.16 2.476 0.379	47.63 2.530 0.384	47.02 2.559 0.389	45.94 2.577 0.398	45.02 2.585 0.405
6.3096 − 03	−2.2	16.42 1.334 0.771	26.52 1.621 0.596	43.71 2.164 0.410	46.17 2.324 0.392	46.78 2.417 0.389	46.56 2.483 0.391	46.03 2.520 0.395	45.03 2.561 0.404	44.08 2.572 0.412
7.9433 − 03	−2.1	14.38 1.276 0.812	23.62 1.542 0.641	40.88 2.066 0.434	44.10 2.237 0.408	45.12 2.344 0.400	45.32 2.427 0.399	45.04 2.473 0.402	44.11 2.536 0.411	43.16 2.560 0.420
1.0000 − 02	−2.0	12.53 1.224 0.850	20.96 1.467 0.687	37.89 1.972 0.461	41.73 2.142 0.427	43.27 2.259 0.414	43.83 2.354 0.410	43.86 2.413 0.411	43.26 2.487 0.417	42.31 2.534 0.426
1.2589 − 02	−1.9	10.87 1.179 0.883	18.52 1.397 0.733	34.81 1.880 0.493	39.08 2.044 0.450	41.17 2.168 0.432	42.13 2.265 0.424	42.50 2.334 0.421	42.33 2.430 0.423	41.61 2.495 0.431
1.5849 − 02	−1.8	9.373 1.140 0.911	16.31 1.332 0.777	31.67 1.792 0.529	36.25 1.951 0.478	38.78 2.072 0.453	40.19 2.169 0.441	40.97 2.239 0.434	41.21 2.353 0.432	40.86 2.440 0.437
1.9953 − 02	−1.7	8.031 1.107 0.935	14.30 1.274 0.818	28.65 1.708 0.568	33.29 1.861 0.510	36.21 1.974 0.478	37.95 2.067 0.461	39.10 2.142 0.451	39.85 2.261 0.444	39.86 2.362 0.444
2.5119 − 02	−1.6	6.833 1.079 0.954	12.48 1.222 0.855	25.78 1.626 0.610	30.37 1.773 0.546	33.48 1.882 0.508	35.51 1.971 0.486	36.89 2.043 0.472	38.26 2.168 0.459	38.68 2.266 0.455
3.1623 − 02	−1.5	5.771 1.057 0.968	10.83 1.176 0.888	23.09 1.546 0.653	27.53 1.689 0.584	30.63 1.794 0.542	32.90 1.880 0.515	34.45 1.952 0.498	36.34 2.072 0.478	37.20 2.168 0.469
3.9811 − 02	−1.4	4.839 1.040 0.979	9.349 1.137 0.916	20.58 1.471 0.697	24.80 1.607 0.626	27.85 1.710 0.580	30.14 1.792 0.549	31.83 1.864 0.528	34.20 1.973 0.501	35.39 2.064 0.488
5.0119 − 02	−1.3	4.027 1.028 0.986	8.015 1.104 0.939	18.28 1.400 0.741	22.24 1.528 0.668	25.18 1.628 0.620	27.46 1.708 0.586	29.22 1.776 0.562	31.85 1.883 0.529	33.37 1.971 0.510
6.3096 − 02	−1.2	3.329 1.018 0.991	6.822 1.077 0.957	16.15 1.335 0.783	19.84 1.454 0.712	22.65 1.549 0.662	24.86 1.626 0.626	26.66 1.692 0.598	29.32 1.795 0.561	31.15 1.881 0.538
7.9433 − 02	−1.1	2.734 1.012 0.995	5.764 1.055 0.971	14.21 1.275 0.824	17.63 1.384 0.754	20.27 1.473 0.704	22.41 1.546 0.666	24.15 1.609 0.637	26.82 1.713 0.596	28.73 1.793 0.569
1.0000 − 01	−1.0	2.232 1.007 0.997	4.833 1.038 0.981	12.43 1.222 0.860	15.58 1.320 0.796	18.06 1.402 0.747	20.07 1.471 0.708	21.77 1.530 0.678	24.39 1.631 0.634	26.38 1.710 0.604
1.2589 − 01	−0.9	1.813 1.004 0.998	4.022 1.026 0.988	10.82 1.176 0.892	13.71 1.262 0.835	16.00 1.336 0.788	17.90 1.400 0.750	19.50 1.456 0.710	22.08 1.551 0.673	24.03 1.628 0.641
1.5849 − 01	−0.8	1.466 1.003 0.999	3.324 1.017 0.993	9.353 1.136 0.920	11.99 1.211 0.870	14.11 1.277 0.827	15.87 1.335 0.791	17.38 1.387 0.761	19.85 1.475 0.714	21.78 1.548 0.679
1.9953 − 01	−0.7	1.182 1.002 1.000	2.729 1.010 0.996	8.032 1.102 0.943	10.43 1.166 0.901	12.37 1.223 0.863	14.00 1.275 0.829	15.42 1.322 0.801	17.76 1.404 0.754	19.61 1.473 0.718
2.5119 − 01	−0.6	0.950 1.001 1.000	2.227 1.006 0.997	6.846 1.075 0.960	9.009 1.127 0.927	10.78 1.177 0.895	12.28 1.222 0.865	13.59 1.264 0.838	15.79 1.338 0.793	17.57 1.402 0.758
3.1623 − 01	−0.5	0.761 1.001 1.000	1.808 1.004 0.999	5.791 1.053 0.974	7.727 1.095 0.948	9.333 1.136 0.922	10.71 1.175 0.896	11.92 1.212 0.872	13.96 1.279 0.831	15.64 1.337 0.797
3.9811 − 01	−0.4	0.609 1.001 1.001	1.462 1.002 0.999	4.860 1.036 0.983	6.578 1.069 0.965	8.024 1.102 0.944	9.276 1.135 0.923	10.38 1.167 0.903	12.27 1.225 0.866	13.84 1.277 0.834
5.0119 − 01	−0.3	0.487 1.001 1.001	1.178 1.002 1.000	4.047 1.024 0.990	5.555 1.048 0.977	6.846 1.075 0.962	7.976 1.102 0.945	8.982 1.128 0.929	10.72 1.178 0.897	12.18 1.224 0.868
6.3096 − 01	−0.2	0.388 1.001 1.001	0.946 1.001 1.000	3.345 1.015 0.994	4.655 1.032 0.986	5.795 1.053 0.975	6.806 1.074 0.962	7.714 1.095 0.949	9.295 1.137 0.923	10.64 1.177 0.899
7.9433 − 01	−0.1	0.310 1.001 1.001	0.758 1.001 1.000	2.747 1.009 0.996	3.870 1.021 0.991	4.866 1.036 0.984	5.761 1.052 0.975	6.573 1.069 0.966	8.002 1.103 0.945	9.232 1.136 0.925
1.0000 + 00	−0.0	0.247 1.001 1.001	0.607 1.001 1.001	2.242 1.005 0.998	3.194 1.013 0.995	4.054 1.023 0.990	4.838 1.035 0.985	5.556 1.048 0.978	6.835 1.075 0.963	7.950 1.102 0.946
1.2589 + 00	0.1	0.196 1.001 1.001	0.485 1.001 1.001	1.820 1.003 0.999	2.618 1.008 0.997	3.353 1.015 0.994	4.030 1.023 0.991	4.658 1.032 0.986	5.792 1.053 0.975	6.792 1.074 0.963
1.5849 + 00	0.2	0.156 1.001 1.001	0.387 1.001 1.001	1.471 1.002 1.000	2.134 1.004 0.998	2.753 1.009 0.997	3.332 1.014 0.995	3.875 1.021 0.992	4.867 1.036 0.985	5.755 1.052 0.976
1.9953 + 00	0.3	0.124 1.000 1.000	0.309 1.000 1.000	1.185 1.001 1.000	1.731 1.003 0.999	2.247 1.005 0.998	2.736 1.009 0.997	3.199 1.013 0.995	4.057 1.023 0.991	4.836 1.035 0.985

Table IX continued; neutron energy 15 MeV

η/rad^{-1}	lg η	$\sigma_0 = 2\ \mu m^2$ $d = 1.60\ \mu m$	$\sigma_0 = 5\ \mu m^2$ $d = 2.52\ \mu m$	$\sigma_0 = 20\ \mu m^2$ $d = 5.05\ \mu m$	$\sigma_0 = 30\ \mu m^2$ $d = 6.18\ \mu m$	$\sigma_0 = 40\ \mu m^2$ $d = 7.14\ \mu m$	$\sigma_0 = 50\ \mu m^2$ $d = 7.98\ \mu m$	$\sigma_0 = 60\ \mu m^2$ $d = 8.74\ \mu m$	$\sigma_0 = 80\ \mu m^2$ $d = 10.1\ \mu m$	$\sigma_0 = 100\ \mu m^2$ $d = 11.3\ \mu m$
zero		98.82 2.539 0.207	93.92 2.545 0.217	81.90 2.570 0.246	77.41 2.581 0.258	74.08 2.590 0.269	71.46 2.595 0.278	69.31 2.600 0.285	65.96 2.607 0.298	63.42 2.610 0.308
1.0000 − 04	−4.0	89.10 2.409 0.227	89.93 2.500 0.225	81.17 2.558 0.247	76.98 2.573 0.259	73.79 2.583 0.270	71.25 2.589 0.278	69.15 2.595 0.285	65.85 2.603 0.298	63.34 2.606 0.308
1.2589 − 04	−3.9	86.41 2.391 0.233	88.96 2.486 0.227	80.98 2.555 0.248	76.87 2.571 0.260	73.72 2.581 0.270	71.20 2.588 0.278	69.11 2.594 0.286	65.83 2.602 0.298	63.32 2.606 0.308
1.5849 − 04	−3.8	83.42 2.369 0.240	87.81 2.469 0.230	80.76 2.551 0.248	76.75 2.569 0.260	73.64 2.579 0.270	71.14 2.587 0.278	69.06 2.593 0.286	65.80 2.601 0.298	63.30 2.605 0.308
1.9953 − 04	−3.7	80.06 2.339 0.249	86.33 2.448 0.234	80.45 2.546 0.249	76.51 2.565 0.261	73.51 2.577 0.270	71.05 2.585 0.279	69.00 2.592 0.286	65.75 2.600 0.298	63.27 2.604 0.308
2.5119 − 04	−3.6	76.34 2.302 0.259	84.64 2.425 0.238	80.08 2.541 0.250	76.35 2.562 0.261	73.37 2.574 0.271	70.94 2.583 0.279	68.91 2.590 0.286	65.70 2.599 0.298	63.23 2.603 0.308
3.1623 − 04	−3.5	72.27 2.257 0.272	82.54 2.398 0.243	79.63 2.533 0.251	76.09 2.557 0.262	73.19 2.571 0.271	70.81 2.580 0.279	68.81 2.587 0.286	65.63 2.597 0.298	63.18 2.602 0.308
3.9811 − 04	−3.4	67.84 2.203 0.287	79.98 2.369 0.250	79.06 2.522 0.253	75.75 2.551 0.263	72.96 2.566 0.272	70.65 2.576 0.280	68.69 2.584 0.287	65.55 2.595 0.299	63.12 2.600 0.308
5.0119 − 04	−3.3	63.14 2.142 0.305	76.96 2.339 0.258	78.38 2.506 0.255	75.34 2.542 0.264	72.68 2.560 0.273	70.44 2.571 0.280	68.53 2.580 0.287	65.44 2.592 0.299	63.05 2.598 0.308
6.3096 − 04	−3.2	58.24 2.075 0.327	73.42 2.304 0.268	77.52 2.487 0.258	74.84 2.530 0.266	72.33 2.552 0.274	70.18 2.566 0.281	68.33 2.576 0.288	65.31 2.588 0.299	62.95 2.595 0.308
7.9433 − 04	−3.1	53.23 2.003 0.352	69.68 2.260 0.281	76.33 2.465 0.261	74.17 2.513 0.268	71.89 2.540 0.275	69.86 2.558 0.282	68.08 2.570 0.289	65.14 2.584 0.300	62.83 2.592 0.309
1.0000 − 03	−3.0	48.20 1.928 0.381	65.61 2.207 0.296	74.91 2.443 0.266	73.31 2.493 0.271	71.37 2.521 0.277	69.46 2.545 0.284	67.77 2.560 0.290	64.93 2.578 0.301	62.68 2.587 0.309
1.2589 − 03	−2.9	43.29 1.852 0.415	61.28 2.146 0.314	73.19 2.419 0.271	72.15 2.470 0.275	70.69 2.500 0.280	68.98 2.525 0.286	67.41 2.547 0.291	64.68 2.569 0.302	62.49 2.582 0.310
1.5849 − 03	−2.8	38.57 1.776 0.452	56.73 2.077 0.335	71.17 2.391 0.277	70.80 2.447 0.279	69.74 2.476 0.283	68.35 2.502 0.288	66.89 2.527 0.294	64.39 2.555 0.303	62.26 2.571 0.312
1.9953 − 03	−2.7	34.15 1.701 0.493	52.02 2.005 0.359	68.89 2.362 0.285	69.17 2.421 0.285	68.42 2.455 0.288	67.45 2.477 0.292	66.17 2.505 0.297	63.92 2.534 0.305	62.00 2.558 0.313
2.5119 − 03	−2.6	30.06 1.628 0.536	47.25 1.929 0.388	66.26 2.328 0.294	67.32 2.393 0.291	66.84 2.435 0.294	66.18 2.456 0.297	65.23 2.480 0.301	63.34 2.510 0.308	61.58 2.535 0.316
3.1623 − 03	−2.5	26.33 1.557 0.583	42.54 1.852 0.421	63.33 2.282 0.305	65.19 2.363 0.299	65.06 2.412 0.300	64.65 2.437 0.302	63.98 2.459 0.306	62.54 2.485 0.312	60.98 2.511 0.318
3.9811 − 03	−2.4	22.97 1.490 0.630	37.98 1.776 0.458	60.16 2.230 0.319	62.74 2.329 0.308	63.15 2.385 0.307	62.93 2.416 0.309	62.50 2.437 0.311	61.41 2.463 0.317	60.18 2.486 0.322
5.0119 − 03	−2.3	19.96 1.426 0.678	33.69 1.701 0.499	56.76 2.167 0.335	60.06 2.282 0.319	61.06 2.352 0.315	61.13 2.390 0.316	60.90 2.414 0.318	60.03 2.445 0.323	59.10 2.466 0.327
6.3096 − 03	−2.2	17.30 1.365 0.726	29.70 1.628 0.542	53.12 2.094 0.354	57.17 2.228 0.333	58.81 2.305 0.325	59.17 2.358 0.324	59.16 2.388 0.325	58.53 2.426 0.329	57.72 2.448 0.333
7.9433 − 03	−2.1	14.94 1.309 0.771	26.06 1.557 0.588	49.21 2.017 0.376	54.06 2.159 0.348	56.20 2.250 0.337	57.10 2.311 0.333	57.22 2.354 0.333	56.90 2.404 0.336	56.27 2.430 0.340
1.0000 − 02	−2.0	12.87 1.258 0.813	22.76 1.489 0.636	45.09 1.938 0.403	50.71 2.081 0.367	53.43 2.182 0.352	54.68 2.255 0.345	55.24 2.307 0.343	55.22 2.376 0.343	54.76 2.410 0.347
1.2589 − 02	−1.9	11.04 1.211 0.851	19.81 1.425 0.684	40.93 1.857 0.435	47.09 2.001 0.390	50.42 2.104 0.369	52.12 2.186 0.359	52.96 2.251 0.354	53.56 2.332 0.352	53.17 2.383 0.354
1.5849 − 02	−1.8	9.436 1.170 0.884	17.18 1.364 0.731	36.75 1.780 0.470	43.21 1.920 0.417	47.08 2.023 0.390	49.32 2.106 0.376	50.51 2.178 0.368	51.60 2.277 0.362	51.70 2.338 0.362
1.9953 − 02	−1.7	8.028 1.134 0.912	14.86 1.308 0.776	32.79 1.705 0.510	39.25 1.840 0.449	43.42 1.943 0.415	46.16 2.025 0.396	47.82 2.096 0.385	49.51 2.207 0.374	49.98 2.283 0.372
2.5119 − 02	−1.6	6.796 1.103 0.936	12.81 1.256 0.818	29.06 1.632 0.552	35.25 1.765 0.485	39.64 1.862 0.446	42.68 1.943 0.421	44.80 2.011 0.406	47.19 2.123 0.389	48.08 2.210 0.383
3.1623 − 02	−1.5	5.721 1.078 0.954	11.00 1.209 0.856	25.59 1.561 0.597	31.45 1.690 0.525	35.76 1.785 0.480	39.03 1.861 0.451	41.41 1.925 0.431	44.47 2.037 0.408	45.96 2.125 0.397
3.9811 − 02	−1.4	4.788 1.057 0.969	9.406 1.168 0.889	22.45 1.493 0.643	27.88 1.618 0.567	32.04 1.709 0.518	35.24 1.784 0.485	37.86 1.843 0.461	41.33 1.951 0.431	43.43 2.039 0.415
5.0119 − 02	−1.3	3.982 1.040 0.979	8.008 1.132 0.917	19.60 1.428 0.690	24.58 1.547 0.612	28.51 1.636 0.559	31.61 1.709 0.523	34.17 1.768 0.496	38.01 1.867 0.460	40.46 1.952 0.438
6.3096 − 02	−1.2	3.292 1.028 0.987	6.782 1.101 0.940	17.06 1.366 0.737	21.58 1.479 0.658	25.20 1.564 0.603	28.17 1.635 0.564	30.63 1.693 0.534	34.45 1.790 0.493	37.29 1.867 0.467
7.9433 − 02	−1.1	2.706 1.018 0.992	5.711 1.075 0.958	14.79 1.309 0.781	18.86 1.414 0.704	22.17 1.496 0.648	24.93 1.564 0.607	27.27 1.622 0.575	31.03 1.715 0.530	33.85 1.790 0.499
1.0000 − 01	−1.0	2.212 1.012 0.995	4.780 1.054 0.971	12.77 1.256 0.823	16.41 1.353 0.750	19.41 1.430 0.694	21.96 1.494 0.652	24.15 1.550 0.618	27.76 1.641 0.569	30.54 1.713 0.535
1.2589 − 01	−0.9	1.799 1.008 0.997	3.976 1.038 0.981	10.99 1.209 0.860	14.23 1.296 0.793	16.93 1.368 0.740	19.25 1.429 0.697	21.27 1.482 0.663	24.66 1.569 0.611	27.37 1.640 0.575
1.5849 − 01	−0.8	1.457 1.005 0.999	3.286 1.026 0.988	9.419 1.167 0.893	12.30 1.245 0.834	14.71 1.310 0.784	16.80 1.366 0.743	18.64 1.416 0.709	21.78 1.499 0.655	24.35 1.568 0.616
1.9953 − 01	−0.7	1.176 1.003 0.999	2.700 1.017 0.993	8.030 1.130 0.921	10.58 1.198 0.870	12.73 1.257 0.825	14.60 1.309 0.786	16.27 1.355 0.753	19.14 1.433 0.700	21.54 1.498 0.659
2.5119 − 01	−0.6	0.946 1.002 1.000	2.207 1.011 0.996	6.809 1.099 0.943	9.058 1.157 0.902	10.97 1.209 0.862	12.64 1.256 0.827	14.14 1.298 0.796	16.74 1.370 0.744	18.95 1.432 0.703
3.1623 − 01	−0.5	0.759 1.001 1.000	1.794 1.007 0.998	5.740 1.073 0.961	7.718 1.122 0.928	9.407 1.167 0.895	10.90 1.208 0.864	12.24 1.246 0.836	14.58 1.312 0.787	16.59 1.369 0.747
3.9811 − 01	−0.4	0.607 1.001 1.000	1.453 1.004 0.999	4.808 1.052 0.974	6.538 1.092 0.949	8.029 1.130 0.923	9.350 1.166 0.897	10.54 1.199 0.872	12.65 1.259 0.828	14.46 1.311 0.790
5.0119 − 01	−0.3	0.486 1.001 1.001	1.172 1.003 1.000	4.001 1.036 0.984	5.505 1.067 0.965	6.814 1.098 0.945	7.982 1.129 0.924	9.041 1.158 0.903	10.92 1.210 0.864	12.55 1.258 0.830
6.3096 − 01	−0.2	0.388 1.001 1.001	0.942 1.002 1.000	3.309 1.024 0.990	4.605 1.047 0.977	5.748 1.072 0.962	6.775 1.097 0.946	7.711 1.122 0.929	9.379 1.167 0.897	10.84 1.209 0.866
7.9433 − 01	−0.1	0.309 1.001 1.001	0.756 1.001 1.000	2.719 1.016 0.994	3.827 1.032 0.986	4.817 1.051 0.975	5.715 1.071 0.963	6.539 1.091 0.950	8.015 1.130 0.924	9.317 1.166 0.898
1.0000 + 00	−0.0	0.246 1.001 1.001	0.605 1.001 1.001	2.222 1.010 0.997	3.160 1.021 0.991	4.011 1.035 0.984	4.790 1.051 0.976	5.510 1.067 0.967	6.809 1.098 0.946	7.963 1.129 0.925
1.2589 + 00	0.1	0.196 1.001 1.001	0.484 1.001 1.001	1.806 1.006 0.998	2.593 1.013 0.995	3.317 1.023 0.991	3.988 1.035 0.985	4.612 1.047 0.978	5.749 1.072 0.964	6.767 1.097 0.947
1.5849 + 00	0.2	0.156 1.001 1.001	0.386 1.001 1.000	1.462 1.003 0.999	2.116 1.008 0.997	2.727 1.015 0.995	3.298 1.023 0.991	3.835 1.032 0.987	4.822 1.051 0.976	5.714 1.071 0.964
1.9953 + 00	0.3	0.124 1.000 1.000	0.308 1.000 1.000	1.179 1.002 0.999	1.719 1.005 0.998	2.228 1.009 0.997	2.710 1.014 0.995	3.167 1.021 0.992	4.017 1.035 0.985	4.792 1.050 0.977

Table IX; continued; neutron energy 20 MeV

η/rad^{-1}	lg η	$\sigma_0 = 2\ \mu m^2$ $d = 1.60\ \mu m$	$\sigma_0 = 5\ \mu m^2$ $d = 2.52\ \mu m$	$\sigma_0 = 20\ \mu m^2$ $d = 5.05\ \mu m$	$\sigma_0 = 30\ \mu m^2$ $d = 6.18\ \mu m$	$\sigma_0 = 40\ \mu m^2$ $d = 7.14\ \mu m$	$\sigma_0 = 50\ \mu m^2$ $d = 7.98\ \mu m$	$\sigma_0 = 60\ \mu m^2$ $d = 8.74\ \mu m$	$\sigma_0 = 80\ \mu m^2$ $d = 10.1\ \mu m$	$\sigma_0 = 1000\ \mu m^2$ $d = 11.3\ \mu m$
zero		115.4 2.339 0.169	110.0 2.344 0.177	96.42 2.372 0.199	91.01 2.388 0.209	86.88 2.401 0.218	83.59 2.411 0.226	80.85 2.420 0.233	76.53 2.433 0.244	73.21 2.443 0.254
1.0000−04	−4.0	101.9 2.245 0.190	104.4 2.313 0.185	95.34 2.364 0.201	90.38 2.382 0.211	86.46 2.396 0.219	83.28 2.407 0.226	80.61 2.416 0.233	76.37 2.430 0.245	73.09 2.440 0.254
1.2589−04	−3.9	98.43 2.232 0.195	103.0 2.304 0.187	95.07 2.361 0.201	90.22 2.381 0.211	86.35 2.395 0.219	83.20 2.406 0.227	80.55 2.415 0.233	76.33 2.429 0.245	73.06 2.439 20.54
1.5849−04	−3.8	94.51 2.216 0.202	101.4 2.293 0.190	94.75 2.359 0.202	90.03 2.379 0.211	86.23 2.394 0.219	83.11 2.405 0.227	80.48 2.415 0.233	76.29 2.429 0.245	73.03 2.439 0.254
1.9953−04	−3.7	90.15 2.193 0.211	99.44 2.279 0.193	94.30 2.355 0.203	89.77 2.377 0.212	86.05 2.392 0.220	82.98 2.404 0.227	80.38 2.413 0.234	76.23 2.428 0.245	72.99 2.438 0.254
2.5119−04	−3.6	85.30 2.166 0.221	97.18 2.260 0.198	93.76 2.351 0.204	89.45 2.374 0.212	85.84 2.390 0.220	82.83 2.402 0.227	80.26 2.412 0.234	76.15 2.427 0.245	72.93 2.438 0.255
3.1623−04	−3.5	80.05 2.133 0.234	94.40 2.239 0.203	93.10 2.346 0.205	89.06 2.370 0.213	85.57 2.387 0.221	82.63 2.400 0.228	80.11 2.410 0.234	76.05 2.425 0.245	72.87 2.437 0.255
3.9811−04	−3.4	74.44 2.093 0.249	91.03 2.217 0.209	92.28 2.339 0.207	88.57 2.365 0.214	85.24 2.383 0.222	82.39 2.397 0.228	79.93 2.408 0.235	75.93 2.424 0.246	72.78 2.435 0.255
5.0119−04	−3.3	68.57 2.047 0.267	87.02 2.193 0.218	91.27 2.330 0.209	87.96 2.360 0.216	84.83 2.379 0.223	82.08 2.393 0.229	79.69 2.405 0.235	75.78 2.421 0.246	72.67 2.433 0.255
6.3096−04	−3.2	62.59 1.996 0.288	82.38 2.167 0.228	90.04 2.319 0.211	87.21 2.352 0.217	84.31 2.374 0.224	81.71 2.389 0.230	79.40 2.401 0.236	75.58 2.419 0.247	72.53 2.431 0.256
7.9433−04	−3.1	56.59 1.940 0.313	77.48 2.135 0.240	88.47 2.303 0.215	86.30 2.342 0.219	83.67 2.367 0.225	81.23 2.384 0.231	79.04 2.396 0.237	75.34 2.415 0.248	72.36 2.428 0.256
1.0000−03	−3.0	50.73 1.880 0.342	72.24 2.096 0.255	86.69 2.283 0.219	85.14 2.329 0.222	82.88 2.356 0.227	80.65 2.376 0.233	78.58 2.391 0.238	75.04 2.411 0.248	72.14 2.425 0.257
1.2589−03	−2.9	45.12 1.818 0.374	66.73 2.050 0.273	84.54 2.261 0.223	83.64 2.311 0.225	81.93 2.343 0.230	79.92 2.365 0.235	78.02 2.382 0.239	74.67 2.406 0.250	71.87 2.420 0.258
1.5849−03	−2.8	39.85 1.755 0.411	61.08 1.998 0.294	81.91 2.236 0.229	81.96 2.290 0.230	80.74 2.326 0.233	79.07 2.350 0.237	77.34 2.371 0.241	74.20 2.397 0.251	71.54 2.415 0.259
1.9953−03	−2.7	34.96 1.692 0.450	55.40 1.941 0.319	78.77 2.210 0.237	79.95 2.267 0.234	79.16 2.307 0.237	77.94 2.331 0.240	76.46 2.356 0.244	73.65 2.386 0.253	71.12 2.407 0.260
2.5119−03	−2.6	30.51 1.629 0.493	49.79 1.882 0.347	75.14 2.184 0.246	77.50 2.241 0.241	77.31 2.287 0.242	76.45 2.312 0.244	75.34 2.338 0.247	72.91 2.371 0.255	70.63 2.395 0.262
3.1623−03	−2.5	26.51 1.567 0.538	44.39 1.819 0.379	71.08 2.153 0.258	74.61 2.214 0.248	75.08 2.263 0.248	74.70 2.292 0.249	73.88 2.317 0.252	71.95 2.351 0.258	69.96 2.378 0.264
3.9811−03	−2.4	22.94 1.506 0.586	39.29 1.756 0.416	66.83 2.114 0.272	71.23 2.187 0.258	72.49 2.236 0.255	72.60 2.270 0.255	72.18 2.294 0.256	70.69 2.332 0.261	69.07 2.358 0.267
5.0119−03	−2.3	19.79 1.447 0.634	34.52 1.692 0.455	62.28 2.068 0.289	67.46 2.154 0.269	69.54 2.208 0.263	70.18 2.244 0.262	70.15 2.270 0.262	69.20 2.312 0.265	67.93 2.339 0.270
6.3096−03	−2.2	17.03 1.391 0.682	30.17 1.629 0.498	57.54 2.015 0.309	63.53 2.113 0.284	66.31 2.174 0.273	67.44 2.215 0.270	67.84 2.243 0.269	67.42 2.289 0.271	66.44 2.320 0.275
7.9433−03	−2.1	14.61 1.338 0.728	26.25 1.567 0.543	52.67 1.955 0.332	59.32 2.065 0.300	62.72 2.134 0.287	64.42 2.181 0.280	65.16 2.214 0.278	65.40 2.263 0.277	64.69 2.300 0.280
1.0000−02	−2.0	12.51 1.288 0.773	22.74 1.506 0.590	47.73 1.893 0.359	54.90 2.009 0.320	58.87 2.086 0.302	61.07 2.139 0.293	62.26 2.180 0.288	63.04 2.234 0.286	62.77 2.275 0.287
1.2589−02	−1.9	10.69 1.242 0.814	19.64 1.447 0.638	42.89 1.828 0.390	50.33 1.947 0.344	54.80 2.031 0.321	57.46 2.091 0.308	59.05 2.136 0.301	60.51 2.201 0.295	60.59 2.246 0.295
1.5849−02	−1.8	9.103 1.200 0.851	16.91 1.390 0.686	38.20 1.762 0.425	45.69 1.882 0.371	50.54 1.967 0.343	53.63 2.035 0.326	55.60 2.086 0.316	57.72 2.158 0.307	58.33 2.210 0.304
1.9953−02	−1.7	7.729 1.162 0.883	14.53 1.337 0.733	33.74 1.698 0.463	41.12 1.816 0.402	46.10 1.902 0.368	49.58 1.970 0.348	51.90 2.027 0.335	54.65 2.108 0.321	55.89 2.167 0.316
2.5119−02	−1.6	6.539 1.129 0.911	12.45 1.287 0.777	29.62 1.634 0.505	36.66 1.748 0.437	41.69 1.834 0.398	45.33 1.903 0.373	47.99 1.961 0.357	51.36 2.051 0.338	53.06 2.116 0.328
3.1623−02	−1.5	5.508 1.100 0.934	10.64 1.240 0.818	25.87 1.571 0.550	32.39 1.683 0.476	37.33 1.767 0.432	41.08 1.834 0.403	43.91 1.892 0.382	47.81 1.985 0.358	50.01 2.058 0.345
3.9811−02	−1.4	4.617 1.076 0.953	9.068 1.198 0.855	22.49 1.510 0.596	28.44 1.619 0.518	33.09 1.701 0.470	36.84 1.766 0.436	39.81 1.822 0.412	43.99 1.916 0.382	46.70 1.990 0.364
5.0119−02	−1.3	3.850 1.056 0.967	7.703 1.160 0.888	19.48 1.450 0.644	24.84 1.558 0.563	29.16 1.637 0.511	32.69 1.700 0.474	35.71 1.752 0.446	40.10 1.847 0.410	43.08 1.919 0.388
6.3096−02	−1.2	3.193 1.040 0.978	6.519 1.125 0.915	16.81 1.393 0.691	21.60 1.496 0.610	25.54 1.574 0.555	28.83 1.637 0.515	31.68 1.686 0.483	36.16 1.774 0.442	39.38 1.847 0.415
7.9433−02	−1.1	2.634 1.028 0.986	5.493 1.098 0.938	14.47 1.338 0.737	18.72 1.437 0.657	22.26 1.512 0.600	25.27 1.574 0.558	27.93 1.623 0.525	32.22 1.706 0.478	35.58 1.773 0.447
1.0000−01	−1.0	2.161 1.019 0.991	4.606 1.074 0.956	12.43 1.287 0.782	16.16 1.380 0.704	19.32 1.452 0.647	22.05 1.512 0.604	24.47 1.560 0.568	28.52 1.642 0.518	31.74 1.706 0.483
1.2589−01	−0.9	1.764 1.012 0.995	3.841 1.054 0.970	10.64 1.240 0.823	13.92 1.326 0.750	16.71 1.395 0.694	19.17 1.451 0.650	21.35 1.499 0.614	25.08 1.578 0.561	28.13 1.641 0.523
1.5849−01	−0.8	1.433 1.008 0.997	3.186 1.038 0.980	9.082 1.197 0.860	11.95 1.276 0.793	14.41 1.340 0.740	16.59 1.393 0.696	18.55 1.439 0.660	21.94 1.515 0.605	24.77 1.578 0.565
1.9953−01	−0.7	1.160 1.005 0.998	2.628 1.026 0.987	7.726 1.159 0.892	10.23 1.229 0.833	12.39 1.288 0.784	14.31 1.338 0.742	16.05 1.382 0.707	19.10 1.455 0.651	21.70 1.515 0.609
2.5119−01	−0.6	0.935 1.003 0.999	2.156 1.018 0.992	6.546 1.125 0.920	8.731 1.187 0.869	10.62 1.240 0.825	12.31 1.287 0.786	13.85 1.328 0.752	16.56 1.397 0.697	18.92 1.454 0.654
3.1623−01	−0.5	0.752 1.002 1.000	1.759 1.011 0.995	5.521 1.096 0.942	7.424 1.150 0.900	9.077 1.197 0.862	10.56 1.239 0.827	11.91 1.277 0.795	14.31 1.341 0.743	16.42 1.396 0.700
3.9811−01	−0.4	0.603 1.001 1.000	1.429 1.007 0.997	4.632 1.071 0.959	6.286 1.117 0.927	7.728 1.158 0.894	9.024 1.196 0.864	10.21 1.230 0.835	12.33 1.290 0.786	14.20 1.340 0.745
5.0119−01	−0.3	0.483 1.001 1.000	1.156 1.005 0.999	3.866 1.052 0.973	5.297 1.089 0.948	6.553 1.124 0.921	7.685 1.157 0.896	8.722 1.187 0.871	10.59 1.241 0.827	12.24 1.288 0.789
6.3096−01	−0.2	0.386 1.001 1.001	0.932 1.003 0.999	3.207 1.036 0.982	4.440 1.066 0.964	5.531 1.095 0.944	6.517 1.123 0.923	7.424 1.149 0.902	9.059 1.197 0.864	10.52 1.240 0.829
7.9433−01	−0.1	0.308 1.001 1.001	0.749 1.002 1.000	2.646 1.025 0.989	3.701 1.047 0.976	4.643 1.071 0.961	5.500 1.094 0.945	6.291 1.116 0.928	7.722 1.158 0.896	9.002 1.196 0.865
1.0000+00	−0.0	0.246 1.001 1.001	0.601 1.001 1.000	2.170 1.016 0.993	3.067 1.033 0.985	3.876 1.051 0.974	4.618 1.070 0.962	5.305 1.088 0.949	6.554 1.124 0.923	7.675 1.157 0.897
1.2589+00	0.1	0.196 1.001 1.001	0.481 1.001 1.000	1.771 1.010 0.996	2.526 1.022 0.991	3.217 1.035 0.983	3.855 1.050 0.975	4.450 1.065 0.965	5.536 1.094 0.945	6.515 1.123 0.924
1.5849+00	0.2	0.156 1.001 1.001	0.384 1.001 1.000	1.438 1.006 0.998	2.070 1.014 0.994	2.654 1.024 0.990	3.199 1.035 0.984	3.710 1.046 0.977	4.651 1.070 0.962	5.504 1.093 0.946
1.9953+00	0.3	0.124 1.000 1.000	0.307 1.000 1.000	1.164 1.004 0.999	1.687 1.009 0.997	2.178 1.015 0.994	2.639 1.023 0.990	3.076 1.032 0.986	3.885 1.050 0.975	4.625 1.069 0.963

Table X. Parameter k_3 in dependence on k_2 and E

n = 3

E	$k_2 =$ 0.150	0.175	0.200	0.225	0.250	0.275	0.300	0.325	0.350	0.400	0.450	0.500	0.550	0.600	0.650	lg E
1.2							0.618	0.632	0.639							0.079
1.5							0.618	0.636	0.646							0.176
2.0							0.616	0.638	0.653	0.665						0.301
2.5								0.638	0.656	0.674						0.398
3.0								0.637	0.657	0.680						0.477
4								0.634	0.658	0.688	0.700					0.602
5									0.657	0.692	0.708					0.699
6									0.655	0.695	0.714					0.778
8									0.651	0.697	0.722	0.734				0.903
10										0.698	0.728	0.741				1.000
12										0.698	0.731	0.747				1.079
15										0.696	0.734	0.753				1.176
20										0.694	0.737	0.761	0.772			1.301
25										0.691	0.738	0.765	0.779			1.398
30										0.688	0.738	0.768	0.784			1.477
40											0.737	0.772	0.790			1.602
50											0.736	0.774	0.795	0.806		1.699
60											0.734	0.776	0.799	0.810		1.778
80											0.731	0.777	0.803	0.817		1.903
100											0.728	0.777	0.806	0.821		2.000
120											0.725	0.776	0.808	0.824		2.079
150												0.775	0.810	0.828		2.176
200												0.773	0.812	0.833	0.843	2.301
250												0.771	0.813	0.836	0.847	2.398
300												0.769	0.813	0.838	0.850	2.477
400												0.766	0.813	0.841	0.854	2.602
500												0.763	0.813	0.843	0.857	2.699
600													0.812	0.844	0.859	2.778
800													0.811	0.846	0.863	2.903
1000													0.809	0.847	0.865	3.000

n = 4

E	$k_2 =$ 0.150	0.175	0.200	0.225	0.250	0.275	0.300	0.325	0.350	0.400	0.450	0.500	0.550	0.600	0.650	lg E
1.2					0.600	0.618	0.630	0.636	0.640							0.079
1.5					0.599	0.619	0.634	0.643	0.648							0.176
2.0					0.598	0.620	0.637	0.650	0.657	0.665						0.301
2.5					0.596	0.620	0.639	0.654	0.664	0.674						0.398
3.0						0.619	0.640	0.656	0.668	0.681						0.477
4						0.618	0.641	0.659	0.673	0.691	0.700					0.602
5						0.616	0.640	0.661	0.677	0.697	0.709					0.699
6						0.614	0.640	0.661	0.679	0.702	0.715					0.778
8							0.638	0.662	0.681	0.709	0.724	0.734				0.903
10							0.637	0.661	0.683	0.713	0.731	0.742				1.000
12							0.635	0.661	0.683	0.716	0.735	0.747				1.079
15							0.632	0.659	0.683	0.719	0.741	0.754				1.176
20								0.657	0.682	0.721	0.747	0.762	0.772			1.301
25								0.655	0.681	0.723	0.751	0.768	0.779			1.398
30								0.654	0.680	0.724	0.754	0.772	0.784			1.477
40								0.650	0.678	0.725	0.757	0.778	0.791			1.602
50									0.676	0.725	0.760	0.782	0.796	0.806		1.699
60									0.675	0.725	0.761	0.785	0.800	0.810		1.778
80									0.671	0.724	0.763	0.790	0.806	0.817		1.903
100									0.669	0.723	0.764	0.792	0.810	0.821		2.000
120										0.722	0.765	0.795	0.813	0.825		2.079
150										0.720	0.765	0.797	0.817	0.829		2.176
200										0.718	0.765	0.799	0.821	0.834	0.843	2.301
250										0.716	0.765	0.801	0.824	0.838	0.847	2.398
300										0.714	0.765	0.802	0.826	0.840	0.850	2.477
400										0.711	0.764	0.803	0.829	0.844	0.854	2.602
500										0.709	0.762	0.803	0.831	0.847	0.857	2.699
600										0.707	0.761	0.803	0.832	0.849	0.860	2.778
800											0.760	0.804	0.834	0.853	0.864	2.903
1000											0.758	0.803	0.835	0.855	0.866	3.000

Table X continued

n = 5

E \ k_2 =	0.150	0.175	0.200	0.225	0.250	0.275	0.300	0.325	0.350	0.400	0.450	0.500	0.550	0.600	0.650	lg E
1.2				0.593	0.612	0.625	0.633	0.637	0.640							0.079
1.5				0.593	0.613	0.628	0.638	0.645	0.649							0.176
2.0				0.592	0.614	0.631	0.644	0.653	0.659	0.665						0.301
2.5				0.591	0.614	0.633	0.648	0.658	0.666	0.675						0.398
3.0				0.590	0.614	0.634	0.650	0.662	0.671	0.682						0.477
4				0.588	0.613	0.635	0.653	0.667	0.678	0.692	0.700					0.602
5					0.613	0.636	0.655	0.671	0.683	0.699	0.709					0.699
6					0.612	0.636	0.656	0.673	0.686	0.704	0.715					0.778
8					0.610	0.635	0.657	0.676	0.691	0.712	0.725	0.734				0.903
10					0.608	0.634	0.658	0.677	0.693	0.717	0.732	0.742				1.000
12					0.607	0.634	0.658	0.678	0.695	0.721	0.737	0.748				1.079
15					0.605	0.632	0.657	0.679	0.697	0.725	0.743	0.754				1.176
20						0.631	0.657	0.679	0.699	0.730	0.749	0.763	0.772			1.301
25						0.629	0.656	0.680	0.700	0.733	0.754	0.769	0.779			1.398
30						0.627	0.655	0.679	0.701	0.735	0.758	0.773	0.784			1.477
40						0.625	0.653	0.679	0.701	0.738	0.763	0.780	0.791			1.602
50						0.623	0.651	0.678	0.701	0.740	0.766	0.784	0.797	0.806		1.699
60							0.650	0.677	0.701	0.741	0.769	0.788	0.801	0.810		1.778
80							0.648	0.675	0.701	0.742	0.773	0.793	0.807	0.817		1.903
100							0.646	0.674	0.700	0.743	0.775	0.796	0.811	0.821		2.000
120							0.644	0.673	0.699	0.744	0.777	0.799	0.814	0.825		2.079
150							0.642	0.671	0.698	0.744	0.779	0.802	0.818	0.829		2.176
200								0.669	0.696	0.744	0.781	0.806	0.823	0.834	0.843	2.301
250								0.667	0.695	0.744	0.782	0.808	0.826	0.838	0.847	2.398
300								0.665	0.694	0.744	0.783	0.810	0.829	0.841	0.850	2.477
400								0.662	0.691	0.743	0.784	0.813	0.832	0.845	0.854	2.602
500								0.660	0.690	0.742	0.784	0.814	0.835	0.848	0.858	2.699
600									0.688	0.742	0.785	0.816	0.837	0.851	0.860	2.778
800									0.685	0.740	0.785	0.817	0.840	0.854	0.864	2.903
1000									0.683	0.739	0.785	0.819	0.842	0.857	0.867	3.000

n = 6

E \ k_2 =	0.150	0.175	0.200	0.225	0.250	0.275	0.300	0.325	0.350	0.400	0.450	0.500	0.550	0.600	0.650	lg E
1.2			0.582	0.602	0.617	0.628	0.634	0.638	0.640							0.079
1.5			0.582	0.603	0.620	0.632	0.640	0.646	0.649							0.176
2.0			0.581	0.604	0.622	0.637	0.647	0.654	0.659	0.665						0.301
2.5			0.580	0.604	0.624	0.640	0.652	0.661	0.667	0.675						0.398
3.0			0.579	0.604	0.625	0.642	0.655	0.665	0.672	0.682						0.477
4			0.577	0.603	0.626	0.644	0.659	0.671	0.680	0.692	0.700					0.602
5				0.602	0.626	0.646	0.662	0.675	0.685	0.699	0.709					0.699
6				0.602	0.626	0.647	0.664	0.678	0.689	0.705	0.715					0.778
8				0.600	0.626	0.648	0.667	0.682	0.695	0.713	0.725	0.734				0.903
10				0.599	0.625	0.648	0.668	0.685	0.699	0.719	0.732	0.742				1.000
12				0.598	0.624	0.648	0.668	0.687	0.701	0.723	0.737	0.748				1.079
15				0.596	0.624	0.648	0.670	0.689	0.704	0.728	0.743	0.755				1.176
20				0.595	0.622	0.648	0.671	0.691	0.708	0.733	0.751	0.763	0.772			1.301
25					0.621	0.647	0.671	0.692	0.710	0.737	0.756	0.769	0.779			1.398
30					0.620	0.647	0.671	0.693	0.711	0.740	0.760	0.774	0.784			1.477
40					0.618	0.646	0.671	0.693	0.713	0.744	0.765	0.780	0.791			1.602
50					0.617	0.645	0.670	0.694	0.714	0.746	0.769	0.785	0.797	0.806		1.699
60					0.615	0.644	0.670	0.694	0.715	0.748	0.772	0.789	0.801	0.810		1.778
80					0.613	0.642	0.669	0.694	0.716	0.751	0.777	0.794	0.807	0.817		1.903
100						0.640	0.668	0.693	0.716	0.753	0.780	0.798	0.812	0.822		2.000
120						0.639	0.667	0.693	0.716	0.754	0.782	0.801	0.815	0.825		2.079
150						0.638	0.666	0.692	0.716	0.756	0.784	0.805	0.819	0.829		2.176
200						0.635	0.664	0.691	0.716	0.757	0.787	0.809	0.824	0.835	0.843	2.301
250						0.634	0.663	0.690	0.716	0.758	0.789	0.812	0.827	0.838	0.847	2.398
300						0.632	0.662	0.690	0.715	0.758	0.791	0.814	0.830	0.841	0.850	2.477
400							0.660	0.688	0.714	0.759	0.793	0.817	0.834	0.846	0.854	2.602
500							0.658	0.687	0.713	0.759	0.794	0.819	0.837	0.849	0.858	2.699
600							0.657	0.686	0.713	0.759	0.795	0.821	0.839	0.851	0.860	2.778
800							0.654	0.684	0.711	0.759	0.797	0.823	0.842	0.855	0.864	2.903
1000							0.652	0.682	0.710	0.759	0.797	0.825	0.844	0.857	0.867	3.000

Table X continued

$n = 7$

E	$k_2 =$ 0.150	0.175	0.200	0.225	0.250	0.275	0.300	0.325	0.350	0.400	0.450	0.500	0.550	0.600	0.650	lg E
1.2		0.568	0.590	0.608	0.621	0.630	0.635	0.638	0.640							0.079
1.5		0.567	0.590	0.610	0.624	0.635	0.642	0.646	0.649							0.176
2.0		0.566	0.590	0.611	0.628	0.640	0.649	0.655	0.660	0.665						0.301
2.5		0.565	0.590	0.612	0.630	0.644	0.654	0.662	0.667	0.675						0.398
3.0			0.590	0.612	0.631	0.646	0.658	0.667	0.673	0.682						0.477
4			0.589	0.613	0.633	0.650	0.663	0.673	0.681	0.692	0.700					0.602
5			0.588	0.613	0.634	0.652	0.667	0.678	0.687	0.700	0.709					0.699
6			0.587	0.613	0.635	0.654	0.669	0.681	0.691	0.706	0.715					0.778
8			0.586	0.612	0.635	0.656	0.672	0.686	0.697	0.714	0.725	0.734				0.903
10			0.585	0.612	0.636	0.657	0.675	0.690	0.702	0.720	0.732	0.742				1.000
12			0.584	0.611	0.636	0.657	0.676	0.692	0.705	0.724	0.738	0.748				1.079
15			0.583	0.610	0.636	0.658	0.678	0.694	0.708	0.729	0.744	0.755				1.176
20				0.609	0.635	0.659	0.679	0.697	0.712	0.735	0.751	0.763	0.772			1.301
25				0.608	0.635	0.659	0.680	0.699	0.715	0.739	0.757	0.769	0.779			1.398
30				0.607	0.634	0.659	0.681	0.700	0.717	0.743	0.761	0.774	0.784			1.477
40				0.605	0.633	0.659	0.682	0.702	0.720	0.747	0.767	0.781	0.791			1.602
50				0.604	0.632	0.658	0.682	0.703	0.721	0.750	0.771	0.786	0.797	0.806		1.699
60				0.603	0.631	0.658	0.682	0.704	0.723	0.753	0.774	0.789	0.801	0.810		1.778
80				0.601	0.630	0.657	0.682	0.705	0.724	0.756	0.779	0.795	0.807	0.817		1.903
100				0.600	0.629	0.656	0.682	0.705	0.726	0.758	0.782	0.799	0.812	0.822		2.000
120					0.628	0.656	0.682	0.705	0.726	0.760	0.785	0.802	0.815	0.825		2.079
150					0.626	0.655	0.681	0.705	0.727	0.762	0.788	0.806	0.819	0.830		2.176
200					0.625	0.653	0.681	0.705	0.728	0.764	0.791	0.810	0.824	0.835	0.843	2.301
250					0.623	0.652	0.680	0.705	0.728	0.766	0.794	0.813	0.828	0.838	0.847	2.398
300					0.622	0.651	0.679	0.705	0.728	0.767	0.795	0.816	0.831	0.841	0.850	2.477
400					0.620	0.650	0.678	0.704	0.728	0.768	0.798	0.819	0.835	0.846	0.854	2.602
500					0.619	0.648	0.677	0.704	0.728	0.769	0.800	0.822	0.838	0.849	0.858	2.699
600						0.647	0.676	0.703	0.728	0.770	0.801	0.824	0.840	0.851	0.860	2.778
800						0.645	0.675	0.702	0.728	0.770	0.803	0.827	0.843	0.855	0.864	2.903
1000						0.644	0.673	0.701	0.727	0.771	0.804	0.829	0.846	0.858	0.867	3.000

$n = 8$

E	$k_2 =$ 0.150	0.175	0.200	0.225	0.250	0.275	0.300	0.325	0.350	0.400	0.450	0.500	0.550	0.600	0.650	lg E
1.2		0.575	0.596	0.612	0.623	0.631	0.636	0.638	0.640							0.079
1.5		0.575	0.596	0.614	0.627	0.636	0.642	0.646	0.649							0.176
2.0		0.574	0.597	0.616	0.631	0.642	0.650	0.656	0.660	0.665						0.301
2.5		0.573	0.597	0.617	0.634	0.646	0.656	0.663	0.668	0.675						0.398
3.0		0.573	0.597	0.618	0.636	0.649	0.660	0.668	0.674	0.682						0.477
4		0.572	0.597	0.619	0.638	0.653	0.665	0.675	0.682	0.693	0.700					0.602
5		0.571	0.597	0.620	0.640	0.656	0.669	0.680	0.688	0.700	0.709					0.699
6		0.570	0.596	0.620	0.641	0.658	0.672	0.684	0.693	0.706	0.715					0.778
8		0.569	0.596	0.620	0.642	0.661	0.676	0.689	0.699	0.714	0.725	0.734				0.903
10			0.595	0.620	0.643	0.662	0.679	0.693	0.704	0.720	0.732	0.742				1.000
12			0.594	0.620	0.643	0.664	0.681	0.695	0.707	0.725	0.738	0.748				1.079
15			0.594	0.620	0.644	0.665	0.683	0.698	0.711	0.730	0.744	0.755				1.176
20			0.592	0.619	0.644	0.666	0.685	0.702	0.715	0.737	0.752	0.763	0.772			1.301
25			0.591	0.619	0.644	0.667	0.687	0.704	0.719	0.741	0.757	0.769	0.779			1.398
30			0.591	0.618	0.644	0.667	0.688	0.706	0.721	0.744	0.761	0.774	0.784			1.477
40			0.589	0.617	0.644	0.668	0.689	0.708	0.724	0.749	0.768	0.781	0.792			1.602
50			0.588	0.616	0.643	0.668	0.690	0.710	0.726	0.753	0.772	0.786	0.797	0.806		1.699
60			0.587	0.616	0.643	0.668	0.691	0.711	0.728	0.756	0.775	0.790	0.801	0.810		1.778
80				0.614	0.642	0.668	0.691	0.712	0.730	0.759	0.780	0.796	0.807	0.817		1.903
100				0.613	0.641	0.668	0.692	0.713	0.732	0.762	0.784	0.800	0.812	0.822		2.000
120				0.612	0.641	0.667	0.692	0.714	0.733	0.764	0.787	0.803	0.815	0.825		2.079
150				0.611	0.640	0.667	0.692	0.714	0.734	0.766	0.790	0.807	0.820	0.830		2.176
200				0.610	0.639	0.666	0.692	0.715	0.735	0.769	0.794	0.811	0.825	0.835	0.843	2.301
250				0.608	0.638	0.666	0.692	0.715	0.736	0.771	0.796	0.815	0.828	0.839	0.847	2.398
300				0.607	0.637	0.665	0.691	0.715	0.737	0.772	0.798	0.817	0.831	0.842	0.850	2.477
400				0.606	0.635	0.664	0.691	0.716	0.738	0.774	0.801	0.821	0.835	0.846	0.854	2.602
500				0.604	0.634	0.663	0.690	0.715	0.738	0.775	0.803	0.824	0.838	0.849	0.858	2.699
600					0.633	0.662	0.690	0.715	0.738	0.776	0.805	0.826	0.841	0.852	0.860	2.778
800					0.631	0.661	0.689	0.715	0.738	0.778	0.807	0.829	0.844	0.855	0.864	2.903
1000					0.630	0.660	0.688	0.715	0.738	0.779	0.809	0.831	0.847	0.858	0.867	3.000

Table X continued

E	$k_2 =$ 0.150	0.175	0.200	0.225	0.250	0.275	0.300	0.325	0.350	0.400	0.450	0.500	0.550	0.600	0.650	lg E
1.2	0.562	0.585	0.603	0.617	0.626	0.632	0.636	0.639	0.640							0.079
1.5	0.562	0.585	0.604	0.619	0.630	0.638	0.643	0.647	0.649							0.176
2.0	0.561	0.585	0.606	0.623	0.635	0.645	0.652	0.657	0.660	0.665						0.301
2.5	0.560	0.585	0.607	0.625	0.639	0.650	0.658	0.664	0.668	0.675						0.398
3.0	0.560	0.585	0.607	0.626	0.641	0.653	0.662	0.669	0.674	0.682						0.477
4	0.559	0.585	0.608	0.628	0.645	0.658	0.668	0.677	0.683	0.693	0.700		n = 10			0.602
5	0.558	0.584	0.609	0.629	0.647	0.661	0.673	0.682	0.689	0.700	0.709					0.699
6	0.558	0.584	0.609	0.630	0.649	0.664	0.676	0.686	0.694	0.706	0.715					0.778
8		0.584	0.609	0.631	0.651	0.667	0.681	0.692	0.701	0.715	0.725	0.734				0.903
10		0.583	0.609	0.632	0.652	0.670	0.684	0.696	0.706	0.721	0.733	0.742				1.000
12		0.582	0.609	0.633	0.654	0.672	0.687	0.699	0.710	0.726	0.738	0.748				1.079
15		0.582	0.608	0.633	0.655	0.674	0.689	0.703	0.714	0.732	0.745	0.755				1.176
20		0.581	0.608	0.633	0.656	0.676	0.693	0.707	0.719	0.738	0.752	0.763	0.772			1.301
25		0.580	0.608	0.633	0.657	0.677	0.695	0.710	0.723	0.743	0.758	0.769	0.779			1.398
30		0.579	0.607	0.633	0.657	0.678	0.697	0.712	0.726	0.747	0.762	0.774	0.784			1.477
40		0.578	0.606	0.633	0.658	0.680	0.699	0.715	0.730	0.752	0.769	0.781	0.792			1.602
50		0.578	0.606	0.633	0.658	0.681	0.700	0.718	0.732	0.756	0.773	0.786	0.797	0.806		1.699
60		0.577	0.605	0.633	0.658	0.681	0.702	0.719	0.735	0.759	0.777	0.790	0.801	0.810		1.778
80		0.576	0.604	0.632	0.658	0.682	0.703	0.722	0.738	0.763	0.782	0.796	0.808	0.817		1.903
100		0.575	0.604	0.632	0.658	0.683	0.704	0.723	0.740	0.766	0.786	0.801	0.812	0.822		2.000
120			0.603	0.631	0.658	0.683	0.705	0.724	0.741	0.769	0.789	0.804	0.816	0.825		2.079
150			0.602	0.631	0.658	0.683	0.706	0.726	0.743	0.771	0.792	0.808	0.820	0.830		2.176
200			0.601	0.630	0.658	0.683	0.706	0.727	0.745	0.775	0.796	0.813	0.825	0.835	0.843	2.301
250			0.600	0.629	0.657	0.683	0.707	0.728	0.747	0.777	0.799	0.816	0.829	0.839	0.847	2.398
300			0.599	0.628	0.657	0.683	0.707	0.729	0.748	0.779	0.802	0.819	0.832	0.842	0.850	2.477
400			0.598	0.627	0.656	0.683	0.708	0.730	0.749	0.781	0.805	0.823	0.836	0.846	0.854	2.602
500			0.597	0.627	0.656	0.683	0.708	0.730	0.750	0.783	0.808	0.826	0.839	0.849	0.858	2.699
600			0.596	0.626	0.655	0.683	0.708	0.731	0.751	0.785	0.810	0.828	0.841	0.852	0.860	2.778
800			0.595	0.625	0.654	0.682	0.708	0.731	0.752	0.787	0.812	0.831	0.845	0.856	0.864	2.903
1000			0.594	0.624	0.653	0.682	0.708	0.732	0.753	0.788	0.814	0.834	0.848	0.859	0.867	3.000

Table XI. Relation of parameters in the special case $n = 2$

E	k_2	k_3	β	E	k_2	k_3	β	E	k_2	k_3	β
1.2	0.380	0.641	5.000+0	12	0.529	0.753	9.091−2	120	0.637	0.831	8.403−3
1.5	0.395	0.652	2.000+0	15	0.541	0.762	7.143−2	150	0.645	0.837	6.711−3
2.0	0.415	0.667	1.000+0	20	0.557	0.774	5.263−2	200	0.655	0.844	5.025−3
2.5	0.430	0.678	6.667−1	25	0.568	0.782	4.167−2	250	0.662	0.849	4.016−3
3.0	0.442	0.687	5.000−1	30	0.577	0.789	3.448−2	300	0.668	0.852	3.344−3
4.0	0.461	0.701	3.333−1	40	0.591	0.799	2.564−2	400	0.676	0.858	2.506−3
5.0	0.476	0.712	2.500−1	50	0.601	0.806	2.041−2	500	0.683	0.862	2.004−3
6.0	0.487	0.721	2.000−1	60	0.609	0.812	1.695−2	600	0.688	0.866	1.669−3
8.0	0.505	0.735	1.429−1	80	0.621	0.820	1.266−2	800	0.695	0.870	1.252−3
10.0	0.518	0.745	1.111−1	100	0.630	0.826	1.010−2	1000	0.701	0.874	1.001−3

Table XII. Parameters β and γ in dependence on k_2 and E

E	$k_2 =$ 0.150	0.175	0.200	0.225	0.250	0.275	0.300	0.325	0.350	0.400	0.450	0.500	0.550	0.600	0.650	lg E
1.2							6.93+0	6.03+0	5.48+0							0.079
1.5							3.12+0	2.60+0	2.32+0							0.176
2.0							1.98+0	1.46+0	1.27+0	1.05+0						0.301
2.5								1.09+0	9.15−1	7.37−1						0.398
3.0								9.25−1	7.40−1	5.80−1						0.477
4								8.37−1	5.70−1	4.21−1	3.46−1		n = 3			0.602
5									4.91−1	3.41−1	2.74−1					0.699
6									4.54−1	2.92−1	2.31−1		β			0.778
8									4.89−1	2.36−1	1.79−1	1.46−1				0.903
10										2.05−1	1.50−1	1.20−1				1.000
12										1.86−1	1.31−1	1.03−1				1.079
15										1.69−1	1.12−1	8.58−2				1.176
20										1.57−1	9.31−2	6.90−2	5.43−2			1.301
25										1.60−1	8.19−2	5.87−2	4.54−2			1.398
30										1.90−1	7.45−2	5.18−2	3.94−2			1.477
40											6.56−2	4.29−2	3.17−2			1.602
50											6.08−2	3.73−2	2.70−2	2.05−2		1.699
60											5.82−2	3.35−2	2.37−2	1.78−2		1.778
80											5.72−2	2.86−2	1.94−2	1.43−2		1.903
100											6.13−2	2.56−2	1.67−2	1.21−2		2.000
120											7.65−2	2.36−2	1.49−2	1.05−2		2.079
150												2.16−2	1.29−2	8.94−3		2.176
200												1.97−2	1.08−2	7.27−3	5.18−3	2.301
250												1.89−2	9.52−3	6.21−3	4.35−3	2.398
300												1.87−2	8.61−3	5.47−3	3.78−3	2.477
400												1.98−2	7.42−3	4.49−3	3.03−3	2.602
500												2.36−2	6.67−3	3.87−3	2.56−3	2.699
600													6.16−3	3.43−3	2.23−3	2.778
800													5.54−3	2.85−3	1.80−3	2.903
1000													5.19−3	2.48−3	1.53−3	3.000
1.2							1.96+0	4.66+0	1.13+1							0.079
1.5							1.37+0	3.63+0	7.89+0							0.176
2.0							5.43−1	2.65+0	5.71+0	4.00+1						0.301
2.5								1.98+0	4.62+0	2.15+1						0.398
3.0								1.45+0	3.89+0	1.61+1			n = 3			0.477
4								4.54−1	2.89+0	1.18+1	9.75+1		γ			0.602
5									2.15+0	9.82+0	4.64+1					0.699
6									1.52+0	8.60+0	3.39+1					0.778
8									2.56−1	7.03+0	2.49+1	4.08+2				0.903
10										5.95+0	2.10+1	1.22+2				1.000
12										5.11+0	1.87+1	8.24+1				1.079
15										4.08+0	1.66+1	6.18+1				1.176
20										2.71+0	1.43+1	4.89+1	6.24+2			1.301
25										1.56+0	1.27+1	4.30+1	2.56+2			1.398
30										3.82−1	1.14+1	3.95+1	1.84+2			1.477

40							9.43+0	3.53+1	1.36+2			1.602
50							7.83+0	3.26+1	1.17+2	8.21+3		1.699
60							6.48+0	3.05+1	1.08+2	1.15+3		1.778
80							4.24+0	2.75+1	9.73+1	5.61+2		1.903
100							2.42+0	2.51+1	9.17+1	4.34+2		2.000
120							7.39−1	2.30+1	8.80+1	3.80+2		2.079
150								2.04+1	8.42+1	3.39+2		2.176
200								1.66+1	7.97+1	3.10+2	6.23+3	2.301
250								1.35+1	7.63+1	2.96+2	2.75+3	2.398
300								1.08+1	7.34+1	2.89+2	2.04+3	2.477
400								6.43+0	6.83+1	2.81+2	1.57+3	2.602
500								3.03+0	6.38+1	2.78+2	1.41+3	2.699
600									5.96+1	2.76+2	1.33+3	2.778
800									5.21+1	2.74+2	1.26+3	2.903
1000									4.53+1	2.72+2	1.24+3	3.000

n = 4

β

1.2	9.22+0	7.53+0	6.60+0	5.96+0	5.47+0							0.079
1.5	4.22+0	3.31+0	2.84+0	2.54+0	2.31+0							0.176
2.0	2.63+0	1.90+0	1.58+0	1.39+0	1.25+0	1.05+0						0.301
2.5	2.25+0	1.43+0	1.16+0	1.00+0	8.90−1	7.35−1						0.398
3.0		1.20+0	9.48−1	8.05−1	7.09−1	5.78−1						0.477
4		9.90−1	7.34−1	6.07−1	5.24−1	4.17−1	3.46−1					0.602
5		9.07−1	6.28−1	5.05−1	4.29−1	3.35−1	2.74−1					0.699
6		8.97−1	5.65−1	4.43−1	3.71−1	2.84−1	2.30−1					0.778
8			4.98−1	3.72−1	3.03−1	2.25−1	1.78−1	1.46−1				0.903
10			4.69−1	3.32−1	2.64−1	1.90−1	1.48−1	1.20−1				1.000
12			4.60−1	3.07−1	2.38−1	1.67−1	1.28−1	1.02−1				1.079
15			4.78−1	2.84−1	2.13−1	1.45−1	1.09−1	8.55−2				1.176
20				2.65−1	1.88−1	1.22−1	8.89−2	6.85−2	5.43−2			1.301
25				2.60−1	1.74−1	1.07−1	7.66−2	5.81−2	4.54−2			1.398
30				2.63−1	1.65−1	9.75−2	6.81−2	5.09−2	3.94−2			1.477
40				2.98−1	1.56−1	8.49−2	5.71−2	4.17−2	3.16−2			1.602
50					1.52−1	7.69−2	5.02−2	3.59−2	2.68−2	2.05−2		1.699
60					1.53−1	7.14−2	4.53−2	3.18−2	2.35−2	1.78−2		1.778
80					1.62−1	6.44−2	3.88−2	2.65−2	1.91−2	1.42−2		1.903
100					1.86−1	6.01−2	3.47−2	2.31−2	1.64−2	1.20−2		2.000
120						5.73−2	3.17−2	2.07−2	1.45−2	1.05−2		2.079
150						5.46−2	2.86−2	1.81−2	1.24−2	8.90−3		2.176
200						5.24−2	2.53−2	1.54−2	1.03−2	7.21−3	5.18−3	2.301
250						5.18−2	2.31−2	1.36−2	8.90−3	6.14−3	4.35−3	2.398
300						5.22−2	2.16−2	1.23−2	7.92−3	5.39−3	3.78−3	2.477
400						5.48−2	1.96−2	1.06−2	6.60−3	4.40−3	3.03−3	2.602
500						5.97−2	1.83−2	9.46−3	5.75−3	3.76−3	2.55−3	2.699
600						6.73−2	1.75−2	8.65−3	5.14−3	3.32−3	2.22−3	2.778
800							1.65−2	7.56−3	4.33−3	2.72−3	1.79−3	2.903
1000							1.59−2	6.84−3	3.79−3	2.34−3	1.52−3	3.000

Table XII continued

E	$k_2 =$ 0.150	0.175	0.200	0.225	0.250	0.275	0.300	0.325	0.350	0.400	0.450	0.500	0.550	0.600	0.650	lg E
1.2					1.21+0	2.59+0	4.63+0	8.37+0	1.81+1							0.079
1.5					9.53−1	2.28+0	4.11+0	7.10+0	1.33+1							0.176
2.0					6.06−1	1.91+0	3.58+0	6.06+0	1.04+1	6.15+1			n = 4			0.301
2.5					2.79−1	1.63+0	3.22+0	5.47+0	9.06+0	3.40+1						0.398
3.0						1.41+0	2.95+0	5.08+0	8.30+0	2.61+1			γ			0.477
4						1.03+0	2.55+0	4.54+0	7.40+0	2.02+1	1.49+2					0.602
5						7.23−1	2.23+0	4.16+0	6.85+0	1.77+1	7.24+1					0.699
6						4.35−1	1.97+0	3.86+0	6.46+0	1.64+1	5.42+1					0.778
8							1.55+0	3.40+0	5.89+0	1.48+1	4.13+1	6.16+2				0.903
10							1.21+0	3.04+0	5.48+0	1.39+1	3.62+1	1.88+2				1.000
12							9.13−1	2.74+0	5.14+0	1.33+1	3.35+1	1.29+2				1.079
15							5.13−1	2.36+0	4.73+0	1.27+1	3.12+1	9.93+1				1.176
20								1.84+0	4.18+0	1.20+1	2.92+1	8.14+1	9.45+2			1.301
25								1.43+0	3.74+0	1.15+1	2.81+1	7.40+1	3.94+2			1.398
30								1.07+0	3.37+0	1.11+1	2.75+1	7.01+1	2.88+2			1.477
40								4.49−1	2.75+0	1.04+1	2.66+1	6.62+1	2.19+2			1.602
50									2.26+0	9.88+0	2.61+1	6.45+1	1.94+2	1.23+4		1.699
60									1.83+0	9.41+0	2.57+1	6.36+1	1.82+2	1.74+3		1.778
80									1.14+0	8.62+0	2.51+1	6.28+1	1.71+2	8.71+2		1.903
100									5.60−1	7.96+0	2.46+1	6.27+1	1.67+2	6.87+2		2.000
120										7.38+0	2.42+1	6.28+1	1.65+2	6.10+2		2.079
150										6.64+0	2.36+1	6.30+1	1.65+2	5.57+2		2.176
200										5.61+0	2.28+1	6.35+1	1.67+2	5.24+2	9.43+3	2.301
250										4.77+0	2.20+1	6.40+1	1.70+2	5.14+2	4.21+3	2.398
300										4.06+0	2.13+1	6.44+1	1.73+2	5.12+2	3.16+3	2.477
400										2.92+0	2.00+1	6.49+1	1.79+2	5.19+2	2.49+3	2.602
500										2.03+0	1.89+1	6.52+1	1.84+2	5.31+2	2.27+3	2.699
600										1.31+0	1.79+1	6.53+1	1.88+2	5.44+2	2.17+3	2.778
800											1.60+1	6.53+1	1.96+2	5.70+2	2.11+3	2.903
1000											1.45+1	6.50+1	2.03+2	5.95+2	2.12+3	3.000
1.2				1.06+1	8.50+0	7.33+0	6.53+0	5.94+0	5.47+0							0.079
1.5				4.82+0	3.74+0	3.17+0	2.80+0	2.52+0	2.30+0							0.176
2.0				2.91+0	2.15+0	1.78+0	1.54+0	1.37+0	1.24+0	1.05+0						0.301
2.5				2.30+0	1.62+0	1.31+0	1.12+0	9.85−1	8.84−1	7.35−1						0.398
3.0				2.03+0	1.35+0	1.07+0	9.03−1	7.88−1	7.02−1	5.77−1						0.477
4				1.89+0	1.08+0	8.25−1	6.82−1	5.86−1	5.16−1	4.16−1	3.46−1		n = 5			0.602
5					9.48−1	7.01−1	5.68−1	4.82−1	4.20−1	3.33−1	2.74−1					0.699
6					8.73−1	6.25−1	4.98−1	4.18−1	3.61−1	2.82−1	2.30−1		β			0.778
8					8.00−1	5.37−1	4.15−1	3.41−1	2.90−1	2.22−1	1.78−1	1.46−1				0.903
10					7.79−1	4.88−1	3.68−1	2.97−1	2.49−1	1.87−1	1.48−1	1.20−1				1.000
12					7.93−1	4.57−1	3.36−1	2.68−1	2.22−1	1.64−1	1.28−1	1.02−1				1.079
15					9.36−1	4.29−1	3.05−1	2.38−1	1.95−1	1.41−1	1.08−1	8.55−2				1.176
20						4.08−1	2.74−1	2.08−1	1.66−1	1.17−1	8.79−2	6.83−2	5.43−2			1.301
25						4.02−1	2.56−1	1.89−1	1.49−1	1.02−1	7.55−2	5.79−2	4.54−2			1.398
30						4.05−1	2.45−1	1.76−1	1.37−1	9.20−2	6.69−2	5.07−2	3.93−2			1.477

40			4.37−1	2.31−1	1.60−1	1.21−1	7.85−2	5.57−2	4.14−2	3.16−2			1.602
50			5.62−1	2.26−1	1.50−1	1.11−1	6.98−2	4.86−2	3.55−2	2.68−2	2.05−2		1.699
60				2.24−1	1.43−1	1.04−1	6.37−2	4.36−2	3.15−2	2.34−2	1.78−2		1.778
80				2.27−1	1.35−1	9.44−2	5.55−2	3.69−2	2.61−2	1.91−2	1.42−2		1.903
100				2.37−1	1.31−1	8.85−2	5.01−2	3.26−2	2.26−2	1.63−2	1.20−2		2.000
120				2.53−1	1.29−1	8.45−2	4.63−2	2.95−2	2.02−2	1.44−2	1.05−2		2.079
150				2.99−1	1.28−1	8.04−2	4.22−2	2.62−2	1.76−2	1.23−2	8.88−3		2.176
200					1.30−1	7.65−2	3.77−2	2.26−2	1.48−2	1.02−2	7.19−3	5.18−3	2.301
250					1.34−1	7.43−2	3.48−2	2.02−2	1.30−2	8.77−3	6.12−3	4.35−3	2.398
300					1.41−2	7.32−2	3.27−2	1.85−2	1.17−2	7.78−3	5.37−3	3.78−3	2.477
400					1.60−1	7.27−2	2.98−2	1.62−2	9.89−3	6.46−3	4.37−3	3.03−3	2.602
500					1.98−1	7.35−2	2.79−2	1.46−2	8.73−3	5.60−3	3.74−3	2.55−3	2.699
600						7.51−2	2.66−2	1.35−2	7.89−3	4.99−3	3.29−3	2.22−3	2.778
800						7.99−2	2.49−2	1.19−2	6.75−3	4.17−3	2.69−3	1.79−3	2.903
1000						8.65−2	2.38−2	1.09−2	6.00−3	3.63−3	2.31−3	1.51−3	3.000

n = 5

ν

1.2	1.15+0	2.28+0	3.83+0	6.24+0	1.07+1	2.26+1							0.079
1.5	9.96−1	2.10+0	3.55+0	5.70+0	9.29+0	1.68+1							0.176
2.0	7.89−1	1.87+0	3.25+0	5.18+0	8.13+0	1.33+1	7.57+1						0.301
2.5	6.22−1	1.70+0	3.04+0	4.87+0	7.53+0	1.19+1	4.23+1						0.398
3.0	4.76−1	1.56+0	2.88+0	4.64+0	7.15+0	1.10+1	3.28+1						0.477
4	2.12−1	1.33+0	2.63+0	4.33+0	6.68+0	1.01+1	2.57+1	1.83+2					0.602
5		1.15+0	2.44+0	4.11+0	6.37+0	9.60+0	2.28+1	8.97+1					0.699
6		1.00+0	2.28+0	3.93+0	6.16+0	9.27+0	2.13+1	6.75+1					0.778
8		7.52−1	2.03+0	3.66+0	5.84+0	8.83+0	1.97+1	5.21+1	7.54+2				0.903
10		5.48−1	1.82+0	3.45+0	5.61+0	8.55+0	1.89+1	4.61+1	2.31+2				1.000
12		3.68−1	1.65+0	3.27+0	5.42+0	8.34+0	1.84+1	4.31+1	1.60+2				1.079
15		7.21−2	1.44+0	3.04+0	5.19+0	8.10+0	1.79+1	4.06+1	1.24+2				1.176
20			1.15+0	2.74+0	4.89+0	7.80+0	1.75+1	3.87+1	1.03+2	1.16+3			1.301
25			9.20−1	2.50+0	4.64+0	7.57+0	1.73+1	3.78+1	9.43+1	4.86+2			1.398
30			7.27−1	2.30+0	4.43+0	7.37+0	1.71+1	3.74+1	9.00+1	3.56+2			1.477
40			4.01−1	1.96+0	4.09+0	7.05+0	1.69+1	3.72+1	8.62+1	2.74+2			1.602
50			4.93−2	1.69+0	3.80+0	6.78+0	1.68+1	3.72+1	8.49+1	2.44+2	1.51+4		1.699
60				1.47+0	3.56+0	6.54+0	1.67+1	3.74+1	8.46+1	2.30+2	2.14+3		1.778
80				1.11+0	3.16+0	6.14+0	1.65+1	3.78+1	8.52+1	2.19+2	1.08+3		1.903
100				8.18−1	2.83+0	5.81+0	1.64+1	3.82+1	8.63+1	2.16+2	8.54+2		2.000
120				5.76−1	2.55+0	5.52+0	1.62+1	3.86+1	8.76+1	2.15+2	7.62+2		2.079
150				2.50−1	2.21+0	5.14+0	1.60+1	3.91+1	8.96+1	2.17+2	7.01+2		2.176
200					1.76+0	4.63+0	1.57+1	3.98+1	9.27+1	2.23+2	6.65+2	1.16+4	2.301
250					1.40+0	4.21+0	1.54+1	4.04+1	9.55+1	2.30+2	6.57+2	5.19+3	2.398
300					1.11+0	3.85+0	1.52+1	4.09+1	9.81+1	2.36+2	6.59+2	3.90+3	2.477
400					6.49−1	3.26+0	1.46+1	4.16+1	1.02+2	2.48+2	6.75+2	3.10+3	2.602
500					2.63−1	2.80+0	1.42+1	4.21+1	1.06+2	2.59+2	6.96+2	2.84+3	2.699
600						2.42+0	1.37+1	4.24+1	1.09+2	2.69+2	7.18+2	2.73+3	2.778
800						1.82+0	1.30+1	4.29+1	1.14+2	2.87+2	7.62+2	2.68+3	2.903
1000						1.36+0	1.23+1	4.32+1	1.19+2	3.03+2	8.03+2	2.70+3	3.000

Table XII continued

	k_z = 0.150	0.175	0.200	0.225	0.250	0.275	0.300	0.325	0.350	0.400	0.450	0.500	0.550	0.600	0.650	
E																lg E
1.2			1.30+1	9.88+0	8.28+0	7.25+0	6.51+0	5.93+0	5.47+0							0.079
1.5			5.98+0	4.39+0	3.61+0	3.13+0	2.78+0	2.52+0	2.30+0							0.176
2.0			3.66+0	2.55+0	2.04+0	1.74+0	1.53+0	1.37+0	1.24+0	1.05+0						0.301
2.5			2.92+0	1.93+0	1.51+0	1.27+0	1.10+0	9.79−1	8.81−1	7.35−1						0.398
3.0			2.58+0	1.61+0	1.24+0	1.03+0	8.86−1	7.81−1	6.99−1	5.77−1						0.477
4			2.37+0	1.30+0	9.66−1	7.84−1	6.64−1	5.78−1	5.12−1	4.15−1	3.46−1			n = 6		0.602
5				1.14+0	8.23−1	6.56−1	5.50−1	4.74−1	4.16−1	3.33−1	2.74−1					0.699
6				1.05+0	7.36−1	5.78−1	4.78−1	4.09−1	3.56−1	2.82−1	2.30−1			β		0.778
8				9.49−1	6.33−1	4.84−1	3.93−1	3.31−1	2.85−1	2.21−1	1.78−1	1.46−1				0.903
10				9.04−1	5.74−1	4.29−1	3.44−1	2.86−1	2.44−1	1.86−1	1.47−1	1.19−1				1.000
12				8.88−1	5.37−1	3.93−1	3.10−1	2.56−1	2.16−1	1.63−1	1.28−1	1.02−1				1.079
15				8.97−1	5.00−1	3.57−1	2.77−1	2.25−1	1.88−1	1.39−1	1.08−1	8.54−2				1.176
20				1.02+0	4.67−1	3.20−1	2.42−1	1.93−1	1.60−1	1.15−1	8.75−2	6.83−2	5.43−2			1.301
25					4.50−1	2.97−1	2.21−1	1.73−1	1.41−1	1.00−1	7.50−2	5.78−2	4.54−2			1.398
30					4.41−1	2.82−1	2.06−1	1.60−1	1.29−1	9.00−2	6.64−2	5.06−2	3.93−2			1.477
40					4.38−1	2.63−1	1.86−1	1.41−1	1.12−1	7.62−2	5.51−2	4.13−2	3.16−2			1.602
50					4.46−1	2.52−1	1.73−1	1.29−1	1.01−1	6.74−2	4.79−2	3.54−2	2.67−2	2.05−2		1.699
60					4.63−1	2.46−1	1.65−1	1.21−1	9.35−2	6.11−2	4.29−2	3.13−2	2.34−2	1.78−2		1.778
80					5.31−1	2.39−1	1.54−1	1.10−1	8.31−2	5.27−2	3.62−2	2.59−2	1.90−2	1.42−2		1.903
100						2.37−1	1.47−1	1.02−1	7.63−2	4.72−2	3.18−2	2.24−2	1.63−2	1.20−2		2.000
120						2.37−1	1.42−1	9.72−2	7.14−2	4.33−2	2.87−2	2.00−2	1.43−2	1.05−2		2.079
150						2.42−1	1.38−1	9.17−2	6.61−2	3.90−2	2.53−2	1.74−2	1.23−2	8.88−3		2.176
200						2.54−1	1.34−1	8.59−2	6.03−2	3.43−2	2.17−2	1.46−2	1.01−2	7.18−3	5.18−3	2.301
250						2.72−1	1.32−1	8.22−2	5.64−2	3.11−2	1.93−2	1.27−2	8.71−3	6.11−3	4.35−3	2.398
300						2.98−1	1.32−1	7.96−2	5.37−2	2.89−2	1.75−2	1.14−2	7.72−3	5.36−3	3.78−3	2.477
400							1.33−1	7.65−2	4.99−2	2.57−2	1.52−2	9.64−3	6.40−3	4.36−3	3.03−3	2.602
500							1.37−1	7.47−2	4.74−2	2.36−2	1.36−2	8.47−3	5.54−3	3.72−3	2.55−3	2.699
600							1.41−1	7.37−2	4.57−2	2.20−2	1.24−2	7.63−3	4.93−3	3.28−3	2.22−3	2.778
800							1.51−1	7.30−2	4.34−2	1.99−2	1.08−2	6.49−3	4.10−3	2.68−3	1.79−3	2.903
1000							1.65−1	7.33−2	4.20−2	1.84−2	9.77−3	5.73−3	3.57−3	2.30−3	1.51−3	3.000
1.2			8.42−1	1.78+0	2.99+0	4.70+0	7.41+0	1.25+1	2.59+1							0.079
1.5			7.31−1	1.66+0	2.83+0	4.44+0	6.83+0	1.09+1	1.94+1							0.176
2.0			5.83−1	1.51+0	2.65+0	4.17+0	6.32+0	9.64+0	1.55+1	8.64+1						0.301
2.5			4.63−1	1.39+0	2.52+0	3.99+0	6.02+0	9.02+0	1.39+1	4.84+1						0.398
3.0			3.61−1	1.29+0	2.41+0	3.86+0	5.82+0	8.64+0	1.30+1	3.77+1				n = 6		0.477
4			1.81−1	1.13+0	2.24+0	3.66+0	5.56+0	8.19+0	1.21+1	2.98+1	2.08+2					0.602
5				1.00+0	2.11+0	3.52+0	5.39+0	7.93+0	1.16+1	2.66+1	1.03+2			γ		0.699
6				8.96−1	2.00+0	3.41+0	5.26+0	7.76+0	1.13+1	2.49+1	7.75+1					0.778
8				7.26−1	1.82+0	3.22+0	5.06+0	7.53+0	1.09+1	2.32+1	6.01+1	8.58+2				0.903
10				5.91−1	1.68+0	3.08+0	4.92+0	7.38+0	1.07+1	2.24+1	5.35+1	2.64+2				1.000
12				4.76−1	1.56+0	2.96+0	4.80+0	7.26+0	1.06+1	2.20+1	5.01+1	1.83+2				1.079
15				3.30−1	1.41+0	2.81+0	4.66+0	7.13+0	1.05+1	2.17+1	4.75+1	1.42+2				1.176
20				1.02−1	1.22+0	2.61+0	4.47+0	6.96+0	1.03+1	2.14+1	4.56+1	1.19+2	1.32+3			1.301
25					1.06+0	2.44+0	4.31+0	6.82+0	1.02+1	2.14+1	4.49+1	1.09+2	5.55+2			1.398
30					9.34−1	2.30+0	4.17+0	6.71+0	1.02+1	2.14+1	4.47+1	1.05+2	4.08+2			1.477

40				7.28−1	2.08+0	3.95+0	6.52+0	1.00+1	2.16+1	4.49+1	1.01+2	3.14+2			1.602
50				5.65−1	1.90+0	3.77+0	6.37+0	9.95+0	2.17+1	4.53+1	1.00+2	2.82+2	1.71+4		1.699
60				4.28−1	1.75+0	3.61+0	6.23+0	9.87+0	2.19+1	4.58+1	1.00+2	2.67+2	2.44+		1.778
80				1.91−1	1.50+0	3.35+0	6.00+0	9.72+0	2.21+1	4.69+1	1.02+2	2.55+2	1.23+		1.903
100					1.31+0	3.13+0	5.80+0	9.59+0	2.24+1	4.80+1	1.04+2	2.52+2	9.79+		2.000
120					1.15+0	2.95+0	5.62+0	9.47+0	2.26+1	4.89+1	1.06+2	2.52+2	8.76+2		2.079
150					9.53−1	2.73+0	5.40+0	9.30+0	2.28+1	5.02+1	1.09+2	2.56+2	8.08+2		2.176
200					7.00−1	2.42+0	5.08+0	9.05+0	2.31+1	5.20+1	1.14+2	2.64+2	7.69+2	1.31+4	2.301
250					5.02−1	2.18+0	4.82+0	8.84+0	2.34+1	5.36+1	1.18+2	2.74+2	7.63+2	5.92+3	2.398
300					3.35−1	1.97+0	4.59+0	8.64+0	2.35+1	5.49+1	1.22+2	2.83+2	7.68+2	4.46+3	2.477
400						1.65+0	4.21+0	8.29+0	2.38+1	5.71+1	1.29+2	2.99+2	7.90+2	3.55+3	2.602
500						1.40+0	3.91+0	7.99+0	2.39+1	5.89+1	1.35+2	3.15+2	8.18+2	3.26+3	2.699
600						1.20+0	3.65+0	7.73+0	2.40+1	6.04+1	1.41+2	3.29+2	8.47+2	3.14+3	2.778
800						8.83−1	3.23+0	7.28+0	2.41+1	6.29+1	1.50+2	3.54+2	9.04+2	3.09+3	2.903
1000						6.42−1	2.89+0	6.89+0	2.41+1	6.48+1	1.57+2	3.75+2	9.57+2	3.13+3	3.000

n = 7

β

1.2	1.77+1	1 20+1	9.58+0	8.17+0	7.21+0	6.49+0	5.93+0	5.46+0							0.079
1.5	8.40+0	5.39+0	4.22+0	3.55+0	3.10+0	2.77+0	2.51+0	2.30+0							0.176
2.0	5.41+0	3.18+0	2.42+0	2.00+0	1.72+0	1.52+0	1.36+0	1.24+0	1.05+0						0.301
2.5	4.60+0	2.43+0	1.80+0	1.47+0	1.25+0	1.09+0	9.75−1	8.80−1	7.35−1						0.398
3.0		2.06+0	1.49+0	1.20+0	1.01+0	8.78−1	7.77−1	6.97−1	5.76−1						0.477
4		1.68+0	1.17+0	9.21−1	7.64−1	6.56−1	5.74−1	5.10−1	4.15−1	3.46−1					0.602
5		1.50+0	1.01+0	7.77−1	6.36−1	5.40−1	4.69−1	4.14−1	3.32−1	2.74−1					0.699
6		1.39+0	9.06−1	6.87−1	5.57−1	4.69−1	4.04−1	3.54−1	2.81−1	2.30−1					0.778
8		1.28+0	7.87−1	5.81−1	4.62−1	3.83−1	3.26−1	2.83−1	2.21−1	1.77−1	1.46−1				0.903
10		1.24+0	7.19−1	5.18−1	4.06−1	3.32−1	2.80−1	2.41−1	1.85−1	1.47−1	1.19−1				1.000
12		1.23+0	6.75−1	4.77−1	3.68−1	2.99−1	2.50−1	2.13−1	1.62−1	1.27−1	1.02−1				1.079
15		1.28+0	6.33−1	4.35−1	3.30−1	2.65−1	2.19−1	1.85−1	1.39−1	1.08−1	8.54−2				1.176
20			5.93−1	3.92−1	2.91−1	2.29−1	1.87−1	1.56−1	1.14−1	8.73−2	6.82−2	5.43−2			1.301
25			5.72−1	3.66−1	2.66−1	2.06−1	1.66−1	1.38−1	9.93−2	7.47−2	5.77−2	4.54−2			1.398
30			5.61−1	3.48−1	2.49−1	1.91−1	1.52−1	1.25−1	8.89−2	6.61−2	5.05−2	3.93−2			1.477
40			5.56−1	3.26−1	2.26−1	1.70−1	1.33−1	1.08−1	7.51−2	5.48−2	4.12−2	3.16−2			1.602
50			5.62−1	3.12−1	2.12−1	1.56−1	1.21−1	9.69−2	6.62−2	4.76−2	3.53−2	2.67−2	2.05−2		1.699
60			5.76−1	3.04−1	2.02−1	1.47−1	1.12−1	8.91−2	5.99−2	4.25−2	3.12−2	2.34−2	1.78−2		1.778
80			6.26−1	2.94−1	1.88−1	1.33−1	1.00−1	7.84−2	5.14−2	3.58−2	2.58−2	1.90−2	1.42−2		1.903
100			7.43−1	2.89−1	1.79−1	1.25−1	9.25−2	7.13−2	4.58−2	3.14−2	2.23−2	1.62−2	1.20−2		2.000
120				2.88−1	1.74−1	1.19−1	8.68−2	6.62−2	4.18−2	2.83−2	1.99−2	1.43−2	1.05−2		2.079
150				2.89−1	1.68−1	1.12−1	8.06−2	6.07−2	3.75−2	2.49−2	1.72−2	1.23−2	8.87−3		2.176
200				2.95−1	1.62−1	1.05−1	7.37−2	5.45−2	3.27−2	2.12−2	1.44−2	1.01−2	7.18−3	5.18−3	2.301
250				3.05−1	1.59−1	1.00−1	6.91−2	5.03−2	2.95−2	1.88−2	1.26−2	8.68−3	6.10−3	4.35−3	2.398
300				3.18−1	1.57−1	9.69−2	6.58−2	4.73−2	2.72−2	1.71−2	1.13−2	7.69−3	5.35−3	3.78−3	2.477
400				3.51−1	1.56−1	9.26−2	6.12−2	4.31−2	2.40−2	1.47−2	9.51−3	6.36−3	4.35−3	3.03−3	2.602
500				4.04−1	1.57−1	8.99−2	5.81−2	4.02−2	2.18−2	1.31−2	8.34−3	5.50−3	3.72−3	2.55−3	2.699
600					1.58−1	8.82−2	5.59−2	3.81−2	2.02−2	1.19−2	7.50−3	4.89−3	3.27−3	2.22−3	2.778
800					1.63−1	8.62−2	5.29−2	3.52−2	1.80−2	1.04−2	6.36−3	4.07−3	2.67−3	1.79−3	2.903
1000					1.69−1	8.52−2	5.09−2	3.32−2	1.65−2	9.28−3	5.60−3	3.53−3	2.29−3	1.51−3	3.000

Table XII continued

E	$k_2 =$ 0.150	0.175	0.200	0.225	0.250	0.275	0.300	0.325	0.350	0.400	0.450	0.500	0.550	0.600	0.650	lg E
1.2		4.56−1	1.26+0	2.24+0	3.53+0	5.38+0	8.33+0	1.39+1	2.86+1							0.079
1.5		3.67−1	1.17+0	2.14+0	3.39+0	5.12+0	7.73+0	1.22+1	2.14+1							0.176
2.0		2.48−1	1.05+0	2.01+0	3.23+0	4.87+0	7.20+0	1.08+1	1.73+1	9.50+1						0.301
2.5		1.43−1	9.63−1	1.91+0	3.12+0	4.71+0	6.91+0	1.02+1	1.55+1	5.34+1						0.398
3.0			8.88−1	1.83+0	3.03+0	4.60+0	6.73+0	9.80+0	1.46+1	4.16+1				n = 7		0.477
4			7.67−1	1.71+0	2.90+0	4.44+0	6.50+0	9.37+0	1.36+1	3.30+1	2.28+2			γ		0.602
5			6.72−1	1.61+0	2.80+0	4.33+0	6.36+0	9.14+0	1.31+1	2.96+1	1.13+2					0.699
6			5.93−1	1.53+0	2.71+0	4.24+0	6.26+0	9.00+0	1.28+1	2.78+1	8.54+1					0.778
8			4.66−1	1.39+0	2.58+0	4.11+0	6.13+0	8.83+0	1.25+1	2.61+1	6.65+1	9.41+2				0.903
10			3.65−1	1.29+0	2.47+0	4.01+0	6.03+0	8.73+0	1.24+1	2.53+1	5.93+1	2.90+2				1.000
12			2.80−1	1.20+0	2.38+0	3.93+0	5.96+0	8.67+0	1.23+1	2.49+1	5.58+1	2.02+2				1.079
15			1.66−1	1.09+0	2.27+0	3.82+0	5.88+0	8.60+0	1.23+1	2.46+1	5.30+1	1.57+2				1.176
20				9.46−1	2.12+0	3.68+0	5.76+0	8.53+0	1.23+1	2.45+1	5.11+1	1.31+2	1.45+3			1.301
25				8.33−1	2.00+0	3.57+0	5.67+0	8.48+0	1.23+1	2.46+1	5.05+1	1.21+2	6.10+2			1.398
30				7.40−1	1.89+0	3.47+0	5.59+0	8.45+0	1.23+1	2.48+1	5.05+1	1.17+2	4.49+2			1.477
40				5.93−1	1.73+0	3.31+0	5.46+0	8.38+0	1.23+1	2.51+1	5.09+1	1.13+2	3.47+2			1.602
50				4.77−1	1.60+0	3.17+0	5.35+0	8.33+0	1.24+1	2.55+1	5.16+1	1.12+2	3.12+2	1.88+4		1.699
60				3.81−1	1.49+0	3.06+0	5.26+0	8.28+0	1.24+1	2.58+1	5.24+1	1.12+2	2.95+2	2.67+3		1.778
80				2.22−1	1.31+0	2.87+0	5.10+0	8.20+0	1.25+1	2.65+1	5.41+1	1.14+2	2.83+2	1.35+3		1.903
100				7.01−2	1.17+0	2.72+0	4.96+0	8.12+0	1.25+1	2.70+1	5.56+1	1.17+2	2.80+2	1.08+3		2.000
120					1.06+0	2.59+0	4.84+0	8.04+0	1.25+1	2.75+1	5.70+1	1.20+2	2.82+2	9.67+2		2.079
150					9.20−1	2.43+0	4.68+0	7.94+0	1.26+1	2.81+1	5.88+1	1.24+2	2.86+2	8.93+2		2.176
200					7.40−1	2.21+0	4.46+0	7.78+0	1.26+1	2.89+1	6.15+1	1.31+2	2.97+2	8.53+2	1.44+4	2.301
250					6.02−1	2.04+0	4.28+0	7.65+0	1.25+1	2.95+1	6.38+1	1.36+2	3.08+2	8.47+2	6.50+3	2.398
300					4.89−1	1.89+0	4.13+0	7.52+0	1.25+1	3.00+1	6.58+1	1.41+2	3.19+2	8.54+2	4.91+3	2.477
400					3.10−1	1.66+0	3.87+0	7.30+0	1.25+1	3.09+1	6.92+1	1.51+2	3.40+2	8.82+2	3.92+3	2.602
500					1.60−1	1.49+0	3.66+0	7.11+0	1.24+1	3.16+1	7.20+1	1.58+2	3.58+2	9.15+2	3.60+3	2.699
600						1.34+0	3.48+0	6.94+0	1.23+1	3.21+1	7.44+1	1.65+2	3.75+2	9.49+2	3.47+3	2.778
800						1.11+0	3.18+0	6.65+0	1.21+1	3.30+1	7.84+1	1.77+2	4.06+2	1.02+3	3.43+3	2.903
1000						9.36−1	2.95+0	6.40+0	1.20+1	3.36+1	8.17+1	1.87+2	4.32+2	1.08+3	3.48+3	3.000
1.2		1.57+1	1.15+1	9.41+0	8.10+0	7.18+0	6.48+0	5.92+0	5.46+0							0.079
1.5		7.18+0	5.13+0	4.13+0	3.51+0	3.09+0	2.76+0	2.51+0	2.30+0							0.176
2.0		4.34+0	2.98+0	2.35+0	1.97+0	1.71+0	1.51+0	1.36+0	1.24+0	1.05+0						0.301
2.5		3.39+0	2.25+0	1.74+0	1.44+0	1.24+0	1.09+0	9.73−1	8.79−1	7.35−1						0.398
3.0		2.93+0	1.88+0	1.43+0	1.17+0	9.99−1	8.72−1	7.75−1	6.96−1	5.76−1						0.477
4		2.47+0	1.50+0	1.11+0	8.96−1	7.53−1	6.50−1	5.72−1	5.09−1	4.15−1	3.46−1		n = 8			0.602
5		2.27+0	1.30+0	9.48−1	7.52−1	6.25−1	5.35−1	4.66−1	4.12−1	3.32−1	2.74−1					0.699
6		2.18+0	1.18+0	8.45−1	6.62−1	5.45−1	4.63−1	4.01−1	3.53−1	2.81−1	2.30−1		β			0.778
8		2.19+0	1.04+0	7.22−1	5.55−1	4.49−1	3.77−1	3.23−1	2.81−1	2.20−1	1.77−1	1.46−1				0.903
10			9.66−1	6.50−1	4.91−1	3.93−1	3.26−1	2.77−1	2.39−1	1.85−1	1.47−1	1.19−1				1.000
12			9.17−1	6.03−1	4.49−1	3.55−1	2.92−1	2.46−1	2.12−1	1.62−1	1.27−1	1.02−1				1.079
15			8.70−1	5.54−1	4.05−1	3.17−1	2.58−1	2.15−1	1.83−1	1.38−1	1.07−1	8.54−2				1.176
20			8.30−1	5.05−1	3.60−1	2.76−1	2.22−1	1.83−1	1.54−1	1.14−1	8.71−2	6.82−2	5.43−2			1.301
25			8.14−1	4.75−1	3.32−1	2.51−1	1.99−1	1.62−1	1.36−1	9.87−2	7.46−2	5.77−2	4.54−2			1.398
30			8.09−1	4.55−1	3.12−1	2.33−1	1.83−1	1.48−1	1.23−1	8.82−2	6.59−2	5.05−2	3.93−2			1.477

40		8.21−1	4.30−1	2.86−1	2.09−1	1.61−1	1.29−1	1.06−1	7.44−2	5.46−2	4.11−2	3.16−2			1.602
50		8.55−1	4.15−1	2.69−1	1.94−1	1.47−1	1.17−1	9.45−2	6.54−2	4.74−2	3.52−2	2.67−2	2.05−2		1.699
60		9.12−1	4.06−1	2.58−1	1.83−1	1.37−1	1.08−1	8.66−2	5.91−2	4.23−2	3.11−2	2.34−2	1.78−2		1.778
80			3.96−1	2.42−1	1.68−1	1.24−1	9.54−2	7.58−2	5.06−2	3.55−2	2.57−2	1.90−2	1.42−2		1.903
100			3.92−1	2.32−1	1.58−1	1.15−1	8.74−2	6.86−2	4.50−2	3.11−2	2.22−2	1.62−2	1.20−2		2.000
120			3.91−1	2.26−1	1.50−1	1.08−1	8.15−2	6.34−2	4.10−2	2.80−2	1.98−2	1.43−2	1.05−2		2.079
150			3.94−1	2.19−1	1.43−1	1.01−1	7.51−2	5.78−2	3.67−2	2.46−2	1.72−2	1.22−2	8.87−3		2.176
200			4.05−1	2.12−1	1.34−1	9.30−2	6.79−2	5.15−2	3.19−2	2.10−2	1.44−2	1.01−2	7.18−3	5.18−3	2.301
250			4.20−1	2.08−1	1.29−1	8.76−2	6.31−2	4.73−2	2.86−2	1.86−2	1.25−2	8.66−3	6.10−3	4.35−3	2.398
300			4.38−1	2.07−1	1.25−1	8.37−2	5.95−2	4.41−2	2.63−2	1.68−2	1.12−2	7.67−3	5.35−3	3.78−3	2.477
400			4.88−1	2.05−1	1.20−1	7.82−2	5.46−2	3.98−2	2.31−2	1.44−2	9.43−3	6.34−3	4.35−3	3.03−3	2.602
500			5.76−1	2.06−1	1.17−1	7.46−2	5.12−2	3.68−2	2.09−2	1.28−2	8.26−3	5.48−3	3.71−3	2.55−3	2.699
600				2.08−1	1.14−1	7.19−2	4.87−2	3.46−2	1.93−2	1.17−2	7.42−3	4.87−3	3.26−3	2.22−3	2.778
800				2.13−1	1.12−1	6.82−2	4.52−2	3.15−2	1.71−2	1.01−2	6.28−3	4.05−3	2.67−3	1.79−3	2.903
1000				2.19−1	1.10−1	6.58−2	4.28−2	2.94−2	1.55−2	9.01−3	5.52−3	3.51−3	2.28−3	1.51−3	3.000

n = 8

γ

1.2	7.58−1	1.57+0	2.60+0	3.97+0	5.94+0	9.09+0	1.51+1	3.09+1							0.079
1.5	6.87−1	1.50+0	2.51+0	3.83+0	5.68+0	8.46+0	1.32+1	2.32+1							0.176
2.0	5.95−1	1.40+0	2.40+0	3.69+0	5.43+0	7.93+0	1.18+1	1.87+1	1.02+2						0.301
2.5	5.23−1	1.33+0	2.32+0	3.60+0	5.29+0	7.64+0	1.11+1	1.69+1	5.75+1						0.398
3.0	4.64−1	1.26+0	2.26+0	3.52+0	5.19+0	7.47+0	1.08+1	1.59+1	4.49+1						0.477
4	3.68−1	1.16+0	2.15+0	3.42+0	5.06+0	7.26+0	1.03+1	1.49+1	3.57+1	2.45+2					0.602
5	2.93−1	1.08+0	2.07+0	3.34+0	4.98+0	7.15+0	1.01+1	1.44+1	3.21+1	1.21+2					0.699
6	2.29−1	1.02+0	2.01+0	3.27+0	4.91+0	7.07+0	1.00+1	1.41+1	3.02+1	9.20+1					0.778
8	1.18−1	9.14−1	1.90+0	3.17+0	4.82+0	6.98+0	9.88+0	1.39+1	2.84+1	7.18+1	1.01+3				0.903
10		8.32−1	1.81+0	3.09+0	4.75+0	6.93+0	9.82+0	1.38+1	2.76+1	6.42+1	3.12+2				1.000
12		7.64−1	1.74+0	3.02+0	4.70+0	6.89+0	9.80+0	1.37+1	2.72+1	6.04+1	2.17+2				1.079
15		6.80−1	1.65+0	2.94+0	4.63+0	6.85+0	9.79+0	1.38+1	2.70+1	5.76+1	1.69+2				1.176
20		5.72−1	1.53+0	2.82+0	4.54+0	6.80+0	9.80+0	1.38+1	2.70+1	5.57+1	1.42+2	1.55+3			1.301
25		4.88−1	1.44+0	2.73+0	4.46+0	6.76+0	9.82+0	1.39+1	2.72+1	5.52+1	1.31+2	6.56+2			1.398
30		4.18−1	1.36+0	2.65+0	4.40+0	6.73+0	9.84+0	1.40+1	2.75+1	5.52+1	1.26+2	4.83+2			1.477
40		3.08−1	1.23+0	2.52+0	4.29+0	6.68+0	9.88+0	1.42+1	2.81+1	5.59+1	1.22+2	3.74+2			1.602
50		2.20−1	1.13+0	2.42+0	4.20+0	6.63+0	9.91+0	1.43+1	2.86+1	5.68+1	1.22+2	3.36+2	2.02+4		1.699
60		1.42−1	1.05+0	2.33+0	4.13+0	6.59+0	9.94+0	1.45+1	2.91+1	5.79+1	1.22+2	3.19+2	2.87+3		1.778
80			9.21−1	2.18+0	3.99+0	6.52+0	9.98+0	1.47+1	3.00+1	5.99+1	1.25+2	3.06+2	1.46+3		1.903
100			8.20−1	2.07+0	3.88+0	6.45+0	1.00+1	1.49+1	3.08+1	6.18+1	1.28+2	3.04+2	1.16+3		2.000
120			7.37−1	1.97+0	3.79+0	6.39+0	1.00+1	1.50+1	3.14+1	6.35+1	1.32+2	3.06+2	1.04+3		2.079
150			6.36−1	1.85+0	3.67+0	6.30+0	1.00+1	1.52+1	3.23+1	6.59+1	1.37+2	3.12+2	9.64+2		2.176
200			5.08−1	1.69+0	3.50+0	6.18+0	1.00+1	1.54+1	3.35+1	6.92+1	1.44+2	3.24+2	9.22+2	1.55+4	2.301
250			4.08−1	1.56+0	3.36+0	6.07+0	1.00+1	1.56+1	3.45+1	7.21+1	1.51+2	3.37+2	9.17+2	6.98+3	2.398
300			3.27−1	1.46+0	3.25+0	5.97+0	9.97+0	1.57+1	3.53+1	7.46+1	1.57+2	3.49+2	9.26+2	5.28+3	2.477
400			1.96−1	1.29+0	3.06+0	5.81+0	9.92+0	1.59+1	3.67+1	7.90+1	1.68+2	3.73+2	9.57+2	4.22+3	2.602
500			7.63−2	1.16+0	2.90+0	5.66+0	9.86+0	1.60+1	3.78+1	8.26+1	1.77+2	3.94+2	9.95+2	3.88+3	2.699
600				1.06+0	2.77+0	5.54+0	9.80+0	1.61+1	3.88+1	8.57+1	1.85+2	4.14+2	1.03+3	3.75+3	2.778
800				8.93−1	2.56+0	5.33+0	9.67+0	1.63+1	4.03+1	9.10+1	2.00+2	4.48+2	1.11+3	3.70+3	2.903
1000				7.68−1	2.39+0	5.15+0	9.56+0	1.63+1	4.15+1	9.55+1	2.12+2	4.79+2	1.18+3	3.76+3	3.000

Table XII continued

n = 10

β

E	k_2 = 0.150	0.175	0.200	0.225	0.250	0.275	0.300	0.325	0.350	0.400	0.450	0.500	0.550	0.600	0.650	lg E
1.2	2.14+1	1.41+1	1.10+1	9.23+0	8.03+0	7.15+0	6.47+0	5.92+0	5.46+0							0.079
1.5	9.92+0	6.35+0	4.87+0	4.03+0	3.47+0	3.07+0	2.76+0	2.51+0	2.30+0							0.176
2.0	6.10+0	3.73+0	2.79+0	2.27+0	1.94+0	1.69+0	1.51+0	1.36+0	1.24+0	1.05+0						0.301
2.5	4.83+0	2.83+0	2.09+0	1.68+0	1.41+0	1.22+0	1.08+0	9.70−1	8.78−1	7.34−1						0.398
3.0	4.21+0	2.38+0	1.72+0	1.37+0	1.15+0	9.86−1	8.66−1	7.72−1	6.95−1	5.76−1						0.477
4	3.63+0	1.91+0	1.35+0	1.06+0	8.69−1	7.40−1	6.44−1	5.68−1	5.08−1	4.15−1	3.46−1					0.602
5	3.40+0	1.67+0	1.16+0	8.90−1	7.26−1	6.12−1	5.28−1	4.63−1	4.11−1	3.32−1	2.74−1					0.699
6	3.34+0	1.52+0	1.03+0	7.87−1	6.36−1	5.32−1	4.56−1	3.97−1	3.51−1	2.80−1	2.30−1					0.778
8		1.35+0	8.89−1	6.63−1	5.28−1	4.36−1	3.70−1	3.19−1	2.79−1	2.20−1	1.77−1	1.46−1				0.903
10		1.25+0	8.03−1	5.90−1	4.64−1	3.79−1	3.19−1	2.73−1	2.37−1	1.84−1	1.47−1	1.19−1				1.000
12		1.19+0	7.46−1	5.40−1	4.20−1	3.41−1	2.85−1	2.42−1	2.09−1	1.61−1	1.27−1	1.02−1				1.079
15		1.12+0	6.87−1	4.90−1	3.76−1	3.02−1	2.50−1	2.11−1	1.81−1	1.37−1	1.07−1	8.53−2				1.176
20		1.07+0	6.27−1	4.37−1	3.30−1	2.62−1	2.14−1	1.79−1	1.52−1	1.13−1	8.69−2	6.82−2	5.43−2			1.301
25		1.04+0	5.90−1	4.04−1	3.01−1	2.36−1	1.91−1	1.58−1	1.33−1	9.79−2	7.44−2	5.76−2	4.54−2			1.398
30		1.02+0	5.64−1	3.80−1	2.80−1	2.17−1	1.75−1	1.44−1	1.20−1	8.74−2	6.57−2	5.04−2	3.93−2			1.477
40		1.02+0	5.31−1	3.49−1	2.52−1	1.93−1	1.53−1	1.24−1	1.03−1	7.35−2	5.43−2	4.11−2	3.16−2			1.602
50		1.03+0	5.10−1	3.28−1	2.34−1	1.77−1	1.39−1	1.12−1	9.18−2	6.46−2	4.71−2	3.52−2	2.67−2	2.05−2		1.699
60		1.05+0	4.96−1	3.14−1	2.21−1	1.65−1	1.28−1	1.03−1	8.38−2	5.82−2	4.20−2	3.11−2	2.34−2	1.78−2		1.778
80		1.13+0	4.79−1	2.94−1	2.03−1	1.49−1	1.14−1	9.03−2	7.29−2	4.97−2	3.52−2	2.56−2	1.90−2	1.42−2		1.903
100		1.29+0	4.69−1	2.81−1	1.91−1	1.38−1	1.05−1	8.21−2	6.57−2	4.41−2	3.08−2	2.21−2	1.62−2	1.20−2		2.000
120			4.63−1	2.72−1	1.82−1	1.31−1	9.82−2	7.62−2	6.05−2	4.01−2	2.77−2	1.97−2	1.43−2	1.05−2		2.079
150			4.59−1	2.62−1	1.72−1	1.22−1	9.07−2	6.96−2	5.48−2	3.57−2	2.43−2	1.71−2	1.22−2	8.87−3		2.176
200			4.58−1	2.52−1	1.62−1	1.12−1	8.22−2	6.23−2	4.84−2	3.09−2	2.07−2	1.43−2	1.00−2	7.17−3	5.18−3	2.301
250			4.60−1	2.46−1	1.55−1	1.06−1	7.65−2	5.73−2	4.41−2	2.77−2	1.82−2	1.24−2	8.64−3	6.10−3	4.35−3	2.398
300			4.65−1	2.41−1	1.49−1	1.01−1	7.22−2	5.37−2	4.10−2	2.53−2	1.65−2	1.11−2	7.64−3	5.34−3	3.78−3	2.477
400			4.78−1	2.36−1	1.42−1	9.43−2	6.63−2	4.85−2	3.66−2	2.21−2	1.41−2	9.33−3	6.32−3	4.34−3	3.02−3	2.602
500			4.93−1	2.33−1	1.37−1	8.96−2	6.22−2	4.50−2	3.35−2	1.99−2	1.25−2	8.17−3	5.46−3	3.71−3	2.55−3	2.699
600			5.10−1	2.31−1	1.34−1	8.62−2	5.91−2	4.23−2	3.13−2	1.83−2	1.14−2	7.33−3	4.84−3	3.26−3	2.22−3	2.778
800			5.50−1	2.30−1	1.29−1	8.14−2	5.48−2	3.86−2	2.82−2	1.61−2	9.78−3	6.19−3	4.02−3	2.66−3	1.78−3	2.903
1000			6.02−1	2.30−1	1.26−1	7.81−2	5.18−2	3.61−2	2.60−2	1.46−2	8.71−3	5.43−3	3.49−3	2.28−3	1.51−3	3.000

Table XII continued

n = 10 γ

E	$k_2 =$ 0.150	0.175	0.200	0.225	0.250	0.275	0.300	0.325	0.350	0.400	0.450	0.500	0.550	0.600	0.650	lg E
1.2	4.74−1	1.18+0	2.04+0	3.15+0	4.64+0	6.81+0	1.03+1	1.69+1	3.44+1							0.079
1.5	4.24−1	1.12+0	1.99+0	3.08+0	4.53+0	6.55+0	9.62+0	1.49+1	2.59+1							0.176
2.0	3.59−1	1.05+0	1.92+0	3.00+0	4.41+0	6.32+0	9.07+0	1.34+1	2.10+1	1.13+2						0.301
2.5	3.09−1	1.00+0	1.86+0	2.94+0	4.34+0	6.19+0	8.79+0	1.27+1	1.90+1	6.40+1						0.398
3.0	2.67−1	9.55−1	1.81+0	2.90+0	4.28+0	6.12+0	8.63+0	1.23+1	1.80+1	5.02+1						0.477
4	2.01−1	8.83−1	1.74+0	2.83+0	4.22+0	6.03+0	8.46+0	1.19+1	1.69+1	4.00+1	2.73+2					0.602
5	1.48−1	8.27−1	1.68+0	2.77+0	4.17+0	5.98+0	8.38+0	1.17+1	1.64+1	3.60+1	1.35+2					0.699
6	1.01−1	7.80−1	1.64+0	2.73+0	4.13+0	5.95+0	8.34+0	1.16+1	1.62+1	3.40+1	1.03+2					0.778
8		7.07−1	1.56+0	2.66+0	4.08+0	5.91+0	8.31+0	1.15+1	1.60+1	3.21+1	8.03+1	1.12+3				0.903
10		6.50−1	1.50+0	2.61+0	4.04+0	5.89+0	8.32+0	1.15+1	1.59+1	3.13+1	7.19+1	3.47+2				1.000
12		6.03−1	1.45+0	2.56+0	4.01+0	5.88+0	8.33+0	1.16+1	1.60+1	3.10+1	6.79+1	2.42+2				1.079
15		5.46−1	1.38+0	2.50+0	3.96+0	5.87+0	8.36+0	1.16+1	1.61+1	3.08+1	6.48+1	1.89+2				1.176
20		4.72−1	1.30+0	2.42+0	3.91+0	5.86+0	8.41+0	1.18+1	1.63+1	3.10+1	6.29+1	1.59+2	1.73+3			1.301
25		4.15−1	1.23+0	2.36+0	3.86+0	5.84+0	8.45+0	1.19+1	1.65+1	3.14+1	6.25+1	1.47+2	7.29+2			1.398
30		3.69−1	1.18+0	2.30+0	3.82+0	5.84+0	8.49+0	1.20+1	1.67+1	3.18+1	6.27+1	1.42+2	5.38+2			1.477
40		2.97−1	1.09+0	2.21+0	3.75+0	5.82+0	8.56+0	1.22+1	1.71+1	3.27+1	6.38+1	1.38+2	4.17+2			1.602
50		2.40−1	1.02+0	2.14+0	3.70+0	5.80+0	8.61+0	1.24+1	1.74+1	3.35+1	6.51+1	1.38+2	3.76+2	2.24+4		1.699
60		1.94−1	9.65−1	2.08+0	3.65+0	5.78+0	8.66+0	1.25+1	1.77+1	3.42+1	6.65+1	1.39+2	3.57+2	3.19+3		1.778
80		1.17−1	8.75−1	1.98+0	3.56+0	5.75+0	8.73+0	1.27+1	1.81+1	3.55+1	6.91+1	1.42+2	3.44+2	1.62+3		1.903
100		4.30−2	8.06−1	1.90+0	3.49+0	5.72+0	8.78+0	1.29+1	1.85+1	3.66+1	7.16+1	1.46+2	3.42+2	1.29+3		2.000
120			7.50−1	1.84+0	3.43+0	5.69+0	8.81+0	1.31+1	1.89+1	3.76+1	7.39+1	1.50+2	3.45+2	1.16+3		2.079
150			6.81−1	1.75+0	3.36+0	5.65+0	8.85+0	1.33+1	1.93+1	3.89+1	7.69+1	1.57+2	3.52+2	1.08+3		2.176
200			5.93−1	1.64+0	3.25+0	5.59+0	8.90+0	1.35+1	1.99+1	4.07+1	8.14+1	1.66+2	3.67+2	1.03+3	1.72+4	2.301
250			5.26−1	1.56+0	3.16+0	5.53+0	8.93+0	1.37+1	2.03+1	4.23+1	8.52+1	1.74+2	3.83+2	1.03+3	7.76+3	2.398
300			4.72−1	1.49+0	3.09+0	5.48+0	8.95+0	1.39+1	2.07+1	4.36+1	8.85+1	1.82+2	3.98+2	1.04+3	5.87+3	2.477
400			3.88−1	1.37+0	2.97+0	5.40+0	8.96+0	1.41+1	2.13+1	4.58+1	9.43+1	1.95+2	4.26+2	1.08+3	4.70+3	2.602
500			3.23−1	1.29+0	2.87+0	5.32+0	8.97+0	1.43+1	2.18+1	4.76+1	9.92+1	2.07+2	4.51+2	1.12+3	4.33+3	2.699
600			2.71−1	1.21+0	2.78+0	5.25+0	8.96+0	1.44+1	2.22+1	4.91+1	1.04+2	2.17+2	4.75+2	1.17+3	4.19+3	2.778
800			1.89−1	1.10+0	2.64+0	5.13+0	8.94+0	1.46+1	2.29+1	5.17+1	1.11+2	2.35+2	5.16+2	1.26+3	4.15+3	2.903
1000			1.22−1	1.01+0	2.53+0	5.03+0	8.92+0	1.48+1	2.34+1	5.38+1	1.17+2	2.51+2	5.58+2	1.34+3	4.22+3	3.000

Table XIII. Parameters lg E_1 and Δk_3 in dependence on the expression $(k_2^* - k_2)$ lg E

lg E_1	k_2^* = 0.150	0.175	0.200	0.225	0.250	0.275	0.300	0.325	0.350	0.400	0.450	0.500	0.550	0.600	0.650	Δk_3
	← $(k_2^* - k_2)$ lg E →															
0.02	−0.004	−0.004	−0.003	−0.003	−0.002	−0.002	−0.001	−0.001	−0.000	0.001	0.002	0.003	0.004	0.005	0.006	0.634
0.04	−0.009	−0.008	−0.007	−0.006	−0.005	−0.004	−0.003	−0.002	−0.001	0.001	0.003	0.005	0.007	0.009	0.011	0.637
0.06	−0.014	−0.012	−0.011	−0.009	−0.008	−0.006	−0.005	−0.003	−0.002	0.001	0.004	0.007	0.010	0.013	0.016	0.639
0.08	−0.018	−0.016	−0.014	−0-012	−0.010	−0.008	−0.006	−0.004	−0.002	0.002	0.006	0.010	0.014	0.018	0.022	0.641
0.10	−0.023	−0.021	−0.018	−0.016	−0.013	−0.011	−0.008	−0.006	−0.003	0.002	0.007	0.012	0.017	0.022	0.027	0.643
0.12	−0.028	−0.025	−0.022	−0.019	−0.016	−0.013	−0.010	−0.007	−0.004	0.002	0.008	0.014	0.020	0.026	0.032	0.646
0.14	−0.034	−0.030	−0.027	−0.023	−0.020	−0.016	−0.013	−0.009	−0.006	0.001	0.008	0.015	0.022	0.029	0.036	0.648
0.16	−0.039	−0.035	−0.031	−0.027	−0.023	−0.019	−0.015	−0.011	−0.007	0.001	0.009	0.017	0.025	0.033	0.041	0.650
0.18	−0.044	−0.040	−0.035	−0.031	−0.026	−0.022	−0.017	−0.013	−0.008	0.001	0.010	0.019	0.028	0.037	0.046	0.653
0.20	−0.050	−0.045	−0.040	−0.035	−0.030	−0.025	−0.020	−0.015	−0.010	0.000	0.010	0.020	0.030	0.040	0.050	0.655
0.22	−0.056	−0.050	−0.045	−0.039	−0.034	−0.028	−0.023	−0.017	−0.012	−0.001	0.010	0.021	0.032	0.043	0.054	0.657
0.24	−0.061	−0.055	−0.049	−0.043	−0.037	−0.031	−0.025	−0.019	−0.013	−0.001	0.011	0.023	0.035	0.047	0.059	0.660
0.26	−0.067	−0.061	−0.054	−0.048	−0.041	−0.035	−0.028	−0.022	−0.015	−0.002	0.011	0.024	0.037	0.050	0.063	0.662
0.28	−0.073	−0.066	−0.059	−0.052	−0.045	−0.038	−0.031	−0.024	−0.017	−0.003	0.011	0.025	0.039	0.053	0.067	0.664
0.30	−0.079	−0.072	−0.064	−0.057	−0.049	−0.042	−0.034	−0.027	−0.019	−0.004	0.011	0.026	0.041	0.056	0.071	0.667
0.32	−0.086	−0.078	−0.070	−0.062	−0.054	−0.046	−0.038	−0.030	−0.022	−0.006	0.010	0.026	0.042	0.058	0.074	0.669
0.34	−0.092	−0.084	−0.075	−0.067	−0.058	−0.050	−0.041	−0.033	−0.024	−0.007	0.010	0.027	0.044	0.061	0.078	0.671
0.36	−0.099	−0.090	−0.081	−0.072	−0.063	−0.054	−0.045	−0.036	−0.027	−0.009	0.009	0.027	0.045	0.063	0.081	0.674
0.38	−0.105	−0.096	−0.086	−0.077	−0.067	−0.058	−0.048	−0.039	−0.029	−0.010	0.009	0.028	0.047	0.066	0.085	0.676
0.40	−0.112	−0.102	−0.092	−0.082	−0.072	−0.062	−0.052	−0.042	−0.032	−0.012	0.008	0.028	0.048	0.068	0.088	0.678
0.45	−0.130	−0.118	−0.107	−0.096	−0.085	−0.073	−0.062	−0.051	−0.040	−0.017	0.005	0.028	0.050	0.073	0.095	0.684
0.50	−0.148	−0.135	−0.123	−0.110	−0.098	−0.085	−0.073	−0.060	−0.048	−0.023	0.002	0.027	0.052	0.077	0.102	0.690
0.55	−0.167	−0.153	−0.139	−0.126	−0.112	−0.098	−0.084	−0.071	−0.057	−0.029	−0.002	0.026	0.053	0.081	0.108	0.696
0.60	−0.187	−0.172	−0.157	−0.142	−0.127	−0.112	−0.097	−0.082	−0.067	−0.037	−0.007	0.023	0.053	0.083	0.113	0.701
0.65	−0.207	−0.191	−0.175	−0.158	−0.142	−0.126	−0.110	−0.093	−0.077	−0.045	−0.012	0.020	0.053	0.085	0.118	0.707
0.70	−0.228	−0.211	−0.193	−0.176	−0.158	−0.141	−0.123	−0.106	−0.088	−0.053	−0.018	0.017	0.052	0.087	0.122	0.712
0.75	−0.250	−0.231	−0.212	−0.194	−0.175	−0.156	−0.137	−0.119	−0.100	−0.062	−0.025	0.013	0.050	0.088	0.125	0.718
0.80	−0.272	−0.252	−0.232	−0.212	−0.192	−0.172	−0.152	−0.132	−0.112	−0.072	−0.032	0.008	0.048	0.088	0.128	0.724
0.85	−0.295	−0.274	−0.253	−0.232	−0.210	−0.189	−0.168	−0.147	−0.125	−0.083	−0.040	0.002	0.045	0.087	0.130	0.729
0.90	−0.319	−0.297	−0.274	−0.252	−0.229	−0.207	−0.184	−0.162	−0.139	−0.094	−0.049	−0.004	0.041	0.086	0.131	0.734
0.95	−0.343	−0.320	−0.296	−0.272	−0.248	−0.225	−0.201	−0.177	−0.153	−0.106	−0.058	−0.011	0.037	0.084	0.132	0.739
1.00	−0.368	−0.343	−0.318	−0.293	−0.268	−0.243	−0.218	−0.193	−0.168	−0.118	−0.068	−0.018	0.032	0.082	0.132	0.745
1.05	−0.394	−0.368	−0.341	−0.315	−0.289	−0.263	−0.236	−0.210	−0.184	−0.131	−0.079	−0.026	0.026	0.079	0.131	0.750
1.10	−0.420	−0.392	−0.365	−0.337	−0.310	−0.282	−0.255	−0.227	−0.200	−0.145	−0.090	−0.035	0.020	0.075	0.130	0.755
1.15	−0.446	−0.418	−0.389	−0.360	−0.331	−0.303	−0.274	−0.245	−0.216	−0.159	−0.101	−0.044	0.014	0.071	0.129	0.759
1.20	−0.473	−0.443	−0.413	−0.383	−0.353	−0.323	−0.293	−0.263	−0.233	−0.173	−0.113	−0.053	0.007	0.067	0.127	0.764
1.25	−0.501	−0.470	−0.438	−0.407	−0.376	−0.345	−0.313	−0.282	−0.251	−0.188	−0.126	−0.063	−0.001	0.062	0.124	0.769
1.30	−0.529	−0.496	−0.464	−0.431	−0.399	−0.366	−0.334	−0.301	−0.269	−0.204	−0.139	−0.074	−0.009	0.056	0.121	0.773
1.35	−0.557	−0.523	−0.490	−0.456	−0.422	−0.388	−0.355	−0.321	−0.287	−0.220	−0.152	−0.085	−0.017	0.050	0.118	0.778
1.40	−0.586	−0.551	−0.516	−0.481	−0.446	−0.411	−0.376	−0.341	−0.306	−0.236	−0.166	−0.096	−0.026	0.044	0.114	0.782
1.45	−0.615	−0.579	−0.543	−0.507	−0.470	−0.434	−0.398	−0.362	−0.325	−0.253	−0.180	−0.108	−0.035	0.037	0.110	0.786
1.50	−0.645	−0.607	−0.570	−0.532	−0.495	−0.457	−0.420	−0.382	−0.345	0.270	−0.195	−0.120	−0.045	0.030	0.105	0.791
1.55	−0.675	−0.636	−0.597	−0.559	−0.520	−0.481	−0.442	−0.404	−0.365	−0.287	−0.210	−0.132	−0.055	0.023	0.100	0.795
1.60	−0.705	−0.665	−0.625	−0.585	−0.545	−0.505	−0.465	−0.425	−0.385	−0.305	−0.225	−0.145	−0.065	0.015	0.095	0.799
1.65	−0.736	−0.695	−0.654	−0.612	−0.571	−0.530	−0.489	−0.447	−0.406	−0.324	−0.241	−0.159	−0.076	0.006	0.089	0.802

1.70	−0.767	−0.725	−0.682	−0.640	−0.597	−0.555	−0.512	−0.470	−0.427	−0.342	−0.257	−0.172	−0.087	−0.002	0.083	0.806
1.75	−0.799	−0.755	−0.711	−0.667	−0.624	−0.580	−0.536	−0.492	−0.449	−0.361	−0.274	−0.186	−0.099	−0.011	0.076	0.810
1.80	−0.830	−0.785	−0.740	−0.695	−0.650	−0.605	−0.560	−0.515	−0.470	−0.380	−0.290	−0.200	−0.110	−0.020	0.070	0.813
1.85	−0.862	−0.816	−0.770	−0.723	−0.677	−0.631	−0.585	−0.538	−0.492	−0.400	−0.307	−0.215	−0.122	−0.030	0.063	0.817
1.90	−0.894	−0.847	−0.799	−0.752	−0.704	−0.657	−0.609	−0.562	−0.514	−0.419	−0.324	−0.229	−0.134	−0.039	0.056	0.820
1.95	−0.927	−0.878	−0.829	−0.781	−0.732	−0.683	−0.634	−0.586	−0.537	−0.439	−0.342	−0.244	−0.147	−0.049	0.048	0.823
2.00	−0.960	−0.910	−0.860	−0.810	−0.760	−0.710	−0.660	−0.610	−0.560	−0.460	−0.360	−0.260	−0.160	−0.060	0.040	0.826
2.05	−0.993	−0.941	−0.890	−0.839	−0.788	−0.736	−0.685	−0.634	−0.583	−0.480	−0.378	−0.275	−0.173	−0.070	0.032	0.830
2.10	−1.026	−0.973	−0.921	−0.868	−0.816	−0.763	−0.711	−0.658	−0.606	−0.501	−0.396	−0.291	−0.186	−0.081	0.024	0.832
2.15	−1.059	−1.006	−0.952	−0.898	−0.844	−0.791	−0.737	−0.683	−0.629	−0.522	−0.414	−0.307	−0.199	−0.092	0.016	0.835
2.20	−1.093	−1.038	−0.983	−0.928	−0.873	−0.818	−0.763	−0.708	−0.653	−0.543	−0.433	−0.323	−0.213	−0.103	0.007	0.838
2.25	−1.127	−1.071	−1.014	−0.958	−0.902	−0.846	−0.789	−0.733	−0.677	−0.564	−0.452	−0.339	−0.227	−0.114	−0.002	0.841
2.30	−1.161	−1.103	−1.046	−0.988	−0.931	−0.873	−0.816	−0.758	−0.701	−0.586	−0.471	−0.356	−0.241	−0.126	−0.011	0.844
2.35	−1.195	−1.136	−1.078	−1.019	−0.960	−0.901	−0.843	−0.784	−0.725	−0.608	−0.490	−0.373	−0.255	−0.138	−0.020	0.846
2.40	−1.229	−1.169	−1.109	−1.049	−0.989	−0.929	−0.869	−0.809	−0.749	−0.629	−0.509	−0.389	−0.269	−0.149	−0.029	0.849
2.45	−1.264	−1.203	−1.141	−1.080	−1.019	−0.958	−0.896	−0.835	−0.774	−0.651	−0.529	−0.406	−0.284	−0.161	−0.039	0.851
2.50	−1.299	−1.236	−1.174	−1.111	−1.049	−0.986	−0.924	−0.861	−0.799	−0.674	−0.549	−0.424	−0.299	−0.174	−0.049	0.854
2.55	−1.334	−1.270	−1.206	−1.142	−1.079	−1.015	−0.951	−0.887	−0.824	−0.696	−0.569	−0.441	−0.314	−0.186	−0.059	0.856
2.60	−1.369	−1.304	−1.239	−1.174	−1.109	−1.044	−0.979	−0.914	−0.849	−0.719	−0.589	−0.459	−0.329	−0.199	−0.069	0.858
2.65	−1.404	−1.337	−1.271	−1.205	−1.139	−1.072	−1.006	−0.940	−0.874	−0.741	−0.609	−0.476	−0.344	−0.211	−0.079	0.860

References

ABEL, H., REGEL, K., ROSSBACH, H., EICHHORN, K., and HABERER, G., 1966, Kernenergie, **9**, 57.

AHNSTRÖM, G., and EDVARDSSON, K. A., 1974, Int. J. Radiat. Biol., **26**, 493.

ALEXANDER, P., DEAN, C. J., HAMILTON, L. D. G., LETT, J. T., and PARKINS, G., 1965, M. D. Anderson Hospital, 18th Annual Symposium, Houston, Texas (Austin: The University of Texas Press), page 241.

ALPER, T., 1956, Radiat. Res., **5**, 573.

ALPER, T., 1973, Brit. Med. Bulletin, **29**, 3.

ALPER, T., and BRYANT, P. E., 1974, Int. J. Radiat. Biol., **26**, 203.

ALPER, T., 1979, Cellular Radiobiology (Cambridge: University Press).

ANDERSON, L., 1973, Phys. Med. Biol., **18**, 779.

ANGLETON, G. M., 1967, Health Phys., **13**, 45.

ANTOKU, S., 1975, Int. J. Radiat. Biol., **27**, 287.

AUXIER, J. A., SNYDER, W. S., and JONES, T. D., 1968, Radiation Dosimetry, Vol. I, edited by F. H. Attix and W. C. Roesch (London, New York: Academic Press Inc.), page 275.

BACH, R. L., and CASWELL, R. S., 1968, Radiat. Res., **35**, 1.

BAILY, N. A., STEIGERWALT, J. E., and HILBERT, J. W., 1972, Radiat. Res., **49**, 26.

BALDYTCHEV, A. S., MALINOVSKI, O. V., POSTNIKOV, L. N., and SHEIKINA, T. A., 1973, Radiobiologia, **XIII**, 206, (russ.).

BARENDSEN, G. W., BEUSKER, T. L. J., VERGROESEN, A. J., and BUDKE, L., 1960, Radiat. Res., **13**, 841.

BARENDSEN, G. W., WALTER, H. M. D FOWLER, J. F., and BEWLEY, D. K., 1963 Radiat. Res., **18**, 106.

BARENDSEN, G. W., 1964, Int. J. Radiat. Biol., **8**, 453.

BARENDSEN, G. W., and WALTER, H. M. D., 1964, Radiat. Res., **21**, 314.

BARENDSEN, G. W., 1966, Biophysical Aspects of Radiation Quality, report of a panel (Vienna: I.A.E.A.), page 44.

BARENDSEN, G. W., KOOT, C. J., VAN KERSEN, G. R., BEWLEY, D. K., FIELD, S. B., and PARNELL, C. J., 1966, Int. J. Radiat. Biol., **10**, 317.

BARENDSEN, G. W., 1967, Theoretical and Experimental Biophysics — a Series of Advances, edited by A. Cole (New York: Marcel Dekker), Vol. 1, page 167.

BARENDSEN, G. W., 1968, Biophysical Aspects of Radiation Quality, Second Panel Report (Vienna: I.A.E.A.), page 63.

BARENDSEN, G. W., and BROERSE, J. J., 1969, Europ. J. Cancer, **5**, 373.

BARENDSEN, G. W., 1974a, Europ. J. Cancer, **10**, 269.

BARENDSEN, G. W., 1974b, Fourth Symposium on Microdosimetry, edited by J. Booz, H. G. Ebert, R. Eickel, and A. Waker (Luxembourg: Euratom), page 235.

BARKAS, W. H., 1963, Nuclear Research Emulsion, Vol. I (London, New York: Academic Press Inc.), page 371.

BARSUKOV, V. S., and MALINOVSKY, O. V., 1974, Citologia, **XVI**, 1397, (russ.).

Bateman, J. L., Rossi, H. H., Kellerer, A. M., Robinson, C. V., and Bond, V. P., 1972, Radiat. Res., **51**, 381.

Baverstock, K. F., and Burns, W. G., 1976, Nature, **260**, 316.

Bazazianc, N. O., Zabrodskaia, A. S., Laritcha, A. F., and Nicolaev, M. N., 1972, Jadernye Konstanty, tchast I (Moscow) — punched card set.

Beekman, W. J., 1949, Physica, **15**, 327.

Bender, M. A., and Gooch, P. C., 1962, Int. J. Radiat. Biol., **5**, 133.

Bengtsson, L. G., 1972, Third Symposium on Microdosimetry, edited by H. G. Ebert (Luxembourg: Euratom), page 483.

Bengtsson, L. G., and Lindborg, L., 1974, Fourth Symposium on Microdosimetry, edited by J. Booz, H. G. Ebert, R. Eickel, and A. Waker (Luxembourg: Euratom), page 823.

Berry, R. J., and Andrews, J. R., 1963, Brit. J. Radiol., **36**, 49.

Berry, R. J., Bewley, D. K., and Parnell, C. J., 1965, Brit. J. Radiol., **38**, 613.

Berry, R. J., and Ellis, R. E., 1966, Nature, **211**, 267.

Berry, R. J., 1970, Radiat. Res., **44**, 237.

Berry, R. J., 1971, Europ. J. Cancer, **7**, 145.

Berry, R. J., Oliver, G. D., George, D. L., Werner, J., and Brennan, J. T., 1973, Brit. J. Radiol., **46**, 996.

Berry, R. J., 1974, Biological Effects of Neutron Irradiation, proceedings of a symposium (Vienna: I.A.E.A.), page 257.

Bewley, D. K., and Hornsey, S., 1964, Biological Effects of Neutron and Proton Irradiation, Vol. 2 (Vienna: I.A.E.A.).

Bewley, D. K., 1968, Biophysical Aspects of Radiation Quality, Second Panel Report (Vienna: I.A.E.A.), page 65.

Bewley, D. K., McNally, N. J., and Page, B. C., 1974, Radiat, Res., **58**, 111.

Biavati, M. H., Rosenzweig, W., Rossi, H. H., and Miyanaga, I., 1963, Radiat. Res., **19**, 512.

Biavati, M. H., and Rossi, H. H., 1965, Annual Report on Research Project, AEC, NYO-2740-2, Columbia University.

Biavati, M. H., and Boer, E., 1966, Annual Report on Research Project, AEC, NYO-2740-3, Columbia University, page 87.

Bichsel, H., 1968, Radiation Dosimetry, Vol. I, edited by F. H. Attix and W. C. Roesch (London, New York: Academic Press Inc.), page 157.

Bichsel, H., 1974, Fourth Symposium on Microdosimetry, edited by J. Booz, H. G. Ebert, R. Eickel, and A. Waker (Luxembourg: Euratom), page 1015.

Biermann, A., 1963, Bulletin of Mathem. Bioph., **25**, 273 and 367.

Bird, R. P., and Burki, H. J., 1975, Int. J. Radiat. Biol., **27**, 105.

Boag, J. W., 1954, Radiat. Res., **1**, 323.

Booz, J., 1970, Second Symposium on Microdosimetry, edited by H. G. Ebert (Brussels: Euratom), page 737.

Booz, J., Oldenburg, U., and Coppola, M., 1972a, First Symposium on Neutron Dosimetry in Biology and Medicine, edited by G. Burger and H. Schraube (Luxembourg: Euratom), page 117.

Booz, J., Giglio, C., Waker, A., and Gaggero G., 1972b, Third Symposium on Microdosimetry, edited by H. G. Ebert (Luxembourg: Euratom), page 833.

Booz, J., and Coppola, M., 1974, Fourth Symposium on Microdosimetry, edited by J. Booz, H. G. Ebert, R. Eickel, and A. Waker (Luxembourg: Euratom), page 983.

Booz, J., 1974, Biological Effects of Neutron Irradiation (Vienna: I.A.E.A.), page 119.

Booz, J., 1976, Fifth Symposium on Microdosimetry, edited by J. Booz, H. G. Ebert, and B. G. R. Smith (Luxembourg: Euratom), page 311.

Bopp, A., and Hagen, U., 1970, BBA, **209**, 320.

Braby, L. A., and Ellett, W. H., 1971, Report AD 731709, Oregon State University.

Brenot, J., Chemtob, M., Chmelevsky, D., Fache, P., Parmentier, N., Soulie, R., Biola, M. T., Haag, J., Le Go, R., Bourguignon, M., Courant, D., Dacher, J., and Ducatez, G., 1974, Fourth Symposium on Microdosimetry, edited by J. Booz, H. G. Ebert, R. Eickel, and A. Waker (Luxembourg: Euratom), page 545.

Broerse, J. J., and Barendsen, G. W., 1966, Nature, **212**, 722.

Broerse, J. J., Barendsen, G. W., and Van Kersen, G. R., 1968, Int. J. Radiat. Biol., **13**, 559.

Broerse, J. J., Engels, A. C., Lelieveld, P., Van Putten, L. M., Duncan, W., Greene, D., Massey, J. B., Gilbert, C. W., Hendry, J. H., and Howard, A., 1971, Int. J. Radiat. Biol., **19**, 101.

Broerse, J. J., and Barendsen, G. W., 1972, Radiobiological Applications of Neutron Irradiation, proceedings of a panel (Vienna: I.A.E.A.), page 135.

Broerse, J. J., and Barendsen, G. W., 1973, Curr. Top. Radiat. Res. Quarterly, **8**, 305.

Broerse, J. J., Reinhold, H. S., Buisman, G. H., and Engels, A. C., 1973, Radiat. Res., **56**, 180.

BROERSE, J. J., and VAN OOSTEROM, P. M., 1974, Biological Effects of Neutron Irradiation, proceedings of a symposium (Vienna: I.A.E.A.), page 275.

BROERSE, J. J., BARENDSEN, G. W., GAISER, J. F., and ZOETELIEF, J., 1977, Radiobiological Research and Radiotherapy, Vol. 2, proceedings of a symposium (Vienna: I.A.E.A.), page 19.

BRUCE, W. R., PEARSON, M. L., and FREEDHOFF, H. S., 1963, Radiat. Res., **19**, 606.

BRUNINX, E., and CROMBEEN, J., 1969, Int. J. Appl. Rad. Isot., **20**, 255.

BRUSTAD, T., 1962, Advan. Biol. Med. Phys., **8**, 161.

BRYANT, P. E., 1973, Int. J. Radiat. Biol., **23**, 217.

BULL, S. R., SMITH, L. A., HERBAUT, Y., ROUX, J. C., and LEROUX, J. B., 1976, Fifth Symposium on Microdosimetry, edited by J. Booz, H. G. Ebert, and B. G. R. Smith (Luxembourg: Euratom), page 249.

BURCH, P. R. J., 1955, Radiat. Res., **3**, 361.

BURCH, P. R. J., 1957a, Brit. J. Radiol., **30**, 524.

BURCH, P. R. J., 1957b, Radiat. Res., **6**, 289.

BURCH, P. R. J., 1970, Second Symposium on Microdosimetry, edited by H. G. Ebert (Brussels: Euratom), page 685.

BURGER, G., and BROERSE, J. J., 1974, Biological Effects of Neutron Irradiation, proceedings of a symposium (Vienna: I.A.E.A.), page 3.

BURLIN, T. E., BENSTOCK, D. M. J., and HADDOW, L. M., 1972, Third Symposium on Microdosimetry, edited by H. G. Ebert (Luxembourg: Euratom), page 657.

BURRELL, A. D., FELDSCHREIBER, P., and DEAN, C. J., 1971, BBA, **247**, 38.

CAIRNS, J., 1963, Cold Spring Harb. Symp. Quant. Biol., **28**, 43.

CANADA, T. R., MCDONALD, J. C., FREED, B. R., KUO, T. Y. T., and LAUGHLIN, J. S., 1974, Radiat. Res., **60**, 205.

CASE, K. M., DE HOFFMANN, F., and PLACZEK, G., 1953, Introduction to the Theory of Neutron Diffusion (Los Alamos: U.S. Atomic Energy Comm.), page 17.

CASWELL, R. S., 1966, Radiat. Res., **27**, 92.

CASWELL, R. S., and COYNE, J. J., 1972, Radiat. Res., **52**, 448.

CASWELL, R. S., and COYNE, J. J., 1974, Fourth Symposium on Microdosimetry, edited by J. Booz, H. G. Ebert, R. Eickel, and A. Waker (Luxembourg: Euratom), page 967.

CASWELL, R. S., and COYNE, J. J., 1976, Fifth Symposium on Microdosimetry, edited by J. Booz, H. G. Ebert, and B. G. R. Smith (Luxembourg: Euratom), page 97.

CHADWICK, K. H., and LEENHOUTS, H. P., 1973, Phys. Med. Biol., **18**, 78.

CHAPMAN, J. D., REUVERS, A. P., DOERN, S. D., GILLESPIE, C. J., and DUGLE, D. L., 1976, Fifth Symposium on Microdosimetry, edited by J. Booz, H. G. Ebert, and B. G. R. Smith (Luxembourg: Euratom), page 775.

CHMELEVSKY, D., CHEMTOB, M., and PARMENTIER, N. C., 1972, Third Symposium on Microdosimetry, edited by H. G. Ebert (Luxembourg: Euratom), page 317.

CHMELEVSKY, D., and KELLERER, A. M., 1977, Rad. and Environm. Biophys., **14**, 123.

CHRISTENSEN, R. C., TOBIAS, C. A., and TAYLOR, W. D., 1972, Int. J. Radiat. Biol., **22**, 457.

COHEN, G., and EISENBERG, H., 1968, Biopolymers, **6**, 1077.

COLE, A., 1969, Radiat. Res., **38**, 7.

COLE, A., COOPER, W. G., SHONKA, F., CORRY, P. M., HUMPHREY, R. M., and ANSEVIN, A. T., 1974, Radiat. Res., **60**, 1.

COLE, A., SHONKA, F., CORRY, P. M., and COOPER, W. G., 1975, Molecular Mechanisms for Repair of DNA, edited by R. B. Setlow and P. C. Hanawalt (New York).

COMBECHER, D., 1978, Third Symposium on Neutron Dosimetry in Biology and Medicine, edited by G. Burger and H. G. Ebert (Brussels, Luxembourg: Euratom), page 97.

COMBECHER, D., KOLLERBAUR, J., LEUTHOLD, G., PARETZKE, H. G., and BURGER, G., 1978, Sixth Symposium on Microdosimetry, edited by J. Booz and H. G. Ebert (London: Harwood Academic Publishers Ltd.), page 295.

COPPOLA, M., and BOOZ, J., 1973, Biophysik, **9**, 225.

COPPOLA, M., PIRRWITZ, D., and BOOZ, J., 1974, Fourth Symposium on Microdosimetry, edited by J. Booz, H. G. Ebert, R. Eickel, and A. Waker (Luxembourg: Euratom), page 1001.

COPPOLA, M., and BOOZ, J., 1975, Rad. and Environm. Biophys., **12**, 157.

COPPOLA, M., EICKEL, R., FITZGERALD, M., PIRRWITZ, D., PORRO, F., and BOOZ, J., 1976, Fifth Symposium on Microdosimetry, edited by J. Booz, H. G. Ebert, and B. G. R. Smith (Luxembourg: Euratom), page 377.

CORMACK, D. V., and JOHNS, H. E., 1952, Brit. J. Radiol., **25**, 369.

CORMACK, D. V., 1968, Biophysical Aspects of Radiation Quality, Second Panel Report (Vienna: I.A.E.A.), page 105.

CORRY, P. M., and Cole, A., 1973, Nature New Biology, **245**, 100.

CRAGGS, J. D., and MASSEY, H. S. W., 1959, Encyclopedia of Physics, Vol XXXVII/1, edited by S. Flügge (Berlin, Göttingen, Heidelberg: Springer), page 314.

VAN DAM, J., BILLIET, G., WAMBERSIE, A., and DUTREIX, A., 1979, High LET Radiations in Clinical Radiotherapy, proceedings of a meeting, edited by G. W. Barendsen, J. J. Broerse, and K. Breur (Europ. J. Cancer suppl.; Oxford, New York, Toronto, Sydney, Paris, Frankfurt: Pergamon Press), page 220.

DATTA, R., COLE, A., and ROBINSON, S., 1976, Radiat. Res., **65**, 139.

DAVIDS, J. A. G., 1972, Radiobiological Applications of Neutron Irradiation, Proceedings of a Panel (Vienna: I.A.E.A.), page 27.

DEERING, R. A., and RICE, R., 1962, Radiat. Res., **17**, 774.

DENEKAMP, J., EMERY, E. W., and FIELD, S. B., 1971, Radiat. Res., **45**, 80.

DENEKAMP, J., and FIELD, S. B., 1974, Europ. J. Cancer, **10**, 241.

DENNIS, J. A., 1972, Third Symposium on Microdosimetry, edited by H. G. Ebert (Luxembourg: Euratom), page 403.

DENNIS, J. A., 1974, Fourth Symposium on Microdosimetry, edited by J. Booz, H. G. Ebert, R. Eickel, and A. Waker (Luxembourg: Euratom), page 980.

DEWEY, W. C., NOEL, J. S., and DETTOR, C. M., 1972, Radiat. Res., **52**, 373.

DICELLO, J. F., 1974, Annual Report on Research Project, C00-3243-3, Columbia University, page 78.

DITTRICH, W., 1961, Z. Naturforschung, **16b**, 398.

DIXON, B., 1968, Thesis, London University.

DOUGLAS, B. G., and FOWLER, J. F., 1976, Radiat. Res., **66**, 401.

DOUTHAT, D. A., 1975, Radiat. Res., **61**, 1.

DUGLE, D. L., GILLESPIE, C. J., and CHAPMAN, J. D., 1976, Proc. Nat. Acad. Sci. USA, **73**, 809.

DUNCAN, W., GREENE, D., HOWARD, A., and MASSEY, J. B., 1969, Int. J. Radiat. Biol., **15**, 397.

DURAND, R. E., and SUTHERLAND, R. M., 1973, Cancer Research, **33**, 213.

DVORAK, R. F., 1975, Annual Report on Research Project, Columbia University.

EDWARDS, A. A., 1978, Third Symposium on Neutron Dosimetry in Biology and Medicine, edited by G. Burger and H. G. Ebert (Brussels, Luxembourg: Euratom), page 535.

EICHHORN, H. J., and LESSEL, A., 1975, Grundlagen der Neutronentherapie, edited by A. Graffi, Th. Matthes, and E. Magdon (Berlin: Akademie-Verlag), page 121.

EICKEL, R., 1975, Thesis, University Saarbrücken.

ELKIND, M. M., 1971, Biophys. J., **11**, 502.

ELKIND, M. M., 1974, Proceedings of the 5th Intern. Congr. Radiat. Res. (Washington), page 1349.

ELKIND, M. M., and BEN-HUR, E., 1974, Fourth Symposium on Microdosimetry, edited by J. Booz, H. G. Ebert, R. Eickel, and A. Waker (Luxembourg: Euratom), page 1.

ELKIND, M. M., 1977, Radiat. Res., **71**, 9.

ELLETT, W. H., and BRABY, L. A., 1972, Third Symposium on Microdosimetry, edited by H. G. Ebert (Luxembourg: Euratom), page 471.

EMERY, E. W., DENEKAMP, J., BALL, M. M., and FIELD, S. B., 1970, Radiat. Res., **41**, 450.

EVANS, R. D., 1958, Encyclopedia of Physics, Vol. XXXIV, edited by S. Flügge (Berlin, Göttingen, Heidelberg: Springer), page 218.

EVANS, R. D., 1968, Radiation Dosimetry, Vol. I, edited by F. H. Attix and W. C. Roesch (London, New York: Academic Press Inc.), page 93.

FANO, U., 1947, Phys. Rev., **72**, 26.

FIELD, S. B., and HORNSEY, S., 1974a, Proceedings of the 5th Intern. Congr. Radiat. Res. (Washington), page 1125.

FIELD, S. B., and HORNSEY, S., 1974b, Europ. J. Cancer, **10**, 621.

FIELDEN, E. M., LILLICRAP, S. C., and ROBINS, A. B., 1971, Radiat. Res., **48**, 421.

FOWLER, J. F., 1966, Biophysical Aspects of Radiation Quality, Report of a Panel (Vienna: I.A.E.A.), page 63.

FOWLER, J. F., 1968, Microdosimetry, Proceedings of a Conference, edited by H. G. Ebert (Brussels: Euratom), page 93.

FOWLER, J. F., 1974, Proceedings of the 5th Intern. Congr. Radiat. Res. (Washington), page 1040.

FOWLER, J. F., DENEKAMP, J., DELAPEYRE, C., HARRIS, S. R., and SHELDON, P. W., 1974, Int. J. Radiat. Biol., **25**, 213.

FOWLER, J. F., 1977, Radiobiological Research and Radiotherapy, Vol. 1, proceedings of a symposium (Vienna: I.A.E.A.), page 157.

FRANKENBERG, D., 1976, Fifth Symposium on Microdosimetry, edited by J. Booz, H. G. Ebert, and B. G. R. Smith (Luxembourg: Euratom), page 833.

FRANKENBERG-SCHWAGER, M., FRANKENBERG, D., BLÖCHER, D., and ADAMCZYK, C., 1979, Int. J. Radiat. Biol., **36**, 261.

FREIFELDER, D., and TRUMBO, B., 1969, Biopolymers, **7**, 681.

FURUNO, I., YADA, Y., MATSUDAIRA, H., and MARUYAMA, T., 1979, Int. J. Radiat. Biol., **36**, 639.

GABBARD, F., BICHSEL, H., and BONNER, T.W., 1959, Nucl. Phys., **14**, 277.

GARCIA, J. D., 1969, Phys. Rev., **177**, 223.

GLASS, W. A., and SAMSKY, D. N., 1967, Radiat. Res., **32**, 138.

GLASS, W. A., 1968, Microdosimetry, Proceedings of a Conference, edited by H. G. Ebert (Brussels: Euratom), page 189.

GLASS, W. A., TOBUREN, L. H., WILSON, W. E., and ROESCH, W. C., 1972, Third Symposium on Microdosimetry, edited by H. G. Ebert (Luxembourg: Euratom), page 71.

GLOCKER, R., 1932, Z. Phys., **77**, 653.

GOODHEAD, D. T., 1977, Int. J. Radiat. Biol., **32**, 43.

GOODMAN, L. J., 1978, Third Symposium on Neutron Dosimetry in Biology and Medicine, edited by G. Burger and H. G. Ebert (Brussels, Luxembourg: Euratom), page 61.

GRAGG, R. L., HUMPHREY, R. M., and MEYN, R. E., 1976, Radiat. Res., **65**, 71.

GRAHN, D., AINSWORTH, E. J., WILLIAMSON, F. S., and FRY, R. J. M., 1972, Radiobiological Applications of Neutron Irradiation, Proceedings of a Panel (Vienna: I.A.E.A.), page 211.

GREEN, A. E. S., and SAWADA, T., 1972, J. Atmosph. Terr. Phys., **34**, 1719.

GREENE, D., and MAJOR, D., 1971, Proceedings of the Second Congress of the European Association of Radiobiology (Amsterdam), page 463.

GROSS, W., BIAVATI, B. J., and ROSSI, H. H., 1970, Second Symposium on Microdosimetry, edited by H. G. Ebert (Brussels: Euratom), page 249.

GROSS, W., and RODGERS, R. C., 1972, Third Symposium on Microdosimetry, edited by H. G. Ebert (Luxembourg: Euratom), page 873.

GUBIN, A. T., 1974, Voprosy microdosimetrii, edited by V. I. Ivanova (Moscow: Atomizdat), page 31, (russ.).

GUBIN, A. T., KOVALEV, E. E., and SAKOVICH, V. A., 1974, Voprosy microdosimetrii, edited by V. I. Ivanova (Moscow: Atomizdat), page 36, (russ.).

GUBIN, A. T., and SAKOVICH, V. A., 1974, Voprosy microdosimetrii, edited by V. I. Ivanova (Moscow: Atomizdat), page 43, (russ.).

GUBIN, A. T., KOVALEV, E. E., and SAKOVICH, V. A., 1976, Radiobiologia, XVI, 942 (russ.); and personal communication.

GUBIN, A. T., KOVALEV, E. E., and SAKOVICH, V. A., 1977, Radiobiologia, **XVII**, 550, (russ.).

GUICHARD, M., GUEULETTE, J., WAMBERSIE, A., and MALAISE, E. P., 1979, High LET Radiations in Clinical Radiotherapy, proceedings of a meeting, edited by G. W. Barendsen, J. J. Broerse, and K. Breur (Europ. J. Cancer suppl.; Oxford, New York, Toronto, Sydney, Paris, Frankfurt: Pergamon Press), page 198.

GÜNTHER, K., and ROSSBACH, H., 1966, studia biophysica, **1**, 25.

GÜNTHER, K., 1968a, Grundzüge einer phänomenologischen Theorie der biologischen Strahlenwirkung (Berlin: Thesis at Deutsche Akademie der Wissenschaften, No. 2/68).

GÜNTHER, K., 1968b, studia biophysica, **8**, 45.

GÜNTHER, K., 1969, studia biophysica, **18**, 253.

GÜNTHER, K., and SCHULZ, W., 1972, studia biophysica, **34**, 165.

GÜNTHER, K., and LEISTNER, W., 1975, Grundlagen der Neutronentherapie, edited by A. Graffi, Th. Matthes, and E. Magdon (Berlin: Akademie-Verlag), page 57.

GÜNTHER, K., 1976, Int. J. Radiat. Biol., **30**, 495.

GÜNTHER, K., SCHULZ, W., and LEISTNER, W., 1977a, studia biophysica, **61**, 163.

GÜNTHER, K., KÖRNER, I. J., and MALZ, W., 1977b, studia biophysica, **64**, 65.

HALL, E. J., GROSS, W., DVORAK, R. F., KELLERER, A. M., and ROSSI, H. H., 1972, Radiat. Res., **52**, 88.

HALL, E. J., ROSSI, H. H., KELLERER, A. M., GOODMAN, L., and MARINO, S., 1973, Radiat. Res., **54**, 431.

HALL, E. J., NOVAK, J. K., KELLERER, A. M., ROSSI, H. H., MARINO, S., and GOODMAN, L. J., 1975a, Radiat. Res., **64**, 245.

HALL, E. J., ROIZIN-TOWLE, L., THEUS, R. B., and AUGUST, L. S., 1975b, Radiology, **117**, 173.

HALL, E. J., GEARD, C. R., POVLAS, S., and ASTOR, M., 1977, Brit. J. Radiol., **50**, 679.

HAMM, R. N., WRIGHT, H. A., RITCHIE, R. H., TURNER, J. E., and TURNER, T. P., 1976, Fifth Symposium on Microdosimetry, edited by J. Booz, H. G. Ebert, and B. G. R. Smith (Luxembourg: Euratom), page 1037.

HAMM, R. N., WRIGHT, H. A., KATZ, R., TURNER, J. E., and RITCHIE, R. H., 1978, Phys. Med. Biol., **23**, 1149.

HANDSCHACK, W., and MALZ, W., 1979, studia biophysica, **76**, 13.

HANDSCHACK, W., and MALZ, W., 1980, studia biophysica, **80**, 141.

HANNAN, W. J., PORTER, D., LAWSON, R. C., and RAILTON, R., 1973, Phys. Med. Biol., **18**, 808.

HARDER, D., 1966, Biophysical Aspects of Radiation Quality, report of a panel (Vienna: I.A.E.A.), page 140.

HARDER, D., and SCHULZ, H. J., 1968, Microdosimetry, proceedings of a conference, edited by H. G. Ebert (Brussels: Euratom), page 385.

HARTWIG, M., 1977, studia biophysica, **61**, 223.

HARTWIG, M., 1978, Acta Biol. Med. Germ., **37**, 421.

HENDRY, J. H., and HOWARD, A., 1971, Int. J. Radiat. Biol., **19**, 51.

HENDRY, J. H., GILBERT, C. W., and HOWARD, A., 1975, Radiat. Res., **61**, 504.

HETHCOTE, H. W., MCLARTY, J. W., and THAMES, H. D., 1976, Radiat. Res., **67**, 387.

HOFER, K. G., KEOUGH, G., and SMITH, J. M., 1977, Curr. Top. Radiat. Res. Quarterly, **12**, 335.

HOGEWEG, B., and BARENDSEN, G. W., 1972, Third Symposium on Microdosimetry, edited by H. G. Ebert (Luxembourg: Euratom), page 857.

HOLT, P. D., 1974, Fourth Symposium on Microdosimetry, edited by J. Booz, H. G. Ebert, R. Eickel, and A. Waker (Luxembourg: Euratom), page 353.

HOPKINS, J. C., 1971, Nucl. Data, **9A**, 137.

HORNSEY, S., HOWARD-FLANDERS, P., and MOORE, D., 1960, Int. J. Radiat. Biol., **2**, 37.

HORNSEY, S., 1970, Radiology, **97**, 649.

HORNSEY, S., ANDREOZZI, U., and MYERS, R., 1977, Radiobiological Research and Radiotherapy, Vol. 2, proceedings of a symposium (Vienna: I.A.E.A.), page 49.

HOWARD-FLANDERS, P., 1958, Advan. Biol. Med. Phys., **6**, 533.

HUG, O., and KELLERER, A. M., 1966, Stochastik der Strahlenwirkung (Berlin, Heidelberg, New York: Springer).

HUTCHINSON, F., and POLLARD, E., 1961, Target Theory and Radiation Effects on Biological Molecules. In: Errera Forrsberg Mechanisms in Radiobiology (London: Academic Press), page 71.

ICRP, 1966, Recommendations of the International Commission on Radiation Protection 1965. ICRP Publication 9 (Oxford: Pergamon Press).

ICRU Report 10b, 1964, Physical Aspects of Irradiation, edited by International Commission on Radiation Units and Measurements (Washington).

ICRU Report 16, 1970, Linear Energy Transfer, edited by International Commission on Radiation Units and Measurements (Washington).

ICRU Report 19, 1971, Radiation Quantities and Units, edited by International Commission on Radiation Units and Measurements (Washington).

ISHIWARI, R., YAMASHITA, S., YUASA, K., and MIYAKE, J., 1956, Phys. Soc. Japan, **11**, 337.

JAIN, V. K., and POHLIT, W., Biophysik, **8**, 254.

JESSE, W. P., 1961, Phys. Rev., **122**, 1195.

KALOS, M., and GOLDSTEIN, H., 1956, U.S. Atomic Energy Commission Report, NDA 12–16 (microcard).

KAMPF, G., REGEL, K., EICHHORN, K., and ABEL, H., 1977a, studia biophysica, **61**, 53.

KAMPF, G., TOLKENDORF, E., REGEL, K., and ABEL, H., 1977b, studia biophysica, **62**, 17.

KAMPF, G., and REGEL, K., 1978, studia biophysica, **73**, 207.

KAMPF, G., 1979, personal communication.

KAPPOS, A., and POHLIT, W., 1972, Int. J. Radiat. Biol., **22**, 51.

KAPULTSEVICH, Y. G., and KABAKOVA, N. M., 1971, Radiobiologica, **XI**, 401, (russ.).

KAPULTSEVICH, Y. G., PARFENOV, E. N., and PETROSIAN, L. P., 1971, Radiobiologia, **XI**, 413, (russ.).

KAPULTSEVICH, Y. G., and PARFENOV, E. N., 1972, Radiobiologia, **XII**, 708, (russ.).

KATZ, R., ACKERSON, B., HOMAYOONFAR, M., and SHARMA, S. C., 1971, Radiat. Res., **47**, 402.

KATZ, R., and SHARMA, S. C., 1973, Nucl. Instrum. Meth., **111**, 93.

KATZ, R., and SHARMA, S. C., 1974a, Phys. Med. Biol. **19**, 413.

KATZ, R., and SHARMA, S. C., 1974b, Int. J. Radiat. Biol., **26**, 143.

KATZ, R., and SHARMA, S. C., 1974c, Fourth Symposium on Microdosimetry, edited by J. Booz, H. G. Ebert, R. Eickel, and A. Waker (Luxembourg: Euratom), page 367.

KATZ, R., 1974, Fourth Symposium on Microdosimetry, edited by J. Booz, H. G. Ebert, R. Eickel, and A. Waker (Luxembourg: Euratom), page 377 (discussion).

KATZ, R., and SHARMA, S. C., 1975, Phys. Med. Biol., **20**, 410.

KATZ, R., FULLERTON, B. G., ROTH, R. A., and SHARMA, S. C., 1976, Health Phys., **30**, 148.

KATZ, R., 1978, Phys. Med. Biol., **23**, 909.

KEBARLE, P., and GODBOLE, E. W., 1962, J. Chem. Phys., **36**, 302.

KEMBER, N. F., 1967, Brit. J. Radiol., **40**, 496.

KELLERER, A. M., and HUG, O., 1963, Biophysik, **1**, 33.

KELLERER, A. M., 1968a, Mikrodosimetrie, Grundlagen einer Theorie der Strahlenqualität, GSF-Bericht B 1.

Kellerer, A. M., 1968b, Biophysical Aspects of Radiation Quality, Second Panel Report (Vienna: I.A.E.A.), page 89.

Kellerer, A. M., 1970, Second Symposium on Microdosimetry, edited by H. G. Ebert (Brussels: Euratom), page 107.

Kellerer, A. M., and Rossi, H. H., 1970, Second Symposium on Microdosimetry, edited by H. G. Ebert (Brussels: Euratom), page 843.

Kellerer, A. M., and Rossi, H. H., 1971, Radiat. Res., **47**, 15.

Kellerer, A. M., 1971a, Radiat. Res., **47**, 359.

Kellerer, A. M., 1971b, Radiat. Res., **47**, 377.

Kellerer, A. M., 1971c, Radiat. Res., **48**, 216.

Kellerer, A. M., and Rossi, H. H., 1972, Curr. Top. Radiat. Res. Quarterly, **8**, 85.

Kellerer, A. M., 1972a, Third Symposium on Microdosimetry, edited by H. G. Ebert (Luxembourg: Euratom), page 807.

Kellerer, A. M., 1972b, Third Symposium on Microdosimetry, edited by H. G. Ebert (Luxembourg: Euratom), page 869.

Kellerer, A. M., and Rossi, H. H., 1974, Fourth Symposium on Microdosimetry, edited by J. Booz, H. G. Ebert, R. Eickel, and A. Waker (Luxembourg: Euratom), page 331.

Kellerer, A. M., and Chmelevsky, D., 1975, Rad. and Environm. Biophys., **12**, 61, 205, and 321.

Kellerer, A. M., 1976, Fifth Symposium on Microdosimetry, edited by J. Booz, H. G. Ebert, and B. G. R. Smith (Luxembourg: Euratom), page 409.

Kellerer, A. M., Hall, E. J., Rossi, H. H., and Teedla, P., 1976, Radiat. Res., **65**, 172.

Kellerer, A. M., 1978, Kerntechnik, **20**, 301.

Kellerer, A. M., and Rossi, H. H., 1978, Radiat. Res., **75**, 471.

Kiefer, L. J., and Dunn, G. H., 1966, Rev. Mod. Phys., **38**, 1.

Kliauga, P., and Rossi, H. H., 1976, Fifth Symposium on Microdosimetry, edited by J. Booz, H. G. Ebert, and B. G. R. Smith (Luxembourg: Euratom), page 127.

Koch, C. J., and Painter, R. B., 1975, Radiat. Res,, **64**, 256.

Kogelnik, H. D., and Withers, H. R., 1978, Radiologia clin., **47**, 362.

Körner, I. J., and Malz, W., 1973, studia biophysica, **41**, **11**.

Körner, I. J., Malz, W., and Günther, K., 1974, studia biophysica, **47**, 127.

Körner, I. J., Günther, K., and Malz, W., 1978, studia biophysica, **70**, 175.

Korogodin, V. I., 1966, Problemy postradiatsionogo vosstanovlenia (Moscow: Atomizdat), (russ.).

Koval, T. M., Hart, R. W., Myser, W. C., and Hink, W. F., 1979, Int. J. Radiat. Biol., **35**, 183.

Krüger, H., and Graul, E. H., 1970, Second Symposium on Microdosimetry, edited by H. G. Ebert (Brussels: Euratom), page 813.

Kunz, W., and Schintlmeister, J., 1965, Tabellen der Atomkerne, Teil II Kernreaktionen, Bd. 1 (Berlin: Akademie-Verlag).

Kunze, P., 1959, Grimsehl, Lehrbuch der Physik, Vol. IV, edited by W. Schallreuter (Leipzig: Teubner), page 61.

Kutcher, G. J., and Green, A. E. S., 1976, Radiat. Res., **67**, 408.

Laird, C. D., 1971, Chromosoma (Berl.), **32**, 378.

Landau, L., 1944, J. Phys. USSR, 8, 201.

De Langguth, E. N., 1968, Biophysik, **5**, 32.

Laurie, J., Orr, J. S., and Foster, C. J., 1972, Brit. J. Radiol., **45**, 362.

Lea, D. E., 1946, Actions of Radiations on Living Cells (Cambridge: The University Press); second edition 1956.

Leenhouts, H. P., and Chadwick, K. H., 1974, Fourth Symposium on Microdosimetry, edited by J. Booz, H. G. Ebert, R. Eickel, and A. Waker (Luxembourg: Euratom), page 381.

Leenhouts, H. P., and Chadwick, K. H., 1976, Fifth Symposium on Microdosimetry, edited by J. Booz, H. G. Ebert, and B. G. R. Smith (Luxembourg: Euratom), page 289.

Lehmann, A. R., and Ormerod, M. G., 1970a, BBA, **204**, 128.

Lehmann, A. R., and Ormerod, M. G., 1970b, BBA, **217**, 268.

Lehmann, A. R., and Stevens, S., 1977, BBA, **474**, 49.

Lennartz, M., Coquerelle, T., and Hagen, U., 1973, Int. J. Radiat. Biol., **24**, 621.

Lennartz, M., Coquerelle, T., Bopp, A., and Hagen, U., 1975, Int. J. Radiat. Biol., **27**, 577.

Lett, J. T., Parkins, G., Alexander, P., and Ormerod, M. G., 1964, Nature, **203**, 593.

Lindhard, L., Scharff, M., and Schiott, E. H., 1963, Mat. Fys. Med. Kong. Danske Vidensk. Selsk., **33**, No. 14.

Littbrand, B., and Révész, L., 1969, Brit. J. Radiol., **42**, 914.

Lücke-Huhle, Ch., 1975, Int. J. Radiat. Biol., **27**, 1.

Lücke-Huhle, Ch., Dertinger, H., Schlag, H., and Weibezahn, K. F., 1977, Radiobiological Research and Radiotherapy, Vol. 1, proceedings of a symposium (Vienna: I.A.E.A.), page 279.

MASUDA, K., 1971, Int. J. Radiat. Biol., **20**, 85.

MASUDA, K., and WITHERS, H. R., 1976, Brit. J. Radiol., **49**, 351.

MCGINNIES, R. T., 1959, National Bureau of Standards Circular 597 (Washington).

MCGRATH, R. A., and WILLIAMS, R. W., 1966, Nature, **212**, 534.

MCNALLY, N. J., and BEWLEY, D. K., 1969, Brit. J. Radiol., **42**, 289.

MENZEL, H. G., and BOOZ, J., 1976, Fifth Symposium on Microdosimetry, edited by J. Booz, H. G. Ebert, and B. G. R. Smith (Luxembourg: Euratom), page 61.

MENZEL, H. G., WAKER, A. J., and HARTMANN, G., 1976, Fifth Symposium on Microdosimetry, edited by J. Booz, H. G. Ebert, and B. G. R. Smith (Luxembourg: Euratom), page 591.

MICHAELS, H. B., and HUNT, J. W., 1978, Radiat. Res., **74**, 23.

MILLAR, B. C., FILDEN, E. M., and MILLAR, J. L., 1978, Int. J. Radiat. Biol., **33**, 599.

MILLER, J. H., and GREEN, A. E. S., 1973, Radiat. Res., **54**, 343.

MODIG, H. G., EDGREN, M., and RÉVÉSZ, L., 1974, Int. J. Radiat. Biol., **26**, 341.

MØLLER, C., 1931, Z. Physik, **70**, 686.

MORTIMER, R., BRUSTAD, T., and CORMACK, D. V., 1965, Radiat. Res., **26**, 465.

MUNSON, R. J., NEARY, G. J., BRIDGES, B. A., and PRESTON, R. J., 1967, Int. J. Radiat. Biol., **13**, 205.

MURTHY, M. S. S., REDDY, N. M. S., RAO, B. S., SUBRAHMANYAM, P., and MADHVANATH, U., 1976, Int. J. Radiat. Biol., **29**, 85.

MYERS, I. T., 1968, Radiation Dosimetry, Vol. I, edited by F. H. Attix and W. C. Roesch (London, New York: Academic Press Inc.), page 317.

NAKAI, S., and MORTIMER, R., 1967, Radiat. Res. Suppl., **7**, 172.

NEARY, G. J., 1965, Int. J. Radiat. Biol., **9**, 477.

NEARY, G. J., PRESTON, R. J., and SAVAGE, J. R. K., 1967, Int. J. Radiat. Biol., **12**, 317.

NEARY, G. J., 1970, Second Symposium on Microdosimetry, edited by H. G. Ebert (Brussels: Euratom), page 51.

NEUFELD, J., and HAMM, R. N., 1976, Fifth Symposium on Microdosimetry, edited by J. Booz, H. G. Ebert, and B. G. R. Smith (Luxembourg: Euratom), page 653.

NIAS, A. H. W., GREENE, D., FOX, M., and THOMAS, R. L., 1967, Int. J. Radiat. Biol., **13**, 449.

NIAS, A. H. W., GREENE, D., and MAJOR, D., 1971, Int. J. Radiat. Biol., **20**, 145.

NORTHCLIFFE, L. C., 1964, Annual Review of Nuclear Science, **13**, 67.

OBATUROV, G. M., 1979, Radiobiologia, **XIX**, 163, (russ.).

ODA, N., NUMAKUNAI, T., and OHTANI, S., 1971, Biophysical Aspects of Radiation Quality (Vienna: I.A.E.A.), page 99.

OLDENBURG, U., and BOOZ, J., 1970, Second Symposium on Microdosimetry, edited by H. G. Ebert (Brussels: Euratom), page 269.

OLDENBURG, U., and BOOZ, J., 1972a, Mass Stopping Power and Pathlength of Neutron Produced Recoils in Tissue and Tissue Equivalent Materials (Ispra: Euratom document EUR 4786e).

OLDENBURG, U., and BOOZ, J., 1972b, Third Symposium on Microdosimetry, edited by H. G. Ebert (Luxembourg: Euratom), page 511.

ÖRE, A., and LARSEN, A., 1964, Radiat. Res., **21**, 331.

ORMEROD, M. G., and LEHMANN, A. R., 1971, BBA, **228**, 331.

PALCIC, B., and SKARSGARD, L. D., 1972, Int. J. Radiat. Biol., **21**, 417.

PARETZKE, H. G., LEUTHOLD, G., BURGER, G., and JACOBI, W., 1974, Fourth Symposium on Microdosimetry, edited by J. Booz, H. G. Ebert, R. Eickel, and A. Waker (Luxembourg: Euratom), page 123.

PARETZKE, H. G., 1974, Fourth Symposium on Microdosimetry, edited by J. Booz, H. G. Ebert, R. Eickel, and A. Waker (Luxembourg: Euratom), page 141.

PARETZKE, H. G., 1978, Sixth Symposium on Microdosimetry, edited by J. Booz and H. G. Ebert (London: Harwood Academic Publishers Ltd.), page 925.

PATAU, J. P., BLANC, D., MATHIEU, J., and MASON, G., 1970, Second Symposium on Microdosimetry, edited by H. G. Ebert (Brussels: Euratom), page 401.

PATTERSON, M. C., ROOZEN, K. J., and SETLOW, R. B., 1973, Int. J. Radiat. Biol., **23**, 495.

PAYNE, M. G., and GARRETT, W. R., 1975, Radiat. Res., **63**, 201.

PETERSON, L. R., 1969, Phys. Rev. A, **187**, 105.

PFISTER, G., HEHN, G., and PRILLINGER, G., 1977, Radiobiological Research and Radiotherapy, Vol. II, proceedings of a symposium (Vienna: I.A.E.A.), page 101.

PHILLIPS, T. L., 1968, Radiology, **91**, 529.

PHILLIPS, T. L., BARSCHALL, H. H., GOLDBERG, E., FU, K., and ROWE, J., 1974, Europ. J. Cancer, **10**, 287.

PINTO, R. T., *et al.*, 1974, Genetics and Biochemistry of the Cell (Leningrad), page 209, (russ.).

POHLIT, W., ILIAKIS, G., and BERTSCHE, U., 1979, High LET Radiations in Clinical Radiotherapy, proceedings of a meeting, edited by G. W. Barendsen, J. J. Broerse, and K. Broer (Europ. J. Cancer suppl.; Oxford, New York, Toronto, Sydney, Paris, Frankfurt: Pergamon Press), page 200.

POLLARD, E. C., GUILD, W. R., HUTCHINSON, F., and SETLOW, R. B., 1955, Progr. Biophys., **5**, 72.

POTTEN, C. S., and CHASE, H. B., 1970, Radiat. Res., **42**, 205.

POWERS, E. L., LYMAN, J. T., and TOBIAS, C. A., 1968, Int. J. Radiat. Biol., **14**, 313.

PUCK, T. T., MARCUS, P. I., and CIERCIURA, S. J., 1956, J. Exptl. Med., **103**, 273.

VAN PUTTEN, L. M., LELIEVELD, P., and BROERSE, J. J., 1970, Proceedings of Symposium on Time and Dose Relationships in Radiation Biology as Applied to Radiotherapy, Brookhaven Report BNL 50203, page 162.

VAN PUTTEN, L. M., LELIEVELD, P., and BROERSE, J. J., 1971, Europ. J. Cancer, **7**, 153.

RAILTON, R., LAWSON, R. C., PORTER, D., and HANNAN, W. J., 1973, Int. J. Radiat. Biol., **23**, 509.

RAILTON, R., PORTER, D., LAWSON, R. C., and HANNAN, W. J., 1974, Int. J. Radiat. Biol., **25**, 121.

RAILTON, R., LAWSON, R. C., and PORTER, D., 1975, Int. J. Radiat. Biol., **27**, 75.

RAJU, M. R., GNANAPURANI, M., STACKLER, B., MARTINS, B. I., MADHVANATH, U., HOWARD, J., LYMAN, J. T., and MORTIMER, R. K., 1971, Radiat. Res., **47**, 635.

RAJU, M. R., and JETT, J. H., 1974, Radiat. Res., **60**, 473.

RAJU, M. R., TOBEY, R. A., JETT, J. H., and WALTERS, R. A., 1975, Radiat. Res., **63**, 422.

RANDOLPH, M. L., 1957, Radiat. Res., **7**, 47.

RAUTH, A. M., and SIMPSON, J. A., 1964, Radiat. Res., **22**, 643.

REGEL, K., 1969, Zur Bestimmung von Qualitätsfaktoren und Dosisäquivalenten in Neutronen-Gamma-Strahlungsfeldern auf der Grundlage des LET (Thesis), (Rossendorf near Dresden: ZfK — 175 report).

REGEL, K., GÜNTHER, K., and KAMPF, G., 1983, Rad. Enivronm. Biophys., **21**, 175.

RITTER, M. A., CLEAVER, J. E., and TOBIAS, C. A., 1977, Nature, **266**, 653.

RODGERS, R., DICELLO, J. F., and GROSS, W., 1972, U.S. Atomic Energy Commission Report C00-3243-1, page 52.

RODGERS, R. C., and GROSS, W., 1974, Fourth Symposium on Microdosimetry, edited by J. Booz, H. G. Ebert, R. Eickel, and A. Waker (Luxembourg: Euratom), page 1027.

ROHRIG, N., and COLVETT, R. D., 1978, Radiat. Res., **76**, 225.

ROHRLICH, F., and CARLSON, B. C., 1954, Phys. Rev. **93**, 38.

ROOTS, R., and SMITH, K. C., 1974, Int. J. Radiat. Biol., **26**, 467.

ROSSBACH, H., 1967, Die lineare Energieübertragung (LET) als Grundlage zur Interpretation strahlenbiologischer Experimente (Thesis), (Rossendorf near Dresden: ZfK — 128 report).

ROSSI, H. H., 1959, Radiat. Res., **10**, 522.

ROSSI, H. H., 1960, Radiat. Res. Suppl., **2**, 290.

ROSSI, H. H., BATEMAN, J. L., BOND, V. P., GOODMAN, L. J., and STICKLEY, E. E., 1960, Radiat. Res., **13**, 503.

ROSSI, H. H., BIAVATI, M. H., and GROSS, W., 1961, Radiat. Res., **15**, 431.

ROSSI, H. H., 1964, Annals of the New York Academy of Sciences, 114 Art. 1, 4.

ROSSI, H. H., 1966, Biophysical Aspects of Radiation Quality, report of a panel (Vienna: I.A.E.A.), page 81.

ROSSI, H. H., 1967, Advanc. biol. med. Phys., **11**, 27.

ROSSI, H. H., 1968, Radiation Dosimetry, Vol. 1, edited by F. H. Attix and W. C. Roesch (New York: Academic Press), page 43.

ROSSI, H. H., 1972, Third Symposium on Microdosimetry, edited by H. G. Ebert (Luxembourg: Euratom), page 1.

ROSSI, H. H., 1974, Fourth Symposium on Microdosimetry, edited by J. Booz, H. G. Ebert, R. Eickel, and A. Waker (Luxembourg: Euratom), page 1095.

ROTH, R. A., SHARMA, S. C., and KATZ, R., 1976, Phys. Med. Biol., **21**, 491.

DE RUITER-BOOTSMA, A. L., KRAMER, M. F., DE ROOIJ, D. G., and DAVIDS, J. A. G., 1974, Biological Effects of Neutron Irradiation, proceedings of a symposium (Vienna: I.A.E.A.), page 325.

RYDBERG, B., and JOHANSON, K. J., 1975, Radiat. Res., **64**, 281.

SAYEG, J. A., BIRGE, A. C., BEAM, C. A., and TOBIAS, C. A., 1959, Radiat. Res., **10**, 449.

SCHNEIDER, D. O., and WHITMORE, G. F., 1963, Radiat. Res., **18**, 286.

SCHULZ, H. J., and HARDER, D., 1970, Second Symposium on Microdosimetry, edited by H. G. Ebert (Brussels: Euratom), page 595.

SCHULZ, W., and GÜNTHER, K., 1973, Symposium on Cosmic Biology and Medicine (Berlin), page 105, (russ.).

SETLOW, R. B., and SETLOW, J. K., 1972, Ann. Rev. Biophys. Bioengng., **1**, 293.

SHELLABARGER, C. J., BROWN, R. D., RAO A. R., SHANLEY, J. P., BOND, V. P., KELLERER, A. M., ROSSI, H. H., GOODMAN

L. J., and MILLS, R. E., 1974, Biological Effects of Neutron Irradiation, proc. of a symposium (Vienna: I.A.E.A.), page 391.

SHIPLEY, W. U., ELKIND, M. M., and PRATHER, W. B., 1971, Radiat. Res., **47**, 437.

SINCLAIR, W. K., and MORTON, R. A., 1963, Nature, **199**, 1158.

SINCLAIR, W. K., 1966, Biophysical Aspects of Radiation Quality, report of a panel (Vienna: I.A.E.A.), page 21.

SINCLAIR, W. K., 1968a, Radiat. Res., **33**, 620.

SINCLAIR, W. K., 1968b, Biophysical Aspects of Radiation Quality, Second Panel Report (Vienna: I.A.E.A.), page 39.

SINCLAIR, W. K., 1974, Proceedings of the 5th Intern. Congr. Radiat. Res. (Washington), page 742.

SKARSGARD, L. D., KIHLMAN, B. A., PARKER, L., PUJARA, C. M., and RICHARDSON, S., 1967, Radiat. Res. Suppl., **7**, 208.

SMITH, B. G. R., and BOOZ, J., 1978, Sixth Symposium on Microdosimetry, edited by J. Booz and H. G. Ebert (London: Harwood Academic Publishers Ltd.), page 759.

SOMMERMEYER, K., 1952, Quantenphysik der Strahlenwirkung in Biologie und Medizin (Leipzig: Geest & Portig).

SOMMERMEYER, K. (editor), 1964, Quantenbiologie (Berlin, Göttingen, Heidelberg: Springer-Verlag).

SPENCER, L. V., and FANO, U., 1954, Phys. Rev., 93, 1172.

STOLTERFOHT, N., and JACOBI, W., 1972, Third Symposium on Microdosimetry, edited by H. G. Ebert (Luxembourg: Euratom), page 95.

SZYBALSKI, W., 1962, M. D. Anderson Hospital, 15th Annual Symposium, Houston, Texas (Austin: The University of Texas Press), page 147.

TAKESHITA, K., and SAWADA, S., 1974, Biological Effects of Neutron Irradiation, proceedings of a symposium (Vienna: I.A.E.A.), page 245.

TERASIMA, T., and TOLMACH, L. T., 1963, Biophys. J., **3**, 11.

TERRISSOL, M. A., and PATAU, J. P., 1974, Fourth Symposium on Microdosimetry, edited by J. Booz, H. G. Ebert, R. Eickel, and A. Waker (Luxembourg: Euratom), page 717.

TILL, J. E., and MCCULLOCH, E. A., 1961, Radiat. Res., **14**, 213.

TIMOFÉEFF-RESSOVSKY, N. W., and ZIMMER, K. G., 1947, Das Trefferprinzip in der Biologie (Leipzig: S. Hirzel Verlag).

TIMOFÉEFF-RESSOVSKY, N. W., IVANOV, V. I., and KOROGODIN, V. J., 1968, Primenenie principa popadania v radiobiologii (Moscow: Atomizdat), (russ.).

TOBUREN, L. H., and WILSON, W. E., 1977, J. Chem. Phys., **66**, 5202.

TOBUREN, L. H., and WILSON, W. E., 1979, Proceedings of the Sixth International Congress of Radiation Research, edited by S. Okada, M. Imamura, T. Terashima, and H. Yamaguchi (Tokyo: JARR), page 80.

TODD, P., 1966, Med. Coll. Virginia Quart., **1**, 4.

TODD, P., 1967, Radiat. Res. Suppl., **7**, 196.

TODD, P., 1975, Radiat. Res., **61**, 288.

TOLKENDORF, E., REGEL, K., ERZGRÄBER, G., and ABEL, H., 1978, personal communication.

TROTT, K. R., 1972, Encyclopedia of Medical Radiology, part 3: Radiation Biology, edited by O. Hug and A. Zuppinger (Berlin, Heidelberg, New York: Springer), page 43.

TSCHALÄR, C., 1968, Rutherford Laboratory Report, RHEL/R 164.

TSCHALÄR, C., and MACCABEE, H. D., 1970, Phys. Rev. B, **1**, 2863.

VARMA, M. N., and BAUM, J. W., 1978, Phys. Med. Biol., **23**, 1162.

VAVILOV, P. V., 1957, Zh. Experim. Teor. Fiz., **32**, 920, (russ.).

VEKSHINA, L. K., SAKOVICH, I. S., and SHEVCHENKO, V. A., 1974, Radiobiologia, XIV, 386, (russ.).

VERGROESEN, A. J., BUDKE, L., and VOS, O., 1963, Int. J. Radiat. Biol., **6**, 117.

VICTOREEN, J. A., 1943, J. Appl. Phys., **14**, 95.

VIDENSKY, V. G., PITKEVICH, V. A., and FARNAKEYEV, V. V., 1973, Radiobiologia, XIII, 412, (russ.).

VIDENSKY, V. G., and PITKEVICH, V. A., 1974, Voprosy microdosimetrii, edited by V. I. Ivanova (Moscow: Atomizdat), page 28.

WAIBEL, E., and GROSSWENDT, B., 1978, Radiat. Res., **76**, 241.

WALSKE, M. C., 1952, Phys. Rev., **88**, 1283.

WALSKE, M. C., 1956, Phys. Rev., **101**, 940.

WAMBERSIE, A., and DUTREIX, J., 1974, Europ. J. Cancer, **10**, 235.

WAMBERSIE, A., and DUTREIX, J., 1976, Fifth Symposium on Microdosimetry, edited by J. Booz, H. G. Ebert, and B. G. R. Smith (Luxembourg: Euratom), page 679.

WAMBERSIE, A., GUEULETTE, J., LAUBLIN, G., VAN DAM, J., MEULDERS, J. P., OCTAVE-PRIGNOT, M., and WINANT, M., 1977, Radiobiological Research and Radiotherapy, Vol. 2, proceedings of a symposium (Vienna: I.A.E.A.), page 151.

WARTERS, R. L., and HOFER, K. G., 1977, Radiat. Res., **69**, 348.

WARTERS, R. L., HOFER, K. G., HARRIS, C. R., and SMITH, J. M., 1977, Curr. Top. Radiat. Res. Quarterly, **12**, 389.

WHITMORE, G. F., GULYAS, S., and BOTOND, J., 1965, Proceedings of the 18th Annual Symposium on Fundamental Cancer Research (Baltimore: Williams and Wilkins), page 423.

WHITMORE, G. F., 1972, Los Alamos Scientific Laboratory Report, LA-5180-C, page 44.

WIDERÖE, R., 1967, Biophysik, **3**, 320.

WIDERÖE, R., 1978, Rad. and Environm. Biophys., **15**, 57.

WILSON, C. T. R., 1923, Proc. Roy. Soc., **A 104**, 1 and 192.

WILSON, K. S. J., and EMERY, E. W., 1968, Microdosimetry, Proceedings of a Conference, edited by H. G. Ebert (Brussels: Euratom), page 79.

WILSON, K. S. J., and FIELD, S. B., 1970, Second Symposium on Microdosimetry, edited by H. G. Ebert (Brussels: Euratom), page 137.

WILSON, W. E., and PARETZKE, H. G., 1974, Fourth Symposium on Microdosimetry, edited by J. Booz, H. G. Ebert, R. Eickel, and A. Waker (Luxembourg: Euratom), page 113.

WILSON, W. E., TOBUREN, L. H., and GLASS, W. A., 1975, Radiat. Res., **63**, 387.

WILSON, W. E., TOBUREN, L. H., and PARETZKE, H. G., 1978, Sixth Symposium on Microdosimetry, edited by J. Booz and H. G. Ebert (London: Harwood Academic Publishers Ltd.), page 239.

WINGATE, C. L., and BAUM, J. W., 1970, Doc. No. BNL 14767, Brookhaven.

WINGATE, C. L., and BAUM, J. W., 1976, Radiat. Res., **65**, 1.

WITHERS, H. R., 1967, Brit. J. Radiol., **40**, 187.

WITHERS, H. R., and ELKIND, M. M., 1969, Radiat. Res., **38**, 598.

WITHERS, H. R., and ELKIND, M. M., 1970, Int. J. Radiat. Biol., **17**, 261.

WITHERS, H. R., HUNTER, N., BARKLEY, H. T., and REID, B. O., 1974a, Radiat. Res., **57**, 88.

WITHERS, H. R., CHU, A. M., MASON, K. A'. REID, B. O., BARKLEY, H. T., and SMATHERS, J. B., 1974b, Europ. J. Cancer, **10**, 249.

WITHERS, H. R., and MASON, K., 1974, Cancer, **34**, 896.

WRIGHT, E. A., BEWLEY, D. K., and PARNELL, C. J., 1966, Brit. J. Radiol., **39**, 551.

YAMAMOTO, O., 1967, Nippon Acta Radiol., **26**, 1361.

YOUNG, D. A., and SMITH, K. C., 1976, Radiat. Res., **68**, 148.

ZEISS, G. D., MEATH, W. J., MCDONALD, J. C., and DAWSON, D. J., 1975, Radiat. Res., **63**, 64.

ZEITZ, L., CANADA, T. R., DJORDJEVIC, B., DYMBORT, G., FREEMAN, R., MCDONALD, J. C., O'NEIL, J., and LAUGHLIN, J. S., 1975, Radiat. Res., **63**, 211.

ZUSMAN, D. R., 1972, J. Mol. Biol., **69**, 119.

Subject index

www.ingramcontent.com/pod-product-compliance
Lightning Source LLC
LaVergne TN
LVHW060628300926
849162LV00008B/1737